Studies in Advanced Mathematics

Titles Included in the Series

Steven R. Bell, The Cauchy Transform, Potential Theory, and Conformal Mapping
John J. Benedetto, Harmonic Analysis and Applications
John J. Benedetto and Michael W. Frazier, Wavelets: Mathematics and Applications
Albert Boggess, CR Manifolds and the Tangential Cauchy–Riemann Complex
Goong Chen and Jianxin Zhou, Vibration and Damping in Distributed Systems, Vol. 1: Analysis, Estimation, Attenuation, and Design. Vol. 2: WKB and Wave Methods, Visualization, and Experimentation
Carl C. Cowen and Barbara D. MacCluer, Composition Operators on Spaces of Analytic Functions
John P. D'Angelo, Several Complex Variables and the Geometry of Real Hypersurfaces
Lawrence C. Evans and Ronald F. Gariepy, Measure Theory and Fine Properties of Functions
Gerald B. Folland, A Course in Abstract Harmonic Analysis
José García-Cuerva, Eugenio Hernández, Fernando Soria, and José-Luis Torrea, Fourier Analysis and Partial Differential Equations
Peter B. Gilkey, Invariance Theory, the Heat Equation, and the Atiyah-Singer Index Theorem, 2nd Edition
Alfred Gray, Modern Differential Geometry of Curves and Surfaces with Mathematica, 2nd Edition
Eugenio Hernández and Guido Weiss, A First Course on Wavelets
Steven G. Krantz, Partial Differential Equations and Complex Analysis
Steven G. Krantz, Real Analysis and Foundations
Clark Robinson, Dynamical Systems: Stability, Symbolic Dynamics, and Chaos
John Ryan, Clifford Algebras in Analysis and Related Topics
Xavier Saint Raymond, Elementary Introduction to the Theory of Pseudodifferential Operators
Robert Strichartz, A Guide to Distribution Theory and Fourier Transforms
André Unterberger and Harald Upmeier, Pseudodifferential Analysis on Symmetric Cones
James S. Walker, Fast Fourier Transforms, 2nd Edition
Gilbert G. Walter, Wavelets and Other Orthogonal Systems with Applications
Kehe Zhu, An Introduction to Operator Algebras

DYNAMICAL SYSTEMS

Stability, Symbolic Dynamics, and Chaos

Second Edition

Clark Robinson

CRC Press
Boca Raton London New York Washington, D.C.

Library of Congress Cataloging-in-Publication Data

Robinson, C. (Clark)
Dynamical systems : stability, symbolic dynamics, and chaos /
Clark Robinson. -- [2nd ed.]
p. cm. -- (Studies in advanced mathematics)
Includes bibliographical references (p. -) and index.
ISBN 0-8493-8495-8 (alk. paper)
1. Differentiable dynamical systems. I. Title. II. Series.
QA614.8.R63 1998
515′.352—dc21 98-34470
CIP

Direct all inquiries to CRC Press LLC, 2000 Corporate Blvd., N.W., Boca Raton, Florida 33431.

International Standard Book Number 0-8493-8495-8
Library of Congress Card Number 98-34470
Printed in the United States of America 1 2 3 4 5 6 7 8 9 0
Printed on acid-free paper

Preface to the First Edition

In recent years, Dynamical Systems has had many applications in science and engineering, some of which have gone under the related headings of chaos theory or nonlinear analysis. Behind these applications there lies a rich mathematical subject which we treat in this book. This subject centers on the orbits of iteration of a (nonlinear) function or of the solutions of (nonlinear) ordinary differential equations. In particular, we are interested in the properties which persist under nonlinear change of coordinates. As such, we are interested in the geometric or topological aspects of the orbits or solutions more than an explicit formula for an orbit (which may not be available in any case). However, as becomes clear in the treatment in this book, there are many properties of a particular solution or the whole system which can be measured by some quantity. Also, although the subject has a geometric or topological flavor, analytic analysis plays an important role (e.g., the local analysis near a fixed point and the stable manifold theory).

There have been several books and monographs on the subject of Dynamical Systems. There are several distinctive aspects which together make this book unique.

First of all, this book treats the subject from a mathematical perspective with the proofs of most of the results included: the only proofs which are omitted either (i) are left to the reader, (ii) are too technically difficult to include in an introductory book, even at the graduate level, or (iii) concern a topic which is only included as a bridge between the material covered in the book and commonly encountered concepts in Dynamical Systems. (Much of the material concerning measures, Liapunov exponents, and fractal dimension is of this latter category.) Although it has a mathematical perspective, readers who are more interested in applied or computational aspects of the subject should find the explicit statements of the results helpful even if they do not concern themselves with the details of the proofs. In particular, the inclusion of explicit formulas for the various bifurcations should be very useful.

Second, this book is meant to be a graduate textbook and not just a reference book or monograph on the subject. This aspect of the book is reflected in the way the background materials are carefully reviewed as we use them. (The particular prerequisites from undergraduate mathematics are discussed below.) The ideas are introduced through examples and at a level which is accessible to a beginning graduate student. Many exercises are included to help the student learn the meaning of the theorems and master the techniques of the proofs and topic under consideration. Since the exercises are not usually just routine applications of theorems but involve similar proofs and or calculations, they are best assigned in groups, weekly or biweekly. For this reason, they are grouped at the end of each chapter rather than in the individual section.

Third, the scope of the book is on the scale of a year long graduate course and is designed to be used in such a graduate level mathematics course in Dynamical Systems. This means that the book is not comprehensive or exhaustive but tries to treat the core concepts thoroughly and treat others enough so the reader will be prepared to read further in Dynamical Systems without a complete mathematical treatment. In fact, this book grew out of a graduate course that I taught at Northwestern University many times between the early 1970s and the present. To the material that I covered in that course, I have added a few other topics: some of which my colleagues treat when they teach the course, others round out the treatment of a topic covered earlier in the book (e.g., Chapter XII*), and others just give greater flexibility to possible courses using this book. Details on which sections form the core of the book are discussed in Section 1.4.

The perspective of the book is centered on multidimensional systems of real variables. Chapters II and III concern functions of one real variable, but this is done mainly because this makes the treatment simpler analytically than that given later in higher dimensions: there are not any (or many) aspects introduced which are unique to one dimension. Some results are proved so they apply in Banach spaces or even complete metric but most of the results are developed in finite dimensions. In particular, no direct connection with partial differential equations or delay equations is given. The fact that the book concerns functions of real rather than complex variables explains why topics such as the Julia set, Mandelbrot set, and Measurable Riemann Mapping Theorem are not treated.

This book treats the dynamics of both iteration of functions and solutions of ordinary differential equations. Many of the concepts are first introduced for iteration of functions where the geometry is simpler, but an attempt has been made to interpret these results for differential equations. A proof of the existence and continuity of solutions with respect to initial conditions is also included to establish the beginnings of this aspect of the subject.

Although there is much overlap in this book and one on ordinary differential equations, the emphasis is different. The dynamical systems approach centers more on properties of the whole system or subsets of the system rather than individual solutions. Even the more local theory in Chapters IV – VII deals with characterizing types of solutions under various hypotheses. Chapters VIII and X deal more directly with more global aspects: Chapter VIII centers on various examples and Chapter X gives the global theory.

Finally, within the various types of Dynamical Systems, this book is most concerned with hyperbolic systems: this focus is most prominent in Chapters VIII, X, XI, and XII. However, an attempt has been made to make this book valuable to people interested in various aspects of Dynamical Systems.

The specific prerequisites include undergraduate analysis (including the Implicit Function Theorem), linear algebra (including the Jordan canonical form), and point set topology (including Cantor sets). For the analysis, one of the following books should be sufficient background: Apostol (1974), Marsden (1974), or Rudin (1964). For the linear algebra, one of the following books should be sufficient background: Hoffman and Kunze (1961) or Hartley and Hawkes (1970). For the point set topology, one of the following books should be sufficient background: Croom (1989), Hocking and Young (1961), or Munkres (1975). What is needed from these other subjects is an ability to use these tools; knowing a proof of the Implicit Function Theorem does not particularly help someone know how to use it. For this reason, we carefully discuss the way these tools are used just before we use them. (See the sections on the Calculus Prerequisites, Cantor Sets, Real Jordan Canonical Form, Differentiation in Higher Dimensions, Implicit Function Theorem, Inverse Function Theorem, Contraction Mapping Theorem, and Definition of a Manifold.) After using these tools in Dynamical Systems, the reader should gain a much better understanding of the importance of these "undergraduate" subjects. The terminology and ideas from differential topology or differential geometry

* Chapter numbers have been revised to agree with those in the current edition.

are also used, including that of a tangent vector, the tangent bundle, and a manifold. However, most surfaces or manifolds are either Euclidean space, tori, or graphs of functions so these ideas should not be too intimidating. Although someone pursuing Dynamical Systems further should learn manifold theory, I have tried to make this book accessible to someone without prior background in this subject. Thus, the prerequisites for this book are really undergraduate analysis, linear algebra, and point set topology and not advanced graduate work. However, the reader should be warned that most beginning graduate students do not find the material at all trivial. The main complicating aspect seems to be the use of a large variety of methods and approaches. The unifying feature is not the methods used but the type of questions which we are trying to answer. By having patience and reviewing the mathematics from other subjects as they are used, the reader should find the material accessible and rich in content, both mathematical and for applications.

The main topic of the book is the dynamics induced by iteration of a (nonlinear) function or by the solutions of (nonlinear) ordinary differential equations. In the usual undergraduate mathematics courses, some properties of solutions of differential equations are considered but more attention is paid to the specific form of the solution. In connection with functions, they are graphed and their minima and maxima are found, but the iterates of a function are not often considered. To iterate a function we repeatedly have the same function act on a point and its images. Thus, for a function f with initial condition x_0, we consider $x_1 = f(x_0)$, and then $x_n = f(x_{n-1})$ for $n \geq 1$. We are interested in finding the qualitative features and long time limiting behavior of a typical orbit, for either an ordinary differential equation or the iterates of a function. Certainly, fixed points or periodic points are important, but sometimes the orbit moves densely through a complicated set such as a Cantor set. We want to understand and bring a structure to this seemingly random behavior. It is often expressed by saying, "we want to bring order out of chaos." One way of finding this structure is via the tool of *symbolic dynamics.* If there is a real valued function f and a sequence of intervals J_i such that the image of J_i by f covers J_{i+1}, $f(J_i) \supset J_{i+1}$, then it is possible to show that there is a point x whose orbit passes through this sequence of intervals, $f^i(x) \in J_i$. Labels for the intervals then can be used as symbols, hence the name of symbolic dynamics for this approach.

Another important concept is that of structural stability. Some types of systems (iterated functions or ordinary differential equations) have dynamics which are equivalent (topologically conjugate) to that of any of its perturbations. Such a system is called *structurally stable.* Finally, the term *chaos* is given a special meaning and interpretation. There is no one set definition of a chaotic system, but we discuss various ideas and measurements related to chaotic dynamics. One of the ironies is that some chaotic systems are also structurally stable.

Chapter I gives a more detailed introduction into the main ideas that are treated in the book by means of examples of functions and differential equations. Suffice it to say here that these three ideas, symbolic dynamics, structural stability, and chaos, form the central part of the approach to Dynamical Systems presented in this book.

In the year-long graduate course at Northwestern, we cover the the material in Chapter II, Sharkovskii's Theorem and Subshifts of Finite Type from Chapter III, Chapter IV except the Perron-Frobenius Theorem, Chapter V except some of the material on periodic orbits for planar differential equations (and sometime the proof of the Stable Manifold Theorem is omitted), a selection of examples from Chapter VIII, and most of Chapter X. In a given year, other selected topics are usually added from among the following: Chapter VII on bifurcations, the material on topological entropy in Chapter IX, and the Kupka-Smale Theorem. A course which did not emphasize the global hyperbolic theory as much could be obtained by skipping Chapter X and treating additional topics, e.g., Chapter VII or more on the measurements of chaos. Section 1.4 discusses the content of the different chapters and possible selections of sections or topics for a course using this book.

There are several other books which give introductions into other aspects or approaches to Dynamical Systems. For other graduate level mathematical introductions to Dynamical Systems, see Devaney (1989), Irwin (1980), Nitecki (1970), and Palis and de Melo (1982). For a more comprehensive treatment of Dynamical Systems, see Katok and Hasselblatt (1994). Some books which emphasize the dynamics of iteration of a function of one variable are Alsedà, Llibre, and Misiurewicz (1993), Block and Coppel (1992), and de Melo and Van Strien (1993). Carleson and Gamelin (1993) gives an introduction to the dynamics of functions of a complex variable. Chow and Hale (1982) gives a more thorough treatment of the bifurcation aspects of Dynamical Systems. The article by Boyle (1993) gives a more thorough introduction into symbolic dynamics as a separate subject and not just how it is used to analyze diffeomorphisms or vector fields. Some books which concentrate on Hamiltonian dynamics are Abraham and Marsden (1978), Arnold (1978), and Meyer and Hall (1992). For an introduction to applications of Dynamical Systems, see Guckenheimer and Holmes (1983), Hirsch and Smale (1974), Wiggins (1990, 1988), and Ott (1993). For applications to ecology, see Hirsch (1982, 1985, 1988, 1990), Hofbauer and Sigmund (1988), Hoppensteadt (1982), May (1975), and Waltman (1983). There are many books written on Dynamical Systems by people in fields outside mathematics, including Lichtenberg and Lieberman (1983), Marek and Schreiber (1991), and Rasband (1990).

I have tried very hard to give references to original papers. However, there are many researchers working in Dynamical Systems and I am not always aware of (or remember) contributions by various people to which I should give credit. I apologize for my omissions. I am sure there are many. I hope the references that I have given will help the reader start finding the related work in the literature.

When referring to a theorem in the same chapter, we use the number as it appears in the statement, e.g., Theorem 2.2 which is the second theorem of the second section of the current chapter. If we are referring to Theorem 2.2 from Chapter VI in a chapter other other than Chapter VI, we refer to it as Theorem VI.2.2 to indicate it comes from a different chapter.

There are not any specific references in this book to using a computer to simulate a dynamical system. However, the reader would benefit greatly by seeing the dynamics as it unfolds by such simulation. The reader can either write a program for him or herself or use several of the computer packages available. On an IBM Personal Computer, I have used the program *Phaser* which comes with the book by Koçak (1989). The program *Dynamics* by Yorke (1990) runs on both IBM Personal Computers and Unix/X11 machines. There are several other programs for IBM Personal Computers but I have not used them myself. Also, the program *DSTool* by J. Guckenheimer, M. R. Myers, F. J. Wicklin, and P. A. Worfolk runs on Unix/X11 machines. Many of the programming languages come with a good enough graphics library that it is not difficult to write one's own specialized program. However, for the X-Window environment on a Unix computer, I found the *VOGLE* library (C graphics C functions) a very helpful asset to write my own programs. There are several programs for the Macintosh, including *MacMath* by Hubbard and West (1992), but I have not used them.

Over the years, I have had many useful conversations with colleagues at Northwestern University and from elsewhere, especially people attending the Midwest Dynamical Systems Seminars. Those colleagues in Dynamical Systems at Northwestern University include Keith Burns, John Franks, Don Saari, Robert Williams, and many postdoctoral instructors and visitors. Those attending the Midwest Dynamical Systems Seminars are too numerous to list, but surely Charles Conley is one who bears mentioning and will long be remembered by many of us. I also owe a great debt to the people who taught me about Dynamical Systems, including Morris Hirsch, Charles Pugh, and Steve Smale. The perspective on Dynamical Systems which I learned from them is still very evident in the selection and treatment of topics in this book.

I would also like to thank the many people who found typographical errors, conceptual errors, or points that needed to be clarified in earlier drafts of this book. I would especially like to thank Keith Burns, Beverly Diamond, Roger Kraft, and Ming-Chia Li. Keith Burns taught out of a preliminary version and made many suggestions for improvements, clarifications, and changed arguments; Beverly Diamond made many suggestions for improvements in grammar and other editing matters; Roger Kraft made both mathematical and typographical corrections; in addition to noting out typographical errors, Ming-Chia Li pointed out aspects which needed clarifying.

This text was typeset using AMS-TeX. The figures were produced using DsTool, Xfig, Maple, and Vogle graphics C Library. I would like to thank Len Evens who supplied me with some macros which were used with AMS-TeX to produce the chapter and section titles and numbers, and the index and table of contents.

I was supported by several National Science Foundation grants during the years this book was written.

Clark Robinson
Department of Mathematics
Northwestern University
Evanston, Illinois 60208
clark@math.nwu.edu

Preface to the Second Edition

The second edition of this book has provided the opportunity for correcting many minor typographical errors or mistakes. Needless to say, the basic approach and content of the book have stayed the same. The discussion of the saddle node bifurcation has been rewritten using notation to make it easier to understand. In an attempt to expand the comparison of results for diffeomorphisms and flows, I have added a section on the horseshoe for a flow with a transverse homoclinic point. This section makes explicit the meaning and interpretation of a horseshoe in the case of a flow instead of a diffeomorphism. Another subsection on horseshoes for nontransverse homoclinic points indicates some recent extensions to the understanding of how horseshoes arise. Also added is a section proving the ergodicity of a hyperbolic toral automorphism. This proof is fairly simple but introduces an important technique which is used to prove ergodicity in other situations. Finally, a new chapter on Hamiltonian systems has been added. This chapter treats mainly local properties near fixed points, but should prove of interest to some of the readers.

Future typographical errors and additions will be posted on a website at Northwestern <http://math.nwu.edu/~clark>. I would appreciate being informed of further typographical errors or suggestions by email.

Clark Robinson
Summer 1998

How many are your works, O Lord!
In wisdom you made them all;
the earth is full of your creatures.
–Psalm 104:24

Contents

* Core Sections

CONTENTS

* Core Sections

* Core Sections

CONTENTS

CHAPTER I
Introduction

The main goal of the study of Dynamical Systems is to understand the long term behavior of states in a system for which there is a deterministic rule for how a state evolves. The systems often involve several variables and are usually nonlinear. In a variety of settings, very complicated behavior is observed even though the equations themselves are not very complicated (only "slightly nonlinear"). Thus, the simple algebraic form of the equations does not mean that the dynamical behavior is simple: in fact, it can be very complicated or even "chaotic." Another aspect of the chaotic nature of the system is the feature of "sensitive dependence on initial conditions." If the initial conditions are only approximately specified, then the evolution of the state may be very different. This feature leads to another difficulty in using approximate, or even real, solutions to predict future states based on present knowledge. To develop an understanding of these aspects of chaotic dynamics, we want to find situations which exhibit this behavior and yet for which we can still understand the important features of how a solution evolves with time.

Sometimes we cannot follow a particular solution with complete certainty because there is round-off error in the calculations or we are using some numerical scheme to find the solution. We are interested to know whether the approximate solution we calculate is related to a true solution of the exact equations. In some of the chaotic systems, we can understand how an ensemble of different initial conditions evolves, and prove that the approximate solution traced by a numerical scheme is shadowed by a true solution with some nearby initial conditions. If the system models the weather, people may not be content to know the range of possible outcomes of the weather that could develop from the known precision of the previous conditions, or to know that a small change of the previous conditions would have produced the weather which had been predicted. However, even for a subject like weather, for which quantitative as well as qualitative predictions are important, it is still useful to understand what factors can lead to instabilities in the evolution of the state of the system. It is now realized that no new better simulation of weather on more accurate computers of the future will be able to predict the weather more than about fourteen days ahead, because of the very nonlinear nature of the evolution of the state of the weather. This type of knowledge can itself be useful.

We now proceed to discuss these ideas in a little more detail in terms of specific equations, some of which arise from modeling different physical situations.

First, a comment about notation. Throughout the book, points and vectors in $\mathbb{R}^n$, a manifold, or a metric space are written in bold face, e.g., $\mathbf{x}$. Points (and vectors) on the line or the circle are denoted without bold face, e.g., x. Also, the components of a point or vector in $\mathbb{R}^n$ are written without bold face, e.g., $\mathbf{x} = (x_1, \ldots, x_n)$. In a few places, it seems strange to use different notation for a point in the line from a point in $\mathbb{R}^n$ for $n \geq 2$, but it seems like the best choice to distinguish a vector from a number.

1.1 Population Growth Models, One Population

In calculus, the differential equation

$$\dot{x} = ax$$

is studied, where $\dot{x} = \dfrac{dx}{dt}$. This equation can be used to model many different situations. In particular, when $a > 0$, it can model the growth of a population with unlimited resources or the effect of continuously compounding interest. When $a < 0$, it can model radioactive decay. For this simple equation, an explicit expression for the solutions can be given,

$$x(t) = x_0 e^{at},$$

where x_0 is the value of x when $t = 0$. If $a > 0$, the solution tends to infinity as t goes to infinity, while the solution tends to zero if $a < 0$. The behavior of the solutions is very simple since the equation is linear. See Figure 1.1.

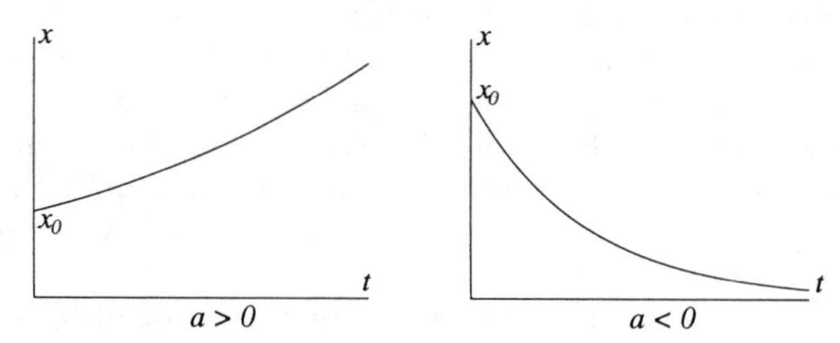

FIGURE 1.1. Plot of Solutions as a Function of Time: for $a > 0$ and $a < 0$

Often, we do not plot the solution as a function of t, but merely plot the solution in the space of possible values of "x", the *phase portrait.* The solution curves are labeled with arrows to indicate the direction that the variable changes as t increases. (This contains more information when more than one variable is involved.) See Figure 1.2. This type of phase space analysis can often yield qualitatively important information, even when an explicit representation of the solution cannot be obtained.

FIGURE 1.2. Phase Portrait: for $a > 0$ and $a < 0$

A more sophisticated population growth model involves a crowding factor. For this equation, it is not assumed that the growth rate, $\dot{x}/x$, is a constant, but that this quantity decreases as x increases, $\dot{x}/x = a - bx$ with $a, b > 0$. This equation can be used to model the growth of a population when the resources are limited. Thus, we get the equation

$$\dot{x} = (a - bx)x.$$

This equation is sometimes called the *logistic model for population growth.* This equation can be solved explicitly (by separation of variables and partial fractions), yielding the solution

$$x(t) = \frac{ax_0}{bx_0 + (a - bx_0)e^{-at}}.$$

As can be seen from the differential equation or from the solution, if x_0 equals either 0 or a/b, then $\dot{x} = 0$, and the solution $x(t)$ is constant in time. Such a solution is also called a *steady state solution* or a *fixed point solution.* If x_0 lies between 0 and a/b, then $\dot{x} > 0$, and the solution continues to increase with time but can never quite reach a/b. By a simple argument about monotone solutions or by using the exact form of the solution, it can be seen that for these initial conditions, $x(t)$ tends to a/b as t goes to infinity. Similarly, if $x_0 > a/b$, then $\dot{x} < 0$ and the solution $x(t)$ monotonically decreases toward a/b as t goes to infinity. See Figure 1.3. (Figure 1.3A shows the graphs of four solutions and Figure 1.3B shows the phase portrait of all solutions.) Thus, for any initial condition $x_0 > 0$, the solution $x(t)$ tends to the quantity a/b as t goes to infinity. Thus, a/b is the long term limit state for any positive initial condition.

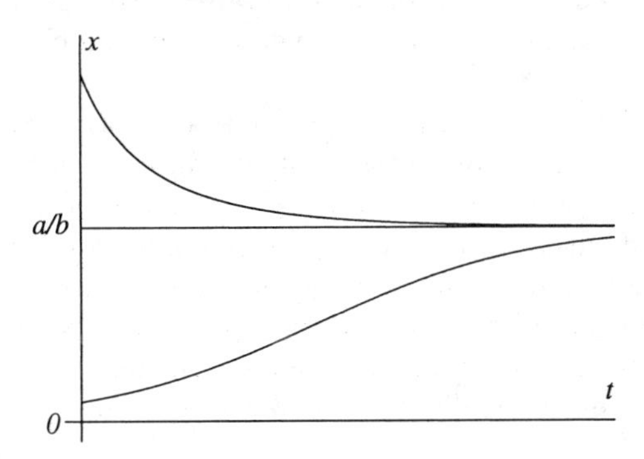

FIGURE 1.3A. Logistic Equation: Solutions as a Function of Time

FIGURE 1.3B. Logistic Equation: Phase Portrait

For differential equations on the real line (one real variable), the dynamics are always about this simple. As t goes to infinity, all solutions must tend to either a steady state solution (a fixed point) or $\pm\infty$.

1.2 Iteration of Real Valued Functions as Dynamical Systems

As mentioned above, the differential equation $\dot{x} = ax$ can be used to model the growth of capital when interest is compounded continuously. If however the interest is compounded at set time intervals (daily, monthly, or yearly) and added to the capital, then the amount of money at the n-th time interval in terms of the previous time is given by

$$x_n = \lambda x_{n-1},$$

where $\lambda > 1$. (Here, λ is equal to one plus the interest rate.) This equation could also be thought to model the growth of a population which reproduces at fixed time intervals (e.g., always in the spring) rather than continuously. Letting $f_\lambda(x) = \lambda x$, we get that $x_n = f_\lambda(x_{n-1})$. We also write $f_\lambda^2(x_0)$ for $f_\lambda \circ f_\lambda(x_0)$ and $f_\lambda^n(x_0) = f_\lambda \circ f_\lambda^{n-1}(x_0)$. For the second iterate, $x_2 = f_\lambda^2(x_0) = f_\lambda(x_1) = \lambda x_1 = \lambda f_\lambda(x_0) = \lambda(\lambda x_0) = \lambda^2 x_0$, and by induction on n,

$$\begin{aligned} x_n &= f_\lambda(x_{n-1}) \\ &= \lambda x_{n-1} \\ &= \lambda(\lambda^{n-1}x_0) \\ &= \lambda^n x_0. \end{aligned}$$

Thus, if $\lambda > 1$, $x_n = f_\lambda^n(x_0)$ tends to infinity as n goes to infinity. If $0 < \lambda < 1$ (some kind of penalty situation rather than adding interest to the capital), then x_n tends to 0 as n goes to infinity. The behavior of these iterates is very similar to the solutions of the linear differential equation. Again, the dynamics are simple, because this function is linear.

In the above situation, repeatedly crediting interest corresponded to iteration of a simple function, f_λ. We put in an initial value of x_0, and then calculate the value of x at later "times" n by the rule $x_n = f_\lambda(x_{n-1})$. This use of a real valued function is much different from the usual idea of merely graphing it. We continue to develop this idea in this introductory chapter and subsequent chapters.

As a next model, we consider the growth of a population at discrete time periods n where there is crowding or competing for resources. In comparison with the continuous differential equations, we get the discrete difference equations $x_n = F_\mu(x_{n-1})$ with

$$F_\mu(x) = \mu x(1-x).$$

(We have scaled the variables so the two constants a and b are equal, and use the single parameter μ.) Robert May (1975) noted that this simple model for discrete time intervals does not necessarily lead to a system which tends toward a steady state population. We study this example extensively in Chapter II, but at the moment we just remark that for $3 < \mu < 1 + \sqrt{6} \approx 3.449$, most initial values of x_0 with $0 < x_0 < 1$ (including $x_0 = 0.5$) do not tend toward a steady state solution, but tend to an orbit of period two: $b = F_\mu(a)$ and $a = F_\mu(b)$. For μ slightly larger than $1 + \sqrt{6}$, repeated iteration of x_0 by F_μ tends to a orbit of period four. For $\mu = 3.73$, the dynamics are even more complicated, and the iterates of an initial x_0 do not seem to tend to any period motion but move chaotically within the interval $(0, 1)$.

For $\mu > 4$, the set of points which stay in the interval $[0, 1]$ for all forward iterates of $F_\mu(x)$ turns out to be a Cantor set. The dynamics of iteration of points in this Cantor set can be analyzed and are very chaotic. The analysis of the dynamics on the Cantor set is by means of "symbolic dynamics." At any "time" n, the iterate x_n is either in the left half of the Cantor set or the right half. This distinction can be made by assigning a symbol (a 1 or a 2) for each time n. It is shown that this crude specification of the location for all times exactly determines the location at the initial state. By specifying a sequence of symbols, an orbit with a certain type of behavior can be shown to exist. Thus, although the dynamics of individual orbits are chaotic, the dynamics of all orbits can be completely understood and described by symbolic dynamics.

This example for a one-dimensional map has many of the properties of what is called a *horseshoe* for a two-dimensional map. In two dimensions, there is again an invariant Cantor set and the dynamics are determined by symbolic dynamics. Many equations used to model phenomenon are shown to be chaotic by proving the existence of a horseshoe. The horseshoe in two and higher dimensions is treated in Section 8.3. Subsections 7.3.1 – 4 treat various ways the horseshoe occurs for various functions or differential equations.

There are several lessons which the dynamics of iterates of the function F_μ teaches us. First, the simple algebraic nature of the function does not insure simple dynamics, and in fact its iterates exhibit chaotic properties. Second, for iteration of a function, the evolution of the state is deterministic. The parameter is fixed and there is a set rule for determining the next state from the previous one, so the evolution of the state from one state to the next one is very predictable. Still, by taking many iterates, the state can exhibit erratic behavior. The last lesson to learn from this example is that all erratic behavior is not always caused by changes in internal forces or stochastic effects, but can also result from the nonlinear nature of the deterministic system itself.

Chapters II and III treat the dynamics of iteration of real valued functions. We do not study the implications for modeling physical situations, but do study the mathematical aspects of the subject. The quadratic example is studied extensively because of the wide variety of types of dynamics it exhibits. Another thing to note about Chapters II and III is that only simple analytic tools are used to prove quite sophisticated results: the Mean Value Theorem, the Intermediate Value Theorem, differentiation of real valued functions, and a few tools from Point Set Topology. The fact that the mathematical tools used are simple does not mean that Chapters II and III are trivial. In fact, many of the key ideas of the book are introduced in this low dimensional setting. Thus, the difficulty of the material in Chapters II and III comes from the range of ideas and the development of the machinery and not the technical nature of the arguments.

1.3 Higher Dimensional Systems

Solutions of differential equations in two variables cannot exhibit chaos in the sense we are using the term. Thus, if $\mathbf{x} \in \mathbb{R}^2$ and $f(\mathbf{x})$ is a given function with values in $\mathbb{R}^2$ (a vector field on the plane), then

$$\dot{\mathbf{x}} = f(\mathbf{x})$$

is an ordinary differential equation in two variables. The Poincaré-Bendixson Theorem states that if $\mathbf{x}(t)$ is a solution which stays bounded as t goes to infinity, then either (i) $\mathbf{x}(t)$ tends to a periodic solution, or (ii) $\mathbf{x}(t)$ repeatedly passes near the same fixed point. In fact, in the second case, the motion $\mathbf{x}(t)$ still cannot be very complicated. Thus, chaos does not occur for differential equations in the plane.

The existence of periodic behavior itself is sometimes interesting. One example where it has been shown that there is an attracting periodic solution is the Van der Pol equations

$$\begin{aligned}\dot{x} &= y - x^3 + x\\ \dot{y} &= -x.\end{aligned}$$

These equations were originally introduced to model a "self exciting" electric circuit. If (x_0, y_0) is any initial condition other than $(0,0)$ (no matter how small), then the solution $(x(t), y(t))$ tends to the periodic motion. Thus, the system excites itself to this periodic motion and, other than transitory effects, the natural motion of a solution is the single periodic motion.

There are many other special equations which model population growth, nonlinear oscillators, and many other physical situations. We do not deal with the modeling aspect of the subject, but develop the mathematical tools to study such equations.

Iterates of functions from $\mathbb{R}^2$ to $\mathbb{R}^2$ can be used to model populations with more than one generation or stages in life (as well as many other situations). The iterates of such functions can exhibit all the complexities of the quadratic map on the real line. In two variables, the map can even be invertible (which the quadratic map on the real line is not) and still exhibit chaos. The simplest algebraic form of such a map is the Hénon map,

$$\begin{pmatrix} x_1 \\ y_1 \end{pmatrix} = F_{A,B}\begin{pmatrix} x \\ y \end{pmatrix} = \begin{pmatrix} A - By - x^2 \\ x \end{pmatrix}.$$

This map has two parameters A and B. It was introduced by Hénon (1976), not as a model of any particular physical situation, but as a map with a simple algebraic form which could easily be studied by means of computer simulation. He found that his map exhibited a "horseshoe" and a "strange attractor" for different parameter values. For $A = 5$ and $B = 0.3$, this map has an invariant Cantor set in the plane, a "horseshoe," and the dynamics of iteration of points in this Cantor set are chaotic. (This can be proved rigorously.) For $A = 1.4$ and $B = -0.3$, it can be proved that there is a trapping region in which points which start in this region remain bounded for all further forward iteration. There is then an "invariant set," Λ, of points which remain in the trapping region for both forward and backward iteration. Computer simulation indicates that this map is chaotic on Λ, and Λ cannot be broken up into smaller dynamically independent pieces ($F_{A,B}$ appears to be *topologically transitive* on Λ). See Figure 3.1. For this reason, this invariant set is called a strange attractor. The fact that $F_{1.4,-0.3}$ is transitive on Λ has not yet been proved rigorously, but much of the structure of the dynamics of this map is well understood.

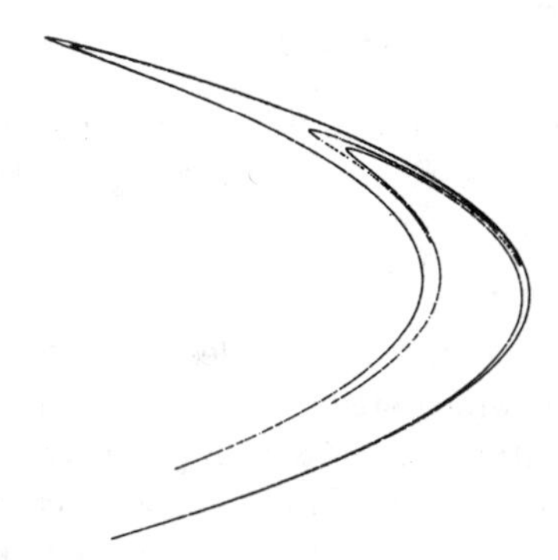

FIGURE 3.1. Hénon Attractor

For larger values of A, $A = 5$ and $B = -0.3$, the Hénon map has an invariant Cantor set Λ. (The set Λ is homeomorphic to the cross product of the usual one-dimensional Cantor set C with itself, $\Lambda \approx C \times C$.) Points not on Λ become unbounded under either forward or backward iteration. To a point on Λ there corresponds a unique string of symbols which exactly determine the orbit: in this sense, the dynamics on Λ are equivalent to the dynamics given by the symbolic dynamics. This type of invariant set is called a "horseshoe." It has many of the properties of the map F_μ on the line discussed above. We study the horseshoe and the Hénon map for these parameter values more in Chapter VIII.

Another way that maps can arise is by considering forced differential equations. For example, we can add a forcing term to the Van der Pol equation and get

$$\begin{aligned} \dot{x} &= y - x^3 + x + g(t) \\ \dot{y} &= -x \end{aligned}$$

where $g(t)$ is a T-periodic function of t, e.g., $g(t) = \cos(\omega t)$ for some fixed ω. The solutions of such nonautonomous equations can cross each other as t evolves, but by following the solutions through a complete period, we can get a well defined map

$$\begin{pmatrix} x_1 \\ y_1 \end{pmatrix} = \begin{pmatrix} x(T) \\ y(T) \end{pmatrix}.$$

Following the solution for multiples of the period yields higher iterates of the map,

$$\begin{pmatrix} x_n \\ y_n \end{pmatrix} = \begin{pmatrix} x(nT) \\ y(nT) \end{pmatrix}.$$

This period map (a special case of what is called a Poincaré map) can exhibit the type of chaotic behavior which we have been discussing. These differential equations can be thought of as equations in three variables, x, y, and t with $dt/dt = 1$. Because one of the variables is time, some people speak of these as differential equations in two and a half dimensions. Thus, differential equations in two and a half dimensions (or three) can exhibit chaos. The advantage of considering iteration of maps first in this book is that this same phenomenon can be observed in lower dimensions; in fact, in one dimension for noninvertible maps.

A type of forced van der Pol type equation was one of the first equations for which "random" or "chaotic" behavior was observed. Cartwright and Littlewood (1945) first discovered this and later Levinson (1949) gave a much simpler analysis of the situation. Much more recently, Levi (1981) gave a more complete analysis and showed how a horseshoe occurs in this situation.

Another problem which has been shown to have a horseshoe is the motion of three point masses moving under Newtonian attraction. Sitnikov (1960) studied the situation of the motion of three masses where the third mass m_3 moves on the z-axis and the first two masses are equal, $m_1 = m_2$, their positions remain symmetric with respect to the z-axis, and they move in an elliptical orbit. The simplest description is for $m_3 = 0$ where the first two masses affect the motion of the third but not vice versa. Then, m_1 and m_2 move in an elliptical orbit which is nearly circular in the (x, y)-plane. The third body oscillates up and down the z-axis. The motion of the first two bodies can be solved independently of the motion of the third mass. Thus, the forces of masses m_1 and m_2 on m_3 can be thought of as a time dependent effect on its motion. Since their motion is periodic, the effect is a time periodic term in the equations for the motion of m_3. The state of the third mass is determined by its height up the z-axis and its velocity. Thus, the equations are time periodic in two variables. By means of a calculation which has become called the Melnikov method, it is possible to show there is a horseshoe for this motion, i.e., there is an invariant Cantor set for the period map and the behavior of an orbit with initial conditions in this Cantor set can be prescribed by sequences of symbols. One way of interpreting the symbolic dynamics for the Cantor set is that it is possible to specify any sequence of integers n_j (as long as all the $n_j \geq N_0$ for some N_0), and then there is an orbit for which the first two masses make n_j revolutions between the j-th and the $j+1$ return of the third mass. Since the sequence is arbitrary, knowing the length of time it took for the third mass to return the last time gives no information about how long it will take to return the next time. This unpredictability, or chaotic nature, of the motion even for very deterministic motion is one of the characteristics of the type of situations we consider. In fact, the number of revolutions can be set to infinity. If $n_{j_0} = \infty$ for some $j_0 < 0$ and all the n_j for $j > j_0$ are bounded, then the orbit is a capture orbit, i.e., for this orbit the first two masses rotate about their center of mass and the third mass comes in from infinity and is captured in bounded motion as time goes to plus infinity. Similarly, if $n_{j_0} = \infty$ for some $j_0 > 0$ and the n_j for $j < j_0$ are bounded, then there is an ejection orbit. The symbols can be thought of as a very inexact determination of the state of the system at a given time. By prescribing this inexact information for all time, the exact present state of the system is determined; it is shown that there is a unique initial condition which goes through this sequence of rough prescriptions of states at the future times. Thus, the existence of a motion of a prescribed nature can be shown by means of such symbolic dynamics. Because of its usefulness in determining types of behavior, symbolic dynamics is one of the fundamental tools we use in our study of Dynamical Systems. In addition to Sitnikov, this situation was studied extensively by Alekseev (1968a, 1968b, 1969). Also see the book by Moser (1973) and the expository paper by Alekseev (1981).

A set of differential equations which have been much discussed are the Lorenz equations given by

$$\begin{aligned}\dot{x} &= -10x + 10y\\ \dot{y} &= 28x - y - xz\\ \dot{z} &= -\frac{8}{3}z + xy.\end{aligned}$$

These equations were introduced by Lorenz (1963) as a very rough model of the fluid flow of the atmosphere (weather). He studied these equations by means of computer simulation and observed chaotic behavior. After much investigation, we understand the features which are causing the chaos in these equations and have a good geometric model for their behavior, but no one has analytically been able to verify that these particular equations satisfy the conditions of the geometric model.

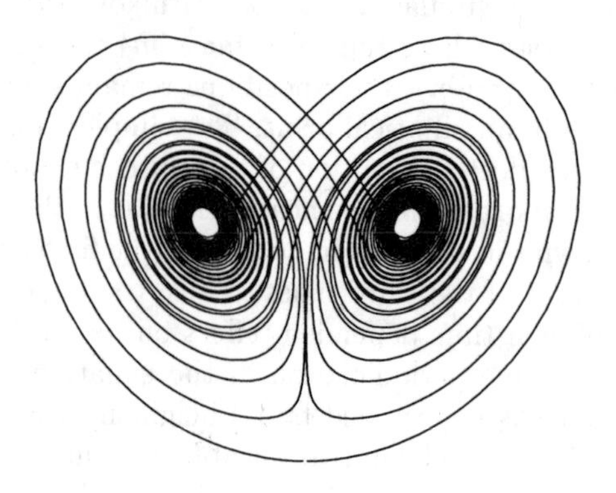

FIGURE 3.2. Lorenz Attractor

So far, we have emphasized that differential equations and iterations of functions can exhibit chaos. Another important idea is determining when two systems have the same dynamics, i.e., the two systems are *topologically conjugate.* A topological conjugacy can be thought of as a continuous change of coordinates. (When this idea is introduced, we explain why we cannot usually find differentiable change of coordinates, i.e., differentiable conjugacies.) Two systems which are topologically conjugate have the same long term dynamical behavior: both are chaotic or not, both have the same type of periodic motions, both are "transitive" or not, etc. Thus, topological conjugacy is an important concept when we classify systems up to equivalent dynamical behavior.

With this idea of topological conjugacy in place, we then want to define what it means for a system to have the same dynamics as all nearby systems in some suitable space of dynamical systems. What is of interest in this consideration is not the stability of a particular solution of the system, but the structure of the dynamics of the whole system. For this reason, Smale (1965) introduced the idea of structural stability based on earlier ideas of Andronov and Pontryagin (1937). A system f is called ***structurally stable*** provided that every system g which is near enough to f is topologically conjugate to f. Thus, the dynamics of f are robust under perturbations. Some systems can be both structurally stable and chaotic, but there certainly are others with only a finite number of periodic orbits which are structurally stable (Morse-Smale systems).

The main analytic feature, which allows us to show that a system is either chaotic or structurally stable, is the existence of a "hyperbolic structure." For these systems, the attention is focused first on the points which have some kind of recurrent behavior, the set of "chain recurrent points." The rough idea is that a system has a hyperbolic structure if each point in this chain recurrent set has a splitting into directions which are contracting and those which are expanding. (For maps on the real line, there is only one direction and it is either everywhere contracting and the motion is periodic, or everywhere expanding and the motion can be chaotic.) The idea of a hyperbolic structure is a generalization of the notion of splitting into various eigenspaces for a single linear map. The splitting at the different points has to be compatible: there needs to be a compatibility along an orbit and also as the point varies with the invariant set. Thus, some infinitesimal displacements (tangent vectors) at a point in the chain recurrent set are contracted and others are expanded. The existence of such a structure is crucial to be able either to apply the mechanism of symbolic dynamics or to show a system is structurally stable. Because these topics are the focus of this book, we mainly treat systems with such a hyperbolic structure on the set of all chain recurrent points. These ideas can be used in other situations where merely an invariant subset has a hyperbolic structure. The question of the existence of capture for three point masses is such a system with a hyperbolic structure on only an invariant subset of states. We do show that a homoclinic orbit (a orbit which tends to the same fixed point as t tends to both $\pm\infty$ but has other behavior between) is a situation where the proper assumption leads to a subsystem with a "horseshoe," i.e., an invariant subsystem with chaotic dynamics. These form an important category of situations where it is possible to prove that there is chaotic dynamics.

1.4 Outline of the Topics of the Chapters

In order to introduce the main perspective early in a situation where the analytic and geometric aspects are simpler, we first consider the iteration of functions of a single real variable. In this setting, we can show how to use sequences of intervals whose images cover the next interval to determine orbits (symbolic dynamics), and how some functions have dynamics which are equivalent to that of any of its perturbations (structural stability). These two ideas which are introduced in Chapter II are central to the approach to Dynamical Systems which we present. In fact, most of Chapter II is used heavily in the rest of the book with the exception of the rotation number of a homeomorphism on the circle, and the rotation number is an important idea in Dynamical Systems.

Chapter III treats topics related to iteration of a function of one variable, and in particular a situation which leads to complicated dynamics. One example of such dynamics is the invariant Cantor set for the quadratic map which we obtain in Chapter II. However, in Chapter III, we connect these ideas with the notion of chaos. We also separate these sections from Chapter II because they can easily be postponed until later in the book. In fact, Sharkovskii's Theorem is only used to motivate subshifts of finite type. In turn, subshifts of finite type are not used again until Chapter VIII when we discuss horseshoes and toral automorphisms in higher dimensions. The material on chaos, Liapunov exponents, period doubling cascade, and the zeta function is not used but is included since these concepts often occur in the literature on Dynamical Systems.

After Chapter III, we turn to higher dimensions for iteration of functions and solutions of ordinary differential equations. In this setting, the analysis near a fixed point is much more analytically complicated than the one-dimensional situation treated in Chapter II. As a first step in this analysis, Chapter IV deals with the dynamics of linear systems, both for iteration of functions and for differential equations. The Perron-Frobenius Theorem could easily be skipped in this chapter as well as some of the detail on solutions of linear differential equations. Chapter V treats the local dynamics of a nonlinear system near a fixed point as a perturbation of the linear system. If the linearization has only contracting and expanding directions and no neutral directions (is hyperbolic), then the dynamics of the linearization determines the local dynamics of the nonlinear system. The proof of these results uses the method of finding a contraction mapping: a mapping from a space of functions to itself is constructed, which is shown to be a contraction mapping and whose fixed point is a conjugacy between the linear and nonlinear system. In addition to determining the local behavior near a periodic point, this chapter introduces some of the methods of proof which are used for the more global results in later chapters. The method of proof of the conjugacy to the linearized system (the Hartman-Grobman Theorem) can be used to show that linear Anosov diffeomorphisms are structurally stable. The proof of the existence of the stable manifold for a fixed point can be modified to prove the existence of stable manifolds for a "hyperbolic invariant set." Thus, Chapter V is also an introduction into the analytic methods for multidimensional dynamical systems. Certainly, some of the material could be skipped or treated more quickly: for example, the proof of the existence of solutions of differential equations, the subsection on the Van der Pol equations, the Poincaré-Bendixson Theorem, the proof of the Stable Manifold Theorem, and the Center Manifold Theorem could be skipped. This chapter is not dependent on Chapter III. It also does not use the material on the invariant Cantor set from Chapter II.

Chapter VI gives a brief introduction into Hamiltonian system. We concentrate mainly on properties near a fixed point. Later in Chapter VIII, we return to Hamiltonian systems when we show how the Melnikov Method can be used to show the existence of horseshoes for time dependent Hamiltonian systems.

In Chapter VII, we treat the local bifurcation of fixed and periodic points, i.e., how the fixed points and their stability vary as a parameter is varied. We do not give a thorough treatment but give the three basic and generic bifurcations for one parameter families of functions. Chow and Hale (1982) has a much more thorough treatment of this aspect of Dynamical Systems. The Implicit Function Theorem is heavily used to prove these theorems. The section on the one-dimensional saddle-node and period doubling bifurcations depend only on Chapter II and could be covered at that time. The rest of the sections depend on Chapter IV. Chapter VII is not used extensively in later parts of the book but there are references to some of these results.

In Chapter VIII, we return to more complicated invariant sets. We give a number of examples and introduce the basic ideas which can be used to understand the structure of this type of dynamics. The unifying feature of these examples is their hyperbolic nature: they have some directions which are contracted and other directions which are expanded. A rigorous expression of these ideas requires some concepts from differential topology or differential geometry. We introduce the ideas necessary and mainly discuss the situation in Euclidean space or on tori where the need for machinery is minimized. This chapter depends on the treatment of the invariant Cantor set of Chapter II (which could be combined with the section in Chapter VII). Also, the main ideas of local dynamics from Chapter V are important for a clear understanding of this material. The key sections are those on the Birkhoff Transitivity Theorem, the Geometric Model Horseshoe, the

Horseshoe from a homoclinic point, attractors, the Solenoid, and Morse-Smale Systems. Section 8.5.1 on Markov Partitions for Hyperbolic Toral Anosov Automorphisms also makes a good connection back to the material on symbolic dynamics but is only needed for Section 10.6 on the Markov Partition for a Hyperbolic Invariant Set. Many of the other examples are interesting but not essential for the later chapters.

Chapter IX returns to the theme of Chapter III on measurements of chaos. All of these concepts appear widely in the literature on Dynamical Systems. We mainly introduce the definitions and main ideas without proofs, except for the material on topological entropy. None of this material is used in an essential way in the rest of the book.

Chapter X gives the theory of the analysis of these hyperbolic systems and their structural stability. This chapter uses heavily the ideas from the Stable Manifold Theorem in Chapter V. The examples of Chapter VIII give substance to the general definition of this chapter. Conley's Fundamental Theorem can be treated by itself at the very beginning, but the ideas of a Liapunov function in Chapter V and a gradient flow given in Chapter VIII are helpful for motivation of the ideas. The more global theorems of this chapter are interesting and important, but much of the richness of Dynamical Systems is revealed in the examples and more local theory of the previous chapters.

Chapter XI gives some generic properties, i.e., properties which are true for most systems in the sense of Baire category. This chapter also is the first place where perturbations of systems are treated to any extent. In the earlier chapters, we addressed conditions which assure that the dynamics cannot be changed by any perturbation. In this chapter, we address the question of what types of perturbations do cause changes in dynamics. In that sense, this chapter relates to Chapter VII. Although the definition of transverse intersection is given earlier in the book, some of the proofs require more theorems from transversality theory. Therefore, we include a section which states the needed theorems and gives references. In terms of generic properties, we prove that most systems have hyperbolic periodic points. We also give some necessary conditions for structural stability and a counter example to the density of structurally stable systems. This last example can be considered as an open set of diffeomorphisms, each of which can be made to undergo a bifurcation by means of a correctly chosen perturbation.

Finally, Chapter XII treats some additional topics in stable manifold theory, mainly concerned with smoothness of the manifolds. In addition to stating some general theorems, we prove that the hyperbolic splitting for an Anosov diffeomorphism on $\mathbb{T}^2$ is differentiable. We also return to prove the differentiability of the center manifold. Certainly, this last topic could be treated right after Chapter V.

CHAPTER II

One-Dimensional Dynamics by Iteration

In this chapter we introduce the concepts of Dynamical Systems through iteration of functions of a single real variable. The main idea is to understand what the orbit of a point is like when iterated repeatedly by the same function: $x_1 = f(x_0)$ and $x_n = f(x_{n-1})$ for $n \geq 1$. In one dimension, the graph of the function can be used quite easily to analyze the iterates of a point by a function. Also, because of the restriction to one dimension, the concepts of a periodic orbit being attracting or repelling become a simple consequence of the Mean Value Theorem. Two other important ideas in Dynamical Systems are the notion of topological conjugacy and symbolic dynamics. Two functions f and g are said to be topologically conjugate if there is a homeomorphism h with $g(x) = h \circ f \circ h^{-1}(x)$. In one dimension, we are able to prove quite easily that simple examples such as $f(x) = 2x$ and $g(x) = 3x$ are topologically conjugate. For the quadratic family, $F_\mu(x) = \mu x(1-x)$, we can show that if $\mu, \mu' > 4$, then F_μ and $F_{\mu'}$ are topologically conjugate even though both maps have infinitely many periodic points. We also use the quadratic map to introduce the ideas of symbolic dynamics. The idea is that we label two intervals I_1 and I_2 and are able to show that sequences of these two intervals are in one-to-one correspondence with orbits of the map under iteration. The final section of the chapter concerns iteration of homeomorphisms of the circle. In this setting, the average rotation of a point under repeated iteration, the rotation number, determines whether the map has periodic points or not.

In later chapters we return to study periodic orbits, the possibility of topological conjugacy, and related topics for iteration of functions in higher dimensions. We also apply these notions to solutions of differential equations. The main reason to treat the case of iteration of one variable first is that it reduces the topological complications while introducing the new concepts of dynamical systems.

2.1 Calculus Prerequisites

In this short section, we recall three results from calculus which we use extensively in the material on iterating a real valued function of one real variable. As we proceed through the material on dynamical systems, we discuss further results from calculus which we need. In particular, material on derivatives in higher dimensions and the Implicit Function Theorem is given later.

Critical Points

The points where a function attains its maximum and minimum are important in studying graphs. They are also important in determining the dynamics of iteration of the function. Therefore, we review the definition and terminology of points where the derivative is zero.

Assume f is differentiable. A point a is called a *critical point* of f provided $f'(a) = 0$. It is called a *nondegenerate critical point* provided $f'(a) = 0$ and $f''(a) \neq 0$. For example, if $F_\mu(x) = \mu x(1-x)$, then $x = 1/2$ is a nondegenerate critical point. The point 0 is a degenerate fixed point for the function $f(x) = x^3$.

Intermediate Value Theorem

Assume J is an interval and $f : J \to \mathbb{R}$ is a continuous function. Further assume that $a, b \in J$ with $a < b$, and z is a value between $f(a)$ and $f(b)$. Then, there is a (at least one) c with $a < c < b$ such that $f(c) = z$. This property says that f attains all the values between $f(a)$ and $f(b)$. This is essentially the result that $f([a, b])$ is connected. This result is also called the *Theorem of Darboux.*

Another consequence of continuity

Assume $f : J \to \mathbb{R}$ is a continuous function and $c_1 < f(a) < c_2$ for a in J. Then, there is an open set U about a in J such that $c_1 < f(x) < c_2$ for all $x \in U$.

Mean Value Theorem

Assume J is an interval and $f : J \to \mathbb{R}$ is a continuous function. Further assume that $a, b \in J$ with $a < b$ and that f is differentiable on (a, b). Then, there exists a c with $a < c < b$ such that

$$\frac{f(b) - f(a)}{b - a} = f'(c).$$

This says that there is a point c for which the slope of the tangent line at c equals the slope of the secant line from $(a, f(a))$ to $(b, f(b))$. This result is also called the *Theorem of Lagrange.*

The Chain Rule

The chain rule concerns the derivative of a composition of functions. If $f, g : \mathbb{R} \to \mathbb{R}$ are two functions, then we write $f \circ g(x)$ for $f(g(x))$, i.e., for the composition of f with g. If both f and g are differentiable, then

$$\begin{aligned}(f \circ g)'(x) &= f'(g(x))g'(x) \qquad \text{or}\\ &= f'(y)g'(x),\end{aligned}$$

where $y = g(x)$. The point is that the derivative of f must be taken at the correct point.

Composition of functions is at the heart of the dynamics of iteration of functions: taking a point $x_0 \in \mathbb{R}$, we want to find $x_1 = f(x_0)$, $x_2 = f(x_1)$, and $x_n = f(x_{n-1})$. Thus, $x_n = f \circ \cdots \circ f(x_0)$, where we take the composition of f with itself n times. We write

$$\begin{aligned}f^2(x) &= f \circ f(x) && \text{and by induction}\\ f^n(x) &= f \circ f^{n-1}(x) && \text{for } n \geq 1.\end{aligned}$$

Note that this is composition of the function and not squaring of the formula which defines f. Thus, if $f(x) = x(1 - x)$, then

$$f^2(x) = f(x(1 - x)) = x(1 - x)[1 - x(1 - x)].$$

For higher iterates, it quickly gets impossible to write out the formula for $f^n(x)$. In this context, the chain rule to the composition of a function f with itself n times can be written as

$$(f^n)'(x_0) = f'(x_{n-1}) \cdots f'(x_0),$$

where $x_j = f^j(x_0)$. In particular, if $f(x) = x(1-x)$, $x_0 = 1/3$, and $n = 3$, then $x_1 = f(1/3) = 2/9$, $x_2 = f^2(1/3) = f(2/9) = 14/81$. The derivative is $f'(x) = 1 - 2x$, and

$$\begin{aligned}(f^3)'(1/3) &= f'(14/81)f'(2/9)f'(1/3)\\ &= (1 - 2\frac{14}{81})(1 - 2\frac{2}{9})(1 - 2\frac{1}{3})\\ &= \frac{53}{81}\cdot\frac{5}{9}\cdot\frac{1}{3}.\end{aligned}$$

The point is that we do not need to compute $f^3(x)$ or $(f^3)'(x)$ explicitly.

If f is invertible, then f^{-1} is the inverse of f, $f^{-2}(x) = (f^{-1})^2(x)$, and $f^{-n}(x) = (f^{-1})^n(x)$ for $-n < 0$. We also write f^0 for the identity, $f^0(x) = x$. Using the chain rule, it can be shown that the derivative of the inverse is the reciprocal of the derivative of the function,

$$(f^{-1})'(x) = \frac{1}{f'(x_{-1})},$$

where $x_{-1} = f^{-1}(x)$.

Terminology about types of functions

Throughout this book, we use some terminology about functions and the extent of their differentiability. Let J be an open subset of $\mathbb{R}$ (possibly all of $\mathbb{R}$), and let $f : J \to \mathbb{R}$ be a function. If f is continuous, we say that *f is C^0*. If f is differentiable at each point of J and both f and f' are continuous, then f is said to be *continuously differentiable* or a *C^1 function.* Given $r \geq 1$, if f together with $f^{(j)}$ are continuous functions for $1 \leq j \leq r$, then f is said to be *r-times continuously differentiable* or a *C^r function.*

A function $f : X \to Y$ between metric spaces X and Y is called a *homeomorphism* provided it is (i) one to one, (ii) onto, (iii) continuous, and (iv) its inverse $f^{-1} : Y \to X$ is continuous. Finally, for J an open subset of $\mathbb{R}$, a function $f : J \to K \subset \mathbb{R}$ is called a *C^r-diffeomorphism from J to K* provided f is a C^r-homeomorphism from J onto K with a C^r inverse $f^{-1} : K \to J$.

2.2 Periodic Points

Throughout this section $f : \mathbb{R} \to \mathbb{R}$ is continuous. Sometimes f is only defined on an interval in $\mathbb{R}$. We add the assumption that f is C^1 or C^2 whenever we take first or second derivatives of the function.

In this section, we discuss the existence of a fixed point x with $f(x) = x$ or periodic point with $f^n(x) = x$ for some $n > 1$. The fixed points or points of low period can be found by solving the equations $f(x) = x$ and $f^n(x) = x$. For higher periods, it is not practical to solve these equations. Below, we discuss an analysis using the graph of f which is more useful for determining points of higher period.

Definition. A point a is a *periodic point of period n* provided $f^n(a) = a$ and $f^j(a) \neq a$ for $0 < j < n$. (Note that n is the *least* period.) If a has period one, then it is called a *fixed point.* If a is a point of period n, then the forward orbit of a, $\mathcal{O}^+(a)$, is called a *periodic orbit.* The notation we use for all points fixed by f^n (not all of least period n) is

$$\begin{aligned}\text{Per}(n, f) &= \{x : f^n(x) = x\} \qquad \text{and}\\ \text{Fix}(f) &= \text{Per}(1, f).\end{aligned}$$

Finally, a point a is *eventually periodic of period n* provided there exists an $m > 0$ such that $f^{m+n}(a) = f^m(a)$, so $f^{j+n}(a) = f^j(a)$ for $j \geq m$, and $f^m(a)$ is a periodic point.

Example 2.1. Let $f(x) = x^3 - x$. The fixed points satisfy the equation $x^3 - x = x$, so are the points $x = 0, \pm\sqrt{2}$. The points $x = \pm 1$ have their first iterates go to the fixed point 0, $f(\pm 1) = 0$, so ± 1 are eventually fixed.

As mentioned in the last section, the critical points satisfy $f'(x) = 3x^2 - 1 = 0$. So the only critical points are $x = \pm 1/\sqrt{3}$. The second derivative is $f''(x) = 6x$, which is nonzero at $\pm 1/\sqrt{3}$, so the critical points are nondegenerate.

As the above example illustrates, to find the fixed points it is enough to solve the equation $f(x) = x$. To find the periodic points of higher period is harder. Of course, theoretically it is possible to find the expression for the higher iterate $f^n(x)$ and then solve $f^n(x) = x$. For n much larger than four, this becomes algebraically very complicated even in the simplest examples. Exercise 2.8 asks you to find the points of period two for the map $F_\mu(x) = \mu x(1 - x)$. Sometimes it is possible to find points of higher period by understanding the graph of f^n. See Exercises 2.6 and 2.7. Also, computer iteration can be used to find the points of higher period. As mentioned in the preface, the programs *Phaser* by Koçak (1989) and *MacMath* by Hubbard and West (1992) are two programs which can carry out such iteration.

Definition. For a continuous function f, the *forward orbit of a point a* is the set $\mathcal{O}^+(a) = \{f^k(a) : k \geq 0\}$. If f is invertible, then the *backward orbit* is defined using negative iterates: $\mathcal{O}^-(a) = \{f^k(a) : k \leq 0\}$. The *(whole) orbit of a point a* is the set $\mathcal{O}(a) = \{f^k(a) : -\infty < k < \infty\}$. If f is not invertible, then we sometimes make choices and construct $x_{-1}, x_{-2}, \ldots$ where $f(x_{-n}) = x_{-n+1}$ or $x_{-n} \in f^{-1}(x_{-n+1})$. (For a noninvertible map, $f^{-1}(y) = \{x : f(x) = y\}$.)

Before discussing how to find the forward orbit using the graph of the function, we give some more fundamental definitions connected with convergence and stability of periodic points.

Definition. A point q is *forward asymptotic to p* provided $|f^j(q) - f^j(p)|$ goes to zero as j goes to infinity. If p is periodic of period n, then q is asymptotic to p if $|f^{jn}(q) - p|$ goes to zero as j goes to infinity. The *stable set of p* is defined to be

$$W^s(p) = \{q : q \text{ is forward asymptotic to } p\}.$$

If f is invertible, then a point q is said to be *backward asymptotic to p* provided $|f^j(q) - f^j(p)|$ goes to zero as j goes to minus infinity. If f is not invertible, then a point q is said to be *backward asymptotic to p* provided there are sequences p_{-j} and q_{-j} with $p_0 = p$, $q_0 = q$, $f(p_{-j}) = p_{-j+1}$, $f(q_{-j}) = q_{-j+1}$, and $|q_{-j} - p_{-j}|$ goes to zero as j goes to infinity. In either case, the *unstable set of p* is defined to be

$$W^u(p) = \{q : q \text{ is backward asymptotic to } p\}.$$

Definition. A point p is *Liapunov stable* (L-stable) provided given any $\epsilon > 0$ there is a $\delta > 0$ such that if $|x - p| < \delta$ then $|f^j(x) - f^j(p)| < \epsilon$ for all $j \geq 0$. This says that for x near enough to p the orbit of x stays near the orbit of p. A point p is *asymptotically stable* provided it is L-stable and $W^s(p)$ contains a neighborhood of p. If p is a periodic point which is asymptotically stable, it is also called an *attracting periodic point* or a *periodic sink*. If p is a periodic point for which $W^u(p)$ is a neighborhood of p, it is called a *repelling periodic point* or *periodic source*.

Example 2.2. Let $f(x) = x^3$. The fixed points satisfy $f(x) = x^3 = x$, so $x = 0, \pm 1$. Note that the fixed points correspond to points where the graph of f, $\{(x, f(x)\}$, intersects the diagonal $\{(x,x)\}$. See Figure 2.1. The graph of f is monotone. On $(0,1)$ the graph of f lies below the diagonal and $f(x) < x$. Thus, for $x \in (0,1)$, $x > f(x) > f^2(x) > \cdots f^n(x) > 0$. Because this sequence is monotone, it must converge to a fixed point and so to 0. (The fact that a bounded monotone sequence of points on an orbit must converge to a fixed point is left to Exercise 2.2.) Thus, $(0,1) \subset W^s(0)$. For backward iterates, $x < f^{-1}(x) < 1$ for $x \in (0,1)$. As j goes to minus infinity, $f^j(x)$ is monotonically increasing to 1, so $(0,1) \subset W^u(1)$.

A similar analysis shows that $(-1,0) \subset W^s(0)$ and $(-1,0) \subset W^u(-1)$ since for $x \in (-1,0)$,

$$x < f(x) < f^2(x) < \cdots < f^n(x) < 0$$
$$-1 < f^{-n}(x) < \cdots < f^{-2}(x) < f^{-1}(x) < x.$$

Because $W^s(0)$ contains a neighborhood of 0, the fixed point 0 is attracting.

For $x > 1$, $f^j(x)$ is monotonically increasing. If this forward orbit were bounded it would have to converge to a fixed point. Since there is no fixed point larger than 1, the orbit must go to infinity as j goes to infinity. As j goes to minus infinity, $f^j(x)$ is monotonically decreasing to 1 so $W^u(1) \supset (1,\infty)$.

Again for $x < -1$, $f^j(x)$ is monotonically decreasing to $-\infty$ as j goes to infinity, and $f^j(x)$ is monotonically increasing as j goes to minus infinity, $W^u(-1) \supset (-\infty,-1)$. Because we have analyzed the iterates of all points in the line, the following list summarizes the stable and unstable sets:

$$\begin{aligned} W^s(0) &= (-1,1) \\ W^u(0) &= \{0\} \\ W^s(\pm 1) &= \{\pm 1\} \\ W^u(1) &= (0,\infty) \\ W^u(-1) &= (-\infty,0). \end{aligned}$$

Because $W^u(1)$ is a neighborhood of 1 (and $W^s(1)$ is not), the iterates of points near 1 move away and the fixed point 1 is repelling (unstable). Similarly, -1 is repelling.

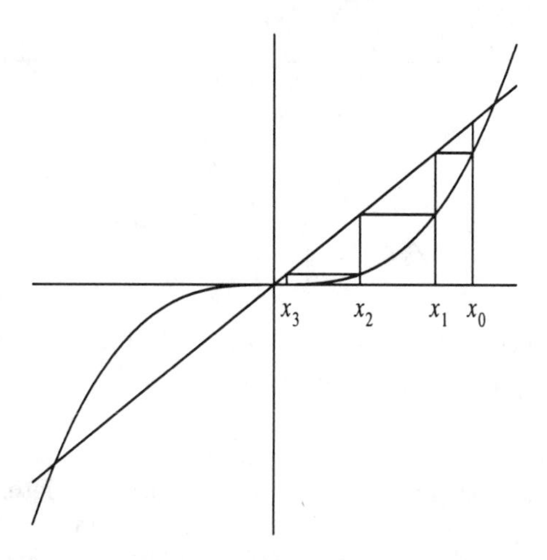

FIGURE 2.1. Graphical Analysis for Example 2.2

To understand the orbit more fully we use the graph of f. We give a general description, but the above example is useful in illustrating the construction. Take a point x_0. Draw a vertical line segment from (x_0, x_0) to the point on the graph $(x_0, f(x_0))$. Then, draw the horizontal line segment from this point on the graph over to the point on the diagonal $(f(x_0), f(x_0))$. Graphically this shows how we map from the point x_0 to $x_1 = f(x_0)$. Repeating this process with line segments from (x_1, x_1) to $(x_1, f(x_1))$ and then from $(x_1, f(x_1))$ to $(f(x_1), f(x_1))$ determines the next point $x_2 = f(x_1)$. Repeating this process determines the orbit. Figure 2.1 draws the graphical representation for the iterates of a point $0 < x_0 < 1$ for Example 2.2. In the figure, we extend the vertical lines down to the axis to indicate the points x_n more clearly. Because of the appearance of this figure, this process is often called the *stair step method* or *cobweb plot* to determine the phase portrait. The reader might want to draw the graphical analysis of orbits for Example 2.2 with $x_0 > 1$, $-1 < x_0 < 0$, and $x_0 < -1$.

The graphical analysis can easily be adapted to find inverse iterates when the function is one to one (monotonic). In this case, take the point x and draw the horizontal line segment from this point (x, x) on the diagonal to the point on the graph: since the function is monotonic, there is only one such point and it is $(f^{-1}(x), x)$. Now, draw the vertical line segment from $(f^{-1}(x), x)$ to the diagonal $(f^{-1}(x), f^{-1}(x))$, or to the axis $(f^{-1}(x), 0)$. In Figure 2.1, applying this process to x_3, we get $x_2 = f^{-1}(x_3)$, $x_1 = f^{-2}(x_3)$, and $x_0 = f^{-3}(x_3)$. This process can either be thought of as (i) reversing the steps to find the forward iterate or (ii) thinking of x as a function of y (interchanging the roles of x and y), which is the way the inverse function is explained in a calculus course (although we did not redraw the the graph with the x-axis up).

If the graph is not monotonic, in applying the graphical analysis to the inverse there may be more than one point on the graph which can be reached by a horizontal line segment from a point (x, x) on the diagonal. Each of these multiple points gives a possible inverse image of x. See Figure 2.2 for $f(x) = -x + x^3$, $x_0 = 0.231$, and

$$f^{-1}(x_0) = \{x_{-1}^{(1)} \approx -0.854, x_{-1}^{(2)} \approx -0.246, \text{ and } x_{-1}^{(3)} = 1.1\}.$$

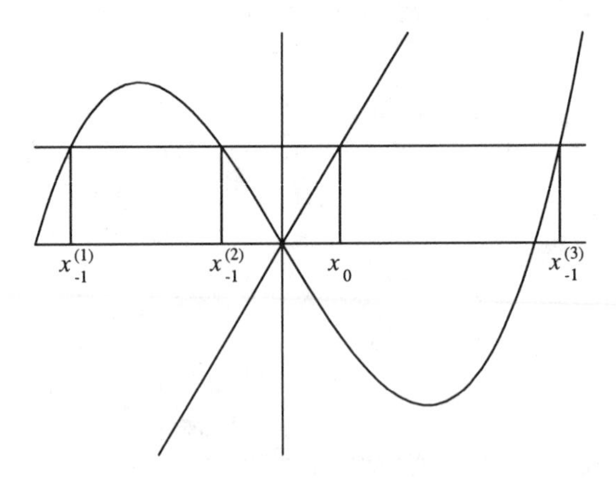

FIGURE 2.2. Function $f(x) = -x+x^3$, $x_0 = 0.231$, and Preimage $f^{-1}(x_0) = \{x_{-1}^{(1)} \approx -0.854, x_{-1}^{(2)} \approx -0.246, \text{ and } x_{-1}^{(3)} = 1.1\}$

The above process describes how to calculate the forward and backward orbit of points. As in Example 2.2, this information can be used to determine whether a fixed point is attracting or repelling. Because this process is involved, it is useful to have a criterion for a periodic orbit to be attracting which only uses the derivative of the function at points along the periodic orbit. The next theorem gives just such a criterion.

Theorem 2.1. *Assume $f : \mathbb{R} \to \mathbb{R}$ is a C^1 function. (a) Assume that p is a fixed point with $|f'(p)| < 1$. Then, p is an attracting fixed point (or asymptotically stable or a sink), i.e., $W^s(p)$ contains a neighborhood of p.*

(b) Assume that p is a periodic point of period n with $|(f^n)'(p)| < 1$. Then, p is an attracting periodic point.

REMARK 2.1. By the Chain Rule, the derivative in part (b) can be calculated as the product of the derivative of f along the orbit:

$$|(f^n)'(p)| = |f'(p_{n-1})| \cdots |f'(p_1)|\,|f'(p_0)|$$

where $p_j = f^j(p)$.

PROOF. (a) Because $|f'(p)| < 1$, there is an interval $[p-\epsilon, p+\epsilon]$ and a λ with $0 < \lambda < 1$ such that $|f'(x)| \leq \lambda$ for $x \in [p-\epsilon, p+\epsilon]$. Then, by the Mean Value Theorem, for $x \in [p-\epsilon, p+\epsilon]$, there is a z between x and p with

$$|f(x) - p| = |f(x) - f(p)| = |f'(z)| \cdot |x - p| \leq \lambda |x - p| < |x - p|.$$

Thus, $f(x)$ is closer to p than x, so $f(x) \in [p-\epsilon, p+\epsilon]$, and we can repeat the argument. By induction

$$|f^j(x) - p| \leq \lambda^j |x - p|.$$

This argument shows that $f^j(x) \in [p-\epsilon, p+\epsilon]$ for all $j \geq 0$, proving that p is L-stable. Because $\lambda^j |x-p|$ goes to zero, $f^j(x)$ converges to p as j goes to infinity, proving that p is attracting.

The above argument can be understood using graphical analysis. There are two cases: $0 < f'(p) < 1$ and $-1 < f'(p) < 0$. (The case $f'(p) = 0$ can be analyzed in a similar manner, but the exact argument depends on the form of the graph of f.) In the first case with $0 < f'(p) < 1$, for $x > p$, $x > f(x) > f^2(x) > \cdots > p$ as can be seen from the graph. Thus, $f^j(x)$ converges to p from above. See Figure 2.1. Similarly, if $x < p$, $f^j(x)$ converges to p from below.

Now, consider the case with $-1 < f'(p) < 0$. If $x > p$, then $f(x) < p$ and $f^2(x) > p$. Because $0 < (f^2)'(p) < 1$, $p < f^2(x) < x$. Thus, $f^j(x)$ converges to p also. See Figure 2.3.

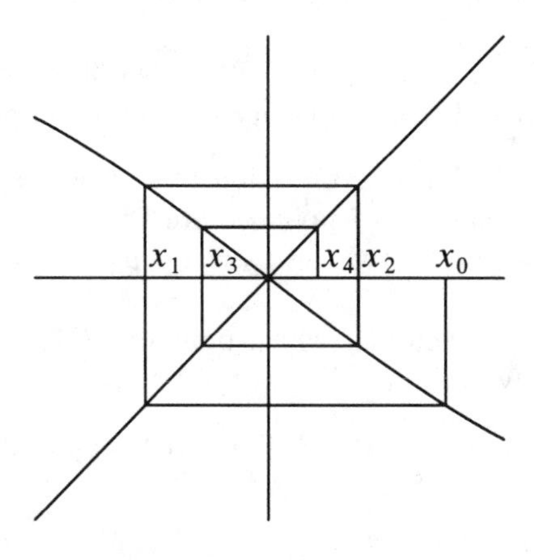

FIGURE 2.3. Near an Attracting Fixed Point with Negative Derivative

(b) For the periodic case, consider $g = f^n$. Then, $g(p) = p$ and $|g'(p)| < 1$. By part (a), $g^j(x)$ converges to p for x near p. By continuity of f^i for $1 \leq i < n$, since any j can be written as $i + kn$ with $0 \leq i < n$, it follows that $|f^j(x) - f^j(p)|$ goes to zero as j goes to infinity for all j and not just for multiples of n. □

The following theorem gives the comparable criterion for a periodic point to be repelling.

Theorem 2.2. *Assume $f : \mathbb{R} \to \mathbb{R}$ is a C^1 function. Assume that p is a periodic point of period n with $|(f^n)'(p)| > 1$. Then, p is repelling. Moreover, for all sufficiently small intervals I about p and $x \in I \setminus \{p\}$ there is a $k = k_x$ such that $f^{kn}(x) \notin I$. This says that all points near p go away from p under iterates of f^n.*

We leave the proof as an exercise. (Exercise 2.3)

Example 2.2 (revisited). Notice that for $f(x) = x^3$, $f'(x) = 3x^2$, $f'(0) = 0$, and $f'(\pm 1) = 3$. By Theorems 2.1 and 2.2, 0 is attracting and ± 1 is repelling. This conclusion is the same one we obtained by directly analyzing the orbits of points. Notice that it is much quicker to get the fact that these points are attracting and repelling using the theorems, i.e., using the criterion on the derivative. However, also notice that the theorems only tell us what happens to orbits near the fixed points, while our earlier analysis of this example analyzed the iterates of all points.

2.2.1 Fixed Points for the Quadratic Family

In this subsection we consider the family of quadratic maps $F_\mu(x) = \mu x(1 - x)$ for $\mu > 0$ and mainly for $\mu > 1$. We first find the fixed points.

Proposition 2.3. *The fixed points of the family of quadratic amps F_μ are 0 and $p_\mu = \dfrac{\mu - 1}{\mu}$. The fixed point 0 is attracting for $0 < \mu < 1$ and repelling for $\mu > 1$. The fixed point p_μ is attracting for $1 < \mu < 3$ and repelling for $0 < \mu < 1$ and $3 < \mu$. The only critical point is $x = 1/2$ which is nondegenerate.*

PROOF. The fixed points satisfy $x = \mu x - \mu x^2$, so are the points $x = 0$ and $x = p_\mu \equiv \frac{\mu-1}{\mu}$ as claimed.

To determine their stability, note that $F'_\mu(x) = \mu - 2\mu x$. Thus, $|F'_\mu(0)| = |\mu|$, so 0 is attracting for $0 < \mu < 1$ and repelling for $\mu > 1$. On the other hand, $|F'_\mu(p_\mu)| = |\mu - 2(\mu - 1)| = |2 - \mu|$. Thus, p_μ is attracting for $1 < \mu < 3$ and repelling for $0 < \mu < 1$ and $3 < \mu$.

The critical points satisfy $F'_\mu(x) = \mu(1 - 2x) = 0$, so the only critical point is $x = 1/2$. Finally, $F''_\mu(x) = -2\mu$ so $x = 1/2$ is a nondegenerate critical point. □

We leave to Exercise 2.8 the determination of the points of period two and their stability. As for the eventually fixed points, note that $F_\mu(1) = 0$, so 1 is eventually fixed. By symmetry of the graph, if we let $\hat{p}_\mu = 1 - p_\mu = 1/\mu$, then $F_\mu(\hat{p}_\mu) = p_\mu$, so $\hat{p}_\mu$ is also eventually fixed.

The following proposition indicates which points of F_μ go to infinity, and so which other points are potentially periodic.

Proposition 2.4. *Assume $\mu > 1$. If $x \notin [0, 1]$, then $F^j_\mu(x)$ goes to minus infinity as j goes to infinity.*

PROOF. For $x < 0$, $F'_\mu(x) = \mu - 2\mu x > 1$. Thus, for $x_0 < 0$, $0 > x_0 > F_\mu(x_0) > F^2_\mu(x_0) > \cdots > F^j_\mu(x_0)$ is decreasing. If this orbit were bounded, it would have to converge to a fixed point which would be a negative point. Since no such fixed point exists, $F_\mu^j(x_0)$ goes to minus infinity.

If $x_0 > 1$, then $F_\mu(x_0) < 0$, so $F_\mu^j(x_0) = F_\mu^{j-1} \circ F_\mu(x_0)$ goes to minus infinity. □

The next proposition shows that all the points in $(0,1)$ converge to the fixed point p_μ for the range of μ for which p_μ is attracting. The solution to Exercise 2.8 shows that this proposition is false for $\mu > 3$. However, for $3 < \mu < \mu_1$, most points in $(0,1)$ are asymptotic to an orbit of period two. For $\mu_1 < \mu < \mu_2$, most points in $(0,1)$ are asymptotic to an orbit of period four. This continues and there are μ_n such that for $\mu_{n-1} < \mu < \mu_n$, it can be shown that most points in $(0,1)$ are asymptotic to an orbit of period 2^n. (Such a proof cannot be done directly by calculating f^{2^n}.) The μ_n converge to μ_∞, and for $\mu > \mu_\infty$ it is not always the case that most points in $(0,1)$ are asymptotic to a periodic orbit. In Section 2.4, we see that for $\mu > 4$, there are many points in $(0,1)$ which are not asymptotic to a periodic orbit. In Section 3.4, we return to a further discussion of this period doubling cascade.

Proposition 2.5. *Assume* $1 < \mu < 3$. *If* $x \in (0,1)$, *then* $F_\mu^j(x)$ *converges to* p_μ *as* j *goes to infinity. Thus,* $W^s(p_\mu) = (0,1)$.

PROOF. (a) First consider $1 < \mu \le 2$. The maximum of the graph occurs at $x = 1/2$. For this range of parameters, $F_\mu(1/2) = \mu/4 \le 1/2$. Using the graph it is then clear that $p_\mu \le 1/2$. The function is thus monotonically increasing on $(0, p_\mu)$ and the graph lies above the diagonal. Thus, for $x_0 \in (0, p_\mu)$, $F_\mu^j(x_0)$ is a monotonically increasing sequence which must converge to the fixed point p_μ. See Figure 2.4. Similarly, on the interval $(p_\mu, 1/2]$ the function is monotonically increasing and the graph lies below the diagonal. Thus, for $y_0 \in (p_\mu, 1/2]$, $F_\mu^j(y_0)$ monotonically decreases to p_μ. Finally, for $x_0 \in (1/2, 1)$, $F_\mu(x_0) \in (0, 1/2)$, so $F_\mu^j(x_0)$ converges to p_μ. This completes the proof for this range of parameters.

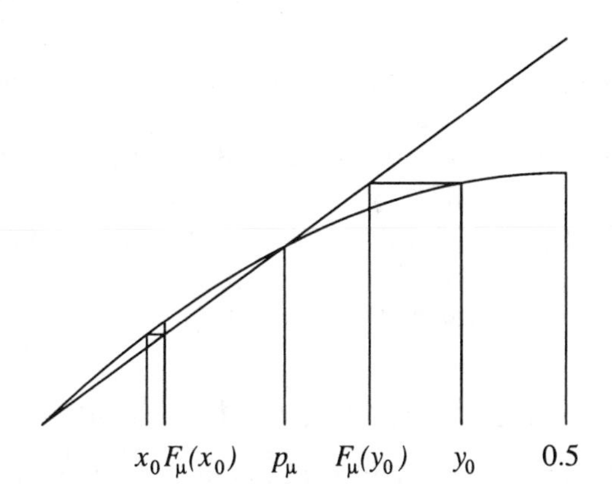

FIGURE 2.4. Iteration of x_0 and y_0 for $1 < \mu < 2$, $0 < x_0 < p_\mu$, and $p_\mu < y_0 \le 0.5$

(b) Now, assume that $2 < \mu < 3$. Note that $p_\mu > 1/2$.

(i) Consider the interval $[1/2, p_\mu]$. Because F_μ^2 is monotone on $[1/2, p_\mu]$, to find the image, it is enough to determine the iterates of the end points:

$$\begin{aligned} F_\mu^2([1/2, p_\mu]) &= F_\mu([p_\mu, \mu/4]) \\ &= [\mu(\frac{\mu}{4})(1 - \frac{\mu}{4}), p_\mu]. \end{aligned}$$

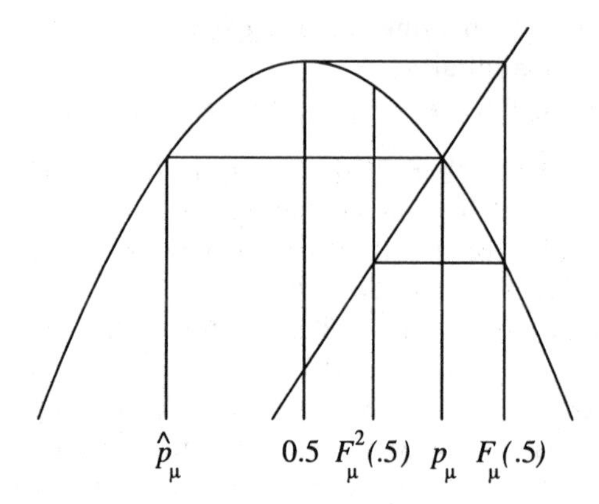

FIGURE 2.5. Iteration of $x = 0.5$ for $2 < \mu < 3$

We want to show that this image is contained in $[1/2, p_\mu]$, $\mu(\mu/4)(1 - \mu/4) > 1/2$, or $0 > \mu^3 - 4\mu^2 + 8 = (\mu - 2)(\mu^2 - 2\mu - 4)$. The roots of $\mu^2 - 2\mu - 4$ are $1 \pm \sqrt{5}$ so this factor is negative for $\mu < 3$. The first factor $\mu - 2 > 0$, so the product is negative as desired. Thus, we have shown that $F^2_\mu(1/2) = \mu(\mu/4)(1 - \mu/4) > 1/2$ and $F^2_\mu([1/2, p_\mu]) \subset [1/2, p_\mu]$. See Figure 2.5. A direct calculation shows that for $2 < \mu < 3$ the only fixed points of F^2_μ are those for F_μ, i.e., 0 and p_μ. (See Exercise 2.8.) Since $F^2_\mu(1/2)$ is above the diagonal, it follows that $F^2_\mu(x)$ is above the diagonal and $x < F^2_\mu(x) < p_\mu$ for $1/2 \le x < p_\mu$. Therefore, all the points in the interval $[1/2, p_\mu]$ converge to p_μ under iteration by F^2_μ. Since $|F'_\mu(p_\mu)| < 1$, it follows that all these points converge to p_μ under iteration by F_μ as well.

(ii) Next, let $\hat{p}_\mu = 1/\mu < 1/2$ as above, so $F_\mu(\hat{p}_\mu) = p_\mu$, $F_\mu([\hat{p}_\mu, 1/2]) = F_\mu([1/2, p_\mu])$, and $F^2_\mu([\hat{p}_\mu, 1/2]) \subset [1/2, p_\mu]$. Thus, all the points in $[\hat{p}_\mu, 1/2]$ also converge to p_μ by the results of the previous case.

(iii) Now, consider $x_0 \in (0, \hat{p}_\mu)$. The function F_μ is monotonically increasing on this interval and the graph lies above the diagonal. Thus, $F^j_\mu(x_0)$ is monotonically increasing as long as the iterates stay in this interval. Because $F_\mu(\hat{p}_\mu) = p_\mu$, the first time that an iterate $F^j_\mu(x_0)$ leaves the interval $(0, \hat{p}_\mu)$, it must land in $[\hat{p}_\mu, p_\mu]$, i.e., $F^k_\mu(x_0) \in [\hat{p}_\mu, p_\mu]$ for some $k > 0$. Then, $F^{k+j}_\mu(x_0)$ converges to p_μ as j goes to infinity.

(iv) Finally, if $x_0 \in (p_\mu, 1)$, then $F_\mu(x_0) \in (0, p_\mu)$, so further iterates converge to p_μ. Combining the cases we have proved the proposition. □

2.3 Limit Sets and Recurrence for Maps

We have defined periodic points and found points of low periods. In the examples we study with complicated dynamics, there are points each of which is not periodic but whose orbit keep returning near where it started. Orbits with such properties are said to have a kind of recurrence. In this section, we introduce several such concepts. Although we give a few examples in this section, the concepts should become clearer as they are used throughout the rest of the book. In this chapter, we mainly use the concepts of the α-limit and ω-limit sets of a point, a nonwandering point, and an invariant set. The other concepts could be postponed until they are used.

A related concept is that of convergence of an orbit to another. We have already defined what it means for q to be forward asymptotic to p. At various times in our study of dynamical systems we need a more general concept: we give this in terms of the α-limit and ω-limit sets of a point which are introduced in this section. These concepts are also used to define the points with a kind of recurrence called the limit set.

We give these definitions in this section for a continuous map $f : X \to X$ where X is a complete metric space with metric d. Several of the definitions use only the forward iterates of f and make sense even if f is not invertible. We do not distinguish these cases but just assume f is a homeomorphism throughout and let the reader determine which definitions make sense for noninvertible maps. In the next section, we return to the study of the quadratic map which gives an example of several of the types of recurrence defined in this section.

Definition. A point $\mathbf{y}$ is an *ω-limit point of* $\mathbf{x}$ *for* f provided there exists a sequence of n_k going to infinity as k goes to infinity such that

$$\lim_{k\to\infty} d(f^{n_k}(\mathbf{x}), \mathbf{y}) = 0.$$

The set of all ω-limit points of $\mathbf{x}$ for f is called the *ω-limit set of* $\mathbf{x}$ and is denoted by $\omega(\mathbf{x})$ or $\omega(\mathbf{x}, f)$. Other books also use the notation of $L_\omega(\mathbf{x}, f)$ or $L^+(\mathbf{x}, f)$. If the map f is invertible, then the *α-limit set of* $\mathbf{x}$ *for* f is defined the same way but with n_k going to minus infinity. If f is not invertible, then it is necessary to make a choice of preimages. The set of all such points is denoted by $\alpha(\mathbf{x})$ or $\alpha(\mathbf{x}, f)$. Again, other books also use the notation of $L_\alpha(\mathbf{x}, f)$ or $L^-(\mathbf{x}, f)$.

Example 3.1. If $f^n(\mathbf{x}) = \mathbf{x}$ is a periodic point, then the ω-limit set equals the orbit, $\omega(\mathbf{x}) = \mathcal{O}(\mathbf{x}) = \{\mathbf{x}, f(\mathbf{x}), \dots, f^{n-1}(\mathbf{x})\}$. In this case, the ω-limit set is finite, and not connected if the period n is greater than 1. If f is invertible, then also $\alpha(\mathbf{x}) = \mathcal{O}(\mathbf{x})$, and in any case $\mathcal{O}(\mathbf{x}) \subset \alpha(\mathbf{x})$.

Example 3.2. Section 2.8 on diffeomorphisms of the circle gives a more complete discussion of the following example on the circle. Let $\rho \notin \mathbb{Q}$. If $\bar{\tau}_\rho(x) = x + \rho$ is considered a map on the reals, then it induces a map $\tau_\rho : S^1 \to S^1$ of the circle by taking points modulus one. For any $x \in S^1$, it is shown in Section 2.8 that the forward and backward orbits of x, $\mathcal{O}^+(x)$ and $\mathcal{O}^-(x)$, are each dense in S^1, and also that $\omega(x) = \alpha(x) = S^1$.

Next we define various types of invariance for a subset.

Definition. A subset $S \subset X$ is said to be *positively invariant* provided $f(\mathbf{x}) \in S$ for all $\mathbf{x} \in S$, i.e., $f(S) \subset S$. A subset $S \subset X$ is said to be *negatively invariant* provided $f^{-1}(S) \subset S$. Finally, a subset $S \subset X$ is said to be *invariant* provided $f(S) = S$. Thus, if S is invariant, then the image of S is both into and onto S but we do not require that S is negatively invariant. If f is invertible (a homeomorphism) and S is an invariant subset for f, then the conditions that $f(S) = S$ and that f one to one imply that S is negatively invariant. Notice that a periodic orbit is always an invariant set. We show below that any $\omega(\mathbf{x})$ is always positively invariant and often invariant.

Example 3.3. Let $F_5(x) = 5x(1 - x)$ on $\mathbb{R}$ (which is not invertible). We show in the next section that F_5 has an invariant Cantor set Λ. In Section 2.5, we show that there are points x^* with $\omega(x^*) = \Lambda$ and $\mathcal{O}^+(x^*)$ dense in Λ.

The following theorem gives many of the basic properties of the limit set of a point.

Theorem 3.1. *Let $f : X \to X$ be a continuous map on a complete metric space X.*

(a) For any $\mathbf{x}$, $\omega(\mathbf{x}) = \bigcap_{N\geq 0} \mathrm{cl}(\bigcup_{n\geq N}\{f^n(\mathbf{x})\})$. *If f is invertible, then* $\alpha(\mathbf{x}) = \bigcap_{N\leq 0} \mathrm{cl}(\bigcup_{n\leq N}\{f^n(\mathbf{x})\})$.

(b) If $f^j(\mathbf{x}) = \mathbf{y}$ *for an integer j, then* $\omega(\mathbf{x}) = \omega(\mathbf{y})$. *Also,* $\alpha(\mathbf{x}) = \alpha(\mathbf{y})$ *if f is invertible.*

(c) For any $\mathbf{x}$, $\omega(\mathbf{x})$ *is closed and positively invariant. If (i)* $\mathcal{O}^+(\mathbf{x})$ *is contained in some compact subset of* X *or (ii)* f *is one to one, then* $\omega(\mathbf{x})$ *is invariant. Similarly, if* f *is invertible, then* $\alpha(\mathbf{x})$ *is closed and invariant, so it is both positively and negatively invariant.*

(d) If $\mathcal{O}^+(\mathbf{x})$ *is contained in some compact subset of* X *(e.g., the forward orbit is bounded in some Euclidean space), then* $\omega(\mathbf{x})$ *is nonempty and compact, and the distance* $d(f^n(\mathbf{x}), \omega(\mathbf{x}))$ *goes to 0 as* n *goes to infinity. Similarly, if* $\mathcal{O}^-(\mathbf{x})$ *is contained in some compact subset of* X, *then* $\alpha(\mathbf{x})$ *is nonempty and compact, and* $d(f^n(\mathbf{x}), \alpha(\mathbf{x}))$ *goes to 0 as* n *goes to minus infinity.*

(e) If $D \subset X$ *is closed and positively invariant, and* $\mathbf{x} \in D$, *then* $\omega(\mathbf{x}) \subset D$. *Similarly, if* f *is invertible and* D *is negatively invariant and* $\mathbf{x} \in D$, *then* $\alpha(\mathbf{x}) \subset D$.

(f) If $\mathbf{y} \in \omega(\mathbf{x})$, *then* $\omega(\mathbf{y}) \subset \omega(\mathbf{x})$, *and (if* f *is invertible) then* $\alpha(\mathbf{y}) \subset \omega(\mathbf{x})$. *Similarly, if* f *is invertible and* $\mathbf{y} \in \alpha(\mathbf{x})$, *then* $\alpha(\mathbf{y}) \subset \alpha(\mathbf{x})$ *and* $\omega(\mathbf{y}) \subset \alpha(\mathbf{x})$.

PROOF. (a) Let $\mathbf{y} \in \omega(\mathbf{x})$. Then, $\mathbf{y} \in \text{cl}(\bigcup_{n \geq N}\{f^n(\mathbf{x})\})$ by the definition of ω-limit set. Therefore, $\mathbf{y} \in \bigcap_{N\geq 0} \text{cl}(\bigcup_{n\geq N}\{f^n(\mathbf{x})\})$. This proves one inclusion. Now, assume $\mathbf{y} \in \bigcap_{N\geq 0} \text{cl}(\bigcup_{n\geq N}\{f^n(\mathbf{x})\})$. Then, for any N, $\mathbf{y} \in \text{cl}(\bigcup_{n\geq N}\{f^n(\mathbf{x})\})$. Now, for each N, take $k_N > k_{N-1}$ with $k_N \geq N$ and $d(f^{k_N}(\mathbf{x}), \mathbf{y}) < \frac{1}{N}$. Then, $d(f^{k_N}(\mathbf{x}), \mathbf{y})$ goes to 0 as N goes to infinity and $\mathbf{y} \in \omega(\mathbf{x})$. This proves the other inclusion, so the sets are equal.

(b) This is clear by the group property of iteration.

(c) For any $\mathbf{x}$, $\omega(\mathbf{x})$ is closed because it is the intersection of closed sets by part (a). To see that it is positively invariant, let $\mathbf{y} \in \omega(\mathbf{x})$. Then, there is a subsequence n_k with $d(f^{n_k}(\mathbf{x}), \mathbf{y})$ going to 0 as k goes to infinity. For any fixed $j \in \mathbb{N}$, $d(f^{n_k+j}(\mathbf{x}), f^j(\mathbf{y}))$ goes to 0 by the continuity of f. Therefore, $f^j(\mathbf{y}) \in \omega(\mathbf{x})$. This proves that $\omega(\mathbf{x})$ is positively invariant. If f is invertible, the above argument works for any $j \in \mathbb{Z}$, proving that $\omega(\mathbf{x})$ is invariant.

If f is not invertible but $\mathcal{O}^+(\mathbf{x})$ is contained in a compact subset, then we argue as follows. Let the subsequence n_k be as above with $d(f^{n_k}(\mathbf{x}), \mathbf{y})$ going to 0 as k goes to infinity. At least a subsequence of the points $f^{n_k-1}(\mathbf{x})$ converge to a point $\mathbf{z}$ which also must be in $\omega(\mathbf{x})$. By continuity, $f(\mathbf{z}) = \mathbf{y}$. This proves that $\omega(\mathbf{x})$ is invariant.

(d) If $\mathcal{O}^+(\mathbf{x})$ is contained in some compact subset of X, then

$$\text{cl}(\bigcup_{n\geq N}\{f^n(\mathbf{x})\})$$

is compact, so

$$\omega(\mathbf{x}) = \bigcap_{N\geq 0} \text{cl}(\bigcup_{n\geq N}\{f^n(\mathbf{x})\})$$

is compact. Also, $\omega(\mathbf{x})$ is the intersection of nested nonempty sets and so is nonempty. Finally, assume $d(f^n(\mathbf{x}), \omega(\mathbf{x}))$ does not go to 0. Then, there exists some $\delta > 0$ and a subsequence of iterates n_k with n_k going to infinity such that $d(f^{n_j}(\mathbf{x}), \omega(\mathbf{x})) \geq \delta$. The points $f^{n_j}(\mathbf{x})$ are bounded, so they have a limit point $\mathbf{z}$ outside of $\omega(\mathbf{x})$, contradicting the definition of $\omega(x)$. Thus, the distance must go to 0 as n goes to infinity.

(e) This follows from either part (a) or the following argument. Let $\mathbf{x} \in D$. By the invariance of D, $f^n(\mathbf{x}) \in D$. Because D is closed, all the limit points of $f^n(\mathbf{x})$ must be in D, proving that $\omega(\mathbf{x}), \alpha(\mathbf{x}) \subset D$.

(f) Let $\mathbf{y} \in \omega(\mathbf{x})$. The set $\omega(\mathbf{x})$ is positively invariant, so by part (e), $\omega(\mathbf{y}) \subset \omega(\mathbf{x})$. A similar argument applies to the α limit sets. □

We now define one type of invariant set which cannot be dynamically broken into smaller pieces: a minimal set. Then, the next proposition characterizes a minimal set in terms of the ω-limit sets of points. With this characterization we can give a simple example of a minimal set.

Definition. A set S is a *minimal set* for f provided (i) S is a closed, nonempty, invariant set and (ii) if B is a closed, nonempty, invariant subset of S, then $B = S$. Clearly, any periodic orbit is a minimal set.

Proposition 3.2. *Let X be a metric space, $f : X \to X$ a continuous map, and $S \subset X$ a nonempty compact subset. Then, S is a minimal set if and only if $\omega(\mathbf{x}) = S$ for all $\mathbf{x} \in S$.*

PROOF. First assume S is minimal. Since S is invariant, for any $\mathbf{x} \in S$, $\omega(\mathbf{x}) \subset S$. Since S is compact, $\omega(\mathbf{x})$ is a nonempty invariant subset of S. Because S is minimal, $\omega(\mathbf{x}) = S$. This proves one direction of the implication of the result.

Now, assume $\omega(\mathbf{x}) = S$ for all $\mathbf{x} \in S$. Assume $\emptyset \neq B \subset S$ is closed and invariant. If $\mathbf{x} \in B$, then $S = \omega(\mathbf{x}) \subset B \subset S$ so $B = S$. □

Example 3.2 (revisited). Let τ_ρ be the map of the circle S^1 as before with $\rho \notin \mathbb{Q}$. The whole circle S^1 is a minimal set for τ_ρ because $\omega(x) = S^1$ for all points. See Section 2.8. Maps of the circle are the principal examples in which we consider minimal sets in this book, although there are other types of maps with minimal sets.

Now, we can give the basic definitions of different types of recurrence.

Definition. Using the definition of ω-limit set and α-limit set, we say that $\mathbf{x}$ is *ω-recurrent* provided $\mathbf{x} \in \omega(\mathbf{x})$, and that $\mathbf{x}$ is *α-recurrent* provided $\mathbf{x} \in \alpha(\mathbf{x})$. The closure of the set of all ω-recurrent points is called the *Birkhoff center*,

$$\mathcal{B}(f) = \mathrm{cl}\{\mathbf{x} : \mathbf{x} \text{ is } \omega\text{-recurrent }\}.$$

The closure of the set of all the ω-limit sets is called the *limit set of f*,

$$L(f) = \mathrm{cl}\left(\bigcup\{\omega(\mathbf{x}, f) : \mathbf{x} \in X\}\right).$$

Definition. For a map $f : X \to X$, a point $\mathbf{p}$ is called *nonwandering* provided for every neighborhood U of $\mathbf{p}$ there is an integer $n > 0$ such that $f^n(U) \cap U \neq \emptyset$. Thus, there is a point $\mathbf{q} \in U$ with $f^n(\mathbf{q}) \in U$. The set of all nonwandering points for f is called the *nonwandering set* and is denoted by $\Omega(f)$.

Definition. An *ϵ-chain of length n from $\mathbf{x}$ to $\mathbf{y}$* for a map f is a sequence $\{\mathbf{x} = \mathbf{x}_0, \ldots, \mathbf{x}_n = \mathbf{y}\}$ such that for all $1 \le j \le n$, $d(f(\mathbf{x}_{j-1}), \mathbf{x}_j) < \epsilon$.

Definition. Let $Y \subset X$. The *ϵ-chain limit set of Y for f* is the set

$$\begin{aligned}\Omega_\epsilon^+(Y, f) = \{\mathbf{x} \in X : \text{ for all } n \ge 1, \text{ there is an } \mathbf{y} \in Y \text{ and an} \\ \epsilon\text{-chain from } \mathbf{y} \text{ to } \mathbf{x} \text{ of length greater than } n\}.\end{aligned}$$

Then, the *forward chain limit set of Y* is the set

$$\Omega^+(Y, f) = \bigcap_{\epsilon > 0} \Omega_\epsilon^+(Y, f).$$

(This set is the analogous object to the ω-limit set of a point but for chains.) If Y is the whole space X, we usually write $\Omega^+(f)$. Similarly,

$$\begin{aligned}\Omega_\epsilon^-(Y, f) = \{\mathbf{y} \in X : \text{ for all } n \ge 1, \text{ there is an } \mathbf{x} \in Y \text{ and an} \\ \epsilon\text{-chain from } \mathbf{y} \text{ to } \mathbf{x} \text{ of length greater than } n\},\end{aligned}$$

and the *backward chain limit set of* Y is the set

$$\Omega^-(Y, f) = \bigcap_{\epsilon>0} \Omega_\epsilon^-(Y, f).$$

(This latter set is like the α-limit set of a point.) Finally, the *chain recurrent set of* f is the set

$$\begin{aligned}\mathcal{R}(f) &= \{\mathbf{x} : \text{there is an } \epsilon\text{-chain from } \mathbf{x} \text{ to } \mathbf{x} \text{ for all } \epsilon > 0\} \\ &= \{\mathbf{x} : \mathbf{x} \in \Omega^+(\mathbf{x}, f)\} = \{\mathbf{x} : \mathbf{x} \in \Omega^-(\mathbf{x}, f)\}.\end{aligned}$$

It takes a little bit of work to see that $\mathbf{x} \in \Omega^+(\mathbf{x}, f)$ is equivalent to there being an ϵ-chain from $\mathbf{x}$ to itself for all ϵ (without assuming the chain is arbitrarily long).

Definition. We define a relation $\sim$ on $\mathcal{R}(f)$ by $\mathbf{x} \sim \mathbf{y}$ if $\mathbf{y} \in \Omega^+(\mathbf{x})$ and $\mathbf{x} \in \Omega^+(\mathbf{y})$. Two such points are called *chain equivalent.* It is clear that this is an equivalence relation. The equivalence classes are called the *chain components* of $\mathcal{R}$. (For flows, these chain components are actually the connected components of $\mathcal{R}$, which is the reason for the name.) If f has a single chain component on an invariant set Λ, then we say that f is *chain transitive on* Λ.

With these definitions, the reader can easily check that

$$\mathcal{B}(f) \subset L(f) \subset \Omega(f) \subset \mathcal{R}(f).$$

In this chapter, we mainly use the concept of the nonwandering set and the α- and ω-limit sets of points. In Chapters VIII, X, and XI, the chain recurrent set is used extensively.

2.4 Invariant Cantor Sets for the Quadratic Family

In this section, we start our study of maps with more complicated dynamics by studying the quadratic family, $F_\mu(x) = \mu x(1-x)$. For suitable choices of the parameter μ, the complicated dynamics of F_μ is not spread out over the whole line but is concentrated on an invariant Cantor set. Cantor sets occur often in maps with complicated or "chaotic" dynamics throughout this book. The use of the quadratic map on the line allows us to introduce such complicated dynamics in a relatively simple setting and give a rather complete analysis. After determining the invariant Cantor set in this section, the next section contains a discussion of symbolic dynamics. By introducing symbols to describe the location of a point, the dynamics of a point in the Cantor set can be determined by means of a sequence of these symbols. Because many different patterns of symbols can be written down, points with many different types of dynamics can be shown to exist.

We start in Subsection 2.4.1 by giving the properties which characterize a Cantor set and reviewing the construction of the middle-α Cantor set in the line. After this treatment, we show how a Cantor set arises for the quadratic map in Subsection 2.4.2.

2.4.1 Middle Cantor Sets

Definitions. Let X be a topological space and $S \subset X$ a subset. The set S is *nowhere dense* provided the interior of the closure of S is the empty set, $\text{int}(\text{cl}(S)) = \emptyset$. The set S is *totally disconnected* provided the connected components are single points. In the real line, a closed set is nowhere dense if and only if it is totally disconnected. In other spaces, these two concepts are different. For example, a curve in the plane is nowhere dense but it is not totally disconnected. The set S is *perfect* provided it is closed and every point p in S is the limit of points $q_n \in S$ with $q_n \neq p$.

A set S is called a *Cantor set* provided it is (i) totally disconnected, (ii) perfect, and (iii) compact.

We write $L(K)$ for the length of an interval K.

Construction of a Middle-α Cantor Set

Let $0 < \alpha < 1$ and $\beta > 0$ be such that $\alpha + 2\beta = 1$. Note that $0 < \beta < \frac{1}{2}$. Let $S_0 = I = [0, 1]$. Start by removing the middle open interval of length α:

$$\begin{aligned} G &= (\beta, 1 - \beta) \qquad \text{and} \\ S_1 &= I \setminus G. \end{aligned}$$

Notice that G is the middle open interval of I which makes up a proportion α of the whole interval. Then,

$$S_1 = J_0 \cup J_2$$

is the union of 2 closed intervals each of length β, $L(J_j) = \beta$ for $j = 0, 2$. We label the intervals so that J_0 is to the left of J_2. (We use 0 and 2 rather than 0 and 1 or 1 and 2 because of the connection we make below with the representation of numbers in base 3.) For $j = 0, 2$, let G_j be the middle open interval which makes up a proportion α of J_j, so $L(G_j) = \alpha\beta$. Let $J_{j,0}$ be the left component of $J_j \setminus G_j$ and $J_{j,2}$ be the right component. For $j_1, j_2 = 0$ or 2, the length of the component J_{j_1,j_2} is β^2, $L(J_{j_1,j_2}) = L(J_{j_1})(1 - \alpha)/2 = \beta^2$. Let

$$S_2 = S_1 \setminus (G_0 \cup G_2) = \bigcup_{j_1, j_2 = 0,2} J_{j_1,j_2}.$$

This set, S_2, is the union of 2^2 closed intervals, each of length β^2. By induction, assume S_{n-1} is the union of 2^{n-1} closed intervals each of length β^{n-1}, and labeled as $J_{j_1,\ldots,j_{n-1}}$ for all combinations of $j_k = 0$ or 2. For each of these intervals, let $G_{j_1,\ldots,j_{n-1}}$ be the middle open interval which makes up a proportion α of the interval $J_{j_1,\ldots,j_{n-1}}$. Thus, $L(G_{j_1,\ldots,j_{n-1}}) = \alpha L(J_{j_1,\ldots,j_{n-1}}) = \alpha\beta^{n-1}$. Let

$$J_{j_1,\ldots,j_{n-1}} \setminus G_{j_1,\ldots,j_{n-1}} = J_{j_1,\ldots,j_{n-1},0} \cup J_{j_1,\ldots,j_{n-1},2},$$

where $J_{j_1,\ldots,j_{n-1},0}$ is to the left of $J_{j_1,\ldots,j_{n-1},2}$. Each of these components has length β^n, $L(J_{j_1,\ldots,j_{n-1},j_n}) = L(J_{j_1,\ldots,j_{n-1}})(1 - \alpha)/2 = \beta^n$. Let

$$\begin{aligned} S_n &= S_{n-1} \setminus \bigcup_{j_1,\ldots,j_{n-1}=0,2} G_{j_1,\ldots,j_{n-1}} \\ &= \bigcup_{j_1,\ldots,j_n=0,2} J_{j_1,\ldots,j_n}. \end{aligned}$$

Since each interval $J_{j_1,\ldots,j_{n-1}}$ of S_{n-1} yields two closed intervals of S_n, S_n has $2(2^{n-1}) = 2^n$ closed intervals. This completes the induction step of the construction.

Finally, let

$$C = \bigcap_{n=0}^{\infty} S_n.$$

We check that C satisfies the properties of a Cantor set in the following claims.

Claim 1. *The set C is nowhere dense, so totally disconnected.*

PROOF. The intervals which make up S_n have length β^n. Therefore, for any point $p \in C$, there is a point q within a distance β^n which is not in S_n and so not in C. Therefore, C has empty interior and is nowhere dense. □

Claim 2. *The set C is perfect.*

PROOF. The sets S_n are closed, and C is the intersection of these sets, so C is closed.

Let $p \in C$ and j a positive integer. Take n such that $\beta^n < 2^{-j}$ and let K be the component of S_n containing p. Then, $K \cap S_{n+1}$ is made up of two intervals. Let q_j be one of the endpoints of the interval of $K \cap S_{n+1}$ not containing p. Then, (i) $q_j \neq p$, (ii) $|p - q_j| < \beta^n < 2^{-j}$, and (iii) $q \in C$ (since all the end points of the S_i are in C). This gives a sequence of points $q_j \in C$, $q_j \neq p$, and q_j converges to p. □

REMARK 4.1. The total length of S_n is $2^n\beta^n = (1-\alpha)^n$. Since $2\beta < 1$, the total length of S_n goes to zero as n goes to infinity. Therefore, all these middle-α Cantor sets have measure zero.

There are Cantor sets with positive measure, but they are not formed by taking a uniform proportion of the intervals out at each step. Let $\alpha_n > 0$ be numbers such that $\prod_{n=1}^{\infty}(1 - \alpha_n) > 0$, i.e., $\sum_{n=1}^{\infty} \alpha_n < \infty$. At the n-th stage, remove the middle-α_n from each of the previous intervals to form the set S_n. Then, the total length of S_n is $1 - \alpha_n$ times the total length of S_{n-1}. By induction, the total length of S_n is $\prod_{j=1}^{n}(1 - \alpha_j)$. As n goes to infinity, it follows that $C = \bigcap_{n=1}^{\infty} S_n$ has measure $\prod_{j=1}^{\infty}(1 - \alpha_j) > 0$.

REMARK 4.2. In the construction, it is not important that each component of S_n has length exactly β^n. Let L_n be the maximum length of a component in S_n,

$$L_n = \max\{L(K) : K \text{ is a component of } S_n\}.$$

What is important is that L_n goes to zero as n goes to infinity. This property implies that C is nowhere dense. If the sum of the lengths of the components of S_n does not go to zero, then C has positive measure. Also, to prove that C is perfect, it is not necessary that for each component K of S_n that $K \cap S_{n+1}$ has two components; what is necessary is that for each component K of S_n that there is a $k \geq n + 1$ such that $K \cap S_k$ has at least two components.

REMARK 4.3. We show that the quadratic map has an invariant set which is a Cantor set in the line but which is not a middle-α Cantor set. Later in the book we see many other invariant sets which are Cantor sets, or locally the cross product of a Cantor set and a disk (in some dimension).

Middle-Third Cantor Set and Ternary Expansion

Any number x in the interval $[0, 1]$ can be written in ternary expansion as

$$x = \sum_{k=1}^{\infty} \frac{j_k}{3^k}.$$

Most numbers have a unique expansion but

$$\frac{1}{3} = \sum_{k=2}^{\infty} \frac{2}{3^k};$$

In fact, for any n and $j_1, \ldots, j_n$,

$$\left(\sum_{k=1}^{n-1} \frac{j_k}{3^k}\right) + \frac{j_n + 1}{3^n} = \sum_{k=1}^{n} \frac{j_k}{3^k} + \sum_{k=n+1}^{\infty} \frac{2}{3^k}.$$

Thus, repeating 2's in a ternary expansion plays the same role as repeating 9's in decimal expansions, and the nonunique representations given above are the only ones that occur.

To make the comparison between the ternary expansion and the middle-third Cantor set, we want to avoid the use of 1's as much as possible. For that reason, we make the following conventions on the coefficients $\mathbf{j} = (j_1, j_2, \ldots)$ which we use in the cases where there is a nonunique representation: (i) we use the $\mathbf{j}$ for which there is an n with $j_n = 2$ and $j_k = 0$ for all $k > n$, but do not use the $\mathbf{j}$ for which there is an n with $j_n = 1$ and $j_k = 2$ for all $k > n$; (ii) we use the $\mathbf{j}$ for which there is an n with $j_n = 0$ and $j_k = 2$ for all $k > n$, but do not use the $\mathbf{j}$ for which there is an n with $j_n = 1$ and $j_k = 0$ for all $k > n$; and finally, (iii) we use $\sum_{k=1}^{\infty} 2 \cdot 3^{-k}$ to represent the number 1. With these restrictions, the representation is unique.

Next we consider the set of numbers which use only 0's and 2's as coefficients in their ternary expansion:

$$C_0 = \{\sum_{n=1}^{\infty} \frac{j_n}{3^n} : j_n = 0 \text{ or } 2\}.$$

Note for points in C_0, there is a unique ternary expansion even without any restrictions. (Their expansions use only 0's and 2's and automatically obey the above rules for choices of the representation.)

We want to represent C_0 as the intersection of sets. Define

$$S'_n = \{\sum_{k=1}^{\infty} \frac{j_k}{3^k} : j_k = 0 \text{ or } 2 \text{ for } 1 \le k \le n \text{ and } j_k = 0, 1, \text{ or } 2 \text{ for } k > n\}.$$

We want to show that $S'_n = S_n$ by induction on n, where the sets S_n are those given above for the middle-third Cantor set. First, note that S'_1 contains all numbers except those which are $1/3 + y$, where y has an (ternary) expansion in 3^{-k} for $k > 1$; thus, $0 < y < 1/3$. The two endpoints are contained in S'_1 because $1/3 = \sum_{k=2}^{\infty} 2 \cdot 3^{-k}$ and $2/3$ can be represented with $j_k = 0$ for $k \ge 2$. (We do not use expansions whose coefficients end with a 1 followed by repeated 0's or a 1 followed by repeated 2's.) Therefore, $S'_1 = S_1$. Also note that the left ends of the two intervals in S'_1 have ternary representations which end in all 0's, and the right endpoints end in all 2's. Now, assume by induction that $S'_{n-1} = S_{n-1}$, and that all the left endpoints in S'_{n-1} have ternary expansions whose coefficients are $j_k = 0$ for $k \ge n$ and the right endpoints have ternary expansions whose coefficients are $j_k = 2$ for $k \ge n$. Let $J_{j_1,\ldots,j_{n-1}}$ be an interval in $S_{n-1} = S'_{n-1}$. Since its left endpoint has a ternary expansion with $j_k = 0$ for $k \ge n$, $J_{j_1,\ldots,j_{n-1}} \setminus S'_n$ is the set of points

$$\sum_{k=1}^{n-1} \frac{j_k}{3^k} + \frac{1}{3^n} + y,$$

where $0 < y < 3^{-n}$. (Again, the open interval is removed because we do not use expansions which end in 1 followed by repeated 0's or 1 followed by repeated 2's.) Therefore, $S'_n \cap J_{j_1,\ldots,j_{n-1}} = S_n \cap J_{j_1,\ldots,j_{n-1}}$ for any $J_{j_1,\ldots,j_{n-1}}$, and so $S'_n = S_n$. Also note that the left endpoints of the intervals in S'_n can be represented by expansions that end in repeated 0's and the right endpoints by expansions which end in repeated 2's. This completes the proof by induction that $S'_n = S_n$ for all n.

Now, letting C be the middle-third Cantor set,

$$\begin{aligned} C &= \bigcap_{n=0}^{\infty} S_n \\ &= \bigcap_{n=0}^{\infty} S'_n \\ &= C_0. \end{aligned}$$

Thus, the middle-third Cantor set consists of those points whose ternary expansion contains all 0's or 2's and no 1's.

Finally, we note a connection between the points in C which are not endpoints and their ternary expansions. The endpoints of S_n for some n are those points in C which are the endpoint of an open interval K, where $K \subset \mathbb{R} \setminus C$. They are also the points which end in repeated 0's or in repeated 2's. Because there are points in C with ternary expansions which have both $j_k = 0$ for arbitrary large k and $j_{k'} = 2$ for other arbitrary large k', there are points which are not the endpoints of any of the S'_n. These are points which have points arbitrarily close which are not in C, but they are also accumulated on by points of C from both sides. Note that one such point that is not an endpoint is

$$\sum_{j=1}^{\infty} \frac{2}{9^j} = \frac{2}{9}\left(\frac{1}{1-\frac{1}{9}}\right) = \frac{1}{4}.$$

The set of endpoints of the any middle-α Cantor set is countable, as can be seen by listing the endpoints of all the S_n. However, the Cantor set C is uncountable. In fact, any perfect set in the line is uncountable. However, for the middle-third Cantor set, this can be seen even more directly. There is a map $\pi : C \to \mathbb{R}$ using the ternary expansion with only 0's and 2's by $\pi(\sum_{n \geq 1} j_n 3^{-n}) = \sum_{n \geq 1} (j_n/2) 3^{-n}$. This takes a point in the Cantor set expressed in a ternary expansion to a point in $[0,1]$ using the corresponding binary expansion. This map is onto $[0,1]$. Since the unit interval is uncountable, the Cantor set C must be uncountable.

2.4.2 Construction of the Invariant Cantor Set

Remember that $F_\mu(x) = \mu x(1-x)$ is the quadratic map, and $I = [0,1]$. We showed before that if $x \notin I$, then $F_\mu(x)$ goes to minus infinity as n goes to infinity. Therefore, we want to find the x-values such that $F_\mu^n(x) \in I$ for all $n \in \mathbb{N}$. Here, and in the rest of this book, $\mathbb{N}$ is the set of nonnegative integers,

$$\mathbb{N} = \{n \in \mathbb{Z} : n \geq 0\}.$$

The maximum of $F_\mu(x)$ occurs at the critical point $x = 1/2$ where $F_\mu(x)$ takes the value $\mu/4$. We consider values of the parameter μ for which this value is greater than 1, $\mu > 4$, so $F_\mu(I)$ covers I and $F_\mu^{-1}(I) \cap I = F_\mu^{-1}(I)$ is the union of two intervals which we label I_1 and I_2,

$$F_\mu^{-1}(I) \cap I = I_1 \cup I_2.$$

(See Figure 4.1.) Later, we take $\mu > 2 + \sqrt{5}$ which insures that $|F'_\mu(x)| > 1$ for $x \in F_\mu^{-1}(I)$. This bound on the derivative makes some calculations easier.

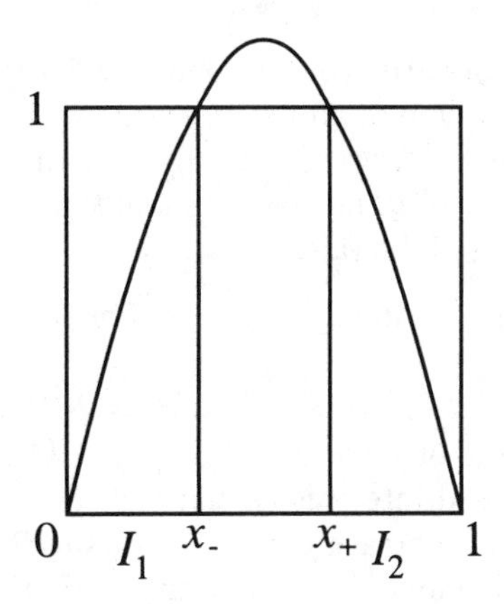

FIGURE 4.1. Intervals for the Quadratic Family

Theorem 4.1. *Assume $\mu > 4$, and let $\Lambda_\mu = \{x : F_\mu^n(x) \in I$ for all $n \geq 0\}$. Then, Λ_μ is a Cantor set.*

We introduce the following notation which is used throughout the proof:

$$I_{i_0,\ldots,i_{n-1}} = \bigcap_{k=0}^{n-1} F_\mu^{-k}(I_{i_k}) = \{x : F_\mu^k(x) \in I_{i_k} \text{ for } 0 \leq k \leq n-1\},$$

where $i_k = 1$ or 2, and

$$S_n = \bigcap_{k=0}^{n} F_\mu^{-k}(I) = \bigcap_{k=0}^{n-1} F_\mu^{-k}(I_1 \cup I_2) = \bigcup_{i_0,i_1,\ldots,i_{n-1}=1,2} I_{i_0,i_1,\ldots,i_{n-1}}.$$

For the proof that Λ_μ is a Cantor set, it is not necessary to label the components of the set S_n so carefully. However, in a later subsection, we use the labeling to prove that the map F_μ restricted to Λ_μ is "topologically conjugate" to a map on a space of symbols. (A topological conjugacy is a homeomorphism which takes orbits of one map to orbits of another map.) We introduce the notation here to avoid proving twice the lemmas used in both proofs. The conjugacy of F_μ restricted to Λ_μ with a map on a space of symbols is interesting because it allows us to prove that the periodic points of F_μ are dense in Λ_μ and that F_μ has a dense orbit in Λ_μ.

We start the proof with the following lemma.

Lemma 4.2. *Assume $\mu > 4$. For all $n \in \mathbb{N}$, the following statements are true.*

(a) For any choice of the labeling with $i_0, \ldots, i_{n-1} \in \{1,2\}$, $I_{i_0,\ldots,i_{n-2}} \cap S_n = I_{i_0,\ldots,i_{n-2},1} \cup I_{i_0,\ldots,i_{n-2},2}$ is the union of two nonempty disjoint closed intervals (which are subsets of $I_{i_0,\ldots,i_{n-2}}$).

(b) For two distinct choices of the labeling $(i_0, \ldots, i_{n-1}) \neq (i'_0, \ldots, i'_{n-1})$, $I_{i_0,\ldots,i_{n-1}} \cap I_{i'_0,\ldots,i'_{n-1}} = \emptyset$, so S_n is the union of 2^n disjoint intervals.

(c) The map F_μ takes the component $I_{i_0,\ldots,i_{n-1}}$ of S_n homeomorphically onto the component $I_{i_1,\ldots,i_{n-1}}$ of S_{n-1}.

REMARK 4.4. Notice from the definition of $I_{i_0,\ldots,i_{n-1}}$,

$$I_{i_0,\ldots,i_{n-1}} = \bigcap_{k=0}^{n-1} F_\mu^{-k}(I_{i_k}) = \{x : F_\mu^k(x) \in I_{i_k} \text{ for } 0 \leq k \leq n-1\},$$

so it is characterized by the forward orbit of points. In Exercise 2.18, the reader is asked to determine the order on the line of all the intervals with three labels, I_{i_0,i_1,i_2}.

In terms of images of these intervals, part (c) says that $F_\mu(I_{i_0,\dots,i_{n-1}}) = I_{i_1,\dots,i_{n-1}}$, where the first label is dropped. The reader should check why this is true from the above characterization of these intervals.

PROOF. We prove the lemma by induction on n. For $n = 0$, $S_0 = I = [0, 1]$ and there is nothing to check.

For $n = 1$, let $G_1 = \{x \in I : F_\mu(x) > 1\}$ and $S_1 = F_\mu^{-1}(I) \cap I = I \setminus G_1$. Then, G_1 is an open interval in the middle of I because $F_\mu(1/2) > 1$, and $S_1 = I \cap S_1$ is the union of two nonempty disjoint closed intervals as claimed, $I_1 \cup I_2$ with $I_{i_0} \subset I$ for $i_0 = 1, 2$. The map F_μ is monotonically increasing on $[0, 1/2]$, so F_μ maps I_1 homeomorphically onto $I = S_0$. Similarly, F_μ is monotonically decreasing on $[1/2, 1]$, so F_μ also maps I_2 homeomorphically onto $I = S_0$. This verifies the induction hypothesis for $n = 1$.

Although we do not need this step, take $n = 2$. The map F_μ is monotone on each of the separate intervals I_1 and I_2. For $i_0 = 1$ or 2, $F_\mu(I_{i_0}) \supset I_1 \cup I_2$, so $S_2 \cap I_{i_0} = I_{i_0} \cap F_\mu^{-1}(S_1)$ is the union of two intervals, $I_{i_0,1} \cup I_{i_0,2}$, where $F_\mu(I_{i_0,k}) = I_k$, and F_μ is a homeomorphism from I_{i_0,i_1} onto I_{i_1}. Since there are four intervals I_{i_0,i_1}, this verifies all three conditions of the induction hypothesis.

Now, assume the lemma is true for n and we verify it for $n+1$. Let $I_{i_0,\dots,i_{n-1}}$ be a component of S_n. Then, $F_\mu(I_{i_0,\dots,i_{n-1}}) = I_{i_1,\dots,i_{n-1}}$ is a component of S_{n-1}, and $I_{i_1,\dots,i_{n-1}} \supset I_{i_1,\dots,i_{n-1}} \cap S_n = I_{i_1,\dots,i_{n-1},1} \cup I_{i_1,\dots,i_{n-1},2}$. Therefore,

$$\begin{aligned} I_{i_0,\dots,i_{n-1}} \cap S_{n+1} &= F_\mu^{-1}(S_n) \cap I_{i_0,\dots,i_{n-1}} \\ &= F_\mu^{-1}(S_n \cap I_{i_1,\dots,i_{n-1}}) \cap I_{i_0} \\ &= [F_\mu^{-1}(I_{i_1,\dots,i_{n-1},1}) \cup F_\mu^{-1}(I_{i_1,\dots,i_{n-1},2})] \cap I_{i_0} \end{aligned}$$

is the union of two nonempty disjoint closed intervals, giving condition (a). Since there are 2^n choices of the index for $I_{i_0,\dots,i_{n-1}}$, S_{n+1} is the union of $2(2^n) = 2^{n+1}$ intervals, giving condition (b). The map F_μ is monotone on $I_{i_0,\dots,i_{n-1},j}$, so it maps homeomorphically onto $I_{i_1,\dots,i_{n-1},j}$, giving conditions (c). This completes the verification of the induction step, and so verifies the lemma. □

With this lemma, some of the properties of a Cantor set can be verified. Since each of the sets S_n is closed, $\Lambda = \bigcap_{n=0}^{\infty} S_n$ is closed. If the lengths of the intervals in S_n go to zero, then Λ is perfect as before. If the lengths of the intervals do not go to zero, then the intersection contains an interval and it is perfect. In any case, Λ is perfect. However, to prove that Λ is nowhere dense, we need to show that the maximal length of an interval in S_n goes to zero. This fact is easier to prove if $|F'_\mu(x)| > 1$ for all $x \in I_1 \cup I_2$. This condition on the derivative is true if and only if $\mu > 2 + \sqrt{5}$. Therefore, in the rest of this section, we assume $\mu > 2 + \sqrt{5}$ and return to the case for $4 < \mu \le 2 + \sqrt{5}$ in the next subsection.

Lemma 4.3. *The absolute value of the derivative is greater than 1 for all points in $I_1 \cup I_2 = S_1$, $|F'_\mu(x)| > 1$ for all $x \in S_1$, if and only if $\mu > 2 + \sqrt{5}$.*

PROOF. The derivative of F_μ is given by $F'_\mu(x) = \mu - 2\mu x$. Also, $F''_\mu(x) = -2\mu < 0$, so the smallest value of $|F'_\mu(x)|$ on I_j occurs where $F_\mu(x) = 1$. Solving $1 = F_\mu(x) = \mu x - \mu x^2$, we get $x_\pm = [\mu \pm (\mu^2 - 4\mu)^{1/2}]/(2\mu)$. But for these points,

$$|F'_\mu(x_\pm)| = \left|\mu - 2\mu\left(\frac{\mu \pm (\mu^2 - 4\mu)^{1/2}}{2\mu}\right)\right| = |\mp (\mu^2 - 4\mu)^{1/2}| = (\mu^2 - 4\mu)^{1/2}.$$

Therefore, we need $|F'_\mu(x_\pm)| = (\mu^2 - 4\mu)^{1/2} > 1$, or $\mu^2 - 4\mu - 1 > 0$. The left side equals zero when $\mu = 2 \pm \sqrt{5}$, and it can be checked to be greater than zero for $\mu > 2 + \sqrt{5}$. (The second value of μ arose because we squared the equation when we solved it.) This proves the lemma. □

The next lemma proves a bound on the maximal length of a component of S_n in terms of a bound on the derivative. If $\mu > 2 + \sqrt{5}$, then this bound goes to zero as n goes to infinity. For an interval K, let $L(K)$ be its length.

Lemma 4.4. *Let $\lambda = \lambda_\mu = \inf\{|F'_\mu(x)| : x \in I_1 \cup I_2\}$. Then, the length of any component $I_{i_0,\ldots,i_{n-1}}$ is bounded by λ^{-n}, $L(I_{i_0,\ldots,i_{n-1}}) \le \lambda^{-n}$ for all possible choices of the labeling.*

PROOF. We prove the lemma by induction. Take $n = 1$. Let $I_{i_0} = [a, b]$. Since $F_\mu(I_{i_0}) = [0, 1]$, $\{F_\mu(a), F_\mu(b)\} = \{0, 1\}$, i.e., endpoints go to endpoints. By the Mean Value Theorem, there is some $c \in [a, b]$ for which $F_\mu(b) - F_\mu(a) = F'_\mu(c)(b - a)$. Then, $1 = |F_\mu(b) - F_\mu(a)| = |F'_\mu(c)| \cdot |b - a| \ge \lambda L(I_{i_0})$. Therefore, $L(I_{i_0}) \le \lambda^{-1}$. This proves the induction step for $n = 1$.

Assume the result is true for $n - 1$. Take a component $I_{i_0,\ldots,i_{n-1}}$ of S_n. Then, the image, $F_\mu(I_{i_0,\ldots,i_{n-1}}) = I_{i_1,\ldots,i_{n-1}}$, is a component of S_{n-1}, and by induction, $L(I_{i_1,\ldots,i_{n-1}}) \le \lambda^{-(n-1)}$. As above by the Mean Value Theorem, there is a $c \in I_{i_0,\ldots,i_{n-1}} = [a, b]$ with $F_\mu(b) - F_\mu(a) = F'_\mu(c)(b - a)$, so

$$
\begin{aligned}
L(I_{i_1,\ldots,i_{n-1}}) &= |F_\mu(b) - F_\mu(a)| \\
&= |F'_\mu(c)(b - a)| \\
&\ge \lambda |b - a| \\
&= \lambda L(I_{i_0,\ldots,i_{n-1}}),
\end{aligned}
$$

and $L(I_{i_0,\ldots,i_{n-1}}) \le \lambda^{-1} L(I_{i_1,\ldots,i_{n-1}}) \le \lambda^{-n}$. This completes the proof of the induction step and the lemma. □

PROOF OF THEOREM 4.1. Assume that $\mu > 2 + \sqrt{5}$. We mentioned above that the S_n are closed, so Λ_μ is closed. We have shown that the length of the components of S_n are shorter than λ^{-n}, which goes to zero with n. The proof that Λ_μ is perfect and nowhere dense is the same as for the Middle-α-Cantor set. Take $p \in \Lambda_\mu$. For $j \in \mathbb{N}$, take n such that $\lambda^{-n} < 2^{-j}$. Then, $p \in I_{i_0,\ldots,i_{n-1}}$ for some choice of the component of S_n. Then, $I_{i_0,\ldots,i_{n-1}} \cap S_{n+1} = I_{i_0,\ldots,i_{n-1},1} \cup I_{i_0,\ldots,i_{n-1},2}$ is the union of two intervals. Take $y_j \in I_{i_0,\ldots,i_{n-1}} \setminus S_{n+1}$ in the gap, and q_j an endpoint of $I_{i_0,\ldots,i_{n-1},i_n}$ where $I_{i_0,\ldots,i_{n-1},i_n}$ is chosen so that $p \notin I_{i_0,\ldots,i_{n-1},i_n}$. Then, y_j is not in S_{n+1} and so is not in Λ_μ. The y_j converge to p, so this shows that Λ_μ is nowhere dense. Also, $q_j \ne p$ and $q_j \in \Lambda_\mu$ since it is an endpoint. The q_j converge to p proving that Λ_μ is perfect. This completes the verification that Λ_μ is a Cantor set for $\mu > 2 + \sqrt{5}$. □

2.4.3 The Invariant Cantor Set for $\mu > 4$

The proof of Theorem 4.1 for all μ with $4 < \mu$ goes back to the work of Fatou and Julia on complex functions. (This is the theorem that if all the critical points of a polynomial have orbits which go to infinity, then the Julia set is totally disconnected.) See Blanchard (1984) or Carleson and Gamelin (1993). The first proof using strictly real variables is found in Henry (1973). (This proof does not prove that $|F^n_\mu(x)| \ge C\lambda^n$ for $x \in \Lambda_\mu$.) The proof given below is mainly given in terms of real variables, but we use Schwarz Lemma of complex variables to prove the key estimate.

For values of μ with $4 < \mu < 2+\sqrt{5}$, there are points $x \in I \cap F_\mu^{-1}(I)$ with $|F'_\mu(x)| < 1$ and other points with $|F'_\mu(x)| > 1$. Thus, in terms of the usual length on the line, we do not have a $\lambda > 1$ such that $L(I_{i_0,\dots,i_n}) \le \lambda^{-1} L(I_{i_1,\dots,i_n})$ for all the subintervals $I_{i_0,\dots,i_n}$.

There are several ways around this difficulty. One method is to look at higher iterates of F_μ and prove there is a $C \ge 0$ and $\lambda > 1$ such that $|(F_\mu^k)'(x)| \ge C\lambda^k$ for all $k \ge 0$ and $x \in I_1 \cup I_2$. By taking an $m \ge 1$ with $C\lambda^m = \lambda' > 1$, we have that F_μ^m is an expansion on $I_1 \cup I_2$, and $L(I_{i_0,\dots,i_n}) \le (\lambda')^{-1} L(I_{i_m,\dots,i_n})$. This inequality can then be used to prove that Λ_μ is nowhere dense. Guckenheimer (1979) and van Strien (1981) have a proof in this spirit using the fact that F_μ has negative "Schwarzian derivative." Also see Misiurewicz (1981) and de Melo and van Strien (1993). Newhouse (1979) has a proof for two-dimensional maps which can be adapted to prove this result.

Rather than give the details of this proof, we give an alternative proof which proceeds by defining a new length on the interval $[0,1]$. This idea is essentially present in the complex variable proof mentioned above.

Definition. Assume that $\rho(x) > 0$ is a continuous density function on an interval K. If $x, y \in K$, then define *the ρ-distance from x to y* by

$$d_\rho(x,y) = \Big|\int_x^y \rho(t)\,dt\Big|.$$

It is easy to check that this is a metric on K: $d_\rho(x,y) > 0$ for $x \ne y$, $d_\rho(x,x) = 0$, $d_\rho(x,y) = d_\rho(y,x)$, and $d_\rho(x,z) \le d_\rho(x,y) + d_\rho(y,z)$. Let $L_\rho(J)$ be the length of an interval J in terms of the distance d_ρ. We think of $\rho(x)$ defining a length or norm of a vector (infinitesimal displacement) at x by $|v|_{\rho,x} = |v|\rho(x)$, where $|v|$ is the usual length of a vector.

REMARK 4.5. Assume K is an interval and $\rho : K \to \mathbb{R}^+$ is a positive density function for which there exist positive constants C_1 and C_2 with $C_1 \le \rho(x) \le C_2$ for all x in K. It can be seen easily by applying estimates to the integral that

$$C_1|x-y| \le d_\rho(x,y) \le C_2|x-y|$$

for any two points $x, y \in K$. Therefore, if we can show that ρ-lengths of a nested set of intervals go to zero, then the usual lengths also go to zero.

The following lemma indicates the property which we want $\rho(x)$ to have.

Lemma 4.5. *Let K be an interval and $\rho : K \to \mathbb{R}^+$ be a positive density function. Let $f : \mathbb{R} \to \mathbb{R}$ be a C^1 function. Assume that there is a $\lambda > 0$ such that*

$$\frac{\rho(f(x))|f'(x)|}{\rho(x)} = \frac{|f'(x)v|_{\rho,f(x)}}{|v|_{\rho,x}} \ge \lambda$$

for $x \in J$, where J is a subinterval of K with $f(J) \subset K$. Then, for this subinterval J, $L_\rho(f(J)) \ge \lambda L_\rho(J)$.

REMARK 4.6. Since the derivative of f is always nonzero on the interval J, f is monotone on J.

PROOF. The following estimate proves the lemma:

$$\begin{aligned} L_\rho(f(J)) &= \left|\int_{t\in f(J)} \rho(t)\,dt\right| \\ &= \left|\int_{s\in J} \rho(f(s))f'(s)\,ds\right| \qquad \text{(where } t = f(s)\text{)} \\ &\geq \int_{s\in J} \left|\frac{\rho(f(s))f'(s)}{\rho(s)}\right| \rho(s)\,ds \\ &\geq \int_{s\in J} \lambda\rho(s)\,ds \\ &= \lambda L_\rho(J). \end{aligned}$$

□

The following proposition relates the condition given in terms of varying norm $|\ |_{\rho,x}$ in the last lemma to a condition for the standard norm.

Proposition 4.6. *Let $f : \mathbb{R} \to \mathbb{R}$ be a C^1 function. Let K be an interval and $\rho : K \to \mathbb{R}^+$ be a positive density function for which there exist positive constants C_1 and C_2 with $C_1 \leq \rho(x) \leq C_2$ for all x in K. Assume that there is a $\lambda > 0$ such that*

$$|f'(x)v|_{\rho,f(x)} \geq \lambda|v|_{\rho,x}$$

provided $x, f(x) \in K$. Then, taking the positive constant $C = C_1/C_2 \leq 1$, the derivative of the n-th iterate of f satisfies

$$|(f^n)'(x)v| \geq C\lambda^n|v|$$

for all $n \geq 0$ and all $x \in J$ in terms of the usual absolute value.

REMARK 4.7. Thus, if f is immediately expanding in terms of some different norm $|\ |_{\rho,x}$, it is eventually expanding in terms of the Euclidean norm.

PROOF. The proof follows directly because the two norms are uniformly equivalent:

$$\begin{aligned} |(f^n)'(x)v| &\geq C_2^{-1}|(f^n)'(x)v|_{\rho,f^n(x)} \\ &\geq C_2^{-1}\lambda^n|v|_{\rho,x} \\ &\geq C_2^{-1}C_1\lambda^n|v|. \end{aligned}$$

□

To use Lemma 4.5, we are free to define the density function $\rho(x)$. We want a choice that satisfies the assumptions of the above lemma. The metric ρ we use is related to the Schwarz Lemma in Complex Variables.

Schwarz Lemma 4.7. *Let $D = \{z \in \mathbb{C} : |z| < 1\}$ be the open disk in the complex plane. Assume $f : D \to D$ is complex analytic with $f(0) = 0$ and $f(D) \neq D$ (not onto). Then, $|f'(0)| < 1$.*

For a proof, see Theorem 15.1.1 in Hille (1962).

Corollary 4.8. *Assume $f : D \to D$ is complex analytic and $f(D) \neq D$ (not onto). Let $g(z) = 1 - |z|^2$ and*

$$\rho(z) = \frac{1}{g(z)}.$$

Then,

$$\frac{\rho(f(z))|f'(z)|}{\rho(z)} = \frac{|f'(z)v|_{\rho,f(z)}}{|v|_{\rho,z}} < 1$$

for all $z \in D$.

REMARK 4.8. This norm $|\cdot|_{\rho,z}$ is called the Poincaré norm and it induces a metric (distance between points) called the Poincaré metric. The unit disk with the Poincaré metric is an example of a non-Euclidean metric on a surface with negative curvature.

PROOF. This theorem is also proved in Hille (1962), Theorem 15.1.3, but is stated in terms of the distance between points and not the length of vectors.

We also give a sketch of the proof using the Schwarz Lemma. Fix $z_0 \in D$ and let $w_0 = f(z_0)$. For $j = 1, 2$, there are fractional linear transformations T_j preserving D,

$$T_j(z) = \frac{a_j - z}{1 - \bar{a}_j z},$$

with $|a_j| < 1$ such that $T_1(0) = z_0$ and $T_2(w_0) = 0$. Thus, $T_2 \circ f \circ T_1(0) = 0$. The fractional linear transformations preserve the length of vectors in terms of $|\cdot|_{\rho,z}$, so

$$\begin{aligned} 1 &= \frac{|T_2'(w_0)|_{\rho,0}}{|1|_{\rho,w_0}} \\ &= \frac{|T_1'(0)|_{\rho,z_0}}{|1|_{\rho,0}}. \end{aligned}$$

Thus, by the Schwarz Lemma,

$$\begin{aligned} 1 > |(T_2 \circ f \circ T_1)'(0)| &= |T_2(w_0)| \cdot |f'(z_0)| \cdot |T_1(0)| \\ &= \frac{|T_2'(w_0)|_{\rho,0}}{\rho(0)} \cdot \frac{|f'(z_0)|_{\rho,w_0}}{\rho(w_0)} \cdot \frac{|T_1'(0)|_{\rho,z_0}}{\rho(z_0)} \\ &= \frac{|T_2'(w_0)|_{\rho,0}}{|1|_{\rho,0}} \cdot \frac{|f'(z_0)|_{\rho,w_0}}{|1|_{\rho,w_0}} \cdot \frac{|T_1'(0)|_{\rho,z_0}}{|1|_{\rho,z_0}} \\ &= \frac{|T_2'(w_0)|_{\rho,0}}{|1|_{\rho,w_0}} \cdot \frac{|f'(z_0)|_{\rho,w_0}}{|1|_{\rho,z_0}} \cdot \frac{|T_1'(0)|_{\rho,z_0}}{|1|_{\rho,0}} \\ &= \frac{|f'(z_0)|_{\rho,w_0}}{|1|_{\rho,z_0}}. \end{aligned}$$

□

In Corollary 4.8, the absolute value of the derivative is less than 1, while in the following lemma, it is greater than 1. The reason for this difference is that in Corollary 4.8, f maps D into D, while $F_\mu(I)$ covers I. In Corollary 4.8, the unit disk is centered at the origin. For the map F_μ, the corresponding interval we use is centered at $1/2$, so a change of variables (given in the proof) modifies the Poincaré norm to give the one stated in the following lemma.

Lemma 4.9. *Let $\rho(x) = [x(1-x)]^{-1}$ on $(0,1)$. (This density is singular at $x = 0, 1$.) Then, for $\mu > 4$ and $x, F_\mu(x) \in (0,1)$,*

$$\frac{\rho(F_\mu(x))|F'_\mu(x)|}{\rho(x)} = \frac{|F'_\mu(x)|_{\rho,F_\mu(x)}}{|1|_{\rho,x}} > 1.$$

PROOF. We consider F_μ as a map of a complex variable, $F_\mu(z) = \mu z(1-z)$. Let $\mathcal{D}_{1/2}$ be the disk of radius $1/2$ centered at the point $1/2$. Notice that F_μ takes circles of radius r about $1/2$ onto circles of radius μr^2 about $\mu/4$: $F_\mu(1/2 + re^{i\theta}) = \mu/4 - \mu r^2 e^{i2\theta}$. In particular, F_μ takes the circle of radius $1/2$ onto the circle of radius $\mu/4$ about $\mu/4$. Thus, for $\mu > 4$, F_μ takes the circle of radius $1/2$ around the outside of the disk $\mathcal{D}_{1/2}$. Also, $z = 1/2$ is the critical point for F_μ (the point where the derivative is zero), and for $\mu > 4$, F_μ takes this critical point outside $\mathcal{D}_{1/2}$: $F_{1/2}(1/2) = \mu/4$ is real and greater than 1. Therefore, $F_\mu(\mathcal{D}_{1/2})$ covers $\mathcal{D}_{1/2}$ twice, and F_μ^{-1} has two branches of the inverse taking $\mathcal{D}_{1/2}$ into itself and each being one to one. (The branches correspond to the inverse of F_μ on the two intervals I_1 and I_2 when F_μ is restricted to the real variable x.) Because there are two branches of the inverse, neither is onto all of $\mathcal{D}_{1/2}$. By Corollary 4.8, each of these inverses is a contraction in terms of the Poincaré metric on $\mathcal{D}_{1/2}$, so F_μ is an expansion for points z with $z, F_\mu(z) \in \mathcal{D}_{1/2}$.

To complete the proof, we need to determine what the Poincaré norm is on this disk, which is not the unit disk centered at the origin. The map $h(z) = 2z - 1$ takes $\mathcal{D}_{1/2}$ onto the unit disk D centered at the origin. If $\rho(\zeta) = (1 - \zeta\bar{\zeta})^{-1}$ is the usual Poincaré norm on D, then $\rho \circ h(z) = [4z\bar{z} - 2z - 2\bar{z}]^{-1}$. For real z, this gives a constant multiple of the norm stated in the lemma. However, the correct way to use the map h to "pull back" the length of vectors is not to just look at this composition but to define $|v|_{*,z} = |h'(z)v|_{\rho,h(z)}$. Since $h'(z) \equiv 2$, this induces the norm $|v|_{*,z} = |v|[2z\bar{z} - z - \bar{z}]^{-1}$, which for real z gives $|v|_{*,x} = |v|2^{-1}[x(1-x)]^{-1}$. If we take two times this norm everywhere, we get the norm stated in the lemma and it will also be expanded by F'_μ.
□

The norm in the previous lemma is singular at 0 and 1. To get rid of this difficulty, we modify it slightly to make the norm singular at the points $-\epsilon$ and $1 + \epsilon$ which are outside of the interval $[0, 1]$. We accomplish this change by mapping the disk of radius $1/2 + \epsilon$ about $1/2$ onto the unit disk rather than the disk of radius $1/2$.

Proposition 4.10. *Assume $4 < \mu$. Let $\rho(x) = [(x+\epsilon)(1+\epsilon-x)]^{-1}$ on $[0,1]$. Then, for $0 < \epsilon < \mu/4 - 1$ and $x, F_\mu(x) \in [0,1]$,*

$$\frac{|F'_\mu(x)|_{\rho,F_\mu(x)}}{|1|_{\rho,x}} = \frac{\rho(F_\mu(x))|F'_\mu(x)|}{\rho(x)} > 1.$$

REMARK 4.9. Notice that this density is nonsingular on $[0, 1]$.

REMARK 4.10. This Proposition can be proved by a direct calculation considering only real x rather than proving it as a corollary of the Schwarz Lemma. The difficulty is that this argument is somewhat involved and needs to consider various intervals to prove the inequality. Exercise 2.14 asks such a direct verification to be carried out for a different density function for which the calculation is simpler and for which there is no easy way to use a complex variable argument.

REMARK 4.11. The first iterate of F_μ stretches lengths in terms of this new metric, so that we are able to take $C = 1$ and $\lambda > 1$ in the inequality $|(F_\mu^k)'(x)|_{\rho,F_\mu^k(x)} \geq C\lambda^k$ which defines a hyperbolic set.

PROOF. We want to modify the proof for Lemma 4.9 by taking $h(z)$ so that it takes the disk of radius $1+\epsilon$ centered at $1/2$ onto the standard unit disk D:

$$h(z) = 2(1+2\epsilon)^{-1}(z - 1/2).$$

Using this h, a direct calculation shows that the induced norm is as follows:

$$|v|_{*,z} = \frac{(1+2\epsilon)2^{-1}}{(x+\epsilon)(1+\epsilon-x)}.$$

Again, this norm is a constant scalar multiple of the one stated in the Proposition.

To prove the proposition, we only need to show that (i) F_μ takes the critical point $z = 1/2$ outside the disk of radius $1/2+\epsilon$ centered at $1/2$, $\mathcal{D}_{1/2+\epsilon}$, and (ii) $F_\mu(\mathcal{D}_{1/2+\epsilon})$ covers $\mathcal{D}_{1/2+\epsilon}$ twice. In connection with the first condition, we saw in the proof of Lemma 4.9 that F_μ mapped the critical point $x = 1/2$ to the point $\mu/4$, so we need $\epsilon > 0$ such that $\mu/4 > 1+\epsilon$ or $\epsilon < \mu/4 - 1$, which we assumed. For the second condition, the proof of Lemma 4.9 showed that F_μ mapped circles of radius r centered at $x = 1/2$ onto circles of radius μr^2 centered at $\mu/4$. Thus, we need $\mu(1/2+\epsilon)^2 > \mu/4 + \epsilon$, which is always true for $\epsilon > 0$ and $\mu > 4$. Choosing $0 < \epsilon < \mu/4 - 1$, we get the result. □

For $\mu > 4$, this last proposition proves that the ρ-length of the intervals $I_{i_0,\dots,i_{n-1}}$ are bounded by λ^{-n}. Applying Remark 4.5, we get the Euclidean length is bounded by $C\lambda^{-n}$ for some $C > 0$. This proves Theorem 4.1 for all these values of the parameter.

2.5 Symbolic Dynamics for the Quadratic Map

In this section, we show there is a way to represent the dynamics of F_μ on Λ by a map on a symbol space made up by points which are sequences of 1's and 2's. The map on the symbol space is called the symbolic representation of the map and is said to give the *symbolic dynamics* for the map.

Definition. Let $\mathbb{N} = \{0, 1, 2, \dots\}$ be the nonnegative integers as always. For p an integer with $2 \leq p$, let $\{1, 2, \dots, p\}^{\mathbb{N}}$ be the space of functions from $\mathbb{N}$ into the set $\{1, 2 \dots, p\}$. We also write this space as $p^{\mathbb{N}}$ or Σ_p^+ to shorten the notation. We define a metric on Σ_p^+ by

$$d(\mathbf{s}, \mathbf{t}) = \sum_{k=0}^{\infty} \frac{\delta(s_k, t_k)}{3^k}$$

for $\mathbf{s} = (s_0, s_1, \dots)$ and $\mathbf{t} = (t_0, t_1, \dots)$, where

$$\delta(i, j) = \begin{cases} 0 & \text{if } i = j \\ 1 & \text{if } i \neq j. \end{cases}$$

Exercise 2.15 is designed to clarify the topology induced by this metric. (Many authors use 2^{-k} in the definition of the metric, but we use 3^{-k} which makes what are called cylinder sets into balls in terms of this metric. See Exercises 2.15–16.) Finally, we define a *shift map* on Σ_p^+ by $\sigma(\mathbf{s}) = \mathbf{t}$ where $t_k = s_{k+1}$, i.e., $\sigma(s_0, s_1, \dots) = (s_1, s_2, \dots)$. The reader can check that σ is continuous with the above metric. The space Σ_p^+ with the shift map σ, (Σ_p^+, σ), is called the *symbol space on p symbols* or the *full (one-sided) p-shift space.* (Later, when we are studying diffeomorphisms in dimensions greater than one, we discuss the two-sided p-shift where $\mathbf{s} : \mathbb{Z} \to \{1, 2 \dots, p\}$.)

Next, we define the map which takes the points in Λ_μ to points in Σ_2^+.

Definition. Define $h : \Lambda_\mu \to \Sigma_2^+$ by $h(x) = \mathbf{j} = (j_0, j_1, \ldots)$, where $F_\mu^k(x) \in I_{j_k}$. Thus, $x \in F_\mu^{-k}(I_{j_k})$ for all k, so for any n, $x \in \bigcap_{k=0}^n F_\mu^{-k}(I_{j_k}) = I_{j_0, j_1, \ldots, j_n}$, which is a component of S_{n+1}. The map h is called the *itinerary map.*

Theorem 5.1. *Let $\mu > 4$ for the quadratic map F_μ defined above. Then, the itinerary map $h : \Lambda_\mu \to \Sigma_2^+$ defined above is a homeomorphism from Λ_μ to Σ_2^+ such that $h \circ F_\mu = \sigma \circ h$.*

PROOF. First we check the condition that $\sigma \circ h = h \circ F_\mu$. Let $x \in \Lambda_\mu$, $\mathbf{s} = h(x)$, and $\mathbf{t} = h(F_\mu(x))$. Then, $F_\mu^k(F_\mu(x)) \in I_{t_k}$ and $F_\mu^k(F_\mu(x)) = F_\mu^{k+1}(x) \in I_{s_{k+1}}$. So $t_k = s_{k+1}$, $\mathbf{t} = \sigma(\mathbf{s})$, and $h(F_\mu(x)) = \sigma(h(x))$.

Next we check that h is onto. Let $\mathbf{s} \in \Sigma_2^+$. As in the earlier theorem, the intersections $I_{s_0, \ldots, s_n} = \bigcap_{k=0}^n F_\mu^{-k}(I_{s_k})$ are nonempty intervals and are nested as n increases. Therefore, there is a $x_0 \in \bigcap_{n=0}^\infty I_{s_0, \ldots, s_n} = \bigcap_{k=0}^\infty F_\mu^{-k}(I_{s_k})$. If $x \in I_{s_0, \ldots, s_n}$, then for $0 \le k \le n$, $x \in F_\mu^{-k}(I_{s_k})$, so $F_\mu^k(x) \in I_{s_k}$. Therefore, $F_\mu^k(x_0) \in I_{s_k}$ for $0 \le k < \infty$ and $h(x_0) = \mathbf{s}$. This proves that h is onto.

We give two proofs of the fact that h is one to one to illustrate different ideas. In both these proofs, we assume $\mu > 2 + \sqrt{5}$ for simplicity. Assume that $\mathbf{s} = h(x) = h(y)$. By the above argument $F_\mu^k(x), F_\mu^k(y) \in I_{s_k}$ for all $0 \le k$, so $x, y \in I_{s_0, \ldots, s_n}$ for all n. Lemma 4.4 proved a bound on the length of $I_{s_0, \ldots, s_n}$, $L(I_{s_0, \ldots, s_n}) \le \lambda^{-(n+1)} L([0,1])$, so $|x - y| \le \lambda^{-(n+1)}$ for all n, and $x = y$.

As a second proof that h is one to one, we use the ideas of expansiveness. If $z \in I_1 \cup I_2$, then $|F_\mu'(z)| \ge \lambda$. Therefore, if $z_1, z_2 \in I_j$, then for some $z' \in I_j$, $|F_\mu(z_1) - F_\mu(z_2)| = |F_\mu'(z')| \cdot |z_1 - z_2| \ge \lambda \cdot |z_1 - z_2|$. If $\mathbf{s} = h(x) = h(y)$, then $F_\mu^k(x), F_\mu^k(y) \in I_{s_k}$ for all $0 \le k$, so $|F_\mu^n(x) - F_\mu^n(y)| \ge \lambda |F_\mu^{n-1}(x) - F_\mu^{n-1}(y)| \ge \lambda^n |x - y|$. If $x \ne y$, then for some n, $\lambda^n |x - y| > L(I_{s_n})$ and $F_\mu^n(x)$ and $F_\mu^n(y)$ cannot be in the same interval. This contradiction proves that h is one to one.

Last, we need to check that h is continuous. Take $x \in \Lambda_\mu$ and $\mathbf{s} = h(x)$. Let $\epsilon > 0$. Pick an n such that $3^{-n} < \epsilon$. For $\delta > 0$ small enough, if $y \in \Lambda_\mu$ and $|y - x| < \delta$, then $y \in I_{s_0, \ldots, s_n}$. For such a $y \in \Lambda_\mu$ with $|y - x| < \delta$, let $\mathbf{t} = h(y)$. Then, $t_k = s_k$ for $0 \le k \le n$. Therefore, $d(h(x), h(y)) \le \sum_{k=n+1}^\infty 3^{-k} = 3^{-(n+1)}[1 - (1/3)]^{-1} = 3^{-n} 2^{-1} < \epsilon$. This proves the continuity of h.

Because the sets Λ_μ and Σ_2^+ are compact and h is a one to one continuous map, it follows that h is a homeomorphism. This completes the proof of the theorem. □

REMARK 5.1. Given two maps $f : X \to X$ and $g : Y \to Y$, a homeomorphism $h : X \to Y$ such that $h \circ f = g \circ h$ is called a *topological conjugacy.* See the next section for further discussion.

The same proof given above (with only minor changes) proves the following theorem about an arbitrary function and p intervals.

Theorem 5.2. *Let $f : \mathbb{R} \to \mathbb{R}$ be a C^1 function and $I_1, \ldots, I_p$ be p disjoint closed bounded intervals with $p \ge 2$. Let $\mathcal{I} = \bigcup_{j=1}^p I_j$. Assume that $f(I_j) \supset \mathcal{I}$ for $1 \le j \le p$. Also assume there is a $\lambda > 1$ such that $|f'(x)| \ge \lambda$ for $x \in \mathcal{I} \cap f^{-1}(\mathcal{I})$, Let $\Lambda = \bigcap_{k=0}^\infty f^{-k}(\mathcal{I})$. Then, Λ is a Cantor set. Define $h : \Lambda \to \Sigma_p^+$ by $h(x) = \mathbf{s}$, where $f^k(x) \in I_{s_k}$. Then, h is a topological conjugacy from f on Λ to σ on Σ_p^+.*

We now use the above result on the conjugacy to prove facts about the periodic points of the map on the line. Before we state the theorem, we give the definition of one more property which is preserved by the conjugacy.

Definition. A map $f : X \to X$ is *(topologically) transitive* on an invariant set Y provided the forward orbit of some some point p is dense in Y. The Birkhoff Transitivity

Theorem proves that a map f is transitive on Y if and only if, given any two open sets U and V in Y, there is a positive integer n such that $f^n(U) \cap V \neq \emptyset$. (See Theorem VIII.2.1.) This property indicates that f mixes up the points of Y and the set is one piece dynamically. A stronger condition is as follows: A map $f : X \to X$ is called *topologically mixing* on an invariant set Y provided for any pair of open sets U and V there is a positive integer n_0 such that $f^n(U) \cap V \neq \emptyset$ for all $n \geq n_0$. Thus, if f is topologically mixing, the iterates $f^n(U)$ intersect V for all sufficiently large values of n.

Theorem 5.3. *Let $f : \mathbb{R} \to \mathbb{R}$ be a C^1 function and $I_1, \ldots, I_p$ be p disjoint closed bounded intervals with $p \geq 2$. Let $\mathcal{I} = \bigcup_{j=1}^{p} I_j$. Assume that $f(I_j) \supset \mathcal{I}$ for $1 \leq j \leq p$. Also assume there is a $\lambda > 1$ such that $|f'(x)| \geq \lambda$ for $x \in \mathcal{I} \cap f^{-1}(\mathcal{I})$. Let $\Lambda = \bigcap_{k=0}^{\infty} f^{-k}(\mathcal{I})$. Then,*

(a) *the cardinality of the number of periodic points is given by $\#(\text{Fix}(f^n|\Lambda) = p^n$,*
(b) *$\text{Per}(f|\Lambda)$ are dense in Λ, and*
(c) *f is transitive on Λ. In fact, f is topologically mixing on Λ.*

REMARK 5.2. We leave to Exercise 2.17 the proof that σ_p is topologically mixing on Σ_p^+ and so $f|\Lambda$ is topologically mixing.

PROOF. The map f restricted to Λ is conjugate to σ on Σ_p^+, so it is enough to prove these facts for σ. (This uses the fact that a conjugacy takes a periodic orbit of period n to a periodic orbit of period n. See Exercise 2.26.)

(a) Given n, there are p^n blocks of length n made up with letters in $\{1, \ldots, p\}$. If $\mathbf{b} = b_0 \cdots b_{n-1}$ is one of these blocks, let $\bar{\mathbf{b}}$ be the string which repeats the block $\mathbf{b}$, $\bar{\mathbf{b}} = b_0 \cdots b_{n-1} b_0 \cdots b_{n-1} \cdots$ or we could write $\bar{\mathbf{b}} = \mathbf{bb}\cdots$. Then, $\sigma^n(\bar{\mathbf{b}}) = \bar{\mathbf{b}}$ so $\bar{\mathbf{b}} \in \text{Per}(n, \sigma)$. On the other hand, if $\sigma^n(\mathbf{s}) = \mathbf{s}$, then $s_{n+j} = s_j$ for all j and $\mathbf{s} = \bar{\mathbf{b}}$, where $\mathbf{b} = s_0 \cdots s_{n-1}$. This shows that the points that are fixed by σ^n are exactly those given by one of the $\bar{\mathbf{b}}$ where $\mathbf{b}$ is a block of length n. There are p^n distinct blocks of length n, so we have proved the claim.

(b) Let $\mathbf{s} \in \Sigma_p^+$ and $\epsilon > 0$. Take n such that $3^{1-n}2^{-1} < \epsilon$. Let $\mathbf{b} = s_0 \cdots s_{n-1}$ and $\mathbf{t} = \bar{\mathbf{b}}$. Then, $\mathbf{t} \in \text{Per}(n, \sigma)$ and

$$\begin{aligned} d(\mathbf{s}, \mathbf{t}) &= \sum_{j=n}^{\infty} \frac{\delta(s_j, t_j)}{3^j} \\ &\leq \sum_{j=n}^{\infty} 3^{-j} = 3^{1-n}2^{-1} < \epsilon. \end{aligned}$$

Since ϵ is arbitrary, this proves that there is a periodic point within ϵ of $\mathbf{s}$, and so the periodic points are dense in Σ_p^+.

(c) To prove there is a point with a dense orbit, we describe such a point for $p = 2$. (An obvious change gives the general case.) Let $\mathbf{t}$ be a sequence which first lists all the blocks of length 1, then lists all the blocks of length 2, and continuing, lists all the blocks of length n for each successive n:

$$\mathbf{t} = 1, 2; (1,1), (2,1), (1,2), (2,2); (1,1,1), (2,1,1), \ldots.$$

(The use of parentheses, commas, and semicolons is merely to clarify the blocks making up the string and has no real meaning.) Then, given any $\mathbf{s} \in \Sigma_2^+$ and k, there is an n such that $\sigma^n(\mathbf{t})$ and $\mathbf{s}$ agree in the first k places. Therefore, $d(\sigma^n(\mathbf{t}), \mathbf{s}) \leq 3^{1-k}2^{-1}$.

Since k is arbitrary, the orbit of $\mathbf{t}$ gets arbitrarily near $\mathbf{s}$. Since $\mathbf{s}$ is arbitrary, the orbit of $\mathbf{t}$ is dense in Σ_2^+.

The fact that f is topologically mixing follows from the fact that given one of the intervals $I_{s_0,\ldots,s_n}$, $f^{n+1}(I_{s_0,\ldots,s_n}) \supset \mathcal{I}$. We leave the details to the reader. This completes the proof of the theorem. □

REMARK 5.3. It is no accident that we proved that the map f is transitive by proving it is conjugate to a shift map and then verified the property for the shift map. It is nearly impossible to specify a point which has a dense orbit for a nonlinear map. However, the very nature of the coding of the shift space allows us to write down a point with a dense orbit for the shift map. Then, the conjugacy proves that the nonlinear map inherits this property even though we cannot write down the point with the dense orbit.

2.6 Conjugacy and Structural Stability

In the last section we showed that the quadratic map on its invariant Cantor set is topologically conjugate to the shift map. We used the conjugacy to determine the number of periodic points and to prove that the quadratic map is topologically transitive on its invariant Cantor set. In this section we continue our discussion of topological conjugacy: we apply it to simple maps of the type considered in Section 2.2 and get conjugacies on intervals in the line (and not just between two Cantor sets). These examples also illustrate some of the properties which topological conjugacies preserve and those which they do not. In the next section we return to the quadratic map and obtain a conjugacy between two nearby quadratic maps on the whole real line.

When constructing the conjugacy for the quadratic map, we did not give any general motivation. We start with such motivation now before stating various conditions which we can place on the conjugacy. The concept of conjugacy arises in many subjects of mathematics. In Linear Algebra, the natural concept is linear conjugacy. Thus, if $\mathbf{x}_1 = A\mathbf{x}$ is a linear map and $\mathbf{x} = C\mathbf{y}$ is a linear change of coordinates for which C has an inverse, then the map on the $\mathbf{y}$-variables is given as follows: $\mathbf{y}_1 = C^{-1}\mathbf{x}_1 = C^{-1}A\mathbf{x} = C^{-1}AC\mathbf{y}$. Thus, the matrix for the map in the $\mathbf{y}$-variables is $C^{-1}AC$. As long as the two maps are defined on the same space (e.g., some $\mathbb{R}^n$), a conjugacy can be considered a change of coordinates of the variables on the space on which the function acts. In Dynamical Systems, we consider a conjugacy (or change of coordinates) which is continuous with a continuous inverse, i.e., a homeomorphism. We could also consider a conjugacy of two functions for which the change of coordinates h is differentiable with a differentiable inverse, i.e., h is a diffeomorphism. A third alternative is a conjugacy where the change of coordinates is an affine function, $h : \mathbb{R} \to \mathbb{R}$ is an affine map, $h(x) = ax + b$. We discuss below why we usually are only able to find a conjugacy by a homeomorphism and not a diffeomorphism. (In the last section, it would not be possible to require that $h : \Lambda_\mu \to \Sigma_2^+$ is differentiable because Σ_2^+ is not a Euclidean space or a manifold.)

Definition. Let $f : X \to X$ and $g : Y \to Y$ be two maps. A map $h : X \to Y$ is called a *topological semi-conjugacy from f to g* provided (i) h is continuous, (ii) h is onto, and (iii) $h \circ f = g \circ h$. We also say that f is *topologically semi-conjugate to g by h*. The map h is called a *topological conjugacy* if it is a semi-conjugacy and (iv) h is one to one and has a continuous inverse (so h is a homeomorphism). We also say that f and g are *topologically conjugate by h*, or sometimes we just say that f and g are *conjugate*.

Definition. To define a differentiable conjugacy, we restrict to functions on the line where we know what we mean by differentiable. Let $J, K \subset \mathbb{R}$ be intervals. Assume $f : J \to J$ and $g : K \to K$ are two C^r- maps for some $r \geq 1$. A map $h : J \to K$ is called a *C^r-conjugacy from f to g* provided (i) h is a C^r-diffeomorphism (h is onto, one to one, C^r, with a C^r inverse) and (ii) $h \circ f = g \circ h$. We also say that f and g are *C^r-conjugate by h*, or *differentiably conjugate*. If the conjugacy h is affine, $h(x) = ax + b$, then we say that f and g are *affinely conjugate*.

We defer to Exercise 2.26 to show that a topological conjugacy takes periodic orbits of one map to periodic orbits of the same period of the other map. Thus, two conjugate maps "have the same dynamics."

Example 6.1. In this first example, we show how to match up two families of quadratic maps: F_μ and a second family g_a. In this case, the conjugacy can even be affine. This is a partial verification of the fact that there is only "one family of quadratic functions" up to affine change of coordinates. Define

$$g_a(y) = ay^2 - 1.$$

A conjugacy must match up the fixed points. The fixed points of F_μ are 0 and $p_\mu = 1 - 1/\mu$. Those of g_a are $y^\pm = [1 \pm (1+4a)^{1/2}]/2a$. Also note that (i) $g_a(-y^+) = y^+$ and $F_\mu(1) = 0$, and (ii) the critical points of g_a and F_μ are 0 and 1/2, respectively. Assume $x = h(y) = my + b$ is the change of coordinates. Because $-y^+ < y^- < y^+$ and $0 < p_\mu < 1$, we must have $h(-y^+) = 1$, $h(y^-) = p_\mu$, $h(y^+) = 0$, and $h(0) = 1/2$. Substituting in h, we get the equations

$$\begin{aligned} m(-y^+) + b &= 1 \\ my^- + b &= 1 - \frac{1}{\mu} \\ my^+ + b &= 0 \\ m \cdot 0 + b &= \frac{1}{2}. \end{aligned}$$

From the last equation, we get that $b = 1/2$. Subtracting the first equation from the second (and using the form of $y^\pm$), we get that $m(1/a) = -1/\mu$, or $m = -a/\mu$. Substituting these values in the third equation we get $-[1 + (1+4a)^{1/2}]/[2\mu] = -1/2$, $\mu = 1 + (1+4a)^{1/2}$, or $4a = \mu^2 - 2\mu$. These last two expressions give necessary conditions for the maps to be conjugate:

$$\begin{aligned} \mu &= 1 + (1+4a)^{1/2} && \text{or} \\ a &= \frac{\mu^2 - 2\mu}{4}, && \text{and} \\ h(y) &= \frac{1}{2} - \frac{ay}{\mu}. \end{aligned}$$

Once we have found these conditions, we can verify directly that this h indeed does work:

$$\begin{aligned} F_\mu \circ h(y) &= F_\mu(-ay/\mu + 1/2) \\ &= \mu(-ay/\mu + 1/2)(ay/\mu + 1/2) \\ &= \frac{\mu}{4} - \frac{a^2y^2}{\mu}, \qquad \text{while} \\ h \circ g_a(y) &= h(ay^2 - 1) \\ &= -\frac{a}{\mu}(ay^2 - 1) + \frac{1}{2} \\ &= -\frac{a^2y^2}{\mu} + \frac{a}{\mu} + \frac{1}{2}. \end{aligned}$$

These two quantities are equal because $4a = \mu^2 - 2\mu$. This shows that these two functions are affinely conjugate when the parameters are correctly related.

Example 6.2. Let

$$D(z) = 2z \mod 2$$

be the *doubling map* on the circle $S \equiv \{z \bmod 2\}$,

$$T(y) = \begin{cases} 2y & \text{if } 0 \leq y \leq 1/2 \\ 2(1-y) & \text{if } 1/2 \leq y \leq 1 \end{cases}$$

be the *tent map*, $g_2(y) = 2y^2 - 1$ be the quadratic map of the last example, and $F_\mu(x) = \mu x(1-x)$ be the standard family of quadratic maps. The tent map is also called the *rooftop map*. The doubling map is also called the *squaring map* because in complex notation it can be written as $D(z) = z^2$ on complex numbers z with $|z| = 1$.

We claim that (i) D is topologically semi-conjugate to both $T|[0,1]$ and $g_2|[-1,1]$, and that (ii) $T|[0,1]$, $g_2|[-1,1]$, and $F_4|[0,1]$ are all topologically conjugate.

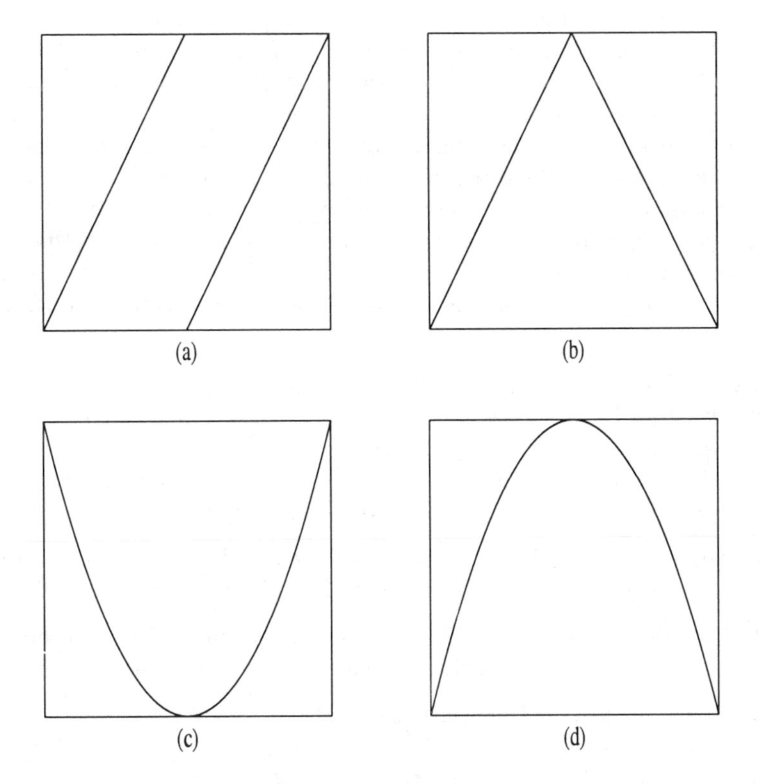

FIGURE 6.1. The Graphs of (a) D on S^1 (or $[0,1]$), (b) T on $[0,1]$, (c) g_2 on $[-1,1]$, and (d) F_4 on $[0,1]$

We first compare D and T. If we consider z is equivalent to $-z$ (or $2-z$) on the circle $S \equiv \{z \bmod 2\}$, this induces a two-to-one projection $\rho : S \to [0,1]$ given by $\rho(z) = |z|$ for $-1 \leq z \leq 1$. Also, if $1 \leq |z| \leq 2$, then $\rho(z) = 2 - |z|$. Then, for $-1/2 \leq z \leq 1/2$, $T \circ \rho(z) = T(|z|) = 2|z|$, and $\rho \circ D(z) = \rho(2|z|) = 2|z|$ because $0 \leq 2|z| \leq 1$. On the other hand, for $1/2 \leq |z| \leq 1$, $T \circ \rho(z) = T(|z|) = 2 - 2|z|$ and $\rho \circ D(z) = \rho(2z) = 2 - 2|z|$ because $1 \leq 2|z| \leq 2$. Therefore, for any z, $T \circ \rho(z) = \rho \circ D(z)$. This proves that D and T are semi-conjugate.

To construct the semi-conjugacy from D to g_2, we consider $z \in S$ to be the point on the circle with angle πz radians and let $x = h(z) = \cos(\pi z)$ be the map to the x coordinate, $h : S \to [-1, 1]$. Then, $h \circ D(z) = \cos(2z) = 2\cos^2(z) - 1 = 2h(z)^2 - 1 = g_2(h(z))$. This proves that D is topologically semi-conjugate to g_2.

Lastly, note that $\rho(z) = \rho(z')$ if and only if z is equal to $\pm\, z'$ modulo 2. This is also true for h, so $k(y) = h \circ \rho^{-1}(y) = \cos(\pi y)$ induces a topological conjugacy from T to g_2, $k : [0, 1] \to [-1, 1]$. Notice that k is differentiable everywhere, but that k^{-1} is not differentiable at $\pm\, 1$.

Also note, by Example 6.1, $g_2|[-1, 1]$ is conjugate to $F_4|[0, 1]$, so combining $T|[0, 1]$ is conjugate to $F_4|[0, 1]$ by $H(y) = 1/2 - (1/2)\cos(\pi y) = \sin^2(\pi y/2)$. Again, H is differentiable everywhere on $[0, 1]$ but H^{-1} is not differentiable at 0 and 1.

In the above examples, some of the conjugacies were affine and so differentiable everywhere, and other conjugacies were differentiable but had inverses which were not differentiable at the endpoints. In Dynamical Systems, we usually consider the notion of topological conjugacy and not C^r-conjugacy. The question arises, why do we only require that the conjugacy h be a homeomorphism and not a diffeomorphism? As the situation of the last section illustrates, sometimes a topological conjugacy is all that exists and it is still useful to match the trajectories of the two maps. The following proposition gives another reason: the assumption that f and g are differentiably conjugate puts a very restrictive condition on the relationship between the derivatives of f and g at their respective periodic points. After this proposition, we define another concept (structural stability) which requires a map f to be topologically conjugate to all "small perturbations" g. After this definition, we give a specific example of a map f that is topologically conjugate to all perturbations, and we also verify that the conjugacy cannot be differentiable for a particular choice of g. This example thus gives a second justification that topological conjugacy is the natural concept in Dynamical Systems.

Proposition 6.1. *Assume $f, g : \mathbb{R} \to \mathbb{R}$ are C^1.*

(a) Assume that f and g are C^1-conjugate by h. Assume x_0 is a n-periodic point for f, and $y_0 = h(x_0)$. Then, $(f^n)'(x_0) = (g^n)'(y_0)$.

(b) Assume f has a point x_0 of period n and assume every n-periodic point y_0 for g has $(f^n)'(x_0) \neq (g^n)'(y_0)$. Then, f and g are not C^1-conjugate.

REMARK 6.1. Since most small perturbations of a map f change the derivative at periodic points, it is very difficult for two maps to be differentiably conjugate.

REMARK 6.2. In higher dimensions, the corresponding result is that the matrices of partial derivatives of f and g are linearly conjugate and so have the same eigenvalues.

PROOF. Clearly part (b) follows from part (a).

To prove part (a), we assume that $h \circ f(x) = g \circ h(x)$, so $h \circ f^n(x) = g^n \circ h(x)$ and $f^n(x) = h^{-1} \circ g^n \circ h(x)$. Taking the derivative at x_0, we get that

$$\begin{aligned}(f^n)'(x_0) &= (h^{-1})'_{(g^n \circ h(x_0))}(g^n)'_{(h(x_0))}h'_{(x_0)} \\ &= (h^{-1})'_{(y_0)}(g^n)'_{(y_0)}h'_{(x_0)} \\ &= (g^n)'_{(y_0)}.\end{aligned}$$

(We moved the point of evaluating the derivatives to subscripts to make the product easier to read.) This proves part (a). □

We next want to discuss maps which are conjugate to all perturbations. Such maps have the same dynamics as all small perturbations and so the structure of the dynamics of the whole map is stable. For this reason, such maps are called structurally stable. To make this definition precise, we need to define what we mean for two functions to be close, i.e., we need to define the distance between two functions.

Definition. Let $r \geq 0$ be an integer. Let $f, g : \mathbb{R} \to \mathbb{R}$ be C^r functions and $J \subset \mathbb{R}$ be an interval (usually closed and bounded). Define the *C^r-distance from f to g* by

$$d_{r,J}(f,g) = \sup\{|f(x) - g(x)|,\ |f'(x) - g'(x)|,\ \cdots, |f^{(r)}(x) - g^{(r)}(x)| : x \in J\}.$$

Obviously, f and g do not need to be defined on the whole real line but only need their domains to include the interval J (unless $J = \mathbb{R}$).

Definition. Assume $r \geq 1$. Let $f : \mathbb{R} \to \mathbb{R}$ be a C^r function. A function f is C^r *structurally stable* provided there exists an $\epsilon > 0$ such that f is conjugate to g on all of $\mathbb{R}$ whenever $g : \mathbb{R} \to \mathbb{R}$ is a C^r function with $d_{r,\mathbb{R}}(f,g) < \epsilon$. A function f is said to be *structurally stable* provided it is C^1 structurally stable.

Example 6.3. Take $f(x) = 3x$ and $g(x) = 2x$. We want to construct a conjugacy h with $h \circ f(x) = g \circ h(x)$, or $h(x) = g^{-1} \circ h \circ f(x)$. If h exists, then $h(x) = g^{-j} \circ h \circ f^j(x)$ for any $j \geq 0$. However, we also have $h(x) = g \circ h \circ f^{-1}(x)$ so $h(x) = g^{-j} \circ h \circ f^j(x)$ for any $-\infty < j < \infty$. Both maps have 0 as a fixed point. As stated before, we need that $h(0) = 0$. The image of 1 by h, $h(1)$, is fairly arbitrary, but we take $h(1) = 1$ and $h(-1) = -1$. Then, $h(3) = h \circ f(1) = g \circ h(1) = g(1) = 2$. The definition of h for x between 1 and 3 is also arbitrary as long as it is monotone, so we let $h_0 : [1,3] \to [1,2]$ be a C^1 map with (i) $h_0(1) = 1$, (ii) $h_0(3) = 2$, (iii) $h_0'(1) = 1$, and (iv) $h_0'(3) = 2/3$. Similarly define $h_0 : [-3,-1] \to [-2,-1]$.

Once h is determined on $[-3,-1] \cup [1,3]$ by $h(x) = h_0(x)$, this determines it for all nonzero x as follows. Take $x \neq 0$. There is a unique $j = j(x) \in \mathbb{Z}$ such that $f^j(x) = 3^j x \in (-3,-1] \cup [1,3)$. Define $h(x) = g^{-j} \circ h_0 \circ f^j(x)$. This defines h for all $x \in \mathbb{R}$. By construction, h is a conjugacy.

In the above construction, the orbit of every $x \neq 0$ goes through the union of the intervals $J = (-3,-1] \cup [1,3)$, and there is no proper subset of J for which this is true. Because of this property, the set J, is called a *fundamental domain for the unstable set of* 0. Often it is preferred to take a fundamental domain as closed so the characterization is changed as follows: the pair of closed intervals $J = [-3,-1] \cup [1,3] = \text{cl}((-3,-1] \cup [1,3))$ is called a *fundamental domain for the unstable set of* 0 because the orbit of every $x \neq 0$ goes through J and there is no proper closed subset of J for which this is true.

Note for those x approaching 1 from below, $h'(x) = (3/2)h_0'(3x)$. The limit of this expression as x approaches 1 is equal to $1 = h_0'(1)$:

$$\begin{aligned}
\lim_{\substack{x \to 1 \\ x < 1}} h'(x) &= \lim_{\substack{x \to 1 \\ x < 1}} \left(\frac{3}{2}\right) h_0'(3x) \\
&= \left(\frac{3}{2}\right)\left(\frac{2}{3}\right) \\
&= 1 \\
&= \lim_{\substack{x \to 1 \\ x > 1}} h_0(x).
\end{aligned}$$

Thus, this extension is differentiable at $x = 1$. Similarly, it is differentiable at 3 and all other points $\pm\, 3^j$. Thus, this extension is C^1 for $x \neq 0$.

Now, let $\{x_i\}$ be an arbitrary sequence with $x_i \neq 0$ that approaches 0 as i goes to infinity. Then, there is a $j = j(i)$ such that $z_i \equiv f^{j(i)}(x_i) \in (-3,-1] \cup [1,3)$. It must be that $j(i)$ goes to infinity as i goes to infinity. Then, $h(x_i) = g^{-j(i)} h_0(z_i) = 2^{-j(i)} h_0(z_i)$ and this must converge to 0 as i goes to infinity. Thus, h is continuous at $x = 0$.

Finally, we want to show that h is not differentiable at 0. Now, let $x_i = 3^{-i}$. Then, x_i approaches 0 as i goes to infinity. Then, $h(x_i) = 2^{-i} h_0(3^i x_i) = 2^{-i} h_0(1) = 2^{-i}$ also approaches 0. However,

$$\begin{aligned}\lim_{i\to\infty} \frac{h(x_i) - h(0)}{x_i - 0} &= \lim_{i\to\infty} \frac{2^{-i}}{3^{-i}} \\ &= \lim_{i\to\infty} \left(\frac{3}{2}\right)^i \\ &= \infty.\end{aligned}$$

Thus, h is not differentiable at 0. Another way to see that h is not continuously differentiable is to show that the derivative of h at the points x_i, $h'(x_i)$, goes to infinity as i goes to infinity:

$$\begin{aligned}\lim_{i\to\infty} h'(x_i) &= \lim_{i\to\infty} (g^{-i})'(1) h'(1) (f^i)'(x_i) \\ &= \lim_{i\to\infty} 2^{-i} \cdot 1 \cdot 3^i \\ &= \infty.\end{aligned}$$

Notice that if g were any map with $g'(x) \geq \lambda > 1$ for all x, then g has a single fixed point and the same proof shows that f is conjugate to g. Thus, f is C^1 structurally stable. Also notice that if $g(x) = (3 + \epsilon)x$, then $d_0(f, g) = \infty$, but the derivatives of f and g are close for all x. Thus, when we consider the perturbations of a function which are small in terms of the distance d_1 on all of $\mathbb{R}$ (which is noncompact), this is very restrictive. Often, we can allow the C^0 size of the perturbation to be larger near $\pm\infty$ if the derivatives are controlled.

Example 6.4. In this example, we consider $f(x) = -x/3$. There are two differences from the previous example: the origin is attracting and not repelling, and the map switches points from one side of the fixed point to the other. However, if we let $J = [1, f^{-2}(1)) = [1, 9)$, then every $x \neq 0$ has a unique $j \in \mathbb{Z}$ such that $f^j(x) \in [1, 9)$. Thus, for a fundamental domain, we do not need to take an interval for negative x as well as positive x; $[1, 9]$ is the complete fundamental domain for the stable set of 0. With this change, the proof that f is structurally stable is also the same. We leave the details to the reader.

Note that the fact that the fundamental domain for $f(x) = -x/3$ is one interval, while that for $g(x) = x/3$ is the union of two intervals implies that these two maps cannot be topologically conjugate. In fact, f is orientation reversing on $\mathbb{R}$ while g is orientation preserving. Two maps which are topologically conjugate must either both be orientation preserving or orientation reversing. This is why f and g cannot be topologically conjugate, even though both have a unique fixed point whose basin of attraction is the whole real line.

Example 6.5. Let $f(x) = x^3/2 - x/2$. This example has fixed points at $x = 0, \pm\sqrt{3}$, and critical points at $\pm 1/\sqrt{3}$. The fixed point $x = 0$ is attracting and the two fixed points $x = \pm\sqrt{3}$ are repelling. The main changes from the previous examples are the existence of several fixed points and the existence of critical points.

If g is any small C^1 perturbation, it will have three fixed points. However, to insure that g has exactly two critical points, we must take g C^2 near f. We take such a g and let p_j for $j = -1, 0, 1$ be the fixed points and $c^\pm$ be the two critical points, with $p_{-1} < c^- < p_0 < c^+ < p_1$, $p_0 < g(c^-) < c^+$, and $c^- < g(c^+) < p_0$.

Notice however that the map f has $f'(x) \geq 4 > 1$ for all x with $x \geq \sqrt{3}$ or $x < -\sqrt{3}$. Thus, we take (partial) fundamental domains for the unstable sets of $\pm\sqrt{3}$ given by $[-f(-4), -4] \cup [4, f(4)]$. We can construct a conjugacy h on $(-\infty, -\sqrt{3}] \cup [\sqrt{3}, \infty)$ from f to g with $h((-\infty, -\sqrt{3}]) = (-\infty, -p_{-1}]$ and $h([\sqrt{3}, \infty)) = [p_1, \infty)$.

Now, we consider the points between $\pm\sqrt{3}$. The point $1/\sqrt{3}$ is a critical point which is a minimum of the function f. Thus, there are points on either side of $1/\sqrt{3}$ on which f takes the same value. Let x_i and y_i sequences of points converging to $1/\sqrt{3}$ with $x_i > 1/\sqrt{3}$, $y_i < 1/\sqrt{3}$ and $f(x_i) = f(y_i)$. If h is a conjugacy from f to g, then we need that $g \circ h(x_i) = h \circ f(x_i) = h \circ f(y_i) = g \circ h(y_i)$. Thus, the points $h(x_i)$ and $h(y_i)$ need to be points which have the same image by g. Thus, $h(1/\sqrt{3})$ has two distinct points arbitrarily near which are taken to the same point by g, and $h(1/\sqrt{3})$ must be the critical point for g, c^+. Similarly, we need $h(-1/\sqrt{3}) = c^-$.

Once we know that h must take the critical point of f to the critical point of g, the rest of the construction of the conjugacy is straightforward. The map f is monotone for x with $-1/\sqrt{3} \leq x \leq 1/\sqrt{3}$. We construct a conjugacy here to the perturbation g much as in the last example using the fundamental domain of the stable set of 0 given by $[f^2(1/\sqrt{3}), 1/\sqrt{3}]$, making sure that the image of the critical point by h_0, $h_0 \circ f(-1/\sqrt{3})$, is $g(c^-)$. Once the conjugacy h_0 is defined on this fundamental domain, extend it to $[-1/\sqrt{3}, 1/\sqrt{3}]$ as before, with $h[-1/\sqrt{3}, 1/\sqrt{3}] = [c^-, c^+]$. (Notice that $f(1/\sqrt{3}) > -1/\sqrt{3}$ and $f(-1/\sqrt{3}) < 1/\sqrt{3}$.)

Next, we need to extend h to the interval $[1/\sqrt{3}, \sqrt{3})$. Let f_+ be the restriction of f to this interval and g_+ be the restriction of g to $[c^+, p_1)$. Now, for $x \in (1/\sqrt{3}, \sqrt{3})$, there is a smallest $j > 0$ with $f_+^j(x) \in [-1/\sqrt{3}, 1/\sqrt{3}]$. Define $h(x) = (g_+)^{-j} \circ h \circ f_+^j(x)$. Make a similar construction for $x \in (-\sqrt{3}, -1/\sqrt{3}]$: $h(x) = (g_-)^{-j} \circ h \circ f_-^j(x)$, where g_- is the restriction of g to $(p_{-1}, c^-]$ and f_- is the restriction of f to $(-\sqrt{3}, -1/\sqrt{3}]$. This extension makes h defined and continuous on the whole real line. This completes the necessary modifications. (The reader may want to check some of the claims we made in the construction.)

2.7 Conjugacy and Structural Stability of the Quadratic Map

In the preface we stated that we wanted to determine which systems were dynamically equivalent to any of its perturbation. In this section we prove that this is the case for the quadratic maps with invariant Cantor sets: any small perturbation g of F_μ is conjugate to F_μ.

Before we prove that a conjugacy exists on the whole real line, we first show that one exists between the nonwandering set of F_μ and the nonwandering set of g. Because the nonwandering sets are contained in a compact interval, J, we only require that the two functions are close on this interval J. To make the statement clearer, we introduce notation for the restriction of the functions to this interval. Let f_J be the restriction of f to J. Then, the nonwandering set of f_J, $\Omega(f_J)$, are the points z such that for any open neighborhood U there is a point $y \in U$ with $f^k(y) \in U$ and $f^j(y) \in J$ for $1 \leq j \leq k$.

We start by giving the definition of two functions being conjugate on their nonwandering sets. We also use the definitions of two functions being close in terms of their derivatives which is introduced in the last section. Sometimes we only require that these values are close on an bounded interval.

Definition. Let $f : \mathbb{R} \to \mathbb{R}$ be a C^r function. As a modification of the concept of structural stability, we consider a map h that only conjugates f restricted to its nonwandering set with g restricted to its nonwandering set. A function f is C^r *Ω-stable on J* provided there exists an $\epsilon > 0$ such that f restricted to $\Omega(f_J)$ is topologically conjugate to g restricted to $\Omega(g_J)$ whenever $g : J \to \mathbb{R}$ is a C^r function with $d_{r,J}(f, g) < \epsilon$.

Theorem 7.1. *Let $F_\mu(x) = \mu x(1 - x)$ as before with $\mu > 4$.*
(a) Then, F_μ is C^1 Ω-stable on $[-2, 2]$, and
(b) F_μ is C^2 structurally stable on $\mathbb{R}$.

REMARK 7.1. By stating the theorem for part (a) the way we do, it implies that F_μ is Ω-conjugate to $F_{\mu'}$ for $|\mu - \mu'|$ small.

PROOF OF THEOREM 7.1(a). We restrict our proof to the case when $\mu > 2 + \sqrt{5}$ but the proof can be modified for $\mu > 4$ using the metric introduced above.

Let $I_\delta = [-\delta, 1 + \delta]$ for $\delta > 0$. For $\delta > 0$ small enough, if $z \in [I_\delta \cap F_\mu^{-1}(I_\delta)] \cup [-2, 0] \cup [1, 2]$, then $|F'_\mu(z)| > \lambda > 1$. Since $F'_\mu(1/2) = 0$, the assumptions imply that $F_\mu(1/2) > 1 + \delta$.

Take $\epsilon > 0$ small enough so that if $d_{1,[-2,2]}(f, g) < \epsilon$, then $g(-\delta) < -\delta$, $g(1+\delta) < -\delta$, $g(1/2) > 1 + \delta$, and $|g'(z)| > \lambda$ for $z \in [I_\delta \cap g^{-1}(I_\delta)] \cup [-2, 0] \cup [1, 2]$. Fix such a map g. These conditions imply that $g|I_\delta$ covers I_δ twice. The next lemma shows that the nonwandering set of g is contained in I_δ.

Lemma 7.2. *If $g^k(x) \in [-2, 2]$ for all $k \geq 0$, then $g^k(x) \in I_\delta$ for all $k \geq 0$. Therefore, $\Omega(g|[-2, 2]) \subset \bigcap_{k=0}^{\infty} g^{-k}(I_\delta)$.*

PROOF. If $g^k(x) \notin I_\delta$, then $g^{k+1}(x) < \max\{g(1 + \delta), g(-\delta)\} < -\delta$. If also $g^{k+1}(x) \in [-2, -\delta]$, then

$$\begin{aligned} g^{k+2}(x) - (-\delta) &< g^{k+2}(x) - g(-\delta) \\ &< \lambda[g^{k+1}(x) - (-\delta)]. \end{aligned}$$

Therefore, $g^{k+2}(x)$ stays less than $-\delta$. As long as $g^{k+j}(x)$ stays in $[-2, -\delta]$,

$$g^{k+j}(x) - (-\delta) < \lambda^{j-1}[g^{k+1}(x) - (-\delta)].$$

Since $\lambda > 1$, this inequality can only last for a finite number of iterates, and $g^{k+j}(x) < -2$ for some $j \geq 0$. Thus, if $g^k(x) \in [-2, 2]$ for all $k \geq 0$, then $g^k(x) \in I_\delta$ for all $k \geq 0$. The statement about the nonwandering set follows from the first statement. □

Lemma 7.3. *Let $\Lambda_g = \bigcap_{k=0}^{\infty} g^{-k}(I_\delta)$. Then, $g|\Lambda_g$ is topologically conjugate to σ on Σ_2^2.*

PROOF. Because $g|I_\delta$ covers I_δ twice, $g^{-1}(I_\delta) \cap I_\delta$ is the union of two intervals I_1^g and I_2^g. By the Mean Value Theorem, the length of each of these intervals is less than λ^{-1} times the length of I_δ. Also, g is monotone on each I_j with derivative greater than λ in absolute value. Let $I_{j,k}^g = g^{-1}(I_k) \cap I_j$. Then, $S_1^g = \bigcap_{i=0}^{2} g^{-i}(I_\delta)$ is the union of the four intervals $I_{j,k}^g$. By the Mean Value Theorem, each of these intervals has length bounded as follows: $L(I_{j,k}) \leq \lambda^{-1} L(I_k) \leq \lambda^{-2} L(I_\delta)$. By induction, $S_n^g = \bigcap_{i=0}^{n} g^{-i}(I_\delta)$ is the union of 2^n intervals of length less than or equal to $\lambda^{-n} L(I_\delta)$. Exactly as in the earlier proof, $\Lambda_g = \bigcap_{k=0}^{\infty} g^{-k}(I_\delta)$ is a Cantor set, and $g|\Lambda_g$ is topologically conjugate to σ on Σ_2^2. □

To finish the proof of the theorem, note that $g|\Lambda_g$ is topologically conjugate to σ on Σ_2^2, which in turn is conjugate to $F_\mu|\Lambda_\mu$. Because topological conjugacy is a transitive relation, we have proved part (a) of the theorem. □

PROOF OF THEOREM 7.1(b). In this part, we let $h : \Lambda_\mu \to \Lambda_g$ be the conjugacy shown to exist by part (a). For part (b), we need to define a conjugacy off the Cantor set Λ_μ. Just as in the simpler example, Example 6.5, which has only a finite number of periodic points, a conjugacy of F_μ and a perturbation on all of $\mathbb{R}$ must take the critical point c_μ of F_μ to a critical point c_g of g. Thus, g must have a unique critical point. In order to be able to know that g has a unique critical point, we must restrict ourselves to g which are C^2 near F_μ.

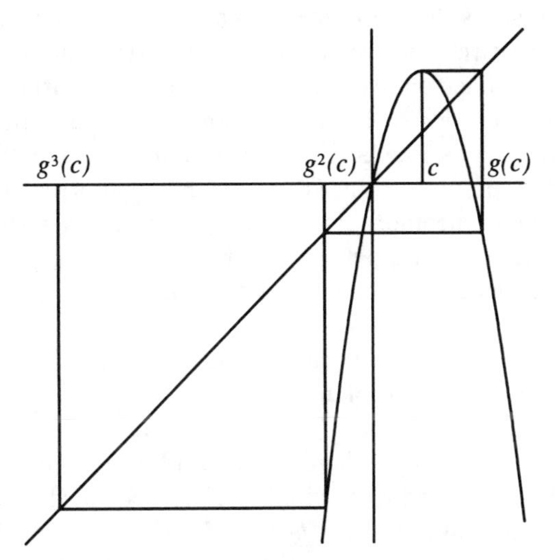

FIGURE 7.1. Orbit of the Critical Point

Note that $F_\mu(c_\mu) > 1$, and $0 > F_\mu^2(c_\mu) > F_\mu^3(c_\mu)$. Also, if g is C^2 near enough to F_μ, then $g(c_g)$ is greater than the points in Λ_g, and $g^2(c_g) > g^3(c_g)$ and both points are less than the points in Λ_g. The conjugacy we construct must take $[F_\mu^3(c_\mu), F_\mu^2(c_\mu)]$ to $[g^3(c_g), g^2(c_g)]$; these two intervals are the *fundamental domains* for F_μ and g, respectively. (These intervals are fundamental domains of the stable set of infinity or the unstable set of the invariant Cantor sets.) Let h_0 be such a map. (It could be taken to be linear on this interval.) The map F_μ is monotone (so has an inverse) when restricted to $x \leq 1/2$. Let $F_{\mu-}$ be this restriction, and $F_{\mu+}$ be the restriction to $x \geq 1/2$. Similarly, let g_- be the restriction of g to points $y \leq c_g$, and g_+ be the restriction to points $y \geq c_g$. Then, g_- and g_+ are each monotone with inverses. As in Examples 6.1 – 6.3, h_0 can be extended to points $x < 0$ by

$$h(x) = g_-^{-j} \circ h_0 \circ F_{\mu-}^j(x),$$

where $F_{\mu-}^j(x) \in [F_\mu^3(c_\mu), F_\mu^2(c_\mu)]$. This extension is continuous, as a map from $\Lambda_\mu \cup \{x : x < 0\}$ to $\Lambda_g \cup \{y : y < z \text{ for all } z \in \Lambda_g\}$.

Next, $F_{\mu+}\{x : x > 1\} = \{x : x < 0\}$, and h is already defined on $\{x : x < 0\}$. Also, $g_+\{y : y > z \text{ for all } z \in \Lambda_g\} = \{y : y < z \text{ for all } z \in \Lambda_g\}$. Thus, we can extend h to $\{x : x > 1\}$ by

$$h(x) = g_+^{-1} \circ h \circ F_{\mu+}(x).$$

With this definition, the map is still continuous and a conjugacy where it is defined. Also, $h(F_\mu(1/2)) = g_+^{-1} \circ h \circ F_\mu^2(1/2) = g_+^{-1} \circ g^2(c_g) = g(c_g)$.

Next, if $G_{1,1,\mu}$ is the gap at the first level for F_μ, and $G_{1,1,g}$ the gap at the first level for g, then the equation

$$h(x) = \begin{cases} g_+^{-1} \circ h \circ F_\mu(x) & \text{for } x \in G_{1,1,\mu} \text{ and } x > 1/2 \\ g_-^{-1} \circ h \circ F_\mu(x) & \text{for } x \in G_{1,1,\mu} \text{ and } x < 1/2 \end{cases}$$

extends h continuously from $G_{1,1,\mu}$ to $G_{1,1,g}$. As above, it can be checked that $h(1/2) = c_g$. We continue inductively to the gaps at n^{th} level, $G_{j,n,\mu}$ for F_μ and $G_{j,n,g}$ for g. For each j, we can extend h from $G_{j,n,\mu}$ to $G_{j,n,g}$ by using the appropriate branch of the inverse of g, g_+^{-1} or g_-^{-1}. We leave the details to the reader. □

2.8 Homeomorphisms of the Circle

In this section, we discuss the periodic orbits and recurrence of orientation preserving homeomorphisms of the circle. By restricting to invertible maps and by exploiting the fact that the circle comes back on itself, we are able to give a rather complete description of what dynamics can occur for such a homeomorphism. An important quantity which makes this determination possible is the rotation number. This number measures the average amount that a point is rotated by the homeomorphism. When this is the rational number p/q the homeomorphism has periodic points with period q. When this number is irrational, there are no periodic orbits. With the assumption that the map is a C^2 diffeomorphism with irrational rotation number, it follows that every orbit is dense in the circle.

We denote by S^1 the unit circle. It can be thought of as either (i) the real numbers modulo 1 or (ii) the points in the plane at a distance one from the origin. In terms of the second way of thinking about S^1, we often identify $\mathbb{R}^2$ with the complex plane $\mathbb{C}$ and so $S^1 = \{z \in \mathbb{C} : |z| = 1\}$. There is also a covering space projection π from $\mathbb{R}$ onto S^1. In terms of the first way of thinking about S^1, $\pi(t) = t \bmod 1$. In terms of the second way of thinking about S^1, $\pi(t) = e^{2\pi i t}$. (Note we used π for both the map and the number $3.14\cdots$, but throughout the rest of this section, π usually denotes the map.) Thus, π can be thought of as taking an angle measurement to a point on the circle.

Throughout this section, $f : S^1 \to S^1$ is assumed to be an orientation preserving homeomorphism. Given such a map, there is a (nonunique) map $F : \mathbb{R} \to \mathbb{R}$ which is called a *lift of* f such that $\pi \circ F = f \circ \pi$. A lift, F, of f satisfies (i) F is monotonically increasing and (ii) $F(t+1) = F(t) + 1$ for all t, so $(F - id)$ has period 1. For example, if f_λ is the rotation by λ on S^1 (or by $2\pi\lambda$ radians), then $F_\lambda(t) = t + \lambda$ is a lift. But for any integer k, $\tilde{F}_\lambda(t) = k + t + \lambda$ is also a lift. In fact, for any homeomorphism f, if F_1 and F_2 are two lifts, then there is an integer k such that $F_2(t) = F_1(t) + k$ for all t. (For each t, $\pi \circ F_2(t) = f \circ \pi(t) = \pi \circ F_1(t)$ so there is an integer k_t with $F_2(t) = F_1(t) + k_t$. Since everything is continuous and the integers are discrete, k_t is independent of t.)

The aim of the section is to define an invariant of f called the rotation number and prove that it can be used to determine whether f has any periodic points or not. The rotation number is a measure of the average amount of rotation of a point along an orbit.

Definition. We start by defining a number for a lift F of f. Let

$$\rho_0(F,t) = \lim_{n\to\infty} \frac{F^n(t) - t}{n}.$$

We show below that this limit exists and is independent of t, and so we denote it by $\rho_0(F)$. We also show that if F_1 and F_2 are two lifts, then $\rho_0(F_1,t) - \rho_0(F_2,t)$ is an integer, so

$$\rho(f) = \rho_0(F,t) \bmod 1$$

is well defined. The number $\rho(f)$ is called the *rotation number of* f.

REMARK 8.1. The definition of rotation number easily implies that $\rho_0(F^k) = k\rho_0(F)$ and $\rho(f^k) = k\rho(f) \bmod 1$.

Example 8.1. Let f_λ be the rotation by λ on S^1 with lift $F_\lambda(t) = t + \lambda$. This map is called a *rigid rotation of* S^1 *by* λ. Since $F_\lambda^n(t) = t+n\lambda$, it is easy to see that $\rho_0(F,t) = \lambda$ for all t and $\rho(f) = \lambda \bmod 1$. In this example every point is rotated by exactly λ so the rotation number should be λ.

In this example we can see the connection between the rationality of the rotation number and the existence of a periodic orbit. Assume $\lambda = p/q$ is rational, i.e., f_λ is a rational rotation. Then, $F_\lambda^q(t) = t + q\lambda = t + p$. Therefore, every point is periodic with period q.

Now, assume that λ is irrational. For f_λ to have a point x of period q, it is necessary for $F_\lambda^q(x) = x + q\lambda = x + k$ for some integer k. Thus, we need $\lambda = k/q$, which is impossible because λ is irrational. Thus, f_λ has no periodic points in this case. Below, Theorem 8.3 shows that a every point in S^1 has a dense orbit when λ is irrational.

We start by proving that an orientation preserving homeomorphism of the circle has a rotation number which is independent of the point.

Theorem 8.1. *Let $f : S^1 \to S^1$ be an orientation preserving homeomorphism with lift F. Then,*

(a) for $t \in \mathbb{R}$, the limit defining $\rho_0(F,t)$ exists and is independent of t,
(b) if $\rho(f) = \rho_0(F,t)$ mod 1, then this is independent of the lift F, and
(c) $\rho(f)$ depends continuously on f.

PROOF. (a) Take any two points $t, s \in \mathbb{R}$. There is an integer ℓ such that $t \le s + \ell < t + 1$. By the monotonicity of F, $F(t) \le F(s+\ell) < F(t+1) = F(t) + 1$, and by induction, $F^p(t) \le F^p(s+\ell) < F^p(t) + 1$. Subtracting $t+1$,

$$\begin{aligned} F^p(t) - t - 1 &\le F^p(s+\ell) - t - 1 \\ &< F^p(s+\ell) - s - \ell \\ &< F^p(t+1) - t \\ &= F^p(t) - t + 1. \end{aligned}$$

Since $F^p(s+\ell) - s - \ell = F^p(s) - s$,

$$F^p(t) - t - 1 < F^p(s) - s < F^p(t) - t + 1. \tag{$*$}$$

Writing $F^{p+n}(t) - t = F^p(F^n(t)) - F^n(t) + F^n(t) - t$ and applying $(*)$ with $s = F^n(t)$,

$$F^{p+n}(t) - t < F^p(t) - t + 1 + F^n(t) - t. \tag{$**$}$$

Applying $(**)$ with $n = p$,

$$F^{2p}(t) - t < 2[F^p(t) - t] + 1,$$

and by induction

$$F^{np}(t) - t < k[F^p(t) - t] + k - 1. \tag{$***$}$$

For any $n \ge p$, write $n = kp + i$ where $0 \le i < p$. Then, by $(**)$ and $(***)$,

$$\begin{aligned} F^n(t) - t &= F^{kp+i}(t) - t \\ &< F^{kp}(t) - t + F^i(t) - t + 1 \\ &< k[F^p(t) - t] + F^i(t) - t + k. \end{aligned}$$

Dividing by n and using the fact that $n \geq kp$,

$$\frac{F^n(t)-t}{n} < \frac{k[F^p(t)-t]}{kp} + \frac{F^i(t)-t}{n} + \frac{k}{kp}.$$

In the same way, using the inequality $F^n(t) - t - 1 < F^n(s) - s$ and the fact that $(k+1)p > n$, we get

$$\frac{k[F^p(t)-t]}{(k+1)p} + \frac{F^i(t)-t}{n} - \frac{k}{(k+1)p} < \frac{F^n(t)-t}{n}.$$

Using the last two inequalities and letting n (and so k) go to infinity with p fixed,

$$\frac{F^p(t)-t}{p} - \frac{1}{p} \leq \limsup_{n\to\infty} \frac{F^n(t)-t}{n} \leq \frac{F^p(t)-t}{p} + \frac{1}{p}.$$

Notice that this last set of inequalities shows that the limsup is finite. Now, letting p go to infinity,

$$\limsup_{n\to\infty} \frac{F^n(t)-t}{n} \leq \liminf_{p\to\infty} \frac{F^p(t)-t}{p}.$$

Thus, we have proved that the limit defining the rotation number $\rho_0(F,t)$ exists and is finite for any $t \in \mathbb{R}$.

Inequality (*) above shows that for two different points $t, s \in \mathbb{R}$,

$$\frac{F^p(t)-t-1}{p} < \frac{F^p(s)-s}{p} < \frac{F^p(t)-t+1}{p}.$$

Since the rotation numbers for both t and s exist, it easily follows that $\rho_0(F,t) = \rho_0(F,s)$ for any two points $t, s \in \mathbb{R}$. Because we have shown that the rotation number is independent of the point, we write $\rho_0(F)$ from now on.

(b) Assume F_1 and F_2 are two lifts of f. We noted above that there is an integer k independent of t such that $F_2(t) = F_1(t) + k$. By induction, $F_2^n(t) = F_1^n(t) + nk$ for any positive integer n. Therefore,

$$\begin{aligned} \rho_0(F_2) &= \lim_{n\to\infty} \frac{F_2^n(t)-t}{n} \\ &= \lim_{n\to\infty} \left(\frac{F_1^n(t)-t}{n} + \frac{nk}{n}\right) \\ &= \rho_0(F_1) + k. \end{aligned}$$

Thus, $\rho_0(F_2) = \rho_0(F_1) \bmod 1$ as claimed.

(c) Let $\epsilon > 0$. Choose an integer $n > 0$ such that $2/n < \epsilon$. Let F be a lift of f. There is an integer p such that $p \leq F^n(0) < p+1$. It then follows that $p-1 < F^n(t) - t < p+1$ for all t (possibly replacing p). For g near enough to f in terms of the C^0 topology, a lift G of g can be chosen so that $p-1 < G^n(t) - t < p+1$. (Note that n is fixed.) The nk-th iterate of 0 can be written as

$$\begin{aligned} F^{nk}(0) &= F^{nk}(0) - 0 \\ &= \sum_{j=0}^{k-1} F^n \circ F^{jn}(0) - F^{jn}(0) \end{aligned}$$

which is greater than $k(p-1)$ and less than $k(p+1)$. Therefore,

$$k(p-1) < F^{nk}(0) < k(p+1),$$

and we also get that

$$k(p-1) < G^{nk}(0) < k(p+1).$$

Because the rotation numbers for F and G exist, they can be calculated by subsequences, so $\rho_0(F) = \lim_{k\to\infty} F^{kn}(0)/kn$ and $\rho_0(G) = \lim_{k\to\infty} G^{kn}(0)/kn$. The above inequalities easily imply that

$$\frac{p-1}{n} < \rho_0(F) < \frac{p+1}{n} \qquad \text{and}$$
$$\frac{p-1}{n} < \rho_0(G) < \frac{p+1}{n}.$$

Therefore, $|\rho_0(F) - \rho_0(G)| < 2/n < \epsilon$. This completes the proof of part (c) and the theorem. □

We leave to the exercises to show that if f has a periodic point, then $\rho(f)$ is rational. The following proposition proves that $\rho(f)$ being rational is also a sufficient condition for f to have a periodic point.

Theorem 8.2. *The rotation number $\rho(f)$ is rational if and only if f has a periodic point. In fact, $\rho(f) = p/q$ if and only if f has a point of period q. (Here, p/q is assumed to be in reduced form with p and q integers and q positive.)*

PROOF. Exercise 2.31 proves that if f has a periodic point, then $\rho(f)$ is rational.

Conversely, assume that $\rho(f) = p/q$ is rational and expressed in lowest terms. Let $\hat{F}$ be a lift of f. By the definitions of $\rho(f)$ and $\rho_0(\hat{F})$, there is an integer k with $\rho_0(\hat{F}) = (p/q) + k$. Then, $F(t) = \hat{F}(t) - k$ is another lift of f and $\rho_0(F) = p/q$. Also, $\rho_0(F^q - p) = \rho_0(F^q) - p = q\rho_0(F) - p = 0$.

Let $G(t) = F^q(t) - p$. We need to show that G has a fixed point on $\mathbb{R}$ so f has a point of period q on S^1. There are three cases: (i) $G(0) = 0$, (ii) $G(0) > 0$, and (iii) $G(0) < 0$. In the first case, 0 is a fixed point, and we are done.

In case (ii), because G is increasing, $0 < G(0) < G^2(0) < \cdots G^n(0) < \cdots$. There are two subcases. First assume that $0 < G^n(0) < 1$ for all n. Because the orbit is monotonically increasing, $G^n(0)$ converges to some point x_0. As we have mentioned before, it follows by continuity that $G(x_0) = x_0$ and G has a fixed point.

As a second subcase, assume there is a $k > 0$ such that $G^k(0) > 1$. Then, $G^{2k}(0) = G^k \circ G^k(0) > G^k(1)$ by the monotonicity of G^k. But $G^k(1) = G^k(0) + 1 > 2$, so $G^{2k}(0) > 2$. By induction, $G^{jk}(0) > j$. Therefore, $G^{jk}(0)/jk > 1/k$ and $\rho_0(G) \geq 1/k$. This contradiction shows that this second subcase cannot occur and case (ii) implies there is a fixed point.

In case (iii), when $G(0) < 0$, $0 > G(0) > G^2(0) > \cdots$. By reasoning as in case (ii), there must be a fixed point. This finishes the proof of the theorem. □

Example 8.2. Let f be the map on S^1 whose lift F is given by

$$F(t) = t + \epsilon \sin(2\pi nt)$$

for n a positive integer and $0 < \epsilon < 1/(2\pi n)$. Then, x is a fixed point of f if there is an integer p with $F(x) = x + \epsilon \sin(2\pi nx) = x + p$, or $\epsilon \sin(2\pi nx) = p$. Since $\epsilon < 1$ it follows that $p = 0$. The solutions are $x = j/2n$ for $j = 0, \ldots, 2n-1$. The derivative

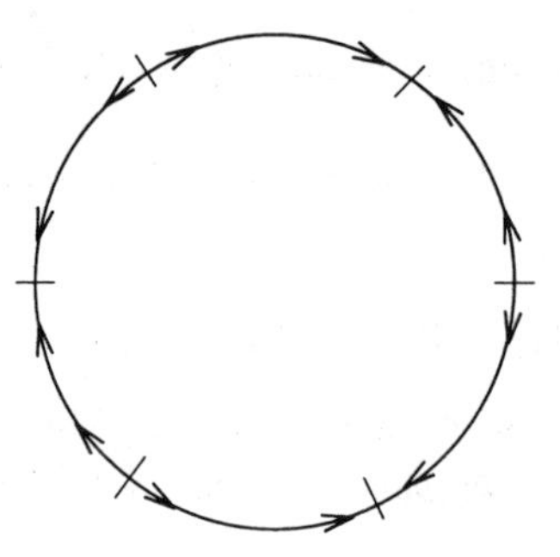

FIGURE 8.1. Example 8.2

$F'(x) = 1 + \epsilon 2\pi n \cos(2\pi n x)$, so $F'(j/2n) = 1 + \epsilon 2\pi n > 1$ for j even and $0 < F'(j/2n) = 1 - \epsilon 2\pi n < 1$ for j odd. Therefore, the points $x = j/2n$ are fixed point sources for j even and fixed point sinks for j odd. See Figure 8.1.

Example 8.3. Let f be the map on S^1 whose lift F is given by

$$F(t) = t + 1/n + \epsilon \sin(2\pi n t)$$

for n a positive integer and $0 < \epsilon < 1/(2\pi n)$. Let $x_j = j/2n$. Then $F(x_j) = x_{j+2}$, so $\{x_j : j \text{ is even}\}$ is one periodic orbit of period n and $\{x_j : j \text{ is odd}\}$ is another periodic orbit of period n. The derivative of F^n at x_0 satisfies

$$(F^n)'(x_0) = F'(x_{2n-2}) \cdots F'(x_0) = (1 + \epsilon 2\pi n)^n > 1,$$

so the first orbit is a source. Similarly,

$$(F^n)'(x_1) = F'(x_{2n-1}) \cdots F'(x_1) = (1 - \epsilon 2\pi n)^n < 1,$$

so the second orbit is a sink. It can be checked that any other point y (i) is not periodic, (ii) has $\alpha(y) = \mathcal{O}(x_0)$, and (iii) has $\omega(y) = \mathcal{O}(x_1)$.

Having looked at some examples with rational rotation number, we return to rotations on the circle with irrational rotation number. We prove that every point for this map has a dense orbit.

Theorem 8.3. *Let f_λ be a rotation on S^1 as defined above. Assume λ is irrational. Then, f_λ has no periodic points and every point in S^1 has a dense orbit in S^1. Thus, for any $x \in S^1$, $\omega(x) = S^1$ and S^1 is a minimal set for f_λ.*

PROOF. As we noted in Example 8.1, the lift F_λ is given by $F_\lambda(t) = t + \lambda$, $\rho_0(F_\lambda) = \lambda$, and $\rho(f_\lambda) = \lambda \bmod 1$. By Example 8.1 or Theorem 8.2, f_λ has no periodic points because λ is irrational.

Now, we turn to showing that all orbits are dense. Since there are no periodic points, all the points $F_\lambda^j(x)$ are distinct modulo one. (If $F_\lambda^n(x) = F_\lambda^m(x) \bmod 1$ for $n \neq m$, then $x + n\lambda = x + m\lambda + k$ for an integer k. Then, $\lambda = k/(n-m)$ is rational.) The set $f_\lambda^j(x)$ must have a limit point in S^1. Thus, given $\epsilon > 0$, there exist integers $n \neq m$ and k such that $|F_\lambda^n(x) - F_\lambda^m(x) - k| < \epsilon$, or in S^1 $d(f_\lambda^n(x), f_\lambda^m(x)) < \epsilon$. The lift F_λ preserves lengths, so letting $q = n - m$, $d(f^q(x), x) < \epsilon$. Then, $d(f^{2q}(x), f^q(x)) < \epsilon$, $d(f^{3q}(x), f^{2q}(x)) < \epsilon$, ... $d(f^{(j+1)q}(x), f^{jq}(x)) < \epsilon$. These intervals eventually cover S^1, so the orbit of x is ϵ-dense in S^1. Because $\epsilon > 0$ is arbitrary, the orbit of x is dense in S^1, $\omega(x) = S^1$, and S^1 is a minimal set for f_λ. □

Theorem 8.4. *Assume $f : S^1 \to S^1$ is a continuous orientation preserving homeomorphism and $\rho(f)$ is irrational. Then, the following are true:*

(a) *$\omega(x)$ is independent of x,*
(b) *$\omega(x)$ is a minimal set, and*
(c) *$\omega(x)$ is either (i) all of S^1 or (ii) a Cantor subset of S^1.*

We leave the proof of this result to Exercise 2.28.

Theorem 8.5. *Assume $f : S^1 \to S^1$ is a continuous orientation preserving homeomorphism and $\tau = \rho(f)$ is irrational.*

(a) Then, f is semi-conjugate to the rigid rotation map with rotation τ, f_τ. The semi-conjugacy takes the orbits of f to orbits of f_τ, is at most two to one on $\omega(x, f)$, and preserves orientation.

(b) If $\omega(x, f) = S^1$, then f is conjugate to f_τ.

(c) If $\omega(x, f) \neq S^1$, then the semi-conjugacy h from f to f_τ collapses the closure of each open interval I in the complement of $\omega(x)$ to a point.

PROOF. Let F be a lift of f. The next lemma proves that the order of orbits on the lift is the same as the order of the lift F_τ of the rigid rotation.

Lemma 8.6. *Choose the lift F so that $\rho_0(F) = \tau$, with τ irrational. Let $t \in \mathbb{R}$ and let $k, m, n, q \in \mathbb{Z}$. Then, (i) $F^n(t) + m < F^k(t) + q$ for some t if and only if (ii) $F^n(t) + m < F^k(t) + q$ for all t if and only if (iii) $n\tau + m < k\tau + q$ if and only if (iv) $F_\tau^n(0) + m < F_\tau^k(0) + q$ where F_τ is the translation by τ.*

PROOF. The equivalence of conditions (iii) and (iv) is clear.

Because the rotation number is irrational, the order of $F^n(t) + m$ and $F^k(t) + q$ in the line is independent of t, so condition (i) is equivalent to condition (ii).

In showing that condition (ii) implies (iii), replacing t by $F^{-k}(t)$ in condition (ii) gives that $F^{n-k}(t) - t < q - m$ for all t. This in turn implies that

$$\begin{aligned}\tau &= \rho_0(F) \\ &= \lim_{j\to\infty} \sum_{i=1}^{j} \frac{F^{i(n-k)}(t) - F^{(i-1)(n-k)}(t)}{j(n-k)} \\ &\leq \frac{q-m}{n-k}.\end{aligned}$$

The inequality must be strict because the rotation number is irrational, so $n\tau + m < k\tau + q$ which is condition (iii).

Assuming condition (iii), we get that $F^{n-k}(t) - t < q - m$ for some t since the rotation number is less than $(q - m)/(n - k)$. Again, the sign of this inequality must be the same for all t because the rotation number is irrational. Substituting $F^k(t)$ for t, we get $F^n(t) - F^k(t) < q - m$ for all t, or condition (ii). □

Let $x_0 \in \omega(x, f)$ and $t_0 \in \mathbb{R}$ with $\pi(t_0) = x_0$. Let $\mathbb{B} = \{F^n(t_0) + m : n, m \in \mathbb{Z}\} \subset \mathbb{R}$. Let $\mathcal{A} = \{F_\tau^n(0) + m : n, m \in \mathbb{Z}\}$. Then, the map $\bar{h} : \mathbb{B} \to \mathcal{A}$ defined by $\bar{h}(F^n(t_0) + m) = F_\tau^n(0) + m$ is order preserving by Lemma 8.6. Also, $\bar{h}(t+1) = \bar{h}(t) + 1$ by the construction. Because of the order preserving property, $\bar{h}$ has a continuous extension to the closure, $\bar{h} : \text{cl}(\mathbb{B}) \to \text{cl}(\mathcal{A}) = \mathbb{R}$. The extension is also order preserving and satisfies $\bar{h}(t + 1) = \bar{h}(t) + 1$. The order preserving map $\bar{h} : \text{cl}(\mathbb{B}) \to \mathbb{R}$ induces a map $h : \omega(x, f) \to S^1$. This map h is semi-conjugacy of $f|\omega(x, f)$ to f_τ.

Claim 1. *If* $\text{cl}(\mathbb{B}) = \mathbb{R}$, *then* $\bar{h}$ *is one to one on* $\mathbb{R}$ *and* h *is one to one on* S^1. *Thus, part (b) of the theorem is true.*

PROOF. Assume there are $x_1 < x_2$ with $\bar{h}(x_1) = \bar{h}(x_2)$. Letting $i = 1, 2$, there are two increasing sequences of points $y_{n_j^i} + m_i \in \mathbb{B}$ for $j \geq 1$ such that $y_{n_j^i} + m_i$ converges to x_i from below. (The m_i can be taken independent of j because the points $y_{n_j^i} + m_i$ converge to a single point.) We can assume that $y_{n_j^1} + m_1 < y_{n_k^2} + m_2$ for all j and k by taking a subsequence. Then, $\bar{h}(y_{n_j^1} + m_1) = F_\tau^{n_j^1}(0) + m_1$ and $\bar{h}(y_{n_j^2} + m_2) = F_\tau^{n_j^2}(0) + m_2$ both converge to the same point and from the same side. Thus, the two sequences in the image must be interlaced while those in the domain are not. This contradicts the order preserving property of $\bar{h}$ and proves the claim about $\bar{h}$.

The properties of h in this case follow from those of $\bar{h}$. □

Claim 2. *If* $\text{cl}(\mathbb{B}) \neq \mathbb{R}$, *then* $\bar{h}$ *takes the two end points of an open interval in* $\mathbb{R} \setminus \text{cl}(\mathbb{B})$ *to the same point. Thus,* $\bar{h}$ *has a continuous order preserving extension to* $\mathbb{R}$ *which takes the closure of each open interval in* $\mathbb{R} \setminus \text{cl}(\mathbb{B})$ *to a single point. This map* $\bar{h} : \mathbb{R} \to \mathbb{R}$ *induces a semi-conjugacy* $h : S^1 \to S^1$ *which satisfies the conditions of part (c) of the theorem.*

PROOF. Take an interval (e_1, e_2) in the complement of $\text{cl}(\mathbb{B})$. No points in the orbit of t_0 intersect (e_1, e_2). We need to prove that $\bar{h}(e_1) = \bar{h}(e_2)$. If $\bar{h}(e_1) \neq \bar{h}(e_2)$, then by the order preserving property of $\bar{h}$, the image $\bar{h}(\text{cl}(\mathbb{B}))$ misses the interval $(\bar{h}(e_1), \bar{h}(e_2))$ in $\mathbb{R}$. However, $\bar{h}(\text{cl}(\mathbb{B})) = \text{cl}(\mathcal{A}) = \mathbb{R}$. This contradiction proves that it must be the case that $\bar{h}(e_1) = \bar{h}(e_2)$. Thus, $\bar{h}$ has a continuous order preserving extension from $\mathbb{R}$ to $\mathbb{R}$ with $\bar{h}([e_1, e_2]) = \{\bar{h}(e_1)\}$. The rest of the conclusions of the claim follow. □

By Claims 1 and 2, there is always an order preserving map $\bar{h} : \mathbb{R} \to \mathbb{R}$ with $\bar{h}(t+1) = \bar{h}(t) + 1$ which induces a semi-conjugacy $h : S^1 \to S^1$ from f to f_τ. The two claims prove the desired properties of h. This completes the proof of Theorem 8.5. □

Theorem 8.7 (Denjoy). *Assume that* $f : S^1 \to S^1$ *is a* C^2 *orientation preserving diffeomorphism. Assume that* $\tau = \rho(f)$ *is irrational. Then* f *is transitive, so* f *is topologically conjugate to the rigid rotation* f_τ.

See Nitecki (1971), de Melo (1989), or de Melo and Van Strien (1993) for a proof.

Theorem 8.8 (Denjoy). *Let* τ *be irrational. Then, there exists a* C^1 *orientation preserving diffeomorphism* f *of* S^1 *such that* $\rho(f) = \tau$ *and* $\omega(x) \neq S^1$.

PROOF. The proof consists of placing the open intervals which correspond to the gaps of the Cantor set in the same order as an orbit of f_τ, the rigid rotation.

Let ℓ_n be a sequence of positive real numbers (lengths) indexed by $n \in \mathbb{Z}$ with (i) $\lim_{n \to \pm\infty}(\ell_{n+1}/\ell_n) = 1$, (ii) $\sum_{n=-\infty}^{\infty} \ell_n = 1$, (iii) $\ell_n > \ell_{n+1}$ for $n \geq 0$, (iv) $\ell_n < \ell_{n+1}$ for $n < 0$, and (v) $3\ell_{n+1} - \ell_n > 0$ for $n \geq 0$. For example, $\ell_n = T(|n| + 2)^{-1}(|n| + 3)^{-1}$ works where $T^{-1} = \sum_{n=-\infty}^{\infty}(|n| + 2)^{-1}(|n| + 3)^{-1}$.

Let I_n be an closed interval of length ℓ_n. We place these intervals on the circle in the same order as the order of the orbit $f_\tau^n(0)$. So to place an interval I_n, consider the sum of the lengths of the intervals I_j where $f_\tau^j(0)$ is between $f_\tau^n(0)$ and 0. This determines the placement of I_n. Since the circle has a total length of one, the measure of $S^1 \setminus \bigcup_{n \in \mathbb{Z}} \text{int}(I_n)$ is zero.

The next step is to define f on the union of the I_n. It is necessary and sufficient for $f'(t) = 1$ on the endpoints in order for the map to have a continuous derivative when it is extended to the closure. Assume $I_n = [a_n, b_n]$, so $\ell_n = b_n - a_n$. The integral

$$\int_{a_n}^{b_n} (b_n - t)(t - a_n)dt = \frac{\ell_n^3}{6},$$

so

$$\frac{6(\ell_{n+1} - \ell_n)}{\ell_n^3} \int_{a_n}^{b_n} (b_n - t)(t - a_n)dt = \ell_{n+1} - \ell_n.$$

Therefore, if we define f for $x \in I_n$ by

$$f(x) = a_{n+1} + \int_{a_n}^{x} 1 + \frac{6(\ell_{n+1} - \ell_n)}{\ell_n^3}(b_n - t)(t - a_n)dt,$$

then $f(b_n) = a_{n+1} + \ell_n + \ell_{n+1} - \ell_n = b_{n+1}$. Also, f is differentiable on I_n with

$$f'(x) = 1 + \frac{6(\ell_{n+1} - \ell_n)}{\ell_n^3}(b_n - x)(x - a_n).$$

Thus, $f'(a_n) = 1 = f'(b_n)$. Notice that for $n < 0$, $\ell_{n+1} - \ell_n > 0$, that

$$\begin{aligned} 1 &\le f'(x) \\ &\le 1 + \frac{6(\ell_{n+1} - \ell_n)}{\ell_n^3}\left(\frac{\ell_n}{2}\right)^2 \\ &= \frac{3\ell_{n+1} - \ell_n}{2\ell_n}, \end{aligned}$$

and $(3\ell_{n+1} - \ell_n)/(2\ell_n)$ goes to 1 as n goes to minus infinity. Similarly for $n \ge 0$ and $x \in I_n$,

$$1 \ge f'(x) \ge \frac{3\ell_{n+1} - \ell_n}{2\ell_n} > 0,$$

so $f'(x)$ goes to 1 as n goes to plus infinity uniformly for $x \in I_n$. From these facts, it follows that f is uniformly C^1 on the union of the interiors of the I_n and has a C^1 extension to all of S^1.

The second derivative $f''(x)$ is given by

$$f''(x) = \frac{6(\ell_{n+1} - \ell_n)}{\ell_n^3}[(b_n - x) - (x - a_n)],$$

so (i) $f''((a_n + b_n)/2) = 0$ in the middle of the interval, and (ii) as x goes to a_n, $f''(x)$ converges to

$$\begin{aligned} f''(a_n) &= \frac{6(\ell_{n+1} - \ell_n)}{\ell_n^3}\ell_n \\ &= \frac{6(\ell_{n+1} - \ell_n)}{\ell_n}\ell_n^{-1}, \end{aligned}$$

which is unbounded as $|n|$ goes to infinity. Thus, f is not C^2, and this example of Denjoy does not contradict the theorem of Denjoy.

Let $\Lambda = S^1 \setminus \bigcup_{n \in \mathbb{Z}} \text{int}(I_n)$. This set can be formed by successively removing open intervals. Because an open interval is eventually removed from any of the closed intervals obtained at a finite stage, Λ is a Cantor set.

The orbit of a point $x \in \Lambda$ is dense in Λ since it is like the orbit of 0 for f_τ by Theorem 8.5. Thus, $\omega(x) = \Lambda$. If $x \in \text{int}(I_n)$, then there is a smaller interval U whose closure is contained in $\text{int}(I_n)$. Since the interval I_n never returns to I_n but wanders among the other I_j, $x \notin \omega(x)$. Also $x \notin \Omega(f)$, the nonwandering set of f. This proves that $\Lambda = \omega(x) = \Omega(f)$ for any $x \in S^1$. This completes the proof. □

Recent work has dealt with the existence of a differentiable conjugacy between a diffeomorphism f with irrational rotation number τ and f_τ. Arnold, Moser, and Herman have obtained results. See de Melo (1989) or the revised version de Melo and Van Strien (1993) for a discussion of these results and references.

2.9 Exercises

Periodic Points

2.1. A homeomorphism f of $\mathbb{R}$ is *(strictly) monotonically increasing* provided $x < y$ implies that $f(x) < f(y)$. It is *(strictly) monotonically decreasing* if $x < y$ implies that $f(x) > f(y)$.
  (a) Prove that any homeomorphism f of $\mathbb{R}$ is either monotonically increasing or monotonically decreasing.
  (b) Prove that a homeomorphism f of $\mathbb{R}$ can never have periodic points whose least period is greater than 2.

2.2. Let $f : \mathbb{R} \to \mathbb{R}$ be continuous. Assume for one point x_0 the orbit $f^j(x_0)$ is a monotone sequence and is bounded. Prove that $f^j(x_0)$ converges to a fixed point.

2.3. Prove Theorem 2.2.

2.4. Let

$$T(x) = \begin{cases} 2x & \text{for } x \le 1/2 \\ 2 - 2x & \text{for } x \ge 1/2. \end{cases}$$

be the *tent map.*
  (a) Sketch the graph on $I = [0, 1]$ of T, T^2, and (a representative graph of) T^n for $n > 2$.
  (b) Use the graph of T^n to conclude that T has exactly 2^n points of period n. (These points do not necessarily have least period n but are fixed by T^n.)
  (c) Prove that the set of all periodic points of T is dense in $[0, 1]$.
  (d) Find the number of points with least period 1, 2, and 4. Also find the number of distinct orbits with least period 1, 2, and 4.

2.5. Let $F_4(x) = 4x(1 - x)$ on $\mathbb{R}$.
  (a) Make a rough sketch of the graph of $F_4^n(x)$ for $n > 2$.
  (b) Use the graph of F_4^n to conclude that F_4^n has exactly 2^n fixed points. (These points do not necessarily have least period n but are fixed by F_4^n.)

2.6. Let T be the tent map defined above. Prove that $x \in [0, 1]$ is eventually periodic if and only if x is rational. The following steps give an outline of the proof.
  (a) Assume x is eventually periodic. Prove that x is rational. Hint: $T^n(x) = 2\,i_n \pm 2^n\,x$, where i_n is an integer. ($i_1 = 0$ or 1)
  (b) Assume $x = 2k/p \in [0, 1]$, where p and $2k$ are relatively prime integers (so p is odd). Show that x is eventually periodic. Hint: Consider the set of all points of the form $2j/p$.
  (c) Assume $x = k/(2^j p)$ with $j \ge 0$, where k and $2^j p$ are relatively prime integers and p is odd. Show that x is eventually periodic. Hint: $T(k/(2^j p)) = k/(2^{j-1}p)$ or $(p - k)/(2^{j-1}p)$.

2.7. Let T be the tent map defined above. Prove that $x \in [0,1]$ is periodic if and only if $x = 2\,k/p$, where p is an odd integer and k is an integer. The following steps give an outline of the proof.

(a) Assume x is periodic. Prove that $x = 2\,k/p$, where p is an odd integer. Hint: $T^n(x) = 2\,i_n \pm 2^n\,x$, where i_n is an integer.

(b) Let p be an odd integer. Prove that T is one to one on the set

$$\{2\,k/p \in [0,1] : k \text{ is an integer }\}.$$

2.8. Consider the quadratic map F_μ for parameter values $\mu > 1$.

(a) Find the points of period two and determine their stability. (Indicate for which parameter values they exist and the stability for different parameter values.)

(b) Let $\mu > 1$ be in the range of parameters for which the orbit of period 2 is attracting. Let $0 < x < 1$. Prove that either (i) there is an integer $k \geq 0$ such that $F_\mu^k(x) = p_\mu$, where p_μ is the fixed point, or (ii) the $\omega(x)$ is the orbit of period 2.

2.9. Let $f(x) = x^3 - \dfrac{5}{4}x$.

(a) Find the fixed points, $\{0, \pm\,p\}$, and determine their stability.

(b) Find the critical points and show they are nondegenerate. Label the critical points $\pm\,c$. Draw the graph of f.

(c) Find all points which satisfy $f(x) = -x$. Use this information to find an orbit of period 2, $\{\pm\,q\}$.

(d) Show that f^2 is monotone on $[-c,c]$, where $\pm\,c$ are the critical points. Use this information to prove that $W^s(\mathcal{O}(q)) \supset [-c,0) \cup (0,c]$. Then, prove that $W^s(\mathcal{O}(q)) = (-p,0) \cup (0,p) \setminus \mathcal{O}^-(0)$, where $\pm\,p$ are the two other fixed points besides 0 and $\mathcal{O}^-(0)$ is the set of all points which eventually map to 0.

Limit Sets

2.10. Let $f : X \to X$ be a continuous map on a metric space. Let $\mathbf{p} \in X$.

(a) Prove that $\mathrm{cl}(\mathcal{O}^+(\mathbf{p})) = \mathcal{O}^+(\mathbf{p}) \cup \omega(\mathbf{p})$.

(b) Prove that if $\omega(\mathbf{p}) = \emptyset$, then $\mathcal{O}^+(\mathbf{p})$ is a closed subset of X.

(c) Assume that $\mathcal{O}^+(\mathbf{p})$ is a compact subset of X. Prove that $\mathcal{O}^+(\mathbf{p}) \supset \omega(\mathbf{p})$.

Invariant Cantor Sets

2.11. Let $I = [0,1]$, and

$$T_s(x) = \begin{cases} sx, & \text{for } x \leq 1/2 \\ s(1-x), & \text{for } x \geq 1/2. \end{cases}$$

This is a *generalized tent map*. Prove that for $s > 2$, $\Lambda = \bigcap\{T_s^{-k}(I) : 0 \leq k < \infty\}$ is the middle-α Cantor set for some α.

2.12. Let $f_\lambda(x) = x^3 - \lambda x$.

(a) Find the fixed points of f_λ and determine their stability for $\lambda > 0$.

(b) Let $I_\lambda = [-(\lambda+1)^{1/2}, (\lambda+1)^{1/2}]$. For $\lambda > 0$, prove that a point x with $x \notin I_\lambda$ has $f_\lambda^n(x) \to +\infty$ or $f_\lambda^n(x) \to -\infty$.

(c) Consider $\lambda = 8$. Show that

$$S_n = \bigcap_{k=0}^{n} f_8^{-k}(I_\lambda)$$

contains 3^n intervals. Show that

$$\Lambda = \bigcap_{n=0}^{\infty} S_n$$

is a Cantor set (i.e., perfect and nowhere dense). Hint: If $x \in S_1$ show that $|f_8'(x)| > 1$.

2.13. Use Lemma 4.4 to show that if $\mu > 2(1+\sqrt{2})$, then Λ_μ has measure zero. Hint: If $\mu > 2(1+\sqrt{2})$, then $|F_\mu'(x)| > 2$ on $I_1 \cup I_2$. Remark: If $\mu > 4$, then Λ_μ has measure zero, but this fact is harder to prove.

2.14. For $\mu > 4$, show that $F_\mu(x) = \mu x(1-x)$ is expanding by a factor of 2 in terms of the density $\rho(x) = [x(1-x)]^{-1/2}$, i.e., show that $|F_\mu'(x)|\rho(F_\mu(x))/\rho(x) \geq 2$ for $x \in [0,1] \cap F_\mu^{-1}([0,1])$.

Symbolic Dynamics

2.15. Let d be the metric on $\Sigma_2^+ = \{1,2\}^{\mathbb{N}}$ defined by

$$d(\mathbf{s},\mathbf{t}) = \sum_{j=0}^{\infty} \frac{|s_j - t_j|}{3^j}.$$

(a) Given $\mathbf{t} \in \Sigma_2^+$ and $n \geq 0$, prove that

$$\{\mathbf{s} \in \Sigma_2^+ : s_j = t_j \text{ for } 0 \leq j \leq n\} = \{\mathbf{s} \in \Sigma_2^+ : d(\mathbf{s},\mathbf{t}) \leq 3^{-n}2^{-1}\}.$$

Also prove that this set is a closed set and an open ball in the above metric.

(b) Given $\mathbf{t} \in \Sigma_2^+$ and $n \geq 1$, prove that

$$\{\mathbf{s} \in \Sigma_2^+ : s_j = t_j \text{ for } 1 \leq j \leq n\}$$

is an open set (but not an open ball). Note that the range of entries of $\mathbf{s}$ starts with $j = 1$ and not 0.

(c) Given $\mathbf{t} \in \Sigma_2^+$ and $m \leq n$, prove that

$$\{\mathbf{s} \in \Sigma_2^+ : s_j = t_j \text{ for } m \leq j \leq n\}$$

is an open set. These sets are called cylinder sets.

(d) Prove that Σ_2^+ with the metric d is a complete metric space.

(e) Prove that Σ_2^+ is compact in terms of the metric d. Remark: The topology induced by d is the same as the one obtained by considering Σ_2^+ as the infinite product of $\{1,2\}$ with itself and using the product topology. Thus, by Tychonoff's Theorem Σ_2^+ is compact. In this exercise the reader is asked to verify this fact directly in this case.

2.16. Let $\lambda > 1$ and ρ_λ be the metric on $\Sigma_2^+ = \{1,2\}^{\mathbb{N}}$ defined by

$$\rho_\lambda(\mathbf{s},\mathbf{t}) = \sum_{j=0}^{\infty} \frac{|s_j - t_j|}{\lambda^j}.$$

Let d be the metric of the previous problem and Section 2.5.

(a) Prove that the identity map, *id*, is a homeomorphism from Σ_2^+ to itself when the domain is given the metric d and the range is given the metric ρ_λ. This shows that the two metrics have the same open sets and so induce the same topology for any $\lambda > 1$.

(b) Given $\mathbf{t} \in \Sigma_2^+$ and $n \geq 0$, prove that

$$\{\mathbf{s} \in \Sigma_2^+ : s_j = t_j \text{ for } 0 \leq j \leq n\}$$

is an open ball in terms of the metric ρ_λ if and only if $\lambda > 2$. Note that these sets are open sets in terms terms of the metric ρ_λ for any $\lambda > 1$ by part (a).

2.17. Prove that the full p-shift is topologically mixing, i.e., σ_p is topologically mixing on Σ_p^+ for any positive integer $p \geq 2$.

2.18. Let $I = [0,1]$, $F_5(x) = 5x(1-x)$, and

$$g(x) = \begin{cases} 4x & \text{for } 0 \leq x \leq 0.5 \\ 4x - 2 & \text{for } 0.5 < x \leq 1. \end{cases}$$

The map g is called the *Baker's map.*

(a) Let $F_5^{-1}(I) \cap I = I_1 \cup I_2$. For a string $s_j \in \{1,2\}$ for $0 \leq j \leq n$, let

$$I_{s_0 \ldots s_n} = \bigcap \{F_5^{-k}(I_{s_k}) : 0 \leq k \leq n\}.$$

For $n = 1, 2, 3$, indicate on a sketch all the possible intervals, $I_{s_0 \ldots s_n}$, for different choices of $s_0 \ldots s_n$. When is the interval $I_{s_0 \ldots s_n 1}$ to the left of $I_{s_0 \ldots s_n 2}$ and when is it to the right? What effect does having a symbol 2 (where the slope is negative on I_2) have on the order of the intervals?

(b) Let $g^{-1}(I) \cap I = J_1 \cup J_2$; for a string $s_j \in \{1,2\}$ for $0 \leq j \leq n$, let

$$J_{s_0 \ldots s_n} = \bigcap \{g^{-k}(J_{s_k}) : 0 \leq k \leq n\}.$$

For $n = 1, 2$, indicate on a sketch all these possible intervals, $J_{s_0 \ldots s_n}$, for different choices of $s_0 \ldots s_n$. What is the difference between the order of the interval in part (b) from those in part (a)?

2.19. Let

$$T(x) = \begin{cases} 2x & \text{for } x \leq 1/2 \\ 2 - 2x & \text{for } x \geq 1/2 \end{cases}$$

be the tent map. Set $I_1 = [0, 1/2]$ and $I_2 = [1/2, 1]$. Define the map $H : \Sigma_2^+ \to [0,1]$ by

$$H(\mathbf{s}) = \bigcap_{k=0}^{\infty} T^{-k}(I_{s_k}).$$

(a) Prove that H is well defined, i.e., that the intersection is a single point.

(b) Prove that H is a semi-conjugacy from σ on Σ_2^+ to T on $[0,1]$.

(c) What sequences correspond to $x = 0, 1, 1/2$, i.e., what are $H^{-1}(0)$, $H^{-1}(1)$, and $H^{-1}(1/2)$?

(d) Prove that H is one to one on most points, and at most two to one, i.e., prove that $H^{-1}(x)$ contains at most two sequences. What sequences go to the same point by H?

(e) Prove that T is topologically transitive.

2.20. Consider F_μ for $\mu > 2 + \sqrt{5}$ with invariant Cantor set Λ_μ. Prove that if $\epsilon > 0$ is small enough, then there is a $\delta > 0$ such that for any δ-chain $\{x_j\}_{j=0}^{\infty}$ with $d(x_j, \Lambda_\mu) < \delta$ for all j there is a unique $x \in \Lambda_\mu$ such that $|x_j - F_\mu^j(x)| < \epsilon$ for all j. (A point x satisfying the conclusion of this exercise is said to *ϵ-shadow the δ-chain* $\{x_j\}$.)

Conjugacy and Structural Stability

2.21. Which of the following are topologically conjugate on all of $\mathbb{R}$? Which are differentiably conjugate? Prove your answer.
 (a) $f(x) = x/2$
 (b) $f(x) = 2x$
 (c) $f(x) = -2x$
 (d) $f(x) = 5x$
 (e) $f(x) = x^3$

2.22. Let $g_a(y) = a\,y^2 - 1$ and $f_c(x) = x^2 - c$. For each parameter value a, find a parameter value c and a linear conjugacy from g_a to f_c.

2.23. Prove that the periodic points of F_4 are dense in $[0,1]$. Hint: Use the conjugacy from the tent map T to F_4 given in Example 6.2, and the fact from Exercise 2.4 that the periodic points for T are dense in $[0,1]$,

2.24. Assume $f : \mathbb{R} \to \mathbb{R}$ has a fixed point at x_0. Find an affine conjugacy h from f to a new function g, where g has a fixed point at 0. Also give the definition of g in terms of the function f. Hint: Think of f as written in terms of coordinates x, $x_1 = f(x)$, g as written in terms of coordinates y, $y_1 = g(y)$, and the affine conjugacy h as transforming the y coordinates into the x coordinates by means of a translation, $x = h(y) = x + b$,

2.25. Prove that $f(x) = x^3 + x/2$ is C^1-structurally stable.

2.26. Assume that $f : X \to X$ and $g : Y \to Y$ are semi-conjugate by means of the map $h : X \to Y$. Let x be a point of least period n. Prove that $h(x)$ is a periodic point for g whose period divides n. Also prove that if h is a conjugacy, then the period of $h(x)$ is n.

Homeomorphisms of the Circle

2.27. Let $f, g : S^1 \to S^1$ be two orientation preserving homeomorphisms that are topologically conjugate by an orientation preserving homeomorphism. Prove that they have the same rotation number, $\rho(f) = \rho(g)$.

2.28. Let $f : S^1 \to S^1$ be an orientation preserving homeomorphism with irrational rotation number.
 (a) Let $I = [f^n(x), f^m(x)]$ for $n < m$ and some $x \in S^1$. Prove that for any $y \in S^1$ there is a positive k such that $f^k(y) \in I$. Hint: Consider $f^{-i(m-n)}(I)$ for consecutive values of i.
 (b) Prove that for any two $x, y \in S^1$ it must be the case that $\omega(x) = \omega(y)$. Conclude that $\omega(x)$ is minimal. Hint: Use part (a) to show that if $z \in \omega(x)$, then $z \in \omega(y)$.
 (c) Take $x \in S^1$ and let $E = \omega(x)$. Prove that E is perfect. Hint: for $z \in E$ show that $z \in \omega(z) = \omega(x)$. What does this imply about how E must accumulate on z?
 (d) For $E = \omega(x)$, prove that E is nowhere dense or $E = S^1$. Hint: Show the boundary of E is invariant. Since E is minimal, what are the possibilities for the boundary of E?

2.29. For the following two functions on the circle, (i) find all the periodic points, (ii) determine the stability of each periodic point, and (iii) describe the phase portrait.

(a) $f(\theta) = \theta + \epsilon \sin(n\theta) \quad \mod 2\pi \quad \text{for} \quad n \geq 1, \quad \epsilon < 1/n.$

(b) $g(\theta) = \theta + (2\pi/n) + \epsilon \sin(n\theta) \quad \mod 2\pi \quad \text{for} \quad n \geq 2, \quad \epsilon < 1/n.$

2.30. This exercise applies the argument on the existence of a rotation number to show that a subadditive sequence has a limit. Suppose that a_n is a sequence of real numbers and c is a fixed real number for which $a_{m+n} \leq a_m + a_n + c$ for all $m, n \in \mathbb{N}$. (In ergodic theory, if $c = 0$, this type of sequence is called *subadditive.*)

(a) Prove that $a_{kp} \leq ka_p + (k-1)c$ for all $k, p \in \mathbb{N}$.

(b) Consider a fixed $p > 0$ and write $n = kp + i$, where $0 \leq i < p$. Prove that

$$\frac{a_n}{n} \leq \frac{a_i}{n} + \frac{a_p + c}{p}.$$

(c) By letting n and p go to infinity in the right order, deduce that

$$\limsup_{n\to\infty} \frac{a_n}{n} \leq \liminf_{p\to\infty} \frac{a_p}{p},$$

and so the limit $\lim_{n\to\infty} a_n/n$ exists. Note that the limit could be $-\infty$.

(d) If $c = 0$ and the sequence is subadditive, prove that

$$\lim_{n\to\infty} \frac{a_n}{n} = \inf_{n\in\mathbb{N}} \frac{a_n}{n}.$$

2.31. Let $f : S^1 \to S^1$ be an orientation preserving homeomorphism with lift $F : \mathbb{R} \to \mathbb{R}$. Assume f has a point x_0 with least period q.

(a) Prove that $F^q(t_0) = t_0 + p$ for some integer p, where $t_0 \in \mathbb{R}$ is a lift of $x_0 \in S^1$.

(b) Prove that $\rho_0(F) = p/q$. Thus, if f has a periodic point, its rotation number is rational.

CHAPTER III
Chaos and Its Measurement

The theme of the chapter is complicated dynamics or chaos of maps on the line. The first section presents a theorem of Sharkovskii; it proves for certain n and k, that the existence of a periodic orbit of period n forces other orbits of period k. This theorem is not exactly about complicated dynamics, but it does show that if a map f on the line has a period which is not equal to a power of 2, then f has infinitely many different periods. The existence of infinitely many different periods is an indication of the complexity of such a map. This theorem is not used later in the book but is of interest in itself, especially since it is proved using mainly the Intermediate Value Theorem and combinatorial bookkeeping. Sharkovskii's Theorem motivates the treatment of subshifts of finite type which is given in the next section. A subshift of finite type is determined by specifying which transitions are allowed between a finite set of states. Given such a system, it is easy to determine the periods which occur and other aspects of the dynamics. These systems are generalizations of the symbolic dynamics which we introduced for the quadratic map in Chapter II. In Chapter VIII and X, we give further examples of nonlinear dynamical systems which are conjugate to subshifts of finite type. Because we can analyze the subshift of finite type, we determine the complexity of the dynamics of the nonlinear map.

The last few sections of this chapter deal with topics related more directly to chaos. We give a couple of alternative definitions of chaos and introduce various properties which chaotic systems tend to possess. We use this context to introduce the concept of a Liapunov exponent for an orbit. This concept generalizes that of the eigenvalue for a periodic orbit, and associates to an orbit a growth rate of the infinitesimal separation of nearby points. This quantity can be defined even when the map does not have a "hyperbolic structure" like the Cantor set for the quadratic map. This quantity is often used to measure chaos in systems which can be simulated on a computer. In Chapter IX, we return to define Liapunov exponents in higher dimensions.

3.1 Sharkovskii's Theorem

In Chapter II we showed that the quadratic map $F_\mu(x) = \mu x(1-x)$ has an invariant Cantor set Λ_μ with points of all periods. In this section we study the following question: if a continuous function $f : \mathbb{R} \to \mathbb{R}$ has a point of period n, does it follow that f must have a point of period k? Stated differently, which periods k are forced by which other periods n? The following theorem is very simple to state and has a relatively simple proof.

Theorem 1.1 (Li and Yorke, 1975). *Assume $f : \mathbb{R} \to \mathbb{R}$ is continuous, and there is a point a such that either (i) $f^3(a) \leq a < f(a) < f^2(a)$ or (ii) $f^3(a) \geq a > f(a) > f^2(a)$. Then, f has points of all periods.*

REMARK 1.1. Note that it follows if f has a point of period three, then it has points of all other periods, hence the title of the Li and Yorke paper, "Period Three Implies Chaos."

REMARK 1.2. There is a more general result of Sharkovskii (1964), which also was proved earlier than the result of Li and Yorke. We prove the simpler result first because the proof is simpler and the lemmas used in the proof of the result of Li and Yorke are needed for the more general result. The treatment of this whole section follows the paper by Block, Guckenheimer, Misiurewicz, and Young (1980). Two good general references for these results and other related one-dimensional results are Alsedà, Llibre, and Misiurewicz (1993) and Block and Coppel (1992).

We assume the first ordering in the theorem with $f^3(a) \leq a < f(a) < f^2(a)$. (The proof for the other ordering is merely obtained by a reflection in the line.) Let $I_1 = [a, f(a)]$ and $I_2 = [f(a), f^2(a)]$. Then, $f(I_1) \supset I_2$ and $f(I_2) \supset I_1 \cup I_2$, as can be seen from the image of the endpoints of the intervals.

Lemma 1.2. *If I and J are closed intervals and $f(I) \supset J$, then there exists a subinterval $K \subset I$ such that $f(K) = J$, $f(\text{int}(K)) = \text{int}(J)$, and $f(\partial K) = \partial J$.*

PROOF. Let $J = [b_1, b_2]$. There exist $a_1, a_2 \in I$ such that $f(a_j) = b_j$. Assume $a_1 < a_2$. (The other case is similar with suprema and infima interchanged.) Let

$$x_1 = \sup\{x : a_1 \leq x \leq a_2 \text{ such that } f(x) = b_1\}.$$

By continuity, $f(x_1) = b_1$. Note that $x_1 < a_2$. Next let

$$x_2 = \inf\{x : x_1 \leq x \leq a_2 \text{ such that } f(x) = b_2\}.$$

Then, $f(x_2) = b_2$. Thus, $f(\{x_1, x_2\}) = \{b_1, b_2\}$. By the definitions of x_1 and x_2, $f((x_1, x_2)) \cap \partial J = \emptyset$. Thus, $f(\text{int}([x_1, x_2])) = \text{int}(J) = (b_1, b_2)$. This proves the lemma. □

Definition. An interval I *f-covers an interval* J provided $f(I) \supset J$. We write $I \to J$.

Lemma 1.3. *(a) Assume that there are two points $a \neq b$ with $f(a) > a$ and $f(b) < b$ and $[a, b]$ is contained in the domain of f. Then, there is a fixed point between a and b.*

(b) If a closed interval I f-covers itself, then f has a fixed point in I.

PROOF. (a) Let $g(x) = f(x) - x$. Then, $g(a) > 0$ and $g(b) < 0$. By the Intermediate Value Theorem there is a point c between a and b where $g(c) = 0$ so $f(c) = c$. This result can also be seen graphically by considering the two cases where (i) $a < b$ and (ii) $a > b$. See Figure 1.1.

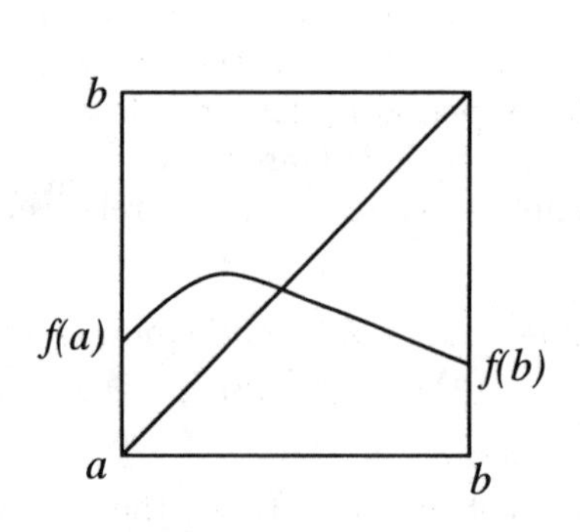

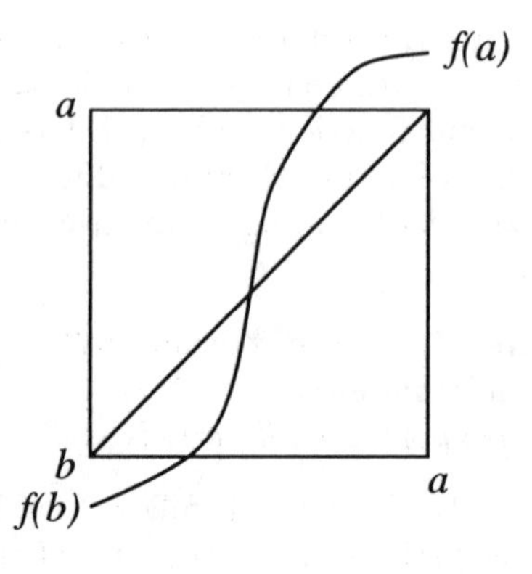

FIGURE 1.1. Fixed Point for Lemma 1.3(a)

(b) By Lemma 1.2, there is an interval $K = [x_1, x_2] \subset I$ with $f(K) = I = [a, b]$. Then, either (i) $f(x_1) = a \leq x_1$ and $f(x_2) = b \geq x_2$, or (ii) $f(x_1) = b > x_1$ and $f(x_2) = a < x_2$. If the equality holds, we are done. Otherwise, part (a) applies to prove there is a fixed point. □

Lemma 1.4. *Assume $J_0 \to J_1 \to \cdots \to J_n = J_0$ is a loop with $f(J_k) \supset J_{k+1}$ for $k = 0, \ldots, n-1$.*

(a) Then, there exists a fixed point x_0 of f^n with $f^k(x_0) \in J_k$ for $k = 0, \ldots, n$.

(b) Further assume that (i) this loop is not a product loop formed by going p times around a shorter loop of length m where $mp = n$, and (ii) $\text{int}(J_i) \cap \text{int}(J_k) = \emptyset$ unless $J_k = J_i$. If the periodic point x_0 of part (a) is in the interior of J_0, then it has least period n.

REMARK 1.3. Note that the loops that we allow can repeat some intervals as, for example, $J_0 \to J_1 \to \cdots \to J_{n-2} \to J_0 \to J_0$, or $J_0 \to J_1 \to J_2 \to J_1 \to J_2 \to J_0$. However, we do not allow a loop such as $J_0 \to J_1 \to J_0 \to J_1$.

PROOF. (a) We give a proof by induction on j. The induction statement is as follows.

(S_j) There exists a subinterval $K_j \subset J_0$ such that for $i = 1, \ldots, j$, $f^i(K_j) \subset J_i$, $f^i(\text{int}(K_j)) \subset \text{int}(J_i)$, and $f^j(K_j) = J_j$.

By Lemma 1.2, the induction hypothesis is true for $j = 1$.

Assume (S_{k-1}) is true. Thus, there exists a K_{k-1}. Then,

$$f^k(K_{k-1}) = f(f^{k-1}(K_{k-1})) = f(J_{k-1}) \supset J_k$$

By Lemma 1.2, there exists a subinterval $K_k \subset K_{k-1}$ such that $f^k(K_k) = J_k$ with $f^k(\text{int}(K_k)) = \text{int}(J_k)$. By the induction assumption (S_{k-1}), the other statements of (S_k) are true.

Now, using the statement (S_n), we have $f^n(K_n) = J_0$. By Lemma 1.3, f^n has a fixed point x_0 in $K_n \subset J_0$. Because $x_0 \in K_n$, $f^i(x_0) \in J_i$ for $i = 0, \ldots, n$. This proves part (a).

For part (b), since $f^n(\text{int}(K_n)) = \text{int}(J_0)$, if $x_0 \in \text{int}(J_0)$, then $x_0 \in \text{int}(K_n)$ and $f^i(x_0) \in \text{int}(J_i)$ for $i = 1, \ldots, n$. Because the loop is not a product, x_0 must have period n. □

PROOF OF THEOREM 1.1.

We assume the first case where $f(a) = b > a$, $f^2(a) = f(b) = c > f(a) = b$, and $f^3(a) = f(c) \leq a$. Let $I_1 = [a, b]$ and $I_2 = [b, c]$. Then, I_1 f-covers I_2 and I_2 f-covers both I_1 and I_2.

First, $F(I_2) \supset I_2$, so there is a fixed point by Lemma 1.3.

Next we show that f has a point of period n for any $n \geq 2$. Take the loop of length n with one interval being I_1 and $n-1$ intervals being repeated copies of I_2: $I_1 \to I_2 \to I_2 \to \cdots \to I_2 \to I_1$. By Lemma 1.4, there exists an $x_0 \in I_1$ such that $f^n(x_0) = x_0$ and $f^j(x_0) \in I_2$ for $j = 1, \ldots, n-1$. If there were a k with $1 \leq k < n$ such that $f^k(x_0) = x_0$, then we would have $x_0 = f^k(x_0) \in I_2$. Thus, we would have $x_0 \in I_1 \cap I_2 = \{b\}$. We now show that $x_0 = b$ is impossible. The argument is slightly different for $n = 2$ and $n \geq 3$. In the case when $n = 2$, $f^2(b) = f^2(x_0) = x_0 = b$, contradicting $f^2(b) = f^3(a) \leq a$. In the case when $n \geq 3$, we must have $f^2(b) = f^2(x_0) \in I_2$, contradicting $f^2(b) = f^3(a) \leq a$. This contradiction shows that $f^j(x_0) \neq x_0$ for $1 \leq j < n$, and x_0 has period n. □

Definition. In order to state the result of Sharkovskii, we need to introduce a new ordering on the positive integers using the symbol $\triangleright$, called the *Sharkovskii ordering*. First, the odd integers greater than one are put in the backward order:

$$3 \triangleright 5 \triangleright 7 \triangleright 9 \triangleright 11 \triangleright \cdots .$$

Next, all the integers which are two times an odd integer are added to the ordering, and then the odd integers times increasing powers of two:

$$3 \triangleright 5 \triangleright 7 \triangleright \cdots \triangleright 2 \cdot 3 \triangleright 2 \cdot 5 \triangleright \cdots \triangleright 2^2 \cdot 3 \triangleright 2^2 \cdot 5 \triangleright \cdots$$
$$\triangleright 2^n \cdot 3 \triangleright 2^n \cdot 5 \triangleright \cdots \triangleright 2^{n+1} \cdot 3 \triangleright 2^{n+1} \cdot 5 \triangleright \cdots.$$

Finally, all the powers of two are added to the ordering in decreasing powers:

$$3 \triangleright 5 \triangleright \cdots \triangleright 2^n \cdot 3 \triangleright 2^n \cdot 5 \triangleright \cdots \triangleright \cdots \triangleright 2^{n+1} \triangleright 2^n \triangleright \cdots \triangleright 2^2 \triangleright 2 \triangleright 1.$$

We have now given an ordering between all positive integers. This ordering seems strange but it turns out to the be ordering which expresses which periods imply which other periods as given in the Theorem of Sharkovskii (Sharkovskii, 1964).

Theorem 1.5 (Sharkovskii). *Let $f : I \subset \mathbb{R} \to \mathbb{R}$ be a continuous function from an interval I into the real line. Assume f has a point of period n and $n \triangleright k$. Then, f has a point of period k. (By period, we mean least period.)*

Until the proof of the theorem is complete, f is assumed to be a continuous function from I to $\mathbb{R}$ as given in the statement. The proof of the theorem involves finding intervals which f-cover each other in certain ways. In order to express these ideas, we introduce the following definition of a type of transition graph.

Definition. Let $\mathcal{A} = \{I_1, \ldots, I_s\}$ be a partition of I into closed intervals with disjoint nonempty interiors. An *transition graph of f for the partition $\mathcal{A}$* is a directed graph with vertices given by the I_j and a directed edge from I_j to I_k if I_j f-covers I_k. See Figures 1.3 and 1.4 for examples.

Example 1.1. Let f have a graph as indicated in Figure 1.2 with three intervals, I_1, I_2, I_3. Then, I_1 f-covers I_2, I_2 f-covers I_1 and I_2, and I_3 f-covers I_1, I_2, and I_3. Thus, the transition graph for the partition $\{I_1, I_2, I_3\}$ is as given in Figure 1.3.

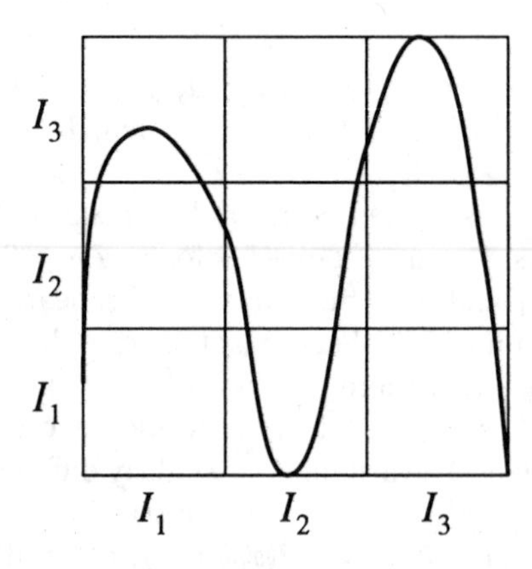

FIGURE 1.2. Graph of the Function in Example 1.1

REMARK 1.4. We first consider the case where n is an odd integer for which (i) $n > 1$ and (ii) f has a point x of period n and f has no points of odd period k with $1 < k < n$ (i.e., $k \triangleright n$). To prove Sharkovskii's Theorem in this case, Peter Stefan had the idea to

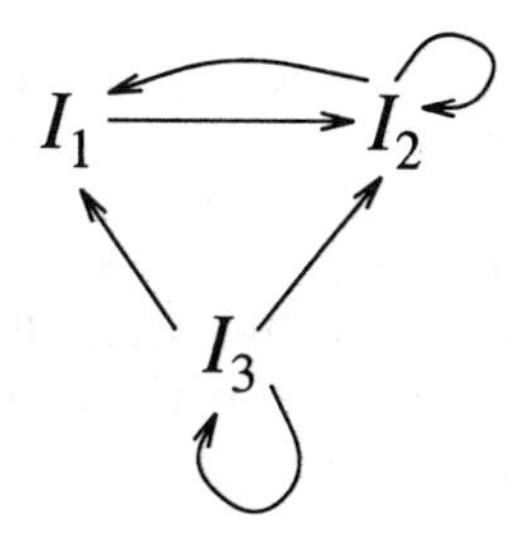

FIGURE 1.3. Transition Graph for the Partition in Example 1.1

prove the existence of an orbit with a special pattern on the line: let x_1 be an n-periodic point such that

$$x_n < x_{n-2} < \cdots < x_3 < x_1 < x_2 < x_4 < \cdots < x_{n-1}$$

where $x_j = f^{j-1}(x_1)$. (The reflection of this ordering is just as good.) A periodic point with such an ordering of its orbit on the line is called a *Stefan cycle*. Lemma 1.6 proves that indeed such an orbit does exist. Given such an orbit, let $I_1 = [x_1, x_2]$, $I_2 = [x_3, x_1]$, $I_3 = [x_2, x_4]$, $I_{2j} = [x_{2j+1}, x_{2j-1}]$, and $I_{2j-1} = [x_{2j-2}, x_{2j}]$ for $j = 2, \ldots, (n-1)/2$. Because of the nature of the orbit, (i) I_1 f-covers I_1 and I_2, (ii) I_j f-covers I_{j+1} for $2 \leq j \leq n-2$, and (iii) I_{n-1} f-covers all the I_j for j odd. Thus, the existence of such a special type of orbit proves that the transition graph of f for the partition $\mathcal{A}$ contains a subgraph of the form given in Figure 1.4. This special transition graph is called a *Stefan transition graph*. Applying the lemmas above to this Stefan transition graph can prove the existence of all the periodic implied by n in the Sharkovskii ordering.

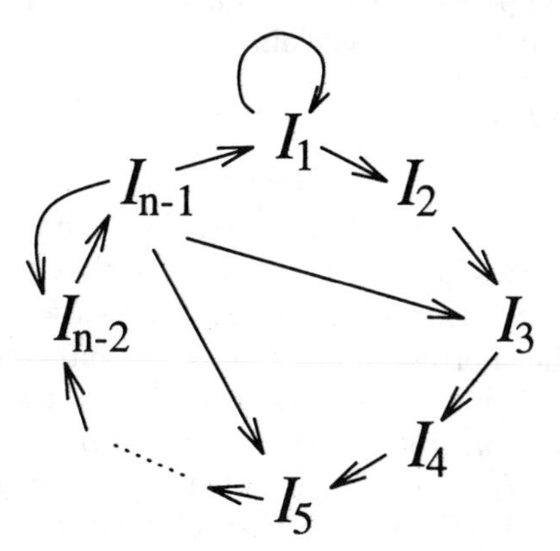

FIGURE 1.4. Transition Subgraph for the Partition in Lemma 1.6

We now turn to the lemma and its proof.

Lemma 1.6. *Assume n is an odd integer with $n > 1$. Assume that f has a point x of period n and f has no points of odd period k with $1 < k < n$ (i.e., $k \triangleright n$.) Let $J = [\min \mathcal{O}(x), \max \mathcal{O}(x)]$. Let $\mathcal{A}$ be the partition of J by the elements of $\mathcal{O}(x)$. Then, the transition graph of f for $\mathcal{A}$ contains a transition subgraph of the following form: The $I_1, \ldots, I_{n-1}$ can be numbered with all the intervals having disjoint interiors such that (i) I_1 f-covers I_1 and I_2, (ii) I_j f-covers I_{j+1} for $2 \leq j \leq n-2$, and (iii) I_{n-1} f-covers all the I_j for j odd. See Figure 1.4.*

PROOF. Let $\mathcal{O}(x) = \{z_1, z_2, \ldots, z_n\}$ where the z_j are ordered as on the line, $z_1 < z_2 < \cdots < z_n\}$. Then, $f(z_n) < z_n$ because $f(z_n)$ is one of the other z_j. Similarly, $f(z_1) > z_1$.

Let $a = \max\{y \in \mathcal{O}(x) : f(y) > y\}$. Then, $a \neq z_n$. Let b be the next larger than a among $\mathcal{O}(x)$ in terms of the ordering of the real line. Let $I_1 = [a, b] \in \mathcal{A}$. We show that this I_1 can be used in the statement of the lemma.

There is a sequence of small steps which we state as claims. First we need to show that I_1 covers itself (Claim 1), and eventually covers all of J (Claim 2). Claim 4 shows that there is a shortest loop with distinct intervals $I_1 \to I_2 \to \cdots \to I_{n-1} \to I_1$. Claims 5 and 6 complete showing that these intervals are situated on the line and behave as claimed.

Claim 1. *The image of I_1 covers itself, $f(I_1) \supset I_1$.*

PROOF. We know that $f(a) > a$ so $f(a) \geq b$. Also, since $b > a$, $f(b) < b$, so $f(b) \leq a$. Therefore, $f(I_1) \supset I_1$, as claimed. □

Claim 2. *The $(n-2)$ image of I_1 covers the whole interval J, $f^{n-2}(I_1) \supset J$.*

PROOF. Since $f(I_1) \supset I_1$, $f^{k+1}(I_1) \supset f^k(I_1)$, so the iterates are nested. The number of points in $\mathcal{O}(x) \setminus \{a, b\}$ is $n-2$, so $z_n \in f^k(I_1)$ for some $0 \leq k \leq n-2$. By the nested property, $z_n \in f^{n-2}(I_1)$. Similarly, $z_1 \in f^{n-2}(I_1)$. Since I_1 is connected, $f^{n-2}(I_1) \supset [z_1, z_n] = J$. □

Claim 3. *There exists a $K_0 \in \mathcal{A}$ with $K_0 \neq I_1$ such that $f(K_0) \supset I_1$.*

PROOF. This proof uses the fact that n is odd, so there are more elements of $\mathcal{O}(x)$ on one side of $\text{int}(I_1)$ than the other. Call $\mathcal{P}$ the elements of $\mathcal{O}(x)$ on the side of $\text{int}(I_1)$ with more elements. There is some $y_1, y_2 \in \mathcal{P}$ with $f(y_1) \in \mathcal{P}$ and $f(y_2) \in \mathcal{O}(x) \setminus \mathcal{P}$. Take adjacent points y_1 and y_2 with iterates as above. Let K_0 be the interval from y_1 to y_2. Then, $f(K_0) \supset I_1$ and $K_0 \neq I_1$, as claimed. □

Claim 4. *There is a loop $I_1 \to I_2 \to \cdots \to I_k \to I_1$ with $I_2 \neq I_1$. The shortest such loop with $k \geq 2$ has $k = n-1$.*

PROOF. Let K_0 be as in Claim 3, so $f(K_0) \supset I_1$. By Claim 2, $f^{n-2}(I_1) \supset K_0$. There are only $n-1$ distinct intervals in $\mathcal{A}$, so there exists such a loop with $2 \leq k \leq n-1$.

Now, assume the smallest k that works satisfies $2 \leq k < n-1$ and we get a contradiction. Since this is the shortest loop, none of the intervals can be repeated or it could be shortened. Either k or $k+1$ is odd. Let $m = k$ or $k+1$ be this odd integer, so $1 < m < n$. Use the loop with m intervals given by $I_1 \to I_2 \to \cdots \to I_k \to I_1$ or $I_1 \to I_2 \to \cdots \to I_k \to I_1 \to I_1$, depending on whether $m = k$ or $m = k+1$. By Lemma 1.4(a), there is a point z with $f^m(z) = z$. The point z cannot be on the boundary of the interval because these points have period n which is greater than m. Thus, z has least period m by Lemma 1.4(b). Since m is odd, this contradicts the assumption on n in the Lemma. This contradiction proves that $k = n-1$. □

For the rest of the proof, we fix $I_1, I_2, \ldots, I_{n-1}$ as given in Claim 4.

Claim 5. *(a) If $f(I_j) \supset I_1$, then $j = 1$ or $n-1$.*
(b) For $j > i+1$, there is no directed edge from I_i to I_j in the transition graph.
(c) The interval I_1 f-covers only I_1 and I_2.

PROOF. Part (a) follows from Claim 4. Parts (b) and (c) follow because the loop is the shortest possible. □

Claim 6. *Either (i) the ordering (in terms of the real line) of the intervals I_j in the loop of Claim 4 is $I_{n-1} \leq I_{n-3} \leq \cdots \leq I_2 \leq I_1 \leq I_3 \leq \cdots \leq I_{n-2}$ and the order of the* *orbit is $f^{n-1}(a) < f^{n-3}(a) < \cdots < f^2(a) < a < f(a) < f^3(a) < \cdots < f^{n-2}(a)$ or (ii) both of these orderings are exactly reversed.*

PROOF. Let $I_1 = [a, b]$. The interval I_1 f-covers only I_1 and I_2, so they must be next to each other. Assume that $I_2 \leq I_1$. (The other possibility gives the reverse order mentioned in the claim.) Then, it must be that $f(a) = b$ and $f(b)$ is the left endpoint of I_2.

Next, $f(\partial I_2) = \partial I_3$. Since one of these endpoints is $f(a) = b$ which is above $\text{int}(I_1)$, both endpoints of I_3 must be above $\text{int}(I_1)$. Also, because of Claim 5a (I_2 does not f-cover I_1) and 5b (I_2 does not f-cover I_j for $j > 3$), I_3 must be adjacent to I_1.

Continue the argument by induction. For $k < n-1$, since I_k does not f-cover I_1 and I_k does not f-cover I_j for $j > k+1$, I_{k+1} must be adjacent to I_{k-1}. This covers all the intervals in the claim.

Note that we have also shown the ordering on the orbit as stated in the claim. □

Claim 7. *The interval I_{n-1} f-covers all the I_j for j odd.*

PROOF. Note that $I_{n-1} = [f^{n-1}(a), f^{n-3}(a)]$. Then, $f(f^{n-1}(a)) = f^n(a) = a$. Also, $f^{n-3}(a) \in I_{n-3}$ so $f(f^{n-3}(a)) = f^{n-2}(a) \in I_{n-2}$ is the far right endpoint of J (the largest element in the orbit $\mathcal{O}(x)$). Thus, $f(I_{n-1}) \supset [a, f^{n-2}(a)] = I_1 \cup I_3 \cup \cdots \cup I_{n-2}$. We have proved the claim. □

All the claims together prove Lemma 1.6. □

Proposition 1.7. *Theorem 1.5 is true if n is odd and maximal in the ordering for which the theorem is true.*

PROOF. Take k with $n \triangleright k$. There are two cases: (a) k is even and $k < n$ and (b) $k > n$ with k either even or odd.

Case a. *The integer k is even and $k < n$.*

PROOF. Consider the loop of length k given by $I_{n-1} \to I_{n-k} \to I_{n-k+1} \to \cdots \to I_{n-1}$. By Lemma 1.4(a) there is a $x_0 \in I_{n-1}$ with $f^k(x_0) = x_0$. The point x_0 cannot be an endpoint because the endpoints have period n. Therefore, x_0 has period k. □

Case b. *The integer $k > n$ with k either even or odd.*

PROOF. Consider the loop of length k given by $I_1 \to I_2 \to \cdots \to I_{n-1} \to I_1 \to I_1 \to \cdots \to I_1$. Again by Lemma 1.4(a), there is a $x_0 \in I_1$ with $f^k(x_0) = x_0$. If $x_0 \in \partial I_1$, then x_0 has period n. Thus, n divides k, so $k \geq 2n \geq n+3$. Also, since $f^n(x_0) \in I_1$, the iterate $f^{n+1}(x_0) \notin I_1$, which contradicts the conclusion of Lemma 1.4(a). Therefore, $x_0 \notin \partial I_1$, and by Lemma 1.4(b), x_0 has period k. This completes the proof of Case b and Proposition 1.7. □

The first step in proving the result for other values of n proves the existence of a point of period two whenever there is a point of even period.

Lemma 1.8. *If f has a point of even period, then it has a point of period two.*

PROOF. Let n be the smallest integer greater than one in the usual ordering of the integers (not the Sharkovskii ordering) such that f has a point of period n. If n is odd, then we are done by Proposition 1.7. Therefore, we can assume that n is even. Let a, $I_1 = [a, b]$, and $J = [\min \mathcal{O}(a), \max \mathcal{O}(a)] = [A, B]$ be as before. In the proof of Lemma 1.6, we only used the fact that n is odd to show that there exists a $K_0 \in \mathcal{A}$ with $K_0 \neq I_1$ and $f(K_0) \supset I_1$.

First assume there is such a K_0. There is a minimal cycle as in Claim 4 with $2 \leq k \leq n-1$. As before, I_k covers all the I_j on the other side. Thus, $I_{n-1} \to I_{n-2} \to I_{n-1}$ is a cycle of length two, and there is a point of period two.

Next assume there is no $K_0 \in \mathcal{A}$ with $K_0 \neq I_1$ and $f(K_0) \supset I_1$. It follows that (i) all the points $x_j \in \mathcal{O}(a)$ with $x_j \leq a$ have $f(x_j) \geq b$ and (ii) all the points $x_j \in \mathcal{O}(a)$ with $x_j \geq b$ have $f(x_j) \leq b$. Since some points in $\mathcal{O}(a)$ are mapped to b and B, both $b, B \in f([A,a])$ and so $f([A,a]) \supset [b,B]$. Similarly, $f([b,B]) \subset [A,a]$. Then, $[A,a] \to [b,B] \to [A,a]$ is a cycle of length two. The intervals are disjoint, so there must be a point of period two. □

The proof of Sharkovskii's Theorem now splits into the following cases.

Case 1: n is odd and maximal in the Sharkovskii ordering and $n \rhd k$.
Case 2: $n = 2^m$ and $n \rhd k$.
Case 3: $n = 2^m p$ with $p > 1$ odd, $m \geq 1$, n is maximal in the Sharkovskii ordering, and $n \rhd k$.

Case 1 is proved above in Proposition 1.7. We split Case 2 up into subcases and prove it next.

Case 2: $n = 2^m$ and $n \rhd k$ so $k = 2^s$ with $0 \leq s < m$.
Case 2a: $s = 0$, i.e., f has a fixed point.
Case 2b: $s = 1$.
Case 2c: $s > 1$.

Proof of Case 2a. We can define a and b as before with $f(a) \geq b$ and $f(b) \leq a$. Therefore, $f([a,b]) \supset [a,b]$ and f has a fixed point. □

Case 2b follows from Lemma 1.8.

Proof of Case 2c. Let $g = f^{k/2} = f^{2^{s-1}}$. The map g has a point of period 2^{m-s+1} with $m - s + 1 \geq 2$. Lemma 1.8 proves that g has a point x_0 of period 2. So $x_0 = g^2(x_0) = f^k(x_0)$ and $x_0 \neq g(x_0) = f^{k/2}(x_0)$. Thus, the period of x_0 for f is 2^t for some $t \leq s$. If $t < s$, then x_0 is fixed by g which is impossible. Therefore, $t = s$ and x_0 is a point of period $2^s = k$. □

We also split Case 3 up into subcases.

Case 3: $n = 2^m p$ with $p > 1$ odd, $m \geq 1$, n is maximal in the Sharkovskii ordering for f, and $n \rhd k$.
Case 3a: $k = 2^s q$ with $s \geq m+1$ and $1 \leq q$ and q odd.
Case 3b: $k = 2^s$ with $s \leq m$.
Case 3c: $k = 2^m q$ with q odd and $q > p$.

We leave the proof of these cases to the exercises. See Exercises 3.2 – 3.4. This completes the proof of Theorem 1.5 (Sharkovskii's Theorem). □

3.1.1 Examples for Sharkovskii's Theorem

There are examples of maps with exactly the orbits implied by the Sharkovskii ordering. First consider the case where the maximal period in the ordering is odd.

Example 1.2. Let $n > 3$ be odd. (If $n = 3$, there are points of all periods and there is nothing to prove.) Let x_1 be a point which is a Stefan cycle for n. Make the graph be piecewise linear connecting the adjacent points $(x_j, f(x_j))$ on the graph by straight line segments. Let $I_1, \ldots, I_{n-1}$ be the intervals as in the proof. See Figure 1.5.

We claim that such a map does not have a point of odd period k with $1 < k < n$. Assume that x is a periodic point with period different than n. If any iterate of x hits one of the endpoints of an I_j, then either x has period n or is not periodic, and so this cannot happen. Thus, $f^j(x) \in \text{int}(I_{i(j)})$ for each j. Because the transition $\mathcal{A}$-graph is

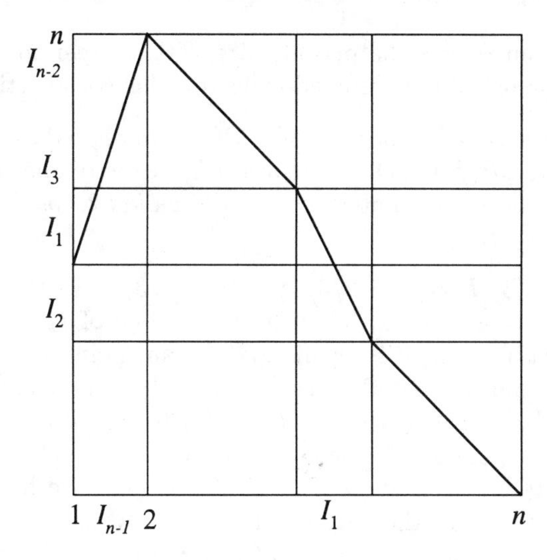

FIGURE 1.5. Example 1.2

exactly the subgraph proved to exist by Lemma 1.6, the length of cycles in the transition graph are exactly those k which are implied by n in the Sharkovskii ordering. Also, the graph over I_1 has slope -2. There is a fixed point in I_1 and all other points must leave I_1 and enter I_2. Thus, all orbits passing through I_1 are either fixed points or have periods at least $n-1$. Other orbits have to have the same period as the period of the cycle of intervals (since the orbit must pass through the interiors). Thus, the possible periods of periodic points are exactly those implied by n in the Sharkovskii ordering. In particular, there are no points of odd period with $1 < k < n$.

Definition. To get other examples with certain periods, we introduce the doubling operator. Let $I = [0, 1]$. Assume $f : I \to I$ is a continuous map. We denote the periods of the orbits of f by $\mathcal{P}(f)$. Now, we define the *double of* f, $\mathcal{D}(f) = g$, by

$$g(x) = \begin{cases} \frac{2}{3} + \frac{1}{3}f(3x) & \text{for } 0 \le x \le \frac{1}{3} \\ [2 + f(1)](\frac{2}{3} - x) & \text{for } \frac{1}{3} \le x \le \frac{2}{3} \\ x - \frac{2}{3} & \text{for } \frac{2}{3} \le x \le 1. \end{cases}$$

See Figure 1.6. It is easily checked that g is continuous.

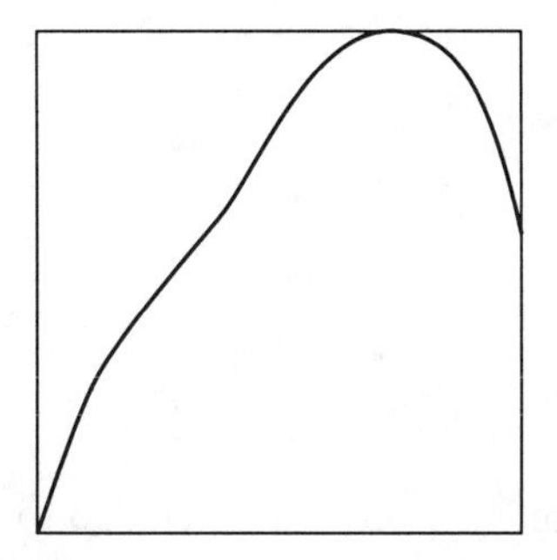

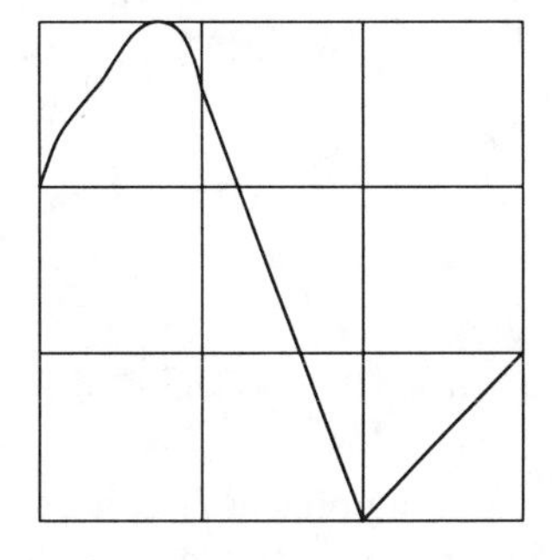

FIGURE 1.6. The Original Map f is in the Left figure and the Double g is in the Right Figure

The next proposition relates the periods of f with the periods of the double of f; this result justifies the use of the name "double" for this construction.

Proposition 1.9. *The set of all periods of g, $\mathcal{P}(g)$, are related to the set of all periods of f, $\mathcal{P}(f)$, by $\mathcal{P}(g) = 2\mathcal{P}(f) \cup \{1\}$. Moreover g has exactly one repelling fixed point, and for each n g has the same number of orbits of period $2n$ as f has of period n and their stability is the same.*

PROOF. Let $I_1 = [0, 1/3]$, $I_2 = (1/3, 2/3)$, and $I_3 = [2/3, 1]$. Because $g(I_2) \supset I_2$, g has a fixed point x_1 in I_2. Because the absolute value of the slope of g in I_2 is at least 2, there is exactly one fixed point in I_2 and it is repelling. Also, any point in I_2 other than the fixed point has an orbit which leaves I_2. Because $[g(I_1) \cup g(I_3)] \cap I_2 = \emptyset$, none of the points in $I_2 \backslash \{x_1\}$ can be periodic. If $x \in I_1$, then $g^2(x) = g(2/3 + f(3x)/3) = f(3x)/3 \in I_1$. Thus, for x to be periodic, its period must be even, $2k$. But by induction, $g^{2k}(x) = f^k(3x)/3$ for $k \geq 1$. Thus, $g^{2k}(x) = x$ if and only if $f^k(3x) = 3x$. We have shown that these periods of g are exactly twice the periods of f. Moreover, since $g'(t) = 1$ on I_3, for a point x of period $2k$ for g, $(g^{2k})'(x) = (f^k)'(3x)$ so the two orbits have the same stability type. The periodic points of g in I_3 are the same because they are on the orbits described above. This proves the proposition. □

Example 1.3. Let $f(x) \equiv 1/3$ for $x \in [0, 1]$. The only periodic point of f is a fixed point. Let $f_1 = \mathcal{D}(f)$. By the above proposition, the periods of f_1, $\mathcal{P}(f_1)$, are $\{1, 2\}$. Also, f_1 has one repelling fixed point and one attracting orbit of period 2. By induction, if $f_n = \mathcal{D}^n(f)$, then the periods of f_n, $\mathcal{P}(f_n)$, are $\{1, 2, \ldots, 2^n\}$, and f_n has one repelling periodic orbit of period 2^j for $1 \leq j < n$ and one attracting periodic orbit of period 2^n. Finally, let $f_\infty(x) = \lim_{n\to\infty} f_n(x)$. We leave to an exercise to prove that f_∞ is continuous and $\mathcal{P}(f_\infty) = \{1, 2, \ldots, 2^n, \ldots\}$, i.e., f_∞ has repelling periodic points of periods 2^n for all n and no other periods. See Exercise 3.8.

We also leave to the exercises the fact that if $n = 2^m p$ for $1 < p$, p odd, and $m \geq 1$, then there is a map f for which $\mathcal{P}(f) = \{k : n \triangleright k\}$. See Exercise 3.5.

3.2 Subshifts of Finite Type

In the proof of Sharkovskii's Theorem, we considered graphs where intervals f-covered each other forming a graph as given in Figure 2.1.

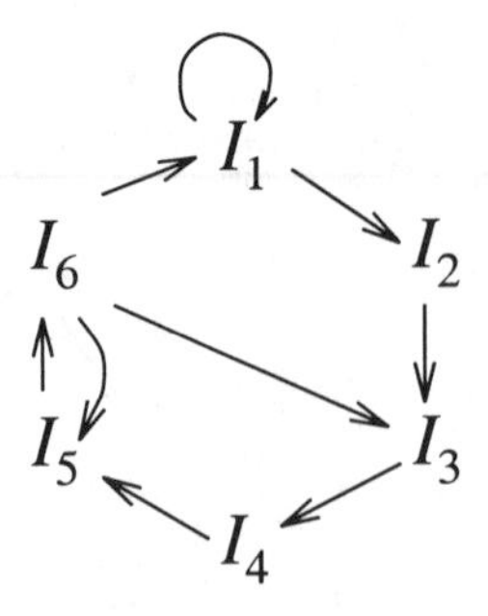

FIGURE 2.1. Graph of Partition in Sharkovskii's Theorem

With this graph, a point in I_1 can go to I_1 or I_2; a point in I_2 can go to I_3; a point in I_3 can go to I_4; a point in I_4 can go to I_5; a point in I_5 can go to I_6; a point in I_6 can go to I_1, I_3, or I_5. Paths in the graph correspond to allowable orbits of points. We can look at only the labeling of the intervals (as we did for the quadratic map $f_\mu(x) = \mu x(1 - x)$

for $\mu > 2 + \sqrt{5}$) and consider sequences $\mathbf{s} = s_0 s_1 s_2 \ldots$ where 1 can be followed by 1 or 2; 2 can only be followed by 3; 3 can only be followed by 4; 4 can only be followed by 5; 5 can only be followed by 6; can 6 can be followed by 1, 3, or 5. All other adjoining combinations are not allowed. Thus, a sequence like $634561123456\ldots$ is allowed.

Definition. Instead of looking at the graph, we can define a *transition matrix* to be a matrix $A = (a_{ij})$ such that (i) $a_{ij} = 0, 1$ for all i and j, (ii) $\sum_j a_{ij} \geq 1$ for all i, and (iii) $\sum_i a_{ij} \geq 1$ for all j. Given a graph of the type in Sharkovskii's Theorem, we can form a transition matrix A by letting $a_{ij} = 0$ if the transition from i to j is not allowed (there is no arrow in the graph from I_i to I_j) and $a_{ij} = 1$ if the transition from i to j is allowed (there is an arrow in the graph from I_i to I_j). The assumption that $\sum_j a_{ij} \geq 1$ for every i means that it is possible to go to some interval from I_i; the assumption that $\sum_i a_{ij} \geq 1$ for every j means that it is possible to get back to I_j from some interval. In the above graph, the transition matrix is

$$A = \begin{pmatrix} 1 & 1 & 0 & 0 & 0 & 0 \\ 0 & 0 & 1 & 0 & 0 & 0 \\ 0 & 0 & 0 & 1 & 0 & 0 \\ 0 & 0 & 0 & 0 & 1 & 0 \\ 0 & 0 & 0 & 0 & 0 & 1 \\ 1 & 0 & 1 & 0 & 1 & 0 \end{pmatrix}$$

Definition. Let Σ_n^+ be the space of all (one-sided) sequences with symbols in the set $\{1, 2, \ldots, n\}$ as defined in Section 2.5, and $\sigma : \Sigma_n^+ \to \Sigma_n^+$ be the shift map given by $\sigma(\mathbf{s}) = \mathbf{t}$ where $t_k = s_{k+1}$. This space has a metric as defined before.

Given an n by n transition matrix A, let

$$\Sigma_A^+ = \{\mathbf{s} \in \Sigma_n^+ : a_{s_k s_{k+1}} = 1 \text{ for } k = 0, 1, 2, \ldots\}.$$

This space Σ_A^+ is made up of the allowable sequences for A. Let $\sigma_A = \sigma | \Sigma_A^+$. The following proposition shows that σ_A acting on a sequence in σ_A gives another sequence in σ_A. The map $\sigma_A : \Sigma_A^+ \to \Sigma_A^+$ is called the *subshift of finite type for the matrix A*. Some other people call this map a *topological Markov chain*. The following proposition also shows that Σ_A^+ is closed.

Proposition 2.1. *(a) The subset Σ_A^+ is closed in Σ_n^+.*
(b) The map σ_A leaves Σ_A^+ invariant, $\sigma_A(\Sigma_A^+) = \Sigma_A^+$.

PROOF. (a) By using cylinder sets, it is easily seen that Σ_A^+ is closed.

(b) If $\mathbf{s} \in \Sigma_A^+$ and $\mathbf{t} = \sigma_A(\mathbf{s})$, then it follows directly that all the transitions in $\mathbf{t}$ are allowed so $\mathbf{t} \in \Sigma_A^+$. On the other hand, if $\mathbf{t} \in \Sigma_A^+$, then there is some s_0 such that $a_{s_0 t_0} = 1$ by the standing assumptions on A. Let $s_k = t_{k-1}$ for $k \geq 1$. Then, $\mathbf{s} \in \Sigma_A^+$ and $\sigma_A(\mathbf{s}) = \mathbf{t}$. □

Definition. In general, a subset $S \subset \Sigma_n^+$ is called a *subshift* provided that it is closed and invariant by the shift map σ. The following example gives a subshift which is not of finite type.

Example 2.1. Let S be the subset of Σ_2^+ consisting of all strings $\mathbf{s}$ such that between any two 2's in the string $\mathbf{s}$, there are an even number of 1's: i.e., if $s_j = 2 = s_k$ with $j < k$, then there are an even number of indices i with $j < i < k$ for which $s_i = 1$. This allows the string $\mathbf{s}$ to start with an odd number of 1's, and $\mathbf{s}$ can have an infinite tail of all 1's or all 2's. A direct check shows that S is closed and invariant under the shift map. Because the number of 1's between two 2's can be an arbitrary even number, S is not a subshift of finite type.

The next thing we want to do is count the number of periodic orbits in Σ_A^+. A fixed point has a repeated symbol, so it is easy to see that the number of fixed points is equal the trace of A. A string which has period k for σ_A keeps repeating the first k symbols that appear in its string, e.g., $124212421242\ldots$ has period 4. Therefore, it is helpful to look at finite strings of symbols which are called *words*. Therefore, 1242 is a word of length 4. Given a transition matrix A, a word $\mathbf{w} = (w_0, \ldots, w_{k-1})$ is called *allowable* provided the transition from w_{j-1} to w_j is allowable for $j = 1, \ldots, k$, i.e., $a_{w_{j-1}, w_j} = 1$ for $j = 1, \ldots, k$. As a first step (the induction step) to determine the number of k-periodic points for σ_A, we prove the following lemma about the number of words of length $k+1$ which start at any symbol i and end at the symbol j.

Lemma 2.2. *Assume that the ij entry of A^k is p, $(A^k)_{ij} = p$. Then, there are p allowable words of length $k+1$, starting at i and ending at j, i.e., words of the form $is_1s_2\ldots s_{k-1}j$.*

PROOF. We prove the result by induction on k. Let $\operatorname{num}(k, i, j)$ be the number of words of length $k+1$ starting at i and ending at j. This result is certainly true for $k = 1$ where $\operatorname{num}(1, i, j)$ is either zero or one depending on whether there is an allowable transition from i to j or not. Now, assume the lemma is true for $k-1$ for all choices of i and j. By matrix multiplication

$$\begin{aligned}(A^k)_{ij} &= \left(A^{k-1}A\right)_{ij} \\ &= \sum_{s_{k-1}} \left(A^{k-1}\right)_{is_{k-1}} a_{s_{k-1}j} \\ &= \sum_{s_{k-1}} \operatorname{num}(k-1, i, s_{k-1}) a_{s_{k-1}j} \\ &= \operatorname{num}(k, i, j).\end{aligned}$$

The last equality is true because if $a_{s_{k-1}j} = 0$, then these words from i to s_{k-1} do not contribute to the count of the words from i to j. On the other hand, if $a_{s_{k-1}j} = 1$, then each of these words contributes one word to the words from i to j. □

Corollary 2.3. *The number of fixed points of σ_A^k is equal to the trace of A^k.*

PROOF. This follows because $\#\operatorname{Fix}(\sigma_A^k|\Sigma_A^+) = \sum_i \operatorname{num}(k, i, i) = \sum_i (A^k)_{ii} = \operatorname{tr}(A^k)$. □

Definition. An n by n matrix of 0's and 1's is called *reducible* provided that there is a pair i, j with $(A^k)_{ij} = 0$ for all $k \geq 1$. An n by n matrix of 0's and 1's is called *irreducible* provided that for each $1 \leq i, j \leq n$ there exists a $k = k(i, j) > 0$ such that $(A^k)_{ij} > 0$, i.e., there is an allowable sequence from i to j for every pair of i and j. The matrix A is called *positive* provided $A_{ij} > 0$ for all i and j and is called *eventually positive* provided there there exists a k which is independent of i and j such that $(A^k)_{ij} > 0$ for all i and j. Thus, both positive and eventually positive matrices are irreducible.

Example 2.2. (a) The following transition matrix is reducible because it is not possible to get from 3 to 1:

$$\begin{pmatrix} 1 & 1 & 0 & 0 \\ 1 & 1 & 1 & 0 \\ 0 & 0 & 1 & 1 \\ 0 & 0 & 1 & 0 \end{pmatrix}$$

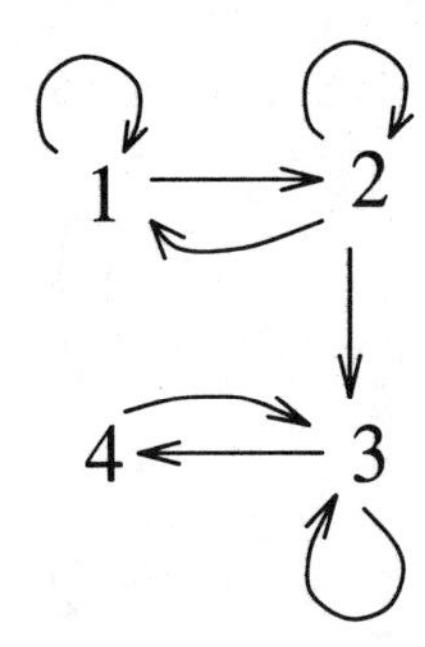

FIGURE 2.2. Graph for Partition in Example 2.2(a)

Its graph is given in Figure 2.2.

(b) The following transition matrix is irreducible:

$$\begin{pmatrix} 1 & 1 & 0 & 0 \\ 1 & 1 & 1 & 0 \\ 1 & 0 & 1 & 1 \\ 0 & 0 & 1 & 0 \end{pmatrix}$$

Its graph is given in Figure 2.3. A calculation shows that A^3 is positive, so A is eventually positive.

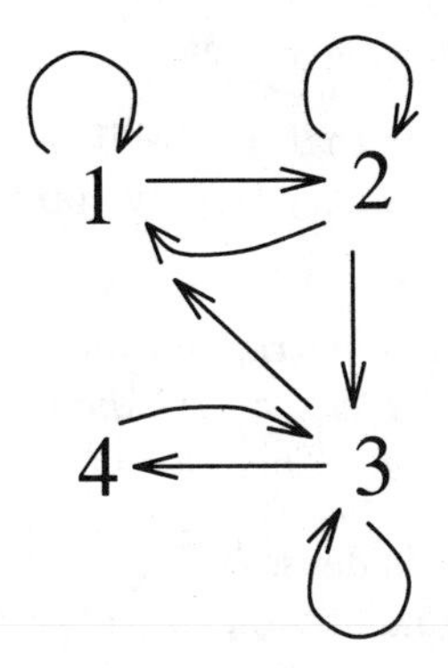

FIGURE 2.3. Graph for Partition in Example 2.2(b)

(c) Let $A = \begin{pmatrix} 1 & 1 \\ 1 & 0 \end{pmatrix}$. Then, $A^2 = \begin{pmatrix} 2 & 1 \\ 1 & 1 \end{pmatrix}$ is positive, so A is eventually positive. Let

$$B = \begin{pmatrix} 0 & I \\ A & 0 \end{pmatrix} = \begin{pmatrix} 0 & 0 & 1 & 0 \\ 0 & 0 & 0 & 1 \\ 1 & 1 & 0 & 0 \\ 1 & 0 & 0 & 0 \end{pmatrix}$$

Then,

$$B^{2k} = \begin{pmatrix} A^k & 0 \\ 0 & A^k \end{pmatrix} \quad \text{and} \quad B^{2k+1} = \begin{pmatrix} 0 & A^k \\ A^{k+1} & 0 \end{pmatrix},$$

so B is irreducible but not eventually periodic.

We often want to exclude the case when A corresponds to a permutation of symbols. A permutation is defined to be a transition matrix where the sum of each row is equal to one and the sum of each column is also equal to one. An example of a permutation is given by

$$A = \begin{pmatrix} 0 & 1 & 0 \\ 0 & 0 & 1 \\ 1 & 0 & 0 \end{pmatrix}.$$

Its graph is given in Figure 2.4.

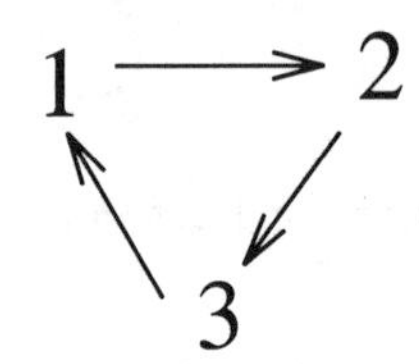

FIGURE 2.4. Graph for Permutation on Three Symbols

The following lemma characterizes permutations in terms of row sums only.

Lemma 2.4. *Let A be a transition matrix. Then, A is a permutation matrix if and only if $\sum_j a_{ij} = 1$ for all i.*

PROOF. If every row sum, $\sum_j a_{ij} = 1$ for all i, then $\sum_{i,j} a_{ij} = n$. Since a transition matrix has $\sum_i a_{ij} \geq 1$ for every j, it follows that $\sum_i a_{ij} = 1$ for every j. This shows A is a permutation matrix. The converse is clear. □

In the next proposition, we show that a subshift is irreducible if and only if the shift map has a dense orbit. As a preliminary step, we prove the following lemma about the dense orbit.

Lemma 2.5. *Let A be a transition matrix. Assume that the point $\mathbf{s}^*$ has a dense orbit in Σ_A^+ for the shift map σ_A and $\mathbf{s}^*$ is not a periodic point. (This last assumption means that A is not a permutation matrix.) Then, for any $k > 1$, $\sigma_A^k(\mathbf{s}^*)$ has a dense orbit.*

PROOF. It is clear that $\sigma_A^{k+j}(\mathbf{s}^*)$ is dense in $\Sigma_A^+ \setminus \{\sigma_A^i(\mathbf{s}^*) : 0 \leq i < k\}$. Thus, we only need to show that this orbit accumulates on $\{\sigma_A^i(\mathbf{s}^*) : 0 \leq i < k\}$. It is clearly sufficient to prove that $\sigma_A^{k+j}(\mathbf{s}^*)$ accumulates on $\mathbf{s}^*$ by taking a higher iterate to get near the other $\sigma_A^i(\mathbf{s}^*)$. Rather than look at $\mathbf{s}^*$, we show that $\mathbf{s}^*$ has a preimage $\mathbf{t}^*$ and that $\sigma_A^{k+j}(\mathbf{s}^*)$ accumulates on $\mathbf{t}^*$.

First, we show that $\mathbf{s}^*$ has a preimage. By assumption (iii) for a transition matrix, there is an element t_0 that can make a transition to $s_0^* \equiv t_1$, so there is a $\mathbf{t}^* \in \sum_A$ with $\sigma_A(\mathbf{t}^*) = \mathbf{s}^*$. If $\mathbf{t}^*$ were on the forward orbit of $\mathbf{s}^*$, then $\mathbf{s}^*$ would be periodic which is not allowed. Therefore, $\mathbf{t}^*$ is not on the forward orbit of $\mathbf{s}^*$, and so $\sigma_A^j(\mathbf{s}^*) \neq \mathbf{t}^*$ for $0 \leq j \leq k$.

Since $\sigma_A^j(\mathbf{s}^*)$ is dense everywhere in Σ_A^+, and $\sigma_A^j(\mathbf{s}^*) \neq \mathbf{t}^*$ for $0 \leq j \leq k$, it follows that $\sigma_A^{k+j}(\mathbf{s}^*)$ must come closer to $\mathbf{t}^*$ than the finite set of points $\{\sigma_A^j(\mathbf{s}^*) : 0 \leq j \leq k\}$. Therefore, $\sigma_A^j \circ \sigma_A^k(\mathbf{s}^*)$ comes arbitrarily near $\mathbf{t}^*$. This completes the proof that the forward orbit of $\sigma_A^k(\mathbf{s}^*)$ is dense in Σ_A^+. □

Proposition 2.6. *Let $A = (a_{i,j})$ be a transition matrix. Then, the following are equivalent:*

(a) *A is irreducible, and*
(b) *σ_A has a dense forward orbit in Σ_A^+.*

PROOF. For a finite word $\mathbf{w}$, let $b(\mathbf{w})$ be the first letter of $\mathbf{w}$ (beginning of $\mathbf{w}$), and $e(\mathbf{w})$ be the last letter of $\mathbf{w}$ (end of $\mathbf{w}$). Given each pair i and j, let $\mathbf{t}_{ij}$ be a choice of the words with $b(\mathbf{t}_{ij}) = i$ and with $e(\mathbf{t}_{ij}) = j$. Such a choice exists because A is irreducible.

First we show that (a) implies (b). We describe the point with a dense orbit, $\mathbf{s}^*$. List all the words of length one with the proper choice of the transition word $\mathbf{t}_{ij}$ between them to make the sequence allowable. Then, list all the allowable words of length two with the proper choice of the transition word $\mathbf{t}_{ij}$ between them to make the sequence allowable. Continue by induction, listing all the allowable words of length n with the proper choice of the transition word $\mathbf{t}_{ij}$ between them to make the sequence allowable. In this way, we construct an infinite allowable sequence which contains all the allowable words of finite length. If $\mathbf{u}$ is a sequence in Σ_A^+ and V is a neighborhood of $\mathbf{u}$, then there is some n such that any sequence which agrees with $\mathbf{u}$ in the first n places is contained in V. Now, for this n, there is a word somewhere in $\mathbf{s}^*$ which agrees with this word of length n in $\mathbf{u}$. Next, there is a k such that $\sigma_A^k(\mathbf{s}^*)$ has this word in the first n places. Thus, $\sigma_A^k(\mathbf{s}^*) \in V$. Since $\mathbf{u}$ and V were arbitrary, this proves that the orbit of $\mathbf{s}^*$ is dense. (The reader might consider the case when A is a permutation matrix separately, but the above proof is also applies to this case.)

Next we show that (b) implies (a). If A is a permutation matrix, then it is clearly the case that A is irreducible. Thus, we can assume that A is not a permutation matrix. Take $\mathbf{s}^* \in \Sigma_A^+$ whose orbit is dense in Σ_A^+. Because the matrix is not a permutation matrix, $\mathbf{s}^*$ cannot be a periodic point.

Take an arbitrary pair i and j. By assumption (ii) for a transition matrix, it is possible to take $\mathbf{a} \in \Sigma_A^+$ such that $a_0 = i$. If a point $\mathbf{t}$ is close enough to $\mathbf{a}$, then $t_0 = a_0$. There is some k_1 such that $\sigma_A^{k_1}(\mathbf{s}^*)$ is within this distance so $\sigma_A^{k_1}(\mathbf{s}^*)_0 = a_0 = i$, and $s_{k_1}^* = a_0 = i$. Thus, i appears in the sequence for $\mathbf{s}^*$.

Similarly, there is a $\mathbf{b} \in \Sigma_A^+$ such that $b_0 = j$. By Lemma 2.5, $\sigma_A^{k_1}(\mathbf{s}^*)$ has a dense forward orbit. The same argument as above shows there is a $k_2 > k_1$ with $\sigma_A^{k_2}(\mathbf{s}^*)_0 = b_0 = j$, and so $s_{k_2}^* = b_0 = j$. Thus, there is an allowable word in $\mathbf{s}^*$ from the k_1^{th} entry to the k_2^{th} entry which goes from i to j, and we can get from i to j for an arbitrary pair i and j. □

By Lemma 2.4, in order to assume that A is not a permutation matrix, it is only necessary to assume that $\sum_j a_{ij} \geq 2$ for some i. We use this assumption to prove that Σ_A^+ is perfect. First, we prove a preliminary lemma.

Lemma 2.7. *Assume that A is an irreducible transition matrix such that $\sum_j a_{i_0 j} \geq 2$ for some i_0. Then for each i, there exists a $k = k(i)$ for which $\sum_j (A^k)_{ij} \geq 2$.*

PROOF. Since A is irreducible, there is a word $\mathbf{w} \in \Sigma_A^+$ such that $b(\mathbf{w}) = i$ and $e(\mathbf{w}) = i_0$. Let the length of $\mathbf{w}$ be k, so there are $k-1$ transitions. Thus, $\text{num}(k-1, i, i_0) \geq 1$. Then, there are least two possible choices after i_0. Thus,

$$\sum_j (A^k)_{ij} \geq \sum_j (A^{k-1})_{i i_0} a_{i_0 j} = \sum_j \text{num}(k-1, i, i_0) a_{i_0 j} \geq 2.$$

□

Proposition 2.8. *Assume that A is an irreducible transition matrix with $\sum_j a_{i_0 j} \geq 2$ for some i_0. Then, Σ_A^+ is perfect.*

PROOF. For each i, there is a $k = k(i)$ such that $\sum_j (A^k)_{ij} \geq 2$. Take an $\mathbf{s} \in \Sigma_A^+$. Take a cylinder set U as a neighborhood, $U = \{\mathbf{t} \in \Sigma_A^+ : t_i = s_i \text{ for } 0 \leq i \leq n\}$. Then, there exists a $k = k(s_n)$ such that $\sum_j (A^k)_{s_n j} \geq 2$. Because there is more than one choice for the transitions from the nth to the $(n+k)^{th}$ entry, there is a $\mathbf{t} \in U$ with $t_{n+m} \neq s_{n+m}$ for some m with $1 \leq m \leq k$. This is true for all $\mathbf{s} \in \Sigma_A^+$ and all cylinder sets, so Σ_A^+ is perfect. □

Proposition 2.9. *Assume that A is an eventually positive transition matrix. Then, σ_A is topologically mixing on Σ_A^+.*

We leave the proof to the exercises. See Exercise 3.14.

Proposition 2.10. *Assume A is a transition matrix. (We do not assume A is irreducible.) Then, the states can be ordered in such a way that A has the following block form:*

$$A = \begin{pmatrix} A_1 & * & * & \cdots & * & * \\ 0 & A_2 & * & \cdots & * & * \\ \vdots & & & & & \vdots \\ 0 & 0 & 0 & \cdots & 0 & A_m \end{pmatrix}$$

where (i) each A_j is irreducible, (ii) the $*$ terms are arbitrary, and (iii) all the terms below the blocks A_j are all 0. Moreover, the nonwandering set $\Omega(\sigma_A) = \Sigma_{A_1}^+ \cup \cdots \cup \Sigma_{A_m}^+$.

We defer the proof to the exercises. See Exercise 3.15.

REMARK 2.1. For an introductory treatment of further topics in symbolic dynamics, see Boyle (1993).

3.3 Zeta Function

Artin and Mazur had the idea to combine the number of periodic points of all periods into a single invariant (Artin and Mazur, 1965). If we list all these numbers, there are countably many invariants. For certain classes of maps, when these numbers are combined together in a certain way, they yield a rational function which has only a finite number of coefficients. Thus, the information given by these countable number of invariants is contained in this finite set of coefficients. We proceed with the formal definitions.

Definition. Let $f : X \to X$ be a map, and $N_k(f) = \#(\mathrm{Per}_k(f)) = \#\,\mathrm{Fix}(f^k)$. The *zeta function for* f is defined to be

$$\zeta_f(t) = \exp\Big(\sum_{k=1}^{\infty} \frac{1}{k} N_k(f) t^k\Big).$$

The zeta function is clearly invariant under topological conjugacy because the number of points of each period is preserved. For more discussion of the zeta function, see Chapter 5 of Franks (1982). In this section, we merely calculate the zeta function for a subshift of finite type. This theorem was originally proved by Bowen and Lanford (1970). In Chapter VIII, we return to prove that the zeta function is a rational function of t for some further types of maps (toral Anosov diffeomorphisms). Before stating the theorem, we give a connection between the determinant, exponential, and trace of a matrix. (The exponential of a matrix is defined by substituting the matrix into the power series for the exponential. It is discussed further in Section 4.3 in the context of solutions of linear differential equations.)

Lemma 3.1 (Liouville's Formula). *Let B be a matrix. Then,*

$$\det(e^B) = e^{\operatorname{tr}(B)}.$$

PROOF. Let $\mathbf{e}_1 \ldots \mathbf{e}_n$ be the standard basis. The following calculation uses the facts that the determinant is alternating in the columns, that $e^{Bt} = (e^{Bt}\mathbf{e}_1, \ldots, e^{Bt}\mathbf{e}_n)$, that $e^{B0} = I$, and that $\frac{d}{dt}e^{Bt} = Be^{Bt}$. Then,

$$\begin{aligned}
\frac{d}{dt}\det(e^{Bt})|_{t=0} &= \sum_j \det(\mathbf{e}_1, \ldots, \mathbf{e}_{j-1}, \frac{d}{dt}e^{Bt}\mathbf{e}_j|_{t=0}, \mathbf{e}_{j+1}, \ldots, \mathbf{e}_n) \\
&= \sum_j \det(\mathbf{e}_1, \ldots, \mathbf{e}_{j-1}, B\mathbf{e}_j, \mathbf{e}_{j+1}, \ldots, \mathbf{e}_n) \\
&= \sum_j \det(\mathbf{e}_1, \ldots, \mathbf{e}_{j-1}, \sum_i b_{ij}\mathbf{e}_i, \mathbf{e}_{j+1}, \ldots, \mathbf{e}_n) \\
&= \sum_{i,j} b_{ij} \det(\mathbf{e}_1, \ldots, \mathbf{e}_{j-1}, \mathbf{e}_i, \mathbf{e}_{j+1}, \ldots, \mathbf{e}_n) \\
&= \sum_j b_{jj} \det(\mathbf{e}_1, \ldots, \mathbf{e}_{j-1}, \mathbf{e}_j, \mathbf{e}_{j+1}, \ldots, \mathbf{e}_n) \\
&= \operatorname{tr}(B).
\end{aligned}$$

For $t = t_0$,

$$\begin{aligned}
\frac{d}{dt}\det(e^{Bt})|_{t=t_0} &= \frac{d}{dt}\det(e^{B(t-t_0)})|_{t=t_0}\det(e^{Bt_0}) \\
&= \operatorname{tr}(B)\det(e^{Bt_0}).
\end{aligned}$$

Solving the scalar differential equation $\frac{d}{dt}\det(e^{Bt}) = \operatorname{tr}(B)\det(e^{Bt})$ with initial condition $\det(e^{B0}) = 1$ gives $\det(e^{Bt}) = e^{\operatorname{tr}(B)t}$. Evaluating this solution at $t = 1$ gives the result. □

Theorem 3.2. *Let $\sigma_A : \Sigma_A^+ \to \Sigma_A^+$ be the subshift of finite type for $A = (a_{ij})$ with $a_{ij} \in \{0, 1\}$ for every pair of i and j. Then, the zeta function of σ_A is rational. Moreover, $\zeta_{\sigma_A}(t) = [\det(I - tA)]^{-1}$.*

PROOF. By Corollary 3.3, $N_k(\sigma_A) = \operatorname{tr}(A^k)$. Therefore, using the linearity of the trace, the power series expansion of the logarithm, and Lemma 3.10, we can make the following calculation.

$$\begin{aligned}
\zeta_{\sigma_A}(t) &= \exp\big(\sum_{k=1}^{\infty} \frac{1}{k}t^k \operatorname{tr}(A^k)\big) \\
&= \exp\big(\operatorname{tr}(\sum_{k=1}^{\infty} \frac{1}{k}t^k A^k)\big) \\
&= \exp\big(\operatorname{tr}(-\log(I - tA))\big) \\
&= \det\big(\exp(\log(I - tA)^{-1})\big) \\
&= \det\big((I - tA)^{-1}\big) \\
&= \big(\det(I - tA)\big)^{-1}.
\end{aligned}$$

This proves the theorem. □

3.4 Period Doubling Cascade

The Sharkovskii Theorem tells us which periods imply which other periods. In particular, if a map $f : \mathbb{R} \to \mathbb{R}$ has finitely many periodic orbits, then all the periods must be powers of 2. For the quadratic family, $F_\mu(x) = \mu x(1-x)$, we saw that it had only fixed points for $0 < \mu \leq 3$, and only fixed points and a point of period 2 for $3 < \mu \leq 1 + \sqrt{6}$. In fact, Douady and Hubbard (1985) proved that for the quadratic family as μ increases new periods are added to the list of periods appearing and never disappear once they have occurred. See de Melo and Van Strien (1993). Let μ_n be the infimum of the parameter values $\mu > 0$ for which F_μ has a point of period 2^n. By the Sharkovskii Theorem, $\mu_n \leq \mu_{n+1}$. Notice that all the $\mu_n < 4$ because F_4 has points of all periods. Let μ_∞ be the limiting value of the μ_n as n goes to infinity. The dynamics for F_{μ_∞} is like the map f_∞ given in Exercise 3.8: there is an invariant set on which F_{μ_∞} acts like an adding machine. This sequence of bifurcations is often called the *period doubling route to chaos.*

At the bifurcation value $\mu_1 = 3$ for the family F_μ, the fixed point p_μ changes from attracting for $1 < \mu < \mu_1$ to repelling for $\mu_1 < \mu$. At $\mu = \mu_1$, $F'_{\mu_1}(p_{\mu_1}) = -1$. For μ slightly larger than μ_1, the 2-periodic orbit $\mathcal{O}(p_{\mu,1})$ is attracting with derivative just less than 1: i.e., $1 > (F^2_\mu)'(p_{\mu,1}) > 0$. In Chapter VII, we study the period doubling bifurcation and show at $\mu = \mu_2$ where the period 4 orbit is created that $(F^2_{\mu_2})'(p_{\mu_2,1}) = -1$. Again, this 2-periodic orbit $\mathcal{O}(p_{\mu,1})$ changes from attracting to repelling as μ moves past μ_2. The period 4 orbit $\mathcal{O}(p_{\mu,2})$ is initially attracting for μ just slightly larger than μ_2 and becomes repelling for $\mu > \mu_3$. This process repeats itself; at $\mu = \mu_n$ the period 2^n orbit $\mathcal{O}(p_{\mu,n})$ is added. This orbit is attracting for $\mu_n < \mu < \mu_{n+1}$ and becomes repelling for $\mu > \mu_{n+1}$.

A natural question to ask is the rate of convergence of the parameter values μ_n to μ_∞. Consider a geometric sequence of numbers, $\lambda_n = C_0 - C_1\lambda^n$, where $0 < \lambda < 1$. For this example, the limiting value $\lambda_\infty = C_0$ and λ (or λ^{-1}) gives the rate of convergence to λ_∞. In general, we want to define a quantity which measures the geometric rate of convergence to the limiting value. Feigenbaum (1978) calculated the rate of convergence by means of the limit

$$\delta = \lim_{n\to\infty} \frac{\mu_n - \mu_{n-1}}{\mu_{n+1} - \mu_n}.$$

This value δ is called the *Feigenbaum constant.* Notice for the sequence $\mu_n = C_0 - C_1\lambda^n$, the value δ would equal λ^{-1}:

$$\begin{aligned}\lim_{n\to\infty} \frac{\mu_n - \mu_{n-1}}{\mu_{n+1} - \mu_n} &= \lim_{n\to\infty} \frac{C_0 - C_1\lambda^n - C_0 + C_1\lambda^{n-1}}{C_0 - C_1\lambda^{n+1} - C_0 + C_1\lambda^n} \\ &= \lambda^{-1}.\end{aligned}$$

Feigenbaum (1978) discovered that this constant is the same for several different families of functions. The value has been calculated to be $\delta = 4.669202\ldots$. Both Feigenbaum (1978) and Coullet and Tresser (1978) suggested using the renormalization method to prove the universality of this constant, i.e., that the constant is the same for any one parameter family of functions which go through the period doubling sequence of bifurcations. Much of this program has now been proved by Feigenbaum, Coullet, Tresser, Collet, Eckmann, Lanford, and others, but there are some mathematical aspects of this program which are still unproven. See de Melo and Van Strien (1993), Collet and Eckmann (1980), and Lanford (1984a, 1984b, 1986).

How can these parameter values μ_n be determined for the family F_μ? We mentioned above that the period 2^n orbit is attracting for $\mu_n < \mu < \mu_{n+1}$. In fact, the critical point $x_0 = 0.5$ must converge to the attracting periodic orbit. See the discussion of negative Schwarzian derivative in Devaney (1989) or de Melo and Van Strien (1993). Thus, to find the attracting periodic orbit, we could iterate the critical point a number of times, say 1000, without recording the iterates, and then record or plot the next 1000 iterates. See Figure 4.1A. Note that for the family F_μ the limiting parameter value $\mu_\infty = 3.5699456$. Therefore, the whole period doubling bifurcation is shown in Figure 4.1B. Since this part of the orbit is near an attracting periodic orbit, we could inspect the orbit to determine the period. By varying μ, we could determine the value of μ when the orbit changed from period 2^{n-1} to 2^n; this value of μ gives μ_n.

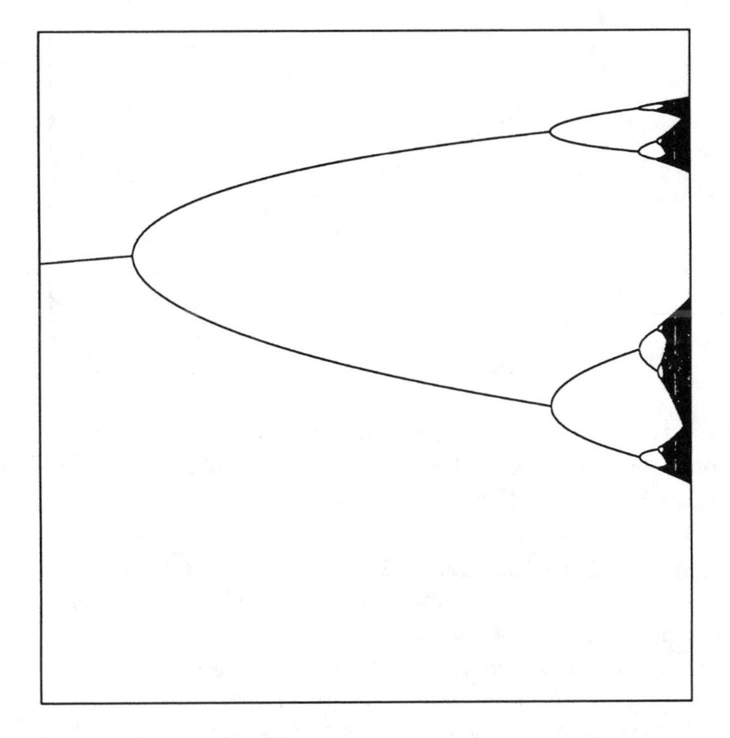

FIGURE 4.1A. The Bifurcation Diagram for the Family F_μ: the Horizontal Direction is the Parameter μ Between 2.9 and 3.6; the Vertical Direction is the Space Variable x Between 0 and 1

A second method to determine the μ_n is to note that the period 2^{n-1} orbit $\mathcal{O}(p_{\mu,n-1})$ becomes unstable at μ_n and

$$(F_{\mu_n}^{2^{n-1}})'(p_{\mu_n,n-1}) = -1.$$

Thus, we could use a numerical scheme (e.g., Newton's method) to search for a point and a parameter value with this property. This search would determine the μ_n.

Finally, there is a third method for determining the rate of convergence given by the Feigenbaum constant by determining slightly different parameter values. We mentioned above that

$$(F_{\mu_n}^{2^n})'(p_{\mu_n,n}) = 1 \qquad \text{and} \qquad (F_{\mu_{n+1}}^{2^n})'(p_{\mu_{n+1},n}) = -1.$$

Between these two parameter values, there is another value μ'_n for which

$$(F_{\mu'_n}^{2^n})'(p_{\mu'_n,n}) = 0,$$

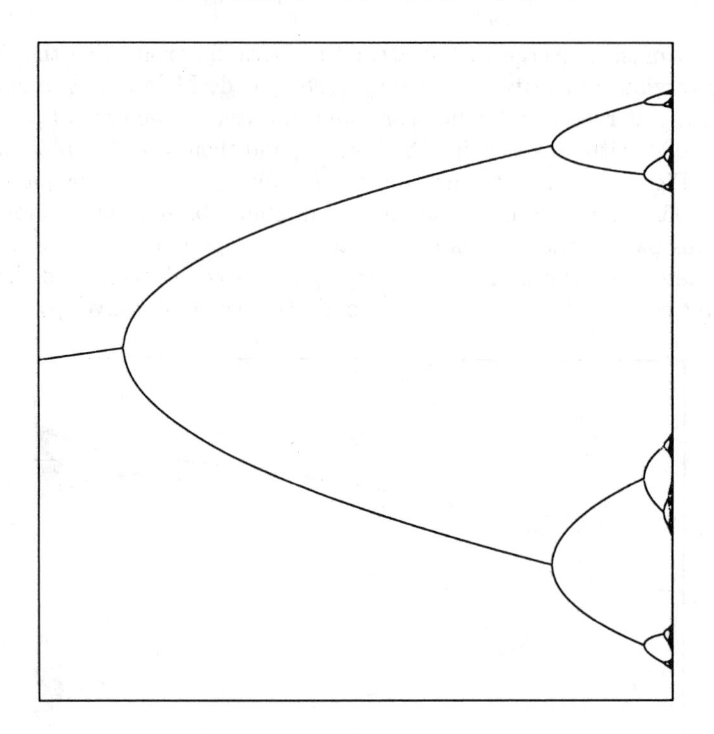

FIGURE 4.1B. The Bifurcation Diagram for the Family F_μ: the Horizontal Direction is the Parameter μ Between 3.54 and 3.5701; the Vertical Direction is the Space Variable x Between 0.47 and 0.57

i.e., the critical point 0.5 has least period 2^n. These parameter values satisfy $\cdots < \mu_n < \mu_n' < \mu_{n+1} < \cdots$. Using these parameter values μ_n' instead of the μ_n gives the same universal constant as the rate of convergence.

For larger values of the parameter μ but with μ still less than 4, an orbit of a point for the quadratic family F_μ seems to be dense in the whole interval $[0,1]$. In fact, Jakobson (1971, 1981) proved that there is a set of parameter values $\mathcal{M} \subset [4-\epsilon, 4]$ such that (i) $\mathcal{M}$ has positive Lebesgue measure (and 4 is a density point) and (ii) for every $\mu \in \mathcal{M}$, F_μ has an invariant measure ν_μ on $[0,1]$ that is absolutely continuous with respect to Lebesgue measure on $[0,1]$. This result implies that most points in $[0,1]$ have orbits which are dense in the interval $[0,1]$ for these parameter values. Many people have written papers on this and related results. See de Melo and Van Strien (1993) for further discussion of this result.

Recently, Benedicks and Carleson (1991) have used results about the transitivity of this one-dimensional family of maps to prove the transitivity of the two-dimensional Hénon family of maps for certain parameter values. We discuss this further in Chapter VIII.

3.5 Chaos

Dynamical systems are often said to exhibit chaos without a precise definition of what this means. In this section, we discuss concepts related to the chaotic nature of maps and give some tentative definitions of a chaotic invariant set.

In Section 2.5, we prove that the quadratic map F_μ, with $\mu > 4$, is transitive on its invariant set Λ_μ. The property of being transitive implies that this set cannot be broken up into two closed disjoint invariant sets. For a set to be called "chaotic," it should be (dynamically) indecomposable in some sense of the word. A weaker notion than transitive, which still includes the some kind of dynamic indecomposability of an invariant set, is that it is chain transitive. See Section 2.3. This latter condition seems more natural than topological transitivity, but it is not as strong and allows certain examples which do not seem chaotic. (Both periodic motion and "quasi-periodic" motion is chain transitive but not very chaotic.) Therefore, in the first definition of a chaotic invariant set, we require the map to be topologically transitive and only give chain transitivity as an alternative assumption.

To define a chaotic invariant set, we also want to add a second assumption which indicates that the dynamics of the map on the invariant set are disorderly, or at least that nearby orbits do not stay near each other under iteration. The following definition of sensitive dependence on initial conditions is one possible such concept. In the next section we define the Liapunov exponents of a map. Another way to express that nearby orbits diverge is that the map has positive Liapunov exponents. See Section 3.6.

Definition. A map f on a metric space X is said to have *sensitive dependence on initial conditions* provided there is an $r > 0$ (independent of the point) such that for each point $\mathbf{x} \in X$ and for each $\epsilon > 0$ there is a point $\mathbf{y} \in X$ with $d(\mathbf{x}, \mathbf{y}) < \epsilon$ and a $k \geq 0$ such that $d(f^k(\mathbf{x}), f^k(\mathbf{y})) \geq r$.

One of the early situations where sensitive dependence was observed was in a set of differential equations in three variables. E. Lorenz was studying the system mentioned in Section 1.3 and discussed in Section 8.10 (Lorenz, 1963). While numerically integrating the equations, he recorded the coordinates of the trajectory to only a three-decimal-place accuracy. After calculating an orbit, he tried to duplicate the latter part of the trajectory by entering as a new initial point $\mathbf{q}$ the coordinates of some point part way through the initial calculation, $\mathbf{p}_{t_0}$. Because the original trajectory had more decimal places stored in memory than he entered the second time by hand, the points $\mathbf{p}_{t_0}$ and $\mathbf{q}$ were not the same but merely nearby points. He observed that the two trajectories, the original trajectory and the one started with the slightly different initial condition, followed each other for a period of time and then diverged from each other rapidly. This divergence is an indication of sensitive dependence on initial conditions of the particular system which he was studying.

Another way that sensitive dependence is manifest is through the round-off errors of the computer. Curry (1979) reports on numerical studies of the Hénon map (for $A = 1.4$ and $B = -0.3$) using two different computers. After 60 iterates, the iterates have nothing to do with each other. Thinking of the numerical orbit on a computer as an ϵ-chain of the function, two different ϵ-chains diverge, giving an indication of sensitive dependence on initial conditions. On the other hand, the plots of the orbits on the two machines seem to fill up the same subset of the plane, giving an indication that the function is topologically transitive (or at least chain transitive) on this invariant set. For further discussion of an attractor for the Hénon map, see Sections 1.3 and 7.9.

The concept of sensitive dependence on initial conditions is closely related to another concept called expansive: a map is expansive provided any two orbits become at least a fixed distance apart.

Definition. A map f on a metric space X is said to be *expansive* provided there is an $r > 0$ (independent of the points) such that for each pair of points $\mathbf{x}, \mathbf{y} \in X$ there is a $k \geq 0$ such that $d(f^k(\mathbf{x}), f^k(\mathbf{y})) \geq r$. If f is a homeomorphism, then in the definition of expansive, we allow $k \in \mathbb{Z}$ and do not require that k is positive, i.e., there is an $r > 0$ such that for each pair of points $\mathbf{x}, \mathbf{y} \in X$, there is a $k \in \mathbb{Z}$ such that $d(f^k(\mathbf{x}), f^k(\mathbf{y})) \geq r$.

If f is expansive and X is a perfect metric space, then it has sensitive dependence on initial conditions. In determining the proper characterization of chaos, the assumption that the map is expansive seems too strong. Therefore, we make the following definitions.

Definition. A map f on a metric space X is said to be *chaotic on an invariant set* Y or exhibits *chaos* provided (i) f is transitive on Y and (ii) f has sensitive dependence on initial conditions on Y.

REMARK 5.1. The use of the term "chaos" was introduced into Dynamical Systems by Li and Yorke (1975). They proved that if a map on the line had a point of period three, then it had points of all periods. They also proved that if a map f on the line has a point of period three, then it has an invariant set S such that

$$\limsup_{n\to\infty} |f^n(p) - f^n(q)| > 0 \qquad \text{and}$$
$$\liminf_{n\to\infty} |f^n(p) - f^n(q)| = 0$$

for every $p, q \in S$ with $p \neq q$. They considered a map with this latter property as chaotic. This property is certainly related to sensitive dependence on initial conditions.

REMARK 5.2. Devaney (1989) gave an explicit definition of a chaotic invariant set in an attempt to clarify the notion of chaos. To our two assumptions, he adds the assumption that the periodic points are dense in Y. Although this property is satisfied by "uniformly hyperbolic" maps like the quadratic map, it does not seem that this condition is at the heart of the idea that the system is chaotic. (This last comment is made even though in the original paper Li and Yorke (1975) proved the existence of periodic points.) Therefore, we leave out conditions about periodic points in our definition of chaos.

The paper of Banks, Brooks, Cairns, Davis, and Stacey (1992) proves that any map which (i) is transitive and (ii) has dense periodic points also must have sensitive dependence on initial conditions. However, as stated above, we consider the conditions that the map (i) is transitive and (ii) has sensitive dependence on initial conditions a more dynamically reasonable choice of conditions in the definition.

REMARK 5.3. As stated above, an alternative definition of a *chaotic invariant set* Y is that (i) f is chain transitive on Y and (ii) f has sensitive dependence on initial conditions on Y.

This definition allows the following example, which does not seem chaotic. Let x and y both be mod 1 variables, so $\{(x, y) : x, y \bmod 1\}$ is the two torus, $\mathbb{T}^2$. Let

$$f(x, y) = (x + y, y)$$

be a shear map. Then, f preserves the y variable. The rotation in the x-direction depends on the y variable. This map is chain transitive on $\mathbb{T}^2$ but not topologically transitive. The controlled nature of the trajectories make it seem non-chaotic.

One way to avoid the above example and still use chain transitivity as the notion of indecomposability is to require that solutions diverge at an exponential rate. This is defined in the next section in terms of the Liapunov exponents. (Also see Section 9.2 for Liapunov exponents in higher dimensions.) Using this concept, we give an alternative definition of a chaotic invariant set: an invariant set Y could be called *chaotic* provided that (i) f is chain transitive on Y and (ii) f has a positive Liapunov exponent on Y. Ruelle (1989a) has a long discussion of a chaotic attractor, in which he includes the requirements that it be irreducible and has a positive Liapunov exponent.

There is another measurement related to chaos which relates to the invariant set for the system. There are various concepts of (fractal) dimensions, including the box dimension, which allow the dimension to be an noninteger value. We give some of these dimensions in Section 9.4 (in higher dimensions where the concepts seem to have their natural setting). For experimental data (without specific equations), Liapunov exponents are not very computable, but the box dimension of the invariant set is computable. Therefore, in the setting of experimental measurements, the box dimension seems like a reasonable measurement of chaos. See the discussion in Chapter 5 of Broer, Dumortier, van Strien, and Takens (1991).

REMARK 5.4. A more mathematical solution to making the notion of chaos precise is in terms of a quantitative measurement of chaos called topological entropy, which is defined in Section 9.1. The topological entropy of a map f is denoted by $h(f)$ and is a number greater than or equal to zero and less than or equal to infinity. This quantity has a complicated definition, but can be thought of as a quantitative measurement of the amount of sensitive dependence on initial conditions of the map. If the nonwandering set of f is a finite number of periodic points, then $h(f) = 0$. In this case, a transitive invariant set is just a single periodic orbit, and this does not have sensitive dependence on initial conditions. If the dynamics of f are complicated, as for the quadratic map F_μ on Λ_μ for $\mu > 4$, then $h(F_\mu) > 0$. Therefore, another characterization of a chaotic invariant set Λ for f might be that $h(f|\Lambda) > 0$. Using this definition, we do not need to add the condition that f is transitive: if $h(f|\Lambda) > 0$, then with mild assumptions there is an invariant subset $\Lambda' \subset \Lambda$ on which f is transitive and for which $h(f|\Lambda') > 0$. (This last statement is not obvious but is true based on results of Chapter X, as well as the more precise definition of topological entropy in Section 9.1.)

REMARK 5.5. Thus, we have given four alternative definitions of a chaotic invariant set. The characterization of chaos in terms of topological entropy is the most satisfactory one from a mathematical perspective but is not very computable in applications (with a computer). The definition in terms of Liapunov exponents is the most computable (possible to estimate) on a computer. The box dimension is most computable for data from experimental work. Thus, there are a number of related important concepts, each of which is important in the appropriate setting. We use the definition of chaos which requires sensitive dependence on initial conditions and topological transitivity for the definition of chaos. The other concepts we refer to by stating the system (i) has positive topological entropy, (ii) has a positive Liapunov exponent, or (iii) has fractional box dimension.

With the above definitions, we can state a result about the quadratic map.

Theorem 5.1. *(a) The shift map σ is chaotic on the full p-shift space, Σ_p^+. In fact, σ is expansive on Σ_p^+.*

(b) For $\mu > 4$, the quadratic map F_μ is chaotic on its invariant Cantor set Λ_μ, i.e., $F_\mu|\Lambda_\mu$ has sensitive dependence on initial conditions and is topological transitivity. In fact, $F_\mu|\Lambda_\mu$ is expansive.

PROOF. We proved in an earlier section that both σ and F_μ are transitive on their respective spaces. (The fact that F_μ is transitive follows from the conjugacy to σ.)

To show that σ is expansive and so has sensitive dependence, let $r = 1$. If $\mathbf{s} \neq \mathbf{t}$ for two points in Σ_p^+, then there is a k such that $s_k \neq t_k$. Then, $\sigma^k(\mathbf{s})$ and $\sigma^k(\mathbf{t})$ differ in the 0-th place and $d(\sigma^k(\mathbf{s}), \sigma^k(\mathbf{t})) \geq 1$. This proves that σ is expansive.

For F_μ, since the itinerary map h is a homeomorphism, if $x, y \in \Lambda_\mu$ are distinct points with $\mathbf{s} = h(x)$ and $\mathbf{t} = h(y)$, then there is a k with $s_k \neq t_k$. Therefore, $F_\mu^k(x)$ and

$F_\mu^k(y)$ are in different intervals I_1 and I_2. Since there is a minimum distance r between these two intervals, $|F_\mu^k(x) - F_\mu^k(y)| \geq r$. This proves that F_μ is expansive. □

REMARK 5.6. The fact that F_μ is expansive on Λ_μ also follows from the following general result that a conjugacy between maps on compact sets preserves expansiveness.

Theorem 5.2. *Let $f : X \to X$ be conjugate to $g : Y \to Y$, where both X and Y are compact. Assume g has sensitive dependence (resp. is expansive) on Y. Then, f has sensitive dependence (resp. is expansive) on X.*

PROOF. Let $r > 0$ be the constant for g for either sensitive dependence or expansiveness. Let $h : X \to Y$ be the conjugacy. By compactness, h is uniformly continuous. Therefore, given the value $r > 0$ as above, there is a $\delta > 0$ such that if $d(\mathbf{p}, \mathbf{q}) < \delta$ in X, then $d(h(\mathbf{p}), h(\mathbf{q})) < r$ in Y. Thus, if $d(h(\mathbf{p}), h(\mathbf{q})) \geq r$ in Y, then $d(\mathbf{p}, \mathbf{q}) \geq \delta$ in X; or denoting the points differently, if $d(\mathbf{p}, \mathbf{q}) \geq r$ in Y, then $d(h^{-1}(\mathbf{p}), (h^{-1}(\mathbf{q})) \geq \delta$ in X.

Now, we check the sensitive dependence case. Let $\mathbf{x} \in X$ and $\epsilon > 0$. Then, there is an $\epsilon' > 0$ such that if $\mathbf{q} \in Y$ is within ϵ' of $\mathbf{y} = h(\mathbf{x})$, then $\mathbf{p} = h^{-1}(\mathbf{q})$ is within ϵ of $\mathbf{x}$. Take such a $\mathbf{q} \in Y$ that is within ϵ' of $\mathbf{y}$ and $k \geq 0$ as given by the condition of sensitive dependence of g at $\mathbf{y}$. Let $\mathbf{p} = h^{-1}(\mathbf{q})$. Then, $d(g^k(\mathbf{y}), g^k(\mathbf{q})) \geq r$, so $d(h^{-1}(g^k(\mathbf{y})), h^{-1}(g^k(\mathbf{q}))) \geq \delta$. But $h^{-1}(g^k(\mathbf{y})) = f^k(h^{-1}(\mathbf{y})) = f^k(\mathbf{x})$ and $h^{-1}(g^k(\mathbf{q})) = f^k(h^{-1}(\mathbf{q})) = f^k(\mathbf{p})$. Therefore, $\mathbf{p}$ is within ϵ of $\mathbf{x}$ and $d(f^k(\mathbf{x}), f^k(\mathbf{q})) \geq \delta$. Thus, the δ from the uniform continuity works as the distance by which nearby points of f move apart in the condition of sensitive dependence. The proof for expansiveness is similar. □

3.6 Liapunov Exponents

In discussing chaos, we referred to Liapunov exponents which measure the (infinitesimal) exponential rate at which nearby orbits are moving apart. In this section we give a precise definition and calculate the exponents in a few examples. In Section 9.2 we return to discuss Liapunov exponents in higher dimensions.

We want to give an expression for determining the growth rate of the derivative of a function $f : \mathbb{R} \to \mathbb{R}$ as the number of iterates increases. If $|(f^n)'(x_0)| \sim L^n$, then $\log(|(f^n)'(x_0)|) \sim \log(L^n) = n \log(L)$, or $(1/n) \log(|(f^n)'(x_0)|) \sim \log(L)$. In the best situations, the limit of this quantity exists as n goes to infinity. If we take the limsup, then it always exists as n goes to infinity.

Definition. Let $f : \mathbb{R} \to \mathbb{R}$ be a C^1 function. For each point x_0, define the *Liapunov exponent of* x_0, $\lambda(x_0)$, as follows:

$$\begin{aligned}\lambda(x_0) &= \limsup_{n\to\infty} \frac{1}{n} \log(|(f^n)'(x_0)|) \\ &= \limsup_{n\to\infty} \frac{1}{n} \sum_{j=0}^{n-1} \log(|f'(x_j)|)\end{aligned}$$

where $x_j = f^j(x_0)$. (The first and second limits are equal by the chain rule.) Note that the right-hand side is an average along an orbit (a time average) of the logarithm of the derivative.

The definition of these exponents as a limsup goes back to the dissertation of Liapunov in 1892. See Liapunov (1907). For a treatment from the point of view of time-dependent linear differential equations, see Cesari (1959) or Hartman (1964). In higher dimensions, the definition is more complicated than the one given above in one dimension. We discuss this situation in Section 9.2. We also discuss in that section the recent theory developed from the work of Oseledec (1968), who showed that the limit exists for almost all points.

We want to give an interpretation of the Liapunov exponent: when it is negative, nearby orbits converge; and when it is positive, nearby orbit diverge. We start by assuming that $\lambda(x_0) < 0$, which implies that $\log |(f^n)'(x_0)| \approx n\,\lambda(x_0)$ or $|(f^n)'(x_0)| \approx e^{n\,\lambda(x_0)} = L(x_0)^n$, where $L(x_0) = e^{\lambda(x_0)} < 1$. By Taylor's expansion,

$$|f^n(x_0 + \delta) - f^n(x_0)| \approx |(f^n)'(x_0)\delta| \approx |\delta|\, L(x_0)^n$$

which goes to zero as n goes to infinity. Thus, the orbits of x_0 and $x_0 + \delta$ converge as n goes to infinity. There is a problem in making this argument rigorous because of the difficulty in interchanging the order of two limits: the limit in taking the derivative (Taylor's expansion) and the limit involved in the Liapunov exponent. Therefore, we often say that if the Liapunov exponent is negative, then orbits of "infinitesimal" displacements converge as n goes to infinity.

Similarly, if $\lambda(x_0) > 0$, then $L(x_0) = e^{\lambda(x_0)} > 1$, and $|f^n(x_0+\delta) - f^n(x_0)| \approx \delta\, L(x_0)^n$. The right-hand side goes to infinity as n goes to infinity, which indicates that the two orbits become far enough apart for the higher-order terms in the Taylor's expansion to become important. Therefore, a positive Liapunov exponent corresponds to sensitive dependence on initial conditions.

Now, we proceed to give three examples where we can calculate or estimate the Liapunov exponents.

Example 6.1. Let

$$T(x) = \begin{cases} 2x & \text{for } 0 \le x \le 0.5 \\ 2(1-x) & \text{for } 0.5 \le x \le 1. \end{cases}$$

be the tent map. If x_0 is such that $x_j = T^j(x_0) = 0.5$ for some j, then $\lambda(x_0)$ is not defined because the derivative is not defined. Such points make up a countable set. For other points $x_0 \in [0,1]$, $|f'(x_j)| = 2$ for all j, so the Liapunov exponent, $\lambda(x_0)$, is $\log(2)$.

Example 6.2. Let $F_\mu(x) = \mu x(1-x)$ for $\mu \ge 2 + \sqrt{5}$. Let Λ_μ be the invariant Cantor set. Then, for $x_0 \in \Lambda_\mu$, $\log(|F_\mu'(x_j)|) \ge \lambda_0 > 0$ for some λ_0. Thus, the average is larger than λ_0, $\lambda(x_0) \ge \lambda_0$. Thus, we may not know an exact value, but it is easy to derive an inequality and know that the exponent is positive.

Before giving the last example, we make some connection between the Liapunov exponent and the space average with respect to an invariant measure. If f has an invariant Borel measure μ with finite total measure and support on a bounded interval, then the Birkhoff Ergodic Theorem (Theorem VIII.2.2) says that the limit of the quantity defining $\lambda(x_0)$ actually exists, and is not just a limsup, for μ-almost all points x_0. In fact, since the measure is a Borel measure and $\log(|f'(x)|)$ is continuous and bounded above, $\lambda(x)$ is a measurable function and

$$\int \lambda(x)\, d\mu(x) = \int \log(|f'(x)|)\, d\mu(x).$$

If f is "ergodic" with respect to μ, then $\lambda(x)$ is constant μ-almost everywhere and

$$\lambda(x) = \frac{1}{|\mu|} \int \log(|f'(x)|)\, d\mu(x) \qquad \mu\text{-almost everywhere},$$

where $|\mu|$ is the total measure of μ. (See Section 8.2 for the definition of ergodic.) This says that the time average of the logarithm of the derivative is equal to the space average (the integral) of the logarithm of the derivative for μ-almost point. The point to understand from this discussion is that if the map preserves a reasonable measure, then the Liapunov exponent is constant almost everywhere.

In higher dimensions, the proof that the appropriate limit exists for almost every x requires a much more complicated ergodic theorem due to Oseledec (1968). See Section 9.2.

Example 6.3. Let $F_4(x) = 4x(1-x)$ be the quadratic map for $\mu = 4$.

If x_0 is such that $x_j = F_4^j(x_0) = 0.5$ for some j, then $\log(|F_4'(x_j)|) = \log(|F_4'(0.5)|) = \log(0) = -\infty$. Therefore, $\lambda(x_0) = -\infty$ for these x_0.

If $x_0 = 0$ or 1, then $\lambda(x_0) = \log(|F_4'(0)|) = \log(4) > 0$.

For points $x_0 \in (0,1)$ for which x_j is never equal to 0 or 1 (and so never equals 0.5), we use the conjugacy of F_4 with the tent map T, $h(y) = \sin^2(\pi y/2)$. (This conjugacy is verified in Example II.6.2.) Note that h is differentiable on $[0,1]$, so there is a $K > 0$ such that $|h'(y)| < K$ for $y \in [0,1]$. Also, $h'(y) > 0$ in the open interval $(0,1)$; so for any (small) $\delta > 0$, there is a bound $K_\delta > 0$ such that $K_\delta < |h'(y)|$ for $h(y) \in [\delta, 1-\delta]$. For x_0 as above,

$$\begin{aligned}
\lambda(x_0) &= \limsup_{n\to\infty} \frac{1}{n} \log(|(F_4^n)'(x_0)|) \\
&= \limsup_{n\to\infty} \frac{1}{n} \log(|(h \circ T^n \circ h^{-1})'(x_0)|) \\
&= \limsup_{n\to\infty} \frac{1}{n} \big(\log(|h'(y_n)|) + \log(|(T^n)'(y_0)|) + \log(|(h^{-1})'(x_0)|) \big) \\
&\le \limsup_{n\to\infty} \frac{1}{n} \big(\log(K) + n\log(2) + \log(|(h^{-1})'(x_0)|) \big) \\
&= \log(2).
\end{aligned}$$

On the other hand, for these x_0, we can pick a sequence of integers n_j going to infinity such that $x_{n_j} \in [\delta, 1-\delta]$. Then, letting $y_0 = h^{-1}(x_0)$ and $y_n = T^n(y_0)$,

$$\begin{aligned}
\lambda(x_0) &\ge \limsup_{j\to\infty} \frac{1}{n_j} \log(|(F_4^{n_j})'(x_0)|) \\
&= \limsup_{j\to\infty} \frac{1}{n_j} \big(\log(|h'(y_{n_j})|) + \log(|(T^{n_j})'(y_0)|) + \log(|(h^{-1})'(x_0)|) \big) \\
&\ge \limsup_{j\to\infty} \frac{1}{n_j} \big(\log(K_\delta) + n_j\log(2) + \log(|(h^{-1})'(x_0)|) \big) \\
&= \log(2).
\end{aligned}$$

Therefore, $\lambda(x_0) = \log(2)$ for all these points. (Note, there are points which repeatedly come near 0.5 but never hit 0.5 for which the limit of the quantity defining the exponent does not exist but only the lim sup.) In particular, the Liapunov exponent is positive for all points whose orbit never hits 0 or 1 (and so never hits 0.5).

Since T preserves Lebesgue measure, the conjugacy also induces an invariant measure μ for F_4; this measure has density function $\pi^{-1}[x(1-x)]^{-1/2}$. Notice the similarity with the density functions we used to prove that F_μ is transitive for $4 < \mu < 2+\sqrt{5}$. By the above argument, $\lambda(x) = \log(2)$ for μ-almost all points. Integrating with respect to this density function gives

$$\begin{aligned}\int_0^1 \log(|F_4'(x)|)\,d\mu(x) &= \int_0^1 \frac{\lambda(x)}{\pi[x(1-x)]^{1/2}}\,dx \\ &= \int_0^1 \frac{\log(2)}{\pi[x(1-x)]^{1/2}}\,dx \\ &= \log(2).\end{aligned}$$

On the other hand,

$$\begin{aligned}\int_0^1 \log(|F_4'(x)|)\,d\mu(x) &= \int_0^1 \frac{\log(|F_4'(x)|)}{\pi[x(1-x)]^{1/2}}\,dx \\ &= \int_0^1 \log(|T'(y)|)\,dy \\ &= \log(2).\end{aligned}$$

These are equal, as the Birkhoff Ergodic Theorem says they must be.

REMARK 6.1. In the last section we mentioned that topological entropy is a measure of complexity of the dynamics of a map. (The formal definition of entropy is given in Section 9.1.) Katok (1980) has proved that if a map preserves a non-atomic (continuous) Borel probability measure μ for which μ-almost all initial conditions have non-zero Liapunov exponents, then the topological entropy is positive, so the map is chaotic. Thus, a good computational criterion for chaos is whether a function has a positive Liapunov exponent for points in a set of positive measure.

3.7 Exercises

Sharkovskii's Theorem

3.1. Let x be a point of period n for f, $f : \mathbb{R} \to \mathbb{R}$ continuous, $\mathcal{O}(x) = \{x_1, \ldots, x_n\}$ be the orbit of x with $x_1 < x_2 < \cdots < x_n$. Let

$$\mathcal{A} = \{[x_1, x_2], [x_2, x_3], \ldots, [x_{n-1}, x_n]\}$$

be the set of intervals induced by the orbit. Assume there exists $I_1 \in \mathcal{A}$ such that $f(I_1) \supset I_1$. Define $J_1 = I_1$, and define inductively

$$J_j = \bigcup\{L \in \mathcal{A} : f(J_{j-1}) \supset L\}$$

for $j = 2, \ldots, n-1$. Further assume that $J_{n-1} \supset K$ for $K \in \mathcal{A}$, where n is the period of x.

(a) Show that there exists $I_{n-2} \in \mathcal{A}$ such that I_{n-2} f-covers K and $I_{n-2} \subset J_{n-2}$.

(b) Show that there exists a sequence $I_j \in \mathcal{A}$ for $j = 1, \ldots, n-1$ with I_1 as above, $I_{n-1} = K$ and such that I_j f-covers I_{j+1} for $j = 1, \ldots, n-2$.

3.2. Assume that $f : \mathbb{R} \to \mathbb{R}$ is continuous and has a point of period n. Assume $n = 2^m p$ with $p > 1$ odd, $m \geq 1$, and n is maximal in the Sharkovskii ordering. Further assume that $k = 2^s q$ with $s \geq m+1$ and $1 \leq q$ and q odd. Prove that f has a point of period k. Thus, prove Case 3a of Sharkovskii's Theorem, page 72.

3.3. Assume that $f : \mathbb{R} \to \mathbb{R}$ is continuous and has a point of period n. Assume $n = 2^m p$ with $p > 1$ odd, $m \geq 1$, and n is maximal in the Sharkovskii ordering. Further assume that $k = 2^s$ with $s \leq m$. Prove that f has a point of period k. Thus, prove Case 3b of Sharkovskii's Theorem, page 72.

3.4. Assume that $f : \mathbb{R} \to \mathbb{R}$ is continuous and has a point of period n. Assume $n = 2^m p$ with $p > 1$ odd, $m \geq 1$, and n is maximal in the Sharkovskii ordering. Further assume that $k = 2^m q$ with q odd and $q > p$. Prove that f has a point of period k. Thus, prove Case 3c of Sharkovskii's Theorem, page 72.

3.5. Let $n = 2^m p$ with $p > 1$ odd, $m \geq 1$. Prove that there is a continuous function $f : [0, 1] \to [0, 1]$ whose periods, $\mathcal{P}(f)$, are exactly the set $\{k : n \triangleright k\}$.

3.6. Construct a map of $\mathbb{R}$ with points of all periods except 3, i.e., construct a map with all periods implied by period 5 from Sharkovskii's Theorem but no other periods. Hint: Take an orbit of period 5, $\{x_1 < x_2 < x_3 < x_4 < x_5\}$, with the order given as in the proof of Sharkovskii's Theorem; let the map be linear on each interval $[x_i, x_{i+1}]$; show that this map works.

3.7. Construct a map with a point of period 10 and all periods implied by 10 by the Sharkovskii ordering but no others. Hint: Take the double of the map in the problem before the last.

3.8. As defined in Example 1.3, let $f_n : [0, 1] \to [0, 1]$ be the function with exactly one point of period 2^i for $0 \leq i \leq n$ and no other periodic points. Define f_∞ by $f_\infty(x) = \lim_{n\to\infty} f_n(x)$.
   (a) Prove that f_∞ is continuous.
   (b) Prove that the periods of f_∞ are exactly $\{2^i : 0 \leq i < \infty\}$.
   (c) Prove that for each n, f_∞ has exactly one periodic orbit of each period 2^n, that it is repelling, and that the points of this orbit lie in the gaps $G_{n,j}$ which define the middle-(1/3) Cantor set.
   (d) Let
$$S_n = [0, 1] \setminus \bigcup_{\substack{1 \leq j \leq 2^{k-1} \\ 1 \leq k \leq n}} G_{k,j}$$
   be the union of the 2^n intervals used to define the middle-(1/3) Cantor set. Prove that $f_\infty(S_n) = S_n$.
   (e) Let $\Lambda = \bigcap_{n \geq 1} S_n$. Prove that Λ is invariant for f_∞.
   (f) Let Σ_2^+ be the set of all sequences of 0's and 1's. Define $A : \Sigma_2^+ \to \Sigma_2^+$ by
$$A(s_0 s_1 s_2 \ldots) = (s_0 s_1 s_2 \ldots) + (1000 \ldots) \bmod 2,$$
   i.e., $(1000\ldots)$ is added to $(s_0 s_1 s_2 \ldots)$ mod 2 with carrying (so $(11\bar{0}) + (1\bar{0}) = (001\bar{0})$). The map A on Σ_2^+ is called the *adding machine*. Define $h : \Lambda \to \Sigma_2^+$ by $h(p) = \mathbf{s}$, where $s_k = 1$ if p belongs to the left hand choice of the interval in S_{n-1}. Prove that h is a topological conjugacy from f_∞ on Λ to A on Σ_2^+.
   (g) Prove that the adding machine A on Σ_2^+ has no periodic points, and every forward orbit is dense in Σ_2^+.

Subshifts of Finite Type

3.9. Give the matrix of the subshift of finite type for the map in Exercise 3.6 and the intervals $[x_i, x_{i+1}]$.

3.10. Let

$$A = \begin{pmatrix} 0 & 1 \\ 1 & 1 \end{pmatrix},$$

and

$$A^n = \begin{pmatrix} a_n & b_n \\ c_n & d_n \end{pmatrix}.$$

(a) For a vector (x_0, y_0), let $(x_n, y_n) = (x_0, y_0)A^n$. With $y_{-1} = x_0$, prove that $x_{n+1} = y_n$, and $y_{n+1} = y_n + y_{n-1}$. (This is a Fibonacci sequence.)

(b) Use the fact that $(1,0)A^n = (a_n, b_n)$ and $(0,1)A^n = (c_n, d_n)$ to prove that $a_n = a_{n-1} + a_{n-2}$ and $d_n = d_{n-1} + d_{n-2}$.

(c) Prove that $\mathrm{tr}(A^n) = \mathrm{tr}(A^{n-1}) + \mathrm{tr}(A^{n-2})$.

3.11. Consider the matrix A given in the last problem. Find all the fixed points of σ_A, σ_A^2, σ_A^3, and σ_A^4. Group the points into orbits and give their least period.

3.12. Let A be an n by n matrix with $a_{ij} \in \{0,1\}$, $\sum_j a_{ij} \geq 1$ for all i, and $\sum_i a_{ij} \geq 1$ for all j. We define i to be equivalent to j, $i \sim j$, if there exist $k = k(i,j) \geq 0$ and $m = m(i,j) \geq 0$ such that $(A^k)_{ij} \neq 0$ and $(A^m)_{ji} \neq 0$. (Because we allow $k = 0 = m$, i is equivalent to itself.) Break $\{1, \ldots, n\}$ into equivalence classes, $\{1, \ldots, n\} = S_1 \cup \cdots \cup S_p$ with $S_i \cap S_j = \emptyset$ for $i \neq j$. Assume that for each equivalence class, S_q, there exists a $i_q \in S_q$ such that $\sum_{j \in S_q} a_{i_q j} \geq 2$. Prove that Σ_A^+ is perfect.

3.13. Let $f : \mathbb{R} \to \mathbb{R}$ be a C^1 function. Assume there are p closed and bounded intervals $I_1, I_2, \ldots, I_p$ and $\lambda > 1$ such that (i) $|f'(x)| \geq \lambda$ for all $x \in \bigcup_{i=1}^p I_i \equiv \mathcal{I}$ and (ii) if $f(I_i) \cap I_j \neq \emptyset$, then $f(I_i) \supset I_j$. Let A be the matrix of the subshift of finite type defined by $a_{ij} = 1$ if $f(I_i) \supset I_j$ and $a_{ij} = 0$ if $f(I_i) \cap I_j = \emptyset$. Further assume that (iii) A is transitive and irreducible. Let $\Lambda = \bigcup_{i=1}^p f^{-i}(\mathcal{I})$. Prove that $f|\Lambda$ is conjugate to the subshift of finite type σ_A on Σ_A^+.

3.14. (This exercise asks you to prove Proposition 2.9.) Let A be an eventually positive transition matrix, with $(A^k)_{ij} \neq 0$ for all i and j.

(a) Prove for $n \geq k$ that $(A^n)_{ij} \neq 0$ for all i and j.

(b) Prove that σ_A is topologically mixing on Σ_A^+.

3.15. (This exercise asks you to prove Proposition 2.10.) Assume A is a transition matrix. (We do not assume A is irreducible.)

(a) Prove that the states can be ordered in such a way that A has the following block form:

$$A = \begin{pmatrix} A_1 & * & * & \cdots & * & * \\ 0 & A_2 & * & \cdots & * & * \\ \vdots & & & & & \vdots \\ 0 & 0 & 0 & \cdots & 0 & A_m \end{pmatrix}$$

where (i) each A_j is irreducible, (ii) the $*$ terms are arbitrary, and (iii) all the terms below the blocks A_j are all 0. Hint: Define an ordering on the states as follows. Call $i \geq j$ provided there is a $k = k(i,j)$ such that $A_{i,j}^k \neq 0$. Call states i and j equivalent provided $i \geq j$ and $j \geq i$. Group together the equivalent state and order all the states in terms of the above ordering.

(b) Prove that the nonwandering set $\Omega(\sigma_A) = \Sigma_{A_1}^+ \cup \cdots \cup \Sigma_{A_m}^+$. Hint: Show that $\Omega(\sigma_{A_j}) = \Sigma_{A_j}^+$ for all j so $\Omega(\sigma_A) \supset \Sigma_{A_1}^+ \cup \cdots \cup \Sigma_{A_m}^+$. Also show that all points in $\Sigma_A^+ \setminus (\Sigma_{A_1}^+ \cup \cdots \cup \Sigma_{A_m}^+)$ are wandering.

3.16. Let Σ_A^+ be a subshift of finite type with metric d defined in Chapter II. Let $\delta = 0.5$. Assume $\{\mathbf{s}^{(j)} \in \Sigma_A^+\}$ is a 0.5-chain for σ_A on Σ_A^+. Explicitly indicate the point $\mathbf{t} \in \Sigma_A^+$ which 0.5-shadows this 0.5-chain.

Zeta Functions

3.17. Let A and B each be square matrices and

$$\zeta(t) = \exp\Big(\sum_{k=1}^{\infty} \frac{\mathrm{tr}(A^k) - \mathrm{tr}(B^k)}{k} t^k\Big).$$

Prove that

$$\zeta(t) = \frac{\det(I - tB)}{\det(I - tA)}.$$

Chaos and Liapunov Exponents

3.18. Prove that F_μ is expansive on $\mathbb{R}$ for $\mu > 4$.

3.19. Consider $f_\mu(x) = \mu x \bmod 1$, for $\mu > 1$.
(a) Calculate the Liapunov exponent.
(b) Prove that $f_\mu(x)$ has sensitive dependence on initial conditions.

3.20. Let $f : \mathbb{R} \to \mathbb{R}$ be C^1. Assume p is a periodic point and $\omega(x_0) = \mathcal{O}(p)$. Prove that the Liapunov exponents of x_0 and p are equal, $\lambda(x_0) = \lambda(p)$.

3.21. Let $F_\mu(x) = \mu\, x\,(1 - x)$ as usual.
(a) For $1 < \mu \leq 3$, find the Liapunov exponents for the different points $x \in [0, 1]$.
(b) For $3 < \mu < 1 + \sqrt{6}$, find the Liapunov exponents for the different points $x \in [0, 1]$. Hint: For these parameter values, there is an attracting orbit of period 2. See Exercise 2.8. Also see Exercise 3.20.

3.22. Let $0 < \alpha < 0.5$, and $f_\alpha : [0, 1] \to [0, 1]$ be defined by

$$f_\alpha(x) = \begin{cases} 2(\alpha - x) & \text{for } 0 \leq x \leq \alpha \\ 2(x - \alpha) & \text{for } \alpha \leq x \leq 0.5 + \alpha \\ 2(0.5 + \alpha - x) + 1 & \text{for } 0.5 + \alpha \leq x \leq 1. \end{cases}$$

(a) Draw the graph of f_α.
(b) Find the intervals on which f_α is transitive and describe its dynamics.

3.23. Let

$$f(x) = \begin{cases} x + 6x^2 & \text{for } 0 \leq x \leq \frac{1}{3} \\ 3(1 - x) & \text{for } \frac{2}{3} \leq x \leq 1, \end{cases}$$

and $f(x) > 1$ for $\frac{1}{3} < x < \frac{2}{3}$. Let $\Lambda = \bigcap_{n \geq 0} f^{-n}([0, 1])$. Put the invariant measure μ on Λ obtained by the conjugacy to the 2-shift, so each of the pieces of $\bigcap_{0 \leq j \leq n} f^{-j}([0, 1])$ has measure 2^{-n}.
(a) Prove that almost every point for μ has positive Liapunov exponent. Hint: Almost every point for μ has $\lim_{n\to\infty} \frac{1}{n} \sum_{j=0}^{n-1} \chi_{[\frac{2}{3},1]}(f^j(x)) = 1/2$, where $\chi_{[\frac{2}{3},1]}$ is the characteristic function for $[\frac{2}{3}, 1]$, i.e., $\chi_{[\frac{2}{3},1]}(t) = 1$ if $t \in [\frac{2}{3}, 1]$ and $\chi_{[\frac{2}{3},1]}(t) = 0$ if $t \notin [\frac{2}{3}, 1]$.
(b) Show that there are an infinite number of points with Liapunov exponent equal to zero. The μ-measure of this set of points is zero because almost every point has a positive Liapunov exponent.

CHAPTER IV
Linear Systems

This chapter begins our study of systems of more than one variable with the consideration of linear systems, both linear ordinary differential equations and linear maps. In the next chapter, we apply these results to the study of nonlinear systems.

Chapters II and III treated only maps in one dimension and not differential equations. In this chapter, we start our consideration of differential equations with the study of linear ordinary differential equations. Most of the results obtained concern linear equations with constant coefficients, e.g., the form of the solutions, the phase portraits, and the topological conjugacy class. These results help in the study of a system of nonlinear differential equations near fixed points. A few results allow the coefficient to depend on time; these are applicable to "linearized behavior" near a general orbit of a system of nonlinear differential equations.

After studying linear ordinary differential equations, we indicate the comparable results for linear maps.

In one dimension, the linear theory is trivial so we did not need any special tools. In several variables, we must use Linear Algebra, including eigenvalues, eigenvectors, and the real Jordan Canonical Form. The first section reviews some of this material, especially the real Jordan Canonical Form.

The reader may already know some of the material of this chapter and can treat it quickly. However, we give a few definitions for general flows which we use later, e.g., topological conjugacy and topological equivalence. These definitions and the material on phase portraits and topological conjugacy may be new and should be mastered before these concepts are applied to nonlinear systems in the next chapter.

4.1 Review: Linear Maps and the Real Jordan Canonical Form

We consider linear maps from $\mathbb{R}^k$ to $\mathbb{R}^n$, which we denote by $\mathbf{L}(\mathbb{R}^k, \mathbb{R}^n)$. Given bases $\{\mathbf{v}^j\}_{j=1}^k$ of $\mathbb{R}^k$ and $\{\mathbf{w}^i\}_{i=1}^n$ of $\mathbb{R}^n$, a linear map $M \in \mathbf{L}(\mathbb{R}^k, \mathbb{R}^n)$ determines an $n \times k$ matrix $A = (a_{i,j})$ by

$$M(\sum_{j=1}^k x_j \mathbf{v}^j) = \sum_{i=1}^n (\sum_{j=1}^k x_j a_{i,j})\mathbf{w}^i.$$

We often identify such a linear map $M \in \mathbf{L}(\mathbb{R}^k, \mathbb{R}^n)$ with this $n \times k$ matrix. With this identification, the linear map is given by

$$A \begin{pmatrix} x_1 \\ \vdots \\ x_k \end{pmatrix} = \begin{pmatrix} y_1 \\ \vdots \\ y_n \end{pmatrix}$$

where the x_j are the coefficients of the basis $\{\mathbf{v}^j\}_{j=1}^k$ and the y_i are the coefficients of the basis $\{\mathbf{w}^i\}_{i=1}^n$, i.e., $y_i = \sum_{j=1}^k a_{i,j} x_j$ as indicated by the first displayed formula above.

The space $\mathbf{L}(\mathbb{R}^k, \mathbb{R}^n)$ is given the *operator norm* (also called the *sup-norm*) defined by

$$||A|| \equiv \sup_{\mathbf{v} \neq \mathbf{0}} \frac{|A\mathbf{v}|}{|\mathbf{v}|},$$

when $A \in \mathbf{L}(\mathbb{R}^k, \mathbb{R}^n)$. This norm $||A||$ measures the maximum stretch of the linear map. Notice that this norm depends on norms on the domain and range space.

We are also often interested in the minimum stretch of the linear map. (This measurement becomes important in some covering estimates of linear or nonlinear maps.) For $A \in \mathbf{L}(\mathbb{R}^k, \mathbb{R}^n)$, we defined the *minimum norm* (or *conorm*) of A by

$$m(A) = \inf_{\mathbf{v} \neq \mathbf{0}} \frac{|A\mathbf{v}|}{|\mathbf{v}|}.$$

The minimum norm is a measure of the minimum expansion of A just as $||A||$ is a measure of the maximum expansion. We are often interested in the case of $A \in \mathbf{L}(\mathbb{R}^n, \mathbb{R}^n)$. If such an A has $m(A) > 0$, then 0 is not an eigenvalue and A is invertible. In turn, if A is invertible, then it is easy to verify that $m(A) = ||A^{-1}||^{-1}$.

For the rest of this section we review the Jordan Canonical Form, and in particular, the real Jordan Canonical Form. Implicitly, we review eigenvalues and eigenvectors.

For the rest of this subsection, A is a $n \times n$ real matrix. First we consider the canonical form over the complex numbers in the case where there is a basis of complex eigenvectors, $\mathbf{v}^1, \ldots, \mathbf{v}^n$. Letting $V = (\mathbf{v}^1 \ldots \mathbf{v}^n)$, then $AV = V\Lambda$ where $\Lambda = \operatorname{diag}(\lambda_1, \ldots, \lambda_n)$. Thus, $V^{-1}AV = \Lambda$ is a diagonal matrix. (If A is symmetric, then the eigenvalues are real and the eigenvectors can always be chosen to be real. Therefore, for a symmetric matrix, there is always a real matrix V which diagonalizes A.)

If an eigenvalue $\lambda_j = \alpha_j + i\beta_j$ is complex, then its eigenvector $\mathbf{v}^j = \mathbf{u}^j + i\mathbf{w}^j$ must also be complex. Since A is real, the complex conjugate $\bar{\lambda}_j = \alpha - i\beta$ is also an eigenvalue and has eigenvector $\bar{\mathbf{v}}^j = \mathbf{u}^j - i\mathbf{w}^j$. Since

$$A(\mathbf{u}^j + i\mathbf{w}^j) = (\alpha_j \mathbf{u}^j - \beta_j \mathbf{w}^j) + i(\beta_j \mathbf{u}^j + \alpha_j \mathbf{w}^j),$$

equating the real and imaginary parts yields

$$\begin{aligned} A\mathbf{u}^j &= \alpha_j \mathbf{u}^j - \beta_j \mathbf{w}^j \qquad \text{and} \\ A\mathbf{w}^j &= \beta_j \mathbf{u}^j + \alpha_j \mathbf{w}^j. \end{aligned}$$

Using the vectors $\mathbf{u}^j$ and $\mathbf{w}^j$ as part of a basis yields a subblock of the matrix in terms of this basis of the form

$$D_j = \begin{pmatrix} \alpha_j & \beta_j \\ -\beta_j & \alpha_j \end{pmatrix}.$$

Thus, if A has a basis of complex eigenvectors, then there is a real basis $\mathbf{z}^1, \ldots, \mathbf{z}^n$ in terms of which $A = \operatorname{diag}(B_1, \ldots, B_q)$ where each of the blocks B_j is either (i) a 1×1 block with real entry λ_k or (ii) B_j is of the form D_j given above.

Next we turn to the case of repeated eigenvalues where the eigenvectors do not span the whole space. If the matrix A has characteristic polynomial $p(x)$, then by substituting A for x we get that $p(A)\mathbf{v} = \mathbf{0}$ for all vectors $\mathbf{v}$. (This is called the Cayley-Hamilton Theorem.) In particular, if $\lambda_1, \ldots, \lambda_{k_0}$ are the distinct eigenvalues with multiplicities $m_1, \ldots, m_{k_0}$, then $S_k = \{\mathbf{v} \in \mathbb{C}^n : (A - \lambda_k I)^{m_k} \mathbf{v} = \mathbf{0}\}$ is a vector space of dimension m_k. Vectors in S_k are called *generalized eigenvectors*.

Now, fix an eigenvalue $\lambda = \lambda_k$ and assume that there is an $m \times m$ Jordan block. This means that there are vectors $\mathbf{v}^1, \ldots, \mathbf{v}^m$ such that $(A-\lambda I)\mathbf{v}^1 = \mathbf{0}$ and $(A-\lambda I)\mathbf{v}^j = \mathbf{v}^{j-1}$ for $2 \le j \le m$. In terms of this (partial) basis, the $m \times m$ (sub)matrix has the form

$$\begin{pmatrix} \lambda & 1 & 0 & \ldots & 0 & 0 \\ 0 & \lambda & 1 & \ldots & 0 & 0 \\ 0 & 0 & \lambda & \ldots & 0 & 0 \\ \vdots & & & \ddots & & \vdots \\ 0 & 0 & 0 & & \lambda & 1 \\ 0 & 0 & 0 & \ldots & 0 & \lambda \end{pmatrix}$$

This gives the Jordan Canonical Form over the complex numbers.

If we use the real and imaginary part of the eigenvectors for the complex eigenvalues to form a basis of real vectors, we get the Real Jordan Canonical Form for A, $A = \text{diag}(B_1, \ldots, B_q)$ where B_j is of one of the following four types:

(i) $B_j = (\lambda_k)$ for some real eigenvalue λ_k,

(ii)

$$B_j = \begin{pmatrix} \lambda & 1 & 0 & \ldots & 0 & 0 \\ 0 & \lambda & 1 & \ldots & 0 & 0 \\ 0 & 0 & \lambda & \ldots & 0 & 0 \\ \vdots & & & \ddots & & \vdots \\ 0 & 0 & 0 & & \lambda & 1 \\ 0 & 0 & 0 & \ldots & 0 & \lambda \end{pmatrix}$$

for some real eigenvalue $\lambda = \lambda_k$,

(iii) $B_j = D_k$ where D_k is a 2×2 matrix with entries α_k and $\pm\beta_k$ as given above for some complex eigenvalue $\lambda_k = \alpha_k + i\beta_k$, or

(iv)

$$B_j = \begin{pmatrix} D_k & I & \ldots & 0 & 0 \\ 0 & D_k & \ldots & 0 & 0 \\ \vdots & & & & \vdots \\ 0 & 0 & \ldots & D_k & I \\ 0 & 0 & \ldots & 0 & D_k \end{pmatrix}$$

where D_k is a 2×2 matrix with entries α_k and $\pm\beta_k$ as given above for some complex eigenvalue $\lambda_k = \alpha_k + i\beta_k$ and I is the 2×2 identity matrix.

See Appendix III of Hirsch and Smale (1974) for more discussion of the Jordan Canonical Form. Also see Gantmacher (1959).

4.2 Linear Differential Equations

The next few sections are concerned with the solutions of linear differential equations. These results are not only interesting for themselves, but are also used in the theory of nonlinear differential equations, e.g., to determine the stability near fixed points. We give a few definitions in the context of a general flow so we can use them throughout the book. For this reason, we give the definition of the flow of a general, possibly nonlinear, differential equation.

Consider the linear equation

$$\frac{d\mathbf{x}}{dt} = A(t)\mathbf{x} \qquad (*)$$

with $\mathbf{x} \in \mathbb{R}^n$ and $A(t) = (a_{ij}(t))$ an $n \times n$ matrix. Often we will take the case where A does not depend on t.

More generally in Chapter V, we consider the ordinary differential equation $\mathbf{x}(0) = \mathbf{x}_0$ and

$$\frac{d}{dt}\mathbf{x} = f(\mathbf{x}) \tag{$\dagger$}$$

where $f : \mathbb{R}^n \to \mathbb{R}^n$ is a function all of whose coordinate functions have continuous partial derivatives. The *flow of the differential equation* is a function $\varphi^t(\mathbf{x}_0)$ for t a real variable and $\mathbf{x}_0 \in \mathbb{R}^n$ such that (i) $\varphi^0(\mathbf{x}_0) = \mathbf{x}_0$ and (ii)

$$\frac{d}{dt}\varphi^t(\mathbf{x}_0) = f(\varphi^t(\mathbf{x}_0))$$

for t for which it is defined, i.e., $\varphi^t(\mathbf{x}_0)$ is the solution curve through $\mathbf{x}_0$ at $t = 0$. Theorem V.3.1 proves that for any $\mathbf{x}_0 \in \mathbb{R}^n$, there is an $\alpha = \alpha(\mathbf{x}_0) > 0$ such that $\varphi^t(\mathbf{x}_0)$ exists and is unique for $|t| < \alpha$.

The time-dependent linear equations $(*)$ can be considered in the form of $(\dagger)$ by forming the equations

$$\begin{aligned} \frac{d\mathbf{x}}{dt} &= A(\tau)\mathbf{x} \\ \frac{d\tau}{dt} &= 1. \end{aligned} \tag{$*$}$$

These equation are time independent (autonomous) and their solutions are essentially the same as $(*)$ since $\tau(t) = \tau_0 + t$. Therefore, the general theory implies that given $\mathbf{x}_0 \in \mathbb{R}^n$, there is a unique solution $\mathbf{x}(t)$ of $(*)$ with $\mathbf{x}(0) = \mathbf{x}_0$ that is defined on some open interval containing 0. If there is a bound on $||A(t)||$ which is independent of t, then it can be shown that $(*)$ has solutions for all t. (See Exercise 5.5.) For the present, we shall take the existence of solutions as known. We shall actually construct solutions for the constant coefficient case (when A is independent of t), giving another proof of the existence in this case. It will also follow in this case that the solutions exist for all t. We will also show that the solutions are unique in this case more simply than in the general theory. For the moment, the following lemma gives the linearity of the set of all solutions.

Lemma 2.1. *If $\mathbf{x} : I \to \mathbb{R}^n$ and $\mathbf{y} : J \to \mathbb{R}^n$ are two solutions of $(*)$ and $a, b \in \mathbb{R}$ are two scalars, then $a\mathbf{x}(t) + b\mathbf{y}(t)$ is a solution of $(*)$ on $I \cap J$.*

REMARK 2.1. Sometimes we allow solutions of $(*)$ with complex values. In this context, Lemma 2.1 is true for $x : I \to \mathbb{C}^n$ and $\mathbf{y} : J \to \mathbb{C}^n$ and $a, b \in \mathbb{C}$. We use this in Lemma 3.2 below.

PROOF. The proof is very direct:

$$\begin{aligned} \frac{d}{dt}[a\mathbf{x}(t) + b\mathbf{y}(t)] &= a\dot{\mathbf{x}}(t) + b\dot{\mathbf{y}}(t) \\ &= aA(t)\mathbf{x}(t) + bA(t)\mathbf{y}(t) \\ &= A(t)[a\mathbf{x}(t) + b\mathbf{y}(t)], \end{aligned}$$

so $a\mathbf{x}(t) + b\mathbf{y}(t)$ is a solution on the common interval of definition. □

Once we have shown that solutions for linear equations with constant coefficients exist for all t, the above lemma shows that the set of solutions of $(*)$ forms a vector space. Since solutions are determined by their initial conditions, it will follow that the dimension of the set of solutions is n. We give the details later.

4.3 Solutions for Constant Coefficients

In this section, we consider the form of the solutions of (*) when A is a constant matrix which is independent of t: $A = (a_{ij})$ and

$$\frac{d\mathbf{x}}{dt} = A\mathbf{x} \tag{**}$$

with $\mathbf{x} \in \mathbb{R}^n$.

This equation is a generalization of the scalar equation $\frac{dy}{dt} = ay$ with $y \in \mathbb{R}$ which has solutions $y = y_0 e^{at}$. In order to motivate the form of solutions we seek, assume there is a basis of vectors $\mathbf{v}^1, \ldots, \mathbf{v}^n$ that puts the matrix A in diagonal form, $\text{diag}(\lambda_1, \ldots, \lambda_n)$. Then in these new coordinates, the differential equation (**) becomes the n scalar equations $\dot{y}_1 = \lambda_1 y_1, \ldots, \dot{y}_n = \lambda_n y_n$. Using the form of the solutions of these scalar equations, we get that the solution

$$\mathbf{y} = \begin{pmatrix} C_1 e^{\lambda_1 t} \\ \vdots \\ C_n e^{\lambda_n t} \end{pmatrix} = C_1 e^{\lambda_1 t}\mathbf{e}^1 + \cdots + C_n e^{\lambda_n t}\mathbf{e}^n$$

where $\mathbf{e}^j$ is the standard unit vector with a one in the jth place and all other coordinates zeroes and $C_j = y_j(0)$. Back in the original basis, this solution is

$$\mathbf{x} = C_1 e^{\lambda_1 t}\mathbf{v}^1 + \cdots + C_n e^{\lambda_n t}\mathbf{v}^n.$$

Rather than write down the details of the way solutions change when we change basis, we use this discussion for motivation and we look for solutions to (**) of the form $e^{\lambda t}\mathbf{v}$. The following proposition gives the conditions that are necessary for this to be a solution.

Before giving the proposition, we need to make one more point. If $\lambda = \alpha + i\beta$ is a complex number with $\alpha, \beta \in \mathbb{R}$, then the exponential of λt can be written in its real and imaginary parts as follows:

$$e^{(\alpha+i\beta)t} = e^{\alpha t}\cos(\beta t) + ie^{\alpha t}\sin(\beta t).$$

This can be seen to be the correct expression by comparing the power series expansion of each side.

Proposition 3.1. *Assume A is a real $n \times n$ constant matrix. The curve $e^{\lambda t}\mathbf{v}$ is a (real) solution of (**) if and only if λ is a (real) eigenvalue of A with (real) eigenvector $\mathbf{v}$.*

PROOF. If $A\mathbf{v} = \lambda\mathbf{v}$, then define $\mathbf{x}(t) = e^{\lambda t}\mathbf{v}$. Then, $\dot{\mathbf{x}}(t) = \lambda e^{\lambda t}\mathbf{v} = e^{\lambda t}A\mathbf{v} = A\mathbf{x}(t)$. Thus, $\mathbf{x}(t)$ is a solution. If λ and $\mathbf{v}$ are real, then clearly the solution $e^{\lambda t}\mathbf{v}$ is real.

Conversely, assume that $e^{\lambda t}\mathbf{v}$ is a solution. Then, $\dot{\mathbf{x}}(t) = \lambda e^{\lambda t}\mathbf{v} = Ae^{\lambda t}\mathbf{v}$, so $\lambda\mathbf{v} = A\mathbf{v}$ and λ is an eigenvalue with eigenvector $\mathbf{v}$. If $\lambda = \alpha + i\beta$ and $\mathbf{v} = \mathbf{u} + i\mathbf{w}$, then $\mathbf{x}(t) = e^{\lambda t}\mathbf{v} = e^{\alpha t}[\cos(\beta t)\mathbf{u} - \sin(\beta t)\mathbf{w}] + ie^{\alpha t}[\sin(\beta t)\mathbf{u} + \cos(\beta t)\mathbf{w}]$. For $\mathbf{x}(t)$ to stay real for all t, it is necessary that $\beta = 0$ and $\mathbf{w} = \mathbf{0}$. Thus, λ and $\mathbf{v}$ must both be real. □

The next step is to find real solutions in the case that the eigenvalues are complex. The following lemma is the main additional step in finding the real form of the solutions in this case.

Lemma 3.2. *Let $A(t)$ be a real matrix which can depend on time. If $\mathbf{z}(t) = \mathbf{x}(t)+i\mathbf{y}(t)$ is a complex solution of $\dot{\mathbf{z}} = A(t)\mathbf{z}$, then $\mathbf{x}(t)$ and $\mathbf{y}(t)$ are each real solutions.*

PROOF. This follows from the linearity of differentiation and matrix multiplication: $\dot{\mathbf{z}}(t) = \dot{\mathbf{x}}(t) + i\dot{\mathbf{y}}(t) = A(t)(\mathbf{x}(t) + i\mathbf{y}(t)) = A(t)\mathbf{x}(t) + iA(t)\mathbf{y}(t)$, so by equating the real and imaginary parts, we get that $\dot{\mathbf{x}}(t) = A(t)\mathbf{x}(t)$ and $\dot{\mathbf{y}}(t) = A(t)\mathbf{y}(t)$, proving the result. □

Combining the lemma with the form of the solutions for complex eigenvalues given in the proof of Proposition 3.2, we get the following result for complex eigenvalues.

Proposition 3.3. *Let A be a real constant matrix with complex eigenvector $\mathbf{v} = \mathbf{u}+i\mathbf{w}$ for the complex eigenvalue $\lambda = \alpha + i\beta$, $\beta \neq 0$. Then, $e^{\alpha t}[\cos(\beta t)\mathbf{u} - \sin(\beta t)\mathbf{w}]$ and $e^{\alpha t}[\sin(\beta t)\mathbf{u} + \cos(\beta t)\mathbf{w}]$ are two real solutions.*

REMARK 3.1. If $\mathbf{v} = \mathbf{u} + i\mathbf{w}$ is the eigenvector for the complex eigenvalue $\lambda = \alpha + i\beta$, then $\mathbf{v} = \mathbf{u} - i\mathbf{w}$ is the eigenvector for the complex conjugate eigenvalue $\bar{\lambda} = \alpha - i\beta$. It can then be checked that these give the same two real solutions of the differential equation as given above.

Next we turn to the case of repeated eigenvalues where the eigenvectors do not span the whole space. We will write out the form of the solutions when the eigenvalues are real (or the complex form of the solutions if a repeated eigenvalue is complex). We will sketch a way of deriving solutions using the exponential of a matrix and then verify that the solutions that we derive are actually valid.

For a matrix B, let e^B be the matrix obtained from the power series:

$$e^B = I + B + \frac{1}{2!}B^2 + \cdots + \frac{1}{k!}B^k + \dots$$

It can be shown that this series converges to a matrix just as for real numbers. However, care must be exercised because it is not always the case that e^{B+C} is equal to $e^B e^C$. This is the case if $BC = CB$ (they commute). In particular,

$$\begin{aligned} e^{At} &= e^{\lambda It+(At-\lambda It)} \\ &= e^{\lambda It}e^{(At-\lambda It)} \\ &= e^{\lambda t}e^{t(A-\lambda I)}. \end{aligned}$$

It can be shown that $e^{At}\mathbf{v}$ is a solution of (**) for any $\mathbf{v}$, but we will give a more specific form of the solutions that does not involve a power series. Given that there are not always as many eigenvectors as the algebraic multiplicity of the eigenvalue, we take a generalized eigenvector $\mathbf{w}$ such that $(A - \lambda I)^{k+1}\mathbf{w} = \mathbf{0}$ but $(A - \lambda I)^k\mathbf{w} \neq \mathbf{0}$. Then, $(A - \lambda I)^j\mathbf{w} = \mathbf{0}$ for $j > k$ and

$$\begin{aligned} e^{At}\mathbf{w} &= e^{\lambda t}e^{t(A-\lambda I)}\mathbf{w} \\ &= e^{\lambda t}\{I\mathbf{w} + t(A - \lambda I)\mathbf{w} + \cdots + \frac{t^k}{k!}(A - \lambda I)^k\mathbf{w} \\ &\qquad + \frac{t^{k+1}}{(k+1)!}(A - \lambda I)^{k+1}\mathbf{w} + \dots\} \\ &= e^{\lambda t}\sum_{j=0}^{k}\frac{t^j}{j!}(A - \lambda I)^j\mathbf{w}. \end{aligned}$$

With the above calculations as motivation, we can check directly that the final form is indeed a solution (without verifying that the series converges).

Proposition 3.4. *If* $(A - \lambda I)^{k+1}\mathbf{w} = \mathbf{0}$, *then*

$$\mathbf{x}(t) = e^{\lambda t} \sum_{j=0}^{k} \frac{t^j}{j!}(A - \lambda I)^j \mathbf{w}$$

is a solution to $(**)$.

PROOF. Let $\mathbf{x}(t)$ be as in the statement. Then, using the fact that $(A - \lambda I)^{k+1}\mathbf{w} = \mathbf{0}$,

$$\begin{aligned}
\dot{\mathbf{x}}(t) &= \lambda e^{\lambda t} \sum_{j=0}^{k} \frac{t^j}{j!}(A - \lambda I)^j \mathbf{w} + e^{\lambda t} \sum_{j=1}^{k} \frac{t^{j-1}}{(j-1)!}(A - \lambda I)^j \mathbf{w} \\
&= \lambda e^{\lambda t} \sum_{j=0}^{k} \frac{t^j}{j!}(A - \lambda I)^j \mathbf{w} + e^{\lambda t} \sum_{j=1}^{k+1} \frac{t^{j-1}}{(j-1)!}(A - \lambda I)^j \mathbf{w} \\
&= e^{\lambda t}\{\lambda \sum_{j=0}^{k} \frac{t^j}{j!}(A - \lambda I)^j \mathbf{w} + (A - \lambda I) \sum_{j=0}^{k} \frac{t^j}{j!}(A - \lambda I)^j \mathbf{w}\} \\
&= A e^{\lambda t} \sum_{j=0}^{k} \frac{t^j}{j!}(A - \lambda I)^j \mathbf{w} \\
&= A\mathbf{x}(t).
\end{aligned}$$

Thus, $\mathbf{x}(t)$ is a solution as claimed. □

We have now given the real form of solutions in all cases (except for repeated complex eigenvalues). Using this, we have existence of solutions and the form of these solutions as given in the following theorem.

Theorem 3.5 (Existence). *Given a real* $n \times n$ *constant matrix* A *and* $\mathbf{x}_0 \in \mathbb{R}^n$, *there is a solution* $\mathbf{x}(t)$ *of* $\dot{\mathbf{x}} = A\mathbf{x}$ *defined for all* t *such that* $\mathbf{x}(0) = \mathbf{x}_0$. *Moreover, each coordinate function of* $\mathbf{x}(t)$ *is a linear combination of functions of the form*

$$t^k e^{\alpha t} \cos(\beta t) \qquad \text{and} \qquad t^k e^{\alpha t} \sin(\beta t)$$

where $\alpha + i\beta$ *is an eigenvalue of* A *and* k *is less than the algebraic multiplicity of the eigenvalue.*

PROOF. The Jordan normal form says that there is always a basis of generalized eigenvectors, $\mathbf{v}^1, \ldots, \mathbf{v}^n$. Let $\mathbf{x}^j(t)$ be the solution with $\mathbf{x}^j(0) = \mathbf{v}^j$ for $j = 1, \ldots, n$. Given any $\mathbf{x}_0 \in \mathbb{R}^n$, solve for $a_1, \ldots, a_n$ such that $\mathbf{x}_0 = \sum_{j=1}^n a_j \mathbf{v}^j$. Then, $\mathbf{x}(t) = \sum_{j=1}^n a_j \mathbf{x}^j(t)$ is a solution with $\mathbf{x}(0) = \mathbf{x}_0$. This shows that there is such a solution, and that it exists for all t. Also, the form of the solutions $\mathbf{x}^j(t)$ found above proves that the coordinate functions are linear combinations of functions of the form stated in the theorem. □

We mentioned above that for any $\mathbf{v}$, $e^{At}\mathbf{v}$ is a solution. Thus, $e^{At}\mathbf{v}$ is the flow for equations $(**)$. If we look only at the matrix e^{At}, it can also be shown that

$$\frac{d}{dt} e^{At} = A e^{At}.$$

Thus, if we allow matrix solutions to the differential equation, e^{At} is such a solution. However, it is not very easy to compute. In Theorem 3.8 below, we show how to use the solutions constructed above to get another matrix solution to the equation. Before giving this construction, we prove directly the uniqueness of solutions using the matrix e^{At} rather than using the general theorem for ordinary differential equations.

Theorem 3.6 (Uniqueness). *Given $\mathbf{x}_0$, there is a unique solution $\mathbf{x}(t)$ to $\dot{\mathbf{x}} = A\mathbf{x}$ with $\mathbf{x}(0) = \mathbf{x}_0$.*

PROOF. Let $\mathbf{x}(t)$ be a solution. Let $\mathbf{y}(t) = e^{-At}\mathbf{x}(t)$. Then,

$$\begin{aligned}\dot{\mathbf{y}}(t) &= -Ae^{-At}\mathbf{x}(t) + e^{-At}\dot{\mathbf{x}}(t)\\ &= -Ae^{-At}\mathbf{x}(t) + e^{-At}A\mathbf{x}(t)\\ &= e^{-At}(-A + A)\mathbf{x}(t)\\ &= \mathbf{0}.\end{aligned}$$

Therefore, $\mathbf{y}(t) \equiv \mathbf{y}(0) = e^{\mathbf{0}}\mathbf{x}(0) = \mathbf{x}_0$, and $e^{-At}\mathbf{x}(t) \equiv \mathbf{x}_0$, so $\mathbf{x}(t) \equiv e^{At}\mathbf{x}_0$. This proves the result. □

REMARK 3.1. The above proof can be modified to apply to the general linear equation (*) by using the fundamental matrix solution introduced below.

Finally, we can prove that the solutions form a vector space of dimension n.

Theorem 3.7. *Given an $n \times n$ constant real matrix A, the set of solutions of (**),*

$$\mathcal{S} = \{\mathbf{x} : \mathbb{R} \to \mathbb{R}^n : \dot{\mathbf{x}}(t) = A\mathbf{x}(t)\},$$

forms a vector space of dimension n.

PROOF. We know from above that solutions exist for all time. By Lemma 2.1, if $\mathbf{x}, \mathbf{y} \in \mathcal{S}$ and $a, b \in \mathbb{R}$, then $a\mathbf{x}(t) + b\mathbf{y}(t) \in \mathcal{S}$. This shows that $\mathcal{S}$ is a vector space.

Let $\mathbf{v}^1, \ldots, \mathbf{v}^n$ be a basis for $\mathbb{R}^n$ (for example, either the standard basis or a basis of generalized eigenvectors of A). Let $\mathbf{x}^j(t)$ be the solution with $\mathbf{x}^j(0) = \mathbf{v}^j$ for $j = 1, \ldots, n$. Given any solution $\mathbf{z} \in \mathcal{S}$, solve for $a_1, \ldots, a_n$ such that $\mathbf{z}(0) = \sum_{j=1}^n a_j\mathbf{v}^j$. Then, both $\mathbf{z}(t)$ and $\sum_{j=1}^n a_j\mathbf{x}^j(t)$ are solutions, and they have the same initial condition at $t = 0$. By uniqueness, they are equal, so $\mathbf{z}(t) = \sum_{j=1}^n a_j\mathbf{x}^j(t)$. This proves that $\{\mathbf{x}^1(t), \ldots, \mathbf{x}^n(t)\}$ span $\mathcal{S}$.

If a linear combination of the solutions $\sum_{j=1}^n a_j\mathbf{x}^j(t) = \mathbf{0}$ for some $a_1, \ldots, a_n$, then by setting $t = 0$ we get that $\sum_{j=1}^n a_j\mathbf{v}^j = \mathbf{0}$. Because the $\mathbf{v}^j$ are independent, it follows that $a_j = 0$ for all j. This proves that the $\mathbf{x}^j(t)$ are independent and so a basis. This shows that S has dimension n. □

Returning to matrix solutions of linear equations, an $n \times n$ matrix $M(t)$ is called a *fundamental matrix solution* of $\dot{\mathbf{x}} = A(t)\mathbf{x}$ provided

$$\frac{d}{dt}M(t) = A(t)M(t)$$

for all t and $M(t_0)$ is nonsingular at one time t_0. The following theorem justifies the assumption the $M(t_0)$ is nonsingular at one time as well as giving an alternative construction of e^{At}.

Theorem 3.8. *Let $A(t)$ be an $n \times n$ real matrix, and consider the linear equation $\dot{\mathbf{x}} = A(t)\mathbf{x}$.*

(a) Assume $\mathbf{x}^1(t), \ldots, \mathbf{x}^n(t)$ are n solutions, and let

$$M(t) = (\mathbf{x}^1(t), \ldots, \mathbf{x}^n(t))$$

be the $n \times n$ matrix formed by putting the vector solutions in as columns. Then, $M(t)$ satisfies the linear equation as a matrix solution:

$$\frac{d}{dt}M(t) = A(t)M(t).$$

(b) If $\mathbf{x}^1(t_0), \ldots, \mathbf{x}^n(t_0)$ are independent for one time t_0, then any solution can be written as $\mathbf{x}(t) = M(t)\mathbf{v}$ for some vector $\mathbf{v}$.

(c) If $M(t)$ is a matrix solution such that $M(t_0)$ is nonsingular at one time t_0, then $M(t)$ is nonsingular at all times.

(d) If $M(t)$ is a matrix solution of (**) with $M(0)$ nonsingular, then

$$e^{At} = M(t)M(0)^{-1}.$$

PROOF. Defining $M(t)$ as stated,

$$\begin{aligned}\frac{d}{dt}M(t) &= (\dot{\mathbf{x}}^1(t), \ldots, \dot{\mathbf{x}}^n(t)) \\ &= (A(t)\mathbf{x}^1(t), \ldots, A(t)\mathbf{x}^n(t)) \\ &= A(t)M(t),\end{aligned}$$

as claimed.

Now, assume the solutions are independent at t_0, and $\mathbf{x}(t)$ is any other solution. Because $M(t_0)$ is nonsingular, it is possible to solve for $\mathbf{v}$ such that $\mathbf{x}(t_0) = M(t_0)\mathbf{v}$. Then, both $\mathbf{x}(t)$ and $M(t)\mathbf{v}$ are solutions that agree at t_0. By the uniqueness of solutions, they are equal for all time.

Part (c) is equivalent to proving that if $M(t)$ is a matrix solution that is singular at one time, then it is singular at all times. Assume that $M(t)$ is such a matrix solution with $M(t_0)$ singular. Thus, the columns of $M(t_0)$ are dependent and there is a nonzero vector $\mathbf{v}$ with $M(t_0)\mathbf{v} = \mathbf{0}$. Then, both $M(t)\mathbf{v}$ and the zero function are solutions that are equal at one time. By uniqueness, they are equal for all time, $M(t)\mathbf{v} = \mathbf{0}$. Thus, $M(t)$ is singular for all time, proving the result.

Now, assume $M(0)$ is nonsingular (or is nonsingular at some time t_0). Then, both $M(t)M(0)^{-1}$ and e^{At} are matrix solutions that are equal at time $t = 0$. By uniqueness of solutions, they are equal for all times. (To reduce it to vector solutions, multiply both matrix solutions by the standard basis elements $\mathbf{e}^j$.) □

REMARK 3.2. Notice that if $M(t)$ is a fundamental matrix solution of $(*)$, then by multiplying on the right by $M(t_0)^{-1}$ we get that $M(t, t_0) = M(t)M(t_0)^{-1}$ is another fundamental matrix solution with $M(t_0, t_0) = I$. This idea is used in the proof of both Theorem 3.8 and 3.9.

To end the section, we restate Liouville's Formula for any fundamental matrix solution of a linear differential equation which depends on time. This result is used when we make the connection between the divergence of a vector field and the way that the flow distorts area.

Theorem 3.9 (Liouville's Formula). *Let $M(t)$ be a fundamental matrix solution of the linear equation (with possibly nonconstant coefficients) $\dot{\mathbf{x}} = A(t)\mathbf{x}$. Then,*

$$\frac{d}{dt}\det\big(M(t)\big) = \det\big(M(t)\big)\,\mathrm{tr}\,\big(A(t)\big) \qquad \textit{and}$$

$$\det\big(M(t)\big) = \det\big(M(0)\big)\,\exp\left(\int_0^t \mathrm{tr}\,\big(A(s)\big)\,ds\right).$$

PROOF. We let $\mathbf{e}^j$ be the standard basis. In the following calculation we use the notation $a_{i,j}(t)$ for the (i,j) entry of $A(t)$. Using the fact that the determinant is multilinear in the columns we get the following:

$$\begin{aligned}
\frac{d}{dt}&\det\big(M(t)M(t_0)^{-1}\big)|_{t=t_0} = \\
&= \frac{d}{dt}\det\big(M(t)M(t_0)^{-1}\mathbf{e}^1,\ldots,M(t)M(t_0)^{-1}\mathbf{e}^n\big)|_{t=t_0} \\
&= \sum_j \det\big(M(t_0)M(t_0)^{-1}\mathbf{e}^1,\ldots,M(t_0)M(t_0)^{-1}\mathbf{e}^{j-1}, \\
&\qquad M'(t_0)M(t_0)^{-1}\mathbf{e}^j, M(t_0)M(t_0)^{-1}\mathbf{e}^{j+1}, \\
&\qquad \ldots, M(t_0)M(t_0)^{-1}\mathbf{e}^n\big) \\
&= \sum_j \det\big(\mathbf{e}^1,\ldots,\mathbf{e}^{j-1},A(t_0)\mathbf{e}^j,\mathbf{e}^{j+1},\ldots,\mathbf{e}^n\big) \\
&= \sum_j \det\big(\mathbf{e}^1,\ldots,\mathbf{e}^{j-1},\sum_i a_{i,j}(t_0)\mathbf{e}^i,\mathbf{e}^{j+1},\ldots,\mathbf{e}^n\big) \\
&= \sum_{i,j} a_{i,j}(t_0)\det(\mathbf{e}^1,\ldots,\mathbf{e}^{j-1},\mathbf{e}^i,\mathbf{e}^{j+1},\ldots,\mathbf{e}^n) \\
&= \sum_j a_{j,j}(t_0)\det(\mathbf{e}^1,\ldots,\mathbf{e}^n) \\
&= \operatorname{tr}\big(A(t_0)\big).
\end{aligned}$$

Therefore, using the multiplicative feature of the determinant,

$$\begin{aligned}
\frac{d}{dt}\det\big(M(t)\big)|_{t=t_0} &= \frac{d}{dt}\det\big(M(t)M(t_0)^{-1}\big)\det\big(M(t_0)\big) \\
&= \det\big(M(t_0)\big)\operatorname{tr}\big(A(t_0)\big).
\end{aligned}$$

Solving this scalar differential equation gives the formula in the theorem relating the quantities $\det\big(M(t)\big)$, $\det\big(M(0)\big)$, and the exponential of $\int_0^t \operatorname{tr}\big(A(s)\big)\,ds$. □

4.4 Phase Portraits

In the last section, we gave the general theory of solutions to linear equations and found the solutions of constant coefficient linear equations. In this section, we give the phase portraits of two-dimensional constant coefficient linear equations and some examples in three dimensions. The *phase portrait* of a differential equation $\dot{\mathbf{x}} = f(\mathbf{x})$ is a drawing of the solution curves with the direction of increasing time indicated. In some abstract sense, the phase portrait is the drawing of all solution curves, but in practice it only includes representative trajectories. The *phase space* is the domain of all $\mathbf{x}$'s considered.

Example 4.1 (A Saddle). Consider

$$\dot{\mathbf{x}} = \begin{pmatrix} 1 & 1 \\ 0 & -2 \end{pmatrix}\mathbf{x}.$$

The general solution is

$$\mathbf{x}(t) = C_1 e^t\begin{pmatrix} 1 \\ 0 \end{pmatrix} + C_2 e^{-2t}\begin{pmatrix} -1 \\ 3 \end{pmatrix}.$$

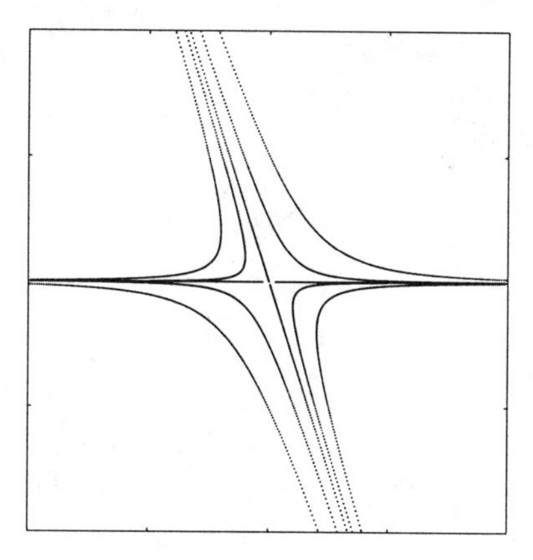

FIGURE 4.1. A Saddle

See Figure 4.1 for the phase portrait. (Solutions along the x-axis are moving away from the origin and those along the line of slope -3 are coming in toward the origin.)

Example 4.2 (A Stable Node). Consider

$$\dot{\mathbf{x}} = \begin{pmatrix} -1 & 0 \\ 0 & -2 \end{pmatrix} \mathbf{x}.$$

The general solution is

$$\mathbf{x}(t) = C_1 e^{-t} \begin{pmatrix} 1 \\ 0 \end{pmatrix} + C_2 e^{-2t} \begin{pmatrix} 0 \\ 1 \end{pmatrix}.$$

See Figure 4.2 for the phase portrait. (All the solutions are coming in toward the origin.)

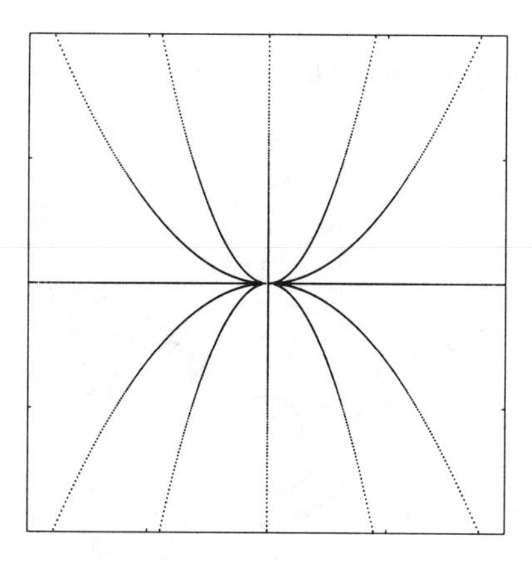

FIGURE 4.2. A Stable Node

Example 4.3 (An Improper Node). Consider

$$\dot{\mathbf{x}} = \begin{pmatrix} -1 & 1 \\ 0 & -1 \end{pmatrix} \mathbf{x}.$$

The general solution is

$$\mathbf{x}(t) = C_1 e^{-t} \begin{pmatrix} 1 \\ 0 \end{pmatrix} + C_2 e^{-t} \begin{pmatrix} t \\ 1 \end{pmatrix}.$$

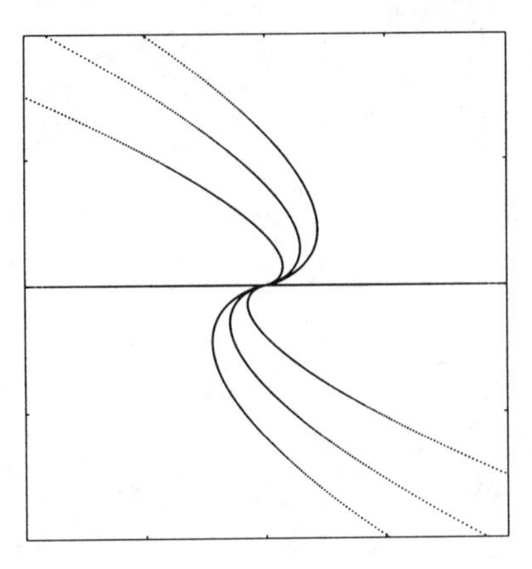

FIGURE 4.3. An Improper Node

See Figure 4.3 for the phase portrait. (All the solutions are coming in toward the origin.)

Example 4.4 (A Center). Consider

$$\dot{\mathbf{x}} = \begin{pmatrix} 0 & -\omega \\ \omega & 0 \end{pmatrix} \mathbf{x}.$$

The eigenvalues are $\pm i\omega$ with eigenvectors $\begin{pmatrix} \pm i \\ 1 \end{pmatrix}$. The general solution is

$$\mathbf{x}(t) = C_1 \begin{pmatrix} -\sin(\omega t) \\ \cos(\omega t) \end{pmatrix} + C_2 \begin{pmatrix} \cos(\omega t) \\ \sin(\omega t) \end{pmatrix}.$$

See Figure 4.4 for the phase portrait. All the solutions are on closed orbits surrounding the origin.

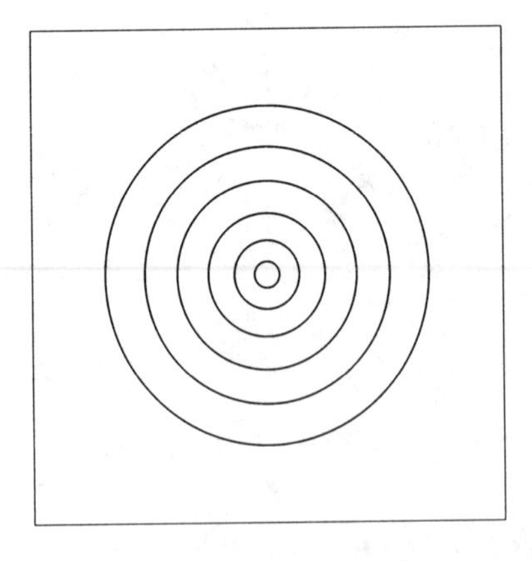

FIGURE 4.4. A Center

Example 4.5 (A Stable Focus). Consider

$$\dot{\mathbf{x}} = \begin{pmatrix} -1 & -2 \\ 2 & -1 \end{pmatrix} \mathbf{x}.$$

The eigenvalues are $-1 \pm 2i$ with eigenvectors $\begin{pmatrix} \pm i \\ 1 \end{pmatrix}$. The general solution is

$$\mathbf{x}(t) = C_1 e^{-t}[\cos(2t)\begin{pmatrix} 0 \\ 1 \end{pmatrix} - \sin(2t)\begin{pmatrix} 1 \\ 0 \end{pmatrix}]$$
$$+ C_2 e^{-t}[\sin(2t)\begin{pmatrix} 0 \\ 1 \end{pmatrix} + \cos(2t)\begin{pmatrix} 1 \\ 0 \end{pmatrix}].$$

See Figure 4.5 for the phase portrait. (All the solutions are coming in toward the origin.)

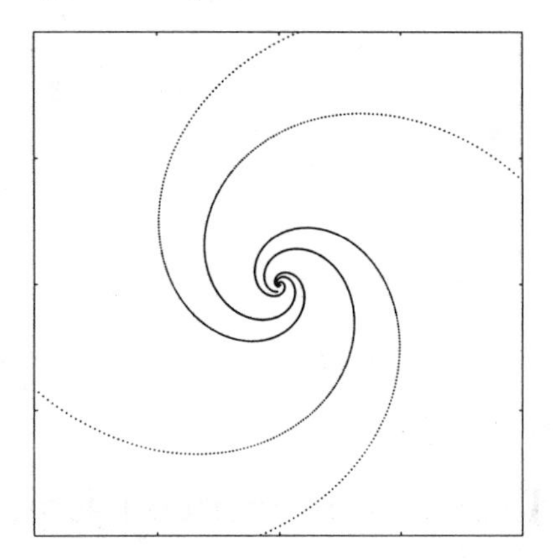

FIGURE 4.5. A Stable Focus

Definition. In general, a linear system is called a *node* provided that every orbit tends to the origin in a definite direction as t goes to infinity (if all the eigenvalues are real and negative), or as t goes to negative infinity (if all the eigenvalues are real and positive). A linear system is called a *proper node* provided it is a node and there is a unique orbit going into (or coming out of) the origin for each direction (all the eigenvalues are equal, real, and nonzero); it is called an *improper node* provided it is a node and there is a direction for which there are more than one orbit going into (or coming out of) the origin in that direction. In this terminology, both Examples 4.2 and 4.3 are improper nodes. A linear system is called a *stable focus* or *stable spiral* (respectively, *unstable focus* or *unstable spiral*) provided the solutions approach the origin as t goes to infinity (respectively, minus infinity) but not from a definite direction.

Example 4.6 (A Three Dimensional Saddle). Consider

$$\dot{\mathbf{x}} = \begin{pmatrix} -1 & -2 & 0 \\ 2 & -1 & 0 \\ 0 & 0 & 1 \end{pmatrix} \mathbf{x}.$$

The eigenvalues are $-1 \pm 2i$ and 1 with eigenvectors $\begin{pmatrix} \pm i \\ 1 \\ 0 \end{pmatrix}$ and $\begin{pmatrix} 0 \\ 0 \\ 1 \end{pmatrix}$. The general solution is

$$\mathbf{x}(t) = C_1 e^{-t}[\cos(2t)\begin{pmatrix} 0 \\ 1 \\ 0 \end{pmatrix} - \sin(2t)\begin{pmatrix} 1 \\ 0 \\ 0 \end{pmatrix}]$$
$$+ C_2 e^{-t}[\sin(2t)\begin{pmatrix} 0 \\ 1 \\ 0 \end{pmatrix} + \cos(2t)\begin{pmatrix} 1 \\ 0 \\ 0 \end{pmatrix}] + C_3 e^{t}\begin{pmatrix} 0 \\ 0 \\ 1 \end{pmatrix}.$$

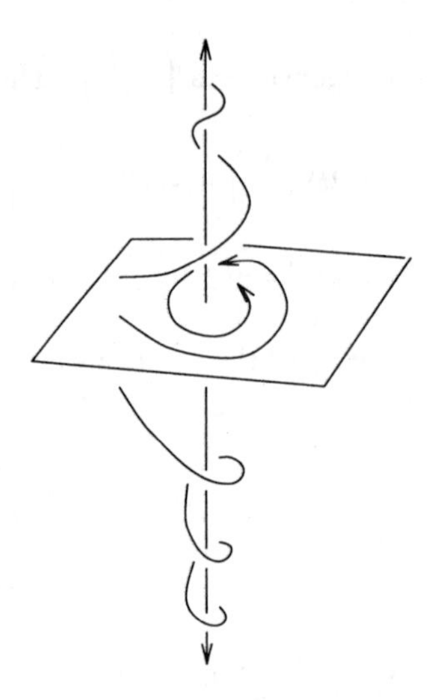

FIGURE 4.6. A Three-Dimensional Saddle

See Figure 4.6 for the phase portrait.

4.5 Contracting Linear Differential Equations

In this section we use the features of the solutions of (**) to characterize the solutions as contracting, expanding, or subexponentially growing. We give the definitions of Liapunov stable and asymptotically stable again in this context. They are essentially the same as we gave for one-dimensional maps.

Definition. The orbit of a point $\mathbf{p}$ is *Liapunov stable* for a flow φ^t provided that given any $\epsilon > 0$, there is a $\delta > 0$ such that if $d(\mathbf{x}, \mathbf{p}) < \delta$, then $d(\varphi^t(\mathbf{x}), \varphi^t(\mathbf{p})) < \epsilon$ for all $t \geq 0$. If $\mathbf{p}$ is a fixed point, then the condition can be written as $d(\varphi^t(\mathbf{x}), \mathbf{p}) < \epsilon$. The orbit of $\mathbf{p}$ is called *asymptotically stable* or *attracting* provided it is Liapunov stable and there is a $\delta_1 > 0$ such that if $d(\mathbf{x}, \mathbf{p}) < \delta_1$, then $d(\varphi^t(\mathbf{x}), \varphi^t(\mathbf{p}))$ goes to zero as t goes to infinity. If $\mathbf{p}$ is a fixed point, then it is asymptotically stable provided it is Liapunov stable and $\delta_1 > 0$ such that if $d(\mathbf{x}, \mathbf{p}) < \delta_1$, then $\omega(\mathbf{x}) = \{\mathbf{p}\}$.

REMARK 5.1. For a linear system, if the origin satisfies the second condition of asymptotic stability, then it is also Liapunov stable. There are nonlinear systems for which this is not the case. See Example V.5.3 for an example of Vinograd (1957).

Another example can be given in polar coordinates with the fixed point at $r = 1$ and $\theta = 0$ by

$$\dot{r} = 1 - r$$
$$\dot{\theta} = \sin(\theta/2).$$

Initial conditions with $r(0)$ near 1, have $r(t)$ tending to 1. Also, $\theta(t)$ tends to 0 modulo 2π. Thus, for any such point, $\omega(r_0, \theta_0) = \{(1,0)\}$. On the other hand, this point is not attracting.

The main theorem of the section proves that if all of the eigenvalues of A have negative real part, then the origin is attracting at an exponential rate determined by the eigenvalues. Thus, one criterion for asymptotic stability of a linear system of differential equations with constant coefficients is that all the eigenvalues of the matrix have negative real part. Before stating the theorem, we give an example which illustrates the fact that for a stable linear system, the usual Euclidean norm of a nonzero solution is not always strictly decreasing.

Example 5.1. Consider the system of linear differential equations given by

$$\begin{aligned}\dot{x} &= -x - y\\ \dot{y} &= 4x - y.\end{aligned}$$

The eigenvalues are $-1 \pm 2i$, and the general solution is given by

$$\begin{pmatrix} x(t) \\ y(t) \end{pmatrix} = e^{-t} \begin{pmatrix} \cos(2t) & -(1/2)\sin(2t) \\ 2\sin(2t) & \cos(2t) \end{pmatrix} \begin{pmatrix} x_0 \\ y_0 \end{pmatrix}.$$

See Figure 5.1 for the phase portrait.

FIGURE 5.1. Phase Portrait for Example 5.1.

The matrix in the above solution, with terms involving $\sin(2t)$ and $\cos(2t)$, has norm less than 2, so

$$\left|\begin{pmatrix} x(t) \\ y(t) \end{pmatrix}\right| \leq 2e^{-t}\left|\begin{pmatrix} x_0 \\ y_0 \end{pmatrix}\right|,$$

which goes to zero exponentially. The time derivative of the Euclidean norm along a nonzero solution is given by

$$\begin{aligned}\frac{d}{dt}(x^2+y^2)^{1/2} &= \frac{1}{2}(x^2+y^2)^{-1/2}(2x\dot{x}+2y\dot{y})\\ &= (x^2+y^2)^{-1/2}(-x^2-y^2+3xy).\end{aligned}$$

Along the line $x = y$, $(d/dt)(x^2+y^2)^{1/2} = 1/\sqrt{2}|x|$ which is positive for $x = y \neq 0$. Therefore, the Euclidean norm is not monotonically decreasing, although it does go to zero at an exponential rate. See Figure 5.2.

In this example, it is possible to take a different norm (the norm in the coordinates which give the Jordan Canonical Form) which is monotonically decreasing along a solution. Let $|\cdot|_*$ be the norm defined by

$$|(x,y)|_* \equiv (4x^2+y^2)^{1/2}.$$

The time derivative of this norm along a nonzero solution is given by

$$\begin{aligned}\frac{d}{dt}|(x,y)|_* &= \frac{1}{2}|(x,y)|_*^{-1}(4\cdot 2x\dot{x}+2y\dot{y})\\ &= |(x,y)|_*^{-1}(-4x^2-y^2)\\ &= -|(x,y)|_*.\end{aligned}$$

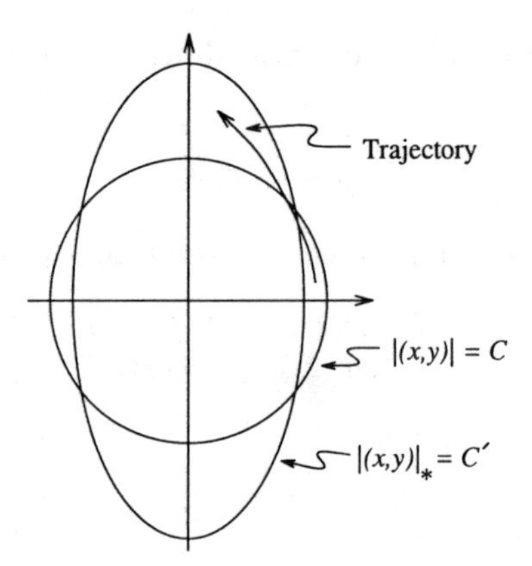

FIGURE 5.2. Solution Curve for Example 5.1 Crossing Level Sets $|(x, y)| = C$ and $|(x, y)|_* = C'$

Solving this linear scalar equation shows that

$$|(x(t), y(t))|_* = e^{-t}|(x_0, y_0)|_*$$

which monotonically and exponentially decreases to zero.

Theorem 5.1. *Let A be an $n \times n$ real matrix, and consider the equation $\dot{\mathbf{x}} = A\mathbf{x}$. The following are equivalent.*

(a) *There is a norm $|\ |_*$ on $\mathbb{R}^n$ and a constant $a > 0$ such that for any initial condition $\mathbf{x} \in \mathbb{R}^n$, the solution satisfies*

$$|e^{At}\mathbf{x}|_* \le e^{-ta}|\mathbf{x}|_* \quad \text{for all } t \ge 0.$$

(b) *For any norm $|\ |'$ on $\mathbb{R}^n$, there exist constants $a > 0$ and $C \ge 1$ such that for any initial condition $\mathbf{x} \in \mathbb{R}^n$, the solution satisfies*

$$|e^{At}\mathbf{x}|' \le Ce^{-ta}|\mathbf{x}|' \quad \text{for all } t \ge 0.$$

(c) *The real parts of all the eigenvalues λ of A are negative, $Re(\lambda) < 0$.*

REMARK 5.2. A norm as in Theorem 5.1(a) is called an *adapted norm.* It is useful to use because solutions immediately contract in terms of this norm as time goes forward. The norm $|\ |_*$ introduced in Example 5.1 is such an adapted norm. The adapted norm for the linear equation is also used to study nonlinear equations near a fixed point.

REMARK 5.3. We give two proofs that condition (c) implies condition (a), i.e., that the condition on the eigenvalues in part (c) implies the existence of a norm in which the differential equation is a contraction as given in part (a). The second (alternative) proof uses the basis which puts the matrix in Jordan canonical form to define the norm. This proof has the advantage that it is constructive; it has the disadvantage that it does not apply in the situation we consider in Chapter VIII of a hyperbolic invariant set (Theorem VIII.1.1). The first proof averages the usual norm along the orbits. It has the advantage that it applies in the more general situation of a hyperbolic invariant set; it has the disadvantage of being more abstract and not easily computable for a particular example.

PROOF. First we show that (a) implies (b). Let $|\ |_*$ be the norm given in (a) and $|\ |'$ be any other norm. Then, there are constants $A_1, A_2 > 0$ such that $A_1|\mathbf{x}|' \le |\mathbf{x}|_* \le A_2|\mathbf{x}|'$. (This is true for any two norms in finite dimensions.) Then, for $t \ge 0$,

$$\begin{aligned}|e^{At}\mathbf{x}|' &\le \frac{1}{A_1}|e^{At}\mathbf{x}|_* \\ &\le \frac{1}{A_1}e^{-at}|\mathbf{x}|_* \\ &\le \frac{A_2}{A_1}e^{-at}|\mathbf{x}|'.\end{aligned}$$

This proves that (b) is true with $C = A_2/A_1$. This completes the proof that (a) implies (b).

Next, we show that (b) implies (c). Suppose (c) is not true: there exists an eigenvalue $\lambda = \alpha + i\beta$ with $\alpha \ge 0$. There is a solution for the corresponding eigenvector of the form $e^{\alpha t}[\sin(\beta t)\mathbf{u} + \cos(\beta t)\mathbf{w}]$, and this solution does not go to zero as $t \to \infty$. This shows that (b) is not true.

Finally, we need to show that (c) implies (a). Let $a > 0$ be chosen so that $Re(\lambda) < -a$ for all the eigenvalues λ of A. Let $\mathbf{v}^1, \dots, \mathbf{v}^n$ be a basis of generalized eigenvectors, and $\mathbf{x}^j(t)$ be the solution with $\mathbf{x}^j(0) = \mathbf{v}^j$. By the form of the solutions given in Theorem 3.5, for each j there is a τ_j such that $|e^{At}\mathbf{v}^j| \le e^{-at}|\mathbf{v}^j|$ for all $t \ge \tau_j$. (Here we use the Euclidean norm, or any other fixed norm.) Then, any $\mathbf{x}$ it can be written as $\mathbf{x} = \sum_{j=1}^n c_j\mathbf{v}^j$, so the solution $e^{At}\mathbf{x} = \sum_{j=1}^n c_j\mathbf{x}^j(t)$. The form of these solutions implies there is a τ which depends on $\mathbf{x}$ such that $|e^{At}\mathbf{x}| \le e^{-at}|\mathbf{x}|$ for all $t \ge \tau$. Using the compactness of $\{\mathbf{x} : |\mathbf{x}| = 1\}$, there is one τ which works for all $\mathbf{x}$ with $|\mathbf{x}| = 1$. By linearity, we get that there is one τ which works for all $\mathbf{x}$. Fix this τ.

What we have is that $e^{At}\mathbf{x}$ is a contraction if we take t larger than τ. We average the norm along the trajectory for times between 0 and τ with a weighting factor e^{at} and show that in terms of this averaged norm, the linear flow is an immediate contraction. To this end, define

$$|\mathbf{x}|_* \equiv \int_0^\tau e^{as}|e^{As}\mathbf{x}|\,ds.$$

We now show that this norm works. Take $t \ge 0$ and write it as $t = n\tau + T$ with $0 \le T < \tau$. In the calculation which follows, we split up the range of s so that $s+t$ runs from $n\tau + T$ to $(n+1)\tau$ and from $(n+1)\tau$ to $(n+1)\tau + T$:

$$\begin{aligned}|e^{At}\mathbf{x}|_* &= \int_0^\tau e^{as}|e^{As}e^{At}\mathbf{x}|\,ds \\ &= \int_0^{\tau-T} e^{as}|e^{An\tau}e^{A(T+s)}\mathbf{x}|\,ds + \int_{\tau-T}^{\tau} e^{as}|e^{A(n+1)\tau}e^{A(T-\tau+s)}\mathbf{x}|\,ds.\end{aligned}$$

Making the substitution $u = T + s$ in the first integral, and $u = T - \tau + s$ in the second integral, and using the estimate above for $|e^{An\tau}\mathbf{x}|$, we get

$$\begin{aligned}|e^{At}\mathbf{x}|_* &\le \int_T^\tau e^{a(u-T-n\tau)}|e^{Au}\mathbf{x}|\,du + \int_0^T e^{a(u+\tau-T-(n+1)\tau)}|e^{Au}\mathbf{x}|\,du \\ &= e^{-at}\int_0^\tau e^{au}|e^{Au}\mathbf{x}|\,du \\ &= e^{-at}|\mathbf{x}|_*.\end{aligned}$$

This proves that (a) holds using the norm $|\ |_*$. □

ALTERNATE PROOF THAT (c) IMPLIES (a). Given $\epsilon > 0$, we can find a basis $\mathcal{B}$ of generalized eigenvectors in terms of which $A = \mathrm{diag}\{A_1, \dots, A_p\}$ where each A_j is one of the following types:

$$\begin{pmatrix} \alpha_j & 0 & \dots & 0 \\ 0 & \alpha_j & \dots & 0 \\ \vdots & & & \vdots \\ 0 & 0 & \dots & \alpha_j \end{pmatrix}, \quad \begin{pmatrix} \alpha_j & \epsilon & \dots & 0 & 0 \\ 0 & \alpha_j & \dots & 0 & 0 \\ \vdots & & & & \vdots \\ 0 & 0 & \dots & \alpha_j & \epsilon \\ 0 & 0 & \dots & 0 & \alpha_j \end{pmatrix},$$

$$\begin{pmatrix} D_j & 0 & \dots & 0 \\ 0 & D_j & \dots & 0 \\ \vdots & & & \vdots \\ 0 & 0 & \dots & D_j \end{pmatrix}, \quad \text{or} \quad \begin{pmatrix} D_j & \epsilon I & \dots & 0 & 0 \\ 0 & D_j & \dots & 0 & 0 \\ \vdots & & & & \vdots \\ 0 & 0 & \dots & D_j & \epsilon I \\ 0 & 0 & \dots & 0 & D_j \end{pmatrix}$$

where

$$D_j = \begin{pmatrix} \alpha_j & -\beta_j \\ \beta_j & \alpha \end{pmatrix}, \quad \text{and} \quad I = \begin{pmatrix} 1 & 0 \\ 0 & 1 \end{pmatrix}.$$

Let $|\ |_{\mathcal{B}}$ be the norm in terms of this basis, i.e., if $\mathbf{x} = \sum_{j=1}^n x_j \mathbf{v}^j$, then $|\mathbf{x}|_{\mathcal{B}} = [\sum_{j=1}^n x_j^2]^{1/2}$.

Using the components in the above basis,

$$\frac{d}{dt}|\mathbf{x}(t)|_{\mathcal{B}} = \frac{\sum_{j=1}^n x_j(t)\dot{x}_j(t)}{|\mathbf{x}(t)|_{\mathcal{B}}} = \frac{\langle \mathbf{x}(t), A\mathbf{x}(t)\rangle_{\mathcal{B}}}{|\mathbf{x}(t)|_{\mathcal{B}}}.$$

Assume that C_1 and C_2 are two constants with $C_1 < Re(\lambda) < C_2$ for all eigenvalues λ of A. It can be shown that if ϵ is small enough in the above Jordan form, then $C_1|\mathbf{x}|_{\mathcal{B}}^2 < \langle \mathbf{x}, A\mathbf{x}\rangle_{\mathcal{B}} < C_2|\mathbf{x}|_{\mathcal{B}}^2$. Therefore,

$$C_1 < \frac{\frac{d}{dt}|\mathbf{x}(t)|_{\mathcal{B}}}{|\mathbf{x}(t)|_{\mathcal{B}}} < C_2,$$

$$C_1 < \frac{d}{dt}\log(|\mathbf{x}(t)|_{\mathcal{B}}) < C_2,$$

$$C_1 t < \log\left(\frac{|\mathbf{x}(t)|_{\mathcal{B}}}{|\mathbf{x}(0)|_{\mathcal{B}}}\right) < C_2 t,$$

and

$$e^{C_1 t}|\mathbf{x}_0|_{\mathcal{B}} < |\mathbf{x}(t)|_{\mathcal{B}} < e^{C_2 t}|\mathbf{x}_0|_{\mathcal{B}}.$$

If all the eigenvalues have negative real part, then taking $a = -C_2 > 0$ above, we get the result claimed. □

From the above theorem, it follows that if the real part of each eigenvalue is negative, then the origin is asymptotically stable for the linear flow. Also note that if for $t \geq 0$, we set $\mathbf{y} = e^{At}\mathbf{x}$ in the first condition of Theorem 5.1, we get that $|\mathbf{y}|_* \leq e^{-tb}|e^{-At}\mathbf{y}|_*$ for all $t \geq 0$, or $e^{b|t|}|\mathbf{y}|_* \leq |e^{At}\mathbf{y}|_*$ for all $t \leq 0$. Thus, going backward in time, the flow is an expansion. For this reason, if all the eigenvalues of A have negative real part, then we say that the differential equation $\dot{\mathbf{x}} = A\mathbf{x}$ has the origin as a *sink*, the origin is *attracting*, or the flow is a *contraction*.

If all the eigenvalues of a matrix A have positive real part, then an analogous result is true and the flow is an expansion for $t \geq 0$ and a contraction for $t \leq 0$. In this case we say that the origin is a *source*, the origin is *repelling*, or that the flow is an *expansion*.

4.6 Hyperbolic Linear Differential Equations

In the previous section, we considered the case when the linear flow $e^{At}\mathbf{x}$ is contracting in all directions or expanding in all directions. In this section we consider the case when it is contracting in some directions and expanding in others. We define the linear differential equation $\dot{\mathbf{x}} = A\mathbf{x}$ to be *hyperbolic* provided all the eigenvalues of A have nonzero real part. If A induces a linear hyperbolic differential equation, and there are at least two eigenvalues of A, λ_u and λ_s, with $Re(\lambda_u) > 0$ and $Re(\lambda_s) < 0$, then the origin is called a (hyperbolic) *saddle* for the differential equation. Thus, in the case of a saddle, some directions expand and some contract. Both the contracting or expanding cases are hyperbolic but are not saddles.

For any linear flow (either hyperbolic or not), we want to characterize the eigenspaces of generalized eigenvectors corresponding to the eigenvalues with positive, zero, and negative real parts. We first introduce some notation and then state the result. Let A be an $n \times n$ matrix. Define the *stable eigenspace*, *unstable eigenspace*, and *center eigenspace* to be

$$\begin{aligned}
\mathbb{E}^s &= \operatorname{span}\{\mathbf{v} : \mathbf{v} \text{ is a generalized eigenvector} \\
&\qquad \text{for an eigenvalue } \lambda \text{ with } Re(\lambda) < 0\}, \\
\mathbb{E}^u &= \operatorname{span}\{\mathbf{v} : \mathbf{v} \text{ is a generalized eigenvector} \\
&\qquad \text{for an eigenvalue } \lambda \text{ with } Re(\lambda) > 0\}, \text{ and} \\
\mathbb{E}^c &= \operatorname{span}\{\mathbf{v} : \mathbf{v} \text{ is a generalized eigenvector} \\
&\qquad \text{for an eigenvalue } \lambda \text{ with } Re(\lambda) = 0\},
\end{aligned}$$

respectively. If A is hyperbolic (so $\mathbb{E}^c = \emptyset$), then the decomposition of $\mathbb{R}^n$ into subspaces given by $\mathbb{R}^n = \mathbb{E}^u \oplus \mathbb{E}^s$ is called a *hyperbolic splitting*. Let

$$\begin{aligned}
V^s = \{\mathbf{v} :\ & \text{there exist } a > 0 \text{ and } C \geq 1 \text{ such that} \\
& |e^{At}\mathbf{v}| \leq Ce^{-at}|\mathbf{v}| \text{ for } t \geq 0\}, \\
V^u = \{\mathbf{v} :\ & \text{there exist } a > 0 \text{ and } C \geq 1 \text{ such that} \\
& |e^{At}\mathbf{v}| \leq Ce^{-a|t|}|\mathbf{v}| \text{ for } t \leq 0\}, \text{ and} \\
V^c = \{\mathbf{v} :\ & \text{for all } a > 0,\ |e^{At}\mathbf{v}|e^{-a|t|} \to 0 \text{ as } t \to \pm\infty\}.
\end{aligned}$$

Thus, the subspace V^s is defined to be all vectors which contract exponentially forward in time; V^c is defined to be the vectors which grow at most subexponentially both forward and backward in time. Finally, notice that V^u is defined as the set of vectors which contract backward in time and not in terms of behavior forward in time. This is done because any vector which is not in $\mathbb{E}^s \oplus \mathbb{E}^c$ expands forward in time, and this is not a subspace and does not characterize V^u. The following theorem shows that the conditions defining the V^σ characterize the subspaces $\mathbb{E}^\sigma$ for $\sigma = s, u, c$.

Theorem 6.1. *Consider the linear differential equation $\dot{\mathbf{x}} = A\mathbf{x}$ with $\mathbf{x}$ in $\mathbb{R}^n$. Let $\mathbb{E}^u$, $\mathbb{E}^c$, $\mathbb{E}^s$, V^u, V^c, and V^s be defined as above. Then, the following are true.*

(a) *The subspaces $\mathbb{E}^\sigma$ are invariant by the flow e^{At} for $\sigma = u, c, s$.*

(b) *The subspace $\mathbb{E}^\sigma = V^\sigma$ for $\sigma = u, c, s$, so $e^{At}|\mathbb{E}^u$ is an exponential expansion, $e^{At}|\mathbb{E}^s$ is an exponential contraction, and $e^{At}|\mathbb{E}^c$ grows subexponentially as $t \to \pm\infty$, and these uniquely characterize the subspaces.*

PROOF. By the form of the solutions given in Theorem 3.5, the subspaces are invariant for e^{At} as claimed.

By Theorem 5.1, $\mathbb{E}^u \subset V^u$ and $\mathbb{E}^s \subset V^s$. On the other hand, if $\mathbf{v} \in V^u \setminus \mathbb{E}^u$, then it has a nonzero component $\mathbf{v}'$ in either $\mathbb{E}^c$ or $\mathbb{E}^s$. By the form of the solutions, $e^{At}\mathbf{v}'$ and so $e^{At}\mathbf{v}$ do not go to zero exponentially as $t \to -\infty$, so $\mathbf{v} \notin V^u$. This contraction proves that $V^u \subset \mathbb{E}^u$ and so $V^u = \mathbb{E}^u$. Similarly, $V^s = \mathbb{E}^s$.

For $\mathbf{v} \in \mathbb{E}^c \setminus \{\mathbf{0}\}$, by the form of the solutions, $e^{At}\mathbf{v}$ has at most polynomial growth as $t \to \pm\infty$, so $\mathbf{v} \in V^u$. As above, if $\mathbf{v} \in V^c \setminus \mathbb{E}^c$, then there is a nonzero component in either $\mathbb{E}^u$ or $\mathbb{E}^s$, and there is exponential growth as t goes to either $+\infty$ or $-\infty$. This contradicts the fact that $\mathbf{v} \in V^c$. This completes the proof that $\mathbb{E}^c = V^c$, and so the theorem. □

Using Theorem 5.1, we can see that if A induces either a purely contracting linear flow or purely expanding linear flow, then nearby B will also induce hyperbolic flows of the same type. For hyperbolic linear flows, the subspaces in the hyperbolic splitting can move, but a small perturbation B of A must remain hyperbolic and must have eigenspaces of the same dimension as those for A. We state this in the following theorem.

Theorem 6.2. *Assume that A induces a hyperbolic flow for the linear differential equation $\dot{\mathbf{x}} = A\mathbf{x}$ for $\mathbf{x} \in \mathbb{R}^n$. If B is an $n \times n$ matrix with entries near enough to A, then $\dot{\mathbf{x}} = B\mathbf{x}$ induces a hyperbolic linear flow with the same dimension splitting $\mathbb{E}^u_B \oplus \mathbb{E}^s_B$ as that for A. Moreover, as A varies continuously to B the subspaces also vary continuously.*

PROOF. Because A is hyperbolic, $\mathbb{R}^n = \mathbb{E}^u \oplus \mathbb{E}^s$ for the eigenspaces of A. Let $p(\lambda)$ be the characteristic polynomial for A. Let γ be a curve in the left half of the complex plane that surrounds all eigenvalues with negative real part for A and is oriented counterclockwise. Let γ' be a curve in the right half of the complex plane that surrounds all eigenvalues with positive real part for A, again with counterclockwise orientation. By residues

$$\dim(\mathbb{E}^s) = \frac{1}{2\pi i}\int_\gamma \frac{p'(z)}{p(z)}\,dz$$

and

$$\dim(\mathbb{E}^u) = \frac{1}{2\pi i}\int_{\gamma'} \frac{p'(z)}{p(z)}\,dz.$$

For B near enough to A, the characteristic polynomial $q(\lambda)$ for B does not vanish on γ or γ'. The two integrals with $p(z)$ replaced by $q(z)$ will count the number of roots for q inside γ and γ', respectively. By continuity of these integrals with respect to changes in p, the number of roots for p and q are the same (since a continuous integer valued function is a constant). In particular, all the roots of $q(z)$ are either inside γ or γ' and none are on the imaginary axis. This proves that B is hyperbolic and the subspaces $\mathbb{E}^s_B$ and $\mathbb{E}^u_B$ have the same dimension as those for A.

Next,

$$P\mathbf{v} = \frac{1}{2\pi i}\int_\gamma (zI - A)^{-1}\mathbf{v}\,dz$$

is a projection from $\mathbb{R}^n$ onto $\mathbb{E}^s$. See Section 148 of Riesz and Nagy (1955). Similarly,

$$Q\mathbf{v} = \frac{1}{2\pi i}\int_\gamma (zI - B)^{-1}\mathbf{v}\,dz$$

is a projection onto the stable eigenspace $\mathbb{E}^s_B$ for B. Again, this integral varies continuously with changes from A to B, so the subspace varies continuously. Similar statements are true for the unstable subspaces. □

4.7 Topologically Conjugate Linear Differential Equations

Definition. We consider two flows to have the same qualitative properties, and so to be topologically similar, if we can match the trajectories of one with the trajectories of the other. There are two ways of doing this depending on whether we demand that the conjugacy match the time parameterization of the two flows or allow a reparameterization. The stronger condition requires that the flows be matched without a reparameterization. We say that two flows φ^t and ψ^t on a space M (M could be $\mathbb{R}^n$ or some manifold) are *topologically conjugate* provided there is a homeomorphism $h : M \to M$ such that $h \circ \varphi^t(\mathbf{x}) = \psi^t \circ h(\mathbf{x})$ for all $\mathbf{x} \in M$ and for all $t \in \mathbb{R}$. Allowing a reparameterization, we say that φ^t and ψ^t are *topologically equivalent* provided there is a homeomorphism $h : M \to M$ such that h takes trajectories of φ^t to trajectories of ψ^t while preserving their orientation. More precisely, φ^t and ψ^t are topologically equivalent if there is a homeomorphism $h : M \to M$ and a (reparameterization) function $\alpha : \mathbb{R} \times M \to \mathbb{R}$ such that $h \circ \varphi^{\alpha(t,\mathbf{x})}(\mathbf{x}) = \psi^t \circ h(\mathbf{x})$ for all $\mathbf{x} \in M$ and for all $t \in \mathbb{R}$, where we assume for each fixed $\mathbf{x}$ that $\alpha(t, \mathbf{x})$ is monotonically increasing in t and is onto all of $\mathbb{R}$. We could also assume the group property on α to insure that it is indeed a reparameterization: $\varphi^{\alpha(t+s,\mathbf{x})}(\mathbf{x}) = \varphi^{\alpha(t,\varphi^{\alpha(s,\mathbf{x})}(\mathbf{x}))} \circ \varphi^{\alpha(s,\mathbf{x})}(\mathbf{x})$.

We use these concepts repeatedly in our study of flows. In most circumstances, it is not possible to preserve the parameterization when we make perturbations. However, the following theorem proves that two linear flows with the same dimensional contracting spaces and the same dimensional expanding spaces are actually conjugate, i.e., it is possible to preserve the parameterization.

Theorem 7.1. *Let A and B be two $n \times n$ real matrices.*

(a) Assume that all the eigenvalues of A and B have negative real part (both are sinks). Then, the two linear flows e^{At} and e^{Bt} are topologically conjugate.

(b) Assume that all the eigenvalues of A and B have nonzero real part and the dimension of the direct sum of all the eigenspaces with negative real part is the same for A and B. (Thus, the dimension of the direct sum of all the eigenspaces with positive real part is the same for A and B.) Then, the two linear flows e^{At} and e^{Bt} are topologically conjugate.

(c) In particular, if all the eigenvalues of A have nonzero real part and B is near enough to A, then the two linear flows e^{At} and e^{Bt} are topologically conjugate.

PROOF. Part (b) follows fairly easily from part (a). The theorem that gives a conjugacy for two systems with all their eigenvalues with negative real part (on the same dimensional space) implies the same result for systems with all their eigenvalues with positive real part (on the same dimensional space). Then, given two hyperbolic linear flows as in part (b), there is a conjugacy between e^{At} and e^{Bt} on the eigenspaces for the eigenvalues with negative real part, and a conjugacy on the eigenspaces for the eigenvalues with positive real part, i.e., there are conjugacies $h_\sigma : \mathbb{E}^\sigma_A \to \mathbb{E}^\sigma_B$ between $e^{At}|\mathbb{E}^\sigma_A$ and $e^{Bt}|\mathbb{E}^\sigma_A$ for $\sigma = u, s$. There are projections $\pi_\sigma : \mathbb{R}^n \to \mathbb{E}^\sigma$ for $\sigma = u, s$, so any $\mathbf{x} \in \mathbb{R}^n$ can be written as $\mathbf{x} = \pi_u(\mathbf{x}) + \pi_s(\mathbf{x})$. The two conjugacies can be combined to give a conjugacy on the total space. Define $h(\mathbf{x}) = h_u(\pi_u(\mathbf{x})) + h_s(\pi_s(\mathbf{x}))$. Using the linearity of the flows, it is easily checked that h is a conjugacy.

Part (c) follows from part (b) because if B is near enough to A, then the dimensions of the splittings for A and B are the same by Theorem 6.2.

It remains to prove part (a). By Theorem 5.1, there exist norms $|\ |_A$ and $|\ |_B$ and constants $a, b > 0$ such that we have the estimates $|e^{At}\mathbf{x}|_A \leq e^{-at}|\mathbf{x}|_A$ and $|e^{Bt}\mathbf{x}|_B \leq$

$e^{-bt}|\mathbf{x}|_B$ for $t \geq 0$ and for any $\mathbf{x}$ in $\mathbb{R}^n$. Running time backward, we get the estimates $|e^{At}\mathbf{x}|_A \geq e^{a|t|}|\mathbf{x}|_A$ and $|e^{Bt}\mathbf{x}|_B \geq e^{b|t|}|\mathbf{x}|_B$ for $t \leq 0$.

We want to match up the trajectories of e^{At} with those for e^{Bt}. Using the above estimates, we see that for each $\mathbf{x} \neq \mathbf{0}$, the trajectory $e^{At}\mathbf{x}$ crosses the unit sphere for $|\ |_A$ exactly once, and each trajectory $e^{Bt}\mathbf{y}$ crosses the unit sphere for $|\ |_B$ exactly once. Let the unit spheres in these two norms be denoted as follows: $\mathcal{S}_A = \{\mathbf{x} : |\mathbf{x}|_A = 1\}$ and $\mathcal{S}_B = \{\mathbf{x} : |\mathbf{x}|_B = 1\}$. These spheres, $\mathcal{S}_A$ and $\mathcal{S}_B$, are called the *fundamental domains* for the two linear flows because of the property that each trajectory of $e^{At}\mathbf{x}$ for $\mathbf{x} \neq \mathbf{0}$ (respectively of $e^{Bt}\mathbf{y}$) crosses $\mathcal{S}_A$ (respectively $\mathcal{S}_B$) exactly once. Therefore, we first of all define a homeomorphism h_0 from $\mathcal{S}_A$ to $\mathcal{S}_B$ by $h_0(\mathbf{x}) = \mathbf{x}/|\mathbf{x}|_B$. (Any homeomorphism would do.) Notice that the inverse of h_0 exists and is given by $h_0^{-1}(\mathbf{y}) = \mathbf{y}/|\mathbf{y}|_A$.

To extend h_0 to all $\mathbb{R}^n$, we need to define the time when the trajectory that starts at $\mathbf{x}$ crosses the unit sphere. Using the above inequalities for the flow e^{At}, it follows that for any $\mathbf{x} \neq \mathbf{0}$, there is a $\tau(\mathbf{x})$ which depends continuously on $\mathbf{x}$ such that $|e^{A\tau(\mathbf{x})}\mathbf{x}|_A = 1$, i.e., $e^{A\tau(\mathbf{x})}\mathbf{x} \in \mathcal{S}_A$. Because of the definition, it follows that $\tau(e^{At}\mathbf{x}) = \tau(\mathbf{x}) - t$.

Now, using this homeomorphism h_0 on the unit sphere and the time $\tau(\mathbf{x})$, we can define a map (homeomorphism) $h : \mathbb{R}^n \to \mathbb{R}^n$ by

$$h(\mathbf{x}) = \begin{cases} e^{-B\tau(\mathbf{x})}h_0(e^{A\tau(\mathbf{x})}\mathbf{x}) & \text{for } \mathbf{x} \neq \mathbf{0}, \text{ and} \\ \mathbf{0} & \text{for } \mathbf{x} = \mathbf{0}. \end{cases}$$

The following calculation shows that h is a conjugacy:

$$\begin{aligned} h(e^{At}\mathbf{x}) &= e^{-B\tau(e^{At}\mathbf{x})}h_0(e^{A\tau(e^{At}\mathbf{x})}e^{At}\mathbf{x}) \\ &= e^{-B(\tau(\mathbf{x})-t)}h_0(e^{A(\tau(\mathbf{x})-t)}e^{At}\mathbf{x}) \\ &= e^{Bt}e^{-B\tau(\mathbf{x})}h_0(e^{A\tau(\mathbf{x})}\mathbf{x}) \\ &= e^{Bt}h(\mathbf{x}). \end{aligned}$$

Because τ and the flows e^{At} and e^{Bt} are continuous, it follows that h is continuous at points $\mathbf{x} \neq \mathbf{0}$. To check continuity at $\mathbf{0}$, notice that if $\mathbf{x}_j$ converges to $\mathbf{0}$, then $\tau_j = \tau(\mathbf{x}_j)$ goes to minus infinity. Letting $\mathbf{y}_j = h_0(e^{A\tau_j}\mathbf{x}_j)$, we have that $|\mathbf{y}_j|_B = 1$. Thus, $|h(\mathbf{x}_j)|_B = |e^{-B\tau_j}\mathbf{y}_j|_B \leq e^{-b|\tau_j|}$ must go to zero. Therefore, $h(\mathbf{x}_j)$ converges to $\mathbf{0} = h(\mathbf{0})$. This proves the continuity at $\mathbf{0}$.

To show that h is one to one, take $\mathbf{x}, \mathbf{y}$ with $h(\mathbf{x}) = h(\mathbf{y})$. If $\mathbf{x} = \mathbf{0}$, then $\mathbf{0} = h(\mathbf{x}) = h(\mathbf{y})$, so $\mathbf{y} = \mathbf{0} = \mathbf{x}$. Now, assume $\mathbf{x} \neq \mathbf{0}$. Then, $h(\mathbf{y}) = h(\mathbf{x}) \neq \mathbf{0}$ so $\mathbf{y} \neq \mathbf{0}$. Letting $\tau = \tau(\mathbf{x})$, $h(e^{A\tau}\mathbf{x}) = e^{B\tau}h(\mathbf{x}) = e^{B\tau}h(\mathbf{y}) = h(e^{A\tau}\mathbf{y})$. This shows that $h(e^{A\tau}\mathbf{y}) = h(e^{A\tau}\mathbf{x}) \in \mathcal{S}_B$, so $e^{A\tau}\mathbf{y} \in \mathcal{S}_A$ and $\tau(\mathbf{y}) = \tau(\mathbf{x})$. Since $h_0(e^{A\tau}\mathbf{x}) = h(e^{A\tau}\mathbf{x}) = h(e^{A\tau}\mathbf{y}) = h_0(e^{A\tau}\mathbf{y})$, and h_0 is one to one, we have $e^{A\tau}\mathbf{x} = e^{A\tau}\mathbf{y}$ and so $\mathbf{x} = \mathbf{y}$. Thus, h is one to one in all cases.

Reversing the roles of A and B in the arguments above, we get that h^{-1} exists (and so h is onto) and is continuous. This completes the proof. □

In contrast to the above results which proved that many different linear contractions are topologically conjugate, there is the following standard result about linear conjugacy which implies that very few different linear differential equations are linearly conjugate.

Theorem 7.2. *Let A and B be two $n \times n$ matrices, and assume that the two flows e^{tA} and e^{tB} are linearly conjugate, i.e., there exists an invertible M with $e^{tB} = Me^{tA}M^{-1}$. Then, A and B have the same eigenvalues.*

PROOF. Differentiating the equality $e^{tB} = Me^{tA}M^{-1}$ with respect to t at $t = 0$, we get that $B = MAM^{-1}$. Then, the characteristic polynomial for B equals that for A:

$$\begin{aligned} p(\lambda) &= \det(B - \lambda I) \\ &= \det(MAM^{-1} - \lambda MIM^{-1}) \\ &= \det(M)\det(A - \lambda I)\det(M^{-1}) \\ &= \det(A - \lambda I). \end{aligned}$$

The fact that the characteristic polynomials are equal implies that they have the same eigenvalues. □

REMARK 7.1. If $B = sA$ for $s > 0$, then certainly the two flows are linearly equivalent. This is the only new feature which the notion of linearly equivalent adds to that of linearly conjugate.

4.8 Nonhomogeneous Equations

In this section we consider nonhomogeneous linear equations. These results are used in the next chapter when studying the stability of fixed points of nonlinear equations and other matters.

The general form of the equations considered is given by

$$\dot{\mathbf{x}} = A(t)\mathbf{x} + \mathbf{g}(t). \tag{NH}$$

Given such an equation, we associate the corresponding homogeneous equation,

$$\dot{\mathbf{x}} = A(t)\mathbf{x}. \tag{H}$$

The following theorem gives the relationship between the solutions of (NH) and those of (H). We leave its proof to the reader.

Theorem 8.1. *(a) If $\mathbf{x}^1(t)$ and $\mathbf{x}^2(t)$ are two solutions of (NH), then $\mathbf{x}^1(t) - \mathbf{x}^2(t)$ is a solution of (H).*

(b) If $\mathbf{x}^n(t)$ is a solution of (NH) and $\mathbf{x}^h(t)$ is a solution of (H), then $\mathbf{x}^n(t) + \mathbf{x}^h(t)$ is a solution of (NH).

(c) If $\mathbf{x}^n(t)$ is a solution of (NH) and $M(t)$ is a fundamental matrix solution of (H), then any solution of (NH) can be written as $\mathbf{x}^n(t) + M(t)\mathbf{v}$.

In terms of the above theorem, we know how to find solutions of the homogeneous equation, at least in the constant coefficient case. We need to find one solution of the nonhomogeneous equation. One way is to look for solutions of the same type as the forcing term. For scalar second-order systems, this method is often called the method of undetermined coefficients. The other method is the method of variation of parameters. The following theorem gives this result for systems. Also, in the scalar equations, there is an arbitrary choice that has to be made. For systems, no such choice is needed: the process is straightforward.

Theorem 8.2 (Variation of Parameters). *Let $M(t)$ be a fundamental matrix solution of the homogeneous equation (H). Then,*

$$\mathbf{x}(t) = M(t)\left(\int_{t_0}^{t} M(s)^{-1}\mathbf{g}(s)\,ds + \mathbf{v}\right)$$

is a solution of the nonhomogeneous equation. If $\mathbf{v}$ *is allowed to vary, then this gives the general solution of the nonhomogeneous equation.*

PROOF. To derive this equation, we look for a solution of the form $\mathbf{x}(t) = M(t)\mathbf{f}(t)$. If this is a solution, then

$$\begin{aligned}\dot{\mathbf{x}}(t) &= A(t)M(t)\mathbf{f}(t) + M(t)\mathbf{f}'(t)\\ &= A(t)\mathbf{x}(t) + M(t)\mathbf{f}'(t).\end{aligned}$$

Since $\mathbf{x}(t)$ is a solution, this has to equal $A(t)\mathbf{x}(t)+\mathbf{g}(t)$, so we need $\mathbf{f}'(t) = M(t)^{-1}\mathbf{g}(t)$. Integrating from t_0 to t, we get

$$\mathbf{f}(t) = \int_{t_0}^{t} M(s)^{-1}\mathbf{g}(s)\,ds + \mathbf{v}$$

for an arbitrary vector $\mathbf{v}$. Substituting this for $\mathbf{f}(t)$, we get the form of $\mathbf{x}(t)$ claimed. The above calculation can be worked backward (or a direct calculation of the derivative of the right-hand side can be made) to show that this indeed gives a solution of the nonhomogeneous equation. The statements about the general solution follow from Theorem 8.1 above. □

4.9 Linear Maps

The results for linear maps are very similar to those for linear differential equations; the groupings of the eigenvalues become those of absolute value bigger than one, equal to one, or less than one rather than those of real part positive, zero, or negative.

Let A be an $n \times n$ real matrix. If $\mathbf{v}$ is an eigenvector for the eigenvalue λ, then $A^n\mathbf{v} = \lambda^n\mathbf{v}$. Thus, if $|\lambda| < 1$, then $|A^n\mathbf{v}| = |\lambda|^n|\mathbf{v}|$, which goes to zero as n goes to infinity. Using the Jordan Canonical Form, we get a result for linear maps that is similar to the one for linear differential equations.

Theorem 9.1. *Let A be an $n \times n$ real matrix and consider the map $A\mathbf{x}$. The following are equivalent.*

(a) There is a norm $|\ |_*$ on $\mathbb{R}^n$ and a constant $0 < \mu < 1$ such that for any initial condition $\mathbf{x} \in \mathbb{R}^n$, the iterates satisfy

$$|A^n\mathbf{x}|_* \le \mu^n|\mathbf{x}|_* \quad \textit{for all } n \ge 0.$$

(b) For any norm $|\ |'$ on $\mathbb{R}^n$ there exist constants $0 < \mu < 1$ and $C \ge 1$ such that for any initial condition $\mathbf{x} \in \mathbb{R}^n$, the iterates satisfy

$$|A^n\mathbf{x}|' \le C\mu^n|\mathbf{x}|' \quad \textit{for all } n \ge 0.$$

(c) All the eigenvalues λ of A satisfy $|\lambda| < 1$.

REMARK 9.1. As for a linear differential equation, a norm as in Theorem 9.1(a) is called an *adapted norm.*

The proof of this theorem is similar to that for linear differential equations with summations replacing integrals, and will be omitted. With this result, a linear map induced by a matrix all of whose eigenvalues have absolute values less than one is said to be a *linear contraction*, and the origin is called a *linear sink* or *attracting fixed point* for this map. If all the eigenvalues have absolute values greater than one, then A is automatically nondegenerate (nonzero determinant) and so $A^n\mathbf{x}$ is an expansion for $n > 0$ and a contraction for $n < 0$. The map induced by A is called a *linear expansion*, and the origin is called a *linear source* or a *repelling fixed point* for this map.

Theorem 9.2. *Assume that B and C are two $n \times n$ matrices which induce invertible linear contractions (or both induce linear expansions), $B\mathbf{x}$ and $C\mathbf{x}$. Further, assume that B and C belong to the same path components of $Gl(n, \mathbb{R})$, the set of invertible $n \times n$ matrices. Then, the linear map $B\mathbf{x}$ is topologically conjugate to the linear map $C\mathbf{x}$.*

REMARK 9.2. The General Linear Group, $Gl(n, \mathbb{R})$, has two path components, those with positive determinant and those with negative determinant. (Compare with Theorem 9.6 at the end of this section.) Thus, if A and B are two elements of $Gl(n, \mathbb{R})$, both of which have positive determinant (both are *orientation preserving*), then A and B are topologically conjugate. Similarly, if both have negative determinant (both are *orientation reversing*), then they are conjugate.

PROOF. The idea of the proof is very similar to that for flows, but the conjugacy on the "fundamental domains" is different.

Let B_t be a curve in $Gl(n, \mathbb{R})$ with $B_0 = C$ and $B_1 = B$.

Take norms $|\ |_B$ and $|\ |_C$ given by Theorem 9.1(a) such that B and C are contractions in terms of these respective norms. Let

$$D_B = \{\mathbf{x} \in \mathbb{R}^n : |\mathbf{x}|_B \leq 1\} \qquad \text{and}$$
$$S_B = \{\mathbf{x} \in \mathbb{R}^n : |\mathbf{x}|_B = 1\},$$

so D_B is the standard unit ball and S_B is the standard unit sphere in $\mathbb{R}^n$ in terms of the norm $|\ |_B$. Similarly, we define D_C and S_C using the norm $|\ |_C$. The following two "annuli"

$$A_B = \text{cl}[D_B \setminus B(D_B)] \quad \text{and}$$
$$A_C = \text{cl}[D_C \setminus C(D_C)]$$

are called *fundamental domains* because, for any $\mathbf{x} \neq \mathbf{0}$, there is a j such that $B^j(\mathbf{x}) \in A_B$, and for most $\mathbf{x} \neq \mathbf{0}$, there is a unique such j.

We need to construct a conjugacy h_0 between the two linear maps on their respective fundamental domains, $h_0 : A_B \to A_C$, such that if $\mathbf{x}, B\mathbf{x} \in A_B$, then $h_0(B\mathbf{x}) = Ch_0(\mathbf{x})$. After constructing h_0 on A_B, we extend it to all of $\mathbb{R}^n$ as we did for differential equations.

The conjugacy h_0 needs to be a homeomorphism between A_B and A_C taking the outer boundary to the outer boundary and the inner boundary to the inner boundary. On the outer boundary, we take h_0 to be essentially the identity (radial projection from S_B to S_C); and on the inner boundary, we take h_0 to be essentially the map CB^{-1} (plus a radial projection onto CS_C). If h_0 has these values on the two boundaries, it is not hard to see that it is a conjugacy on A_B. To construct h_0 with these properties, we separated the adjustment of the radius from the change in the "angle" variable in S^{n-1}. For the radial variable, the radial line segments from the inside boundary to the outside boundary have different lengths for A_B and A_C. In order to adjust this radial component of the points, we first define a map h_B from the standard annulus $[0,1] \times S^{n-1}$ to the fundamental domain A_B. Similarly, we define h_C from $[0,1] \times S^{n-1}$ to A_C. Having adjusted the radial component by means of the maps h_B and h_C, we define a map H from $[0,1] \times S^{n-1}$ to $[0,1] \times S^{n-1}$ using the path from C to B in $Gl(n, \mathbb{R})$. This map H preserves the t value in $[0,1]$ and is the map in S^{n-1} induced by B_tB^{-1}. It is the identity for $t = 1$ and CB^{-1} for $t = 0$; thus, H makes the adjustments of the "angular" component in S^{n-1} in a manner so that $h_0 = h_C \circ H \circ h_B^{-1}$ satisfies the necessary conjugacy equation for points on the outer boundary: if $\mathbf{x} \in S_B$, then $Ch_0(\mathbf{x}) = h_0(B\mathbf{x})$.

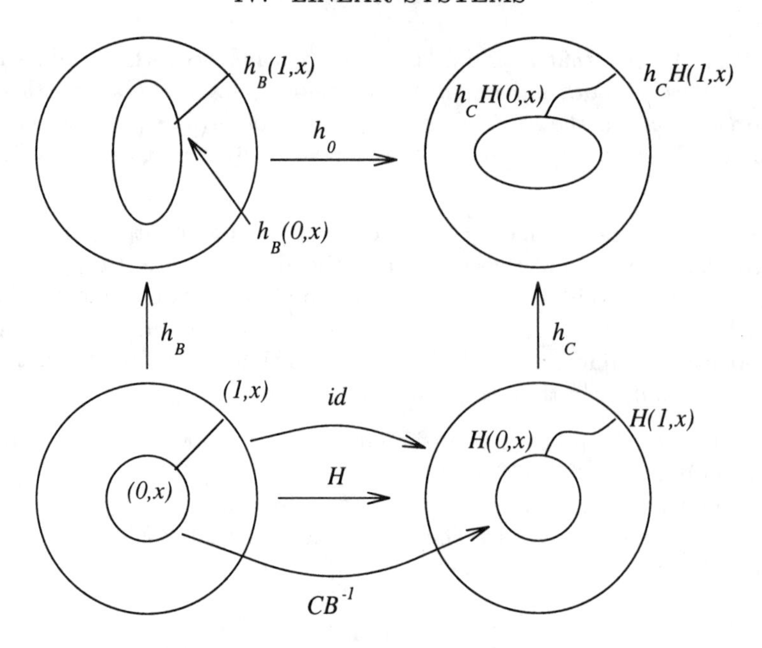

FIGURE 9.1. The Construction of h_0

Beginning the actual constructions, let

$$h_B : [0,1] \times S^{n-1} \to A_B$$

be given by

$$h_B(t, \mathbf{x}) = \tau_B(t, \mathbf{x})\mathbf{x}$$

where τ_B is the affine map in t such that (i) $\tau_B(1, \mathbf{x}) = |\mathbf{x}|_B^{-1}$ so $h_B(1, \mathbf{x}) = \mathbf{x}/|\mathbf{x}|_B \in S_B$ and (ii) $h_B(0, \mathbf{x}) = \tau_B(0, \mathbf{x})\mathbf{x} \in BS_B$. For $\tau_B(0, \mathbf{x})\mathbf{x}$ to be in BS_B, we need $\tau_B(0, \mathbf{x})B^{-1}\mathbf{x} \in S_B$, $\tau_B(0, \mathbf{x})|B^{-1}\mathbf{x}|_B = 1$, or $\tau_B(0, \mathbf{x}) = |B^{-1}\mathbf{x}|_B^{-1}$. Since we choose τ_B to be an affine map in t,

$$\tau_B(t, \mathbf{x}) = \frac{t}{|\mathbf{x}|_B} + \frac{1-t}{|B^{-1}\mathbf{x}|_B}.$$

For any $\mathbf{x} \in S^{n-1}$, $h_B(1, \mathbf{x}) = \mathbf{x}/|\mathbf{x}|_B \in S_B$ is on the outer boundary of A_B, and $h_B(0, \mathbf{x}) = \mathbf{x}/|B^{-1}\mathbf{x}|_B \in BS_B$ is on the inner boundary of A_B. Thus, h_B takes the radial line segment $[0,1] \times \{\mathbf{x}\}$ onto the radial line segment from $\mathbf{x}/|B^{-1}\mathbf{x}|_B$ to $\mathbf{x}$, i.e., the line segment from the inner boundary of A_B to the outer boundary of A_B. For use in verifying the conjugacy condition, note that for any $\mathbf{y} \in S_B$, letting $\mathbf{x} = B\mathbf{y}/|B\mathbf{y}|$,

$$h_B(0, \frac{B\mathbf{y}}{|B\mathbf{y}|}) = \frac{B\mathbf{y}}{|\mathbf{y}|_B} = B\mathbf{y} \qquad \text{or}$$

$$h_B^{-1}(B\mathbf{y}) = (0, \frac{B\mathbf{y}}{|B\mathbf{y}|}).$$

Similarly, we define τ_C and h_C with

$$h_C : [0,1] \times S^{n-1} \to A_C.$$

Converting the formulae for h_B into those for h_C, for any $\mathbf{x} \neq \mathbf{0}$, $\mathbf{x}/|\mathbf{x}| \in S^{n-1}$,

$$h_C(0, \frac{C\mathbf{x}}{|C\mathbf{x}|}) = \frac{C\mathbf{x}}{|\mathbf{x}|_C}, \qquad \text{and}$$

$$h_C(1, \frac{\mathbf{x}}{|\mathbf{x}|}) = \frac{\mathbf{x}}{|\mathbf{x}|_C}.$$

As stated above, next we use the curve B_t in $Gl(n, \mathbb{R})$ with $B_0 = C$ and $B_1 = B$ to define

$$H : [0,1] \times S^{n-1} \to [0,1] \times S^{n-1}$$

in a manner which preserves the "radius" $t \in [0,1]$ and continuously changes the "angular" component in S^{n-1}. In fact, define

$$H(t, \mathbf{x}) = \left(t, \frac{B_t B^{-1}\mathbf{x}}{|B_t B^{-1}\mathbf{x}|}\right).$$

On the outer boundary $\{1\} \times S^{n-1}$, $H(1, \mathbf{x}) = (1, \mathbf{x})$ is the identity. On the inner boundary $\{0\} \times S^{n-1}$,

$$H(0, \mathbf{x}) = \left(0, \frac{CB^{-1}\mathbf{x}}{|CB^{-1}\mathbf{x}|}\right), \qquad \text{so}$$
$$H\left(0, \frac{B\mathbf{x}}{|B\mathbf{x}|}\right) = \left(0, \frac{C\mathbf{x}}{|C\mathbf{x}|}\right),$$

which is essentially the map $\mathbf{x} \mapsto CB^{-1}\mathbf{x}$.

Finally, we combine these maps and define

$$h_0 = h_C \circ H \circ h_B^{-1} : A_B \to A_C.$$

First note that because each of the maps h_C, H, and h_B are one to one and onto, h_0 is also. Next we check that h_0 is a conjugacy whenever $\mathbf{x}, B\mathbf{x} \in A_B$, i.e., for $\mathbf{x} \in S_B$. For these $\mathbf{x} \in S_B$,

$$\begin{aligned}
Ch_0(\mathbf{x}) &= Ch_C \circ H \circ h_B^{-1}(\mathbf{x}) \\
&= Ch_C \circ H\left(1, \frac{\mathbf{x}}{|\mathbf{x}|}\right) \\
&= Ch_C\left(1, \frac{\mathbf{x}}{|\mathbf{x}|}\right) \\
&= \frac{C\mathbf{x}}{|\mathbf{x}|_C}.
\end{aligned}$$

On the other hand,

$$\begin{aligned}
h_0(B\mathbf{x}) &= h_C \circ H \circ h_B^{-1}(B\mathbf{x}) \\
&= h_C \circ H\left(0, \frac{B\mathbf{x}}{|B\mathbf{x}|}\right) \\
&= h_C\left(0, \frac{C\mathbf{x}}{|C\mathbf{x}|}\right) \\
&= \frac{C\mathbf{x}}{|\mathbf{x}|_C} \\
&= Ch_0(\mathbf{x}).
\end{aligned}$$

This checks the conjugacy of h_0 on A_B.

Having defined h_0 from A_B to A_C, we extend it to all of $\mathbb{R}^n$ by

$$h(\mathbf{x}) = \begin{cases} \mathbf{0} & \text{for } \mathbf{x} = \mathbf{0} \\ C^{-j(\mathbf{x})} h_0(B^{j(\mathbf{x})}\mathbf{x}) & \text{for } \mathbf{x} \neq \mathbf{0} \end{cases}$$

where $B^{j(\mathbf{x})}\mathbf{x} \in A_B$. The only difference from the case of the differential equations is that we need to check that h is well defined. However, if both $B^j\mathbf{x}, B^{j+1}\mathbf{x} \in A_B$, then

$$\begin{aligned} C^{-j-1}h_0(B^{j+1}\mathbf{x}) &= C^{-j-1}h_0(BB^j\mathbf{x}) \\ &= C^{-j-1}Ch_0(B^j\mathbf{x}) \\ &= C^{-j}h_0(B^j\mathbf{x}) \end{aligned}$$

from the conjugacy property for h_0 verified above. Thus, h is well defined. The reader should also check that it is continuous at points on $B(S_B)$. The continuity at $\mathbf{0}$ is similar to before: if $\mathbf{x}_i$ converges to zero, then $j(\mathbf{x}_i)$ goes to minus infinity, so $C^{-j(\mathbf{x}_i)}h_0(B^{j(\mathbf{x}_i)}\mathbf{x})$ goes to $\mathbf{0}$. The rest of the proof is similar to that for differential equations. □

Having discussed linear contracting maps and linear expanding maps, we now consider the possibility of both types of eigenvalues. If all the eigenvalues of A have absolute value which is not equal to one, $A\mathbf{x}$ is called a *hyperbolic linear map*. In general, we define the eigenspaces much as before:

$$\begin{aligned} \mathbb{E}^u &= \text{span}\{\mathbf{v} : \mathbf{v} \text{ is a generalized eigenvector} \\ &\qquad\qquad \text{for an eigenvalue } \lambda \text{ with } |\lambda| > 1\}, \\ \mathbb{E}^c &= \text{span}\{\mathbf{v} : \mathbf{v} \text{ is a generalized eigenvector} \\ &\qquad\qquad \text{for an eigenvalue } \lambda \text{ with } |\lambda| = 1\}, \text{ and} \\ \mathbb{E}^s &= \text{span}\{\mathbf{v} : \mathbf{v} \text{ is a generalized eigenvector} \\ &\qquad\qquad \text{for an eigenvalue } \lambda \text{ with } |\lambda| < 1\}. \end{aligned}$$

These subspaces are called the *unstable eigenspace*, *center eigenspace*, and *stable eigenspace*, respectively. Much as before, $\mathbb{E}^s$ is characterized as vectors which exponentially contract as $n \to \infty$, and $\mathbb{E}^u$ is characterized as vectors which exponentially contract as $n \to -\infty$. The center subspace, $\mathbb{E}^c$, is characterized as vectors which grow subexponentially as $n \to \pm\infty$, i.e.,

$$\mathbb{E}^c = \{\mathbf{v} : \text{ for all } 0 < \mu < 1,\ |A^n\mathbf{v}|\mu^{|n|} \to 0 \text{ as } n \to \pm\infty\}.$$

Next we prove the preservation of the hyperbolic nature of a linear map under perturbation, which is analogous to Theorem 6.2 for differential equations. This result is then used to prove that nearby hyperbolic linear maps are topologically conjugate, which is analogous to Theorem 7.1. Let $H(n,\mathbb{R})$ be the set of matrices A in $Gl(n,\mathbb{R})$ such that A induces a hyperbolic linear map, $A\mathbf{x}$.

Theorem 9.3. *Assume that $A \in H(n,\mathbb{R})$. If B is near enough to A, then $B \in H(n,\mathbb{R})$ and the splitting for B, $\mathbb{E}^u_B \oplus \mathbb{E}^s_B$, has subspaces with the same dimensions as those for A. Moreover, as A varies continuously to B, the subspaces also vary continuously.*

PROOF. The proof is the same as that for differential equations. Let $p(\lambda)$ be the characteristic polynomial for A. Let γ be a simple closed curve inside the open unit disk of the complex plane that surrounds all the eigenvalues of A with negative real part and is oriented counterclockwise. Let γ' be a simple closed curve outside the closed unit disk of the complex plane that surrounds all eigenvalues of A with positive real part but does not surround the unit disk, again with counterclockwise orientation. Then,

$$\frac{1}{2\pi i}\int_\gamma \frac{p'(z)}{p(z)}\,dz$$

counts the number of roots of $p(z)$ inside γ and varies continuously with the change from $p(z)$ to the characteristic polynomial $q(z)$ for B. A similar statement holds for the unstable eigenvalues. Thus, for B near enough to A, all the roots for $q(z)$ are either inside γ or γ' and so B is hyperbolic with the same dimensional subspaces as A.

As before, the projections,

$$P\mathbf{v} = \frac{1}{2\pi i}\int_\gamma (zI - C)^{-1}\mathbf{v}\,dz,$$

onto the stable subspace vary continuously with changes of C from A to B, so the subspace $\mathbb{E}^s_B$ varies continuously. See Section 148 of Riesz and Nagy (1955). A similar statement holds for $\mathbb{E}^u_B$. Therefore, the subspaces of B are near those for A. □

Theorem 9.4. *Assume that A is an $n \times n$ matrix which induces a linear hyperbolic map $A\mathbf{x}$. If B is near enough to A, then $B\mathbf{x}$ is topologically conjugate to $A\mathbf{x}$.*

This result follows from the cases of linear contractions and linear expansions (Theorem 9.2), and the fact that the dimension of the splitting does not change (Theorem 9.3) in the same way that it did for differential equations.

In contrast to the above results which proved that many different linear contractions are topologically conjugate, there is the following standard result about linear conjugacy which implies that very few different linear maps are linearly conjugate.

Theorem 9.5. *Let B and C be two invertible matrices in $Gl(n,\mathbb{R})$. Assume B and C are linearly conjugate, i.e., there exists an M in $Gl(n,\mathbb{R})$ with $C = MBM^{-1}$. Then, B and C have the same eigenvalues.*

REMARK 9.3. This result is a standard fact in linear algebra. Compare with the proof of Theorem 7.2. The details are left to the reader.

REMARK 9.4. By the uniqueness of the Jordan Canonical Form, if B and C are linearly conjugate, then they have the same Jordan canonical form if the blocks are ordered in the same way.

Now, we return to the question of the path components of $Gl(n,\mathbb{R})$ and $H(n,\mathbb{R})$. Let $Cont(n,\mathbb{R})$ be the set of matrices $A \in Gl(n,\mathbb{R})$ such that A induces a contracting linear map, $A\mathbf{x}$. Let $Exp(n,\mathbb{R})$ be the set of matrices $A \in Gl(n,\mathbb{R})$ such that A induces a expanding linear map, $A\mathbf{x}$. We have the following result.

Theorem 9.6. *(a) Let $A, B \in Cont(n,\mathbb{R})$. Assume that $\det(A)$ and $\det(B)$ have the same sign. Then, there is a curve $\{A_t \in Cont(n,\mathbb{R}) : 0 \le t \le 1\}$ such that $A_0 = A$ and $A_1 = B$.*

(b) Let $A, B \in H(n,\mathbb{R})$. Let $\mathbb{E}^s_A$ and $\mathbb{E}^u_A$ be the eigenspaces for A, and $\mathbb{E}^s_B$ and $\mathbb{E}^u_B$ be the eigenspaces for B. Assume that (i) the dimension of $\mathbb{E}^s_A$ equals the dimension of $\mathbb{E}^s_B$ (so the dimension of $\mathbb{E}^u_A$ equals the dimension of $\mathbb{E}^u_B$), (ii) the signs of $\det(A|\mathbb{E}^s)$ and $\det(B|\mathbb{E}^s)$ are the same, and (iii) the signs of $\det(A|\mathbb{E}^u)$ and $\det(B|\mathbb{E}^u)$ are the same. Then, there is a curve $\{A_t \in H(n,\mathbb{R}) : 0 \le t \le 1\}$ such that $A_0 = A$ and $A_1 = B$.

REMARK 9.5. For $\sigma = s, u$, $\det(A|\mathbb{E}^\sigma)$ is calculated in terms of a basis of $\mathbb{E}^\sigma$. The condition that $\det(A|\mathbb{E}^\sigma) = \pm 1$ is really the condition that A preserves or reverses the orientation on the invariant subspace $\mathbb{E}^\sigma$.

REMARK 9.6. Because the determinant is a continuous function, if $A, B \in Cont(n,\mathbb{R})$ have $\text{sign}(\det(A)) \neq \text{sign}(\det(B))$, then there cannot be a curve in $Gl(n,\mathbb{R})$, let alone in $Cont(n,\mathbb{R})$, connecting A and B. Thus, the result of this theorem is sharp.

PROOF. We break up the proof into lemmas and leave the proof of one main step to the exercises.

Lemma 9.7. *Let A be the matrix given as follows:*

$$A = \begin{pmatrix} \alpha & -\beta \\ \beta & \alpha \end{pmatrix}.$$

Assume $r = (\alpha^2 + \beta^2)^{1/2} \neq 0$. Let θ be chosen so that $\alpha = r\cos(\theta)$ and $\beta = r\sin(\theta)$ and let $\theta_t = (1-t)\theta$. Define the curve of matrices

$$A_t = \begin{pmatrix} r\cos(\theta_t) & -r\sin(\theta_t) \\ r\sin(\theta_t) & r\cos(\theta_t) \end{pmatrix},$$

for $0 \leq t \leq 1$. Then, (i) $A_0 = A$, (ii) A_1 is a diagonal matrix, and (iii) for $0 \leq t \leq 1$, the eigenvalues of A_t have the same absolute value as those of A, namely r. Thus, we have given a curve of matrices from a block in the Jordan Canonical Form corresponding to a complex eigenvalue to a diagonal block.

The validity of this lemma is obvious.

Lemma 9.8. *Assume A is a real diagonal matrix, $A = \text{diag}(\lambda_1, \ldots, \lambda_n)$, in $H(n, \mathbb{R})$. Let*

$$\mu_j = \begin{cases} 2 & \text{if } 1 < \lambda_j \\ 0.5 & \text{if } 0 < \lambda_j < 1 \\ -0.5 & \text{if } -1 < \lambda_j < 0 \\ -2 & \text{if } \lambda_j < -1. \end{cases}$$

Define $\lambda_{j,t} = (1-t)\lambda_j + t\mu_j$, and $A_t = \text{diag}(\lambda_{1,t}, \ldots, \lambda_{n,t})$ for $0 \leq t \leq 1$. Then, (i) $A_t \in H(n, \mathbb{R})$ for $0 \leq t \leq 1$, (ii) $A_0 = A$, and (iii) A_1 is a diagonal matrix whose entries are ± 2 and ± 0.5.

The verification of the lemma is direct and is left to the reader. See Exercise 4.12(b). Exercises 4.11 and 4.12 ask the reader to prove the following result using Lemma 9.7.

Lemma 9.9. *(a) Assume $A \in Cont(n, \mathbb{R})$. Then, there is a curve*

$$\{A_t \in Cont(n, \mathbb{R}) : 0 \leq t \leq 1\}$$

such that (i) $A_0 = A$ and (ii)

$$A_1 = \begin{cases} \text{diag}(0.5, \ldots, 0.5) & \text{if } \det(A) > 0 \\ \text{diag}(0.5, \ldots, 0.5, -0.5) & \text{if } \det(A) < 0. \end{cases}$$

(b) Assume $A \in H(n, \mathbb{R})$. Then, there is a curve $\{A_t \in H(n, \mathbb{R}) : 0 \leq t \leq 1\}$ such that (i) $A_0 = A$ and (ii)

$$A_1|\mathbb{E}^s = \begin{cases} \text{diag}(0.5, \ldots, 0.5) & \text{if } \det(A|\mathbb{E}^s) > 0 \\ \text{diag}(0.5, \ldots, 0.5, -0.5) & \text{if } \det(A|\mathbb{E}^s) < 0, \end{cases}$$

and

$$A_1|\mathbb{E}^u = \begin{cases} \text{diag}(2, \ldots, 2) & \text{if } \det(A|\mathbb{E}^u) > 0 \\ \text{diag}(2, \ldots, 2, -2) & \text{if } \det(A|\mathbb{E}^u) < 0. \end{cases}$$

PROOF OF THEOREM 9.6. The proofs of parts (a) and (b) are essentially the same so we consider only part (a). By Lemma 9.9, since $A, B \in Cont(n, \mathbb{R})$ and $\det(A)$ and $\det(B)$ have the same sign, there are curves A_t and B_t in $Cont(n, \mathbb{R})$ with $A_0 = A$, $B_0 = B$, and $A_1 = B_1$. We can combine these two curves by defining

$$C_t = \begin{cases} A_{2t} & \text{for } 0 \leq t \leq 0.5 \\ B_{2-2t} & \text{for } 0.5 \leq t \leq 1. \end{cases}$$

Then, $C_0 = A$, $C_1 = B$, and $C_t \in Cont(n, \mathbb{R})$ for $0 \leq t \leq 1$. This completes the proof. □

4.9.1 Perron-Frobenius Theorem

In this subsection we return to considering irreducible matrices. These were defined for transition matrices, but the definition makes sense if all the entries are nonnegative, as given below. The proof of the theorem of this section is not essential elsewhere in this book, although we do refer to the theorem in a few situations.

Definition. An $n \times n$ matrix $A = (a_{ij})$ is called *nonnegative* provided all the entries are nonnegative, $a_{ij} \geq 0$. An $n \times n$ matrix $A = (a_{ij})$ is called *positive* provided all the entries are positive, $a_{ij} > 0$. It is called *eventually positive* provided it is nonnegative and there is an integer $m > 0$ for which A^m is positive. An $n \times n$ matrix $A = (a_{ij})$ is called *reducible* provided that there is a pair i, j with $(A^m)_{ij} = 0$ for all $m \geq 1$. It is called *irreducible* provided that it is not reducible, i.e., for $1 \leq i, j \leq n$ there exists $m = m(i,j) > 0$ such that $(A^m)_{ij} \neq 0$. If A is eventually positive and irreducible, then $(A^j)_{ij} > 0$ for all $j \geq m$.

With these definitions, there is the following result of Perron, see Perron (1907) and Gantmacher (1959).

Theorem 9.10 (Perron-Frobenius). *Assume A is an eventually positive matrix.*

(a) Then, there is a real positive eigenvalue λ_1 which is a simple root of the characteristic equation such that if λ_j is any other eigenvalue (with $j > 1$), then $\lambda_1 > |\lambda_j|$. The eigenvector $\mathbf{v}^1$ for the eigenvalue λ_1 can be chosen with all entries strictly positive, $v_i^1 > 0$ for $1 \leq i \leq n$. In fact, all other eigenvectors, whether for real or complex eigenvalues, have components of both signs. This is true for both eigenvectors on the right and on the left.

(b) If A^m has all positive entries and $\mathbf{x}$ is any unit vector with $x_i \geq 0$ for all i, then $A^j\mathbf{x}/|A^j\mathbf{x}|$ converges to $\mathbf{v}^1/|\mathbf{v}^1|$ as j goes to infinity, and there are positive constants C_1 and C_2 such that $C_1\lambda_1^j \leq |A^j\mathbf{x}| \leq C_2\lambda_1^j$ and $C_1\lambda_1^j \leq (A^j\mathbf{x})_i \leq C_2\lambda_1^j$ for $j \geq m$ and $1 \leq i \leq n$. (Here, $(A^j\mathbf{x})_i$ is the ith component of $A^j\mathbf{x}$.)

REMARK 9.7. Frobenius proved a generalization of this result for irreducible nonnegative matrices (Frobenius, 1912). A good general reference for these results and a proof of the more general result is Gantmacher (1959).

REMARK 9.8. One application of this theorem is to a system with a finite number of states where probabilities of making the transition from one state to another are known. Assume there are n states and p_{ij} is the probability of making the transition from state j to state i for $1 \leq i, j \leq n$. Let $P = (p_{ij})_{1 \leq i,j \leq n}$ be the transition matrix. Since the probability of making the transition from state j to some other state is 1, it is assumed that $\sum_i p_{ij} = 1$. Also, assume $p_{ij} > 0$ for all pairs (i, j), so there is a positive probability of transition from any state to any other state. Note that P preserves the set of all distributions $\mathbf{x}$ for which $\sum_j x_j = 1$, i.e., $\mathbf{x}$ represents the distribution within the finite states $\{1, \ldots, n\}$. Also note that $(1, \ldots, 1)$ is a left eigenvector for the eigenvalue 1, $(1, \ldots, 1)P = (\sum_i p_{i1}, \ldots, \sum_i p_{in}) = (1, \ldots, 1)$. Since all the entries of $(1, \ldots, 1)$ are positive, $\lambda_1 = 1$ is the largest eigenvalue in absolute value by the conclusion of the Perron Theorem. The eigenvalue $\lambda_1 = 1$ also has right eigenvector $\mathbf{s}^*$ with all the $s_j^* > 0$. Also, $\mathbf{s}^*$ can be normalized so that $\sum_j s_j^* = 1$. This vector $\mathbf{s}^*$ represents the final steady state distribution within the states, because if $\mathbf{x}$ is any initial distribution with all the $x_j \geq 0$ and $\sum_j x_j = 1$, then $P^k\mathbf{x}$ converges to $\mathbf{s}^*$ as k goes to infinity because of the inequalities between the eigenvalues and the fact that P preserves vectors with $\sum_j x_j = 1$.

PROOF. By replacing A with A^m, we can assume that A is positive. (Alternatively, we could take only powers A^j for $j \geq m$.)

Let $\{\mathbf{e}^j : 1 \leq j \leq n\}$ be the standard basis of $\mathbb{R}^n$. Let

$$Q = \{\mathbf{x} = \sum_j x_j \mathbf{e}^j : x_j \geq 0 \text{ for all } j\}$$

be the first "quadrant", $S^{n-1} = \{\mathbf{x} : |\mathbf{x}| = 1\}$ be the sphere of all unit vectors, and

$$\Delta = Q \cap S^{n-1}$$

be the simplex of unit vectors in the first quadrant.

The matrix A induces a map, f_A, on S^{n-1} by

$$f_A(\mathbf{x}) = \frac{A\mathbf{x}}{|A\mathbf{x}|}.$$

Because A is positive, $A\mathbf{e}^j = \sum_i a_{ij}\mathbf{e}^i \in \text{int}(Q)$. Applying the map f_A to these unit vectors, $f_A(\Delta) \subset \text{int}(\Delta, S^n)$, where the interior is taken relative to S^n. The simplex Δ is homeomorphic to D^{n-1}, the closed unit disk in $\mathbb{R}^{n-1}$, i.e., Δ is the image of a homeomorphism from D^{n-1} into $\mathbb{R}^n$. By the Brouwer Fixed Point Theorem, f_A must have a fixed point, $\mathbf{v}^1$, in Δ. Then, $A\mathbf{v}^1 = \lambda_1 \mathbf{v}^1$ for some positive real number λ_1. Thus, λ_1 is an eigenvalue with unit eigenvector $\mathbf{v}^1$. (Here, $\mathbf{v}^1$ is a column vector because it is multiplied on the right of A.) The components of $\mathbf{v}^1$ are all positive because $\mathbf{v}^1 \in \text{int}(Q)$.

The above argument can be repeated with the action of A on row vectors where the vector is multiplied on the left. We obtain a left eigenvector $\mathbf{w}^1$ with all positive entries for some real eigenvalue λ^* where $\mathbf{w}^1$ is a row vector, $\mathbf{w}^1 A = \lambda^* \mathbf{w}^1$. Alternatively, apply the above result to A^{tr}. (We do not use the fact that $\lambda^* = \lambda_1$, but this is true. Because both $\mathbf{w}^1$ and $\mathbf{v}^1$ have positive entries, $\mathbf{w}^1\mathbf{v}^1 > 0$ and $\lambda^*\mathbf{w}^1\mathbf{v}^1 = \mathbf{w}^1 A\mathbf{v}^1 = \lambda_1 \mathbf{w}^1\mathbf{v}^1$, so $\lambda^* = \lambda_1$.)

Now, define the $(n-1)$-dimensional subspace W by

$$W = \{\mathbf{x} \in \mathbb{R}^n : \mathbf{w}^1\mathbf{x} = 0\}.$$

Thus, $\mathbf{w}^1$ is the normal (co)vector to this subspace. Any nonzero vector $\mathbf{x} \in W$ has some component positive and other components negative, because all the components of $\mathbf{w}^1$ are positive and $\mathbf{w}^1\mathbf{x} = 0$. It follows that W is invariant by A: if $\mathbf{x} \in W$, then $\mathbf{w}^1(A\mathbf{x}) = (\mathbf{w}^1 A)\mathbf{x} = \lambda^*\mathbf{w}^1\mathbf{x} = 0$ so $A\mathbf{x} \in W$. Thus, A restricted to W has $n-1$ eigenvalues (with multiplicity), $\lambda_2, \ldots, \lambda_n$. We need to prove that $\lambda_1 > |\lambda_j|$ for $j > 1$.

To prove this claim, first take the case where λ_j is real with real eigenvector $\mathbf{v}^j \in W$ of length 1. Since $\mathbf{v}^j \in W$, it must have both positive and negative components as claimed in the theorem. Let V be the two-dimensional subspace spanned by $\mathbf{v}^1$ and $\mathbf{v}^j$. We change to the inner product on V which makes these two vectors into an orthonormal basis. Let S^1 be the unit sphere in V in terms of this new inner product. Any $\mathbf{x} \in S^1$ can be represented by

$$\mathbf{x} = \cos(\varphi)\mathbf{v}^j + \sin(\varphi)\mathbf{v}^1$$

for some φ. Then,

$$A[\cos(\varphi)\mathbf{v}^j + \sin(\varphi)\mathbf{v}^1] = \lambda_j \cos(\varphi)\mathbf{v}^j + \lambda_1 \sin(\varphi)\mathbf{v}^1, \qquad \text{so}$$

$$f_A(\cos(\varphi)\mathbf{v}^j + \sin(\varphi)\mathbf{v}^1) = \frac{\cos(\varphi)\mathbf{v}^j + (\lambda_1/\lambda_j)\sin(\varphi)\mathbf{v}^1}{|\cos(\varphi)\mathbf{v}^j + (\lambda_1/\lambda_j)\sin(\varphi)\mathbf{v}^1|}.$$

There is a φ_0 with $0 < \varphi_0 < \pi/2$, $0 < \cos(\varphi_0) < 1$, such that $\mathbf{x}_0 = \cos(\varphi_0)\mathbf{v}^j + \sin(\varphi_0)\mathbf{v}^1$ is on the boundary of $\Delta = Q \cap S^1$ relative to S^1, $\partial(\Delta, S^1)$. Because A is a positive matrix, $f_A^k \mathbf{x}_0 \in \text{int}(\Delta, S^1)$ for $k \geq 1$. If $|\lambda_j| > \lambda_1$, then by the form of f_A above, $f_A^k(\mathbf{x}_0)$ would converge to $\mathbf{v}^j$ as k goes to infinity, which contradicts the fact that $f_A^k \mathbf{x}_0 \in \text{int}(\Delta, S^1)$. On the other hand, if $|\lambda_j| = \lambda_1$, then $\lambda_j^2 = \lambda_1$ and $f_A^2(\mathbf{x}_0)$ would equal $\mathbf{x}_0$, which again contradicts the fact that $f_A^k \mathbf{x}_0 \in \text{int}(\Delta, S^1)$. Therefore, when λ_j is real, the only possibility left is that $\lambda_1 > |\lambda_j|$.

Next we consider the case of a complex eigenvalue, $\lambda_j = \gamma[\cos(\psi) + i\sin(\psi)]$. Then, there is a a complex eigenvector $\mathbf{v}^j + i\mathbf{w}^j$ with $\mathbf{v}^j, \mathbf{w}^j \in W$ such that $A\mathbf{v}^j = \gamma\cos(\psi)\mathbf{v}^j + \gamma\sin(\psi)\mathbf{w}^j$ and $A\mathbf{w}^j = -\gamma\sin(\psi)\mathbf{v}^j + \gamma\cos(\psi)\mathbf{w}^j$. Since $\mathbf{v}^j, \mathbf{w}^j \in W$, both of these vectors must have both positive and negative components as claimed in the theorem. Let V be the three-dimensional subspace spanned by $\mathbf{v}^1$, $\mathbf{v}^j$, and $\mathbf{w}^j$. We change to the inner product on V which makes these three vectors into an orthonormal basis. Let S^2 be the unit sphere in V in terms of this new inner product. Define

$$\mathbf{x}(\theta) = \cos(\theta)\mathbf{v}^j + \sin(\theta)\mathbf{w}^j,$$

so $\mathbf{x}(\theta) \in S^2$. A direct calculation shows that $A\mathbf{x}(\theta) = \gamma\mathbf{x}(\theta+\psi)$, so $f_A(\mathbf{x}(\theta)) = \mathbf{x}(\theta+\psi)$ and these points move on the unit circle in the plane spanned by $\mathbf{v}^j$ and $\mathbf{w}^j$. Any point in $S^2 \subset V$ can be represented as $\cos(\varphi)\mathbf{x}(\theta) + \sin(\varphi)\mathbf{v}^1$, and

$$A[\cos(\varphi)\mathbf{x}(\theta) + \sin(\varphi)\mathbf{v}^1] = \gamma\cos(\varphi)\mathbf{x}(\theta+\psi) + \lambda_1\sin(\varphi)\mathbf{v}^1, \qquad \text{so}$$

$$f_A(\cos(\varphi)\mathbf{x}(\theta) + \sin(\varphi)\mathbf{v}^1) = \frac{\cos(\varphi)\mathbf{x}(\theta+\psi) + (\lambda_1/\gamma)\sin(\varphi)\mathbf{v}^1}{|\cos(\varphi)\mathbf{x}(\theta+\psi) + (\lambda_1/\gamma)\sin(\varphi)\mathbf{v}^1|}.$$

Assume $|\lambda_j| = \gamma > \lambda_1$. By the form of f_A given above, if $\cos(\varphi) \neq 0$, then $f_A^k(\cos(\varphi)\mathbf{x}(\theta) + \sin(\varphi)\mathbf{v}^1)$ converges to the unit circle in the plane spanned by $\mathbf{v}^j$ and $\mathbf{w}^j$ as k goes to infinity. In particular such a point which starts in the boundary of $\Delta = Q \cap S^2$ relative to S^2, $\partial(\Delta, S^2)$, does not remain in the $\text{int}(\Delta, S^2)$ as k goes to infinity. This contradicts the fact that A is a positive matrix.

If $|\lambda_j| = \gamma = \lambda_1$,

$$f_A(\cos(\varphi)\mathbf{x}(\theta) + \sin(\varphi)\mathbf{v}^1) = \cos(\varphi)\mathbf{x}(\theta+\psi) + \sin(\varphi)\mathbf{v}^1.$$

Therefore, all points on S^2 are recurrent for f_A. (If ψ is a rational multiple of 2π, then all points in S^2 are periodic or f_A; if ψ is an irrational multiple of 2π, then all points in S^2 are recurrent but only $\pm\,\mathbf{v}^1$ are periodic.) In any case, $f_A^k(\partial(\Delta, S^2))$ cannot be contained in $\text{int}(\Delta, S^2)$ for all positive k. Again, this contradicts the fact that A is a positive matrix.

Combining all the cases, we have proved part (a) of the theorem: $\lambda_1 > |\lambda_j|$ for all $j > 1$.

To prove part (b) of the theorem, assume that $\mathbf{x}$ is a unit vector with $x_j \geq 0$ for all j. Let $a = (\mathbf{x}\cdot\mathbf{v}^1)/|\mathbf{v}^1|^2$. Note that $a > 0$ because all the components of $\mathbf{v}^1$ are positive and all those of $\mathbf{x}$ are nonnegative. Then, $\mathbf{x} - a\mathbf{v}^1$ is in the subspace W defined above. Therefore,

$$\mathbf{x} = a\mathbf{v}^1 + \mathbf{v}^*$$

with $\mathbf{v}^* \in W$,

$$\begin{aligned} A^k\mathbf{x} &= A^k a\mathbf{v}^1 + A^k\mathbf{v}^* \\ &= a\lambda_1^k\mathbf{v}^1 + A^k\mathbf{v}^*, \qquad \text{and} \\ \frac{A^k\mathbf{x}}{\lambda_1^k} &= a\mathbf{v}^1 + \frac{A^k\mathbf{v}^*}{\lambda_1^k} \end{aligned}$$

where

$$\lim_{k\to\infty} \frac{|A^k \mathbf{v}^*|}{\lambda_1^k} = 0.$$

From the above convergence, it follows that there are positive constants $0 < C_1 < C_2$ such that

$$C_1 \leq \frac{|A^k \mathbf{x}|}{\lambda_1^k} \leq C_2 \qquad \text{and}$$

$$C_1 \leq \frac{(A^k \mathbf{x})_i}{\lambda_1^k} \leq C_2$$

for $1 \leq i \leq n$. Moreover, $f_A(\mathbf{x})$ converges to $\mathbf{v}^1/|\mathbf{v}^1|$. This completes the proof of the theorem. □

REMARK 9.9. The above proof shows that $\lambda_1 > |\lambda_j|$ for all $j > 1$, so that for any $\mathbf{x} \in \Delta$, $f_A^k(\mathbf{x})$ converges to $\mathbf{v}^1/|\mathbf{v}^1|$ as k goes to infinity. In fact, using the comparison of the eigenvalues it follows that f_A is a contraction on Δ. We did not prove that f_A is a contraction on Δ directly, but merely used the Brouwer Fixed Point Theorem to get the eigenvector for λ_1. After getting the eigenvector for λ_1, we argued that since f_A maps Δ into its interior in S^n, the other eigenvalues cannot be larger than λ_1 in absolute value. Then, as remarked above, the inequalities between the eigenvalues implies that $f_A|\Delta$ is a contraction by general arguments.

REMARK 9.10. The matrix A can have complex eigenvalues and off diagonal terms in its Jordan canonical form, i.e., 1's in its Jordan form. In fact, let $\mathbf{v}^1 = (1, \ldots, 1)^{tr}$,

$$W = \{\mathbf{x} : \mathbf{x} \cdot \mathbf{v}^1 = 0\},$$

and $\mathbf{v}^j$ for $1 < j \leq n$ be any basis for W. Let C be any matrix in terms of the basis $\{\mathbf{v}^2, \ldots, \mathbf{v}^n\}$. Clearly, C can have any Jordan canonical form. Then, consider the linear map L, which preserves W, has the matrix C in terms of the basis $\{\mathbf{v}^2, \ldots, \mathbf{v}^n\}$ on W, and $L(\mathbf{v}^1) = \lambda_1 \mathbf{v}^1$. Then, any of the standard unit vectors can be represented in terms of this basis,

$$\mathbf{e}^j = \sum_i y_{j,i} \mathbf{v}^i.$$

Because $\mathbf{v}^1$ has all positive coefficients, and for $j \geq 2$ the sum of the coefficients of $\mathbf{v}^j$ in terms of the standard basis is zero, it follows that $y_{j,1} > 0$ for all j. Then,

$$\begin{aligned} L(\mathbf{e}^j) &= L(\sum_i y_{j,i} \mathbf{v}^i) \\ &= \lambda_1 y_{j,1} \mathbf{v}^1 + \sum_{i=2}^n y_{j,i} L(\mathbf{v}^i). \end{aligned}$$

Also let $A = (a_{i,j})$ be the matrix of L in terms of the standard basis, so

$$L(\mathbf{e}^j) = \sum_i a_{ij} \mathbf{e}^i.$$

Comparing coefficients, we get that

$$a_{ij} = \lambda_1 y_{j,1} + \sum_{k=2}^n y_{j,k} L(\mathbf{v}^k) \cdot \mathbf{e}^i.$$

Thus, for λ_1 large enough, $a_{ij} > 0$ for all i and j, A is positive, and $L(\mathbf{e}^j)$ is in the interior of Q for all j. This proves that A can have any type of Jordan canonical form for the eigenvalues which correspond to the eigenvectors on W.

4.10 Exercises

Jordan Canonical Form

4.1. Let A be an $n \times n$ matrix, and $\mathbf{v}^1, \dots, \mathbf{v}^n$ be a basis such that $A\mathbf{v}^1 = \lambda\mathbf{v}^1$ and $A\mathbf{v}^j = \lambda\mathbf{v}^j + \mathbf{v}^{j-1}$ for $j = 2, \dots, n$. Given $\epsilon > 0$, find a new basis $\mathbf{w}^j$ such that $A\mathbf{w}^1 = \lambda\mathbf{w}^1$ and $A\mathbf{w}^j = \lambda\mathbf{w}^j + \epsilon\mathbf{w}^{j-1}$ for $j = 2, \dots, n$. Hint: Try $\mathbf{w}^j = s_j\mathbf{v}^j$ for suitable choices of s_j.

Solutions and Phase Portraits for Constant Coefficients

4.2. Find a basis of solutions and draw the phase portrait for $\dot{\mathbf{x}} = A\mathbf{x}$ for each of the following choices of A.

$$\text{(a)} \begin{pmatrix} 3 & 1 \\ 0 & 2 \end{pmatrix}, \quad \text{(b)} \begin{pmatrix} 2 & 0 \\ 1 & 2 \end{pmatrix}, \quad \text{(c)} \begin{pmatrix} 1 & 12 \\ 3 & 1 \end{pmatrix}, \quad \text{(d)} \begin{pmatrix} 5 & 3 \\ -2 & 1 \end{pmatrix},$$

$$\text{(e)} \begin{pmatrix} -2 & 0 & 0 \\ 0 & -1 & 0 \\ 0 & 0 & 3 \end{pmatrix}, \text{(f)} \begin{pmatrix} -2 & 0 & 0 \\ 0 & 1 & -1 \\ 0 & 1 & 1 \end{pmatrix}.$$

4.3. Consider the second-order linear equation given by $\ddot{\mathbf{x}} = A\mathbf{x}$.

(a) Prove that $\mathbf{x}(t) = e^{\lambda t}\mathbf{v}$ is a solution of $\ddot{\mathbf{x}} = A\mathbf{x}$ if and only if $\mu = \lambda^2$ is an eigenvalue of A with eigenvector $\mathbf{v}$.

(b) Find a basis of (four) solutions of $\ddot{\mathbf{x}} = A\mathbf{x}$ for

$$A = \begin{pmatrix} -(k_1 + k_2) & k_2 \\ k_2 & -(k_1 + k_2) \end{pmatrix}.$$

Hyperbolic Linear Differential Equations

4.4. Let $H_{\text{diff}}(n, \mathbb{R})$ be the set of matrices A in $Gl(n, \mathbb{R})$ that induce a hyperbolic linear differential equation, $\dot{x} = A\mathbf{x}$. Assume that $A \in H_{\text{diff}}(n, \mathbb{R})$.

(a) Prove there is a curve $\{A_t \in H_{\text{diff}}(n, \mathbb{R}) : 0 \le t \le 1\}$ such that (i) $A_0 = A$ and (ii) A_1 has no nonzero off diagonal terms in its Jordan canonical form.

(b) Prove there is a curve $\{A_t \in H_{\text{diff}}(n, \mathbb{R}) : 0 \le t \le 1\}$ such that (i) $A_0 = A$ and (ii)

$$A_1|\mathbb{E}^s = \text{diag}(-1, \dots, -1) \qquad \text{and}$$
$$A_1|\mathbb{E}^u = \text{diag}(1, \dots, 1).$$

Conjugacy and Structural Stability

4.5. Consider linear systems of constant coefficients in $\mathbb{R}^2$. Construct explicit conjugacies $h : \mathbb{R}^2 \to \mathbb{R}^2$ between $\dot{\mathbf{x}} = -\mathbf{x}$ and the following other systems:

(a) $\dot{\mathbf{y}} = \begin{pmatrix} -1 & 0 \\ 0 & -2 \end{pmatrix} \mathbf{y}$,

(b) $\dot{\mathbf{y}} = \begin{pmatrix} -2 & 1 \\ 0 & -2 \end{pmatrix} \mathbf{y}$, and

(c) $\dot{\mathbf{y}} = \begin{pmatrix} -2 & 1 \\ 1 & -2 \end{pmatrix} \mathbf{y}$.

4.6. Consider linear systems of constant coefficients in $\mathbb{R}^2$. Construct explicit conjugacies $h : \mathbb{R}^2 \to \mathbb{R}^2$ between

$$\dot{\mathbf{x}} = \begin{pmatrix} -1 & 0 \\ 0 & 1 \end{pmatrix} \mathbf{x} \quad \text{and} \quad \dot{\mathbf{y}} = \begin{pmatrix} 5 & -6 \\ 4 & -6 \end{pmatrix} \mathbf{y}.$$

4.7. Suppose that $A \in GL(\mathbb{R}^n)$ is not hyperbolic. Show that the map $A\mathbf{x}$ is not structurally stable. (In this context, structurally stable would mean that all linear maps which are close in terms of the norm on linear maps are topologically conjugate to $A\mathbf{x}$. Hint: Consider the family of maps $A_r(\mathbf{x}) = rA(\mathbf{x})$ for $r \in (1-\epsilon, 1+\epsilon)$.

4.8. Let $f, g : \mathbb{R}^n \to \mathbb{R}^n$ be defined by $f(\mathbf{x}) = a\mathbf{x}$ and $g(\mathbf{x}) = b\mathbf{x}$, where $1 < a < b$. Find an explicit formula for a conjugacy from the map g to the map f on all of $\mathbb{R}^n$, i.e., a homeomorphism h for which $f \circ h = h \circ g$. Prove that any such h which is differentiable must have all partial derivatives at the origin equal to $\mathbf{0}$, and h^{-1} does not have partial derivatives at $\mathbf{0}$.

Nonhomogeneous Equations

4.9. Prove Theorem 8.1.

Linear Maps

4.10. (This exercise gives a direct calculation of the eventual contraction of a Jordan block with a real eigenvalue absolute value less than 1. Compare with Theorem 9.1.) Assume $0 < |\lambda| < 1$. Assume A is a matrix with $A\mathbf{e}^1 = \lambda\mathbf{e}^1$ and $A\mathbf{e}^k = \lambda\mathbf{e}^k + a\mathbf{e}^{k-1}$ for $1 < k \le m$.

(a) Prove that

$$A^n\mathbf{e}^k = \sum_{j=0}^{k-1} \binom{n}{j} a^j \lambda^{n-j} \mathbf{e}^{k-j}$$

for $n \ge j$.

(b) Prove that $\binom{n}{j} \le n^j/j! \le n^j$.

(c) Prove that $n^j|\lambda^j|$ goes to zero as n goes to infinity.

(d) Prove that $|A^n\mathbf{e}^k|$ goes to zero as n goes to infinity for any $1 \le k \le m$.

4.11. Let $\text{Cont}(n, \mathbb{R})$ be the set of matrices A in $Gl(n, \mathbb{R})$ that induce a contracting linear map, $A\mathbf{x}$.

(a) Assume that $A \in \text{Cont}(n, \mathbb{R})$. Prove there is a curve $\{A_t \in \text{Cont}(n, \mathbb{R}) : 0 \le t \le 1\}$ such that (i) $A_0 = A$ and (ii) A_1 is diagonal with entries equal to either 0.5 or -0.5. Hint: Use Lemma 9.7. Allow for 1's in the off diagonal terms of the Jordan canonical form.

(b) Exhibit a curve from $\text{diag}(-0.5, -0.5)$ to $\text{diag}(0.5, 0.5)$. Hint: The curve of matrices cannot remain diagonal. Use Lemma 9.7.

(c) Assume $A \in \text{Cont}(n, \mathbb{R})$. Prove there is a curve $\{A_t \in \text{Cont}(n, \mathbb{R}) : 0 \le t \le 1\}$ such that (i) $A_0 = A$ and (ii)

$$A_1 = \begin{cases} \text{diag}(0.5, \dots, 0.5) & \text{if } \det(A) > 0 \\ \text{diag}(0.5, \dots, 0.5, -0.5) & \text{if } \det(A) < 0. \end{cases}$$

Hint: Use parts (a) and (b). Notice that this proves Lemma 9.9(a) for contracting linear maps.

4.12. Let $H(n, \mathbb{R})$ be the set of matrices A in $Gl(n, \mathbb{R})$ that induce a hyperbolic linear map, $A\mathbf{x}$.

(a) Assume that $A \in H(n, \mathbb{R})$. Prove there is a curve $\{A_t \in H(n, \mathbb{R}) : 0 \le t \le 1\}$ such that (i) $A_0 = A$ and (ii) A_1 is a real diagonal matrix. Hint: Use Exercise 4.11. Notice that the contracting and expanding subspaces are not necessarily spanned by the a subset of the standard basis.

(b) Assume $A \in H(n, \mathbb{R})$. Prove there is a curve $\{A_t \in H(n, \mathbb{R}) : 0 \le t \le 1\}$ such that (i) $A_0 = A$ and (ii) A_1 is diagonal with entries equal to either 2, 0.5, -0.5, or -2, i.e., prove Lemma 9.8. Notice that 1's are allowed in the off diagonal terms in the Jordan canonical form.

(c) Prove Lemma 9.9(b). More precisely, assume $A \in H(n, \mathbb{R})$. Prove there is a curve $\{A_t \in H(n, \mathbb{R}) : 0 \leq t \leq 1\}$ such that (i) $A_0 = A$ and (ii)

$$A_1|\mathbb{E}^s = \begin{cases} \text{diag}(0.5, \ldots, 0.5) & \text{if } \det(A|\mathbb{E}^s) > 0 \\ \text{diag}(0.5, \ldots, 0.5, -0.5) & \text{if } \det(A|\mathbb{E}^s) < 0, \end{cases}$$

and

$$A_1|\mathbb{E}^u = \begin{cases} \text{diag}(2, \ldots, 2) & \text{if } \det(A|\mathbb{E}^u) > 0 \\ \text{diag}(2, \ldots, 2, -2) & \text{if } \det(A|\mathbb{E}^u) < 0. \end{cases}$$

Hint: Use the previous problem as well as parts (a) and (b). (Note that for $\sigma = s, u$, $\det(A|\mathbb{E}^\sigma)$ is calculated in terms of a basis of $\mathbb{E}^\sigma$. The condition that $\det(A|\mathbb{E}^\sigma)$ is positive or negative is really the condition that A preserves or reverses orientation on the invariant subspace $\mathbb{E}^\sigma$.)

CHAPTER V

Analysis Near Fixed Points and Periodic Orbits

In this chapter we consider solutions of systems of nonlinear differential equations and the iteration of nonlinear functions of more than one variable. We also consider the phase portraits for both types of nonlinear systems. For nonlinear systems, even the analysis near a fixed point is more complicated. Rather than just using inequalities from the Mean Value Theorem, we must use the more complicated linear theory from the last chapter and nonlinear theorems from differential calculus like the Inverse Function Theorem, Implicit Function Theorem, and the Contraction Mapping Theorem. We are able to prove that if the linearization is hyperbolic, then the nonlinear map or differential equation is topologically conjugate to the linearization. This theorem is simple in one dimension, but requires a proof using the Contraction Mapping Theorem in higher dimensions. We also introduce the nonlinear invariant manifold which is tangent to contracting directions of the linearization, called the stable manifold, and the corresponding invariant manifold which is tangent to the expanding directions, called the unstable manifold. The proof that these manifolds exist is again a nontrivial fact which needs an involved proof.

As the above summary and this chapter's title indicate, this chapter concerns only the behavior near a single periodic orbit. In Chapters VIII and X we return to consider more global and complicated dynamics such as those for the quadratic map on the Cantor set and the structural stability of certain classes of examples. In between, Chapter VII is concerned with how periodic points change or bifurcate as a parameter is varied.

The first few sections present a review of differentiation of function between Euclidean spaces as a linear map, and the important theorems from differential calculus in the form we use them: Inverse Function Theorem, Implicit Function Theorem, and the Contraction Mapping Theorem. As a first application of the Contraction Mapping Theorem, we prove the existence of solutions of nonlinear differential equations. After these beginning sections, we begin our study of properties of solutions of nonlinear differential equations and the iteration of nonlinear maps.

5.1 Review: Differentiation in Higher Dimensions: The Derivative as a Linear Map

The general references for this material are Dieudonné (1960), Lang (1968), Marsden (1974), and Smith (1971). Both Dieudonné (1960) and Lang (1968) talk about the derivative of functions between Banach spaces. The third reference, Marsden (1974), is concerned with derivatives between Euclidean spaces, and the approach is more elementary but not as developed. Many other books on real analysis define the total derivative as a linear map.

We are concerned with maps $f : U \subset \mathbb{R}^k \to \mathbb{R}^n$, where U is an open subset of $\mathbb{R}^k$. If all the partial derivatives exist and are continuous, then the map is said to be *continuously differentiable*, C^1. We put the partial derivatives at a point $\mathbf{p}$ into a single matrix, $Df_{\mathbf{p}}$,

$$Df_{\mathbf{p}} = (\frac{\partial f_i}{\partial x_j}(\mathbf{p})).$$

We identify $n \times k$ matrices with linear maps from $\mathbb{R}^k$ to $\mathbb{R}^n$, $\mathbf{L}(\mathbb{R}^k, \mathbb{R}^n)$. Thus, $Df_{\mathbf{p}}$ should actually be thought of as a linear map in $\mathbf{L}(\mathbb{R}^k, \mathbb{R}^n)$. Which coordinate function is used determines the row and which partial derivative is taken determines the column. With this choice of the entries in the matrix, the ith coordinate of the product of this matrix with a vector $\mathbf{v}$ is given by

$$(Df_{\mathbf{p}}\mathbf{v})_i = \sum_j (\frac{\partial f_i}{\partial x_j}(\mathbf{p}))v_j,$$

which is what it should be in terms of the partial derivatives. In fact, if all the partial derivatives exist in U and are continuous, then using the Mean Value Theorem it can be shown that for $\mathbf{p} \in U$,

$$f(\mathbf{x}) = f(\mathbf{p}) + Df_{\mathbf{p}}(\mathbf{x} - \mathbf{p}) + R(\mathbf{x}, \mathbf{p}),$$

where

$$\lim_{\mathbf{x} \to \mathbf{p}} \frac{R(\mathbf{x}, \mathbf{p})}{|\mathbf{x} - \mathbf{p}|} = 0.$$

This latter condition is often expressed by saying that

$$R(\mathbf{x}, \mathbf{p}) = o(|\mathbf{x} - \mathbf{p}|).$$

The fact that $f(\mathbf{x}) = f(\mathbf{p}) + Df_{\mathbf{p}}(\mathbf{x} - \mathbf{p}) + o(|\mathbf{x} - \mathbf{p}|)$ can be taken as the definition of f being differentiable with its derivative being the linear map $Df_{\mathbf{p}}$. If any (matrix or) linear map $A \in \mathbf{L}(\mathbb{R}^k, \mathbb{R}^n)$ exists such that

$$\lim_{\mathbf{x} \to \mathbf{p}} \frac{f(\mathbf{x}) - f(\mathbf{p}) - A(\mathbf{x} - \mathbf{p})}{|\mathbf{x} - \mathbf{p}|} = 0,$$

then f is said to be *differentiable at* $\mathbf{p}$ and A is called the *(Frechet) derivative* at $\mathbf{p}$. The derivative at $\mathbf{p}$ can be shown to be unique, and if the partial derivatives exist, then the above matrix is the unique matrix which gives the derivative. The definition does not give a way to calculate the derivative but the partial derivatives do.

The space $\mathbf{L}(\mathbb{R}^k, \mathbb{R}^n)$ is given the *operator norm* as defined in the last chapter. The derivative is called *continuous* provided the map $Df : U \to \mathbf{L}(\mathbb{R}^k, \mathbb{R}^n)$ is continuous with respect to the Euclidean norm on the domain and the operator norm on the space of linear maps. In fact, if the partial derivatives all exist and are continuous, then the derivative is continuous, and vice versa. Such a map is called *continuously differentiable* or C^1.

If U is a region where $f(\mathbf{x})$ is defined and C^1, then we can let $K = \sup\{\|Df_{\mathbf{x}}\| : \mathbf{x} \in U\}$. By the Mean Value Theorem,

$$|f(\mathbf{x}) - f(\mathbf{y})| \leq K|\mathbf{x} - \mathbf{y}|$$

if the line segment from $\mathbf{x}$ to $\mathbf{y}$ is contained in U. See Dieudonné (1960), Lang (1968), or Marsden (1974). It is possible for a function to have this latter property but not be differentiable. In this case, if there is some $K > 0$ such that $|f(\mathbf{x}) - f(\mathbf{y})| \leq K|\mathbf{x} - \mathbf{y}|$ for all $\mathbf{x}$ and $\mathbf{y}$, then f is called *Lipschitz* with Lipschitz constant K. We write $\text{Lip}(f) = K$

if K is the smallest constant that works. Thus, if f is C^1, then it is Lipschitz. However, the function $f(x) = |x|$ on $\mathbb{R}$ is Lipschitz but not C^1. In the same way, we call f α-*Hölder* for $\alpha > 0$ provided there is a constant $K > 0$ such that $|f(\mathbf{x}) - f(\mathbf{y})| \leq K|\mathbf{x}-\mathbf{y}|^\alpha$ for all $\mathbf{x}$ and $\mathbf{y}$. Thus, if a function is 1-Hölder, then it is Lipschitz. Finally, we call a function $C^{r+\alpha}$ for r a positive integer and $0 < \alpha \leq 1$ if the rth order partial derivatives are α-Hölder.

Given the above definition of the derivative, if $f : U \subset \mathbb{R}^k \to \mathbb{R}^m$ and $g : V \subset \mathbb{R}^m \to \mathbb{R}^n$ are differentiable at $\mathbf{p}$ and $\mathbf{q} = f(\mathbf{p})$, respectively, then the composition, $g \circ f$, is differentiable at $\mathbf{p}$ with derivative $D(g \circ f)_\mathbf{p} = Dg_\mathbf{q} Df_\mathbf{p}$. Thus, the derivative of the composition is the composition of the derivatives. This is called the chain rule. In calculus books this derivative of the composition is often written out as a product of partial derivatives. The reader should satisfy him or herself that these ways of expressing the derivative of the composition are compatible.

Using this approach to understand the second derivative and higher derivatives seems complicated when first encountered. As stated above, the derivative of a map $f : U \subset \mathbb{R}^k \to \mathbb{R}^n$ at a point $\mathbf{p}$ gives a map $Df : U \to \mathbf{L}(\mathbb{R}^k, \mathbb{R}^n)$ as the point $\mathbf{p}$ varies. This second space is itself isomorphic to the Euclidean space $\mathbb{R}^{kn}$. It can be given either the Euclidean norm or the operator norm. (Any two norms on finite-dimensional spaces are equivalent.) The *second derivative* at $\mathbf{p}$, $D^2 f_\mathbf{p}$, is then the derivative of this map and so is an element of $\mathbf{L}(\mathbb{R}^k, \mathbf{L}(\mathbb{R}^k, \mathbb{R}^n))$.

However, $\mathbf{L}(\mathbb{R}^k, \mathbf{L}(\mathbb{R}^k, \mathbb{R}^n))$ is isomorphic to bilinear maps from $\mathbb{R}^k$ to $\mathbb{R}^n$ which we denote by $\mathbf{L}^2(\mathbb{R}^k, \mathbb{R}^n)$. (Note $\mathbf{L}^2(\mathbb{R}^k, \mathbb{R}^n)$ are those maps in $\mathbf{L}(\mathbb{R}^k \times \mathbb{R}^k, \mathbb{R}^n)$ which are linear in each factor of $\mathbb{R}^k$ separately.) An element $B \in \mathbf{L}^2(\mathbb{R}^k, \mathbb{R}^n)$ acts on two vectors in $\mathbb{R}^k$ and gives a vector in $\mathbb{R}^n$. In terms of the standard bases $\{\mathbf{e}^i\}_{i=1}^k$ of $\mathbb{R}^k$ and $\{\mathbf{s}^\ell\}_{\ell=1}^n$ of $\mathbb{R}^n$, two vectors $\mathbf{v}$ and $\mathbf{w}$ can be written as $\mathbf{v} = \sum_i v_i \mathbf{e}^i$ and $\mathbf{w} = \sum_j w_j \mathbf{e}^j$, and

$$\begin{aligned} B(\mathbf{v}, \mathbf{w}) &= \sum_{\ell=1}^{n} \Big(\sum_{1 \leq i,j \leq k} b^\ell_{i,j} v_i w_j \Big) \mathbf{s}^\ell \\ &= \sum_{1 \leq i,j \leq k} \Big(\sum_{\ell=1}^{n} b^\ell_{i,j} \mathbf{s}^\ell \Big) v_i w_j. \end{aligned}$$

Such a bilinear map is *symmetric* provided $B(\mathbf{v}, \mathbf{w}) = B(\mathbf{w}, \mathbf{v})$ for all vectors $\mathbf{v}$ and $\mathbf{w}$. In terms of the entries $b^\ell_{i,j}$, B is symmetric provided $b^\ell_{i,j} = b^\ell_{j,i}$ for all indices i, j, and ℓ. The set of all symmetric bilinear forms from $\mathbb{R}^k$ to $\mathbb{R}^n$ is denoted by $\mathbf{L}^2_s(\mathbb{R}^k, \mathbb{R}^n)$.

Returning to the second derivative, $D^2 f_\mathbf{p}$ acts on two vectors in $\mathbb{R}^k$ and gives a vector in $\mathbb{R}^n$. If $\mathbf{v}$ and $\mathbf{w}$ are expressed in terms of the standard basis $\mathbf{e}^1, \ldots, \mathbf{e}^k$ as $\mathbf{v} = \sum_i v_i \mathbf{e}^i$ and $\mathbf{w} = \sum_j w_j \mathbf{e}^j$, then

$$D^2 f_\mathbf{p}(\mathbf{v}, \mathbf{w}) = \sum_{i,j} \Big(\frac{\partial^2 f}{\partial x_i \partial x_j} \Big)_\mathbf{p} v_i w_j.$$

(Note that each $\Big(\frac{\partial^2 f}{\partial x_i \partial x_j} \Big)_\mathbf{p}$ is a vector.) The fact that the mixed cross partial derivatives are equal,

$$\frac{\partial^2 f}{\partial x_i \partial x_j}(\mathbf{p}) = \frac{\partial^2 f}{\partial x_j \partial x_i}(\mathbf{p}),$$

implies that $D^2 f_\mathbf{p}$ is a symmetric bilinear form, $D^2 f_\mathbf{p}(\mathbf{v}, \mathbf{w}) = D^2 f_\mathbf{p}(\mathbf{w}, \mathbf{v})$ for all $\mathbf{v}$ and $\mathbf{w}$. This process can be continued to define higher derivatives, and the rth derivative at

$\mathbf{p}$ is a symmetric r-linear form, $D^r f_{\mathbf{p}} \in \mathbf{L}_s^r(\mathbb{R}^k, \mathbb{R}^n)$. We write $D^r f_{\mathbf{p}}(\mathbf{v})^r$ to mean that the rth derivative is acting on the same vector $\mathbf{v}$ r times, $D^r f_{\mathbf{p}}(\mathbf{v}, \cdots, \mathbf{v})$. If all the derivatives (or all the partial derivatives) of order $1 \leq j \leq r$ exist and are continuous, then f is said to be *r-continuously differentiable*, or f is C^r.

Just as for C^1 (or as for functions of one real variable), if $f : U \subset \mathbb{R}^k \to \mathbb{R}^m$ and $g : V \subset \mathbb{R}^m \to \mathbb{R}^n$ are C^r with $\mathbf{q} = f(\mathbf{p}) \in V$ for $\mathbf{p} \in U$, then $g \circ f$ is C^r in a neighborhood of $\mathbf{p}$. The formula for the second derivative can be calculated:

$$D^2(g \circ f)_{\mathbf{p}}(\mathbf{v}, \mathbf{w}) = D^2 g_{f(\mathbf{p})}(Df_{\mathbf{p}}\mathbf{v}, Df_{\mathbf{p}}\mathbf{w}) + Dg_{f(\mathbf{p})} D^2 f_{\mathbf{p}}(\mathbf{v}, \mathbf{w}).$$

This is found by differentiating $D(g \circ f)_{\mathbf{p}} = Dg_{f(\mathbf{p})} Df_{\mathbf{p}}$ using the product rule (and the chain rule). There is one term for each place that the variable $\mathbf{p}$ appears in the formula. The result is very similar to functions of one variable, but all the derivatives are matrices which are applied to vectors. The higher derivatives of the composition can be calculated, but the formulas are somewhat complicated. (For explicit formulas see Abraham and Robbin (1967).)

Using these higher derivatives, we can state Taylor's Theorem. If $f : U \subset \mathbb{R}^k \to \mathbb{R}^n$ is C^r, then

$$f(\mathbf{x}) = f(\mathbf{p}) + Df_{\mathbf{p}}(\mathbf{x} - \mathbf{p}) + \frac{1}{2!} D^2 f_{\mathbf{p}}(\mathbf{x} - \mathbf{p})^2 + \cdots + \frac{1}{r!} D^r f_{\mathbf{p}}(\mathbf{x} - \mathbf{p})^r + R(\mathbf{x}, \mathbf{p}),$$

where

$$\lim_{\mathbf{x} \to \mathbf{p}} \frac{R(\mathbf{x}, \mathbf{p})}{|\mathbf{x} - \mathbf{p}|^r} = 0.$$

This last condition is often expressed by saying that $R(\mathbf{x}, \mathbf{p}) = o(|\mathbf{x} - \mathbf{p}|^r)$.

If $f : \mathbb{R}^k \to \mathbb{R}^k$ is C^r, $Df_{\mathbf{p}}$ is a linear isomorphism at each point $\mathbf{p} \in \mathbb{R}^k$, and f is one to one and onto, then f is called a *C^r-diffeomorphism*. The set of C^r-diffeomorphisms on $\mathbb{R}^k$ is denoted by $\text{Diff}^r(\mathbb{R}^k)$. By the Inverse Function Theorem, it follows that f^{-1} is also C^r. (The statement of the Inverse Function Theorem is given in Section 5.2.2.)

We have completed our introduction to the notion of derivatives as linear maps which we need to start studying the dynamics of functions of several variables. Other main topics of the multidimensional differential calculus which we use are the Implicit and Inverse Function Theorems. They are not used immediately. The Implicit Function Theorem is used in discussions of the Poincaré map of a differential equation near a closed orbit and in bifurcation questions. The Inverse Function Theorem is used in the proof of the Stable Manifold Theorem. Because this material can be postponed until it is needed, we put it in a separate section made up of several subsection (Sections 5.2–5.2.2). Section 5.2.3 deals with the related topic of the Contraction Mapping Theorem which is used repeatedly, including in the proof of the existence of solutions for differential equations, Section 5.3.

5.2 Review: The Implicit Function Theorem

In a course on advanced calculus or real analysis, a proof of the Implicit Function Theorem is often given, but often the students do not get a very good idea about how it is used. On the other hand, most calculus students learn to calculate using implicit differentiation, but have no real idea of the significance of the method of calculation. In Dynamical Systems several uses are made of the theorem for bifurcation results and results concerning the Poincaré map. In this section, we want to discuss the idea and meaning of the theorem. The proof will be left to books on real analysis. See Dieudonné (1960), Lang (1968), or Marsden (1974). (Also, the proof of the Hartman–Grobman Theorem later in the chapter solves a functional equation by applying a similar contraction mapping argument.) In Subsection 5.2.2 we give the statement of the Inverse Function Theorem. Finally, in Subsection 5.2.3, we give a statement and proof of the Contraction Mapping Theorem.

We will first give the statement and interpret the result for the case when the function is real valued. In the next Subsection, 5.2.1, we discuss the case when the function is vector valued.

Theorem 2.1 (Implicit Function Theorem). *Assume that $U \subset \mathbb{R}^{n+1}$ is an open set and $F : U \to \mathbb{R}$ is a C^r function for some $r \geq 1$. For $\mathbf{p} \in \mathbb{R}^{n+1}$, we write $\mathbf{p} = (\mathbf{x}, y)$ with $\mathbf{x} \in \mathbb{R}^n$ and $y \in \mathbb{R}$. Assume that $(\mathbf{x}_0, y_0) \in U$ and*

$$\frac{\partial F}{\partial y}(\mathbf{x}_0, y_0) \neq 0. \tag{i}$$

Let $C = F(\mathbf{x}_0, y_0) \in \mathbb{R}$. Then, there are open sets V containing $\mathbf{x}_0$ and W containing y_0 with $V \times W \subset U$, and a C^r function $h : V \to W$ such that

$$h(\mathbf{x}_0) = y_0 \tag{ii}$$

$$F(\mathbf{x}, h(\mathbf{x})) = C \qquad \text{for all } \mathbf{x} \in V. \tag{iii}$$

Further, for each $\mathbf{x} \in V$, $h(\mathbf{x})$ is the unique $y \in W$ such that $F(\mathbf{x}, y) = C$.

This theorem states several things. First, it says that indeed the set of $(\mathbf{x}, y)$ that satisfy $F(\mathbf{x}, y) = C$ can be represented locally as a graph, $y = h(\mathbf{x})$. Next, it says that this function is differentiable. Once we know that it is differentiable, then implicit differentiation gives a method of calculating its derivative: differentiating $F(\mathbf{x}, h(\mathbf{x})) = C$ with respect to x_i and using the chain rule, we get

$$\frac{\partial F}{\partial x_i}(\mathbf{x}, h(\mathbf{x})) + \frac{\partial F}{\partial y}(\mathbf{x}, h(\mathbf{x}))\frac{\partial h}{\partial x_i}(\mathbf{x}) = 0 \qquad \text{so}$$

$$\frac{\partial h}{\partial x_i}(\mathbf{x}) = -\frac{\partial F}{\partial x_i}(\mathbf{x}, h(\mathbf{x}))\left[\frac{\partial F}{\partial y}(\mathbf{x}, h(\mathbf{x}))\right]^{-1}.$$

If we combine these partial derivatives together in the (Frechet) derivative of h, we get

$$\begin{aligned} Dh(\mathbf{x}) &= \left(\frac{\partial h}{\partial x_1}(\mathbf{x}), \dots, \frac{\partial h}{\partial x_n}(\mathbf{x})\right) \\ &= -\left[\frac{\partial F}{\partial y}(\mathbf{x}, h(\mathbf{x}))\right]^{-1}\left(\frac{\partial F}{\partial x_1}(\mathbf{x}, h(\mathbf{x})), \dots, \frac{\partial F}{\partial x_n}(\mathbf{x}, h(\mathbf{x}))\right). \end{aligned}$$

Thus, the assumption in the theorem that $\frac{\partial F}{\partial y}(\mathbf{x}_0, y_0) \neq 0$ is exactly the assumption which makes the method of implicit differentiation valid. Or formally, the assumption that the partial derivative with respect to y is nonzero is the correct assumption to make so that y can be solved in terms of $\mathbf{x}$.

A second more geometric way of understanding the assumption on the partial derivative is in terms of the tangent line. Consider the example of two variables ($n = 1$), $F(x, y) = x^2 + y^2 = 1$. The tangent line at $\mathbf{p}_0 = (x_0, y_0)$ is given by

$$\nabla F_{\mathbf{p}_0} \cdot (x - x_0, y - y_0) = 0.$$

See Figure 2.1. If $\frac{\partial F}{\partial y}(\mathbf{p}_0) \neq 0$, then we can represent this line as a graph of y in terms of x, i.e., we can solve for y in terms of x:

$$(x - x_0)\frac{\partial F}{\partial x}(\mathbf{p}_0) + (y - y_0)\frac{\partial F}{\partial y}(\mathbf{p}_0) = 0,$$

$$(y - y_0) = \frac{-(x - x_0)\dfrac{\partial F}{\partial x}(\mathbf{p}_0)}{\dfrac{\partial F}{\partial y}(\mathbf{p}_0)}.$$

Conversely, if we can solve the tangent line equation to give y as a function of x, then $\frac{\partial F}{\partial y}(\mathbf{p_0}) \neq 0$. For example, at $(x_0, y_0) = (\frac{1}{2}, \frac{\sqrt{3}}{2})$,

$$y - \frac{\sqrt{3}}{2} = \frac{-(x - \frac{1}{2})(2)(\frac{1}{2})}{(2)\frac{\sqrt{3}}{2}} = -\frac{1}{\sqrt{3}}(x - \frac{1}{2})$$

gives the tangent line. The Implicit Function Theorem says that if the tangent line at $\mathbf{p_0}$ can be represented as a graph of y in terms of x, then nearby the nonlinear level set $F(x, y) = C$ can be represented as a graph $y = h(x)$. The same ideas apply for $n > 1$.

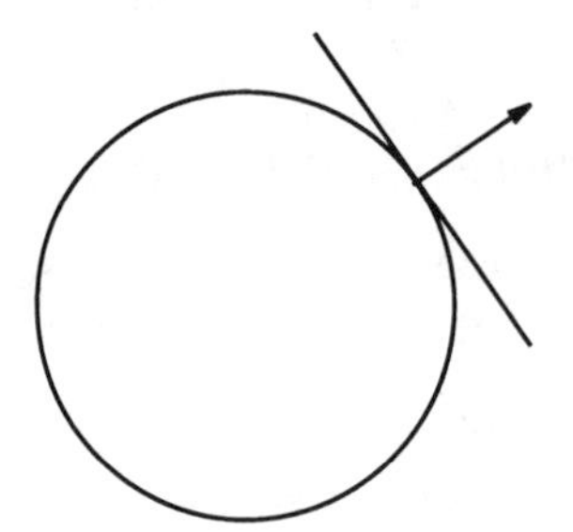

FIGURE 2.1. Gradient and Tangent Line to the Circle

Notice that in the example, $x^2 + y^2 = 1$ at $(1, 0)$, $\frac{\partial F}{\partial y}(1, 0) = 0$, and the method breaks down. The tangent line is vertical so it cannot be represented as a graph over the x variable. Also, near $(1, 0)$, the level set $x^2 + y^2 = 1$ cannot be represented as a graph of y in terms of x. Thus, both the hypothesis and the conclusion fail to be true at this point. However, $\dfrac{\partial F}{\partial x}(1, 0) \neq 0$, so reversing the roles of x and y, it is possible to solve for x in terms of y, i.e., the level set is a graph of x in terms of y.

5.2.1 Higher Dimensional Implicit Function Theorem

Similar results are true for functions that are vector valued as given in the following theorem. (See a calculus book on implicit differentiation involving partial derivatives, e.g., Edwards and Penny (1990), pages 797 and 806.)

Theorem 2.2 (Higher Dimensional Implicit Function Theorem). *Assume that $U \subset \mathbb{R}^n \times \mathbb{R}^k$ is an open set and $F : U \to \mathbb{R}^k$ is a C^r function for some $r \geq 1$. Represent a point $\mathbf{p} \in U$ by $\mathbf{p} = (\mathbf{x}, \mathbf{y})$ with $\mathbf{x} \in \mathbb{R}^n$ and $\mathbf{y} \in \mathbb{R}^k$, and the coordinate functions of F by f_i, $F = (f_1, \dots, f_k)$. Assume that for $(\mathbf{x_0}, \mathbf{y_0}) \in U$*

$$\left(\frac{\partial f_i}{\partial y_j}(\mathbf{x_0}, \mathbf{y_0})\right)_{1 \leq i,j \leq k}$$

is an invertible $k \times k$ matrix (nonzero determinant). Let $C = F(\mathbf{x_0}, \mathbf{y_0}) \in \mathbb{R}^k$. Then, there are open sets V containing $\mathbf{x_0}$ and W containing $\mathbf{y_0}$ with $V \times W \subset U$, and a C^r function $h : V \to W$ such that

$$\begin{aligned} h(\mathbf{x_0}) &= \mathbf{y_0} \\ F(\mathbf{x}, h(\mathbf{x})) &= C \qquad \text{for all } \mathbf{x} \in V. \end{aligned}$$

Further, for each $\mathbf{x} \in V$, $h(\mathbf{x})$ is the unique $\mathbf{y} \in W$ such that $F(\mathbf{x}, \mathbf{y}) = C$.

Just as in two dimensions, once it is known that h is a C^r function, it is possible to solve for the matrix of partial derivatives, $(\frac{\partial h_i}{\partial x_j})_{1\le i\le k, 1\le j\le n}$ in terms of the two matrices of partial derivatives $(\frac{\partial f_\ell}{\partial x_j})_{1\le \ell\le k, 1\le j\le n}$ and $(\frac{\partial f_\ell}{\partial y_i})_{1\le \ell, i\le k}$:

$$(\frac{\partial f_\ell}{\partial x_j})_{1\le \ell\le k, 1\le j\le n} + (\frac{\partial f_\ell}{\partial y_i})_{1\le \ell, i\le k}(\frac{\partial h_i}{\partial x_j})_{1\le i\le k, 1\le j\le n} = 0 \qquad \text{so}$$

$$(\frac{\partial h_i}{\partial x_j})_{1\le i\le k, 1\le j\le n} = -(\frac{\partial f_\ell}{\partial y_i})^{-1}_{1\le \ell, i\le k}(\frac{\partial f_\ell}{\partial x_j})_{1\le \ell\le k, 1\le j\le n}.$$

Or if we use the notation of $Dh_{\mathbf{x}_0}$ for the Frechet derivative of h, $D_{\mathbf{x}}F_{(\mathbf{x}_0,\mathbf{y}_0)}$ for the matrix of partial derivatives with respect to the x_j's, and $D_{\mathbf{y}}F_{(\mathbf{x}_0,\mathbf{y}_0)}$ for the matrix of partial derivatives with respect to the y_k's, then this equation can be written as

$$Dh_{\mathbf{x}_0} = -(D_{\mathbf{y}}F_{(\mathbf{x}_0,\mathbf{y}_0)})^{-1}D_{\mathbf{x}}F_{(\mathbf{x}_0,\mathbf{y}_0)}.$$

Notice that this formula is very similar to that for a real valued function F, where matrices have replaced numbers and the inverse of a matrix has replaced dividing by a number.

Also, the theorem can be interpreted to say that if we can solve the linear (affine) equation

$$F(\mathbf{x}_0,\mathbf{y}_0) + (\frac{\partial f_i}{\partial x_j}(\mathbf{x}_0,\mathbf{y}_0))(\mathbf{x}-\mathbf{x}_0) + (\frac{\partial f_i}{\partial y_j}(\mathbf{x}_0,\mathbf{y}_0))(\mathbf{y}-\mathbf{y}_0) = C$$

for $\mathbf{y}$ in terms of $\mathbf{x}$, then locally near $(\mathbf{x}_0,\mathbf{y}_0)$ we can (theoretically) solve the nonlinear equation $F(\mathbf{x},\mathbf{y}) = C$ for $\mathbf{y}$ in terms of $\mathbf{x}$. This means that if implicit differentiation works, then indeed locally $\mathbf{y}$ is a differentiable function of $\mathbf{x}$. Also, the above linear equation gives the tangent space to the level set $F(\mathbf{x},\mathbf{y}) = C$. If this linear tangent space can be represented as a graph with $\mathbf{y}$ given as a function of $\mathbf{x}$, then the nonlinear equations can also be represented as a nearby graph.

Just as in two dimensions, it might be that at some points

$$(\frac{\partial f_i}{\partial y_j}(\mathbf{x}_0,\mathbf{y}_0))_{1\le i,j\le k}$$

is not invertible. However, if the $k\times(n+k)$ matrix

$$DF_{(\mathbf{x}_0,\mathbf{y}_0)} = (\frac{\partial f_i}{\partial x_j}(\mathbf{x}_0,\mathbf{y}_0), \frac{\partial f_i}{\partial y_j}(\mathbf{x}_0,\mathbf{y}_0)),$$

which is the Frechet derivative at $(\mathbf{x}_0,\mathbf{y}_0)$, has rank k, then it is possible to select k columns that give an invertible submatrix. The theorem then says that the corresponding k variables can be solved near the point in terms of the remaining n variables to give the level set $F(\mathbf{x},\mathbf{y}) = C$ as a graph.

For a more thorough treatment, see Dieudonné (1960) or Lang (1968).

5.2.2 The Inverse Function Theorem

The Inverse Function Theorem can easily be proved from the Implicit Function Theorem and vice versa. However, although both theorems involve the assumption that a matrix of partial derivatives is invertible, they seem very different.

Theorem 2.3 (Inverse Function Theorem). *Assume that $U \subset \mathbb{R}^n$ is an open set and $f : U \to \mathbb{R}^n$ is a C^r function for some $r \geq 1$. Assume that for $\mathbf{x}_0 \in U$, $Df_{\mathbf{x}_0}$ is an invertible linear map (matrix). Then, there exist open sets V containing $\mathbf{x}_0$ and W containing $\mathbf{y}_0 = f(\mathbf{x}_0)$ and a C^r function $g : W \to V$ which is the inverse of f on V: $g \circ f(\mathbf{x}) = \mathbf{x}$ for $\mathbf{x} \in V$ and $f \circ g(\mathbf{y}) = \mathbf{y}$ for $\mathbf{y} \in W$. Further, $Dg_{f(\mathbf{x})} = [Df_{\mathbf{x}}]^{-1}$.*

One way of interpreting this theorem is that if the linearized (affine) equations (function) $\mathbf{y} = A(\mathbf{x}) = f(\mathbf{x}_0) + Df_{\mathbf{x}_0}(\mathbf{x} - \mathbf{x}_0)$ have an inverse, then the nonlinear equations $\mathbf{y} = f(\mathbf{x})$ have an inverse near $\mathbf{x} = \mathbf{x}_0$.

In fact, in the proof of the stable manifold theorem, we need a more precise statement of the neighborhoods on which the function is one to one. We give this statement in the following theorem whose proof we leave to the exercises. See Exercise 5.3. (Also see Hirsch Pugh (1970) and Lang (1968).) The statement of the theorem uses the minimum norm of a linear map which we defined in the last chapter. Also, for $r > 0$, let

$$\bar{B}(\mathbf{x}, r) = \{\mathbf{y} \in \mathbb{R}^n : |\mathbf{y} - \mathbf{x}| \leq r\}$$

be the closed ball of radius r centered at $\mathbf{x}$.

Theorem 2.4 (Covering Estimate). *Assume that $U \subset \mathbb{R}^n$ is an open set and $f : U \to \mathbb{R}^n$ is a C^1 function. Assume that $\mathbf{x}_0 \in U$, and that $L = Df_{\mathbf{x}_0}$ is an invertible linear map with a bounded linear inverse. Let $\mathbf{y}_0 = f(\mathbf{x}_0)$. Take any β with $0 < \beta < 1$. Let $r > 0$ be such that (i) $\bar{B}(\mathbf{x}_0, r) \subset U$ and (ii) $\|L - Df_{\mathbf{x}}\| \leq m(L)(1 - \beta)$ for all $\mathbf{x} \in \bar{B}(\mathbf{x}_0, r)$. Then,*

$$f(\bar{B}(\mathbf{x}_0, r)) \supset \{\mathbf{y}_0\} + L(\bar{B}(\mathbf{0}, \beta r)) \supset \bar{B}(\mathbf{y}_0, m(L)\beta r).$$

Moreover, every point $\mathbf{y} \in \{\mathbf{y}_0\} + L(\bar{B}(\mathbf{0}, \beta r))$ has exactly one preimage $\mathbf{x} \in \bar{B}(\mathbf{x}_0, r)$, $f(\mathbf{x}) = \mathbf{y}$, so the inverse function g is a C^1 function from

$$\{\mathbf{y}_0\} + L(\bar{B}(\mathbf{0}, \beta r)) \supset \bar{B}(\mathbf{y}_0, m(L)\beta r)$$

into $\bar{B}(\mathbf{x}_0, r)$.

As stated above, we leave the proof of the covering estimates to the exercises. See Exercise 5.3.

5.2.3 Contraction Mapping Theorem

In this section we consider maps on some metric space Y which decrease distance, so-called *contraction mappings*. Finding the fixed points of a contraction map can itself be thought of as a problem in the dynamics of iteration. The theorem below shows that such a map g has a unique fixed point. Besides being interesting in itself, this result is used in many of the proofs of other theorems. In various proofs, a map is constructed whose fixed point gives the desired conclusion. For example, later in this chapter we prove that a nonlinear map with a "hyperbolic" fixed point is conjugate to a linear map in a neighborhood of the fixed point (Hartman-Grobman Theorem). This result is proved by constructing a map Θ on a set of functions. The fixed point of Θ turns out to be the conjugacy. A main step in the proof is verifying that Θ is a contraction mapping and concluding that there exists a unique fixed point. Similarly, in the next section we use the contraction mapping method to prove the existence of solutions of differential equations. The contraction mapping method is also used in the proofs of the Implicit and Inverse Function Theorems.

With this motivation, we turn to the statement and proof of the Contraction Mapping Theorem.

Theorem 2.5 (Contraction Mapping Theorem). *Assume Y is a complete metric space with metric d, and $g : Y \to Y$ is a Lipschitz function with Lipschitz constant* $\mathrm{Lip}(g) = \kappa < 1$. *Then, there is a unique fixed point $\mathbf{y}^*$, $g(\mathbf{y}^*) = \mathbf{y}^*$. More specifically, if $\mathbf{y}_0$ is any point in Y, then $\{g^n(\mathbf{y}_0)\}_{n\in\mathbb{N}}$ is a Cauchy sequence and* $d(\mathbf{y}_0, \mathbf{y}^*) \le d(\mathbf{y}_0, g(\mathbf{y}_0))/[1 - \mathrm{Lip}(g)]$.

PROOF. By induction,

$$\begin{aligned} d(g^n(\mathbf{y}_0), g^{n+1}(\mathbf{y}_0)) &\le \kappa d(g^{n-1}(\mathbf{y}_0), g^n(\mathbf{y}_0)) \\ &\le \kappa^n d(\mathbf{y}_0, g(\mathbf{y}_0)). \end{aligned}$$

Then,

$$\begin{aligned} d(g^n(\mathbf{y}_0), g^{n+k}(\mathbf{y}_0)) &\le \sum_{j=0}^{k-1} d(g^{n+j}(\mathbf{y}_0), g^{n+j+1}(\mathbf{y}_0)) \\ &\le \sum_{j=0}^{k-1} \kappa^{n+j} d(\mathbf{y}_0, g(\mathbf{y}_0)) \\ &\le \frac{\kappa^n}{1-\kappa} d(\mathbf{y}_0, g(\mathbf{y}_0)). \end{aligned}$$

This latter inequality proves that the sequence $g^n(\mathbf{y}_0)$ is Cauchy and so converges to a limit point $\mathbf{y}^*$.

If there were two fixed points $\mathbf{y}^*$ and $\mathbf{y}'$, then

$$\begin{aligned} d(\mathbf{y}^*, \mathbf{y}') &= d(g(\mathbf{y}^*), g(\mathbf{y}')) \\ &\le \kappa d(\mathbf{y}^*, \mathbf{y}'), \end{aligned}$$

which is impossible unless $d(\mathbf{y}^*, \mathbf{y}') = 0$ and $\mathbf{y}^* = \mathbf{y}'$. Therefore, the fixed point is unique.

To get the final conclusion, note that

$$\begin{aligned} d(\mathbf{y}_0, \mathbf{y}^*) &= \lim_{n\to\infty} d(\mathbf{y}_0, g^n(\mathbf{y}_0)) \\ &\le \lim_{n\to\infty} \sum_{j=0}^{n-1} d(g^j(\mathbf{y}_0), g^{j+1}(\mathbf{y}_0)) \\ &= \sum_{j=0}^{\infty} d(g^j(\mathbf{y}_0), g^{j+1}(\mathbf{y}_0)) \\ &\le \sum_{j=0}^{\infty} \kappa^j d(\mathbf{y}_0, g(\mathbf{y}_0)) \\ &= \frac{1}{1-\kappa} d(\mathbf{y}_0, g(\mathbf{y}_0)). \end{aligned}$$

This last inequality proves the desired result. □

5.3 Existence of Solutions for Differential Equations

In this section, we start our consideration of nonlinear systems of differential equations. In the last chapter we considered linear differential equations of the form $\dot{\mathbf{x}} = A\mathbf{x}$. (As before, $\dot{\mathbf{x}} = \frac{d}{dt}\mathbf{x}$ is the derivative with respect to time.) A simple example of a nonlinear systems is

$$\begin{aligned}\dot{x} &= y\\ \dot{y} &= -x + x^3.\end{aligned}$$

In general, we consider an equation of the form $\dot{\mathbf{x}} = f(\mathbf{x})$, where $\mathbf{x}$ is some point in an open set U in $\mathbb{R}^n$ (or on some manifold like the torus) and $f : U \subset \mathbb{R}^n \to \mathbb{R}^n$ is a differentiable function. The function f is called a *vector field* because it assigns a vector $f(\mathbf{x})$ to each point in U. We want to consider the solution, $\mathbf{x}(t)$, with $\mathbf{x}(t_0)$ some prescribed value. More precisely, given $\mathbf{x}_0$ and t_0, we want $\mathbf{x}(t)$ to be defined on an open interval of times, I, with $t_0 \in I$, $\mathbf{x}(t_0) = \mathbf{x}_0$, and

$$\frac{d}{dt}\mathbf{x}(t) = f(\mathbf{x}(t)) \tag{†}$$

for t in I. Thus, the tangent vector to the solution curve, $\dot{\mathbf{x}}(t)$, is equal to the vector field f evaluated at the position at this time, $f(\mathbf{x}(t))$.

In Chapter IV, we used the concept of a flow when discussing topologically conjugate flows. Here we reintroduce this notion which is used extensively in the rest of the book and make the connection with the solutions of a nonlinear differential equation. Given the differential equation $\dot{\mathbf{x}} = f(\mathbf{x})$, we let $\varphi^t(\mathbf{x}_0)$ be the solution $\mathbf{x}(t)$ with the given initial condition $\mathbf{x}_0$ at $t = 0$: $\varphi^0(\mathbf{x}_0) = \mathbf{x}_0$ and $\frac{d}{dt}\varphi^t(\mathbf{x}_0) = f(\varphi^t(\mathbf{x}_0))$ for all t for which it is defined. We also sometimes write $\varphi(t, \mathbf{x}_0)$ for $\varphi^t(\mathbf{x}_0)$. (Other books often write $\varphi_t(\mathbf{x}_0)$.) The function $\varphi^t(\mathbf{x}_0)$ is called the *flow* of the differential equation. The function f defining the differential equation is also called the *vector field which generates the flow.*

The first theorem below shows that the solution is uniquely determined by the initial conditions $\mathbf{x}_0$ and the time t; the notation of the flow $\varphi^t(\mathbf{x}_0)$ emphasizes this dependence. Later in the section we show that $\varphi^t(\mathbf{x}_0)$ is a continuous function of the initial condition $\mathbf{x}_0$. The final important property of the flow is the group property: $\varphi^t \circ \varphi^s(\mathbf{x}_0) = \varphi^{t+s}(\mathbf{x}_0)$. Then, $\varphi^{-t} \circ \varphi^t(\mathbf{x}_0) = \varphi^0(\mathbf{x}_0) = \mathbf{x}_0$ so $\varphi^{-t} = (\varphi^t)^{-1}$, and for fixed t, φ^t is a homeomorphism on its domain of definition. (There might be some points for which the solution is not defined up to time t.)

Before verifying these properties for the flow generated by a differentiable vector field, we summarize these important properties of a flow. Since we occasionally refer to flows on metric spaces which are not the solutions of a differential equation on $\mathbb{R}^n$, we state the definition in this context.

Definition. For a metric space X, any continuous map $\varphi : U \subset \mathbb{R} \times X \to X$ defined on an open set $U \supset \{0\} \times X$ is called a *flow* provided (i) it satisfies the group property $\varphi^t \circ \varphi^s(\mathbf{x}_0) = \varphi^{t+s}(\mathbf{x}_0)$ and (ii) for fixed t, φ^t is a homeomorphism on its domain of definition.

We now turn to verifying the properties of the flow generated by a differentiable vector field on $\mathbb{R}^n$.

Theorem 3.1 (Existence and Uniqueness of Solutions of Ordinary Differential Equations). *Let $U \subset \mathbb{R}^n$ be an open set, and $f : U \to \mathbb{R}^n$ be a Lipschitz function (or C^1). Let $\mathbf{x}_0 \in U$ and $t_0 \in \mathbb{R}$. Then, there exists an $\alpha > 0$ and a solution, $\mathbf{x}(t)$, of $\dot{\mathbf{x}} = f(\mathbf{x})$ defined for $t_0 - \alpha < t < t_0 + \alpha$ such that $\mathbf{x}(t_0) = \mathbf{x}_0$. Moreover, if $\mathbf{y}(t)$ is another solution with $\mathbf{y}(t_0) = \mathbf{x}_0$, then $\mathbf{x}(t) = \mathbf{y}(t)$ on their common interval of definition about t_0.*

PROOF OF EXISTENCE. For $\mathbf{x}_0 \in U$ take $b > 0$ such that $B(\mathbf{x}_0, b) \equiv \{\mathbf{x} : |\mathbf{x} - \mathbf{x}_0| \leq b\} \subset U$. The function f is Lipschitz so there is a $K > 0$ and $M > 0$ such that $|f(\mathbf{x}) - f(\mathbf{y})| \leq K|\mathbf{x} - \mathbf{y}|$ and $|f(\mathbf{x})| \leq M$ for all $\mathbf{x}, \mathbf{y} \in B(\mathbf{x}_0, b)$.

If $\mathbf{x}(t)$ is a solution with $\mathbf{x}(t_0) = \mathbf{x}_0$, then

$$\mathbf{x}(t) = \mathbf{x}_0 + \int_{t_0}^{t} \dot{\mathbf{x}}(s)\, ds = \mathbf{x}_0 + \int_{t_0}^{t} f(\mathbf{x}(s))\, ds. \qquad (*)$$

Conversely, if $\mathbf{x} : J \to \mathbb{R}^n$ satisfies $(*)$, then $\mathbf{x}$ is a solution of $\dot{\mathbf{x}} = f(\mathbf{x})$ with $\mathbf{x}(t_0) = \mathbf{x}_0$. Thus, to get solutions to the ordinary differential equation we find solutions of $(*)$.

To this end, for $\mathbf{y} : J \to B(\mathbf{x}_0, b)$, we define

$$\mathcal{F}(\mathbf{y})(t) = \mathbf{x}_0 + \int_{t_0}^{t} f(\mathbf{y}(s))\, ds.$$

The idea is to show $\mathcal{F}$ is a contraction on some function space which has a fixed point which satisfies equation $(*)$, and thus is a solution of the differential equation.

We need to define the function space on which $\mathcal{F}$ acts. First we need to specify the length of the interval J. Take a with $a < \min\{b/M, 1/K\}$ and let $J = [t_0 - a, t_0 + a]$. We are going to consider $\mathcal{F}$ as acting on potential solutions defined for t in J. We take $a < b/M$ so that $\mathcal{F}(\mathbf{y})(t)$ does not leave $B(\mathbf{x}_0, b)$ for t in J. We take $a < 1/K$ so that $\mathcal{F}$ is a contraction by $\lambda = aK$.

We now explicitly define the function space $\mathcal{S}$ of potential solutions on which $\mathcal{F}$ acts. Let

$$\mathcal{S} = \{\mathbf{y} : J \to B(\mathbf{x}_0, b) : \mathbf{y} \text{ is } C^0,\ \mathbf{y}(t_0) = \mathbf{x}_0, \mathrm{Lip}(\mathbf{y}) \leq M\},$$

the space of M-Lipschitz curves that go through $\mathbf{x}_0$ at $t = t_0$ and take their values in $B(\mathbf{x}_0, b)$. We put the C^0*-sup-norm* on $\mathcal{S}$: for $\mathbf{y}, \mathbf{z} \in \mathcal{S}$ we set

$$\|\mathbf{y} - \mathbf{z}\|_0 = \sup\{|\mathbf{y}(t) - \mathbf{z}(t)| : t \in J\}.$$

With this norm, it can be shown that $\mathcal{S}$ is a complete metric space.

We need to show that $\mathcal{F}$ preserves $\mathcal{S}$, i.e., if $\mathbf{y}$ is in $\mathcal{S}$, then $\mathbf{y}_1 = \mathcal{F}(\mathbf{y})$ is also in $\mathcal{S}$. Clearly, $\mathbf{y}_1$ is continuous with $\mathcal{F}(\mathbf{y})(t_0) = \mathbf{x}_0$. Next,

$$\begin{aligned} |\mathbf{y}_1(t_2) - \mathbf{y}_1(t_1)| &= \left| \int_{t_1}^{t_2} f(\mathbf{y}(s))\, ds \right| \leq \left| \int_{t_1}^{t_2} |f(\mathbf{y}(s))|\, ds \right| \\ &\leq \left| \int_{t_1}^{t_2} M\, ds \right| = M|t_2 - t_1|. \end{aligned}$$

This shows that $\mathcal{F}(\mathbf{y})$ is Lipschitz as a function of t with $\mathrm{Lip}(\mathcal{F}(\mathbf{y})) \leq M$. Then, for $|t - t_0| \leq a$, $|\mathbf{y}_1(t) - \mathbf{x}_0| = |\mathbf{y}_1(t) - \mathbf{y}_1(t_0)| \leq M|t - t_0| \leq Ma < b$ so $\mathcal{F}(\mathbf{y})(t)$ takes values in $B(\mathbf{x}_0, b)$ for $|t - t_0| \leq a$. Thus, we have shown that for $\mathbf{y}$ in $\mathcal{S}$, $\mathcal{F}(\mathbf{y})$ is in $\mathcal{S}$, so $\mathcal{F} : \mathcal{S} \to \mathcal{S}$.

Finally, to show that $\mathcal{F}$ is a contraction on $\mathcal{S}$,

$$\begin{aligned}
\|\mathcal{F}(\mathbf{y}) - \mathcal{F}(\mathbf{z})\|_0 &= \sup_{t\in J} |\mathcal{F}(\mathbf{y})(t) - \mathcal{F}(\mathbf{z})(t)| \\
&\leq \sup_{t\in J} \Big| \int_{t_0}^{t} |f(\mathbf{y}(s)) - f(\mathbf{z}(s))| \, ds \Big| \\
&\leq \sup_{t\in J} \Big| \int_{t_0}^{t} K|\mathbf{y}(s) - \mathbf{z}(s)| \, ds \Big| \\
&\leq \sup_{t\in J} \Big| \int_{t_0}^{t} K\|\mathbf{y} - \mathbf{z}\|_0 \, ds \Big| \\
&\leq Ka\|\mathbf{y} - \mathbf{z}\|_0 \\
&\leq \lambda\|\mathbf{y} - \mathbf{z}\|_0.
\end{aligned}$$

Thus, $\mathcal{F}$ is a contraction by λ on $\mathcal{S}$ and so has a unique fixed point $\mathbf{x}^*$ in $\mathcal{S}$. But a fixed point of $\mathcal{F}$ clearly satisfies equation $(*)$. This proves the existence of a solution of the differential equation. □

The uniqueness of the fixed point proves that the solution is unique among the curves which are M-Lipschitz. For the proof of the uniqueness of the solutions among all curves, together with the continuity of solutions on initial conditions, we use the following result which is called Gronwall's Inequality.

Theorem 3.2 (Gronwall's Inequality). *Let $v(t)$ and $g(t)$ be continuous nonnegative scalar functions on (a, b), $a < t_0 < b$, $C \geq 0$, and*

$$v(t) \leq C + \Big| \int_{t_0}^{t} v(s)g(s) \, ds \Big| \qquad \text{for } a < t < b.$$

Then,

$$v(t) \leq C \exp\Big(\Big| \int_{t_0}^{t} g(s) \, ds \Big|\Big).$$

PROOF. First we consider the case for $t_0 \leq t < b$. It is not possible to differentiate inequalities and retain the inequality, so we define

$$U(t) = C + \int_{t_0}^{t} v(s)g(s) \, ds.$$

Then, $v(t) \leq U(t)$ and we can differentiate U. In fact, $U'(t) = v(t)g(t) \leq U(t)g(t)$.
First, if $C > 0$, then $U(t) > 0$ so

$$\begin{aligned}
\frac{U'(t)}{U(t)} &\leq g(t), \\
\log\Big(\frac{U(t)}{U(t_0)}\Big) &\leq \int_{t_0}^{t} g(s) \, ds, \quad \text{and} \\
U(t) &\leq C \exp\Big(\int_{t_0}^{t} g(s) \, ds\Big)
\end{aligned}$$

since $U(t_0) = C$. Using the fact that $v(t) \leq U(t)$, we are done when $C > 0$ and $t_0 \leq t < b$.

If $a < t \leq t_0$ and $C > 0$, then we define

$$U(t) = C + \int_t^{t_0} v(s)g(s)\,ds,$$

so $U'(t) = -v(t)g(t) \geq -U(t)g(t)$. Then,

$$\log\big(\frac{U(t_0)}{U(t)}\big) \geq -\int_t^{t_0} g(s)\,ds, \quad \text{and}$$

$$U(t) \leq C \exp\big(\int_t^{t_0} g(s)\,ds\big).$$

This completes the modifications for $C > 0$ and $a < t \leq t_0$.

Next, if $C = 0$, we can take $C_j > 0$ which converge to zero and have the assumed inequality true for C_j. By the first case,

$$v(t) \leq C_j \exp\big(|\int_{t_0}^t g(s)\,ds\,|\big).$$

Since this last term goes to zero as j goes to infinity, we get that $v(t) \equiv 0$ in this case, which verifies the conclusion. □

We can now give the proof of uniqueness. However, we give the proof of continuity with respect to initial conditions at the same time.

Theorem 3.3 (Continuity with Respect to Initial Conditions). *With the assumptions of Theorem 3.1, the solution $\varphi^t(\mathbf{x}_0)$ depends continuously on the initial condition x_0.*

Proof of Uniqueness and Continuity. Assume that $\mathbf{x}(t)$ and $\mathbf{y}(t)$ are two solutions with $\mathbf{x}(t_0) = \mathbf{x}_0$ and $\mathbf{y}(t_0) = \mathbf{y}_0$. Then,

$$\mathbf{x}(t) - \mathbf{y}(t) = \mathbf{x}_0 - \mathbf{y}_0 + \int_{t_0}^t f(\mathbf{x}(s)) - f(\mathbf{y}(s))\,ds.$$

Let $v(t) = |\mathbf{x}(t) - \mathbf{y}(t)|$, which is nonnegative, and $v(t_0) = |\mathbf{x}_0 - \mathbf{y}_0|$. We get that

$$\begin{aligned} v(t) &\leq v(t_0) + \Big|\int_{t_0}^t |f(\mathbf{x}(s)) - f(\mathbf{y}(s))|\,ds\,\Big| \\ &\leq v(t_0) + \Big|\int_{t_0}^t K\,v(s)\,ds\,\Big|. \end{aligned}$$

So by Gronwall's Inequality, $v(t) \leq v(t_0)e^{K|t-t_0|}$, or $|\mathbf{x}(t) - \mathbf{y}(t)| \leq |\mathbf{x}_0 - \mathbf{y}_0|\,e^{K|t-t_0|}$. This clearly implies the solutions depend continuously on $\mathbf{x}_0$.

Also, if $\mathbf{x}_0 = \mathbf{y}_0$, then we get that $v(t) \equiv 0$, or $\mathbf{x}(t) \equiv \mathbf{y}(t)$. This last statement gives the uniqueness. □

Example 3.1. If the differential equation is not Lipschitz, then it is possible to have nonunique solutions. One example is the equation $\dot{x} = 3\,x^{\frac{2}{3}}$ on the real line. Given any t_0, there is a solution $x(t)$ which equals $(t - t_0)^3$ for $t \geq t_0$ and 0 for $t \leq t_0$. There is yet another solution $z(t) \equiv 0$ for all t. Then, both $x(t_0)$ and $z(t_0)$ equal zero but $x(t)$ and $z(t)$ are not equal on any interval about t_0.

We next discuss the *maximal interval of definition of the solution.* That is, for fixed $\mathbf{x}$, we extend the solution so $\varphi^t(\mathbf{x})$ is defined for $t \in (t_-, t_+)$ but no larger open interval of times t. (Here t_- and t_+ possibly depend on $\mathbf{x}$.) The interval is open by the existence of solutions on a short interval starting at any point. The following example gives an equation for which the solutions are not defined for all time.

Example 3.2. Consider $\dot{x} = x^2$. If $x_0 > 0$, then solving the equation by separation of variables shows that $\varphi^t(x_0) = \dfrac{x_0}{1 - tx_0}$. This solution is defined for $-\infty < t < 1/x_0$, so $t_+ = 1/x_0 < \infty$. What is the interval of definition for $x_0 < 0$?

The following theorem shows that if the solution is bounded, then it is defined for all time: $t_- = -\infty$ and $t_+ = \infty$. Thus, if a solution is not defined for all time, then it must be unbounded (it must leave all compact subsets of U).

Theorem 3.4. *Let $U \subset \mathbb{R}^n$ be an open set and $f : U \to \mathbb{R}^n$ a C^1 function.*

(a) Given $\mathbf{x} \in U$, let (t_-, t_+) be the maximal interval of definition for $\varphi^t(\mathbf{x})$. If $t_+ < \infty$, then given any compact subset $C \subset U$, there is a time t_C with $0 \le t_C < t_+$ such that $\varphi^{t_C}(\mathbf{x}) \notin C$. Similarly, if $t_- > -\infty$, then there is a t_{C-} with $t_- < t_{C-} \le 0$ such that $\varphi^{t_{C-}}(\mathbf{x}) \notin C$.

(b) In particular, if $f : \mathbb{R}^n \to \mathbb{R}^n$ is defined on all of $\mathbb{R}^n$ and $|f(\mathbf{x})|$ is bounded, then the solutions exist for all t.

PROOF. (a) Given a compact C, since f is C^1, there are constants $M > 0$ and $K > 0$ such that $|f(\mathbf{y})| \le M$ and $|f(\mathbf{y}) - f(\mathbf{z})| \le K|\mathbf{y} - \mathbf{z}|$ for $\mathbf{y}, \mathbf{z} \in C$. By the existence proof, as long as the solution $\varphi^t(\mathbf{x})$ remains in C, it is M-Lipschitz with respect to t. Assume that $\varphi^t(\mathbf{x}) \in C$ for $0 \le t < t_+$. Because $\varphi^t(\mathbf{x})$ is M-Lipschitz with respect to t, it has a unique limit point $\varphi^{t_+}(\mathbf{x}) \in C$. By the existence of solutions, there are $\delta > 0$ and a solution defined for $(t_+ - \delta, t_+ + \delta)$ that agrees with $\varphi^{t_+}(\mathbf{x})$ for $t = t_+$. Thus, the interval (t_-, t_+) is not maximal. This contradiction shows that the solution $\varphi^t(\mathbf{x})$ must leave C before t_+.

(b) If $|f(\mathbf{y})| \le M$ for all $\mathbf{y} \in \mathbb{R}^n$, then $|\varphi^t(\mathbf{x}) - \mathbf{x}| \le M|t|$ for all t, and the solution must stay in the ball $B(\mathbf{x}, R)$ for time $|t| \le R/M$. By part (a), it follows that $t_+ = \infty$ and $t_- = -\infty$. □

Given an initial point $\mathbf{x}_0$, let (t_-, t_+) be the maximal interval of definition. The set $\mathcal{O}(\mathbf{x}_0) = \{\varphi^t(\mathbf{x}_0) : t_- < t < t_+\}$ is called the *orbit through* $\mathbf{x}_0$.

Example 3.2 shows that the flow of a differential equation is not necessarily defined for all time. The following theorem shows how a differential equation can be modified to keep the same solution curves in phase space but change the parameterization so each trajectory is defined for all time.

Theorem 3.5 (Reparameterization). *Assume $f : U \subset \mathbb{R}^n \to \mathbb{R}^n$ is a vector field defined on an open set U of $\mathbb{R}^n$ with flow φ^t. Then, there is a C^∞ real valued function $g : U \to (0,1] \subset \mathbb{R}$ such that $F(\mathbf{x}) = g(\mathbf{x})\, f(\mathbf{x})$, $F : U \to \mathbb{R}^n$, has a flow ψ^t defined for all time. Moreover, $\psi^t(\mathbf{x}) = \varphi^{\tau(t,\mathbf{x})}(\mathbf{x})$ where τ satisfies the differential equation $\dot{\tau}(t,\mathbf{x}) = g \circ \varphi^{\tau(t,\mathbf{x})}(\mathbf{x}) > 0$; therefore, ψ^t is a reparameterization of the flow φ^t with the same oriented solution curves.*

PROOF. If $U = \mathbb{R}^n$, we let $g(\mathbf{x}) = [1 + |f(\mathbf{x})|^2]^{-1/2}$ where we use the Euclidean norm so it is differentiable. Then, $F(\mathbf{x}) = g(\mathbf{x}) f(\mathbf{x})$ has $|F(\mathbf{x})| \le 1$ for all $\mathbf{x}$; the solutions exist for all time by Theorem 3.4(b).

If U is not all of $\mathbb{R}^n$, we let $G : U \to (0,1] \subset \mathbb{R}$ be a C^∞ positive function such that (i) $\|DG_{\mathbf{x}}\| \le 1$ and (ii) $|G(\mathbf{x})|$ goes to zero as $\mathbf{x}$ goes to the boundary of U or as $|\mathbf{x}|$ goes to infinity. The function G can be thought of as the square of the distance to the boundary of U. Let $g(\mathbf{x}) = G(\mathbf{x})^2[1 + |f(\mathbf{x})|^2]^{-1/2}$ and $F(\mathbf{x}) = g(\mathbf{x})\, f(\mathbf{x})$. Then, $|F(\mathbf{x})| \le |G(\mathbf{x})|^2 \le 1$. By Theorem 3.4, to show that a solution for F is defined for all time it is enough to show that $G \circ \psi^t(\mathbf{x})$ does not go to zero in finite time, or that $[G \circ \psi^t(\mathbf{x})]^{-1}$ does not go to infinity in finite time. But,

$$\frac{d}{dt}[G \circ \psi^t(\mathbf{x})]^{-1} = -[G \circ \psi^t(\mathbf{x})]^{-2} DG_{\psi^t(\mathbf{x})} F \circ \psi^t(\mathbf{x})$$

and

$$\begin{aligned}[G \circ \psi^t(\mathbf{x})]^{-1} &\le [G(\mathbf{x})]^{-1} + \int_0^{|t|} |G \circ \psi^s(\mathbf{x})|^{-2} \cdot \|DG_{\psi^t(\mathbf{x})}\| \cdot |F \circ \psi^s(\mathbf{x})|\, ds \\ &\le [G(\mathbf{x})]^{-1} + \int_0^{|t|} ds \\ &\le [G(\mathbf{x})]^{-1} + |t|.\end{aligned}$$

Thus, $[G \circ \psi^t(\mathbf{x})]^{-1}$ does not go to infinity in finite time, and the solution $\psi^t(\mathbf{x})$ is defined for all time.

For the reparameterization,

$$\frac{d}{dt}\varphi^{\tau(t,\mathbf{x})}(\mathbf{x}) = f \circ \varphi^{\tau(t,\mathbf{x})}(\mathbf{x})\dot{\tau}(t,\mathbf{x}).$$

Therefore, if τ is defined as the solution of

$$\dot{\tau}(t,\mathbf{x}) = g \circ \varphi^{\tau(t,\mathbf{x})}(\mathbf{x}),$$

then both $\varphi^{\tau(t,\mathbf{x})}(\mathbf{x})$ and $\psi^t(\mathbf{x})$ satisfy the same differential equation, and so are equal by uniqueness of solutions. Because g is strictly positive, $\tau(t,\mathbf{x})$ is a monotonically increasing function of time and the orbits of φ^t and ψ^t have the same orientation. □

The final result of this section gives the group property for the flow.

Theorem 3.6. *The flow φ^t satisfies the following group property. Letting (t_-, t_+) be the maximal interval of definition for the initial condition $\mathbf{x}_0$, then*

$$\varphi^t(\varphi^s(\mathbf{x}_0)) = \varphi^{t+s}(\mathbf{x}_0)$$

for all t and s for which $s, t+s \in (t_-, t_+)$.

PROOF. Let $J(\mathbf{y})$ be the maximal interval of definition for $\varphi^t(\mathbf{y})$. Then, $\varphi^t(\varphi^s(\mathbf{x}))$ is defined for $t \in J(\varphi^s(\mathbf{x}))$. Define the new function

$$\mathbf{y}(\tau) = \begin{cases} \varphi^\tau(\mathbf{x}) & \text{for } 0 \le \tau \le s \\ \varphi^{\tau-s}(\varphi^s(\mathbf{x})) & \text{for } s \le \tau \le s+t. \end{cases}$$

It is easily checked that $\mathbf{y}(\tau)$ is a solution. By uniqueness, it must equal $\varphi^\tau(\mathbf{x})$ for $s \le \tau \le s+t$. Therefore, $s+t \in J(\mathbf{x})$ and $\mathbf{y}(\tau) = \varphi^\tau(\mathbf{x})$ for $0 \le \tau \le s+t$, or $\varphi^t(\varphi^s(\mathbf{x})) = \mathbf{y}(s+t) = \varphi^{t+s}(\mathbf{x})$, verifying the claim of the theorem. □

If the flow is generated by a differentiable vector field, we have proved above that $\varphi^t(\mathbf{x}_0)$ is a continuous function of $\mathbf{x}_0$ and a differentiable function of t. Actually, if the vector field f is C^r with $r \ge 1$, then $\varphi^t(\mathbf{x}_0)$ is C^r jointly in $\mathbf{x}_0$ and t. We do not prove this, but a proof can be found in many graduate level ordinary differential equations books, e.g., Hale (1969) or Hartman (1964). By differentiating the equation $\frac{d}{dt}\varphi^t(\mathbf{x}) = f(\varphi^t(\mathbf{x}))$ with respect to the initial condition $\mathbf{x}$ and interchanging the order of differentiation, we get the equation

$$\frac{d}{dt}D\varphi^t_{\mathbf{x}} = Df_{\varphi^t(\mathbf{x})}D\varphi^t_{\mathbf{x}}$$

or

$$\frac{d}{dt}D\varphi^t_{\mathbf{x}}\mathbf{v} = Df_{\varphi^t(\mathbf{x})}D\varphi^t_{\mathbf{x}}\mathbf{v}$$

for any vector $\mathbf{v}$. Either form of these equations is called the *First Variation Equation.*

5.4 Limit Sets and Recurrence for Flows

In this section, X is a metric space with distance d, and φ^t is a flow on X. In the applications in this book, X is a Euclidean space or manifold and φ^t is the flow of a differentiable vector field f. We start with the definition of a fixed point and a periodic orbit, and then give the definitions and results about limit sets, nonwandering sets, and chain recurrent sets. These theorems for flows are similar to those for maps. (Compare this section with Section 2.3.)

Definition. A point $\mathbf{p}$ is called a *fixed point* for the flow φ^t if $\varphi^t(\mathbf{p}) = \mathbf{p}$ for all t. Sometimes such a point is also called an *equilibrium* or *singular point.* If the flow is obtained as the solutions of a differential equation $\dot{\mathbf{x}} = f(\mathbf{x})$, then a fixed point is a point for which $f(\mathbf{p}) = \mathbf{0}$. A point $\mathbf{p}$ is called a *periodic point* provided there is $T > 0$ such that $\varphi^T(\mathbf{p}) = \mathbf{p}$ and $\varphi^t(\mathbf{p}) \neq \mathbf{p}$ for $0 < t < T$. The time $T > 0$ which satisfies the above conditions is called the *period* of the orbit. (The period is the least period because $\varphi^t(\mathbf{p}) \neq \mathbf{p}$ for $0 < t < T$.) The orbit of such a point $\mathbf{p}$, $\mathcal{O}(\mathbf{p})$, is called a *periodic orbit.* Periodic orbits are also called *closed orbits* since the set of points on the orbit is a closed curve. It easily follows that if $\mathbf{p}$ is a periodic point of period T, and $\mathbf{q} \in \mathcal{O}(\mathbf{p})$, then $\mathbf{q}$ is a periodic point of period T.

Definition. A point $\mathbf{y}$ is an *ω-limit point of* $\mathbf{x}$ *for* φ^t provided there exists a sequence of t_k going to infinity such that $\lim_{k\to\infty} d(\varphi^{t_k}(\mathbf{x}), \mathbf{y}) = 0$. The set of all ω-limit points of $\mathbf{x}$ for φ^t is denoted by $\omega(\mathbf{x})$, $\omega(\mathbf{x}, \varphi^t)$, $L_\omega(\mathbf{x})$, or $L_\omega(\mathbf{x}, \varphi^t)$, and is called the *$\omega$-limit set.* The *$\alpha$-limit set of* $\mathbf{x}$ is defined the same way but with t_k going to minus infinity. The set of all such points is denoted by $\alpha(\mathbf{x})$ or $L_\alpha(\mathbf{x})$ (or with φ^t specified in the notation).

All the basic results for limit sets of maps are also true for flows as indicated in the following theorem. In addition, if the forward orbit of a point $\mathcal{O}^+(\mathbf{x}) = \{\varphi^t(\mathbf{x}) : t \geq 0\}$ is bounded, then $\omega(\mathbf{x})$ is connected. Remember that this is not true for a map as the example of a periodic point shows. (The reason flows are different than maps is that an orbit for a flow is connected but not for a map.)

Theorem 4.1. *Let φ^t be a flow on a metric space X.*

(a) The ω-limit set can be represented in terms of the forward orbit as follows:

$$\omega(\mathbf{x}) = \bigcap_{T\geq 0} \operatorname{cl} \bigcup_{t\geq T} \{\varphi^t(\mathbf{x})\}.$$

(b) If $\varphi^t(\mathbf{x}) = \mathbf{y}$ for a real number t, then $\omega(\mathbf{x}) = \omega(\mathbf{y})$ and $\alpha(\mathbf{x}) = \alpha(\mathbf{y})$.

(c) The limit sets, $\omega(\mathbf{x})$ and $\alpha(\mathbf{x})$, are closed and both positively and negatively invariant (contain complete orbits).

(d) If $\mathcal{O}^+(\mathbf{x})$ is contained in some compact subset of X, then $\omega(\mathbf{x})$ is nonempty, compact, and connected. Further, $d(\varphi^t(\mathbf{x}), \omega(\mathbf{x}))$ goes to zero as t goes to infinity. Similarly, if $\mathcal{O}^-(\mathbf{x})$ is contained in a compact subset of X, then $\alpha(\mathbf{x})$ is nonempty, compact, and connected, and $d(\varphi^t(\mathbf{x}), \omega(\mathbf{x}))$ goes to zero as t goes to minus infinity.

(e) If $D \subset X$ is closed and positively invariant and $\mathbf{x} \in D$, then $\omega(\mathbf{x}) \subset D$.

(f) If $\mathbf{y} \in \omega(\mathbf{x})$, then $\omega(\mathbf{y}) \subset \omega(\mathbf{x})$ and $\alpha(\mathbf{y}) \subset \alpha(\mathbf{x})$.

PROOF. All the proofs are the same as for diffeomorphisms except the new result about $\omega(\mathbf{x})$ being connected in part (d). For a flow, the sets $\bigcup_{t\geq T}\{\varphi^t(\mathbf{x})\}$ are connected, so $\operatorname{cl} \bigcup_{t\geq T}\{\varphi^t(\mathbf{x})\}$ is a nested collection of compact and connected sets, and so

$$\bigcap_{T\geq 0} \operatorname{cl} \Big(\bigcup_{t\geq T} \{\varphi^t(\mathbf{x})\} \Big)$$

is connected. (See Exercise 5.10.) □

In this section, we present some examples of flows in the plane which illustrate some of these possibilities of limit sets. In Section 5.9, we discuss the Poincaré-Bendixson Theorem which gives restriction on the possible limit sets for flows in the plane.

Example 4.1. The condition that the orbit is bounded is necessary to prove that the limit set is connected. This example gives a flow in the plane for which the orbit is unbounded and the limit set is not connected. Let $y_1 < 0 < y_2$ be two fixed values of y and L_j be the horizontal line $\{(x, y_j)\}$ for $j = 1, 2$. Let φ^t be the flow for which $\varphi^t(x, y_1) = (x - t, y_1)$, $\varphi^t(x, y_2) = (x + t, y_2)$, the origin is a fixed point, and the orbits of points $\mathbf{q}$ between the lines $y = y_1$ and $y = y_2$ spiral out limiting on both L_1 and L_2. See Figure 4.1. For $\mathbf{q} = (x, y) \neq \mathbf{0}$ and $y_1 < y < y_2$, $\omega(\mathbf{q}) = \{(x, y) : y = y_1 \text{ or } y = y_2\}$, which is not connected. For these same $\mathbf{q}$, $\alpha(\mathbf{q}) = \mathbf{0}$. For $\mathbf{q} = (x, y_j)$ for $j = 1$ or 2, $\omega(\mathbf{q}) = \alpha(\mathbf{q}) = \emptyset$.

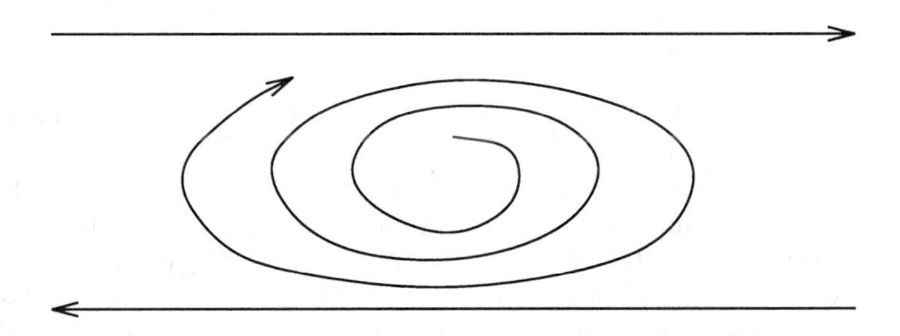

FIGURE 4.1. Example with Disconnected Limit Set

Example 4.2. Consider the equations

$$\dot{x} = -y + \mu x(1 - x^2 - y^2)$$
$$\dot{y} = x + \mu y(1 - x^2 - y^2)$$

for $\mu > 0$. In polar coordinates, this is the system

$$\dot{\theta} = 1$$
$$\dot{r} = \mu r(1 - r^2).$$

The time derivative of r satisfies

$$\dot{r} > 0 \qquad \text{for } 0 < r < 1$$
$$\dot{r} < 0 \qquad \text{for } 1 < r.$$

So, for initial conditions $\mathbf{p} \neq (0, 0)$ (i.e., $r_0 \neq 0$), the trajectory has

$$\omega(\mathbf{p}) = S^1 = \{(x, y) : x^2 + y^2 = 1\}.$$

Thus, solutions forward in time limit on the unique periodic orbit with $r \equiv 1$. Such an orbit with trajectories spiraling toward it from both sides is called a *limit cycle*.

Example 4.3. Consider the equations

$$\begin{aligned}\dot{x} &= y \\ \dot{y} &= x - x^3 - \mu y(2y^2 - 2x^2 + x^4)\end{aligned}$$

for $\mu > 0$. We proceed to explain the phase portrait as given in Figure 4.2. There are three fixed points: $\mathbf{0} = (0,0)$ and $\mathbf{p}^{\pm} = (\pm 1, 0)$. As we show later in the chapter, the two fixed points $\mathbf{p}^{\pm}$ are repelling and the origin is a "saddle fixed point." To aid in the analysis of the phase portrait, consider the real valued (Liapunov) function $L(x,y) = y^2/2 - x^2/2 + x^4/4$. (See Section 5.5.3 for a discussion of Liapunov functions.) The time derivative of L along trajectories can be calculated as follows:

$$\begin{aligned}\dot{L}(x,y) &\equiv \frac{d}{dt} L \circ \varphi^t(x,y)|_{t=0} \\ &= y\dot{y} + (-x + x^3)\dot{x} \\ &= y[x - x^3 - \mu y(2y^2 - 2x^2 + x^4)] + (-x + x^3)y \\ &= -4\mu y^2 L(x,y).\end{aligned}$$

Because $\dot{L}(x,y) \equiv 0$ along the level curve $L^{-1}(0)$, the trajectories are tangent to this level curve, and so the level curve $L^{-1}(0)$ is invariant under the flow. (If $\mu = 0$, then each level curve is invariant for the flow.) The level curve $L^{-1}(0)$ is made up of three trajectories: the fixed point $\mathbf{0}$ and two trajectories $\gamma^+ = \{(x,y) : x > 0 \text{ and } L(x,y) = 0\}$ and $\gamma^- = \{(x,y) : x < 0 \text{ and } L(x,y) = 0\}$. For $\mathbf{q} \in \gamma^{\pm}$, it can be shown that $\omega(\mathbf{q}) = \alpha(\mathbf{q}) = \mathbf{0}$. Because of this property, the trajectories $\gamma^{\pm}$ are called homoclinic connections for the saddle point $\mathbf{0}$. See Figure 4.2.

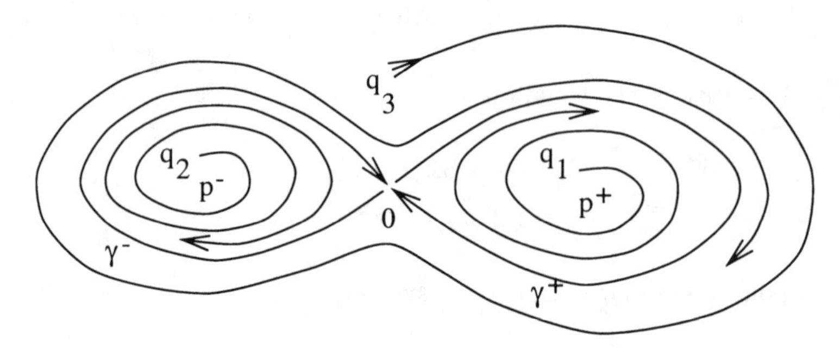

FIGURE 4.2. Example 4.3

For all $\mathbf{q}$ inside γ^+, $\dot{L}(\mathbf{q}) \geq 0$ and $\dot{L}(\mathbf{q}) > 0$ if $\mathbf{q}$ is off the x-axis. Using these facts, it can also be shown that for $\mathbf{q}_1$ inside γ^+ but not equal to $\mathbf{p}^+$, $\omega(\mathbf{q}_1) = \{\mathbf{0}\} \cup \gamma^+$ and $\alpha(\mathbf{q}_1) = \{\mathbf{p}^+\}$. Similarly, for $\mathbf{q}_2$ inside γ^- but not equal to $\mathbf{p}^-$, $\omega(\mathbf{q}_2) = \{\mathbf{0}\} \cup \gamma^-$ and $\alpha(\mathbf{q}_2) = \{\mathbf{p}^-\}$. Finally, for $\mathbf{q}$ outside both γ^+ and γ^-, $\dot{L}(\mathbf{q}) \leq 0$ and $\dot{L}(\mathbf{q}) < 0$ if $\mathbf{q}$ is off the x-axis. Again, it can be shown that for $\mathbf{q}_3$ outside both γ^+ and γ^-, $\omega(\mathbf{q}_3) = \{\mathbf{0}\} \cup \gamma^+ \cup \gamma^-$ and $\alpha(\mathbf{q}_3) = \emptyset$. At the moment we are not worrying about the fact that these particular equations have this phase portrait, but merely that this phase portrait illustrates a flow with different limit sets for different points. In particular, the α-limit set of certain points equals a single fixed point and for other points it is the empty set. The ω-limit set of certain points equals a single fixed point; for other points, it is the fixed point at the origin together with one of the homoclinic connections (γ^+ or γ^-); for still other points, it is the fixed point at the origin together with both the homoclinic connections γ^+ and γ^-.

Definition. Using the definition of ω-limit set and α-limit set, we say that $\mathbf{x}$ is *ω-recurrent* provided $\mathbf{x} \in \omega(\mathbf{x})$ and that $\mathbf{x}$ is *α-recurrent* provided $\mathbf{x} \in \alpha(\mathbf{x})$. The closure of the set of all recurrent points is called the *Birkhoff center*,

$$\mathcal{B}(\varphi^t) = \text{cl}\{\mathbf{x} : \mathbf{x} \text{ is } \omega\text{-recurrent } \}.$$

The closure of the set of all the ω-limit sets is called the *limit set of* φ^t,

$$L(\varphi^t) = \text{cl}\left(\bigcup\{\omega(\mathbf{x}, \varphi^t) : \mathbf{x} \in X\}\right).$$

The only changes that are needed in the definitions of nonwandering and chain recurrent points from those in Section 2.3 is that the times must be large enough.

Definition. For a flow φ^t on X, a point $\mathbf{p}$ is called *nonwandering* provided for every neighborhood U of $\mathbf{p}$ there is a time $t > 1$ such that $\varphi^t(U) \cap U \neq \emptyset$. Thus, there is a point $\mathbf{q} \in U$ with $\varphi^t(\mathbf{q}) \in U$. (Note the times for the flow are real numbers while those for maps are integers.)

Definition. An *ϵ-chain of length T from $\mathbf{x}$ to $\mathbf{y}$* for a flow φ^t is a sequence $\{\mathbf{x} = \mathbf{x}_0, \ldots, \mathbf{x}_n = \mathbf{y}; t_1, \ldots, t_n\}$ with $t_j \geq 1$ for all $1 \leq j \leq n$, $d(\varphi^{t_j}(\mathbf{x}_{j-1}), \mathbf{x}_j) < \epsilon$, and $t_1 + \cdots + t_n = T$.

The definitions of *ϵ-chain limit set*, *chain limit set*, *chain recurrent set*, and *chain components* remain the same. It is not hard to prove that for a flow, the chain components are indeed the connected components of $\mathcal{R}$.

As for maps, the reader can easily check that for a flow φ^t,

$$\text{cl}\bigcup\{\omega(\mathbf{x}, \varphi^t) : \mathbf{x} \in X\} \subset \Omega(\varphi^t) \subset \mathcal{R}(\varphi^t).$$

5.5 Fixed Points for Nonlinear Differential Equations

For linear systems of differential equations, the asymptotic stability is determined by the real parts of the eigenvalues. For a fixed point of a nonlinear system, the real parts of the eigenvalues of the derivative also determine the stability. In the next two subsections we state this result for nonlinear equations and prove it for the case of an attracting fixed point. In Section 5.6, we give the comparable results for maps. In between, in Section 5.5.3, we define a Liapunov function for a nonlinear ordinary differential equation, which can also be used to determine the stability of a fixed point.

Definition. We consider a (nonlinear) differential equation $\dot{\mathbf{x}} = f(\mathbf{x})$ with $\mathbf{x} \in \mathbb{R}^n$. The point $\mathbf{p}$ is a fixed point if $f(\mathbf{p}) = \mathbf{0}$. In this case the flow fixes the point $\mathbf{p}$ for all t, $\varphi^t(\mathbf{p}) = \mathbf{p}$ for all t. To linearize the flow near $\mathbf{p}$, let $A = Df_{\mathbf{p}}$ be the matrix of partial derivatives. Then, the *linearized differential equation at* $\mathbf{p}$ is given by $\dot{\mathbf{x}} = A(\mathbf{x} - \mathbf{p})$.

Let $\mathbf{p}$ be a fixed point for $\dot{\mathbf{x}} = f(\mathbf{x})$. We call the fixed point *hyperbolic* provided $Re(\lambda) \neq 0$ for all the eigenvalues λ of $Df_{\mathbf{p}}$. There are several special cases. A hyperbolic fixed point is called a *sink* or *attracting* provided the real parts of all the eigenvalues of $Df_{\mathbf{p}}$ are negative. This terminology is used because Theorem 5.1 below proves that in this case, not only is the linearized equation asymptotically stable (attracting) but also the nonlinear flow. A hyperbolic fixed point is called a *source* or *repelling*, provided the real parts of all the eigenvalues of $Df_{\mathbf{p}}$ are positive. Finally, a hyperbolic fixed point is called a *saddle* provided it is neither a sink nor a source, so there are two eigenvalues

λ_+ and λ_- with $Re(\lambda_+) > 0$ and $Re(\lambda_-) < 0$ (and the real parts of all the other eigenvalues are nonzero). A nonhyperbolic fixed point is said to have a *center.* (In a more restrictive sense of the word, a *center* is a family of closed orbits surrounding a fixed point as occurs for a linear two-dimensional center or an undamped pendulum.)

We are often interested in the set of all points whose ω-limit set is a fixed point (or in a more complicated set). The *basin of attraction* of a fixed point $\mathbf{p}$ is the set of all points $\mathbf{q}$ such that $\omega(\mathbf{q}) = \{\mathbf{p}\}$, which we denote by $W^s(\mathbf{p})$. Note this concept is most interesting and most often used when $\mathbf{p}$ is a sink.

Theorems 5.1 and 5.2 below prove the stability of a nonlinear sink, and Theorem 5.3 and Corollary 5.4 state these results for nonlinear hyperbolic fixed points. We end this section with several examples.

Example 5.1. Consider the second-order equation $\ddot{\theta} + \sin(\theta) + \delta\,\dot{\theta} = 0$. By letting $x_1 = \theta$ and $x_2 = \dot{\theta}$, this can be written as a first-order system

$$\dot{\mathbf{x}} = \begin{pmatrix} \dot{x}_1 \\ \dot{x}_2 \end{pmatrix} = f(\mathbf{x}) \equiv \begin{pmatrix} x_2 \\ -\sin(x_1) - \delta\, x_2 \end{pmatrix}.$$

The fixed points are $\mathbf{x} = (n\pi, 0)$ for n an integer. The derivative of f is

$$Df_{\mathbf{x}} = \begin{pmatrix} 0 & 1 \\ -\cos(x_1) & -\delta \end{pmatrix}.$$

At $(0,0)$, or $(n\pi, 0)$ for n even, the derivative is given by

$$Df_{(0,0)} = \begin{pmatrix} 0 & 1 \\ -1 & -\delta \end{pmatrix}.$$

The characteristic equation is $\lambda^2 + \delta\,\lambda + 1 = 0$ with eigenvalues $\lambda = (-\delta \pm\, [\delta^2 - 4]^{1/2})/2$. Since $\delta^2 - 4 < \delta^2$, for $\delta > 0$, the real part of both eigenvalues is negative. Thus, $(0,0)$ is a nonlinear sink. Notice that for $0 < \delta < 2$, the eigenvalues are complex; and for $\delta \geq 2$, the eigenvalues are real. For $0 > \delta$, $(0,0)$ is a source.

Next, for the fixed point $(\pi, 0)$, or $(n\pi, 0)$ for n odd,

$$Df_{(\pi,0)} = \begin{pmatrix} 0 & 1 \\ 1 & -\delta \end{pmatrix},$$

and the eigenvalues are $\lambda = (-\delta \pm\, [\delta^2 + 4]^{1/2})/2$. Notice that for $\delta = 0$, the eigenvalues are ± 1. For any δ, the fixed point is a saddle.

Example 5.2. Consider the equation $\ddot{x} + x - x^3 = 0$. The fixed points are at $x = 0$, ± 1 and $\dot{x} = 0$. The derivative of the system of equations is

$$Df_{\mathbf{x}} = \begin{pmatrix} 0 & 1 \\ -1 + 3x^2 & 0 \end{pmatrix}.$$

The eigenvalues for the fixed point $(0,0)$ are $\pm\, i$. Thus, this fixed point is nonhyperbolic; the linearized equation is a center. The fixed points $x = \pm 1$ have eigenvalues $\pm\sqrt{2}$, and thus are saddles.

5.5.1 Nonlinear Sinks

In this section we consider the dynamics near a nonlinear sink. The following theorem proves that the nonlinear flow near the fixed point sink is an exponential contraction if its linearized flow is one, too. Also, because the solutions are bounded for the nonlinear flow, they exist for all $t \geq 0$.

Theorem 5.1. *Let* $\mathbf{p}$ *be a fixed point for the equations* $\dot{\mathbf{x}} = f(\mathbf{x})$ *with* $\mathbf{x} \in \mathbb{R}^n$. *Also assume that there is a constant* $a > 0$ *such that all the eigenvalues* λ *for* $A = Df_{\mathbf{p}}$ *have negative real part with* $Re(\lambda) < -a < 0$. *Then, the following two statements are true.*

(a) There is a norm $|\ |_*$ *on* $\mathbb{R}^n$ *and a neighborhood* $U \subset \mathbb{R}^n$ *of* $\mathbf{p}$ *such that for any initial condition* $\mathbf{x} \in U$, *the solution is defined for all* $t \geq 0$ *and satisfies*

$$|\varphi^t(\mathbf{x}) - \mathbf{p}|_* \leq e^{-ta}|\mathbf{x} - \mathbf{p}|_* \quad \textit{for all } t \geq 0.$$

(b) For any norm $|\ |'$ *on* $\mathbb{R}^n$, *there exist a neighborhood* $U \subset \mathbb{R}^n$ *of* $\mathbf{p}$ *and a constant* $C \geq 1$ *such that for any initial condition* $\mathbf{x} \in U$, *the solution is defined for all* $t \geq 0$ *and satisfies*

$$|\varphi^t(\mathbf{x}) - \mathbf{p}|' \leq Ce^{-ta}|\mathbf{x} - \mathbf{p}|' \quad \textit{for all } t \geq 0.$$

REMARK 5.1. A norm as in part (a) of the theorem is called an *adapted norm.*

REMARK 5.2. This theorem can be proved by using either approach to the proof of Theorem IV.5.1. We present the proof using the averaged norm of the first proof. We leave to the exercises the proof using the Jordan canonical form. See Exercise 5.20.

PROOF. We first do a change of coordinates $\mathbf{y} = \mathbf{x} - \mathbf{p}$ to move the fixed point to the origin. Then,

$$\begin{aligned}\dot{\mathbf{y}} &= \dot{\mathbf{x}} = f(\mathbf{y} + \mathbf{p}) \\ &= A\mathbf{y} + g(\mathbf{y})\end{aligned}$$

where $g(\mathbf{y})$ contains all the terms which are quadratic and higher. Therefore, $g(\mathbf{0}) = \mathbf{0}$ and $Dg_{\mathbf{0}} = \mathbf{0}$. We let $\psi^t(\mathbf{y})$ be the flow in the $\mathbf{y}$ coordinates and $\varphi^t(\mathbf{x})$ be the flow in the $\mathbf{x}$ coordinates. Since $\psi^t(\mathbf{y}_0)$ is the solution of $\dot{\mathbf{y}} = A\mathbf{y} + g(\mathbf{y})$ with $\psi^0(\mathbf{y}_0) = \mathbf{y}_0$, then thinking of $g(\psi^t(\mathbf{y}_0))$ as a known function of t, it is also a solution of the nonhomogeneous linear equation

$$\dot{\mathbf{y}} = A\mathbf{y} + g(\psi^t(\mathbf{y}_0))$$

with $\mathbf{y} = \mathbf{y}_0$ at $t = 0$. Therefore, we can apply the variation of parameters formula to the solution to get

$$\psi^t(\mathbf{y}_0) = e^{At}\mathbf{y}_0 + \int_0^t e^{A(t-s)}g(\psi^s(\mathbf{y}_0))\,ds.$$

We now want to use the estimates for the linear flow to get estimates for the nonlinear flow. We have to introduce a little error because of the nonlinear terms, so we take $\epsilon > 0$ and $b = a + \epsilon$ such that all the eigenvalues λ for A satisfy $Re(\lambda) < -b$. Now, let $|\ |_*$ be any adapted norm for the linear equations which is shown to exist by Theorem IV.5.1, i.e., the norm satisfies $|e^{At}\mathbf{y}_0|_* \leq e^{-ta}|\mathbf{y}_0|_*$ for $t \geq 0$. The nonlinear term g starts with quadratic terms, so there is a $\delta > 0$ such that for $|\mathbf{y}|_* \leq \delta$ we have $|g(\mathbf{y})|_* < \epsilon|\mathbf{y}|_*$.

We let U' be the neighborhood of $\mathbf{y} = \mathbf{0}$ given by $U' = \{\mathbf{y} : |\mathbf{y}|_* \leq \delta\}$. Applying this estimate to the integral above, if $|\psi^s(\mathbf{y}_0)|_* \leq \delta$ for $0 \leq s \leq t$, then

$$\begin{aligned} |\psi^t(\mathbf{y}_0)|_* &\leq |e^{At}\mathbf{y}_0|_* + \int_0^t |e^{A(t-s)}g(\psi^s(\mathbf{y}_0))|_*\, ds \\ &\leq e^{-bt}|\mathbf{y}_0|_* + \int_0^t e^{-b(t-s)}|g(\psi^s(\mathbf{y}_0))|_*\, ds \\ &\leq e^{-bt}|\mathbf{y}_0|_* + \int_0^t e^{bs}e^{-bt}\epsilon|\psi^s(\mathbf{y}_0)|_*\, ds \\ e^{bt}|\psi^t(\mathbf{y}_0)|_* &\leq |\mathbf{y}_0|_* + \int_0^t \epsilon e^{bs}|\psi^s(\mathbf{y}_0)|_*\, ds. \end{aligned}$$

We can now apply Gronwall's Inequality to the function $e^{bt}|\psi^t(\mathbf{y}_0)|_*$ and get

$$\begin{aligned} e^{bt}|\psi^t(\mathbf{y}_0)|_* &\leq |\mathbf{y}_0|_* e^{\epsilon t}, \\ |\psi^t(\mathbf{y}_0)|_* &\leq |\mathbf{y}_0|_* e^{(-b+\epsilon)t} = |\mathbf{y}_0|_* e^{-at}. \end{aligned}$$

It follows that if $\mathbf{y}_0 \in U'$ with $|\mathbf{y}_0|_* \leq \delta$, then the solution will remain in U' for all $t \geq 0$: $|\psi^t(\mathbf{y}_0)|_* \leq |\mathbf{y}_0|_* e^{(-b+\epsilon)t} \leq \delta$. Thus, the solution is defined for all $t \geq 0$, and $|\psi^t(\mathbf{y}_0)|_*$ goes to zero exponentially. In the $\mathbf{x}$ coordinates, let $U = \{\mathbf{x} : |\mathbf{x} - \mathbf{p}|_* \leq \delta\}$. For $\mathbf{x}_0 \in U$, let $\mathbf{y}_0 = \mathbf{x}_0 - \mathbf{p}$ and $\varphi^t(\mathbf{x}_0) = \psi^t(\mathbf{y}_0) + \mathbf{p}$. Thus, the solution $\varphi^t(\mathbf{x}_0)$ is defined for all $t \geq 0$ and $|\varphi^t(\mathbf{x}_0) - \mathbf{p}|_* \leq e^{-at}|\mathbf{x}_0 - \mathbf{p}|_*$, i.e., the solution goes to the fixed point at an exponential rate as claimed.

For any norm $|\ |'$, the result for part (b) follows just as in the linear case by the equivalence of norms. □

The next theorem is a special case of the Hartman-Grobman Theorem which is stated at the beginning of the next subsection. Here we state and prove it for the special case of a fixed point sink. Instead of the linear flow, we consider the affine flow $\dot{\mathbf{x}} = A(\mathbf{x} - \mathbf{p})$ which has solutions $\mathbf{p} + e^{At}(\mathbf{x} - \mathbf{p})$.

Theorem 5.2. *Let* $\mathbf{p}$ *be a fixed point sink for* $\dot{\mathbf{x}} = f(\mathbf{x})$. *Then, the flow* φ^t *of* f *is conjugate in a neighborhood of* p *to the affine flow* $\mathbf{p} + e^{At}(\mathbf{y} - \mathbf{p})$ *where* $A = Df_{\mathbf{p}}$. *More precisely, there are a neighborhood* U *of* $\mathbf{p}$ *and a homeomorphism* $h : U \to U$ *such that* $\varphi^t(h(\mathbf{x})) = h(\mathbf{p} + e^{At}(\mathbf{y} - \mathbf{p}))$ *as long as* $\mathbf{p} + e^{At}(\mathbf{y} - \mathbf{p}) \in U$.

PROOF. The proof is very similar to the proof of Theorem IV.7.1 for linear flows. One of the differences is that because the nonlinear flow and the affine flow are close, it is possible to use the identity map on the small sphere rather than some h_0. Take $0 < a < b$ with $Re(\lambda) < -b$ for all the eigenvalues λ of A. Set $\epsilon = b - a$. Take $\delta > 0$ as in Theorem 5.1 above such that for $|\mathbf{x} - \mathbf{p}|_* \leq \delta$, $|e^{At}(\mathbf{x} - \mathbf{p})|_* \leq e^{-at}|\mathbf{x} - \mathbf{p}|_*$ and $|\varphi^t(\mathbf{x}) - \mathbf{p}|_* \leq e^{-at}|\mathbf{x} - \mathbf{p}|_*$ for all $t \geq 0$.

Let $U = \{\mathbf{x} : |\mathbf{x} - \mathbf{p}|_* \leq \delta\}$. For $\mathbf{x} \in U$ take $\tau(\mathbf{x})$ such that $|e^{A\tau(\mathbf{x})}(\mathbf{x} - \mathbf{p})|_* = \delta$. Define $h : U \to U$ by

$$h(\mathbf{x}) = \begin{cases} \mathbf{p} & \text{for } \mathbf{x} = \mathbf{p} \\ \varphi^{-\tau(\mathbf{x})}(\mathbf{p} + e^{A\tau(\mathbf{x})}(\mathbf{x} - \mathbf{p})) & \text{for } \mathbf{x} \neq \mathbf{p}. \end{cases}$$

The proof that h is a conjugacy is now the same as in Theorem IV.7.1 as long as times are restricted so that $\tau(\mathbf{x}) \leq t < \infty$. □

Example 5.3. Vinograd (1957) gave an example of a nonlinear differential equation with the origin an isolated fixed point which is not Liapunov stable but which has a neighborhood U for every $\mathbf{q} \in U$ has $\omega(\mathbf{q}) = \mathbf{0}$. The equations are

$$\dot{x} = \frac{x^2(y-x)+y^5}{(x^2+y^2)[1+(x^2+y^2)^2]}$$
$$\dot{y} = \frac{y^2(y-2x)}{(x^2+y^2)[1+(x^2+y^2)^2]}.$$

The phase portrait is given in Figure 5.1. See Hahn (1967) page 191 for an analysis of this example.

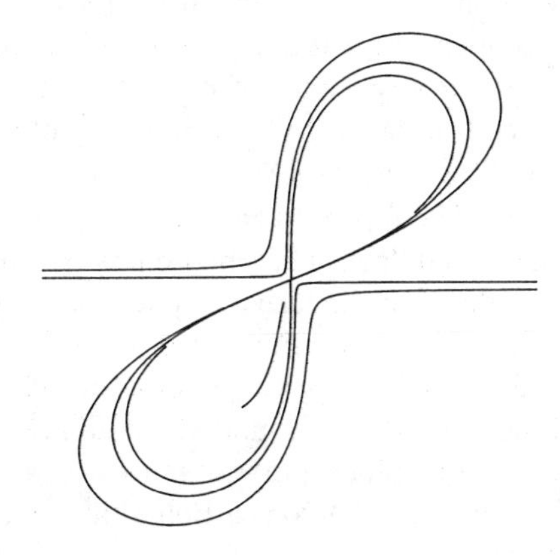

FIGURE 5.1. Phase Portrait for Example 5.3

5.5.2 Nonlinear Hyperbolic Fixed Points

In the last subsection we considered nonlinear sinks. In this subsection we consider hyperbolic fixed points of nonlinear differential equations and state the results linking their stability with the real parts of the eigenvalues of the derivative of the vector field at the fixed point. We give this stability result as a corollary of the Hartman-Grobman Theorem. We defer the proof of the Hartman-Grobman Theorem to Section 5.7.

Theorem 5.3 (Hartman-Grobman). *Let* $\mathbf{p}$ *be a hyperbolic fixed point for* $\dot{\mathbf{x}} = f(\mathbf{x})$. *Then, the flow* φ^t *of* f *is conjugate in a neighborhood of* p *to the affine flow* $\mathbf{p}+e^{At}(\mathbf{y}-\mathbf{p})$ *where* $A = Df_p$. *More precisely, there are a neighborhood* U *of* p *and a homeomorphism* $h : U \to U$ *such that* $\varphi^t(h(\mathbf{x})) = h(\mathbf{p} + e^{At}(\mathbf{y}-\mathbf{p}))$ *as long as* $\mathbf{p} + e^{At}(\mathbf{y}-\mathbf{p}) \in U$.

We delay the proof of this theorem to Section 5.7.2 because it involves much more analysis than the special case of a sink, given in Theorem 5.2. As stated above, the stability of hyperbolic fixed points follows from the Hartman-Grobman Theorem.

Corollary 5.4. *Let* $\mathbf{p}$ *be a hyperbolic fixed point for* $\dot{\mathbf{x}} = f(\mathbf{x})$. *If* $\mathbf{p}$ *is a source or a saddle, then the fixed point* $\mathbf{p}$ *is not Liapunov stable (unstable). If* $\mathbf{p}$ *is a sink, then it is asymptotically stable (attracting).*

The fact that a source is not Liapunov stable follows from a result like Theorem 5.1. It can be proved directly that a saddle is not Liapunov stable as is given in Hirsch and Smale (1974), page 187.

Combining the conjugacy of linear hyperbolic flows given in Theorem IV.7.1 with the Hartman-Grobman Theorem, we can state the following corollary.

Theorem 5.5. *Assume $\dot{\mathbf{x}} = f(\mathbf{x})$ has a hyperbolic fixed point at $\mathbf{p}$, and $\dot{\mathbf{x}} = g(\mathbf{x})$ has a hyperbolic fixed point at $\mathbf{q}$ (where both equations are in $\mathbb{R}^n$). Assume that the number of negative eigenvalues at the two fixed points are equal. Then, there are neighborhoods U of $\mathbf{p}$ and V of $\mathbf{q}$ and a homeomorphism $h : U \to V$ such that h is a conjugacy of the flow of f on U to the flow of g on V.*

REMARK 5.3. Since any two contracting linear flows are topologically conjugate, it is clear that a topological conjugacy does not preserve much of the geometric nature of the phase portrait. By contrast, a differentiable conjugacy does preserve much of this structure.

It is possible to prove that a nonlinear flow is differentiably conjugate to the linear flow near a hyperbolic fixed point if the eigenvalues satisfy a nonresonance condition. Assume that f is C^∞ and the eigenvalues at a hyperbolic fixed point are λ_j for $1 \le j \le n$. Assume that $\lambda_k \neq \sum_{j=1}^n m_j \lambda_j$ for any choice of $m_j \geq 0$ with $\sum_{j=1}^n m_j \geq 2$. Then, the conjugacy h in Theorem 5.3 can be taken to be a diffeomorphism from V onto U. This result was initially proven by Sternberg (1958).

In two dimensions, Hartman (1960) proved that near a hyperbolic fixed point, any C^2 flow is C^1 conjugate to its linearized flow. This theorem is true even when the eigenvalue has multiplicity two. In particular if the linearized equations are diagonal with a negative real eigenvalue, then the solutions for the linearized equations go straight toward the fixed point. The nonlinear flow could have trajectories which spiral as they approach the origin but the increase in the angle is bounded (as follows from the C^1 conjugacy or a direct estimate). Thus, the two phase portraits are similar up to changes which C^1 nonlinear change of coordinates allows. Also see Belitskii (1973) for C^1 conjugacies for the general hyperbolic case.

See the discussion in Hartman (1964) on differentiable conjugacies.

Finally, we remark again that the assumption of a differentiable conjugacy between two flows in neighborhoods of their fixed points implies that they have the same eigenvalues. This follows directly from Theorem IV.7.2 about linear flows.

Theorem 5.6. *Assume $\dot{\mathbf{x}} = f(\mathbf{x})$ has a hyperbolic fixed point at $\mathbf{p}$, and $\dot{\mathbf{x}} = g(\mathbf{x})$ has a hyperbolic fixed point at $\mathbf{q}$ (where both equations are in $\mathbb{R}^n$). Assume that the flows are differentiably conjugate in a neighborhood of these two fixed points. Then, the eigenvalues of the linearization of f at $\mathbf{p}$ equal the eigenvalues of the linearization of g at $\mathbf{q}$.*

5.5.3 Liapunov Functions Near a Fixed Point

In this section we introduce the use of a real valued function which is decreasing along trajectories. For a system of equations for an oscillator, the function is often an energy function. We start by giving a specific example as motivation.

Example 5.4. Consider the equation of a pendulum with friction,

$$\begin{aligned} \dot{x} &= y \\ \dot{y} &= -\sin(x) - \delta y \end{aligned}$$

with $\delta > 0$. The fixed point $(0, 0)$ is attracting (asymptotically stable or a fixed point sink) as we saw in Example 5.1 using the eigenvalues and Theorem 5.1. In this section we will find a second way to see that is true using the "energy function" $L(x, y) = \frac{1}{2}y^2 - \cos(x)$. (The first term is the "kinetic energy" and the second is the "potential energy.") We are interested in the time derivative of the real valued function L along a solution curve $(x(t), y(t))$:

$$\begin{aligned}
\dot{L}(x(t), y(t)) &\equiv \frac{d}{dt}L(x(t), y(t)) \\
&= y(t)\dot{y}(t) + \sin(x(t))\dot{x}(t) \\
&= y(t)[-\sin(x(t)) - \delta y(t)] + \sin(x(t))y(t) \\
&= -\delta y(t)^2 \\
&\leq 0.
\end{aligned}$$

Since this function is nonincreasing, an easy argument proves that the origin is Liapunov stable. (See the proof of Theorem 5.7 below.) Since the function is decreasing most of the time, we can prove that the origin is asymptotically stable. (See Theorem 5.8 below.) These theorems also give an idea of the basin of attraction of this fixed point sink, namely all points with $L(x, y) < L(\pi, 0) = 1$ and $-\pi < x < \pi$.

Functions L as in the above example are called Liapunov functions. We give their important characteristic in the following definition.

Definition. Let $\dot{\mathbf{x}} = f(\mathbf{x})$ be a differential equation on $\mathbb{R}^n$ with flow $\varphi^t(\mathbf{x})$ and a fixed point $\mathbf{p}$. A real valued C^1 function L is called a *weak Liapunov function* for the flow φ^t on an open neighborhood U of $\mathbf{p}$ provided $L(\mathbf{x}) > L(\mathbf{p})$ and $\dot{L}(\mathbf{x}) \equiv \frac{d}{dt}L(\varphi^t(\mathbf{x}))|_{t=0} \leq 0$ for all $\mathbf{x} \in U \setminus \{\mathbf{p}\}$. These conditions imply that $L \circ \varphi^t(\mathbf{x}) \leq L(\mathbf{x})$ for all $\mathbf{x} \in U$ and for all $t \geq 0$ such that $\varphi^s(\mathbf{x}) \in U$ for all $0 \leq s \leq t$. The function L is called a *Liapunov function* on an open neighborhood U of $\mathbf{p}$ (or *strong*, *strict*, or *complete Liapunov function*) provided it is a weak Liapunov function which also satisfies $\dot{L}(\mathbf{x}) < 0$ for all $\mathbf{x} \in U \setminus \{\mathbf{p}\}$.

One feature of this method is that it is possible to calculate $\dot{L}$ without actually knowing the solutions. This makes it possible to verify that a given function is in fact a Liapunov function for a given differential equation in a certain neighborhood of a fixed point. A general reference for this section is Hirsch and Smale (1974), pages 192–199.

Theorem 5.7. *Let $\mathbf{p}$ be a fixed point for $\dot{\mathbf{x}} = F(\mathbf{x})$. Let U be a neighborhood of $\mathbf{p}$ and $L : U \to \mathbb{R}$ a weak Liapunov function on the neighborhood U of $\mathbf{p}$ for the differential equation. Then, $\mathbf{p}$ is Liapunov stable.*

(b) If L is a (strict) Liapunov function on the neighborhood U of $\mathbf{p}$ for the differential equation, then $\mathbf{p}$ is asymptotically stable.

PROOF. To show that $\mathbf{p}$ is Liapunov stable, given any $\epsilon > 0$ we need to find a neighborhood $U_1 \subset B(\mathbf{p}, \epsilon)$ such that a solution starting in U_1 stays in $B(\mathbf{p}, \epsilon)$. First of all, we can assume that ϵ is small enough so that $B(\mathbf{p}, \epsilon) \subset U$. Choose α such that

$$L(\mathbf{p}) < \alpha < \min\{L(\mathbf{x}) : \mathbf{x} \in \partial B(\mathbf{p}, \epsilon)\},$$

and let

$$U_1 = \{\mathbf{x} \in B(\mathbf{p}, \epsilon) : L(\mathbf{x}) < \alpha\}.$$

If $\mathbf{x} \in U_1$, then $L \circ \varphi^t(\mathbf{x}) \leq L(\mathbf{x}) < \alpha$, so the solution cannot cross the boundary of $B(\mathbf{p}, \epsilon)$. Therefore, $\varphi^t(\mathbf{x}) \in B(\mathbf{p}, \epsilon)$ for all $t \geq 0$. This proves that $\mathbf{p}$ is Liapunov stable.

For the second part, take U_1 as above. For $\mathbf{x} \in U_1$, we want to show that $\varphi^t(\mathbf{x})$ goes to $\mathbf{p}$ as t goes to infinity. Let $\mathbf{z}$ be an ω-limit point of $\mathbf{x}$, i.e., there is a subsequence of times $t_n \to \infty$ such that $\varphi^{t_n}(\mathbf{x}) \to \mathbf{z}$. Since the closed ball $\text{cl}(B(\mathbf{p}, \epsilon))$ is compact,

$\mathbf{z} \in \mathrm{cl}(B(\mathbf{p}, \epsilon))$. We claim that $\mathbf{z} = \mathbf{p}$. Assume there is an ω-limit point $\mathbf{z} \neq \mathbf{p}$. By continuity, $L(\varphi^{t_n}(\mathbf{x}))$ converges to $L(\mathbf{z})$. The fact that L is decreasing along the solution implies that $L(\varphi^{t_n}(\mathbf{x}))$ converges to $L(\mathbf{z})$ from above. Since $\mathbf{z} \neq \mathbf{p}$, $L(\varphi^s(\mathbf{z})) < L(\mathbf{z})$ for (small) positive s. By continuity, for $\mathbf{y}$ sufficiently near $\mathbf{z}$, $L(\varphi^s(\mathbf{y})) < L(\mathbf{z})$. Letting $\mathbf{y} = \varphi^{t_n}(\mathbf{x})$, for some large n, we get that $\varphi^{t_n+s}(\mathbf{x}) < L(\mathbf{z})$. Then, for $t_m > t_n + s$, $\varphi^{t_m}(\mathbf{x}) < \varphi^{t_n+s}(\mathbf{x}) < L(\mathbf{z})$. This contradicts the fact that $L(\varphi^{t_n}(\mathbf{x}))$ converges to $L(\mathbf{z})$ from above. Thus, $\mathbf{p}$ is the only ω-limit point of $\mathbf{x}$, so $\mathbf{p}$ is asymptotically stable. □

When applying Theorem 5.7 to the pendulum example above, we see that $(0, 0)$ is Liapunov stable for $\delta \geq 0$. However, $\dot{L}$ is not strictly negative on any deleted neighborhood, so it does not imply that the point is asymptotically stable for $\delta > 0$. To get this result, we need a refinement of the above theorem given next.

Theorem 5.8. *Let $\mathbf{p}$ be a fixed point for $\dot{\mathbf{x}} = F(\mathbf{x})$. Let U be a neighborhood of $\mathbf{p}$ and $L : U \to \mathbb{R}$ a weak Liapunov function on U. Suppose $S \subset U$ is a closed, bounded, positively invariant neighborhood of $\mathbf{p}$. Let*

$$\mathbf{Z} = \{\mathbf{x} \in S : \dot{L}(\mathbf{x}) = 0\}.$$

Further suppose that $\{\mathbf{p}\}$ is the largest positively invariant subset of $\mathbf{Z}$. Then, $\mathbf{p}$ is asymptotically stable, and S is contained in the basin of attraction of $\mathbf{p}$.

PROOF. Take any $\mathbf{x} \in S$. The ω-limit set $\omega(\mathbf{x})$ is nonempty because S is closed, bounded, and positively invariant. Let $\mathbf{z}$ be an ω-limit point. By the argument in Theorem 5.7, $L(\varphi^s(\mathbf{z})) = L(\mathbf{z})$ for all $s > 0$, so $\varphi^s(\mathbf{z}) \in \mathbf{Z}$ for all $s > 0$. Thus, $\mathbf{z}$ must be in a positively invariant subset of $\mathbf{Z}$ and $\mathbf{z} = \mathbf{p}$. □

5.6 Stability of Periodic Points for Nonlinear Maps

In this section we consider a map (usually a diffeomorphism) $f : \mathbb{R}^n \to \mathbb{R}^n$ or $f : M \to M$ where M is a surface or manifold such as $M = \mathbb{T}^n$ or S^n. A linear map is given by $f(\mathbf{x}) = A\mathbf{x}$ where A is an $n \times n$ matrix.

The reader should compare the results of this section with the results for maps on $\mathbb{R}$ given in Chapter II. The conditions on the derivative at the fixed point for one-dimensional maps are replaced by conditions on the eigenvalues of the derivative.

Definition. A point $\mathbf{p}$ is a *fixed point* of f if $f(\mathbf{p}) = \mathbf{p}$. It is a *periodic point* of (least) period k if $f^k(\mathbf{p}) = \mathbf{p}$ but $f^j(\mathbf{p}) \neq \mathbf{p}$ for $0 < j < k$. A periodic point $\mathbf{p}$ is called *Liapunov stable* provided given any $\epsilon > 0$, there exists a $\delta > 0$ such that $|f^j(\mathbf{x}) - f^j(\mathbf{p})| < \epsilon$ for all $|\mathbf{x} - \mathbf{p}| < \delta$ and $j \geq 0$. A periodic point $\mathbf{p}$ is called it asymptotically stable or *attracting* provided it is Liapunov stable and there is a $\delta > 0$ such that if $|\mathbf{x} - \mathbf{p}| < \delta$, then $|f^j(\mathbf{x}) - f^j(\mathbf{p})|$ goes to zero as j goes to infinity.

The *basin of attraction* of an attracting periodic orbit $\mathcal{O}(\mathbf{p})$ is the set of all points $\mathbf{q}$ such that $\omega(\mathbf{p}) = \mathcal{O}(\mathbf{p})$. This basin is denoted by $W^s(\mathcal{O}(\mathbf{p}))$.

The following theorem gives the existence of an adapted metric for maps at an attracting fixed point which is similar to Theorem IV.5.1 for differential equations.

Theorem 6.1. *Let $\mathbf{p}$ be a periodic point for $f : \mathbb{R}^n \to \mathbb{R}^n$ with period k. Assume there is a constant $0 < \mu < 1$ such that all the eigenvalues λ of $Df^k_{\mathbf{p}}$ have $|\lambda| < \mu$.*

(a) Then, there is a norm $|\ |_*$ on $\mathbb{R}^n$ and a neighborhood $U \subset \mathbb{R}^n$ of $\mathbf{p}$ such that for any initial condition $\mathbf{x} \in U$, the iterates satisfy

$$|f^{jk}(\mathbf{x}) - f^{jk}(\mathbf{p})|_* \leq \mu^j |\mathbf{x} - \mathbf{p}|_* \quad \textit{for all } j \geq 0.$$

(b) For any norm $|\ |'$ on $\mathbb{R}^n$, there exist a neighborhood $U \subset \mathbb{R}^n$ of $\mathbf{p}$ and a constant $C \geq 1$ such that for any initial condition $\mathbf{x} \in U$, the iterates satisfy

$$|f^{jk}(\mathbf{x}) - f^{jk}(\mathbf{p})|' \leq C\mu^j |\mathbf{x} - \mathbf{p}|' \quad \text{for all } j \geq 0.$$

REMARK 6.1. As for differential equations, a norm as in part (a) of the theorem is called an *adapted norm.*

REMARK 6.2. The proof is similar to that for flows with integrals replaced by summations and is omitted.

To look at the case that is not a sink, for a linear map with matrix A, we let

$$\begin{aligned}
\mathbb{E}^u &= \text{span}\{\mathbf{v}^u : \mathbf{v}^u \text{ is a generalized eigenvector} \\
&\qquad\qquad \text{for an eigenvalue } \lambda_u \text{ of } A \text{ with } |\lambda_u| > 1\}, \\
\mathbb{E}^s &= \text{span}\{\mathbf{v}^s : \mathbf{v}^s \text{ is a generalized eigenvector} \\
&\qquad\qquad \text{for an eigenvalue } \lambda_s \text{ of } A \text{ with } |\lambda_s| < 1\}, \\
\mathbb{E}^c &= \text{span}\{\mathbf{v}^c : \mathbf{v}^c \text{ is a generalized eigenvector} \\
&\qquad\qquad \text{for an eigenvalue } \lambda_c \text{ of } A \text{ with } |\lambda_c| = 1\}.
\end{aligned}$$

Definition. For a periodic point $\mathbf{p}$ of period k, we consider these spaces with $A = Df^k_{\mathbf{p}}$. The periodic point $\mathbf{p}$ of period k is called *hyperbolic* provided $\mathbb{E}^c = \{\mathbf{0}\}$, i.e., all the eigenvalues λ of $Df^k_{\mathbf{p}}$ satisfy $|\lambda| \neq 1$. A hyperbolic periodic point $\mathbf{p}$ of period k is called a *sink* provided all the eigenvalues of $Df^k_{\mathbf{p}}$ are less than 1 in absolute value, $|\lambda| < 1$, i.e., both $\mathbb{E}^u$, $\mathbb{E}^c = \{\mathbf{0}\}$. Theorem 6.1 above proves that a periodic sink is asymptotically stable (or attracting). In the same way, a hyperbolic periodic point is called a *source* provided all the eigenvalues are greater than one in absolute value, $|\lambda| > 1$, i.e., both $\mathbb{E}^s, \mathbb{E}^c = \{\mathbf{0}\}$. Applying Theorem 6.1 to the inverse, we get that a periodic source is *repelling.* Finally, a hyperbolic periodic point with $\mathbb{E}^u \neq \{\mathbf{0}\}$ and $\mathbb{E}^s \neq \{\mathbf{0}\}$ is a *saddle.*

In the same way that Theorem 5.2 followed from Theorem 5.1, we could prove that a nonlinear sink is topologically conjugate in a neighborhood of the fixed point to the linear map induced by the derivative at the fixed point. The following theorem states the more general Hartman-Grobman Theorem for diffeomorphisms.

Theorem 6.2 (Hartman-Grobman Theorem). *Let $f : \mathbb{R}^n \to \mathbb{R}^n$ be a C^r diffeomorphism with a hyperbolic fixed point $\mathbf{p}$. Then, there exist neighborhoods U of $\mathbf{p}$ and V of $\mathbf{0}$ and a homeomorphism $h : V \to U$ such that $f(h(\mathbf{x})) = h(A\mathbf{x})$ for all $\mathbf{x} \in V$, where $A = Df_{\mathbf{p}}$.*

We delay the proof of this theorem to Section 5.7.1.

REMARK 6.3. Just as for differential equations, it is possible to prove that a nonlinear flow is differentiably conjugate to the linear flow near a hyperbolic fixed point if the eigenvalues satisfy a nonresonance condition. Assume that f is C^∞ and the eigenvalues at a hyperbolic fixed point are λ_j for $1 \leq j \leq n$. Assume that $\lambda_k \neq \prod_{j=1}^n \lambda_j^{m_j}$ for any choice of $m_j \geq 0$ with $\sum_{j=1}^n m_j \geq 2$. Then, it is possible to prove the existence of a conjugacy h as in Theorem 6.2 that is a diffeomorphism from V onto U. This result was initially proven by Sternberg (1958).

In two dimensions, Hartman (1960) proved that near a hyperbolic fixed point, any C^2 diffeomorphism is C^1 conjugate to its linearized map. Thus, the two phase portraits are similar up to changes which C^1 nonlinear change of coordinates allows. Also see Belitskii (1973) for C^1 conjugacies in the general hyperbolic case.

See the discussion in Hartman (1964) on differentiable conjugacies.

As in the case of a flow, the stability of a fixed point follows from the Hartman–Grobman Theorem.

Corollary 6.3. *Let $f : \mathbb{R}^n \to \mathbb{R}^n$ be a C^r diffeomorphism with a hyperbolic fixed point $\mathbf{p}$. If $\mathbf{p}$ is a source or a saddle, then the fixed point $\mathbf{p}$ is not Liapunov stable. If $\mathbf{p}$ is a sink, then it is asymptotically stable.*

If a fixed point is hyperbolic, then the linear part determines the stability type of the fixed point. Small changes in the linear part preserve the same stability type. To end this section we give a result that says that a hyperbolic fixed point persists for small changes in the map. More specifically, we consider a one-parameter family of maps f_μ. We could assume that $\mathbf{x}_0$ is a hyperbolic fixed point for f_{μ_0}. However, it is enough to assume that 1 is not an eigenvalue of $D(f_{\mu_0})_{\mathbf{x}_0}$ in order to show the fixed point persists, as the following theorem shows.

Theorem 6.4. *Let $f_\mu(\mathbf{x})$ be a one-parameter family of differentiable maps with $\mathbf{x} \in \mathbb{R}^n$. Assume that $f_\mu(\mathbf{x})$ is C^1 as a function jointly of μ and $\mathbf{x}$. Assume that $f_{\mu_0}(\mathbf{x}_0) = \mathbf{x}_0$ and 1 is not an eigenvalue of $D(f_{\mu_0})_{\mathbf{x}_0}$. Then, there are (i) an open set U about $\mathbf{x}_0$, (ii) an interval N about μ_0, and (iii) a C^1 function $\mathbf{p} : N \to U$ such that $\mathbf{p}(\mu_0) = \mathbf{x}_0$ and $f_\mu(\mathbf{p}(\mu)) = \mathbf{p}(\mu)$. Moreover, for $\mu \in N$, f_μ has no other fixed points in U other than $\mathbf{p}(\mu)$. Finally,*

$$\mathbf{p}'(\mu) = [D(f_\mu)_{\mathbf{p}(\mu)} - I]^{-1} \frac{\partial f}{\partial \mu}\Big|_{(\mathbf{p}(\mu),\mu)}.$$

PROOF. We want to find points $\mathbf{x}$ such that $f_\mu(\mathbf{x}) = \mathbf{x}$. We define the function $G(\mathbf{x}, \mu) = f_\mu(\mathbf{x}) - \mathbf{x}$ and the condition becomes finding zeros of $G(\cdot, \mu)$. This is set up in the form where the Implicit Function Theorem might apply. Note that $G(\mathbf{x}_0, \mu_0) = \mathbf{0}$ and $\frac{\partial G}{\partial x}(\mathbf{x}_0, \mu_0) = D(f_{\mu_0})_{\mathbf{x}_0} - I$ is invertible since 1 is not an eigenvalue of $D(f_{\mu_0})_{\mathbf{x}_0}$. Therefore, the Implicit Function Theorem does indeed apply to give a C^1 function $\mathbf{p}(\mu)$ such that $G(\mathbf{p}(\mu), \mu) = 0$ for μ in some open interval N about μ_0 and these are the only zeroes in some set $N \times U$ where U is an open neighborhood of $\mathbf{x}_0$. The calculation of the derivative follows by implicit differentiation. □

5.7 Proof of the Hartman-Grobman Theorem

To prove the Hartman-Grobman Theorem for a diffeomorphism in a neighborhood of a fixed point, we first prove a case where the nonlinear map is defined on all of a Banach (or Euclidean) space, and it is a bounded distance away from the linear map on the whole space. Next we apply the global theorem to prove the local Hartman-Grobman Theorem for a diffeomorphism, Theorem 6.2. Finally, we prove the local Hartman-Grobman Theorem for a flow, Theorem 5.3.

To carry out these arguments, we need to work with a few function spaces and with bounded linear maps. For two Banach spaces $\mathbb{E}_1$ and $\mathbb{E}_2$, if $A : \mathbb{E}_1 \to \mathbb{E}_2$ is a linear map we define the *operator norm* or *sup-norm* of A just as in finite dimensions by

$$||A|| = \sup_{\mathbf{v} \neq \mathbf{0}} \frac{|A\mathbf{v}|_2}{|\mathbf{v}|_1}.$$

The linear map A is a bounded linear map provided the norm $||A||$ is finite. (Note that the function A does not take on "bounded" values in $\mathbb{E}_2$.) We let $\mathbf{L}(\mathbb{E}_1, \mathbb{E}_2)$ be the set of *bounded linear maps* from $\mathbb{E}_1$ to $\mathbb{E}_2$ with the norm $\|\cdot\|$. We also use the *minimum norm* which is also defined in the same way as we defined it in Section 4.1 for finite dimensions:

$$m(A) = \inf_{\mathbf{v}\neq\mathbf{0}} \frac{|A\mathbf{v}|_2}{|\mathbf{v}|_1}.$$

If A is invertible, then $m(A) = \|A^{-1}\|^{-1}$.

A function $f : \mathbb{R}^n \to \mathbb{R}^n$ is said to be *bounded* provided there is a uniform $C > 0$ such that $|f(\mathbf{x})| \leq C$ for all $\mathbf{x} \in \mathbb{R}^n$. (Notice the difference from the use of the term for a bounded linear map.) We let $C_b^0(\mathbb{R}^n) = C_b^0(\mathbb{R}^n, \mathbb{R}^n)$ be the space of all bounded continuous maps from $\mathbb{R}^n$ to itself. We put the C^0-sup topology on $C_b^0(\mathbb{R}^n)$,

$$\|v_1 - v_2\|_0 = \sup_{\mathbf{x}\in\mathbb{R}^n} |v_1(\mathbf{x}) - v_2(\mathbf{x})|.$$

With this norm, $C_b^0(\mathbb{R}^n)$ is a complete metric space. See Dieudonné (1960).

For a differentiable map $g : \mathbb{R}^n \to \mathbb{R}^k$, at each point $\mathbf{a} \in \mathbb{R}^n$ the derivative is (a matrix or) a bounded linear map, $Dg_{\mathbf{a}} : \mathbb{R}^n \to \mathbb{R}^k$. We let $C_b^1(\mathbb{R}^n)$ be the set of C^1 functions from $\mathbb{R}^n$ to itself, $g : \mathbb{R}^n \to \mathbb{R}^n$, such that g is in $C_b^0(\mathbb{R}^n)$ and such that there is a uniform bound on the derivatives, i.e., a constant C independent of $\mathbf{a} \in \mathbb{R}^n$ such that $\|Dg_{\mathbf{a}}\| \leq C$.

We want to consider C^1 functions $f : \mathbb{R}^n \to \mathbb{R}^n$ such that $f = A + g$ with $A \in \mathbf{L}(\mathbb{R}^n, \mathbb{R}^n)$ an invertible hyperbolic linear map and $g \in C_b^1(\mathbb{R}^n)$. The global Hartman-Grobman Theorem says that such an f can be conjugated to A by a continuous map $h : \mathbb{R}^n \to \mathbb{R}^n$ with $h = id + v$ and $v \in C_b^0(\mathbb{R}^n)$.

Theorem 7.1. *Let $A \in \mathbf{L}(\mathbb{R}^n, \mathbb{R}^n)$ be an invertible hyperbolic linear map. There exists an $\epsilon > 0$ such that if $g \in C_b^1(\mathbb{R}^n)$ with $\mathrm{Lip}(g) < \epsilon$, then $f = A + g$ is topologically conjugate to A by a map $h = id + v$ with $v \in C_b^0(\mathbb{R}^n)$, and the conjugacy is unique among maps $id + k$ with $k \in C_b^0(\mathbb{R}^n)$. In fact, let $0 < a < 1$ be such that each eigenvalue λ of A has either $|\lambda| < a$ or $|\lambda^{-1}| < a$. If both $\epsilon(1-a)^{-1} < 1$ and $m(A) - \epsilon > 0$, then this ϵ works. (The condition on the eigenvalues can also be expressed by saying that spectrum$(A) \subset \{\lambda : |\lambda| < a$ or $|\lambda^{-1}| < a\}$. The condition that $m(A) - \epsilon > 0$ insures that f is one to one.)*

REMARK 7.1. Palis and de Melo (1982) have two proofs of this theorem. The first on pages 59–63 is similar to that given here. The second on pages 80–88 is a more geometrical proof (using the λ-Lemma). We use this latter type of reasoning later in this book. Irwin (1980), pages 92, 113–114, has a proof similar to the one given here.

MOTIVATION AND OUTLINE OF THE PROOF. Formally, the conjugacy can be proved to exist by the Implicit Function Theorem. For $g \in C_b^1(\mathbb{R}^n)$, we want to find a map $id + v$ which conjugates $A + g$ with A, i.e.,

$$\begin{aligned} (A+g) \circ (id+v) &= (id+v) \circ A, \\ (A+g) \circ (id+v) \circ A^{-1} &= id + v, \\ 0 &= id + v - (A+g) \circ (id+v) \circ A^{-1}, \qquad \text{or} \\ 0 &= v - A \circ v \circ A^{-1} - g \circ (id+v) \circ A^{-1}. \end{aligned}$$

Therefore, we define $\Psi : C_b^1(\mathbb{R}^n) \times C_b^0(\mathbb{R}^n) \to C_b^0(\mathbb{R}^n)$ by

$$\Psi(g, v) = v - A \circ v \circ A^{-1} - g \circ (id+v) \circ A^{-1}.$$

Given g, we want to find a v_g such that $\Psi(g, v_g) = 0$. Such a v_g corresponds to a semi-conjugacy of $A + g$ with A. Notice that $\Psi(0,0) = id - A \circ id \circ A^{-1} = 0$, so we can hope to use the Implicit Function Theorem near $(0,0) \in C_b^1(\mathbb{R}^n) \times C_b^0(\mathbb{R}^n)$ to solve for the v_g with $\Psi(g, v_g) = 0$.

It can be proved (after some work) that Ψ is a C^1 map, and the partial with respect to the second variable is

$$\begin{aligned}(\frac{\partial \Psi}{\partial v})_{(0,0)}\hat{v} &= \hat{v} - A \circ \hat{v} \circ A^{-1} \\ &\equiv (id - A_{\#})\hat{v} \\ &\equiv \mathcal{L}(\hat{v}),\end{aligned}$$

where $A_{\#}\hat{v} = A \circ \hat{v} \circ A^{-1}$. (See Franks (1979), Irwin (1972, 1980).) If we were to show that Ψ is a C^1 map with partial derivative $\mathcal{L}$ and that $\mathcal{L}$ is an isomorphism (a bounded linear map with a bounded linear inverse), then the Implicit Function Theorem would show that we can solve for $v = v_g$ as a function of g such that $\Psi(g, v_g) \equiv 0$. This would prove the theorem.

Instead of verifying that Ψ is C^1 with partial derivative $\mathcal{L}$, we verify that $\mathcal{L}$ is an isomorphism and imitate (or repeat) the proof of the Implicit Function Theorem. In the direct proof of the Implicit Function Theorem (as opposed to the proof using the Inverse Function Theorem), the problem of finding a zero of Ψ is changed into finding a fixed point by considering the function $\Theta : C_b^1(\mathbb{R}^n) \times C_b^0(\mathbb{R}^n) \to C_b^0(\mathbb{R}^n)$ given by

$$\Theta(g, v) = \mathcal{L}^{-1}\{\mathcal{L}v - \Psi(g, v)\}.$$

(The functions g are required to be bounded so that Θ takes its values in $C_b^0(\mathbb{R}^n)$.) After showing that $\mathcal{L}$ is an isomorphism using Lemma 7.2, we prove that if $\|\mathcal{L}\|\ \text{Lip}(g) < 1$, then $\Theta(g, \cdot)$ is a contraction on $C_b^0(\mathbb{R}^n)$ with a unique fixed point v_g. Letting $f = A + g$ and $h_f = id + v_g$, it follows that $h_f = f \circ h_f \circ A^{-1}$ or $h_f \circ A = f \circ h_f$. Thus, h_f is a semiconjugacy. Lemma 7.4 proves that h_f is one to one and Lemma 7.5 proves that h is onto, so h_f is a conjugacy from A to f. This ends the outline of the proof.

Before starting the proof, we give more notation and then prove a preliminary lemma. The matrix A is hyperbolic, so we can define the stable and unstable subspaces as usual:

$$\begin{aligned}\mathbb{E}^u &= \text{span}\{\mathbf{v}^u : \mathbf{v}^u \text{ is a generalized eigenvector} \\ &\qquad \text{for an eigenvalue } \lambda_u \text{ of } A \text{ with } |\lambda_u| > 1\}, \\ \mathbb{E}^s &= \text{span}\{\mathbf{v}^s : \mathbf{v}^s \text{ is a generalized eigenvector} \\ &\qquad \text{for an eigenvalue } \lambda_s \text{ of } A \text{ with } |\lambda_s| < 1\}.\end{aligned}$$

Then, $\mathbb{R}^n = \mathbb{E}^u \oplus \mathbb{E}^s$.

We use this decomposition to decompose the space of continuous bounded functions, $C_b^0(\mathbb{R}^n, \mathbb{R}^n) = C_b^0(\mathbb{R}^n, \mathbb{E}^u) \oplus C_b^0(\mathbb{R}^n, \mathbb{E}^s)$. We put norms on $\mathbb{E}^u$ and $\mathbb{E}^s$ such that the linear map A is a contraction and expansion on the two subspaces: $\|(A|\mathbb{E}^u)^{-1}\| \le a < 1$ and $\|A|\mathbb{E}^s\| \le a < 1$. On $\mathbb{R}^n$, we put the maximum norm of the norms on $\mathbb{E}^u$ and $\mathbb{E}^s$: if $\mathbf{v} = \mathbf{v}^u + \mathbf{v}^s$ with $\mathbf{v}^\sigma \in \mathbb{E}^\sigma$ for $\sigma = u, s$, then

$$|\mathbf{v}| \equiv \max\{|\mathbf{v}^u|, |\mathbf{v}^s|\}.$$

Let

$$A_{\#}v = A \circ v \circ A^{-1}$$

be the map on $C_b^0(\mathbb{R}^n, \mathbb{R}^n)$ as above and

$$\mathcal{L}(v) = (id - A_{\#})v = v - A \circ v \circ A^{-1}.$$

For $\sigma = u, s$, we also let $A_{\#}^{\sigma} = A_{\#}|C_b^0(\mathbb{R}^n, \mathbb{E}^{\sigma})$ and $\mathcal{L}^{\sigma} = \mathcal{L}|C_b^0(\mathbb{R}^n, \mathbb{E}^{\sigma})$. Because we use the norms induced by the maximum norm on $\mathbb{R}^n$, $\|\mathcal{L}\| = \max\{\|\mathcal{L}^u\|, \|\mathcal{L}^s\|\}$.

The first step is to give some results about linear maps on a Banach space. We use these results below, applied to $A_{\#}^{\sigma}$, to prove that $\mathcal{L}$ is an isomorphism.

Lemma 7.2. *Let $\mathbb{E}$ be a Banach space and $G, B \in \mathbf{L}(\mathbb{E}, \mathbb{E})$.*

(a) If $\|G\| \le a < 1$, then $id - G$ is an isomorphism and $\|(id - G)^{-1}\| \le \frac{1}{1-a}$. In fact, the inverse $(id - G)^{-1}$ can be represented by the series $\sum_{j=0}^{\infty} G^j$.

(b) If B is an isomorphism with $\|B^{-1}\| \le a < 1$, then $B - id$ is an isomorphism with $\|(B - id)^{-1}\| \le a/(1 - a)$. Again, this inverse, $(B - id)^{-1}$, can be represented by a power series, $\sum_{j=1}^{\infty} B^{-j}$.

PROOF. To prove (a), given $\mathbf{y}$ we want to find $\mathbf{x}$ such that $\mathbf{x} - G\mathbf{x} = \mathbf{y}$, or $\mathbf{x} = \mathbf{y} + G\mathbf{x}$. We can find this $\mathbf{x}$ as a fixed point of a map, u. Let $u : \mathbb{E} \times \mathbb{E} \to \mathbb{E}$ be given by $u(\mathbf{x}, \mathbf{y}) = \mathbf{y} + G\mathbf{x}$. Then,

$$\begin{aligned} u(\mathbf{x}_1, \mathbf{y}) - u(\mathbf{x}_2, \mathbf{y}) &= G(\mathbf{x}_1 - \mathbf{x}_2), \quad \text{so} \\ |u(\mathbf{x}_1, \mathbf{y}) - u(\mathbf{x}_2, \mathbf{y})| &\le a\,|(\mathbf{x}_1 - \mathbf{x}_2|. \end{aligned}$$

Thus, for $\mathbf{y}$ fixed, $u(\cdot, \mathbf{y})$ is a contraction. By the contraction mapping result, there is a unique fixed point $\mathbf{x_y}$, $\mathbf{x_y} = u(\mathbf{x_y}, \mathbf{y}) = \mathbf{y} + G(\mathbf{x_y})$. Thus, $\mathbf{y} = (id - G)\mathbf{x_y}$. The existence of $\mathbf{x_y}$ shows that $id - G$ is onto. The uniqueness shows that $id - G$ is one to one. To get the bound on the norm of the inverse, notice that if $\mathbf{x} = (id - G)^{-1}\mathbf{y}$, then $\mathbf{x} - G\mathbf{x} = \mathbf{y}$ and

$$\begin{aligned} &|\mathbf{x}| - a|\mathbf{x}| \le |\mathbf{y}|, \\ &|\mathbf{x}| \le \frac{|\mathbf{y}|}{1 - a}, \qquad \text{so} \\ &\frac{|(id - G)^{-1}\mathbf{y}|}{|\mathbf{y}|} \le \frac{1}{1 - a}. \end{aligned}$$

Thus, $(id - G)^{-1}$ is a bounded linear map and $\|(id - G)^{-1}\| \le 1/(1 - a)$.

We will not bother with the details of convergence to show the inverse can be given by the series indicated. However, if the series converges, then it can easily be checked that it is the inverse as follows:

$$\begin{aligned} (id - G)\sum_{j=0}^{\infty} G^j &= \sum_{j=0}^{\infty} G^j - \sum_{j=1}^{\infty} G^j \\ &= id. \end{aligned}$$

Turning to part (b), by part (a), $B^{-1} - id$ is an isomorphism. Since $B - id = B(id - B^{-1})$, it follows that it is also an isomorphism. Its inverse is $(B - id)^{-1} = (id - B^{-1})^{-1}B^{-1}$, so $\|(B - id)^{-1}\| \le (1/(1 - a))a$. We leave to the reader to check the series. This completes the proof of Lemma 7.2. □

PROOF OF THEOREM 7.1. We want to show that each $\mathcal{L}^\sigma = id - A^\sigma_\#$ is invertible. Writing $C^0_b(\mathbb{E}^\sigma)$ for $C^0_b(\mathbb{R}^n, \mathbb{E}^\sigma)$, the norm of $A^\sigma_\#$ is given as follows:

$$\|A^\sigma_\#\| = \sup_{v \in C^0_b(\mathbb{E}^\sigma)\setminus\{0\}} \frac{\|A^\sigma_\# v\|_0}{\|v\|_0}.$$

Then, for $v \in C^0_b(\mathbb{E}^s)$,

$$\begin{aligned}
\|A^s_\# v\|_0 &= \sup_{\mathbf{x}\in\mathbb{R}^n} |Av \circ A^{-1}\mathbf{x}| \\
&= \sup_{\mathbf{y}\in\mathbb{R}^n} |Av(\mathbf{y})| \\
&\le a\|v\|_0.
\end{aligned}$$

Thus, $\|A^s_\#\| \le a$, and by Lemma 7.2(a), $\mathcal{L}^s = id - A^s_\#$ is invertible with $\|(\mathcal{L}^s)^{-1}\| \le 1/(1-a)$. By a similar calculation, $\|(A^u_\#)^{-1}\| \le a$, and by Lemma 7.2(b), $\mathcal{L}^u = id - A^u_\#$ is invertible with $\|(\mathcal{L}^u)^{-1}\| \le a/(1-a)$. Because the norm on $\mathbb{R}^n$ is the maximum of the norms on $\mathbb{E}^u$ and $\mathbb{E}^s$, we get that $\|\mathcal{L}^{-1}\| \le 1/(1-a)$.

We have the map $\Psi(g, v) = v - A_\#(v) - g \circ (id + v) \circ A^{-1}$ and its 'linearization' at $(0, 0)$, $\mathcal{L}v = v - A_\#(v)$. Imitating the proof of the Implicit Function Theorem, we let

$$\begin{aligned}
\Theta(g, v) &= \mathcal{L}^{-1}\{\mathcal{L}v - \Psi(g, v)\} \\
&= \mathcal{L}^{-1}\{v - A_\#(v) - v + A_\#(v) + g \circ (id + v) \circ A^{-1}\} \\
&= \mathcal{L}^{-1}\{g \circ (id + v) \circ A^{-1}\}.
\end{aligned}$$

Thus,

$$\begin{aligned}
&\|\Theta(g, v_1) - \Theta(g, v_2)\|_0 \\
&\quad\le \|\mathcal{L}^{-1}\| \sup_{\mathbf{x}\in\mathbb{R}^n} |g \circ (id + v_1) \circ A^{-1}\mathbf{x} - g \circ (id + v_2) \circ A^{-1}\mathbf{x}| \\
&\quad\le \left(\frac{1}{1-a}\right) \operatorname{Lip}(g) \sup_{\mathbf{y}\in\mathbb{R}^n} |v_1(\mathbf{y}) - v_2(\mathbf{y})| \\
&\quad\le \left(\frac{1}{1-a}\right) \operatorname{Lip}(g) \|v_1 - v_2\|_0.
\end{aligned}$$

For a fixed g with $(\frac{1}{1-a})\operatorname{Lip}(g) < 1$, $\Theta(g, \cdot)$ is a contraction on $C^0_b(\mathbb{R}^n)$. Because $C^0_b(\mathbb{R}^n)$ is a complete metric space, there is a unique fixed point v_g with $\Theta(g, v_g) = v_g$. A direct calculation shows that this is equivalent to $\Psi(g, v_g) = 0$. Letting $f = A + g$ and $h_f = id + v_g$, the fact that $\Psi(g, v_g) = 0$ implies that $h_f = (A+g) \circ h_f \circ A^{-1} = f \circ h_f \circ A^{-1}$, or $h_f \circ A = f \circ h_f$. All that remains is to show that $h = h_f$ is a homeomorphism. Before proving this fact, we show that $f = A + g$ is a diffeomorphism.

Lemma 7.3. *The map f is one to one and onto, so is a diffeomorphism.*

PROOF. Assume that $f(\mathbf{x}) = f(\mathbf{y})$. Then, $0 = f(\mathbf{x}) - f(\mathbf{y}) = A(\mathbf{x} - \mathbf{y}) + g(\mathbf{x}) - g(\mathbf{y})$. Therefore,

$$\begin{aligned}
0 &= |f(\mathbf{x}) - f(\mathbf{y})| \\
&\ge m(A)|\mathbf{x} - \mathbf{y}| - \operatorname{Lip}(g)|\mathbf{x} - \mathbf{y}| \\
&\ge \big(m(A) - \operatorname{Lip}(g)\big)|\mathbf{x} - \mathbf{y}|,
\end{aligned}$$

so $\mathbf{x} = \mathbf{y}$. This shows that f is one to one.

The map f is onto because it is a bounded distance from the linear map A which is onto and one to one. □

Lemma 7.4. *The map $h = h_f$ is one to one.*

PROOF. There are two types of proofs that h is one to one. One uses the uniqueness of the conjugacy h (within maps h for which $h - id$ is bounded) and the fact that we can also solve for a unique k with $A \circ k = k \circ f$. Then, $A \circ k \circ h = k \circ f \circ h = k \circ h \circ A$. Thus, $k \circ h$ conjugates A with itself. By the uniqueness of the maps which conjugate A with itself (within maps for which $h - id$ is bounded), $k \circ h = id$. This proves that h is a one to one, and even that h is a homeomorphism. We do not give the details of this proof.

The second proof uses a property called expansiveness. If $h(\mathbf{x}) = h(\mathbf{y})$, then $h \circ A\mathbf{x} = f \circ h(\mathbf{x}) = f \circ h(\mathbf{y}) = h \circ A\mathbf{y}$. By induction, $h(A^n\mathbf{x}) = h(A^n\mathbf{y})$ for $n \geq 0$. Using the fact that f is invertible and $f^{-1} \circ h = h \circ A^{-1}$, we can also show that $h(A^n\mathbf{x}) = h(A^n\mathbf{y})$ for $n \leq 0$, so for all $n \in \mathbb{Z}$.

Now, we write $\mathbf{x} = \mathbf{x}^u + \mathbf{x}^s$ and $\mathbf{y} = \mathbf{y}^u + \mathbf{y}^s$ with $\mathbf{x}^\sigma, \mathbf{y}^\sigma \in \mathbb{E}^\sigma$. If $\mathbf{x} \neq \mathbf{y}$, then either $\mathbf{x}^u \neq \mathbf{y}^u$ or $\mathbf{x}^s \neq \mathbf{y}^s$. If $\mathbf{x}^u \neq \mathbf{y}^u$, then $|A^j\mathbf{x}^u - A^j\mathbf{y}^u| \geq a^{-j}|\mathbf{x}^u - \mathbf{y}^u|$. Thus, we can take a $j \geq 0$ with $|A^j\mathbf{x}^u - A^j\mathbf{y}^u| \geq 3\|h - id\|_0 > 0$. (If $h = id$, then h is a homeomorphism and we are done.) Then, letting $\mathbf{x}_j = A^j\mathbf{x}$ and $\mathbf{y}_j = A^j\mathbf{y}$, $h(\mathbf{x}_j) - h(\mathbf{y}_j) = \mathbf{x}_j - \mathbf{y}_j + (h - id)(\mathbf{x}_j) - (h - id)(\mathbf{y}_j)$, so $0 = |h(\mathbf{x}_j) - h(\mathbf{y}_j)| \geq |\mathbf{x}_j^u - \mathbf{y}_j^u| - |(h - id)(\mathbf{x}_j)| - |(h - id)(\mathbf{y}_j)| \geq \|h - id\|_0 > 0$. This contradiction shows that it is impossible for $\mathbf{x}^u \neq \mathbf{y}^u$. Similarly, using negative iterates, we can prove that $\mathbf{x}^s = \mathbf{y}^s$. This completes the proof that h is one to one. □

Lemma 7.5. *The map $h = h_f$ is onto, so it is a homeomorphism of $\mathbb{R}^n$.*

PROOF. The proof that h is onto uses the fact that it is a bounded distance from the identity: let $b = \|h - id\|_0$. We use the notation that $B(r) = B(0, r)$ is the open ball centered at the origin of radius r, $\text{cl}(B(r))$ is the closed ball centered at the origin of radius r, and $S(r) = \text{cl}(B(r)) \setminus B(r)$ is the sphere of radius r centered at the origin.

Notice that for $\mathbf{x} \in \text{cl}(B(r))$, $|h(\mathbf{x})| \leq |h(\mathbf{x}) - \mathbf{x}| + |\mathbf{x}| \leq b + r$, so $h(\text{cl}(B(r))) \subset \text{cl}(B(r + b))$. Similarly, for $\mathbf{x} \in S(r)$, $|h(\mathbf{x})| \geq |\mathbf{x}| - |h(\mathbf{x}) - \mathbf{x}| \geq r - b$, so $h(S(r)) \subset \text{cl}(B(r + b)) \setminus B(r - b)$.

Because h is one to one, the Brouwer Invariance of Domain Theorem implies that h takes an open set to an open set; in particular, the images $h(B(r))$ are open. By taking the union, $h(\mathbb{R}^n)$ is open. (See Dugundji (1966) page 359 for the Brouwer Invariance of Domain Theorem.)

One the other hand, we show that the image $h(\mathbb{R}^n)$ is closed. Assume $\mathbf{z}_0 \in \text{cl}(h(\mathbb{R}^n))$. There exists $\mathbf{x}_j \in \mathbb{R}^n$ with $h(\mathbf{x}_j)$ converging to $\mathbf{z}_0$. Thus, $h(\mathbf{x}_j)$ is bounded with $|h(\mathbf{x}_j)| \leq |\mathbf{z}_0| + 1 \equiv R$. Since $R \geq |h(\mathbf{x}_j)| \geq |\mathbf{x}_j| - b$, we get that $|\mathbf{x}_j| \leq R + b$, and the $\mathbf{x}_j$ are bounded. By compactness of $\text{cl}(B(R + b))$, there is a subsequence $\mathbf{x}_{j_i}$ which converges to a point $\mathbf{x}_0 \in \text{cl}(B(R + b))$. By continuity of h, $h(\mathbf{x}_0) = \mathbf{z}_0$. Therefore, $\mathbf{z}_0$ is in the image, and $\text{cl}(h(\mathbb{R}^n)) = h(\mathbb{R}^n)$.

Because $h(\mathbb{R}^n)$ is both open and closed in $\mathbb{R}^n$ and $\mathbb{R}^n$ is connected, $h(\mathbb{R}^n) = \mathbb{R}^n$, i.e., h is onto.

In finite dimensions, a continuous bijection is a homeomorphism. We show this fact explicitly in this situation, i.e., we show that h^{-1} is continuous. Assume $\mathbf{y}_n$ is a sequence of points contained in some $\text{cl}(B(R))$ converging to $\mathbf{y}_\infty$. By the above arguments, there are $\mathbf{x}_n$ and $\mathbf{x}_\infty$ in $\text{cl}(B(R + b))$ such that $h(\mathbf{x}_n) = \mathbf{y}_n$ and $h(\mathbf{x}_\infty) = \mathbf{y}_\infty$. Thus, $h^{-1}(\mathbf{y}_n) = \mathbf{x}_n, h^{-1}(\mathbf{y}_\infty) = \mathbf{x}_\infty \in \text{cl}(B(R + b))$. Assume the $\mathbf{x}_n$ do not converge to $\mathbf{x}_\infty$. Then, there is a subsequence $\mathbf{x}_{n_j}$ converging to $\mathbf{p} \neq \mathbf{x}_\infty$. By continuity of h,

$$h(\mathbf{p}) = \lim_{j \to \infty} h(\mathbf{x}_{n_j}) = \mathbf{y}_\infty = h(\mathbf{x}_\infty).$$

This contradicts the fact that h is one to one. Therefore, $h^{-1}(\mathbf{y}_n)$ must converge to $h^{-1}(\mathbf{y}_\infty)$, proving that h^{-1} is continuous.

The key idea which made the above proof work is that h is proper. A map h is called *proper* provided the inverse images of compact sets are compact. This completes the proof of the lemma and Theorem 7.1.

□

5.7.1 Proof of the Local Theorem

To prove the local version, we need to use what are called *bump functions*. These are functions which make the transition from being identically zero to functions which are identically one. We give the construction in the following lemma.

Lemma 7.6. *Given numbers $0 < a < b$, there is a C^∞ function β on $\mathbb{R}^n$ such that $0 \le \beta(\mathbf{x}) \le 1$ for all $\mathbf{x} \in \mathbb{R}^n$ and*

$$\beta(\mathbf{x}) = \begin{cases} 1 & \text{for } |\mathbf{x}| \le a \\ 0 & \text{for } |\mathbf{x}| \ge b. \end{cases}$$

PROOF. We start by defining a function of a real variable,

$$\alpha(x) = \begin{cases} 0 & \text{for } x \le 0 \\ e^{-1/x} & \text{for } x > 0. \end{cases}$$

A direct check shows that α is C^∞.

Next, for $a < b$, let $\gamma(x) = \alpha(x-a)\alpha(b-x)$. Then, $\gamma(x) \ge 0$ and is greater than zero exactly on the open interval (a, b). Again γ is C^∞.

Now, if we define

$$\delta(x) = \frac{\int_x^b \gamma(s)\,ds}{\int_a^b \gamma(s)\,ds},$$

then $0 \le \delta(x) \le 1$ for all $x \in \mathbb{R}$ and

$$\delta(x) = \begin{cases} 1 & \text{for } x \le a \\ 0 & \text{for } x \ge b. \end{cases}$$

Thus, δ is almost the desired function on the real line.

Lastly, define $\beta(\mathbf{x})$ on $\mathbb{R}^n$ by $\beta(\mathbf{x}) = \delta(|\mathbf{x}|)$. This function has all the desired properties. □

We now use this bump function to construct a map satisfying Theorem 7.1 from a nonlinear map in a neighborhood of a fixed point.

Proposition 7.7. *Let U_0 be an open neighborhood of the origin in $\mathbb{R}^n$. Let $f : U_0 \to \mathbb{R}^n$ be a C^r local diffeomorphism for $r \ge 1$ with $f(\mathbf{0}) = \mathbf{0}$ and $A = Df_{\mathbf{0}}$. Then, given any $\epsilon > 0$, there is a smaller neighborhood $U \subset U_0$ of $\mathbf{0}$ and a C^r extension $\bar{f} : \mathbb{R}^n \to \mathbb{R}^n$ with $\bar{f}|U = f|U$, $(\bar{f} - A) \in C_b^1(\mathbb{R}^n, \mathbb{R}^n)$, and $\text{Lip}(\bar{f} - A) < \epsilon$.*

PROOF. Let $\beta(x)$ be the C^∞ bump function given by Lemma 7.6 on $\mathbb{R}$ with $a = 1$ and $b = 2$. Then, there is a uniform $K \ge 1$ such that $|\beta'(x)| \le K$ for all $x \in \mathbb{R}$.

Let $g = f - A$, so $Dg_{\mathbf{0}} = \mathbf{0}$. Take $r > 0$ such that $\|Dg_{\mathbf{x}}\| < \epsilon/(4K)$ for all $\mathbf{x} \in B(\mathbf{0}, 2r)$ and $B(\mathbf{0}, 2r) \subset U_0$. Finally, let

$$\varphi(\mathbf{x}) = \beta(\frac{|\mathbf{x}|}{r})g(\mathbf{x}) \qquad \text{and}$$
$$\bar{f}(\mathbf{x}) = A\mathbf{x} + \varphi(\mathbf{x}).$$

Clearly, $\bar{f}$ equals f inside $U \equiv B(\mathbf{0}, r)$ and A outside $B(\mathbf{0}, 2r)$. Thus, $\bar{f} \in C_b^1(\mathbb{R}^n, \mathbb{R}^n)$.

All that remains is to check the bound on the Lipschitz constant of φ. Since $\varphi \equiv 0$ outside $B(\mathbf{0}, 2r)$, we can assume $\mathbf{x}, \mathbf{y} \in B(\mathbf{0}, 2r)$ in the following calculation of this Lipschitz constant:

$$\begin{aligned}
|\varphi(\mathbf{x}) - \varphi(\mathbf{y})| &= |\beta(\frac{|\mathbf{x}|}{r})g(\mathbf{x}) - \beta(\frac{|\mathbf{y}|}{r})g(\mathbf{y})| \\
&\le |\beta(\frac{|\mathbf{x}|}{r}) - \beta(\frac{|\mathbf{y}|}{r})| \cdot |g(\mathbf{x})| + |\beta(\frac{|\mathbf{y}|}{r})| \cdot |g(\mathbf{x}) - g(\mathbf{y})| \\
&\le K \cdot \frac{1}{r} \cdot |\mathbf{x} - \mathbf{y}| \cdot (\frac{\epsilon}{4K}) \cdot |\mathbf{x}| + 1 \cdot (\frac{\epsilon}{4K}) \cdot |\mathbf{x} - \mathbf{y}| \\
&\le \epsilon(\frac{1}{2} + \frac{1}{4K})|\mathbf{x} - \mathbf{y}| \\
&\le \epsilon|\mathbf{x} - \mathbf{y}|.
\end{aligned}$$

This completes the proof of the proposition. □

For an arbitrary nonlinear map with a fixed point at $\mathbf{p}$, there is a simple translation that brings the fixed point to the origin. By Proposition 7.7, there is an extension $\bar{f}$ which equals f in a neighborhood of $\mathbf{0}$ and satisfies Theorem 7.1. Then, $\bar{f}$ and A are conjugate on all of $\mathbb{R}^n$. Because $\bar{f}$ and f are equal on a neighborhood of $\mathbf{0}$, f and A are conjugate on a neighborhood of the origin. This shows how the local version of the Hartman-Grobman Theorem for a diffeomorphism, Theorem 6.2, follows from Theorem 7.1.

5.7.2 Proof of the Hartman-Grobman Theorem for Flows

As in the statement of the theorem of the local Hartman-Grobman Theorem for flows near a fixed point, Theorem 5.3, we consider the differential equation $\dot{\mathbf{x}} = f(\mathbf{x})$ in a neighborhood of a hyperbolic fixed point $\mathbf{p}$. By a translation we can take $\mathbf{p} = \mathbf{0}$. Let $B = Df_{\mathbf{0}}$. Using a bump function, we can find an extension $\bar{f}$ which equals f in a neighborhood of $\mathbf{0}$ and equals B outside a larger neighborhood. Let φ^t be the flow for the differential equation $\dot{\mathbf{x}} = \bar{f}(\mathbf{x})$ and e^{tB} the flow for the linear equation $\dot{\mathbf{x}} = B\mathbf{x}$. Note that φ^t is also the flow for f in a neighborhood of $\mathbf{0}$. Take the time one maps $\varphi^1(\mathbf{x})$ and $e^B\mathbf{x}$. By Theorem 7.1 there is a conjugacy h between φ^1 and e^B, $h \circ e^B = \varphi^1 \circ h$, and h is unique among maps for which $h - id$ is bounded. The following lemma shows that h actually conjugates φ^t and e^{tB} for all times t.

Lemma 7.8. *The map h satisfies $\varphi^t \circ h \circ e^{-tB} = h$ for all real t.*

PROOF. Fix a real number t. First, we show that not only h but also $\varphi^t \circ h \circ e^{-tB}$ is a conjugacy between φ^1 and e^B:

$$\begin{aligned}
\varphi^1 \circ \left[\varphi^t \circ h \circ e^{-tB}\right] \circ e^{-B} &= \varphi^t \circ \left[\varphi^1 \circ h \circ e^{-B}\right] \circ e^{-tB} \\
&= \varphi^t \circ h \circ e^{-tB}
\end{aligned}$$

because h is a conjugacy for the time one maps of the flows. The conjugacy h is unique among maps for which $h - id$ is a continuous bounded map. To see that $\varphi^t \circ h \circ e^{-tB} - id$ is bounded, note that

$$\varphi^t \circ h \circ e^{-tB} - id = (\varphi^t - e^{tB}) \circ h \circ e^{-tB} + e^{tB} \circ (h - id) \circ e^{-tB}.$$

The second quantity on the right-hand side is bounded because $h - id$ is bounded. The two flows are equal outside a bounded set so the first quantity on the right-hand side is bounded by compactness. Thus, $\varphi^t \circ h \circ e^{-tB} - id$ is bounded and also is a conjugacy. By the uniqueness of Theorem 7.1, $\varphi^t \circ h \circ e^{-tB} = h$ and we have proved the lemma. □

By the lemma we have a conjugacy of the extensions on all of $\mathbb{R}^n$. By restricting to a neighborhood of the fixed point, we have proved the local Hartman-Grobman Theorem for flows, Theorem 5.3.

The next section discusses the behavior of a flow in a neighborhood of a periodic orbit.

5.8 Periodic Orbits for Flows

We consider periodic orbits in this section and see how the eigenvalues of an appropriate matrix determine the stability. For a different approach than given here to the analysis of orbits in a neighborhood of a periodic orbit, see Hale (1969).

Remember that a point $\mathbf{p}$ is a *periodic point with (least) period* T provided $\varphi^T(\mathbf{p}) = \mathbf{p}$ and $\varphi^t(\mathbf{p}) \neq \mathbf{p}$ for $0 < t < T$. If $\mathbf{p}$ is a periodic point with period T, then the set of points $\mathcal{O}(\mathbf{p}) = \{\varphi^t(\mathbf{p}) : 0 \leq t \leq T\}$ is called a *periodic orbit* or *closed orbit.*

It is important to understand the flow in a neighborhood of a periodic orbit. One way to do this is to follow the nearby trajectories as they make one circuit around the periodic orbit. The following example gives a case where the orbits can be given as explicit functions of time.

Example 8.1. Consider the equations

$$\dot{x} = -y + x(1 - x^2 - y^2)$$
$$\dot{y} = x + y(1 - x^2 - y^2).$$

which in polar coordinates are given by

$$\dot{r} = r(1 - r^2)$$
$$\dot{\theta} = 1.$$

If we look at orbits with $\theta(0) = 0$, then $\theta(2\pi) = 2\pi$. Thus, the solutions return to the surface $\{\theta = 0 \bmod 2\pi\}$ after a time of 2π. The map P which takes the r value at time 0 to the r value at time 2π incorporates the effect of making one circuit around the periodic orbit. This map P is called the *first return map* or *Poincaré map* from the surface $\{\theta = 0 \bmod 2\pi\}$ to itself. The solutions for this system of differential equations, and so also the Poincaré map, can be explicitly calculated, and used to show that $r = 1$ is an attracting periodic orbit. Also, in this case, it can be seen directly that $r = 1$ is an attracting periodic orbit because $\dot{r} > 0$ for $r < 1$ and $\dot{r} < 0$ for $r > 1$.

Definition. In general, let γ be a periodic orbit of period T with $\mathbf{p} \in \gamma$. Then, for some k, the kth coordinate function of the vector field must be nonzero at $\mathbf{p}$, $f_k(\mathbf{p}) \neq 0$. We take the hyperplane $\Sigma = \{\mathbf{x} : x_k = p_k\}$. This hyperplane Σ is called a *cross section* or *transversal* at $\mathbf{p}$. For $\mathbf{x} \in \Sigma$ near $\mathbf{p}$, the flow $\varphi^t(\mathbf{x})$ returns to Σ in time $\tau(\mathbf{x})$, which is about T, as can be seen by the Implicit Function Theorem (as carried out explicitly in the proof of the following theorem.) Let $V \subset \Sigma$ be an open set in Σ on which $\tau(\mathbf{x})$ is a differentiable function. The *first return map* or *Poincaré map* is defined to be $P(\mathbf{x}) = \varphi^{\tau(\mathbf{x})}(\mathbf{x})$ for $\mathbf{x} \in V$.

REMARK 8.1. Example 8.1 gives a trivial example where the Poincaré map can be explicitly calculated. In two subsections below, we determine properties of the Poincaré map in two different situations of differential equations in $\mathbb{R}^2$. These two subsections are not necessary for the rest of the book but give some idea of how the Poincaré map could be determined when the equations cannot be explicitly solved. In particular, we determine the nature of the Poincaré map for the Van der Pol equation, and we determine the derivative of the Poincaré map for equations in two dimensions in terms of an integral of the divergence of the vector field. In the first subsection below, we show how it is possible to take a map and recover a flow which has this map as a Poincaré map. This process is called the suspension of a map.

The proof of the following theorem carries out the construction of the Poincaré map and determines some of its properties in terms of the Implicit Function Theorem. It should also be noted that a cross section can be taken to be a hypersurface which is nonlinear but we do not include the necessary modifications.

Theorem 8.1. *Consider a C^r flow φ^t for $r \geq 1$.*

(a) If $\mathbf{p}$ is on a periodic orbit γ of period T and Σ is a transversal at $\mathbf{p}$, then the first return time $\tau(\mathbf{x})$ is defined in a neighborhood V of $\mathbf{p}$ in Σ and $\tau : V \to \mathbb{R}$ is C^r.

(b) If $P : V \to \Sigma$ given by $P(\mathbf{x}) = \varphi^{\tau(\mathbf{x})}(\mathbf{x})$ is the first return map, then P is C^r.

PROOF. Let $\Sigma = \{\mathbf{x} : x_k = p_k\}$ be the cross section at $\mathbf{p}$ with $f_k(\mathbf{p}) \neq 0$. Define $\psi(\mathbf{x}, t) = \varphi_k^t(\mathbf{x}) - p_k$. Then, $\psi(\mathbf{x}, t) = 0$ if and only if $\varphi^t(\mathbf{x}) \in \Sigma$. Also, $\psi(\mathbf{p}, T) = 0$. Finally, $\frac{\partial}{\partial t}\psi(\mathbf{x}, t) = f_k \circ \varphi^t(\mathbf{x})$ and $\frac{\partial}{\partial t}\psi(\mathbf{p}, t)|_{t=T} = f_k(\mathbf{p}) \neq 0$. By the Implicit Function Theorem, there is a neighborhood V of $\mathbf{p}$ in Σ such that for $\mathbf{x} \in V$, it is possible to solve for t as a function of $\mathbf{x}$, $t = \tau(\mathbf{x})$, such that $\psi(\mathbf{x}, \tau(\mathbf{x})) \equiv 0$. Moreover, $\tau(\mathbf{x})$ is a C^r function of $\mathbf{x}$. (The reader may want to look at the section on the statement of the Implicit Function Theorem.) Once we have that $\tau(\mathbf{x})$ is C^r, it follows that the Poincaré map P is C^r because $\varphi^t(\mathbf{x})$ is jointly C^r in t and $\mathbf{x}$. □

Now, that we have shown that the Poincaré map is differentiable, we can indicate the relationship between the eigenvalues of the derivative of the Poincaré map and the eigenvalues of the derivative of the flow, $D\varphi_q^T$. The first step is given in the following lemma.

Lemma 8.2. *(a) If $\varphi^t(\mathbf{x})$ is a solution of $\dot{\mathbf{x}} = f(\mathbf{x})$, then $D\varphi_x^t f(\mathbf{x}) = f(\varphi^t(\mathbf{x}))$ for any t.*

(b) If γ is a periodic orbit of period T and $\mathbf{p} \in \gamma$, then $D\varphi_{\mathbf{p}}^T$ has 1 as an eigenvalue with eigenvector $f(\mathbf{p})$.

(c) If $\mathbf{p}$ and $\mathbf{q}$ are two points on a T-periodic orbit γ, then the derivatives $D\varphi_{\mathbf{p}}^T$ and $D\varphi_{\mathbf{q}}^T$ are linearly conjugate and so have the same eigenvalues.

PROOF. We have that

$$\begin{aligned} f(\varphi^t(\mathbf{x})) &= \frac{d}{ds}\varphi^s(\mathbf{x})|_{s=t} \\ &= \frac{d}{ds}\varphi^t \circ \varphi^s(\mathbf{x})|_{s=0} \\ &= D\varphi_{\mathbf{x}}^t f(\mathbf{x}). \end{aligned}$$

This proves part (a).

For part (b), $\varphi^T(\mathbf{p}) = \mathbf{p}$ so $f(\mathbf{p}) = f(\varphi^T(\mathbf{p})) = D\varphi_{\mathbf{p}}^T f(\mathbf{p})$. This proves part (b).

For part (c), assume that $\mathbf{q} = \varphi^\tau(\mathbf{p})$. Then,

$$\varphi^T \circ \varphi^\tau(\mathbf{x}) = \varphi^\tau \circ \varphi^T(\mathbf{x}),$$

so taking the spatial derivative (with respect to $\mathbf{x}$) at $\mathbf{p}$ yields

$$D\varphi^T_{\mathbf{q}} D\varphi^\tau_{\mathbf{p}} = D\varphi^\tau_{\mathbf{p}} D\varphi^T_{\mathbf{p}}.$$

Thus, $D\varphi^T_{\mathbf{q}}$ and $D\varphi^T_{\mathbf{p}}$ are linearly conjugate by the linear map $D\varphi^\tau_{\mathbf{p}}$. □

Definition. If γ is a periodic orbit of period T with $\mathbf{p} \in \gamma$, then the above result shows that the eigenvalues of $D\varphi^T_{\mathbf{p}}$ are $1, \lambda_1, \ldots, \lambda_{n-1}$. The $n-1$ eigenvalues $\lambda_1, \ldots, \lambda_{n-1}$ are called the *characteristic multipliers* (or eigenvalues) of the periodic orbit. (Note that Lemma 8.2(c) shows that the characteristic multipliers do not depend on the point $\mathbf{p}$.)

A periodic orbit γ is called *hyperbolic* provided $|\lambda_j| \neq 1$ for all the characteristic multipliers ($1 \leq j \leq n-1$). It is called *attracting*, a *periodic attractor*, or a *periodic sink* provided all the characteristic multipliers are less than 1 in absolute value. It is called *repelling*, a *periodic repeller*, or a *periodic source* provided all the characteristic multipliers are greater than 1 in absolute value. A hyperbolic periodic orbit which is neither a source nor a sink is called a *saddle periodic orbit*.

With these results and definitions, we can now state and prove the following theorem, which says that the characteristic multipliers of the periodic orbit and the eigenvalues of the Poincaré map are the same.

Theorem 8.3. *Let $\mathbf{p}$ be a point on a periodic orbit γ of period T. Then, the characteristic multipliers of the periodic orbit are the same as the eigenvalues of the derivative of the Poincaré map at $\mathbf{p}$.*

PROOF. Let $\pi : \mathbb{R}^n \to \Sigma$ be the projection along $f(\mathbf{p})$. The Poincaré map is $P(\mathbf{x}) = \varphi^{\tau(\mathbf{x})}(\mathbf{x})$ for $\mathbf{x} \in \Sigma$, so $DP_{\mathbf{x}} = D\varphi^{\tau(\mathbf{x})}_{\mathbf{x}}|\Sigma \; + \; f \circ \varphi^{\tau(\mathbf{x})}(\mathbf{x}) D\tau_{\mathbf{x}}|\Sigma$. Now, taking the point $\mathbf{p}$ on γ, $DP_{\mathbf{p}} = \pi D\varphi^{\tau(\mathbf{p})}_{\mathbf{p}}|\Sigma$ because $\pi f(\mathbf{p}) = 0$. Lastly, the characteristic multipliers of the periodic orbit are the eigenvalues of $\pi D\varphi^{\tau(\mathbf{p})}_{\mathbf{p}}|\Sigma$. This can be seen by taking a basis of vectors $\mathbf{v}^1, \ldots, \mathbf{v}^{n-1}$ in Σ and adding $f(p)$ to make a basis of $\mathbb{R}^n$. Then, for some entries C,

$$D\varphi^{\tau(\mathbf{p})}_{\mathbf{p}} = \begin{pmatrix} B & 0 \\ C & 1 \end{pmatrix}, \qquad \text{and}$$

$$\pi D\varphi^{\tau(\mathbf{p})}_{\mathbf{p}}|\Sigma = B.$$

Thus, the eigenvalues of B are the characteristic multipliers. □

Now, we can state and prove the theorem giving sufficient conditions for stability in terms of the characteristic multipliers.

Theorem 8.4. *(a) Let γ be a periodic orbit for which all the characteristic multipliers λ_j satisfy $|\lambda_j| < 1$. Then, γ is asymptotically stable (attracting).*

(b) With the assumptions of part (a), if $\mathbf{q}$ is a point for which $d(\varphi^t(\mathbf{q}), \gamma)$ goes to 0 as t goes to infinity, then there is a point $\mathbf{z} \in \gamma$ such that $d(\varphi^t(\mathbf{q}), \varphi^t(\mathbf{z}))$ goes to 0 as t goes to infinity. This is called the in phase condition, $\mathbf{q}$ is in phase with $\mathbf{z}$.

(c) Let γ be a periodic orbit for which there is at least one characteristic multiplier λ_k with $|\lambda_k| > 1$. Then, γ is not Liapunov stable.

PROOF. Let Σ be a cross section at $\mathbf{p}$, and $P : V \subset \Sigma \to \Sigma$ be the Poincaré map. Then, $P(\mathbf{p}) = \mathbf{p}$ and all of the eigenvalues λ_j of $DP_{\mathbf{p}}$ have $|\lambda_j| < 1$. Thus, $\mathbf{p}$ is an asymptotically stable fixed point for P, and there is a subneighborhood $V_0 \subset V$ of $\mathbf{p}$ such that for $\mathbf{q} \in V_0$, $d(P^n(\mathbf{q}), \mathbf{p})$ goes to 0 as n goes to infinity. The next lemma shows that this property of the map carries over to the flow in a neighborhood of $\mathbf{p}$ in Σ.

Lemma 8.5. *If* $\mathbf{q} \in V_0$, *then* $\lim_{t\to\infty} d(\varphi^t(\mathbf{q}), \gamma) = 0$.

PROOF. For $\mathbf{q} \in V_0$, let $\mathbf{q}_n = P^n(\mathbf{q})$. Then, $\mathbf{q}_n \to \mathbf{p}$ as $n \to \infty$ by the choice of V_0 (and because $\mathbf{p}$ is asymptotically stable). Let $\tau_n = \tau(\mathbf{q}_n)$ be the return times used to define the Poincaré map for these points. Then, τ_n converges to $T = \tau(\mathbf{p})$ because τ is a differentiable (and so continuous) function. Therefore, $\tau_n \le 2T$ for $n \ge N_1$. We need to control the flow of points in times between crossing the transversal, i.e., for times less than $2T$. Given $\epsilon > 0$, there exists a $\delta > 0$ such that if $d(\mathbf{x}, \mathbf{p}) \le \delta$ and $0 \le t \le 2T$, then $\epsilon > d(\varphi^t(\mathbf{x}), \varphi^t(\mathbf{p})) \ge d(\varphi^t(\mathbf{x}), \gamma)$. Now, for the $\mathbf{q}_n$, there is a $N_2 \ge N_1$ such that for $n \ge N_2$, $d(\mathbf{q}_n, \mathbf{p}) \le \delta$, so $d(\varphi^t(\mathbf{q}_n), \gamma) < \epsilon$ for $0 \le t \le \tau_n \le 2T$. Now, we want to see that the flow is near for all sufficiently large times. Let $t_n = \sum_{j=0}^{n-1} \tau_j$, so $\varphi^{t_n}(\mathbf{q}) = \mathbf{q}_n$ converges to $\mathbf{p}$ as n goes to infinity. Thus, for $t \ge t_{N_2}$, there is some $n \ge N_2$ with $t_n \le t < t_n + \tau_n$ and $d(\varphi^t(\mathbf{q}), \gamma) < \epsilon$. Since this is possible for any $\epsilon > 0$, we have proved the lemma. □

The above lemma proves part (a) if we start on the transversal, Σ. Now, let U be an arbitrary neighborhood of γ. Let $V_0 \subset \Sigma$ be as in Lemma 8.5, and take $V_1 \subset V_0$ a smaller neighborhood of $\mathbf{p}$ in Σ such that $\tau(\mathbf{x}) \le 2T$ for $\mathbf{x} \in V_1$, and $U_1 \equiv \{\varphi^t(\mathbf{x}) : \mathbf{x} \in V_1 \text{ and } 0 \le t \le \tau(\mathbf{x})\} \subset U$. Now, take $V_2 \subset V_1 \subset \Sigma$ such that $P^n(V_2) \subset V_1$ for all $n \ge 0$. Finally, let $U_2 \equiv \{\varphi^t(\mathbf{x}) : \mathbf{x} \in V_2 \text{ and } 0 \le t \le \tau(\mathbf{x})\}$. Now, take $\mathbf{q} \in U_2$. Let $\mathbf{q}_0 = \varphi^{-t_0}(\mathbf{q}) \in V_2 \subset V_1$, so $\varphi^t(\mathbf{q}_0) \in U_2 \subset U$ for $0 \le t \le \tau(\mathbf{q}_0) \le 2T$. Thus, $\varphi^t(\mathbf{q}) \in U_1$ for $0 \le t \le \tau(\mathbf{q}_0) - t_0$. Let $\mathbf{q}_n = P^n(\mathbf{q}_0) = \varphi^{t_n}(\mathbf{q})$. Then, $\mathbf{q}_n \in V_1$ and $\varphi^t(\mathbf{q}) \in U_1 \subset U$ for $t_n \le t \le t_{n+1}$. Thus, $\varphi^t(\mathbf{q}) \in U_1 \subset U$ for all $t \ge 0$, and so γ is *orbitally Liapunov stable.* Also, $\mathbf{q}_n = \varphi^{t_n}(\mathbf{q}) \in V_1$ and $d(\varphi^t(\mathbf{q}), \gamma)$ goes to 0 as t goes to infinity by an argument as in Lemma 8.5. This proves part (a).

We want to find a point $\mathbf{z} \in \gamma$ such that $d(\varphi^t(\mathbf{z}), \varphi^t(\mathbf{q}))$ goes to zero as t goes to infinity. It suffices to consider $\mathbf{q} \in V_0 \subset \Sigma$. As above, $\mathbf{q}_n = P^n(\mathbf{q}) = \varphi^{t_n}(\mathbf{q})$ where $t_n = \sum_{j=0}^{n-1} \tau \circ P^j(\mathbf{q})$. We want to show that t_n grows like nT, so we look at the difference, $h_n = t_n - nT = \sum_{j=0}^{n-1}(\tau_j - T)$. We claim that h_n is a Cauchy sequence.

Lemma 8.6. *The numbers* h_n *form a Cauchy sequence.*

PROOF. We need the fact that the derivative of τ is bounded by a constant $C > 0$, $\|D\tau_{\mathbf{x}}\| \le C$, so τ is Lipschitz, i.e., $|\tau(\mathbf{x}) - \tau(\mathbf{y})| \le Cd(\mathbf{x}, \mathbf{y})$ for $\mathbf{x}, \mathbf{y} \in V_0$. Also, there is a $C_1 \ge 1$ and $\nu < 1$ such that $d(\mathbf{q}_j, \mathbf{p}) \le C_1 \nu^j d(\mathbf{q}, \mathbf{p})$ for all $j \ge 0$. Then,

$$\begin{aligned}
|h_{n+k} - h_k| &= \left|\sum_{j=k}^{n+k-1} (\tau(\mathbf{q}_j) - \tau(\mathbf{p}))\right| \qquad (\text{since } T = \tau(\mathbf{p})) \\
&\le \sum_{j=k}^{n+k-1} Cd(\mathbf{q}_j, \mathbf{p}) \\
&\le \sum_{j=k}^{n+k-1} CC_1 \nu^j d(\mathbf{q}, \mathbf{p}) \\
&\le CC_1 d(\mathbf{q}, \mathbf{p}) \frac{\nu^k}{1 - \nu}.
\end{aligned}$$

This last quantity goes to zero as k goes to infinity, so $|h_{n+k} - h_k|$ does also. This proves the lemma. □

Now, since h_n is a Cauchy sequence, it approaches some limit value A. Then, $t_n \approx A + nT$. We use A to define $\mathbf{z}$ as $\mathbf{z} = \varphi^{-A}(\mathbf{p})$. We need to show that this $\mathbf{z}$ works. Let

$h_n = t_n - nT$ or $t_n = h_n + nT$. Then,

$$\begin{aligned} d(\varphi^{t_n}(\mathbf{z}), \varphi^{t_n}(\mathbf{q})) &= d(\varphi^{h_n+nT-A}(\mathbf{p}), P^n(\mathbf{q})) \\ &= d(\varphi^{h_n-A}(\mathbf{p}), P^n(\mathbf{q})) \qquad \text{(since } \varphi^{nT}(\mathbf{p}) = \mathbf{p}) \\ &\le d(\varphi^{h_n-A}(\mathbf{p}), \mathbf{p}) + d(P^n(\mathbf{p}), P^n(\mathbf{q})). \end{aligned}$$

On the last line, the first term goes to zero because $h_n - A$ goes to zero, and the second term goes to zero because $\mathbf{p}$ is asymptotically stable for the map. Therefore, $d(\varphi^t(\mathbf{z}), \varphi^t(\mathbf{q}))$ goes to zero as $t = t_n$ goes to infinity. For arbitrary large t, there is an n with $t_n \le t < t_{n+1}$ and $|t_n - t_{n+1}| < 2T$. Using the uniform continuity of $\varphi^t(\mathbf{x})$ for a bounded time interval, we get that $d(\varphi^t(\mathbf{z}), \varphi^t(\mathbf{q}))$ goes to zero as t goes to infinity. This proves part (b).

The instability in part (c) follows as before from the Hartman–Grobman Theorem.
□

The In Phase Property does not always hold if the attraction to the periodic orbit is not at a geometric rate (given by the eigenvalues), as the following example shows.

Example 8.2. This example is given in polar coordinates. Consider

$$\begin{aligned} \dot{r} &= -(r - r^*)^3 \\ \dot{\theta} &= 1 + r - r^*, \end{aligned}$$

where $r^* > 0$. Then, $r = r^*$ is an attracting periodic orbit. The solutions are given by

$$r(t) = \begin{cases} r^* & \text{for } r_0 = r^* \\ \dfrac{\operatorname{sign}(r_0 - r^*)}{[2t + (r_0 - r^*)^{-2}]^{1/2}} & \text{for } r_0 \ne r^* \end{cases}$$

$$\theta(t) = \begin{cases} \theta_0 + t & \text{for } r_0 = r^* \\ \theta_0 + t + \operatorname{sign}(r_0 - r^*)\{[2t + (r_0 - r^*)^{-2}]^{1/2} \\ \qquad -(r_0 - r^*)^{-1}\} & \text{for } r_0 \ne r^*. \end{cases}$$

Let $r(t)$ and $\theta(t) = \theta_0 + t + [2t + (r_0 - r^*)^{-2}]^{1/2} - (r_0 - r^*)^{-1}$ be a solution with $r_0 > r^*$. We would like to select a solution $\tilde{r}(t) \equiv r^*$ and $\tilde{\theta}(t) = \tilde{\theta}_0 + t$ with $r_0 = r^*$ and so that the distance between the solutions goes to zero. But

$$|\theta(t) - \tilde{\theta}(t)| = |[2t + (r_0 - r^*)^{-2}]^{1/2} - (r_0 - r^*)^{-1} + \theta_0 - \tilde{\theta}_0|$$

does not go to zero for any choice of $\tilde{\theta}_0$. In fact, the difference of the angles goes to infinity before we take everything mod 2π, which means that the solution off the periodic orbit keeps going around the angle direction more rapidly than the solution on the periodic orbit. This proves that the solutions do not approach the periodic orbit in phase with a solution on the periodic orbit.

We now consider a flow near a hyperbolic periodic orbit, γ, of period T. If $\mathbf{p} \in \gamma$, then

$$D\varphi^T_{\mathbf{p}}(\mathbb{E}^u_{\mathbf{p}} \oplus \mathbb{E}^s_{\mathbf{p}}) = \mathbb{E}^u_{\mathbf{p}} \oplus \mathbb{E}^s_{\mathbf{p}}.$$

We can define the splitting along the whole orbit by

$$\mathbb{E}^\sigma_{\mathbf{q}} = D\varphi^t_{\mathbf{p}}(\mathbb{E}^\sigma_{\mathbf{p}})$$

if $\varphi^t(\mathbf{p}) = \mathbf{q}$, and consider the *normal bundle*

$$\mathcal{N} = \{(\mathbf{q}, \mathbf{y}) : \mathbf{q} \in \gamma \text{ and } \mathbf{y} \in \mathbb{E}^u_{\mathbf{q}} \oplus \mathbb{E}^s_{\mathbf{q}}\}.$$

Then we can consider a flow Ψ^t on this normal bundle $\mathcal{N}$ which is linear on the "fibers" $\mathbb{E}^u_{\mathbf{q}} \oplus \mathbb{E}^s_{\mathbf{q}}$,

$$\Psi^t(\mathbf{q}, \mathbf{y}) = (\varphi^t(\mathbf{q}), D\varphi^t_{\mathbf{q}}\mathbf{y}).$$

Note that $\mathbb{E}^u_{\mathbf{q}} \oplus \mathbb{E}^s_{\mathbf{q}}$ is $(n-1)$-dimensional and the set of $\mathbf{q}$ on γ is one-dimensional, so the total space $\mathcal{S}$ is n-dimensional. Thus, it makes sense to say that Ψ^t on this space is conjugate to φ^t on a neighborhood of γ. We can now state the following Hartman–Grobman Theorem near a periodic orbit for a flow.

Theorem 8.7 (Hartman-Grobman). *Let γ be a periodic orbit for the differential equation $\dot{\mathbf{x}} = f(\mathbf{x})$. Then, the flow φ^t of f is topologically equivalent in a neighborhood of γ to the linear bundle flow Ψ^t (defined above) in a neighborhood of $\gamma \times \{\mathbf{0}\}$.*

PROOF. The Poincaré map of Ψ^t is the derivative of the Poincaré map for φ^t. Since $\Psi^T|\mathbb{E}^u_{\mathbf{p}} \oplus \mathbb{E}^s_{\mathbf{p}}$ is hyperbolic, the Poincaré map of φ^t in a neighborhood of $\mathbf{p}$ is conjugate to the Poincaré map of Ψ^t in a neighborhood of $\{\mathbf{0}\}$. From the conjugacy of the Poincaré maps, it follows that the flow φ^t in a neighborhood of γ is topologically equivalent to Ψ^t in a neighborhood of $\gamma \times \{\mathbf{0}\}$. □

REMARK 8.2. The above result about the flow being in phase for a periodic sink can be used to show that the two flows are topologically conjugate in this case. In fact, if the periodic orbit is hyperbolic, then it is always topologically conjugate (and not just equivalent) to the linear bundle flow. The main step is to find a transversal Σ such that for a neighborhood Σ_0 of γ in Σ, $\varphi^T(\Sigma_0) \subset \Sigma$. See Pugh and Shub (1970a) and Irwin (1970a).

5.8.1 The Suspension of a Map

The above discussion took the flow near a periodic orbit and found a map on a one-dimensional lower space. This is a local map defined only at points near the periodic orbit. Sometimes it is possible to make this construction more globally, but we do not investigate this. See Fried (1982). On the other hand, there is a general construction that takes a C^r diffeomorphism on a space and constructs a C^r flow on a space of one higher dimension. (It also works for a homeomorphism to give a continuous flow.) The new space is not a Euclidean space and is often not even the product of a Euclidean space and a circle. In any case this construction is useful in order to understand what flows are possible. The flow is called the *suspension of the map* and is essentially what topologists call the mapping torus. Given a map $f : X \to X$, we consider the space $X \times \mathbb{R}$ with the equivalence relation, $(\mathbf{x}, s+1) \sim (f(\mathbf{x}), s)$. Then, we consider the space

$$\tilde{X} = X \times \mathbb{R}/\sim.$$

To get all points, it is enough to consider $0 \leq s \leq 1$, but the other points are included because they make it clear that the quotient space has a C^r structure if f is C^r. Now, we consider the equations on $X \times \mathbb{R}$ given by

$$\dot{\mathbf{x}} = 0$$
$$\dot{s} = 1.$$

This induces a flow $\bar{\varphi}^t$ on $X \times \mathbb{R}$ which passes to a flow φ^t on the quotient space $\tilde{X}$. Notice that $\bar{\varphi}^1(\mathbf{x}, 0) = (\mathbf{x}, 1) \sim (f(\mathbf{x}), 0)$, so the flow on $\tilde{X}$ indeed has f as its (global) Poincaré map. This ends our discussion of the construction.

5.8.2 An Attracting Periodic Orbit for the Van der Pol Equations

In the introductory chapter, we mentioned that the Van der Pol equations

$$\begin{aligned}\dot{x} &= y - x^3 + x \\ \dot{y} &= -x\end{aligned} \tag{1}$$

have an attracting periodic orbit. In this section we prove this result for a slightly more general set of equations called the *Lienard equations.* The importance of this example is that it gives a set of equations for which it is possible to verify the properties of the Poincaré map and the existence of an attracting periodic orbit for a nontrivial example.

Consider the second-order scalar equation given by

$$\ddot{x} + g(x)\dot{x} + x = 0. \tag{2}$$

If $g(x)$ is zero, this is the linear oscillator. The term involving $\dot{x}$ is a "frictional" term where the friction depends on the position x. In fact, for small x, we are going to take $g(x)$ negative so it is an "anti-frictional" term, while for large x we are going to take $g(x)$ positive so it is a "frictional" term. As mentioned in the introductory chapter, these equations can be used for a certain type of electrical circuit. See Hirsch and Smale (1974). We give the precise assumptions below.

To change the equations into the form we use in our analysis, we do not let $v = \dot{x}$. Instead, we let $f(x)$ be the antiderivative of $g(x)$, $f'(x) = g(x)$ with $f(0) = 0$, and set $y = \dot{x} + f(x)$. Then, equation (2) becomes

$$\begin{aligned}\dot{x} &= y - f(x) \\ \dot{y} &= -x.\end{aligned} \tag{3}$$

This set of equations is called the Lienard equations. Figure 8.1 shows the phase portrait for $f(x) = -x + x^3$.

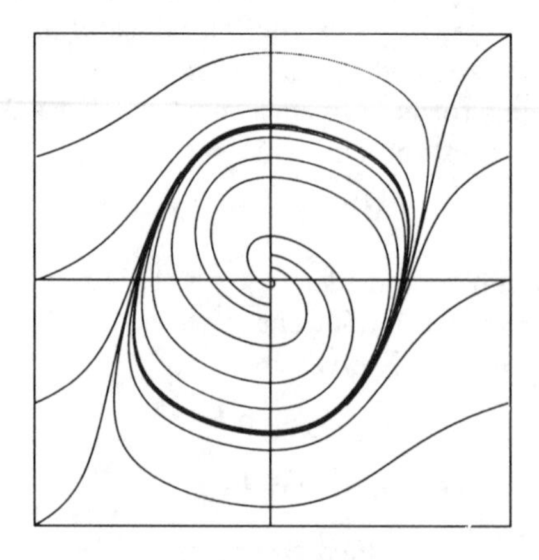

FIGURE 8.1. Phase Portrait of the Lienard Equation for $f(x) = -x + x^3$ with $-2 \leq x \leq 2$ and $-2 \leq y \leq x$

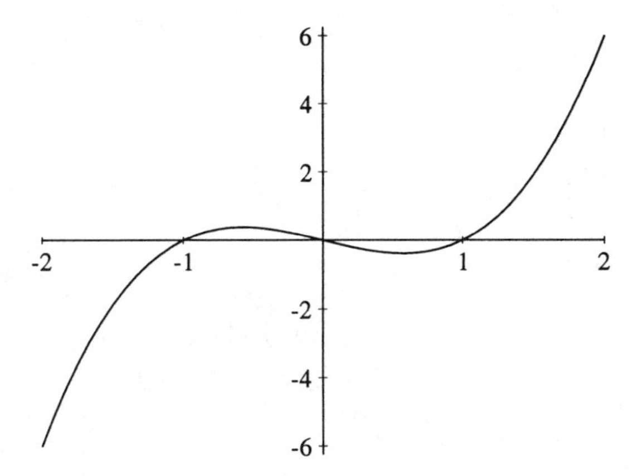

FIGURE 8.2. The Graph of $f(x) = -x + x^3$ for $-2 \leq x \leq 1$

We now proceed to give the assumptions on f that make our analysis work. The model for f is $f(x) = -x + x^3$ which gives the Van der Pol equations (1) given above. Notice for $f(x) = -x + x^3$ that (i) f is an odd function of x, (ii) $f(x) < 0$ for $0 < x < 1$ and $f(x) > 0$ for $x > 1$, (iii) f is a strictly monotone increasing function of x for $x > 1$, and (iv) $f(x)$ goes to infinity as x goes to infinity. See Figure 8.2. We make similar assumptions on the general function which we consider. Assume

(i) f is a C^1 odd function of x,
(ii) there is an $x^* > 0$ such that $f(x) < 0$ for $0 < x < x^*$ and $f(x) > 0$ for $x > x^*$,
(iii) f is a monotone increasing function of x for $x > x^*$, and
(iv) $f(x)$ goes to infinity as x goes to infinity.

With these assumptions, we can state the main result.

Theorem 8.8. *With assumptions (i) – (iv) given above, equations (3) have a unique nontrivial periodic orbit γ. This orbit is globally attracting in the sense that if $\mathbf{q} \neq \mathbf{0}$, then $\omega(\mathbf{q}) = \gamma$.*

PROOF. The first thing to note is that the only fixed point is at the origin. Next, we want to show that nontrivial solutions (solutions for $\mathbf{q} \neq \mathbf{0}$) go around the origin. To check that solutions go around the origin, we look at the curves where $\dot{x} = 0$ and $\dot{y} = 0$. Define

$$\begin{aligned}
\mathbf{v}^+ &= \{(0, y) : y > 0\}, \\
\mathbf{v}^- &= \{(0, y) : y < 0\}, \\
\mathbf{g}^+ &= \{(x, y) : y = f(x), x > 0\}, \qquad \text{and} \\
\mathbf{g}^- &= \{(x, y) : y = f(x), x < 0\}.
\end{aligned}$$

Then, $\dot{x} = 0$ on $\mathbf{g}^+$ and $\mathbf{g}^-$, and $\dot{y} = 0$ on $\mathbf{v}^+$ and $\mathbf{v}^-$. These curves are the isoclines for these equations. Let A be the region between $\mathbf{v}^+$ and $\mathbf{g}^+$, B be the region between $\mathbf{g}^+$ and $\mathbf{v}^-$, C be the region between $\mathbf{v}^-$ and $\mathbf{g}^-$, and D be the region between $\mathbf{g}^-$ and $\mathbf{v}^+$.

Lemma 8.9. *Every nontrivial trajectory moves clockwise around the origin crossing $\mathbf{v}^+$, entering A, crossing $\mathbf{g}^+$, entering B, crossing $\mathbf{v}^-$, entering C, crossing $\mathbf{g}^-$, entering D, and returning to $\mathbf{v}^+$.*

PROOF. Start by assuming that $(x_0, y_0) \in \mathbf{v}^+$. Let $(x(t), y(t))$ be the solution. Then, $\dot{x}(0) > 0$, so the solution enters A and $x_1 = x(t_1) > 0$ for small time $t_1 > 0$. As long as the solution stays in region A, $\dot{x} > 0$, so $x(t) \geq x_1$. In the region

$$\{(x, y) : x \geq x_1, y \leq y_0, (x, y) \in A\},$$

$\dot{y} \leq -x_1 < 0$, so the solution can only exit A by crossing the curve $\mathbf{g}^+$ at some time $t_2 > 0$.

The argument in region B is slightly more delicate. Let $(x_2, y_2) \in \mathbf{g}^+$ be the solution at time t_2. The trajectory enters region B for some time $t_3 > t_2$. Moreover, all along $\mathbf{g}^+$, $\dot{y} < 0$, so once the trajectory leaves a small neighborhood of $\mathbf{g}^+$, it can never return. Therefore, along the trajectory, there is an upper bound on $\dot{x}$, $\dot{x} \leq -a < 0$ for $t > t_3$, as long as it stays in region B, and $x(t) \leq x_3 - a(t - t_3)$ for as long as it stays in region B. The trajectory can leave B only by crossing $\mathbf{v}^-$ or by the solution becoming unbounded. By the above bound, $x(t)$ must become zero at least by time $t = t_3 + x_3/a$. However, in this time interval, $\dot{y} = -x \geq -x_3$, so $y(t) \geq y_3 - x_3(x_3/a)$. Since there is an a priori bound on $y(t)$ on this time interval, the solution must exit by crossing $\mathbf{v}^-$.

By the symmetry of the equations, the arguments in the other regions are similar to those given above. □

Because the solutions travel from $\mathbf{v}^+$ to $\mathbf{v}^-$ and back to $\mathbf{v}^+$, we can define the Poincaré maps $\beta : \mathbf{v}^+ \to \mathbf{v}^-$ and $\sigma : \mathbf{v}^+ \to \mathbf{v}^+$. Because of the symmetry of the equations, if $(x(t), y(t))$ is a solution, then $(-x(t), -y(t))$ is also a solution. Therefore, $\sigma(\mathbf{q}) = -\beta(-\beta(\mathbf{q}))$. Note that $\beta(\mathbf{q}) \in \mathbf{v}^-$, so $-\beta(\mathbf{q}) \in \mathbf{v}^+$, $\beta(-\beta(\mathbf{q})) \in \mathbf{v}^-$, and $-\beta(-\beta(\mathbf{q})) \in \mathbf{v}^+$.

Lemma 8.10. *The following are equivalent:*

(a) $\mathbf{p} \in \mathbf{v}^+$ *is on a periodic orbit,*
(b) $\sigma(\mathbf{p}) = \mathbf{p}$, *and*
(c) $\beta(\mathbf{p}) = -\mathbf{p}$.

PROOF. Note that solutions do not cross themselves and all solutions on $\mathbf{v}^+$ enter A, so σ and β are one to one monotone functions. Therefore, the only way for $\mathbf{p} \in \mathbf{v}^+$ to be a periodic solution is for $\sigma(\mathbf{p}) = \mathbf{p}$. This proves that (a) and (b) are equivalent.

If $\beta(\mathbf{p}) = -\mathbf{p}$, then $\sigma(\mathbf{p}) = -\beta(-\beta(\mathbf{p})) = -\beta(\mathbf{p}) = \mathbf{p}$. On the other hand, if $\beta(\mathbf{p}) < -\mathbf{p}$, then $-\beta(\mathbf{p}) > \mathbf{p}$, $\sigma(\mathbf{p}) = -\beta(-\beta(\mathbf{p})) > -\beta(\mathbf{p}) > \mathbf{p}$, and $\sigma(\mathbf{p}) \neq \mathbf{p}$. The case for $\beta(\mathbf{p}) > -\mathbf{p}$ is similar. □

Thus, to find the periodic orbits, we can use the Poincaré map β. To determine the properties of β, we use a "Liapunov function" $L(x, y) = (1/2)(x^2 + y^2)$. Then, the time derivative along solutions is given by

$$\dot{L}(x, y) = -xf(x),$$

which is not always of one sign. The change of L as solutions move from $\mathbf{v}^+$ to $\mathbf{v}^-$ is given by

$$\delta(\mathbf{p}) \equiv L(\beta(\mathbf{p})) - L(\mathbf{p}) = \int_0^{t^1(\mathbf{p})} \dot{L}(x(t), y(t))\, dt,$$

where $t^1(\mathbf{p})$ is the time to reach $\mathbf{v}^-$. Then, $\beta(\mathbf{p}) = -\mathbf{p}$ if and only if $\delta(\mathbf{p}) = 0$. To calculate δ, we sometimes look at the solutions as functions of x and write $(x, y(x))$, or as functions of y and write $(x(y), y)$. We write $\dfrac{dL}{dx}(x, y)$ for the total derivative along the solution written as a function of x, $(x, y(x))$, and similarly $\dfrac{dL}{dy}(x, y)$ for the total derivative along the solution written as a function of y, $(x(y), y)$. Then,

$$\begin{aligned}\frac{dL}{dx}(x,y) &= \frac{\dot{L}(x,y)}{\dot{x}} \\ &= \frac{-xf(x)}{y-f(x)},\end{aligned}$$

and

$$\begin{aligned}\frac{dL}{dy}(x,y) &= \left(\frac{\partial L}{\partial x}\right)\left(\frac{\dot{x}}{\dot{y}}\right) + \frac{\partial L}{\partial y} \\ &= x\left(\frac{y-f(x)}{-x}\right) + y \\ &= f(x).\end{aligned}$$

Remember that x^* is the value where $f(x^*) = 0$. There is a unique point $\mathbf{p}^* \in \mathbf{v}^+$ that flows to $(x^*, 0)$ when it first reaches $\mathbf{g}^+$. Let $r^* = |\mathbf{p}^*|$. This value is important in determining the properties of the function $\delta(\mathbf{p})$.

Lemma 8.11. *(a) For* $\mathbf{p} \in \mathbf{v}^+$ *with* $0 < |\mathbf{p}| \le r^*$, $\delta(\mathbf{p}) > 0$.
(b) The function $\delta(\mathbf{p})$ *is a monotonically decreasing function of* $|\mathbf{p}|$ *for* $|\mathbf{p}| \ge r^*$.
(c) As $|\mathbf{p}|$ *goes to infinity,* $\delta(\mathbf{p})$ *goes to minus infinity.*

PROOF. If $0 < |\mathbf{p}| \le r^*$, then $x(t) \le x^*$ along the trajectory from $\mathbf{p}$ to $\beta(\mathbf{p})$. Thus, $f(x(t)) \le 0$ (and strictly negative for most times), $\dot{L}(x(t), y(t)) \ge 0$ (and strictly positive for most times), and so $\delta(\mathbf{p}) > 0$. This proves part (a).

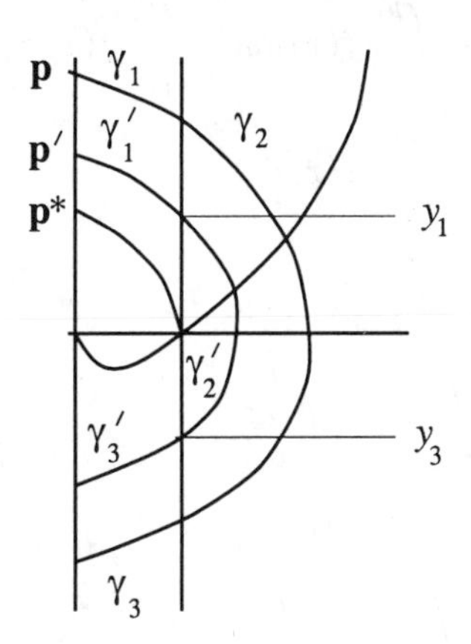

FIGURE 8.3

Now, assume that $|\mathbf{p}| > |\mathbf{p}'| \ge r^*$. Let γ_1 be the part of the trajectory of $\mathbf{p}$ from $\mathbf{v}^+$ until it hits the vertical line $\{(x^*, y)\}$. Let γ_2 be the part of the trajectory from this first time of hitting the vertical line $\{(x^*, y)\}$ until it hits this same vertical line again. Finally, let γ_3 be the part of the trajectory from this second crossing of $\{(x^*, y)\}$ until it reaches $\mathbf{v}^-$. Let γ be the combination of γ_1, γ_2, and γ_3, which is the trajectory from $\mathbf{p}$ to $\beta(\mathbf{p})$. See Figure 8.3. Similarly, let γ_j' be the parts of the trajectory for $\mathbf{p}'$. We need to compare the changes of L along γ_j and γ_j'. For $0 \le x < x^*$, if y and y' are chosen so that $(x, y) \in \gamma_1$ and $(x, y') \in \gamma_1'$, then $y > y'$, $y - f(x) > y' - f(x) > 0$, and

$-xf(x) > 0$, so the changes of L along γ_1 and γ_1' are compared as follows:

$$\begin{aligned}\int_{\gamma_1} \dot{L}(x,y)\,dt &= \int_0^{x^*} \frac{-xf(x)}{y - f(x)}\,dx \\ &< \int_0^{x^*} \frac{-xf(x)}{y' - f(x)}\,dx \\ &= \int_{\gamma_1'} \dot{L}(x,y)\,dt.\end{aligned}$$

Similarly, along γ_3 and γ_3', $y < y' < 0$ but x is decreasing, so

$$\begin{aligned}\int_{\gamma_3} \dot{L}(x,y)\,dt &= -\int_0^{x^*} \frac{-xf(x)}{y - f(x)}\,dx \\ &< -\int_0^{x^*} \frac{-xf(x)}{y' - f(x)}\,dx. \\ &= \int_{\gamma_3'} \dot{L}(x',y')\,dt.\end{aligned}$$

For the comparison of the changes of L along γ_2 and γ_2', we need to further split γ_2 into three subpieces. Let y_1' and y_3' be the y values where γ_1' and γ_3' meet $\{(x^*, y)\}$. Let $\tilde{\gamma}_2$ be the part of γ_2 between y_1' and y_3'. We use this part of the trajectory as a graph over y and notice that y is decreasing. Also let y_1 and y_3 be the y values where γ_1 and γ_3 meet $\{(x^*, y)\}$. Then,

$$\begin{aligned}\int_{\gamma_2} \dot{L}(x,y)\,dt &= -\int_{y_1}^{y_1'} f(x)\,dy - \int_{y_3'}^{y_1'} f(x)\,dy - \int_{y_3}^{y_3'} f(x)\,dy \\ &< -\int_{y_3'}^{y_1'} f(x)\,dy \\ &< -\int_{y_3'}^{y_1'} f(x')\,dy \\ &= \int_{\gamma_2'} \dot{L}(x',y')\,dt.\end{aligned}$$

Note that

$$-\int_{\tilde{\gamma}_2} f(x)\,dy < -\int_{\tilde{\gamma}_2'} f(x)\,dy$$

because the x values on γ_2 are larger than the corresponding x' values on γ_2' so $f(x) > f(x')$ and the integral is more negative. Combining the comparisons of integrals shown above, we get that $\delta(\mathbf{p}) < \delta(\mathbf{p}')$. This proves part (b).

To get that the limit of $\delta(\mathbf{p})$ equals minus infinity, note that as $|\mathbf{p}|$ goes to infinity, the x values on $\tilde{\gamma}_2$ go to infinity, so

$$\delta(\mathbf{p}) < -\int_{y_3'}^{y_1'} f(x)\,dy \to -\infty.$$

□

PROOF OF THEOREM 8.8. We showed above that $\mathbf{p} \in \mathbf{v}^+$ is on a periodic orbit if and only if $\delta(\mathbf{p}) = 0$. Lemma 8.11 proved that $\delta(\mathbf{p})$ is positive for $|\mathbf{p}| \leq r^*$. For $|\mathbf{p}| \geq r^*$, $\delta(\mathbf{p})$ is monotonically decreasing, so it has a unique zero at some $\mathbf{p}_0$ with $|\mathbf{p}_0| > r^*$. Thus, there is a unique periodic orbit.

Further, for any $\mathbf{p}$ with $|\mathbf{p}| > |\mathbf{p}_0|$, $\delta(\mathbf{p}) < 0$. The trajectory for $\mathbf{p}$ can never cross the periodic orbit through $\mathbf{p}_0$, so it stays outside this periodic orbit and $\delta(\sigma^j(\mathbf{p})) < 0$ for $j > 0$. Therefore, $|\sigma^j(\mathbf{p})|$ is monotonically decreasing, and the solution comes inward toward $\mathbf{p}_0$ with each revolution. The limit of the $\delta(\sigma^j(\mathbf{p}))$ must be a fixed point of σ and so must be $\mathbf{p}_0$. Then, the trajectory for $\mathbf{p}$ limits on the periodic orbit for $\mathbf{p}_0$. Similarly, if $|\mathbf{p}| < |\mathbf{p}_0|$, then $\delta(\sigma^j(\mathbf{p})) > 0$, $|\sigma^j(\mathbf{p})|$ is monotonically increasing, and so $\delta(\sigma^j(\mathbf{p}))$ must converge to $\mathbf{p}_0$. This completes the proof of the theorem. □

5.8.3 Poincaré Map for Differential Equations in the Plane

In this subsection, we consider a differential equation in the plane given by

$$\begin{aligned}\dot{x} &= X(x, y)\\ \dot{y} &= Y(x, y).\end{aligned}$$

Let $V(x, y) = \begin{pmatrix} X(x,y) \\ Y(x,y) \end{pmatrix}$ be the corresponding vector field. Denote the divergence of V at $\mathbf{q}$ by $(\operatorname{div} V)(\mathbf{q})$. Assume that Σ is a transversal and $\Sigma' \subset \Sigma$ is an open subset on which the Poincaré map is defined, $P : \Sigma' \to \Sigma$. Since the differential equations are in the plane, Σ is a curve which can be parameterized by $\gamma : I \to \Sigma$ with $\gamma(I') = \Sigma'$ and $|\gamma'(s)| = 1$. Let $V_\perp(\mathbf{q})$ be the scalar component of V perpendicular to the tangent line to Σ at $\mathbf{q}$ given by

$$V_\perp \circ \gamma(s) = \det(\gamma'(s), V \circ \gamma(s)).$$

In the case where Σ is a horizontal line, $\{(x, y^*) : x_1 < x < x_2\}$, then $V_\perp(\mathbf{q}) = Y(\mathbf{q})$. In the case where Σ is a vertical line, $\{(x^*, y) : y_1 < y < y_2\}$, then $V_\perp(\mathbf{q}) = -X(\mathbf{q})$.

As in the general case, let $\tau(\mathbf{q})$ be the return time for $\mathbf{q} \in \Sigma'$, so $P(\mathbf{q}) = \varphi^{\tau(\mathbf{q})}(\mathbf{q})$. With these definitions and notation, we can state the main theorem of this subsection.

Theorem 8.12. *Let $\gamma : I' \to \Sigma'$ be a parameterization of the transversal Σ' as above with $|\gamma'(s)| = 1$. Then, for $s \in I'$,*

$$(P \circ \gamma)'(s) = \frac{V_\perp \circ \gamma(s)}{V_\perp \circ P \circ \gamma(s)} \exp\Big(\int_0^{\tau \circ \gamma(s)} (\operatorname{div} V) \circ \varphi^t \circ \gamma(s)\, dt \Big).$$

In particular, if $P(\mathbf{q}_0) = \mathbf{q}_0$ and $\gamma(s_0) = \mathbf{q}_0$, then

$$(P \circ \gamma)'(s_0) = \exp\Big(\int_0^{\tau(\mathbf{q}_0)} (\operatorname{div} V) \circ \varphi^t(\mathbf{q}_0)\, dt \Big).$$

REMARK 8.3. This theorem is contained in Section 28 of Andronov, Leontovich, Gordon, and Maier (1973). The case where $P(\mathbf{q}_0) = \mathbf{q}_0$ is the one most often used. In the application in the example below, we use the general case.

PROOF. The first variation equation states that

$$\frac{d}{dt} D\varphi^t_{\mathbf{q}} = DV_{\varphi^t(\mathbf{q})} D\varphi^t_{\mathbf{q}}.$$

Since $\det(D\varphi^0_{\mathbf{q}}) = \det(id) = 1$, Liouville's formula for time-dependent linear equations gives that

$$\det(D\varphi^{\tau(\mathbf{q})}_{\mathbf{q}}) = \exp\Big(\int_0^{\tau(\mathbf{q})} (\operatorname{div} V) \circ \varphi^t(\mathbf{q})\, dt\Big).$$

(We leave it as an exercise to prove this time-dependent version of Liouville's formula. See Exercise 5.35.) Notice that the right-hand side of this equality is the integral in the formula for $(P \circ \gamma)'(s)$ as stated in the theorem. Therefore, to complete the proof, we must relate $(P \circ \gamma)'(s)$ with $\det(D\varphi^{\tau\circ\gamma(s)}_{\gamma(s)})$.

Taking the derivative of $P \circ \gamma(s) = \varphi^{\tau\circ\gamma(s)}(\gamma(s))$ with respect to s yields

$$\begin{aligned}(P \circ \gamma)'(s) &= (D\varphi^{\tau\circ\gamma(s)}_{\gamma(s)})\gamma'(s) + (\tau \circ \gamma)'(s)[V \circ \varphi^{\tau\circ\gamma(s)}(\gamma(s))]\\ &= (D\varphi^{\tau\circ\gamma(s)}_{\gamma(s)})\gamma'(s) + (\tau \circ \gamma)'(s)[V \circ P \circ \gamma(s)].\end{aligned}$$

Then,

$$\begin{aligned}&(P \circ \gamma)'(s)[V_\perp \circ P \circ \gamma(s)]\\ &\quad= \det((P \circ \gamma)'(s), V \circ P \circ \gamma(s))\\ &\quad= \det((D\varphi^{\tau\circ\gamma(s)}_{\gamma(s)})\gamma'(s), V \circ P \circ \gamma(s))\\ &\qquad\qquad + \det((\tau \circ \gamma)'(s)[V \circ P \circ \gamma(s)], V \circ P \circ \gamma(s))\\ &\quad= \det((D\varphi^{\tau\circ\gamma(s)}_{\gamma(s)})\gamma'(s), (D\varphi^{\tau\circ\gamma(s)}_{\gamma(s)})V \circ \gamma(s))\\ &\quad= \det(D\varphi^{\tau\circ\gamma(s)}_{\gamma(s)}) \det(\gamma'(s), V \circ \gamma(s))\\ &\quad= \exp\Big(\int_0^{\tau\circ\gamma(s)} (\operatorname{div} V) \circ \varphi^t \circ \gamma(s)\, dt\Big) V_\perp \circ \gamma(s).\end{aligned}$$

Dividing by $V_\perp \circ P \circ \gamma(s)$ gives the desired formula. □

Example 8.3. Consider the Volterra-Lotka equations which model the populations of two species which are predator and prey:

$$\begin{aligned}\dot{x} &= x(A - By)\\ \dot{y} &= y(Cx - D)\end{aligned}$$

with all $A, B, C, D > 0$. There is a unique fixed point in the interior of the first quadrant with $x^* = D/C$ and $y^* = A/B$. Let $\Sigma = \{(x, y^*) : x^* \le x < \infty\}$. By using arguments like those for the Van der Pol equation, it can be shown that every point (x, y^*) of Σ with $x > x^*$ returns to Σ, $P : \Sigma \to \Sigma$. The argument below shows that this map extends differentiably so that $P(x^*) = x^*$. We show that all points in the open first quadrant except (x^*, y^*) lie on periodic orbits by applying Theorem 8.12. This fact is usually verified by finding a real valued function L which is constant on orbits, $\dot{L} \equiv 0$.

Let V be the vector field for the above differential equations. The divergence of V is given by

$$\begin{aligned}(\operatorname{div} V)(x, y) &= (A - By) + (Cx - D)\\ &= \frac{\dot{x}}{x} + \frac{\dot{y}}{y}.\end{aligned}$$

We write $P(x)$ for the x-value of the Poincaré map of the point (x, y^*). The integral in Theorem 8.12 becomes

$$\exp\Big(\int_0^{T(x)} \big(\frac{\dot{x}}{x} + \frac{\dot{y}}{y}\big)\, dt\Big) = \exp\Big(\int_x^{P(x)} \frac{1}{x}\, dx + \int_{y^*}^{y^*} \frac{1}{y}\, dy\Big) = \frac{P(x)}{x}.$$

Applying the formula of Theorem 8.12 yields

$$P'(x) = \frac{Y(x, y^*)}{Y(P(x), y^*)} \cdot \frac{P(x)}{x}.$$

Defining $f(s) = Y(s, y^*)/s$, we get that

$$f \circ P(x) P'(x) = f(x).$$

Integrating from x^* to x yields

$$F \circ P(x) - F \circ P(x^*) = F(x) - F(x^*),$$

where F is the antiderivative of f. Since $P(x^*) = x^*$, this gives

$$F \circ P(x) = F(x).$$

Since $f(s) > 0$ for $s > x^*$, $F(s)$ is strictly monotonically increasing. Therefore, the fact that $F \circ P(x) \equiv F(x)$ implies that $P(x) \equiv x$. This completes the proof that all orbits are periodic.

For other examples applying Theorem 8.12, see Robinson (1985).

5.9 Poincaré-Bendixson Theorem

The Poincaré-Bendixson Theorem is a result about flows in a region in the plane or on the two sphere, S^2. The reason for the restriction to these domains is that it depends on the Jordan Curve Theorem. It also depends on the fact that a transversal is one-dimensional. The conclusion of the theorem is that the ω-limit set of a point is either a closed orbit or contains a fixed point. In order to insure that the ω-limit set is nonempty, we need to assume that the forward orbit is bounded, i.e., $\mathcal{O}^+(p)$ is contained in a compact subset of the domain. See Section 5.4 for examples of limit sets for flows in the plane. For other references with more details on this result, see Hale (1969), Hartman (1964), and Hirsch and Smale (1974).

Theorem 9.1 (Poincaré-Bendixson Theorem). *Let $\mathcal{D}$ be a planar domain, i.e., either a simply connected subset of $\mathbb{R}^2$ or $\mathcal{D} = S^2$. Let $\varphi^t(\mathbf{x})$ be a C^1 flow on $\mathcal{D}$. Let $\mathbf{p} \in \mathcal{D}$ be a point such that $\mathcal{O}^+(\mathbf{p})$ is bounded and $\omega(\mathbf{p})$ does not contain any fixed points. Then, $\omega(\mathbf{p})$ is a periodic orbit. If we replace the assumptions on the forward orbit with the assumption that $\mathcal{O}^-(\mathbf{p})$ is bounded and $\alpha(\mathbf{p})$ does not contain any fixed points, then the conclusion is that $\alpha(\mathbf{p})$ is a periodic orbit.*

In the case where the original point $\mathbf{p}$ is not itself on a closed orbit, the closed orbit is called a limit cycle. Thus, a *limit cycle* is a periodic orbit γ such that there exists a point $\mathbf{p} \notin \gamma$ with $\omega(\mathbf{p}) = \gamma$. When applying the theorem to prove the existence of a periodic orbit, this periodic orbit is usually a limit cycle.

The above theorem can be used to prove the existence of a periodic orbit as the following corollary shows.

Corollary 9.2. *Let φ^t be a C^1 flow on a planar region $\mathcal{D}$ as in Theorem 1. Let $\mathcal{R} \subset \mathcal{D}$ be a positively invariant bounded region such that $\mathcal{R}$ does not contain any fixed points. Then, $\mathcal{R}$ contains a periodic orbit.*

The proof of this corollary follows immediately from the Poincaré-Bendixson Theorem. This corollary can be applied to specific equations as the following example shows.

Example 9.1. Consider the equations

$$\dot{x} = y$$
$$\dot{y} = -x + y(1 - x^2 - 2y^2).$$

Let r be the polar coordinate. Then, along a solution,

$$\begin{aligned} \frac{d}{dt}\left(\frac{r^2}{2}\right) &= x\dot{x} + y\dot{y} \\ &= xy - xy + y^2(1 - x^2 - 2y^2) \\ &= y^2(1 - x^2 - 2y^2). \end{aligned}$$

Then, $\dot{r} \geq 0$ on $r = 2^{-1/2}$ and $\dot{r} \leq 0$ on $r = 1$. Thus, the annulus $\{(x, y) : 1/2 \leq r^2 \leq 1\}$ is positively invariant. It also contains no fixed points. Therefore, this annulus contains a periodic orbit.

Now, we begin the proof of the Poincaré-Bendixson Theorem.

PROOF OF THE POINCARÉ-BENDIXSON THEOREM. Let L_0 be a connected open transversal to the flow, i.e., L_0 is the image of an open interval. Let L be a connected open (sub-)transversal with $\text{cl}(L) \subset L_0$. Let

$$L' = \{\mathbf{x} \in L : \text{ there is a } t_{\mathbf{x}} > 0 \text{ such that } \varphi^{t_{\mathbf{x}}}(\mathbf{x}) \in L\}.$$

We assume that we are dealing with an L for which L' is nonempty. For $\mathbf{x} \in L'$, let $\tau(\mathbf{x}) > 0$ be the smallest time such that $\varphi^t(\mathbf{x}) \in L$. Then, $g(\mathbf{x}) = \varphi^{\tau(\mathbf{x})}(\mathbf{x})$ is the Poincaré map and depends differentiably on $\mathbf{x} \in L'$, $g : L' \to L$. Below we often say we have a transversal L and do not mention L_0 or L'.

Lemma 9.3. *If $g^k(\mathbf{x})$ is defined for $k = 0, \ldots, n$, then $g^k(\mathbf{x})$ is a monotone function of k in terms of the ordering on L_0.*

PROOF. Consider the Jordan curve Γ formed by the trajectory $\{\varphi^t(\mathbf{x}) : 0 \leq t \leq \tau(\mathbf{x})\}$ and the line segment on L' connecting $\mathbf{x}$ and $g(\mathbf{x})$. Then, Γ is the boundary of a subset $D \subset \mathcal{D}$. The trajectories can only leave or enter D across the line segment on L'. Points are either all leaving across this segment or all entering. If they are all entering, then $g^k(\mathbf{x}) \in \text{int}(D)$ for $k \geq 1$. Thus, the entire sequence is on the same side of $\mathbf{x}$ as $g(\mathbf{x})$. The same argument applies to all the iterates. This shows the sequence is monotone. □

Lemma 9.4. *Let L be a transversal as above. Then, $\omega(\mathbf{p}) \cap L$ is at most one point.*

PROOF. Assume $\mathbf{y}_0 \in \omega(\mathbf{p}) \cap L$. Then, $\varphi^t(\mathbf{p})$ gets near $\mathbf{y}_0$ infinitely often. If it gets near enough to L (in a closed neighborhood of $\mathbf{y}_0$), then it has to cross L. (This follows by methods like the proof of the existence of the Poincaré map using the Implicit Function Theorem.) Thus, the trajectory crosses L infinitely often. The sequence of points where it crosses L is monotone by Lemma 9.3. A monotone sequence can accumulate on at most one point, so $L \cap \omega(\mathbf{p}) = \{\mathbf{y}_0\}$. This proves the lemma. □

Lemma 9.5. *Assume $\mathcal{O}(\mathbf{p})$ is bounded, $\omega(\mathbf{p})$ contains no fixed points, and $\mathbf{q} \in \omega(\mathbf{p})$. Then, $\mathbf{q}$ is on a periodic orbit, and $\omega(\mathbf{p}) = \omega(\mathbf{q}) = \mathcal{O}(\mathbf{q})$.*

PROOF. Since $\mathbf{q} \in \omega(\mathbf{p})$, $\omega(\mathbf{q}) \subset \omega(\mathbf{p})$. Take $\mathbf{z} \in \omega(\mathbf{q})$. Then, $\mathbf{z}$ cannot be a fixed point, so we can take a transversal L at $\mathbf{z}$. There is a sequence of times t_n such that $\varphi^{t_n}(\mathbf{q})$ accumulates on $\mathbf{z}$. These times can be chosen so that $\varphi^{t_n}(\mathbf{q}) \in L$; they also must be in $\omega(\mathbf{p})$ by its invariance. Since $\omega(\mathbf{p}) \cap L$ is at most one point, $\varphi^{t_n}(\mathbf{q}) = \varphi^{t_1}(\mathbf{q})$ for all n. Thus, $\varphi^{t_2 - t_1}(\mathbf{q}) = \mathbf{q}$ and $\mathbf{q}$ is a periodic point. But then, $\omega(\mathbf{q}) = \mathcal{O}(\mathbf{q}) = \{\varphi^t(\mathbf{q}) : 0 \leq t \leq \tau\}$, where $\tau = \tau(\mathbf{q})$ is the (minimal) period of $\mathbf{q}$.

It remains to prove that $\omega(\mathbf{p}) = \mathcal{O}(\mathbf{q})$. Take a transversal L at $\mathbf{q}$. There is a sequence of times s_n, going to infinity, such that $\mathbf{p}_n = \varphi^{s_n}(\mathbf{p}) \in L$ and these points converge to $\mathbf{q}$. Because $\mathbf{q}$ is a periodic point, the times can be chosen so that $s_{n+1} - s_n = \tau(\mathbf{p}_n) \approx \tau$, where $\tau = \tau(\mathbf{q})$ is the period of $\mathbf{q}$. These points are monotone on L, so they must accumulate on exactly one point from one side. Because the return times are bounded, it follows that the orbit $\mathcal{O}^+(\mathbf{p})$ can only accumulate on $\mathcal{O}(\mathbf{q})$ and $\omega(\mathbf{p}) = \mathcal{O}(\mathbf{p})$. This completes the proof of the lemma. □

These lemmas complete the proof of the Poincaré-Bendixson Theorem. □

The following generalization of the Poincaré-Bendixson Theorem allows $\omega(\mathbf{p})$ to contain a fixed point. See page 56 of Hale (1969) for a proof.

Theorem 9.6. *Assume that $\mathcal{O}^+(\mathbf{p})$ is contained in a closed bounded subset K of the planar domain $\mathcal{D}$ of the differential equation $\dot{\mathbf{x}} = f(\mathbf{x})$. Assume further that f has only finitely many fixed points in K. Then, one of the following is satisfied:*

(a) *$\omega(\mathbf{p})$ is a periodic orbit,*
(b) *$\omega(\mathbf{p})$ is a single fixed point of f,*
(c) *$\omega(\mathbf{p})$ consists of a finite number of fixed points, $\mathbf{q}_1, \ldots, \mathbf{q}_m$, together with a finite set of orbits $\gamma_1, \ldots, \gamma_n$, such that for each j, $\alpha(\gamma_j)$ is a single fixed point and $\omega(\gamma_j)$ is a single fixed point: $\omega(\mathbf{p}) = \{\mathbf{q}_1, \ldots, \mathbf{q}_m\} \cup \gamma_1 \cup \cdots \cup \gamma_n$, $\alpha(\gamma_j) = \mathbf{q}_{i(j,\alpha)}$, and $\omega(\gamma_j) = \mathbf{q}_{i(j,\omega)}$.*

5.10 Stable Manifold Theorem for a Fixed Point of a Map

In this section we consider a C^k differentiable map on a Banach space, $f : U \subset \mathbb{E} \to \mathbb{E}$. We allow f to be non-invertible and not onto. Assume $\mathbf{p}$ is a fixed point, and let $A = Df_{\mathbf{p}}$ be the derivative at $\mathbf{p}$. The *spectrum* of A is the set of complex numbers for which $A - \lambda I$ is not an isomorphism,

$$\text{spec}(A) = \{\lambda \in C : A - \lambda I \text{ is not an isomorphism}\}.$$

In finite dimensions the spectrum is the same as the set of eigenvalues. A fixed point $\mathbf{p}$ for f is called *hyperbolic* if $\text{spec}(Df_{\mathbf{p}}) \cap \{\alpha : |\alpha| = 1\} = \emptyset$. By standard results in Spectral Theory, there are invariant subspaces $\mathbb{E}^u$ and $\mathbb{E}^s$ corresponding to the part of the spectrum outside and inside the unit circle, and constants $0 < \mu < 1$ and $\lambda > 1$ such that $\mathbb{E} = \mathbb{E}^u \oplus \mathbb{E}^s$, $\text{spec}(Df_{\mathbf{p}}|\mathbb{E}^u) \subset \{\alpha : |\alpha| > \lambda\}$, and $\text{spec}(Df_{\mathbf{p}}|\mathbb{E}^s) \subset \{\alpha : |\alpha| < \mu\}$. In fact, we identify $\mathbb{E}$ with $\mathbb{E}^u \oplus \mathbb{E}^s$, so we write a point $\mathbf{x} \in \mathbb{E}$ as $(\mathbf{x}^u, \mathbf{x}^s)$ where $\mathbf{x}^\sigma \in \mathbb{E}^\sigma$ for $\sigma = u, s$. In finite dimensions, these correspond to the subspaces spanned by the generalized eigenvectors for the eigenvalues of absolute value greater than 1 and less than 1, respectively. Because the spectrum of $\mathbb{E}^u$ is bounded away from 0 (and by the construction of $\mathbb{E}^u$), $Df_{\mathbf{p}}|\mathbb{E}^u$ is an isomorphism on $\mathbb{E}^u$. Further, there is $C > 0$ such that $\|Df^n_{\mathbf{p}}|\mathbb{E}^s\| < C\mu^n$ and $\|Df^{-n}_{\mathbf{p}}|\mathbb{E}^u\| < C\lambda^{-n}$ for $n > 0$. By the usual change in the norm on $\mathbb{E}$, we can take $C = 1$. Such a norm is called an *adapted norm* or *adapted metric*. The subspaces $\mathbb{E}^s$ and $\mathbb{E}^u$ are called the *stable* and *unstable subspaces* for the fixed point $\mathbf{p}$, respectively.

Given a hyperbolic fixed point $\mathbf{p}$ for a C^k map f, and given a neighborhood $U' \subset U$ of $\mathbf{p}$, the *local stable manifold* for $\mathbf{p}$ in the neighborhood U' is defined to be the following set:

$$W^s(\mathbf{p}, U', f) = \{\mathbf{q} \in U' : f^j(\mathbf{q}) \in U' \text{ for } j > 0 \text{ and } d(f^j(\mathbf{q}), \mathbf{p}) \to 0 \text{ as } j \to \infty\}.$$

To define the unstable manifold, we need to look at the past history. Because f is not necessarily invertible, we need a replacement for the backward iterates. We define a *past history of a point* $\mathbf{q}$ to be a sequence of points $\{\mathbf{q}_{-j}\}_{j=0}^{\infty}$ such that $\mathbf{q}_0 = \mathbf{q}$ and $f(\mathbf{q}_{-j-1}) = \mathbf{q}_{-j}$ for $j \geq 0$. The *local unstable manifold* for $\mathbf{p}$ in U' is defined to be the following set:

$$W^u(\mathbf{p}, U', f) = \{\mathbf{q} \in U' : \text{there exists some choice of the past history of } \mathbf{q} \ \{\mathbf{q}_{-j}\}_{j=0}^{\infty} \subset U' \text{ such that } d(\mathbf{q}_{-j}, \mathbf{p}) \to 0 \text{ as } j \to \infty\}.$$

Sometimes we write $W^s_{\text{loc}}(\mathbf{p}, f)$ and $W^u_{\text{loc}}(\mathbf{p}, f)$ to indicate local stable and unstable manifolds for a suitably small but not specified neighborhood U'. We also write $W^s_\epsilon(\mathbf{p}, f)$ for $W^s(\mathbf{p}, B(\mathbf{p}, \epsilon), f)$, and similarly $W^u_\epsilon(\mathbf{p}, f)$ for $W^u(\mathbf{p}, B(\mathbf{p}, \epsilon), f)$.

The following theorem states that these local stable and unstable manifolds are C^k embedded manifolds which can be represented as the graph of a map from a disk in one of the subspaces to the other subspace. The Hartman – Grobman Theorem already proves that the stable and unstable manifolds are topological disks but it does not prove they are differentiable. In fact, the hard part of the proof of the Stable Manifold Theorem is to show that these manifolds are Lipschitz. Once this is known, it can be shown they are differentiable. Again, the Hartman – Grobman Theorem does not prove they are Lipschitz. On the other hand, the Hartman – Grobman Theorem proves that all the orbits near the fixed point behave like the linear map, while the Stable Manifold Theorem only gives information about points on the stable and unstable manifolds.

To represent a closed disk in one of the subspaces, we use the following notation: for any Banach space $\mathcal{E}$ and $\delta > 0$, the closed disk in $\mathcal{E}$ about the origin of radius δ is represented by $\mathcal{E}(\delta) = \{\mathbf{x} \in \mathcal{E} : |\mathbf{x}| \leq \delta\}$.

Theorem 10.1 (Stable Manifold Theorem). *Let $\mathbf{p}$ be a hyperbolic fixed point for a C^k map $f : U \subset \mathbb{E} \to \mathbb{E}$ with $k \geq 1$. We assume that the derivatives are uniformly continuous in terms of the point at which the derivative is taken. Then, there is some neighborhood of $\mathbf{p}$, $U' \subset U$, such that $W^s(\mathbf{p}, U', f)$ and $W^u(\mathbf{p}, U', f)$ are each C^k embedded disks which are tangent to $\mathbb{E}^s$ and $\mathbb{E}^u$, respectively. In fact, considering $\mathbb{E} = \mathbb{E}^u \times \mathbb{E}^s$, there is a small $r > 0$ such that taking $U' \equiv \mathbf{p} + (\mathbb{E}^u(r) \times \mathbb{E}^s(r))$, $W^s(\mathbf{p}, U', f)$ is the graph of a C^k function $\sigma^s : \mathbb{E}^s(r) \to \mathbb{E}^u(r)$ with $\sigma^s(\mathbf{0}) = \mathbf{0}$ and $D\sigma^s_{\mathbf{0}} = \mathbf{0}$:*

$$W^s(\mathbf{p}, U', f) = \{\mathbf{p} + (\sigma^s(\mathbf{y}), \mathbf{y}) : \mathbf{y} \in \mathbb{E}^s(r)\}.$$

Similarly, there is a C^k function $\sigma^u : \mathbb{E}^u(r) \to \mathbb{E}^s(r)$ with $\sigma^u(\mathbf{0}) = \mathbf{0}$ and $D\sigma^u_{\mathbf{0}} = \mathbf{0}$ such that

$$W^u(\mathbf{p}, U', f) = \{\mathbf{p} + (\mathbf{x}, \sigma^u(\mathbf{x})) : \mathbf{x} \in \mathbb{E}^u(r)\}.$$

Moreover, for $r > 0$ small enough and $U' = \mathbf{p} + (\mathbb{E}^u(r) \times \mathbb{E}^s(r))$,

$$\begin{aligned} W^s(\mathbf{p}, U', f) &= \{\mathbf{q} \in U' : f^j(\mathbf{q}) \in U' \text{ for } j \geq 0\} \\ &= \{\mathbf{q} \in U' : f^j(\mathbf{q}) \in U' \text{ for } j \geq 0 \text{ and } d(f^j(\mathbf{q}), \mathbf{p}) \leq \mu^j d(\mathbf{q}, \mathbf{p}) \text{ for all } j \geq 0\}. \end{aligned}$$

This means that every point that is not on $W^s(\mathbf{p}, U', f)$ leaves U' under forward iteration, and that points on $W^s(\mathbf{p}, U', f)$ converge to $\mathbf{p}$ at an exponential rate given by the bound on the stable spectrum. Similarly,

$$\begin{aligned} W^u(\mathbf{p}, U', f) &= \{\mathbf{q} \in U' : \text{there exists some choice of the past history of } \mathbf{q} \\ &\qquad \text{with } \{\mathbf{q}_{-j}\}_{j=0}^{\infty} \subset U'\} \\ &= \{\mathbf{q} \in U' : \text{there exists some choice of the past history of } \mathbf{q} \\ &\qquad \text{with } \{\mathbf{q}_{-j}\}_{j=0}^{\infty} \subset U' \text{ and} \\ &\qquad d(\mathbf{q}_{-j}, \mathbf{p}) \leq \lambda^{-j} d(\mathbf{q}, \mathbf{p}) \text{ for all } j \geq 0\}. \end{aligned}$$

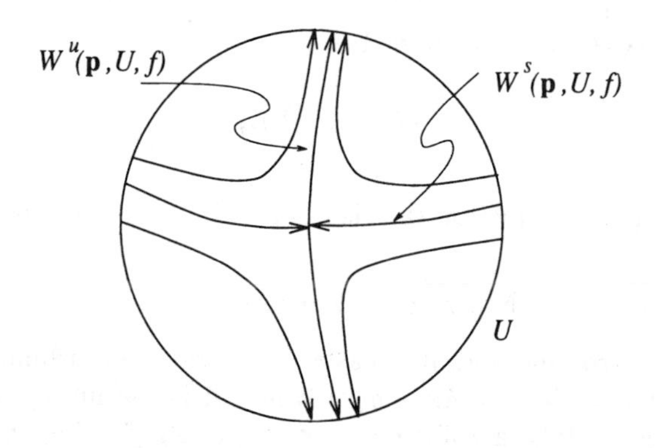

FIGURE 10.1. Stable and Unstable Manifolds in a Neighborhood of the Fixed Point

Once we have local stable and unstable manifolds, then the *(global) unstable manifold* is obtained by

$$W^u(\mathbf{p}, f) = \bigcup_{j \geq 0} f^j W^u(\mathbf{p}, U', f).$$

If f is invertible, then the *(global) stable manifold* is obtained by

$$W^s(\mathbf{p}, f) = \bigcup_{j \geq 0} f^{-j} W^s(\mathbf{p}, U', f).$$

We end this section with a discussion of the types of proofs of the Stable Manifold Theorem. There are two basic types, the Graph Transform Method of Hadamard (1901) and the variation of parameters method of Perron (1929). For historical notes, see Hartman (1964), page 271. In this discussion we take $\mathbf{p} = \mathbf{0}$.

Graph Transform Method of Hadamard

In the Graph Transform Method, the approach is to take a trial function $\sigma : \mathbb{E}^u(r) \to \mathbb{E}^s(r)$ which might possibly give the unstable manifold. A new function $\Gamma(\sigma) : \mathbb{E}^u(r) \to \mathbb{E}^s(r)$ is defined so that

$$\text{graph}[\Gamma(\sigma)] = f(\text{graph}(\sigma)) \cap (\mathbb{E}^u(r) \times \mathbb{E}^s(r)).$$

The set of all such possible trial functions is defined by

$$\Sigma(r, L) = \{\sigma : \mathbb{E}^u(r) \to \mathbb{E}^s(r) : \sigma \text{ is continuous}, \sigma(\mathbf{0}) = \mathbf{0}, \text{Lip}(\sigma) \leq L\}.$$

It is shown that $\Gamma : \Sigma(r, L) \to \Sigma(r, L)$ is a contraction in the C^0-sup topology. It thus has a fixed point function, $\Gamma(\sigma^u) = \sigma^u$, which gives the local unstable manifold as the graph of a Lipschitz function. Because the map Γ is not a contraction on the function space if the function space is given the C^1 or C^k topology, the fixed point needs to be proven to be C^1 (and then C^k) as a second step by different means. Notice that in this method, only one iterate of f is used to define Γ. For this approach, see Hirsch and Pugh (1970), Hirsch, Pugh, and Shub (1977), and Shub (1987). Also see Section 12.1.

In the proof presented in the next section, we use a modification of the Hadamard proof. Let

$$B_N = \bigcap_{j=0}^{N-1} f^j(\mathbb{E}^u(r) \times \mathbb{E}^s(r)).$$

Then, $B_{N+1} = f(B_N) \cap B_0$. It is proved that

$$B_\infty = \bigcap_{j=0}^{\infty} B_j$$

is the graph of a Lipschitz function that is in fact C^1. It is then proved that it is C^k by induction on k.

Variation of Parameters Method of Perron

To explain the Perron method, it is easier to consider an ordinary differential equation. Assume that $\dot{\mathbf{x}} = f(\mathbf{x}) = A\mathbf{x} + g(\mathbf{x})$ where A is the linear part at $\mathbf{0}$, $g(\mathbf{0}) = \mathbf{0}$, and $Dg_{\mathbf{0}} = \mathbf{0}$. Again, take a splitting so $\mathbf{x} = (\mathbf{x}_u, \mathbf{x}_s)^{tr}$. Applying the variation of parameters to the equation $\dot{\mathbf{x}}(t) = A\mathbf{x}(t) + g(\mathbf{x}(t))$, and thinking of the second term as a known nonhomogeneous term, we get

$$\mathbf{x}(t_2) = e^{A(t_2-t_1)}\mathbf{x}(t_1) + \int_{t_1}^{t_2} e^{A(t_2-s)} g(\mathbf{x}(s))\, ds.$$

We then break this equation into the stable and unstable components. The stable component is specified with an initial condition at $t = 0$, $\mathbf{x}_s(0) = \mathbf{a}_s$, so taking $t_2 = t$ and $t_1 = 0$ for the stable component we get

$$\mathbf{x}_s(t) = e^{A_s t}\mathbf{a}_s + \int_0^t e^{A_s(t-s)} g_s(\mathbf{x}(s))\, ds.$$

On the other hand, we cannot specify the unstable component at $t = 0$, but it is determined by the fact that the unstable component goes to zero (or is at least bounded) as t goes to infinity. Therefore, taking $t_2 = t$ and $t_1 = \infty$ for the unstable component and $\mathbf{x}_u(\infty) = \mathbf{0}$, we get

$$\mathbf{x}_u(t) = \mathbf{0} + \int_\infty^t e^{A_u(t-s)} g_u(\mathbf{x}(s))\, ds.$$

Combining the right-hand sides of these two equations, we can get a transformation of trial solutions on the stable manifold. To get a transformation of the whole stable manifold, we look at trial solutions $\mathbf{x}(t, \mathbf{a}_s)$ where $\mathbf{a}_s \in \mathbb{E}^s(r)$ parameterizes the stable component of the initial conditions, and let $\mathcal{F}$ be the function space of such trial sets of solutions on the stable manifold. A transform $T : \mathcal{F} \to \mathcal{F}$ is defined by

$$\begin{aligned}[T(\mathbf{x})](t, \mathbf{a}_s) = e^{A_s t}\mathbf{a}_s &+ \int_0^t e^{A_s(t-s)} g_s(\mathbf{x}(s, \mathbf{a}_s))\, ds \\ &+ \int_\infty^t e^{A_u(t-s)} g_u(\mathbf{x}(s, \mathbf{a}_s))\, ds,\end{aligned}$$

which takes one parameterized set of potential solutions on the stable manifold into another set. This transformation can be shown to be a contraction which has a fixed point which gives the stable manifold. Notice that one iterate of T uses the whole trial orbit of $\mathbf{x}(t, \mathbf{a}_s)$ which is pulled back with the linear flow. For this approach, see Hale (1969), Chow and Hale (1982), or Kelley (1967).

The proof in Irwin's book, Irwin (1980), is a mixture of these methods. For N the natural numbers, let

$$\mathcal{L}(\mathbb{E}(r)) = \{\sigma \in C^0(N, \mathbb{E}(r)) : \sigma(j) \to \mathbf{0} \text{ as } j \to \infty\}.$$

A function is then defined,

$$\mathcal{F} : E^s(r) \times \mathcal{L}(\mathbb{E}(r)) \to \mathcal{L}(\mathbb{E}(r)),$$

by the formulas

$$\begin{aligned}\mathcal{F}(\mathbf{x}_s, \sigma)(0) &= (\mathbf{x}_s, \sigma(0) + A_u^{-1}[\sigma_u(1) - f_u(\sigma(0))]) \in \mathbb{E}^s(r) \times \mathbb{E}^u(r), \\ \mathcal{F}(\mathbf{x}_s, \sigma)(n) &= (f_s(\sigma(n-1)), \sigma(n) + A_u^{-1}[\sigma_u(n+1) - f_u(\sigma(n))]) \text{ for } n > 0.\end{aligned}$$

It is then shown that $\mathcal{F}$ is C^k and a contraction for each fixed $\mathbf{x}_s$, so there exists a C^k function $g : \mathbb{E}^s(r) \to \mathcal{L}(\mathbb{E}(r))$ with $\mathcal{F}(\mathbf{x}_s, g(\mathbf{x}_s)) = g(\mathbf{x}_s)$. Note that for this fixed function, $f(g(\mathbf{x}_s)(n)) = g(\mathbf{x}_s)(n+1)$. Then, $h(\mathbf{x}_s) \equiv g(\mathbf{x}_s)(0)$ is C^k and its image is the stable manifold. For details, see Irwin (1980).

Differentiability of the Stable Manifold

The differentiability of the stable manifold is very important for our applications. There are several methods of proving this property. The main difficulty is that the various transformations are not contractions of the set of possible C^1 manifolds so we cannot directly look at a contraction map on this space. Hirsch and Pugh (1970) used the Fiber Contraction Principle. See Exercise 12.4. We use the method of cones inspired by McGehee (1973). Also compare with the argument in Moser (1973) to show that there is a subshift (horseshoe) as a subsystem of a dynamical system. Both of these references are based on work of Conley. Following these references, we use the cones to not only prove that the stable manifold is Lipschitz but also to show that it is C^1.

Cones are closely related to the idea of hyperbolicity. If $L : \mathbb{R}^n \to \mathbb{R}^n$ is a hyperbolic linear map, and C^u is a cone of vectors roughly in the expanding direction, then $L(C^u) \subset C^u$, i.e., the cone of vectors is invariant by L. (See following subsection for the definition of a *cone of vectors*. Some authors use the name of a *sector* for what we call a cone.) If the exact expanding direction is not known, then it is often easier to find an invariant cone, C^u. Using such a cone, it is possible to show that L is hyperbolic.

The use of cones goes back at least to Alekseev (1968a). Sinai (1968,1970) also used the concept very early. As mentioned above, they are used in McGehee (1973) and Moser (1973). Recent uses of cones include Wojtkowski (1985), Burns and Gerber (1989), and Katok and Burns (1994).

5.10.1 Proof of the Stable Manifold Theorem

This subsection contains the proof of Theorem 10.1, the Stable Manifold Theorem where we take $\mathbf{p} = \mathbf{0}$. The proof presented here follows the method developed by Conley as given in McGehee (1973), but with some modifications which were influenced by Hirsch and Pugh (1970). Thus, it is in the spirit of the Hadamard proof (more than the Perron proof) but has some differences. There is also a lot of similarity to the argument of Conley given in Moser (1973) to show there is a subshift (horseshoe) as a subsystem of a dynamical system. The arguments in both McGehee (1973) and Moser (1973) are presented in the case of two dimensions where both the stable and unstable directions are one-dimensional. Some modifications need to be made to take care of the higher dimensional cases. Because the map is not assumed to be invertible, we need to indicate how we show both the stable and unstable manifolds exist.

OUTLINE OF THE PROOF. To start the proof, we want to show that $W^s_r(\mathbf{0})$ is the graph of a function $\varphi^s : \mathbb{E}^s(r) \to \mathbb{E}^u(r)$ which is Lipschitz with Lipschitz constant less than α (α-Lipschitz) for some fixed α. To verify that $W^s_r(\mathbf{0})$ is a graph, we use the characterization of $W^s_r(\mathbf{0})$ as points whose forward orbit remains in $\mathcal{B}(r) = \mathbb{E}^s(r) \times \mathbb{E}^u(r)$:

$$\begin{aligned} W^s_r(\mathbf{0}) &= \{\mathbf{x} : f^j(\mathbf{x}) \in \mathcal{B}(r) \text{ for all } j \geq 0\} \\ &= \bigcap_{j=0}^{\infty} f^{-j}(\mathcal{B}(r)). \end{aligned}$$

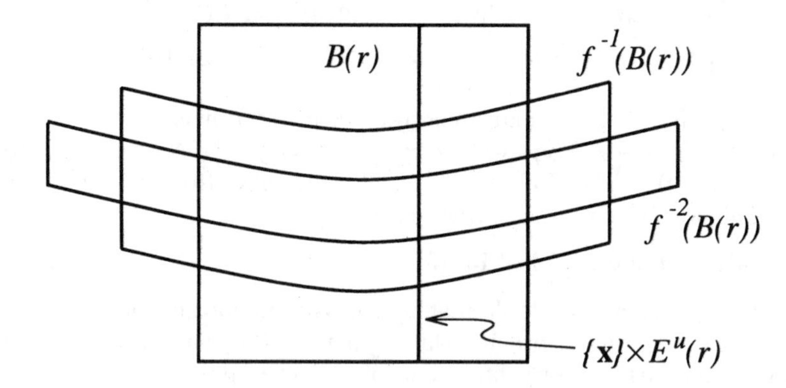

FIGURE 10.2. Intersection of the Boxes $f^{-n}(\mathcal{B}(r))$

See Figure 10.2. (Later in Proposition 10.7, we verify that these points tend to the fixed point.) In order for $W^s_r(\mathbf{0})$ to be a graph, we need for $W^s_r(\mathbf{0}) \cap [\{\mathbf{x}\} \times \mathbb{E}^u(r)]$ to be a single point for each $\mathbf{x} \in \mathbb{E}^s(r)$. The disk $\{\mathbf{x}\} \times \mathbb{E}^u(r)$ is a trivial example of a C^1 graph of a function from $\mathbb{E}^u(r)$ into $\mathbb{E}^s(r)$ with slope less than α^{-1}. We define an *unstable disk* to be the graph of a C^1 function $\varphi : \mathbb{E}^u(r) \to \mathbb{E}^s(r)$ with $\|D\varphi_{\mathbf{y}}\| \leq \alpha^{-1}$ for all $\mathbf{y} \in \mathbb{E}^u(r)$. Similarly, an α-Lipschitz *stable disk* is defined to be the graph of a function $\psi : \mathbb{E}^s(r) \to \mathbb{E}^u(r)$ with $\text{Lip}(\psi) \leq \alpha$. Using the bounds on the terms in the derivative of f, Lemma 10.5 shows that if D^u_0 is an unstable disk, then $D^u_1 = f(D^u_0) \cap \mathcal{B}(r)$ is an unstable disk (with slope less than α^{-1}) and

$$\text{diam}(D^u_0 \cap f^{-1}(D^u_1)) \leq (\lambda - \epsilon\alpha^{-1})^{-1} \,\text{diam}(D^u_0) = (\lambda - \epsilon\alpha^{-1})^{-1} 2r.$$

By induction, $D^u_j = f(D^u_{j-1}) \cap \mathcal{B}(r)$ is an unstable disk, and

$$\text{diam}(\bigcap_{j=0}^{n} f^{-j}(D^u_j)) \leq (\lambda - \epsilon\alpha^{-1})^{-n} 2r.$$

From this it follows that the infinite intersection is a single point which we call $(\mathbf{x}, \varphi^s(\mathbf{x}))$:

$$\begin{aligned} [\{\mathbf{x}\} \times \mathbb{E}^u(r)] \cap W^s_r(\mathbf{0}) &= [\{\mathbf{x}\} \times \mathbb{E}^u(r)] \cap \bigcap_{j=0}^{\infty} f^{-j}(\mathcal{B}(r)) \\ &= \{(\mathbf{x}, \varphi^s(\mathbf{x}))\}. \end{aligned}$$

This shows that $W_r^s(\mathbf{0})$ is a graph.

To show that it is α-Lipschitz, we consider the *stable and unstable cones* which are defined by

$$C^s(\alpha) = \{(\mathbf{v}_s, \mathbf{v}_u) \in \mathbb{E}^s \times \mathbb{E}^u : |\mathbf{v}_u| \le \alpha|\mathbf{v}_s|\} \qquad \text{and}$$
$$C^u(\alpha) = \{(\mathbf{v}_s, \mathbf{v}_u) \in \mathbb{E}^s \times \mathbb{E}^u : |\mathbf{v}_u| \ge \alpha|\mathbf{v}_s|\}.$$

See Figure 10.3. The condition that

$$W_r^s(\mathbf{0}) \cap [\{\mathbf{q}\} + C^u(\alpha)] = \{\mathbf{q}\}$$

for all points $\mathbf{q} \in W_r^s(\mathbf{0})$ is equivalent to the graph being α-Lipschitz. Proposition 10.6 shows that this intersection is indeed a single point by the same argument which shows that $W_r^s(\mathbf{0})$ is a graph.

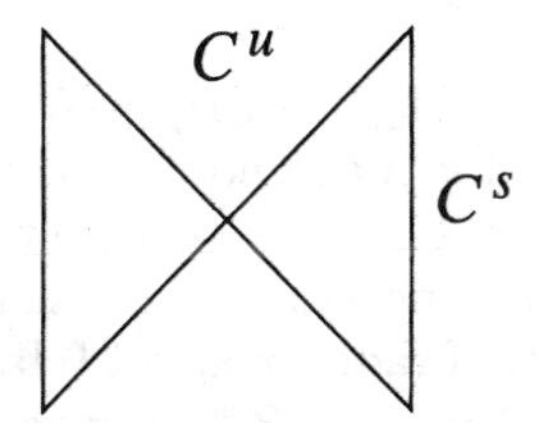

FIGURE 10.3. Stable and Unstable Cones

After we have shown that $W_r^s(\mathbf{0})$ is the graph of a Lipschitz function, Proposition 10.8 shows that it is C^1 and Theorem 10.10 proves that it is C^k by induction on k.

To carry out the above argument, we need some estimates about the effect of the map f on the stable and unstable components. Lemma 10.3 shows that the distance between the unstable components of two points $\mathbf{q}$ and $\mathbf{q}'$ is increased by the action of the nonlinear map f provided the displacement from $\mathbf{q}$ to $\mathbf{q}'$ lies in the unstable cone (is "mainly an unstable displacement"). Before making these nonlinear estimates, we give similar linear estimates in Lemma 10.2: the derivate $Df_{\mathbf{q}}$ preserves the unstable cones and stretches the length of vectors in the unstable cone. We prove this linear case first because it is easier and is used in Proposition 10.8 to prove that $W_r^s(\mathbf{0})$ is C^1.

REMARK 10.1. There are some differences in the proof given here from those in the references cited. In particular, if $\dim(\mathbb{E}^u) \neq 1$, then

$$\partial[\bigcap_{j=0}^{n} f^{-j}(\mathcal{B}(r))] \setminus [\partial\mathbb{E}^s(r) \times \mathbb{E}^u(r)]$$

need not be made up of two graphs over $\mathbb{E}^s(r)$, as is true in Moser (1973) where $\dim(\mathbb{E}^u) = 1$. The fact that f need not be invertible necessitates some care in the proof for the unstable manifold: $\partial[\cap_{j=0}^{n} f^j(\mathcal{B}(r))] \cap [\mathbb{E}^s(r) \times \{\mathbf{y}_0\}]$ need not be in the image by f of $\partial[\cap_{j=0}^{n-1} f^j(\mathcal{B}(r))]$, $f(\partial[\cap_{j=0}^{n-1} f^j(\mathcal{B}(r))])$. In general, greater care needs to be taken in several of the arguments when f is not assumed to be invertible.

To start carrying out the above outline, we first introduce some notation. It is convenient in expressing some of the ideas to use the minimum norm of a linear map which is first introduced in the Section 4.1. We also use the estimates from the Inverse

Function Theorem given in Section 5.2.2. (Those estimates should be reviewed at this time.) Remember that the *minimum norm* of a linear map $A : \mathbb{E} \to \mathbb{E}$ is defined by

$$m(A) = \inf_{\mathbf{v} \neq \mathbf{0}} \frac{|A\mathbf{v}|}{|\mathbf{v}|}.$$

The minimum norm is a measure of the minimum expansion of A just as $\|A\|$ is a measure of the maximum expansion. If A is invertible, then $m(A) = \|A^{-1}\|^{-1}$. For a hyperbolic fixed point $\mathbf{0}$ for which $\|Df_{\mathbf{0}}^{-n}|\mathbb{E}^u\| < C\lambda^{-n}$ for $n > 0$, it follows that $m(Df_{\mathbf{0}}^n|\mathbb{E}^u) > C\lambda^n$.

By taking local coordinates we can assume that the fixed point is $\mathbf{0}$ in the Banach space $\mathbb{E}$ and $\mathbb{E} = \mathbb{E}^s \times \mathbb{E}^u$, where these subspaces come from the hyperbolic splitting at this fixed point. (This cross product is isomorphic to the original Banach space.) Let $\pi_u : \mathbb{E} \to \mathbb{E}^u$ be the projection along $\mathbb{E}^s$ onto $\mathbb{E}^u$, and $\pi_s : \mathbb{E} \to \mathbb{E}^s$ be the projection along $\mathbb{E}^u$ onto $\mathbb{E}^s$. Then, $f(\mathbf{0}) = \mathbf{0}$ and we use the splitting $\mathbb{E}^s \times \mathbb{E}^u$ to write

$$Df_{\mathbf{0}} = \begin{pmatrix} A_{ss} & \mathbf{0} \\ \mathbf{0} & A_{uu} \end{pmatrix},$$

where $A_{ss} = \pi_s Df_{\mathbf{0}}|(\mathbb{E}^s \times \{\mathbf{0}\})$ and $A_{uu} = \pi_u Df_{\mathbf{0}}|(\{\mathbf{0}\} \times \mathbb{E}^u)$. Because the fixed point is hyperbolic, there exist $0 < \mu < 1 < \lambda$ and norms on each subspace $\mathbb{E}^s$ and $\mathbb{E}^u$ so that $m(A_{uu}) > \lambda > 1$ and $\|A_{ss}\| < \mu < 1$. We fix these norms, and on $\mathbb{E}$ we put the norm which is the maximum of the components in the $\mathbb{E}^s$ and $\mathbb{E}^u$ subspaces. In any Banach space $\mathbb{E}$, we denote the closed ball of radius r about $\mathbf{0}$ by $\mathbb{E}(r) = \{\mathbf{x} \in \mathbb{E} : |\mathbf{x}| \leq r\}$. We define a neighborhood of the fixed point in $\mathbb{E}$ by taking the cross product of the closed balls in $\mathbb{E}^u$ and $\mathbb{E}^s$ of radius r (which is also the closed ball in $\mathbb{E}$ because we are using the maximum of the norm in the stable and unstable subspaces):

$$\mathcal{B}(r) = \mathbb{E}^s(r) \times \mathbb{E}^u(r) \subset \mathbb{E}.$$

Then, $f : \mathcal{B}(r) \to \mathbb{E}$, with $f(\mathbf{0}) = \mathbf{0}$.

We fix $\alpha > 0$, which serves as a bound on the slopes of graphs over $\mathbb{E}^s$ into $\mathbb{E}^u$. For all but special cases, we can take $\alpha = 1$. We take $\epsilon > 0$ small enough so that

$$\begin{aligned} &\mu + \epsilon\alpha + \epsilon < 1 \qquad \text{and} \\ &\lambda - \epsilon\alpha^{-1} - 2\epsilon > 1. \end{aligned}$$

(These give bounds on the effects of the off diagonal terms to the expansion and contraction.) Given these α and ϵ, we can find $r > 0$ small enough so that for $\mathbf{q} \in \mathcal{B}(r)$,

$$Df_{\mathbf{q}} = \begin{pmatrix} A_{ss}(\mathbf{q}) & A_{su}(\mathbf{q}) \\ A_{us}(\mathbf{q}) & A_{uu}(\mathbf{q}) \end{pmatrix}$$

with

$$\begin{aligned} \|A_{ss}(\mathbf{q})\| &< \mu, \\ m(A_{uu}(\mathbf{q})) &> \lambda, \\ \|A_{su}(\mathbf{q})\| &< \epsilon, \\ \|A_{us}(\mathbf{q})\| &< \epsilon, \qquad \text{and} \\ \|Df_{\mathbf{q}} - Df_{\mathbf{0}}\| &< \epsilon. \end{aligned}$$

We say that f *satisfies hyperbolic estimates* in a neighborhood in which estimates of the above type are true. This name is justified because these estimates imply that the map contracts displacements which are primarily in the stable direction and expands displacements which are primarily in the unstable direction as is shown in Lemmas 10.3 and 10.4 below.

Having fixed the notation and choice of of ϵ, we start by proving the linear estimates as discussed above.

Lemma 10.2. *For $\mathbf{q} \in \mathcal{B}(r)$, $Df_{\mathbf{q}}C^u(\alpha) \subset C^u(\alpha)$.*

PROOF. Let $\mathbf{v} = (\mathbf{v}_s, \mathbf{v}_u) \in C^u(\alpha) \setminus \{\mathbf{0}\}$, and $\mathbf{q} \in \mathcal{B}(r)$. Then,

$$\begin{aligned}
\frac{|\pi_s Df_{\mathbf{q}}\mathbf{v}|}{|\pi_u Df_{\mathbf{q}}\mathbf{v}|} &= \frac{|A_{su}(\mathbf{q})\mathbf{v}_u + A_{ss}(\mathbf{q})\mathbf{v}_s|}{|A_{uu}(\mathbf{q})\mathbf{v}_u + A_{us}(\mathbf{q})\mathbf{v}_s|} \\
&\le \frac{\|A_{su}(\mathbf{q})\| \cdot |\mathbf{v}_u| + \|A_{ss}(\mathbf{q})\| \cdot |\mathbf{v}_s|}{m(A_{uu}(\mathbf{q}))\,|\mathbf{v}_u| - \|A_{us}(\mathbf{q})\| \cdot |\mathbf{v}_s|} \\
&= \frac{\|A_{ss}(\mathbf{q})\|(|\mathbf{v}_s|/|\mathbf{v}_u) + \|A_{su}(\mathbf{q})\|}{m(A_{uu}(\mathbf{q})) - \|A_{us}(\mathbf{q})\|(|\mathbf{v}_s|/|\mathbf{v}_u|)} \\
&\le \frac{\mu\alpha^{-1} + \epsilon}{\lambda - \epsilon\alpha^{-1}} \le \alpha^{-1}(\frac{\mu + \epsilon\alpha}{\lambda - \epsilon\alpha^{-1}}) < \alpha^{-1}.
\end{aligned}$$

But we chose ϵ and α so that $\mu + \epsilon\alpha < 1$ and $\lambda - \epsilon\alpha^{-1} > 1$. □

Lemma 10.3. *Let $\mathbf{q}, \mathbf{q}' \in \mathcal{B}(r)$ and $\mathbf{q}' \in \{\mathbf{q}\} + C^u(\alpha)$. Then, the following are true:*

(a) $|\pi_s \circ f(\mathbf{q}') - \pi_s \circ f(\mathbf{q})| \le \alpha^{-1}(\mu + \epsilon\alpha)|\pi_u(\mathbf{q}' - \mathbf{q})|$,

(b) $|\pi_u \circ f(\mathbf{q}') - \pi_u \circ f(\mathbf{q})| \ge (\lambda - \epsilon\alpha^{-1})|\pi_u(\mathbf{q}' - \mathbf{q})|$, *and*

(c) $f(\mathbf{q}') \in \{f(\mathbf{q})\} + C^u(\alpha)$.

PROOF. (a) The idea is to calculate how $\pi_s \circ f$ varies along the line segment from $\mathbf{q}$ to $\mathbf{q}'$, $\psi(t) = \mathbf{q} + t(\mathbf{q}' - \mathbf{q})$ for $0 \le t \le 1$. Using the Mean Value Theorem,

$$\begin{aligned}
|\pi_s \circ f(\mathbf{q}') - \pi_s \circ f(\mathbf{q})| &= |\pi_s \circ f \circ \psi(1) - \pi_s \circ f \circ \psi(0)| \\
&\le \sup_{0\le t\le 1} |\frac{d}{dt}\pi_s \circ f \circ \psi(t)| \\
&\le \sup_{0\le t\le 1} \|A_{su}(\psi(t))\| \cdot |\pi_u(\mathbf{q}' - \mathbf{q})| \\
&\qquad + \sup_{0\le t\le 1} \|A_{ss}(\psi(t))\| \cdot |\pi_s(\mathbf{q}' - \mathbf{q})| \\
&\le \{ \sup_{0\le t\le 1} \|A_{su}(\psi(t))\| \\
&\qquad + \alpha^{-1} \sup_{0\le t\le 1} \|A_{ss}(\psi(t))\|\}|\pi_u(\mathbf{q}' - \mathbf{q})| \\
&\le (\epsilon + \alpha^{-1}\mu)|\pi_u(\mathbf{q}' - \mathbf{q})| \\
&= \alpha^{-1}(\mu + \epsilon\alpha)|\pi_u(\mathbf{q}' - \mathbf{q})|.
\end{aligned}$$

This proves part (a).

(b) To get an estimate of minimum expansion, we would like to take the inverse of the function. To get a function from a space to itself, we split the displacement from $\mathbf{q}$ to $\mathbf{q}'$ into the component in the $\mathbb{E}^u$ direction and then the $\mathbb{E}^s$ direction, and apply the Inverse Function Theorem (Theorem 2.4) to the part that concerns the displacement in the $\mathbb{E}^u$ direction.

Let $\mathbf{q} = (\mathbf{q}_s, \mathbf{q}_u)$ and $\mathbf{q}' = (\mathbf{q}'_s, \mathbf{q}'_u)$ for $\mathbf{q}$ and $\mathbf{q}'$ as in the statement, i.e., $\mathbf{q}_\sigma = \pi_\sigma\mathbf{q}$ and $\mathbf{q}'_\sigma = \pi_\sigma\mathbf{q}'$ for $\sigma = u, s$. Let $\mathbf{z} = (\mathbf{q}_s, \mathbf{q}'_u)$, so $\mathbf{q}' - \mathbf{q} = (\mathbf{q}' - \mathbf{z}) + (\mathbf{z} - \mathbf{q})$, with $\mathbf{z} - \mathbf{q} \in \{\mathbf{0}\} \times \mathbb{E}^u$ and $\mathbf{q}' - \mathbf{z} \in \mathbb{E}^s \times \{\mathbf{0}\}$.

For $\mathbf{y} \in \mathbb{E}^u(r)$ and $\mathbf{x} \in \mathbb{E}^s(r)$, define $h(\mathbf{y}) = \pi_u \circ f(\mathbf{q}_s, \mathbf{y})$ and $g(\mathbf{x}) = \pi_u \circ f(\mathbf{x}, \mathbf{0})$. Then, $g(\mathbf{q}_s) = \pi_u \circ f(\mathbf{q}_s, \mathbf{0}) = h(\mathbf{0})$. Also, $Dg_{\mathbf{x}} = A_{us}(\mathbf{x}, \mathbf{0})$, so $\|Dg_{\mathbf{x}}\| \le \epsilon$. By the Mean Value Theorem, $\epsilon r \ge \epsilon|\mathbf{q}_s| \ge |g(\mathbf{q}_s)| = |h(\mathbf{0})|$. Then, $Dh_{\mathbf{y}} = A_{uu}(\mathbf{q}_s, \mathbf{q}_u + \mathbf{y})$, which is an isomorphism with minimum norm greater than λ, and $\|Dh_{\mathbf{y}} - A_{uu}(\mathbf{0}, \mathbf{0})\| < \epsilon$. Applying Theorem 2.4, $h|\mathbb{E}^u(r)$ is onto $\mathbb{E}^u((\lambda - \epsilon)r - |h(\mathbf{0})|) \supset \mathbb{E}^u((\lambda - 2\epsilon)r) \supset \mathbb{E}^u(r)$, and h has a unique inverse on this set. Also, by the Inverse Function Theorem, the derivative of h^{-1} is given by $D(h^{-1})_{h(\mathbf{y})} = (Dh_{\mathbf{y}})^{-1}$, which has norm less than λ^{-1}. Therefore, by the Mean Value Theorem,

$$|h^{-1}(\mathbf{w}_2) - h^{-1}(\mathbf{w}_1)| \leq \lambda^{-1}|\mathbf{w}_2 - \mathbf{w}_1|.$$

Letting $\mathbf{w}_2 = h(\mathbf{q}'_u) = \pi_u \circ f(\mathbf{z})$ and $\mathbf{w}_1 = h(\mathbf{q}_u) = \pi_u \circ f(\mathbf{q})$, so $h^{-1}(\mathbf{w}_2) = \mathbf{q}'_u$ and $h^{-1}(\mathbf{w}_1) = \mathbf{q}_u$, we get

$$\lambda|\pi_u(\mathbf{q}' - \mathbf{q})| \leq |\pi_u \circ f(\mathbf{z}) - \pi_u \circ f(\mathbf{q})|.$$

Turning to the displacement from $\mathbf{z}$ to $\mathbf{q}'$,

$$\begin{aligned} |\pi_u \circ f(\mathbf{q}') - \pi_u \circ f(\mathbf{z})| &\leq \sup \|A_{us}\| \cdot |\pi_s(\mathbf{q}' - \mathbf{q})| \\ &\leq \epsilon|\pi_s(\mathbf{q}' - \mathbf{q})| \\ &\leq \epsilon\alpha^{-1}|\pi_u(\mathbf{q}' - \mathbf{q})|. \end{aligned}$$

Combining,

$$\begin{aligned} |\pi_u \circ f(\mathbf{q}') - \pi_u \circ f(\mathbf{q})| &\geq |\pi_u \circ f(\mathbf{z}) - \pi_u \circ f(\mathbf{q})| \\ &\qquad - |\pi_u \circ f(\mathbf{q}') - \pi_u \circ f(\mathbf{z})| \\ &\geq (\lambda - \epsilon\alpha^{-1})|\pi_u(\mathbf{q}' - \mathbf{q})|. \end{aligned}$$

This completes the proof of part (b).

(c) For $\mathbf{q}$ and $\mathbf{q}'$ in the statement of part (c), to show that $f(\mathbf{q}') \in \{f(\mathbf{q})\} + C^u(\alpha)$, we need to look at the ratio of stable and unstable components. Using part (a) to estimate the numerator and part (b) to estimate the denominator, we get

$$\begin{aligned} \frac{|\pi_s[f(\mathbf{q}') - f(\mathbf{q})]|}{|\pi_u[f(\mathbf{q}') - f(\mathbf{q})]|} &\leq \alpha^{-1}\frac{(\mu + \epsilon\alpha)}{(\lambda - \epsilon\alpha^{-1})} \\ &< \alpha^{-1}. \end{aligned}$$

This last inequality uses the fact that $\mu + \epsilon\alpha < 1$ and $\lambda - \epsilon\alpha^{-1} > 1$. From the above inequality, we get that $f(\mathbf{q}') \in \{f(\mathbf{q})\} + C^u(\alpha)$. □

From this last lemma, we can determine the effect of f^{-1} on displacements within the stable cone.

Lemma 10.4. *Assume that* $f(\mathbf{q}_{-1}) = \mathbf{q}$, $f(\mathbf{q}'_{-1}) = \mathbf{q}'$, $\mathbf{q}' \in \{\mathbf{q}\} + C^s(\alpha)$, *and all the points* $\mathbf{q}, \mathbf{q}', \mathbf{q}_{-1}, \mathbf{q}'_{-1} \in \mathcal{B}(r)$. *Then,*

(a) $\mathbf{q}'_{-1} \in \{\mathbf{q}_{-1}\} + C^s(\alpha)$,
(b) $|\pi_s\mathbf{q}'_{-1} - \pi_s\mathbf{q}_{-1}| \geq (\mu + \epsilon\alpha)^{-1}|\pi_s\mathbf{q}' - \pi_s\mathbf{q}|$.

PROOF. (a) We use the fact that $C^s(\alpha)$ and $C^u(\alpha)$ are complementary cones, i.e.,

$$C^s(\alpha) = \mathbb{E} \setminus \mathrm{int}(C^u(\alpha)).$$

By Lemma 10.3(c), $\mathbf{q}' \notin \{\mathbf{q}\} + C^u(\alpha)$ implies $\mathbf{q}'_{-1} \notin \{\mathbf{q}_{-1}\} + C^u(\alpha)$. Therefore $\mathbf{q}'_{-1}$ is in the complementary cone, $\mathbf{q}'_{-1} \in \{\mathbf{q}_{-1}\} + C^s(\alpha)$.

(b) Following the method of the proof of Lemma 10.3(a), and letting $\psi(t) = \mathbf{q}_{-1} + t(\mathbf{q}'_{-1} - \mathbf{q}_{-1})$,

$$\begin{aligned}
|\pi_s\mathbf{q}' - \pi_s\mathbf{q}| &= |\pi_s \circ f \circ \psi(1) - \pi_s \circ f \circ \psi(0)| \\
&\leq \sup_{0\leq t\leq 1} |\frac{d}{dt}\pi_s \circ f \circ \psi(t)| \\
&\leq \sup_{0\leq t\leq 1} \|A_{ss}(\psi(t))\| \cdot |\pi_s\mathbf{q}'_{-1} - \pi_s\mathbf{q}_{-1}| \\
&\qquad + \sup_{0\leq t\leq 1} \|A_{su}(\psi(t))\| \cdot |\pi_u\mathbf{q}'_{-1} - \pi_u\mathbf{q}_{-1}| \\
&\leq \{ \sup_{0\leq t\leq 1} \|A_{ss}(\psi(t))\| \\
&\qquad + \alpha \sup_{0\leq t\leq 1} \|A_{su}(\psi(t))\|\} \cdot |\pi_s\mathbf{q}'_{-1} - \pi_s\mathbf{q}_{-1}| \\
&\leq (\mu + \alpha\epsilon) \cdot |\pi_s\mathbf{q}'_{-1} - \pi_s\mathbf{q}_{-1}|.
\end{aligned}$$

Dividing both sides of the inequality by the constant, we get the result of part (b). □

The following lemma makes precise the induction process with unstable disks, which was discussed in a descriptive fashion earlier in this section.

Lemma 10.5. *Let D^u_0 be a C^1 unstable disk over $\mathbb{E}^u(r)$.*

(a) Then, $D^u_1 = f(D^u_0) \cap \mathcal{B}(r)$ is an unstable disk over $\mathbb{E}^u(r)$ and

$$\operatorname{diam}[\pi_u(f^{-1}(D^u_1) \cap D^u_0)] \leq (\lambda - \epsilon\alpha^{-1})^{-1}2r.$$

(b) Inductively, let $D^u_n = f(D^u_{n-1}) \cap \mathcal{B}(r)$. Then, D^u_n is an unstable disk for $n \geq 1$ and

$$\operatorname{diam}[\pi_u \bigcap_{j=0}^{n} f^{-j}(D^u_j)] \leq (\lambda - \epsilon\alpha^{-1})^{-n}2r.$$

See Figure 10.4.

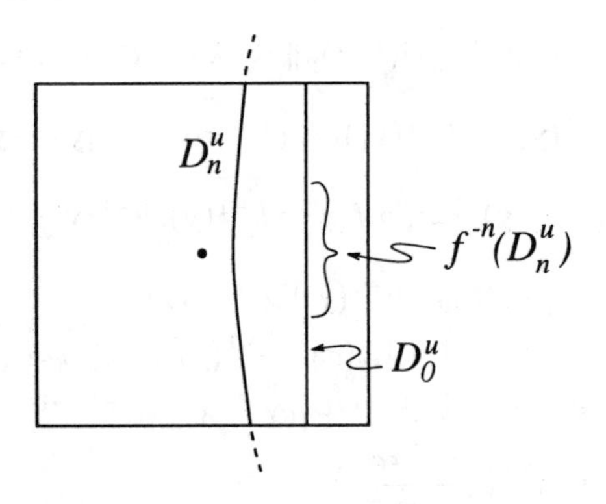

FIGURE 10.4. Disks D^u_n and $f^{-n}(D^u_n)$

PROOF. (a) Since D^u_0 is an unstable disk, it is the graph of a C^1 function $\varphi_0 : \mathbb{E}^u(r) \to \mathbb{E}^s(r)$ with $\|D(\varphi_0)_\mathbf{y}\| \leq \alpha^{-1}$. Let $\sigma_0 : \mathbb{E}^u(r) \to \mathcal{B}(r)$ be the associated function defined by $\sigma_0(\mathbf{y}) = (\varphi_0(\mathbf{y}), \mathbf{y})$. Define $h = \pi_u \circ f \circ \sigma_0$. We need to determine the size of the set covered by h and on which it has a unique inverse. We obtain this by applying Theorem 2.4 for which we need to estimate $\|Dh_{\mathbf{y}} - A_{uu}(\mathbf{0})\|$:

$$\begin{aligned}\|Dh_{\mathbf{y}} - A_{uu}(\mathbf{0})\| &= \|D(\pi_u \circ f \circ \sigma_0)_{\mathbf{y}} - A_{uu}(\mathbf{0})\| \\ &= \|A_{uu}(\sigma_0(\mathbf{y})) + A_{us}(\sigma_0(\mathbf{y}))D(\varphi_0)_{\mathbf{y}} - A_{uu}(\mathbf{0})\| \\ &\leq \|A_{uu}(\sigma_0(\mathbf{y})) - A_{uu}(\mathbf{0})\| + \|A_{us}(\sigma_0(\mathbf{y}))\|\|D(\varphi_0)_{\mathbf{y}}\| \\ &\leq (\epsilon + \epsilon\alpha^{-1}).\end{aligned}$$

By Theorem 2.4, h is a C^1 local diffeomorphism and has an inverse defined on

$$h(\mathbb{E}^u(r)) \supset \{h(\mathbf{0})\} + \mathbb{E}^u((\lambda - \epsilon - \epsilon\alpha^{-1})r) \supset \mathbb{E}^u((\lambda - \epsilon - \epsilon\alpha^{-1})r - |h(\mathbf{0})|).$$

To get the estimate on $|h(\mathbf{0})|$ and so the image under h, notice that $h(\mathbf{0}) = \pi_u \circ f(\varphi_0(\mathbf{0}), \mathbf{0})$ with $|\varphi_0(\mathbf{0})| \leq r$, and $\pi_u \circ f(\mathbf{0}, \mathbf{0}) = \mathbf{0}$. By the Mean Value Theorem applied to $\pi_u \circ f$ and the displacement from $(\mathbf{0}, \mathbf{0})$ to $(\varphi_0(\mathbf{0}), \mathbf{0})$, we get that $|h(\mathbf{0})| \leq \sup(\|A_{us}\|)|\varphi_0(\mathbf{0}) - \mathbf{0}| \leq \epsilon r$. Using this estimate, we get that h has an inverse defined on

$$h(\mathbb{E}^u(r)) \supset \mathbb{E}^u((\lambda - \epsilon\alpha^{-1})r - |h(\mathbf{0})|) \supset \mathbb{E}^u((\lambda - \epsilon\alpha^{-1} - \epsilon)r) \supset \mathbb{E}^u(r).$$

Since $h = \pi_u \circ f \circ \sigma_0$ is a C^1 diffeomorphism which covers $\mathbb{E}^u(r)$, $h^{-1}(\mathbb{E}^u(r)) \subset \mathbb{E}^u(r)$. Therefore, we can define what is called the graph transform of σ_0,

$$\sigma_1 = f \circ \sigma_0 \circ h^{-1}|\mathbb{E}^u(r).$$

Notice that $\pi_u \circ \sigma_1 = \pi_u \circ f \circ \sigma_0 \circ h^{-1} = h \circ h^{-1} = id|\mathbb{E}^u(r)$. Thus, σ_1 is a graph over $\mathbb{E}^u(r)$.

We also need an estimate on $\|D(h^{-1})_{\mathbf{y}}\|$. For $\mathbf{v}_u \in \mathbb{E}^u$,

$$\begin{aligned}|D(\pi_u \circ f \circ \sigma_0)_{\mathbf{y}}\mathbf{v}_u| &= |A_{uu}(\sigma_0(\mathbf{y}))\mathbf{v}_u + A_{us}(\sigma_0(\mathbf{y}))D(\varphi_0)_{\mathbf{y}}\mathbf{v}_u| \\ &\geq |A_{uu}(\sigma_0(\mathbf{y}))\mathbf{v}_u| - |A_{us}(\sigma_0(\mathbf{y}))D(\varphi_0)_{\mathbf{y}}\mathbf{v}_u| \\ &\geq m(A_{uu}(\sigma_0(\mathbf{y})))\,|\mathbf{v}_u| - \|A_{us}(\sigma_0(\mathbf{y}))\| \cdot \|D(\varphi_0)_{\mathbf{y}}\||\mathbf{v}_u| \\ &\geq (\lambda - \epsilon\alpha^{-1})|\mathbf{v}_u|.\end{aligned}$$

Therefore, $m(Dh_{\mathbf{y}}) \geq \lambda - \epsilon\alpha^{-1}$, $\|D(h^{-1})_{\mathbf{y}}\| \leq (\lambda - \epsilon\alpha^{-1})^{-1}$, and

$$|h^{-1}(\mathbf{y}_2) - h^{-1}(\mathbf{y}_1)| \leq (\lambda - \epsilon\alpha^{-1})^{-1}|\mathbf{y}_2 - \mathbf{y}_1|.$$

If we let $\varphi_1(\mathbf{y}) = \pi_s \circ \sigma_1(\mathbf{y}) = \pi_s \circ f(\varphi_0 \circ h^{-1}(\mathbf{y}), h^{-1}(\mathbf{y}))$, then

$$\begin{aligned}\|D(\varphi_1)_{\mathbf{y}}\| &\leq \|A_{su}(\sigma_0 \circ h^{-1}(\mathbf{y}))D(h^{-1})_{\mathbf{y}}\| \\ &\qquad + \|A_{ss}(\sigma_0 \circ h^{-1}(\mathbf{y}))D(\varphi_0)_{h^{-1}(\mathbf{y})}D(h^{-1})_{\mathbf{y}}\| \\ &\leq \epsilon(\lambda - \epsilon\alpha^{-1})^{-1} + \mu\alpha^{-1}(\lambda - \epsilon\alpha^{-1})^{-1} \\ &\leq \alpha^{-1}\left(\frac{\mu + \epsilon\alpha}{\lambda - \epsilon\alpha^{-1}}\right) \\ &< \alpha^{-1}.\end{aligned}$$

Note that for this argument we need that $\mu + \epsilon\alpha < 1$ and $\lambda - \epsilon - \epsilon\alpha^{-1} > 1$. Letting $D_1^u = \sigma_1(\mathbb{E}^u(r))$, we have verified the conclusions of part (a).

Part (b) follows from part (a) by induction on n. □

REMARK 10.2. Lemma 10.5 is the heart of the Hadamard proof of the existence of a Lipschitz unstable manifold. We could define the function space

$$\mathcal{H} = \mathcal{H}(\alpha, r) = \{\sigma : \mathbb{E}^u(r) \to \mathcal{B}(r) : \pi_u \circ \sigma = id,\ \mathrm{Lip}(\pi_s \circ \sigma) \leq \alpha^{-1}\}.$$

By an argument like that in Lemma 10.5, we can show that for any σ in $\mathcal{H}$, $f(\text{image}(\sigma))$ is again the graph of a Lipschitz function in the function space which we denote by $f_{\#}(\sigma)$. This map $f_{\#}$ on the function space $\mathcal{H}$ is called the graph transform. Then, it is possible to show that $f_{\#}$ is a contraction on $\mathcal{H}$ with the C^0-sup topology, and so has a unique fixed section σ^u, for which $\pi_s \circ \sigma^u$ is α^{-1}-Lipschitz. The graph of $\pi_s \circ \sigma^u$ or the image of σ^u is the unstable manifold. For this type of approach, see Hirsch and Pugh (1970), Hirsch, Pugh, and Shub (1977), and Shub (1987). Instead of following that method, we use the above lemma on C^1 functions (which are unstable disks) to prove the existence of a Lipschitz stable manifold.

The following lemma now gives the induction step and the proof that W^u_r is the graph of an α-Lipschitz function. As discussed above, we use Lemma 10.5 to prove that for each $\mathbf{x} \in \mathbb{E}^s(r)$, $W^s_r(\mathbf{0}) \cap (\{\mathbf{x}\} \times \mathbb{E}^u(r))$ is exactly one point, and so $W^s_r(\mathbf{0})$ is a graph. Finally, Lemma 10.3 shows that W^u_r is α-Lipschitz.

Proposition 10.6. *The local stable manifold, $W^s_r(\mathbf{0}) = \bigcap_{j=0}^{\infty} f^{-j}(\mathcal{B}(r))$, is a graph of an α-Lipschitz function $\varphi^s : \mathbb{E}^s(r) \to \mathbb{E}^u(r)$ with $\varphi^s(\mathbf{0}) = \mathbf{0}$.*

PROOF. Let $S_n = \bigcap_{j=0}^{n} f^{-j}(\mathcal{B}(r))$ so $W^s_r(\mathbf{0}) = \bigcap_{n=0}^{\infty} S_n$: the forward orbit of a point in the infinite intersection is contained in $\mathcal{B}(r)$ so it lies on $W^s_r(\mathbf{0})$.

The first step is to use Lemma 10.5 to see that $W^s_r(\mathbf{0}) \cap [\{\mathbf{x}\} \times \mathbb{E}^u(r)]$ is at most one point for each $\mathbf{x} \in \mathbb{E}^s(r)$. Fix $\mathbf{x} \in \mathbb{E}^s(r)$ and let $D^u_0(\mathbf{x}) = \{\mathbf{x}\} \times \mathbb{E}^u(r)$. By Lemma 10.5 and induction on n,

$$D^u_n(\mathbf{x}) = f(D^u_{n-1}(\mathbf{x})) \cap \mathcal{B}(r)$$

is an unstable disk, and

$$[\{\mathbf{x}\} \times \mathbb{E}^u(r)] \cap S_n = \bigcap_{j=0}^{n} f^{-j}(D^u_j(\mathbf{x})) \subset D^u_0(\mathbf{x})$$

is a nested sequence of closed and complete sets whose diameters are bounded above by $(\lambda - \epsilon\alpha^{-1})^{-n} 2r$. Since the diameters of $[\{\mathbf{x}\} \times \mathbb{E}^u(r)] \cap S_n$ go to zero as n goes to infinity, the intersection $[\{\mathbf{x}\} \times \mathbb{E}^u(r)] \cap W^s_r(\mathbf{0})$ is a single point for each $\mathbf{x}$ and $W^s_r(\mathbf{0})$ is a graph over $\mathbb{E}^s$. See Figure 10.5. We show below that the ω-limit set of a point in the intersection defining $W^s_r(\mathbf{0})$ is the fixed point.

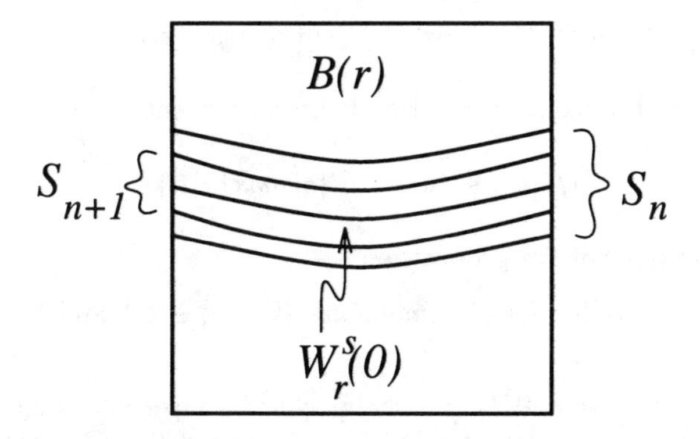

FIGURE 10.5. Sets S_n Converging to $W^s_r(\mathbf{0})$

To show that the graph is Lipschitz, let $\mathbf{q} \in W^s_r(\mathbf{0})$. Assume that

$$\mathbf{q}' \in W^s_r(\mathbf{0}) \cap [\{\mathbf{q}\} + C^u(\alpha)]$$

Because both $\mathbf{q}, \mathbf{q}' \in W^s_r(\mathbf{0})$, $f^j(\mathbf{q}), f^j(\mathbf{q}') \in \mathcal{B}(r)$ for all $j \geq 0$. By induction, applying Lemma 10.3, we get that for all $j \geq 0$,

$$f^j(\mathbf{q}) \in \{f^j(\mathbf{q}')\} + C^u(\alpha)$$

and

$$|\pi_u(f^j(\mathbf{q}) - f^j(\mathbf{q}'))| \geq (\lambda - \epsilon\alpha^{-1})^j |\pi_u(\mathbf{q} - \mathbf{q}')|.$$

If $\mathbf{q}' \neq \mathbf{q}$, then since $(\lambda - \epsilon\alpha^{-1})^j$ goes to infinity either $f^j(\mathbf{q})$ or $f^j(\mathbf{q}')$ must leave $\mathcal{B}(r)$. This contradicts the fact that both $\mathbf{q}$ and $\mathbf{q}'$ are in $W^s_r(\mathbf{0})$. Thus, $\mathbf{q} = \mathbf{q}'$, and there is at most one point in $W^s_r(\mathbf{0}) \cap (\{\mathbf{q}\} + C^u(\alpha))$. Applying this to $\mathbf{x}, \mathbf{y} \in \mathbb{E}^u(r)$ with $\mathbf{x} \neq \mathbf{y}$, since $\sigma^u(\mathbf{x}) = (\mathbf{x}, \varphi^s(\mathbf{x})) \neq (\mathbf{y}, \varphi^s(\mathbf{y})) = \sigma^u(\mathbf{y})$, it follows that $\sigma^u(\mathbf{y}) \notin \{\sigma^u(\mathbf{x})\} + C^u(\alpha)$, $\sigma^u(\mathbf{y}) - \{\sigma^u(\mathbf{x})\} \notin C^u(\alpha)$, and $\sigma^u(\mathbf{y}) - \{\sigma^u(\mathbf{x})\} \in C^s(\alpha)$. Thus, $\text{Lip}(\pi_u \circ \sigma^s) \leq \alpha$. This completes the proof of Proposition 10.6. □

Before we prove that $W^s_r(\mathbf{0})$ is C^1, we check its characterization in terms of the rate of convergence to 0.

Proposition 10.7. *If $\mathbf{q} \in W^s_r(\mathbf{0})$, then $|f^j(\mathbf{q})| \leq \alpha(\mu + \epsilon\alpha)^j |\mathbf{q}|$ for $j \geq 0$, so $f^j(\mathbf{q})$ converges to the fixed point $\mathbf{0}$ at an exponential rate.*

PROOF. Because $W^s_r(\mathbf{0})$ is α-Lipschitz, $|\pi_u[f^j(\mathbf{q}) - \mathbf{0}]| \leq \alpha|\pi_s[f^j(\mathbf{q}) - \mathbf{0}]|$, so the iterates $f^j(\mathbf{q}) \in \{\mathbf{0}\} + C^s(\alpha)$. Applying Lemma 10.3 to $f^j(\mathbf{q})$ and $f^j(\mathbf{0}) = \mathbf{0}$, we get that

$$\begin{aligned}
(\mu + \epsilon\alpha)\, |f^{j-1}(\mathbf{q})| &\geq (\mu + \epsilon\alpha) |\pi_s \circ f^{j-1}(\mathbf{q})| \\
&\geq (\mu + \epsilon\alpha) |\pi_s [f^{j-1}(\mathbf{q}) - f^{j-1}(\mathbf{0})]| \\
&\geq |\pi_s [f^j(\mathbf{q}) - f^j(\mathbf{0})]| \\
&\geq |\pi_s \circ f^j(\mathbf{q})|.
\end{aligned}$$

By induction, we get that

$$(\mu + \epsilon\alpha)^j |\mathbf{q}| \geq |\pi_s \circ f^j(\mathbf{q})|.$$

Then, the unstable component is bounded by α times this amount:

$$|\pi_u \circ f^j(\mathbf{q})| \leq \alpha |\pi_s \circ f^j(\mathbf{q})| \leq \alpha(\mu + \epsilon\alpha)^j |\mathbf{q}|.$$

Because we are using the maximum norm of the components,

$$|f^j(\mathbf{q})| \leq (\mu + \epsilon\alpha)^j |\mathbf{q}| \max\{1, \alpha\}.$$

This completes the proof of the proposition. □

Proposition 10.8. *The local stable manifold, $W^s_r(\mathbf{0})$, is C^1 and is tangent to $\mathbb{E}^s$ at the fixed point $\mathbf{0}$.*

PROOF. We have shown that $W^s_r(\mathbf{0})$ is the graph of a Lipschitz function $\varphi^s : \mathbb{E}^s(r) \to \mathbb{E}^u(r)$ with $\text{Lip}(\varphi^s) \leq \alpha$. We let $\sigma^s : \mathbb{E}^s(r) \to \mathcal{B}(r)$ be defined by $\sigma^s(\mathbf{x}) = (\mathbf{x}, \varphi^s(\mathbf{x}))$. We proceed to show that σ^s and φ^s are C^1.

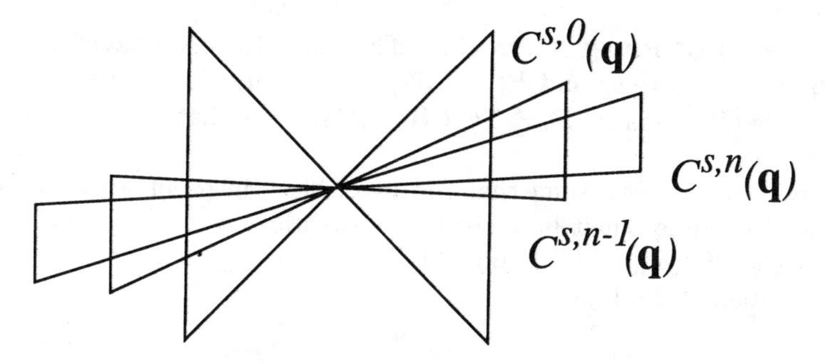

FIGURE 10.6. Nested Cones

For any $\mathbf{q} \in W_r^s(\mathbf{0})$, since $\mathrm{Lip}(\varphi^s) \leq \alpha$, $\sigma^s(\mathbb{E}^s(r)) \subset \{\mathbf{q}\} + C^s(\alpha)$. To get the derivative of φ^s, we use the comparison of the nonlinear action of f on $C^s(\alpha)$ with the linear action of $Df_{\mathbf{q}}$ on $C^s(\alpha)$. Define $C^{s,0}(\mathbf{q}) = C^s(\alpha)$, and for $n \geq 1$,

$$C^{s,n}(\mathbf{q}) = (Df^n_{\mathbf{q}})^{-1}C^s(\alpha).$$

Lemma 10.9. *Let* $\mathbf{q} \in W_r^s(\mathbf{0})$. *Then,*

$$C^{s,n}(\mathbf{q}) \subset C^{s,n-1}(\mathbf{q}) \subset \cdots \subset C^{s,1}(\mathbf{q}) \subset C^{s,0}(\mathbf{q}),$$

and there is a bounded positive angle between vectors in $C^{s,n+1}(\mathbf{q})$ *and the complement of* $C^{s,n}(\mathbf{q})$. *Taking the intersection,* $\bigcap_{n\geq 0} C^{s,n}(\mathbf{q})$ *is a plane* $P_{\mathbf{q}}$ *which is the graph of a linear map* $L_{\mathbf{q}} : \mathbb{E}^s \to \mathbb{E}^u$. *The plane* $P_{\mathbf{q}}$ *depends continuously on* $\mathbf{q}$.

PROOF.

By Lemma 10.2, $Df_{f^j(\mathbf{q})}C^u(\alpha) \subset C^u(\alpha)$, so

$$C^{s,1}(f^j(\mathbf{q})) = [Df_{f^j(\mathbf{q})}]^{-1}C^s(\alpha) \subset C^s(\alpha) = C^{s,0}(f^j(\mathbf{q})).$$

By induction,

$$C^{s,n}(\mathbf{q}) \subset C^{s,n-1}(\mathbf{q}) \subset \cdots \subset C^{s,1}(\mathbf{q}) \subset C^{s,0}(\mathbf{q}).$$

See Figure 10.6. It follows from the argument of Proposition 10.6 applied to this sequence of linear maps rather than one nonlinear map, that $C^{s,n}(\mathbf{q})$ converges to a set that is a graph of a Lipschitz function from $\mathbb{E}^s$ to $\mathbb{E}^u$. (This means that the maximal angle in the cone $C^{s,n}(\mathbf{q})$ converges to zero as n goes to infinity.) Each of the nested cones $C^{s,n}(\mathbf{q})$ contains a plane, so it follows that the intersection contains a plane $P_{\mathbf{q}}$ (which can depend on $\mathbf{q}$). Because the intersection is graph of a single Lipschitz function, the intersection is equal to this plane $P_{\mathbf{q}}$, which is the graph of a linear map $L_{\mathbf{q}} : \mathbb{E}^s \to \mathbb{E}^u$.

In fact, the estimates given in the proof of Lemma 10.2 show that there is a bounded positive angle between vectors in $C^{s,1}(f^n(\mathbf{q}))$ and the complement of $C^{s,0}(f^n(\mathbf{q}))$, which implies that there is a bounded positive angle between vectors in

$$C^{s,n+1}(\mathbf{q}) = D(f^{-n})_{f^n(\mathbf{q})}C^{s,1}(f^n(\mathbf{q}))$$

and the complement of

$$C^{s,n}(\mathbf{q}) = D(f^{-n})_{f^n(\mathbf{q})}C^{s,0}(f^n(\mathbf{q})).$$

The cone $C^{s,n+1}(\mathbf{q}) \setminus \{\mathbf{q}\}$ is in the interior of $C^{s,n}(\mathbf{q})$, so for $\mathbf{q}'$ near enough to $\mathbf{q}$, $P_{\mathbf{q}'} \subset C^{s,n+1}(\mathbf{q}') \subset C^{s,n}(\mathbf{q})$. By the topology on the set of planes, this shows that $P_{\mathbf{q}}$ is a continuous function of $\mathbf{q}$. □

RETURN TO PROOF OF PROPOSITION 10.8: The goal is to prove that φ^s is differentiable at $\mathbf{x}_0 = \pi_s \mathbf{q}$ with $D\varphi^s_{\mathbf{x}_0} = L_{\mathbf{q}}$ and $P_{\mathbf{q}} = T_{\mathbf{q}} W^s_r(\mathbf{0})$. Because $P_{\mathbf{q}}$ varies continuously, this shows that φ^s is C^1. Because $P_{\mathbf{0}} = \mathbb{E}^s \times \{\mathbf{0}\}$, this shows that $W^s_r(\mathbf{0})$ is tangent to $\mathbb{E}^s$ at $\mathbf{0}$.

To simplify the proof, we assume that f is invertible. The proof below can be modified when f is not invertible but it becomes less straightforward: it involves showing that $\sigma^s(\mathbf{x}) - \sigma^s(\mathbf{x}_0) \notin C^{s,n}(\sigma^s(\mathbf{x}_0))$ is impossible.

By the definition of the cones,

$$D(f^{-n-1})_{f^{n+1}(\mathbf{q})} C^{s,0}(f^{n+1}(\mathbf{q})) = C^{s,n+1}(\mathbf{q}).$$

By Lemma 10.9,

$$C^{s,n+1}(\mathbf{q}) \cap \{\mathbf{v} : |\mathbf{v}| = 1\} \subset \operatorname{int} C^{s,n}(\mathbf{q})$$

and there is a positive angle between vectors in $C^{s,n+1}(\mathbf{q})$ and the complement of $C^{s,n}(\mathbf{q})$. In comparing the effect of f^{-n-1} and $D(f^{-n-1})_{f^{n+1}(\mathbf{q})}$, we can only use small displacements; so we define the truncated cones

$$C^{s,n}(f^{n+1}(\mathbf{q}), \delta) = C^{s,n}(f^{n+1}(\mathbf{q})) \cap \pi_s^{-1}(\mathbb{E}^s(\delta)).$$

By the differentiability of f^{-n-1} and the above inclusion for the cones using the derivative of f^{n+1}, there is a $\delta > 0$ such that

$$f^{-n-1}(\{f^{n+1}(\mathbf{q})\} + C^{s,0}(f^{n+1}(\mathbf{q}), \delta)) \subset \{\mathbf{q}\} + C^{s,n}(\mathbf{q}).$$

(The proof of this fact uses the the definition of the derivative and the fact that there is a positive angle between vectors in $C^{s,n+1}(\mathbf{q})$ and the complement of $C^{s,n}(\mathbf{q})$.) By the continuity of σ^s and f^{n+1} and the fact that φ^s is α-Lipschitz, there is a $\eta > 0$ such that if $|\mathbf{x} - \mathbf{x}_0| \leq \eta$, then

$$f^{n+1}(\sigma^s(\mathbf{x})) \in \{f^{n+1}(\sigma^s(\mathbf{x}_0))\} + C^{s,0}(f^{n+1}(\sigma^s(\mathbf{x}_0)), \delta).$$

By the above inclusion using f^{-n-1},

$$\sigma^s(\mathbf{x}) \in \{\sigma^s(\mathbf{x}_0)\} + C^{s,n}(\sigma^s(\mathbf{x}_0)).$$

Thus, we have shown that given an n there is an $\eta = \eta(n) > 0$ such that for $|\mathbf{y} - \mathbf{x}_0| < \eta$,

$$\sigma^s(\mathbf{x}) - \sigma^s(\mathbf{x}_0) \in C^{s,n}(\sigma^s(\mathbf{x}_0)),$$

i.e., we have shown that σ^s satisfies the definition of the derivative at $\mathbf{x}_0$. □

Theorem 10.10. *If f is C^k for $k \geq 1$, then $W^s_r(\mathbf{0})$ is C^k.*

PROOF. If $k = 1$, then this is a restatement of Proposition 10.8. Assume that f is C^k for $k \geq 2$, with $f(\mathbf{0}) = \mathbf{0}$. Define a new function $F(\mathbf{p}, \mathbf{v}) = (f(\mathbf{p}), Df_{\mathbf{p}}\mathbf{v})$. This function F is defined for $(\mathbf{p}, \mathbf{v})$ with $\mathbf{p} \in \mathcal{B}(r)$ and $\mathbf{v}$ is a (tangent) vector at $\mathbf{p}$, $\mathbf{v} \in T_{\mathbf{p}}\mathbb{E} \approx \mathbb{E}$. Then F is C^{k-1}, $F(\mathbf{0}, \mathbf{0}) = (\mathbf{0}, \mathbf{0})$, and

$$DF_{(\mathbf{p},\mathbf{v})}(\dot{\mathbf{p}}, \dot{\mathbf{v}}) = \begin{pmatrix} Df_{\mathbf{p}}\dot{\mathbf{p}} \\ D^2 f_{\mathbf{p}}(\mathbf{v}, \dot{\mathbf{p}}) + Df_{\mathbf{p}}\dot{\mathbf{v}} \end{pmatrix},$$

so

$$DF_{(\mathbf{0},\mathbf{0})}(\dot{\mathbf{p}}, \dot{\mathbf{v}}) = \begin{pmatrix} Df_{\mathbf{0}}\dot{\mathbf{p}} \\ Df_{\mathbf{0}}\dot{\mathbf{v}} \end{pmatrix} = \begin{pmatrix} Df_{\mathbf{0}} & \mathbf{0} \\ \mathbf{0} & Df_{\mathbf{0}} \end{pmatrix} \begin{pmatrix} \dot{\mathbf{p}} \\ \dot{\mathbf{v}} \end{pmatrix}.$$

Thus, $(\mathbf{0}, \mathbf{0})$ is a hyperbolic fixed point for F. By induction, F has a C^{k-1} stable manifold, $W^s_r((\mathbf{0}, \mathbf{0}), F)$. This manifold is characterized as points $(\mathbf{p}, \mathbf{v})$ which have $\mathbf{p}_j \in \mathcal{B}(r)$, and $|\mathbf{v}_j| \leq r$ for all $j \geq 0$, where $(\mathbf{p}_j, \mathbf{v}_j) \equiv F^j(\mathbf{p}, \mathbf{v})$. Thus, $\mathbf{p} \in W^s_r(\mathbf{0}, f)$. In fact, $|\mathbf{v}_j| = |Df^j_p \mathbf{v}|$ goes to zero exponentially as j goes to infinity. But vectors in $T_p W^s_r(\mathbf{0}, f)$ have this property. By the uniqueness of $W^s_r((\mathbf{0}, \mathbf{0}), F)$ and counting the dimensions of $T_p W^s_r(\mathbf{0}, f)$ and $W^s_r((\mathbf{0}, \mathbf{0}), F) \cap (\{\mathbf{p}\} \times T_{\mathbf{p}}\mathbb{E})$,

$$\{\mathbf{v} : (\mathbf{p}, \mathbf{v}) \in W^s_r((\mathbf{0}, \mathbf{0}), F)\} = T_{\mathbf{p}} W^s_r(\mathbf{0}, f).$$

Thus, $TW^s_r(\mathbf{0}, f)$ is C^{k-1} and $W^s_r(\mathbf{0}, f)$ is C^k. □

Theorem 10.11. *If f is C^k for $k \geq 1$, then the local unstable manifold $W^u_r(\mathbf{0}, f)$ is C^k.*

PROOF. If f is invertible, then the existence of the unstable manifold for f follows from that of the stable manifold of f^{-1}. We will now indicate the changes needed to take care of the case when f is not invertible.

We can use Lemmas 10.3 and 10.4 as given. All we need is a replacement for Lemma 10.5.

Lemma 10.12. *Let $D^s = D^s_0$ be an α-Lipschitz stable disk over $\mathbb{E}^s(r)$.*

(a) Then, $D^s_1 = f^{-1}(D^s) \cap \mathcal{B}(r)$ is an α-Lipschitz stable disk over $\mathbb{E}^s(r)$ and

$$\text{diam}[f(D^s_1)] \leq (\mu + \epsilon\alpha)2r.$$

(b) Inductively define $D^s_n = f^{-1}(D^s_{n-1}) \cap \mathcal{B}(r)$. Then, D^s_n is an α-Lipschitz stable disk for $n \geq 1$ and $\text{diam}[f^n(D^s_j)] \leq (\mu + \epsilon\alpha)^n 2r$.

PROOF. The first step is to show that $D^s_1 = f^{-1}(D^s)$ contains exactly one point in each fiber $\{\mathbf{x}\} \times \mathbb{E}^u(r)$. But $\mathbf{p} \in f^{-1}(D^s) \cap [\{\mathbf{x}\} \times \mathbb{E}^u(r)]$ if and only if $f(\mathbf{p}) \in D^s \cap f[\{\mathbf{x}\} \times \mathbb{E}^u(r)]$ and $\mathbf{p} \in \{\mathbf{x}\} \times \mathbb{E}^u(r)$. By Lemma 10.5, the second set is the graph of a C^1 function $\psi : \mathbb{E}^u(r) \to \mathbb{E}^s(r)$ whose derivative has norm slightly less than α^{-1}. There is a unique point of intersection of these two sets. This fact can be seen by considering the composition $\pi_u \circ \sigma(\pi_s \psi)$ which is a contraction (has derivative with norm less than 1). Let $\mathbf{z}$ be this point of intersection: $\{\mathbf{z}\} = D^s \cap f[\{\mathbf{x}\} \times \mathbb{E}^u(r)]$. By the proof of Lemma 10.5, there is a unique $\mathbf{p} \in [\{\mathbf{x}\} \times \mathbb{E}^u(r)]$ with $f(\mathbf{p}) = \mathbf{z}$. Thus, $\mathbf{p} \in f^{-1}(D^s) \cap [\{\mathbf{x}\} \times \mathbb{E}^u(r)]$ is unique. By Lemma 10.4(a), the graph is α-Lipschitz and $\text{diam}[f(D^s_1)] \leq (\mu + \epsilon\alpha)2r$. This indicates the changes in the proof of Lemma 10.5. □

Finally, we want to check the characterization of the points in W^u_r.

Proposition 10.13. *If $\mathbf{p} \in W^u_r$, then for any past history $\mathbf{p}_{-j} \in \mathcal{B}(r)$ with $\mathbf{p}_0 = \mathbf{p}$ and $f(\mathbf{p}_{-j-1}) = \mathbf{p}_{-j}$, it follows that $|\mathbf{p}_{-j}| \leq (\lambda - \epsilon\alpha)^{-j}|\mathbf{p}|$. Also, the past history is unique in W^u_r.*

PROOF. Take $\mathbf{p}_0 = \mathbf{p} \in W^u_r$ and $\mathbf{p}_{-j} \in \mathcal{B}(r)$ a past history. Thus, $\mathbf{p}_0 \in \{\mathbf{0}\} + C^u(\alpha)$. Applying induction and using Lemma 10.3 with $\mathbf{q} = \mathbf{0}$,

$$(\lambda - \epsilon\alpha)^j |\pi_u \mathbf{p}_{-j}| \leq |\pi_u(\mathbf{p} - \mathbf{0})| = |\pi_u \mathbf{p}_0|$$

so

$$|\pi_u \mathbf{p}_{-j}| \leq (\lambda - \epsilon\alpha)^{-j} |\pi_u \mathbf{p}_0|.$$

Thus, $|\pi_u \mathbf{p}_{-j}|$ goes to zero exponentially fast as stated. Since $|\pi_s \mathbf{p}_{-j}| \leq \alpha |\pi_u \mathbf{p}_{-j}|$, the stable component also goes to zero exponentially fast.

To show the uniqueness, assume that $\mathbf{p}_{-j}$ and $\mathbf{q}_{-j}$ are both past histories for $\mathbf{p}$ which remain in $B(r)$ for all $j \geq 0$. Then, $\mathbf{p}_0 = \mathbf{q}_0$, so $\mathbf{q}_0 \in \{\mathbf{p}_0\} + C^s(\alpha)$. By induction and using Lemma 10.4, $\mathbf{q}_{-j} \in \{\mathbf{p}_{-j}\} + C^s(\alpha)$ for $j \geq 0$. Applying the estimate of Lemma 10.4, the only way that both $\mathbf{p}_{-j-k}$ and $\mathbf{q}_{-j-k}$ can remain in $\mathcal{B}(r)$ for all $k \geq 0$ is for $\mathbf{p}_{-j} = \mathbf{q}_{-j}$. This proves uniqueness in W^u_r. □

5.10.2 Center Manifold

In this section, we give the modifications of the Stable Manifold Theorem for the case that allows eigenvalues (spectrum) on the unit circle. We only discuss the case of finite dimensions, although the results in Banach spaces are true with some modifications of the assumptions. First of all, the stable and unstable manifolds exist in this situation as stated in Theorem 10.14. There also exists a manifold which is tangent to the center eigenspace as stated in Theorem 10.15. Theorem 10.15 also gives the existence of so called center-stable and center-unstable manifolds.

Theorem 10.14 (Stable Manifold Theorem). *Let $f : U \subset \mathbb{R}^n \to \mathbb{R}^n$ be a C^k map for $1 \leq k \leq \infty$ with $f(\mathbf{p}) = \mathbf{p}$. Then, $\mathbb{R}^n$ splits into the eigenspaces of $Df_{\mathbf{p}}$, $\mathbb{R}^n = \mathbb{E}^u \oplus \mathbb{E}^c \oplus \mathbb{E}^s$, which correspond to the eigenvalues of $Df_{\mathbf{p}}$ greater than 1, equal to 1, and less than 1. There exists a neighborhood of $\mathbf{p}$, $V \subset U$, such that $W^s(\mathbf{p}, V, f)$ and $W^u(\mathbf{p}, V, f)$ are C^k manifolds tangent to $\mathbb{E}^s$ and $\mathbb{E}^u$, respectively, and are characterized by the exponential rate of convergence of orbits to $\mathbf{p}$ as follows. Assume that $0 < \mu < 1 < \lambda$ and norms on $\mathbb{E}^u$ and $\mathbb{E}^s$ are chosen such that $\|Df_{\mathbf{p}}|\mathbb{E}^s\| < \mu$ and $m(Df_{\mathbf{p}}|\mathbb{E}^u) > \lambda$. Then,*

$$W^s(\mathbf{p}, V, f) = \{\mathbf{q} \in V : d(f^j(\mathbf{q}), \mathbf{p}) \leq \mu^j d(\mathbf{q}, \mathbf{p}) \text{ for all } j \geq 0\}$$

and

$$W^u(\mathbf{p}, V, f) = \{\mathbf{q} \in V : d(\mathbf{q}_{-j}, \mathbf{p}) \leq \lambda^{-j} d(\mathbf{q}, \mathbf{p}) \text{ for all } j \geq 0 \\ \text{where } \{\mathbf{q}_{-j}\}_{j=0}^{\infty} \text{ is some choice of a past history of } \mathbf{q}\}$$

REMARK 10.3. We do not give a proof of this theorem, although methods similar to those for the proof of the earlier Stable Manifold Theorem work with some modification. Most proofs of the Stable Manifold Theorem probably apply, but this theorem is certainly proved in Kelley (1967) and Hirsch, Pugh, and Shub (1977).

There is also a *center-stable manifold*, $W^{cs}(\mathbf{p}, V, f)$, which is tangent to $\mathbb{E}^c \oplus \mathbb{E}^s$, but it is not necessarily unique and is difficult to characterize locally. Also, if f is C^k for $1 \leq k < \infty$, then there is a neighborhood on which the center-stable manifold is C^k. See Section 12.3. However, the neighborhood can depend on k, and so if f is C^∞, there are examples where there is no neighborhood on which the manifold is C^∞. See van Strien (1979), Carr (1981), and Exercise 5.44. Finally, the manifold $W^{cs}(\mathbf{p}, V, f)$ is not strictly invariant, although all points which stay in V for all future iterates are contained in $W^{cs}(\mathbf{p}, V, f)$.

In the same way, there is a *center-unstable manifold*, $W^{cu}(\mathbf{p}, V, f)$, which is characterized by choices of backward iterates.

It is easier to characterize these manifolds by forming an extension, $\bar{f}$, of f to all of $\mathbb{R}^n$ which is close to $f(\mathbf{p}) + Df_{\mathbf{p}}(\mathbf{q} - \mathbf{p})$ on all of $\mathbb{R}^n$. Once the extension is fixed, $W^{cu}(\mathbf{p}, \bar{f})$ and $W^{cs}(\mathbf{p}, \bar{f})$ are unique. By a translation we can assume that $\mathbf{p} = \mathbf{0}$. Let $A = Df_{\mathbf{0}}$. Given $\epsilon > 0$, there is an $r > 0$ and an extension $\bar{f} : \mathbb{R}^n \to \mathbb{R}^n$ that is C^k and such that $\bar{f}|B(\mathbf{0}, r) = f|B(\mathbf{0}, r)$, $\|\bar{f} - A\|_{C^1} < \epsilon$, and $\bar{f} = A$ off $B(\mathbf{0}, 2r)$. With this notation, the manifolds are characterized in the following theorem.

Theorem 10.15 (Center Manifold Theorem). *Let $f : U \subset \mathbb{R}^n \to \mathbb{R}^n$ be a C^k map for $1 \leq k \leq \infty$ with $f(\mathbf{0}) = \mathbf{0}$ and $A = Df_{\mathbf{0}}$. Let k' be chosen to be (i) k if $k < \infty$ and (ii) some integer with $1 \leq k' < \infty$ if $k = \infty$. Assume that $0 < \mu < 1 < \lambda$, and norms on $\mathbb{E}^u$ and $\mathbb{E}^s$ are chosen such that $\|Df_{\mathbf{p}}|\mathbb{E}^s\| < \mu$ and $m(Df_{\mathbf{p}}|\mathbb{E}^u) > \lambda$. Let $\epsilon > 0$ be small enough so that $\|Df_{\mathbf{p}}|\mathbb{E}^s\| < \mu - \epsilon$ and $m(Df_{\mathbf{p}}|\mathbb{E}^u) > \lambda + \epsilon$. Let $r > 0$ and $\bar{f} : \mathbb{R}^n \to \mathbb{R}^n$ be a C^k map with $\bar{f}|B(\mathbf{0}, r) = f|B(\mathbf{0}, r)$, $\|\bar{f} - A\|_{C^1} < \epsilon$, and $\bar{f} = A$ off*

$B(\mathbf{0}, 2r)$. If $r > 0$ is small enough (it can depend on k'), then there exists an invariant $C^{k'}$ center-stable manifold, $W^{cs}(\mathbf{0}, \bar{f})$, which is a graph over $\mathbb{E}^c \oplus \mathbb{E}^s$, which is tangent to $\mathbb{E}^c \oplus \mathbb{E}^s$ at $\mathbf{0}$, and which is characterized as follows:

$$W^{cs}(\mathbf{0}, \bar{f}) = \{\mathbf{q} : d(\bar{f}^j(\mathbf{q}), \mathbf{0})\lambda^{-j} \to \mathbf{0} \text{ as } j \to \infty\}.$$

This means that $\bar{f}^j(\mathbf{q})$ grows more slowly than λ^j. Similarly, there exists an invariant $C^{k'}$ center-unstable manifold, $W^{cu}(\mathbf{0}, \bar{f})$, which is a graph over $\mathbb{E}^u \oplus \mathbb{E}^c$, which is tangent to $\mathbb{E}^u \oplus \mathbb{E}^c$ at $\mathbf{0}$, and which is characterized as follows:

$$W^{cu}(\mathbf{0}, \bar{f}) = \{\mathbf{q} : d(\mathbf{q}_{-j}, \mathbf{0})\mu^j \to \mathbf{0} \text{ as } j \to \infty \text{ where } \{\mathbf{q}_{-j}\}_{j=0}^{\infty} \text{ is some choice of a past history of } \mathbf{q}\}.$$

This means that $\mathbf{q}_{-j}$ grows more slowly than μ^{-j} as $j \to \infty$, or $-j \to -\infty$. Then, the center manifold of the extension is defined as

$$W^c(\mathbf{0}, \bar{f}) = W^{cs}(\mathbf{0}, \bar{f}) \cap W^{cu}(\mathbf{0}, \bar{f}).$$

It is $C^{k'}$ and tangent to $\mathbb{E}^c$. There are also local center-stable, local center-unstable, and local center manifolds of f defined as

$$\begin{aligned} W^{cs}(\mathbf{0}, B(\mathbf{0}, r), f) &= W^{cs}(\mathbf{0}, \bar{f}) \cap B(\mathbf{0}, r), \\ W^{cu}(\mathbf{0}, B(\mathbf{0}, r), f) &= W^{cu}(\mathbf{0}, \bar{f}) \cap B(\mathbf{0}, r), \quad \text{and} \\ W^{c}(\mathbf{0}, B(\mathbf{0}, r), f) &= W^{c}(\mathbf{0}, \bar{f}) \cap B(\mathbf{0}, r), \end{aligned}$$

respectively. These local manifolds depend on the extension; but if $f^j(\mathbf{q}) \in B(\mathbf{0}, r)$ for all $-\infty < j < \infty$, then $\mathbf{q} \in W^c(\mathbf{0}, \bar{f})$ for any extension $\bar{f}$ and $\mathbf{q} \in W^c(\mathbf{0}, B(\mathbf{0}, r), f)$.

REMARK 10.4. The proof that a C^1 center-stable and center-unstable manifolds exist is essentially the same as in the last subsection. In Section 12.3, we return to discuss why it is C^r for $2 \leq r < \infty$. For a more thorough discussion of the Center Manifold Theorem, see Carr (1981). There are proofs in Kelley (1967), Hirsch, Pugh, and Shub (1977), and Chow and Hale (1982).

Example 10.1 (Nonunique Center Manifold). The following example illustrates the fact that the center manifold is not unique. It is attributed to Anosov in Kelley (1967), page 149. Consider the differential equations

$$\begin{aligned} \dot{x} &= x^2 \\ \dot{y} &= -y. \end{aligned}$$

It is easy to see that the origin is a non-hyperbolic fixed point with eigenvalues -1 and 0. The stable manifold of the origin is clearly the y-axis. To determine the various center manifolds, note that $\dfrac{dy}{dx} = \dfrac{-y}{x^2}$ and $y = Ce^{1/x}$ is a solution for any choice of C. Thus, for any choice of C, the graph of the following function gives a center manifold:

$$u(x, C) = \begin{cases} 0 & \text{for } x \geq 0 \\ C\,e^{1/x} & \text{for } x < 0. \end{cases}$$

Note that $u(x, C) \to 0$ as $x \to 0$ with $x < 0$. Thus, $u(x, C)$ is continuous at $x = 0$. With more calculations, it can be shown that $u(x, C)$ is C^∞ at $x = 0$. For any C, the graph of $u(x, C)$ is tangent to the x-axis at $x = 0$ and is invariant. Therefore, for any choice of C, this graph is a center manifold. Thus, far from being unique, there is a one parameter family of center manifolds.

5.10.3 Stable Manifold Theorem for Flows

The statement of the Stable Manifold Theorem for a fixed point of a flow is similar to that for a diffeomorphism. We do not comment further on it except to mention that the sum of the dimensions of the stable and unstable manifolds of a single fixed point equals the total dimension of the ambient space. This is similar to a diffeomorphism but different than the stable and unstable manifolds of a periodic orbit for a flow.

For a hyperbolic periodic orbit γ, it is certainly possible to get the stable manifold of a Poincaré map $P : \Sigma \to \Sigma$, $W^s_\epsilon(\mathbf{p}, P)$, and let

$$W^s_\epsilon(\gamma, \varphi^t) = \bigcup_{t \geq 0} \varphi^t(W^s_\epsilon(\mathbf{p}, P).$$

However, this representation does not tell us much about the geometry of the local stable manifold of the periodic orbit.

To explain this geometry, we need to look at the contracting and expanding splitting along the whole periodic orbit. Let φ^t be a flow on a manifold M for a vector field X and with a periodic orbit γ of period T. For any $\mathbf{q} \in \gamma$, there is a splitting

$$T_\mathbf{q}M = \mathbb{E}^s_\mathbf{q} \oplus \mathbb{E}^u_\mathbf{q} \oplus \langle X(\mathbf{q}) \rangle,$$

where $\langle X(\mathbf{q}) \rangle$ is the span of the vector $X(\mathbf{q})$, and $\mathbb{E}^\sigma_\mathbf{q} = D\varphi^t_\mathbf{p} \mathbb{E}^\sigma_\mathbf{p}$ for $\sigma = s, u$ if $\mathbf{q} = \varphi^t(\mathbf{p})$. Just as we defined the normal bundle for the periodic orbit when discussing the Hartman-Grobman Theorem, we can define the *stable bundle* and *unstable bundle of the periodic orbit* by

$$\mathbb{E}^s_\gamma = \bigcup \{(\mathbf{q}, \mathbf{y}) : \mathbf{q} \in \gamma \text{ and } \mathbf{y} \in \mathbb{E}^s_\mathbf{q}\},$$

and

$$\mathbb{E}^u_\gamma = \bigcup \{(\mathbf{q}, \mathbf{y}) : \mathbf{q} \in \gamma \text{ and } \mathbf{y} \in \mathbb{E}^u_\mathbf{q}\}.$$

Further, for $\sigma = s, u$ and $\epsilon > 0$, let

$$\mathbb{E}^\sigma_\gamma(\epsilon) = \bigcup \{(\mathbf{q}, \mathbf{y}) : \mathbf{q} \in \gamma \text{ and } \mathbf{y} \in \mathbb{E}^\sigma_\mathbf{q}(\epsilon)\}.$$

If the derivative of the flow restricted to the stable bundle, $D\varphi^T_\mathbf{p}|\mathbb{E}^s_\mathbf{p}$, is orientation preserving (has positive determinant), then $\mathbb{E}^s_\gamma(\epsilon)$ is isomorphic to the cross product of γ and $\mathbb{E}^s_\mathbf{p}(\epsilon)$. If $D\varphi^T_\mathbf{p}|\mathbb{E}^s_\mathbf{p}$ is orientation reversing (has negative determinant), then $\mathbb{E}^s_\gamma(\epsilon)$ is not isomorphic to the cross product of γ and $\mathbb{E}^s_\mathbf{p}(\epsilon)$ but is a twisted product. By going around the orbit twice, an oriented basis of $\mathbb{E}^s_\mathbf{p}$ can be brought back to itself (while remaining a basis of the appropriate $\mathbb{E}^s_\mathbf{q}$ the whole way) but not after only once around γ. If $\mathbb{E}^s_\mathbf{p}$ is one-dimensional and the stable bundle is twisted, then $\mathbb{E}^s_\gamma(\epsilon)$ is isomorphic to a Möbius strip. In higher dimensions, it is a corresponding twisted product if the bundle is not oriented.

The local stable manifold of γ, $W^s_\epsilon(\gamma, \varphi^t)$, can be represented as a graph over $\mathbb{E}^s_\gamma(\epsilon)$ for some small $\epsilon > 0$. Thus, if $\mathbb{E}^s_\mathbf{p}$ is one-dimensional, the local stable manifold is either a graph over an untwisted strip (an annulus) or a Möbius strip. Similar statements can be made about the local unstable manifold. This geometric difference of types of local stable and unstable manifolds for a periodic orbit does not arise for fixed points of flows or for diffeomorphisms. The differentiation between $\mathbb{E}^s_\gamma$ being twisted or not is related to Floquet theory for the time periodic linear system of differential equations (first variation equation)

$$\frac{d}{dt}\mathbf{v} = DX_{\varphi^t(\mathbf{p})}\mathbf{v}$$

along the periodic orbit. See Hartman (1964) or Hale (1969).

Notice that the sum of the dimensions of $\mathbb{E}^s_\gamma$ and $\mathbb{E}^u_\gamma$ is equal to the dimension of M plus 1,

$$\dim(\mathbb{E}^s_\gamma) + \dim(\mathbb{E}^u_\gamma) = \dim(M) + 1,$$

since both bundles contain the direction along γ. This is another difference from the case for stable and unstable manifolds of fixed points for flows or for diffeomorphisms. This difference is important when we consider transverse stable and unstable manifolds for flows: for a periodic orbit γ, $W^s(\gamma, \varphi^t)$ can be transverse to $W^u(\gamma, \varphi^t)$ (and so generate a horseshoe); but for a fixed point $\mathbf{p}_0$, $W^s(\mathbf{p}_0, \varphi^t)$ cannot be transverse to $W^u(\mathbf{p}_0, \varphi^t)$ at points away from $\mathbf{p}_0$. See the discussion of flows in Chapters VIII and X.

5.11 The Inclination Lemma

This section concerns a result which follows from the proof of the stable manifold theorem, or at least from similar ideas. It is used in Chapter X to develop the general theory of invariant sets with a hyperbolic structure (some contracting and some expanding directions). Let $\mathbf{p}$ be a hyperbolic fixed point for f. The result concerns the iterates of a disk, of the same dimension as the $W^u(\mathbf{p})$, which crosses $W^s(\mathbf{p})$. We first need to define what we mean by a disk crossing $W^s(\mathbf{p})$ transversally.

Definition. Let D_1 and D_2 be the differentiable images in $\mathbb{R}^n$ of two disks, $D_j = g_j(\mathbb{R}^{n_j}(r_j))$ for $j = 1, 2$, where $g_j : \mathbb{R}^{n_j} \to \mathbb{R}^n$ is C^k and one to one. We say that these two embedded disks are *transverse at* $\mathbf{p}$ provided either $\mathbf{p} \notin D_1 \cap D_2$ or

$$Dg_{1,(\mathbf{p})}\mathbb{R}^{n_1} \oplus Dg_{2,(\mathbf{p})}\mathbb{R}^{n_2} = \mathbb{R}^n.$$

We say that D_1 and D_2 are *transverse* provided they are transverse at all points.

In the definition, note that if D_1 and D_2 are transverse and $n_1 + n_2 < n$, then $D_1 \cap D_2 = \emptyset$. Thus, two transverse lines in $\mathbb{R}^3$ do not intersect. Next, if $n_1 + n_2 = n$, and the disks are transverse, then they intersect in isolated points. (A complete proof of this fact uses the Implicit Function Theorem.) So, a line and a plane which are transverse in $\mathbb{R}^3$ intersect in a single point. If $n_1 + n_2 > n$, then the transverse objects intersect in a curve, surface, or higher dimensional object (of dimension $n_1 + n_2 - n$). So, two planes which are transverse in $\mathbb{R}^3$ intersect in a line. For a more complete treatment of transversality, see Abraham and Robbin (1967), Guillemin and Pollack (1974), or Hirsch (1976).

Now, we can state the Inclination Lemma (or Lambda Theorem).

Theorem 11.1 (Inclination Lemma). *Let $\mathbf{p}$ be a hyperbolic fixed point for a C^k diffeomorphism f. Let $r > 0$ be small enough so that in the neighborhood of $\mathbf{p}$ given by $\{\mathbf{p}\} + (\mathbb{E}^u(r) \times \mathbb{E}^s(r))$, the hyperbolic estimates hold which prove the Stable (and Unstable) Manifold Theorem. Let D^u be an embedded disk of the same dimension as $\mathbb{E}^u$ and such that D^u is transverse to $W^s_r(\mathbf{p}, f)$. Let $D^u_1 = f(D^u) \cap (\mathbb{E}^u(r) \times \mathbb{E}^s(r))$ and $D^u_{n+1} = f(D^u_n) \cap (\mathbb{E}^u(r) \times \mathbb{E}^s(r))$. Then, D^u_n converges to $W^u_r(\mathbf{p}, f)$ in the C^k topology (pointwise and with all its derivatives). See Figure 11.1. In other words, given $\epsilon > 0$, there is n_0 such that for all $n \geq n_0$, D^u_n is within ϵ of $W^u_r(\mathbf{p}, f)$ in terms of the C^k-topology. (The latter condition means that if D^u_n is given as the graph of $\sigma_n : W^u_r(\mathbf{p}, f) \to \mathbb{E}^s$, then σ_n and its first k derivatives are smaller than ϵ.)*

The proof of this theorem follows from the methods of the proof of the Stable Manifold Theorem. See Palis and de Melo (1982) for the details. Palis and de Melo (1982) also contains another (geometric) proof of the Hartman – Grobman Theorem using the Inclination Lemma.

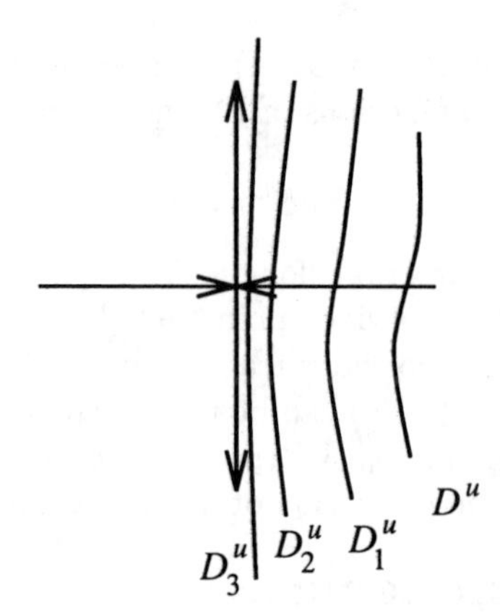

FIGURE 11.1. The Disks D_n^u Converging to the Unstable Manifold

This type of result can also be formulated for an invariant set with a hyperbolic structure which we define in Chapter VIII.

5.12 Exercises

Differentiation

5.1. Assume $f : U \subset \mathbb{R}^k \to \mathbb{R}^m$ is C^1 and $\mathbf{p} \in U$. Show that

$$\frac{\partial f}{\partial x_j}(\mathbf{p}) = Df_{\mathbf{p}}\mathbf{e}^j,$$

where $\mathbf{e}^j = (0, \ldots, 0, 1, 0, \ldots, 0)$ is the standard unit vector with zeroes in all but the jth position.

5.2. Assume $f : U \subset \mathbb{R}^k \to \mathbb{R}^m$ and $g : V \subset \mathbb{R}^m \to \mathbb{R}^n$ are differentiable at $\mathbf{p}$ and $\mathbf{q} = f(\mathbf{p})$, respectively. Using the matrix product write out the (i, j) entry of $Dg_{\mathbf{q}}Df_{\mathbf{p}}$ in terms of partial derivatives. Discuss the relationship between the chain rule in terms of products of matrices and combinations of partial derivatives.

Inverse Function Theorem

5.3. This exercise asks for a proof of the covering estimates of the Inverse Function Theorem stated in Theorem 2.4. Assume that $U \subset \mathbb{R}^n$ is an open set containing $\mathbf{0}$, and that $f : U \to \mathbb{R}^n$ is a C^1 function with $f(\mathbf{0}) = \mathbf{0}$. Assume that $L = Df_{\mathbf{0}}$ is an invertible linear map (so has a bounded linear inverse). Take any β with $0 < \beta < 1$. Let $r > 0$ be such that (i) $\bar{B}(\mathbf{0}, r) \subset U$ and (ii) $\|L - Df_{\mathbf{x}}\| \leq m(L)(1 - \beta)$ for all $\mathbf{x} \in \bar{B}(\mathbf{0}, r)$. Define $\varphi(\mathbf{x}, \mathbf{y}) = \mathbf{y} + \mathbf{x} - L^{-1}f(\mathbf{x})$.

(a) For $\mathbf{y} \in \bar{B}(\mathbf{0}, \beta r)$ fixed, prove that the derivative with respect to $\mathbf{x}$ is given by $D_{\mathbf{x}}(\varphi(\cdot, \mathbf{y}))_{\mathbf{x}} = I - L^{-1}Df_{\mathbf{x}}$ for any $\mathbf{x} \in \bar{B}(\mathbf{0}, r)$, and that $\|D_{\mathbf{x}}(\varphi(\cdot, \mathbf{y}))_{\mathbf{x}}\| \leq 1 - \beta$.

(b) For $\mathbf{y} \in \bar{B}(\mathbf{0}, \beta r)$, prove that

$$|\varphi(\mathbf{x}_1, \mathbf{y}) - \varphi(\mathbf{x}_2, \mathbf{y})| \leq (1 - \beta)|\mathbf{x}_1 - \mathbf{x}_2|$$

for all $\mathbf{x}_1, \mathbf{x}_2 \in \bar{B}(\mathbf{0}, r)$.

(c) For $\mathbf{y} \in \bar{B}(\mathbf{0}, \beta r)$, prove that $\varphi(\cdot, \mathbf{y}) : \bar{B}(\mathbf{0}, r) \to \bar{B}(\mathbf{0}, r)$, and that it has a unique fixed point.

(d) For $\mathbf{y} \in L(\bar{B}(\mathbf{0}, \beta r))$, prove that there is a unique $\mathbf{x} \in \bar{B}(\mathbf{0}, r)$ such that $f(\mathbf{x}) = \mathbf{y}$. Hint: A fixed point $\mathbf{x}$ of $\varphi(\cdot, \mathbf{y})$ corresponds to a point $\mathbf{x}$ such that $f(\mathbf{x}) = L\mathbf{y}$.

(e) Prove that $L(\bar{B}(\mathbf{0}, \beta r)) \supset \bar{B}(\mathbf{0}, m(L)\beta r)$.

Contraction Mapping

5.4. Assume Y is a complete metric space with metric d, Λ is a metric space, and $0 < \kappa < 1$. Assume $g : \Lambda \times Y \to Y$ is uniformly continuous, and for each $\mathbf{x} \in \Lambda$, $g(\mathbf{x}, \cdot)$ is Lipschitz with $\text{Lip}(g(\mathbf{x}, \cdot)) \le \kappa$.

(a) Prove that there is a continuous map $\sigma : \Lambda \to Y$ such that $g(\mathbf{x}, \sigma(\mathbf{x})) = \sigma(\mathbf{x})$ for every $\mathbf{x} \in \Lambda$. Hint: Apply Theorem 2.5 to show that for each $\mathbf{x} \in \Lambda$, there is a $\sigma(\mathbf{x}) \in Y$ such that $g(\mathbf{x}, \sigma(\mathbf{x})) = \sigma(\mathbf{x})$. Then, for $\mathbf{x} \in \Lambda$ near $\mathbf{x}_0 \in \Lambda$, apply the estimate in Theorem 2.5 on the fixed point $\sigma(\mathbf{x})$ to see how near it is to $\sigma(\mathbf{x}_0)$.

(b) Assume that g is uniformly Lipschitz in both variables. Prove that σ is Lipschitz.

(c) Assume that g is C^1 and prove that σ is C^1. Hint: Use that Taylor expansion of g:

$$\begin{aligned} g(x, g(x)) =& g(x_0, \sigma(x_0)) + \frac{\partial g}{\partial x}(x_0, \sigma(x_0))(x - x_0) \\ &+ \frac{\partial g}{\partial y}(x_0, \sigma(x_0))(\sigma(x) - \sigma(x_0)) + o(|x - x_0|, |\sigma(x) - \sigma(x_0)|). \end{aligned}$$

Solutions of Differential Equations

5.5. Assume $A(t)$ is an $n \times n$ matrix with real entries such that $||A(t)||$ is bounded for all t, i.e., there is a $C > 0$ such that $||A(t)|| \le C$ for all t. Prove that the solutions of the linear system of equations $\dot{\mathbf{x}} = A(t)\mathbf{x}$ exist for all t.

5.6. Let C_0, C_1, and C_2 be positive constants. Assume $v(t)$ is a continuous nonnegative real valued function on $\mathbb{R}$ such that

$$v(t) \le C_0 + C_1|t| + C_2\Big|\int_0^t v(s)\,ds\Big|.$$

Prove that

$$v(t) \le C_0 e^{C_2|t|} + \frac{C_1}{C_2}\left[e^{C_2|t|} - 1\right].$$

Hint: Use

$$U(t) = C_0 + C_1 t + C_2 \int_0^t v(s)\,ds.$$

5.7. Let C_1 and C_2 be positive constants. Assume $f : \mathbb{R}^n \to \mathbb{R}^n$ is a C^1 function such that $|f(\mathbf{x})| \le C_1 + C_2|\mathbf{x}|$ for all $\mathbf{x} \in \mathbb{R}^n$. Prove that the solutions of $\dot{\mathbf{x}} = f(\mathbf{x})$ exist for all t. Hint: Use the previous exercise.

5.8. Consider the differential equation depending on a parameter

$$\dot{\mathbf{x}} = f(\mathbf{x}; \mu),$$

where $\mathbf{x}$ is in $\mathbb{R}^n$, μ is in $\mathbb{R}^p$, and f is a C^1 function jointly in $\mathbf{x}$ and μ. Let $\varphi^t(\mathbf{x}; \mu)$ be the solution with $\varphi^0(\mathbf{x}; \mu) = \mathbf{x}$. Prove that $\varphi^t(\mathbf{x}; \mu)$ depends continuously on the parameter μ.

5.9. (Differentiably flow conjugate) Let $f, g : \mathbb{R}^n \to \mathbb{R}^n$ be two vector fields with flows φ^t and ψ^t, respectively.

(a) Assume $h : \mathbb{R}^n \to \mathbb{R}^n$ is differentiable flow conjugacy of the flows φ^t and ψ^t, i.e., $h \circ \varphi^t \circ h^{-1} = \psi^t$ for all t and h is a C^1 diffeomorphism. Prove that

$$Dh_{h^{-1}(\mathbf{x})} f \circ h^{-1}(\mathbf{x}) = g(\mathbf{x}).$$

In differential geometry, the vector field given by $Dh_{h^{-1}(\mathbf{x})} f \circ h^{-1}(\mathbf{x})$ is labeled by $(h_* f)(\mathbf{x})$.

(b) Assume f and g are differentiably flow equivalent, i.e., they satisfy the equation

$$\psi^t(\mathbf{x}) = h \circ \varphi^{\alpha(t,\mathbf{x})} \circ h^{-1}(\mathbf{x})$$

where $h : \mathbb{R}^n \to \mathbb{R}^n$ is diffeomorphism and $\alpha : \mathbb{R} \times \mathbb{R}^n \to \mathbb{R}$ is differentiable with $\frac{d}{dt}\alpha(t, \mathbf{x}) > 0$. Prove that

$$Dh_{h^{-1}(\mathbf{x})} f \circ h^{-1}(\mathbf{x}) = \beta(\mathbf{x})g(\mathbf{x}),$$

where $\beta : \mathbb{R}^n \to (0, \infty)$.

5.10. (Flow Box Coordinates) Consider a differential equation $\dot{\mathbf{x}} = f(\mathbf{x})$ for $f : \mathbb{R}^n \to \mathbb{R}^n$ a C^r vector field for $r \geq 1$. Let $\mathbf{a}$ be a point for which $f(\mathbf{a}) \neq \mathbf{0}$. Prove there exists a C^r change of coordinates $\mathbf{x} = g(\mathbf{y})$ valid near $\mathbf{a}$ in terms of which the differential equations become

$$\dot{y}_1 = 1, \qquad \dot{y}_j = 0 \qquad \text{for } 2 \leq j \leq n.$$

The outline of the proof is as follows.

(a) Let $\mathbf{v} = f(\mathbf{a})$. One of the coordinates $v_k \neq 0$. Renumber the equations so that $k = 1$. Take the hyperplane

$$S = \{(a_1, y_2, \ldots, y_n)\}.$$

Define $g(\mathbf{y}) = \varphi^{y_1}(a_1, y_2, \ldots, y_n)$, where φ^t is the flow of the differential equation. Prove that

$$Dg_{\mathbf{a}} = \begin{pmatrix} v_1 & 0 & \ldots & 0 \\ v_2 & 1 & \ldots & 0 \\ \vdots & \vdots & & \vdots \\ v_n & 0 & \ldots & 1. \end{pmatrix}$$

(b) Using the inverse function theorem, prove that g defines a C^r set of coordinates near $\mathbf{a}$, i.e., g is C^r and has a C^r inverse.

(c) Prove the differential equations in the $\mathbf{y}$ variables is as the form given above.

Limit Sets

5.11. Let X be a metric space, and $S_j \subset X$ for $1 \leq j$ be a sequence of nested, compact, and connected subsets: $S_j \supset S_{j+1}$. Prove that $\bigcap_{j=1}^{\infty} S_j$ is connected.

5.12. Consider the flow $e^{tA}\mathbf{x}$ for the linear differential equation

$$\dot{\mathbf{x}} = A\mathbf{x}$$

with $\mathbf{x} \in \mathbb{R}^n$.

(a) Prove that all the eigenvalues of A have nonzero real part (i.e., the fixed point $\mathbf{0}$ is hyperbolic) if and only if for each $\mathbf{x} \in \mathbb{R}^n$, either $\omega(\mathbf{x}) = \{\mathbf{0}\}$ or $\omega(\mathbf{x}) = \emptyset$.

(b) For $n = 4$, show there is a choice of A and a point $\mathbf{x}$ for which $\omega(\mathbf{x}) \supset \mathcal{O}(\mathbf{x})$ but $\mathcal{O}(\mathbf{x})$ is neither a fixed point nor a periodic orbit.

5.13. Let $\mathbf{\dot{x}} = X(\mathbf{x})$ be a (nonlinear) differential equation in $\mathbb{R}^n$. Assume that a trajectory $\varphi^t(\mathbf{p})$ is bounded and each of the coordinates of the solution is monotonic for $t \geq t_0$, i.e., $\frac{d}{dt}\pi_j\varphi^t(\mathbf{p})$ does not change sign for $t \geq t_0$ and $1 \leq j \leq n$, where $\pi_j : \mathbb{R}^n \to \mathbb{R}$ is the projection of the jth coordinate. Prove that $\omega(\mathbf{p})$ is a single fixed point.

5.14. Let $f : X \to X$ and $g : Y \to Y$ are continuous maps on compact metric spaces. Assume $h : X \to Y$ is a topological conjugacy. Prove that h takes the chain recurrent set of f to the chain recurrent set of g, $h(\mathcal{R}(f)) = \mathcal{R}(g)$, i.e., prove that the chain recurrent set is a topological invariant.

Fixed Points for Differential Equations

5.15. Find the fixed points and classify them for the system of equations

$$\begin{aligned}
\dot{x} &= v, \\
\dot{v} &= -x + \omega\, x^3, \\
\dot{\omega} &= -\omega.
\end{aligned}$$

5.16. Find the fixed points and classify them for the Lorenz system of equations

$$\begin{aligned}
\dot{x} &= -10x + 10y, \\
\dot{y} &= 28x - y - xz, \\
\dot{z} &= \frac{8}{3}z + xy.
\end{aligned}$$

5.17. Consider the equation of a pendulum with friction,

$$\begin{aligned}
\dot{x} &= y \\
\dot{y} &= -\sin(x) - \delta y
\end{aligned}$$

with $\delta > 0$.

(a) Using a Liapunov function, prove that the ω-limit set of any point $\mathbf{q}_0 = (x_0, y_0)$ is a single fixed point.

(b) Let $\mathbf{p}_k = (k\pi, 0)$ be the fixed points. For a fixed point $\mathbf{p}_k$, let

$$W^s(\mathbf{p}_k) = \{\mathbf{q} : \omega(\mathbf{q}) = \mathbf{p}_k\}$$

be the *basin of attraction* of $\mathbf{p}_k$ (or the stable manifold of $\mathbf{p}_k$). Discuss how the basins of attraction for the fixed points are located in $\mathbb{R}^2$. In particular, explain how $W^s(\mathbf{p}_{2k+1})$ separates $W^s(\mathbf{p}_{2k})$ from $W^s(\mathbf{p}_{2k+2})$.

(c) Discuss the difference of the motion of a point $\mathbf{q} = (0, y_0)$ which lies in $W^s(\mathbf{p}_{2k})$ from the motion if it lies in $W^s(\mathbf{p}_{2k+2})$. In particular, how many rotations through multiple of 2π in the x-variable does each forward orbit make?

5.18. Discuss the basins of attraction of the fixed points for the following system of differential equations with $\delta > 0$:

$$\begin{aligned}
\dot{x} &= y \\
\dot{y} &= x - x^3 - \delta\, y.
\end{aligned}$$

5.19. Consider the equations

$$\dot{x} = x(A - x - ay),$$
$$\dot{y} = y(B - bx - y)$$

which model two competing species. Assume $A, B, a, b > 0$, $A > aB$, and $B > bA$.

(a) Find the fixed points. Hint: Consider the isoclines where $\dot{x} = 0$ and $\dot{y} = 0$.

(b) For any point $\mathbf{q}$ in the interior of the first quadrant, prove, using Exercise 5.13, that the $\omega(\mathbf{q})$ is a fixed point with both coordinates positive. Hint: Consider the various regions where $\dot{x}$ and $\dot{y}$ have fixed sign. (The isoclines are the boundaries of these regions.)

5.20. This exercise asks for an alternate proof of Theorem 5.1 (on nonlinear sinks for ordinary differential equations) using the Jordan Canonical Form approach. Assume $\mathbf{0}$ is a fixed point for the equations $\dot{\mathbf{x}} = f(\mathbf{x})$ with $\mathbf{x} \in \mathbb{R}^n$. Also assume that there is a constant $a > 0$ such that all the eigenvalues λ for $A = Df_{\mathbf{0}}$ have negative real part with $Re(\lambda) < -a < 0$.

(a) Let $\langle\ ,\ \rangle_{\mathcal{B}}$ be the inner product and $\|\ \|_{\mathcal{B}}$ be the norm determined by an arbitrary basis $\mathcal{B}$. Show that

$$\frac{d}{dt}\|\mathbf{x}(t)\|_{\mathcal{B}} = \frac{\langle \mathbf{x}(t), f(\mathbf{x}(t))\rangle_{\mathcal{B}}}{\|\mathbf{x}(t)\|_{\mathcal{B}}}.$$

(b) Let $\epsilon > 0$ be small enough so that $Re(\lambda) < -a - \epsilon$ for all the eigenvalues of A. Let $\mathcal{B}$ be the basis used in the alternative proof of the linear sink theorem, so

$$\langle \mathbf{x}, A\mathbf{x}\rangle_{\mathcal{B}} < (-a - \epsilon)\|\mathbf{x}\|_{\mathcal{B}}^2.$$

Prove that if U is a small enough neighborhood of 0, then

$$\langle \mathbf{x}, f(\mathbf{x})\rangle_{\mathcal{B}} < -a\|\mathbf{x}\|_{\mathcal{B}}^2.$$

(c) Prove that for a small enough neighborhood U of 0 and $\mathbf{x}_0 \in U$, the solution $\varphi^t(\mathbf{x}_0)$ satisfies $\|\varphi^t(\mathbf{x}_0)\|_{\mathcal{B}} \le e^{-at}\|\mathbf{x}_0\|_{\mathcal{B}}$.

(d) Prove that $\mathbf{0}$ is a nonlinear sink.

Remark: This exercise proves that $\|\mathbf{x}\|_{\mathcal{B}}$ or $\|\mathbf{x}\|_{\mathcal{B}}^2$ is a Liapunov function in a neighborhood of $\mathbf{0}$.

5.21. Let $f : \mathbb{R}^n \to \mathbb{R}$ be a C^2 function. Let $X(\mathbf{x}) = \nabla f$ be the gradient vector field for f.

(a) If $\mathbf{p}$ is a fixed point for X, prove that the eigenvalues are real.

(b) Prove that $\mathbf{p}$ is a hyperbolic fixed point of X if and only if $Df_{\mathbf{p}} = \mathbf{0}$ and $D^2 f_{\mathbf{p}}(\cdot,\cdot)$ is a nondegenerate bilinear form.

5.22. Assume $\dot{\mathbf{x}} = A\mathbf{x}$ and $\dot{\mathbf{y}} = B\mathbf{y}$ are both hyperbolic linear flows on $\mathbb{R}^n$ and are differentiably conjugate, i.e., there is a C^1 diffeomorphism $h : \mathbb{R}^n \to \mathbb{R}^n$ and a C^1 function $\tau : \mathbb{R} \times \mathbb{R}^n \to \mathbb{R}$ such that

$$h\big(e^{A\tau(t,\mathbf{x})}\mathbf{x}\big) = e^{Bt}h(\mathbf{x}).$$

Prove that the eigenvalues of A are proportional to the eigenvalues of B.

Periodic Points for Maps

5.23. Let $h : \mathbb{R} \to \mathbb{R}$ be given by $h(x) = x^3$. Find a C^∞ diffeomorphism $f : \mathbb{R} \to \mathbb{R}$ such that $h^{-1} \circ f \circ h$ is not differentiable at 1.

5.24. Consider the map given in polar coordinates by $f(r,\theta) = (r^2, \theta - 0.5\sin(\theta))$. This can be considered as a map on the two sphere by adding a fixed point at infinity.

(a) Find the fixed points and classify their stability. (Include the fixed point at infinity).

(b) Find the basins of attractions of all the fixed point sinks, including the fixed point at infinity.

5.25. Let $F_{AB} = (A - By - x^2, x)$ be the Hénon map. Find the fixed points and classify them for different values of the parameters A and B.

5.26. Prove Theorem 6.1 in the case of a fixed point. (This is the theorem that says that a map with a fixed point, all of whose eigenvalues are less than 1 in absolute value, is a contraction.)

5.27. Let f and g_k be diffeomorphisms of $\mathbb{R}$ given by

$$\begin{aligned} f(x) &= x + \frac{1}{2}\sin(x) && \text{and} \\ g_k(x) &= x + \frac{1}{2}h_k(x), && \text{where} \\ h_k(x) &= x - \frac{x^3}{3!} + \frac{x^5}{5!} - \cdots + (-1)^k \frac{x^{2k+1}}{(2k+1)!} && \text{for } k \geq 1. \end{aligned}$$

For $k \geq 1$, prove that f and g are not topologically conjugate.

5.28. Suppose h is a conjugacy between $f : \mathbb{R}^n \to \mathbb{R}^n$ and $g : \mathbb{R}^n \to \mathbb{R}^n$.

(a) Show that $\mathbf{p}$ is a periodic point of f if and only if $h(\mathbf{p})$ is a periodic point of g.

(b) Show that if $f^j(\mathbf{p})$ converges to $\mathbf{q}$ as j goes to infinity, then $g^j(h(\mathbf{p}))$ converges to $h(\mathbf{q})$ as j goes to infinity.

(c) Show that for any $\mathbf{p} \in \mathbb{R}^n$, we have $h(\omega(\mathbf{p}, f)) = \omega(h(\mathbf{p}), g)$, where $\omega(\mathbf{p}, f)$ are the ω-limit sets of $\mathbf{p}$.

5.29. Assume $f : \mathbb{R}^n \to \mathbb{R}^n$ is a C^1 diffeomorphism. Assume $\mathbf{p}$ is a hyperbolic periodic point for f. Given any positive integer n, prove there is a neighborhood U of $\mathbf{p}$ such that any periodic point of f in $U \setminus \{\mathbf{p}\}$ has period greater than n.

Hartman-Grobman Theorem

5.30. This exercise gives another proof of Theorem 7.1. Let $A \in \mathbf{L}(\mathbb{R}^n, \mathbb{R}^n)$ be an invertible hyperbolic linear map. Assume that f is a C^1 diffeomorphism of $\mathbb{R}^n$ with $A \circ f^{-1} - id \in C_b^0(\mathbb{R}^n, \mathbb{R}^n)$. To solve the equation $A \circ h = h \circ f$ for $h = id + k$ with $k \in C_b^0(\mathbb{R}^n, \mathbb{R}^n)$, it is sufficient to solve

$$\begin{aligned} A \circ (id + k) \circ f^{-1} &= id + k, \\ A \circ f^{-1} + A \circ k \circ f^{-1} &= id + k, \\ A \circ f^{-1} - id &= k - A \circ k \circ f^{-1}, && \text{or} \\ \mathcal{L}(k) &= A \circ f^{-1} - id, \end{aligned}$$

where

$$\mathcal{L}(k) = k - A \circ k \circ f^{-1}.$$

(a) Prove that $\mathcal{L}$ is an invertible bounded linear operator on $C_b^0(\mathbb{R}^n, \mathbb{R}^n)$. This includes the fact that $\mathcal{L}$ preserves the space $C_b^0(\mathbb{R}^n, \mathbb{R}^n)$.

(b) Prove that there is a unique solution k_0 to the equation $\mathcal{L}(k) = A \circ f^{-1} - id$.

(c) Let $h = k_0 + id$ where k_0 is the solution obtained in part (b). Prove that h is (i) a homeomorphism, and (ii) a conjugacy between A and f.

5.31. Assume φ^t and ψ^t are two flows on $\mathbb{R}^n$ for which $\mathbf{0}$ is a hyperbolic fixed point sink. Show that there is a conjugacy h in a neighborhood of $\mathbf{0}$ from φ^1 and ψ^1 (the time one maps) that is not a conjugacy of φ^t and ψ^t, and in fact does not take trajectories of φ^t to trajectories of ψ^t.

Periodic Orbits for Flows

5.32. Assume γ is a periodic orbit for the flow φ^t and β is a periodic orbit for the flow ψ^t (both flows in n-dimensional spaces). Let P_φ be the Poincaré map for the flow φ^t for a transversal $\Sigma_{\mathbf{p}}$ at $\mathbf{p} \in \gamma$; and P_ψ be the Poincaré map for the flow ψ^t for a transversal $\Sigma_{\mathbf{q}}$ at $\mathbf{q} \in \beta$. Assume that P_φ in a neighborhood of $\mathbf{p}$ in $\Sigma_{\mathbf{p}}$ is topologically conjugate to P_ψ in a neighborhood of $\mathbf{q}$ in $\Sigma_{\mathbf{q}}$. Prove that the flow φ^t in a neighborhood of γ is topologically equivalent to ψ^t in a neighborhood of β.

5.33. Assume γ is an attracting periodic orbit for the flow φ^t. Prove that φ^t in a neighborhood of γ is topologically conjugate to the linear bundle flow defined in Section 5.8 for Theorem 8.7.

5.34. Assume that two diffeomorphisms f and g are topologically conjugate. Prove that their suspensions are topologically conjugate.

5.35. Let $A(t)$ be a time-dependent n by n curve of matrices. Let $M(t)$ be a fundamental matrix solution for the system of differential equations $\dot{\mathbf{x}} = A(t)\mathbf{x}$. Prove Liouville's Formula:

$$\det(M(t_1)) = \det(M(t_0)) \int_{t_0}^{t_1} \operatorname{tr}(A(t))\, dt.$$

5.36. Consider the Volterra-Lotka equations

$$\begin{aligned} \dot{x} &= x(A - By - ax) = xM(x,y), \\ \dot{y} &= y(Cx - D - by) = yN(x,y) \end{aligned}$$

with all $A, B, C, D, a, b > 0$. These equations model the populations of two species which are predator y and prey x and an increase in either population adversely affects its own growth rate (there is a crowding factor in both equations),

(a) Find conditions on the constants so there is a unique fixed point $\mathbf{p}^* = (x^*, y^*)$ with $x^* > 0$ and $y^* > 0$. Hint: Look at the isoclines where $\dot{x} = 0$ and $\dot{y} = 0$.

(b) Letting $V(x,y)$ be the vector field for these equations, verify that

$$(\operatorname{div} V)(x,y) = \dot{x}/x + \dot{y}/y + xM_x + yN_y$$

where M_x and N_y are the partial derivatives with respect to x and y of the respective functions.

Let $\Sigma = \{(x, y^*) : x \geq x^*\}$ and $P : \Sigma \to \Sigma$ be the Poincaré map. The solution of the rest of this exercise proves that for any $\mathbf{q}$ in the interior of the first quadrant, $\omega(\mathbf{q}) = \{\mathbf{p}^*\}$.

(c) Verify that

$$P'(x) = \frac{N(x,y^*)}{N(P(x),y^*)} \cdot \frac{P(x)}{x} \cdot e^{\mu(x)}$$

for $x > x^*$ where $\mu(x) < 0$.

(d) Define $f(x) = N(x, y^*)/x$ and F the antiderivative of f. Prove that $F \circ P(x) < F(x)$ for $x > x^*$, and that $P^n(x)$ converges to x^* as n goes to infinity.

(e) Conclude that for $\mathbf{q}$ in the interior of the first quadrant, $\omega(\mathbf{q}) = \{\mathbf{p}^*\}$.

5.37. Assume X is a C^1 vector field on $\mathbb{T}^2$ with no fixed points. Prove that there is a closed curve Γ on $\mathbb{T}^2$ which is a transversal to the flow. Further, prove that Γ cannot be contracted to a point in $\mathbb{T}^2$. Hint: Consider the vector field Y on $\mathbb{T}^2$ which is everywhere perpendicular to X, e.g., if $\tilde{X}$ and $\tilde{Y}$ are the lifts of X and Y to $\mathbb{R}^2$, then it is possible to take $\tilde{Y}(x,y) = (X_2(x,y), -X_1(x,y))$ where $\tilde{X}(x,y) = (X_1(x,y), X_2(x,y))$. Take any point $\mathbf{q} \in \mathbb{T}^2$ and consider $\mathbf{p} \in \omega(\mathbf{q}, Y)$. The trajectory of $\mathbf{q}$ for Y repeatedly comes near $\mathbf{p}$. By considering a pair of points on the trajectory, show that it can be modified to make a transversal for the flow of X.

5.38. Consider a vector field X on $\mathbb{T}^2$ whose lift $\tilde{X}$ to $\mathbb{R}^2$ has first component equal to 1, $\tilde{X}(x,y) = (1, X_2(x,y))$. Prove that X has a periodic orbit if and only if the Poincaré map has a rational rotation number.

Poincaré-Bendixson Theorem

5.39. Let $\dot{\mathbf{x}} = V(\mathbf{x})$ be a differential equation in $\mathbb{R}^2$ with only a finite number of fixed points. Assume $\mathbf{p}_0$ is a point whose forward orbit, $\mathcal{O}^+(\mathbf{p}_0)$, is bounded.

(a) Assume $\mathbf{p}_1 \in \omega(\mathbf{p}_0)$ and $\mathbf{p}_2 \in \omega(\mathbf{p}_1)$ with $V(\mathbf{p}_2) \neq \mathbf{0}$. Apply Lemma 9.4 and the argument of Lemma 9.5 to prove that $\mathbf{p}_1$ is on a periodic orbit.

(b) If $\mathbf{p}_1 \in \omega(\mathbf{p}_0)$ is not a periodic orbit, prove that $\alpha(\mathbf{p}_1)$ and $\omega(\mathbf{p}_1)$ are each single fixed points.

(c) If $\omega(\mathbf{p}_0)$ is not a periodic orbit, prove that $\omega(\mathbf{p}_0)$ contains a finite set of fixed points and a finite set of other orbits $\mathcal{O}(\mathbf{q}_i)$ where $\alpha(\mathbf{q}_i)$ and $\omega(\mathbf{q}_i)$ are each single fixed points for each $\mathbf{q}_i$.

5.40. Let $A = S^1 \times [a,b]$ be an annulus with covering space $\tilde{A} = \mathbb{R} \times [a,b]$. (The "angle variable" is not taken modulo 1 in the covering space.) Let X be a vector field on A with lift $\tilde{X} = (\tilde{X}_1, \tilde{X}_2)$ to $\tilde{A}$. Assume that $\tilde{X}_1(x,a) < 0$ and $\tilde{X}_1(x,b) > 0$ for all x, and $\operatorname{div}(X) \equiv 0$, i.e., X is area preserving. Prove that the flow of X has a fixed point in A. Hint: Assume X has no fixed points and let Y be a nonzero vector field which is perpendicular to X everywhere in A. Prove that Y has a periodic orbit γ in A using the Poincaré-Bendixson Theorem. Get a contradiction to the area preserving assumption on X by considering X along γ.

5.41. Consider the differential equations

$$\dot{x} = a - x - \frac{4xy}{1+x^2}$$
$$\dot{y} = bx\left(1 - \frac{y}{1+x^2}\right)$$

for $a, b > 0$.

(a) Show that $x^* = a/5$ and $y^* = 1 + (x^*)^2$ is the only fixed point.

(b) Show that the fixed point is repelling for $b < 3a/5 - 25/a$ and $a > 0$. Hint: Show that $\det(DF_{(x^*,y^*)}) > 0$ and $\operatorname{tr}(DF_{(x^*,y^*)}) > 0$.

(c) Let x_1 be the value of x where the isocline $\{\dot{x} = 0\}$ crosses the x-axis. Let $y_1 = 1 + (x_1)^2$. Prove that the rectangle $\{(x,y) : 0 \le x \le x_1, 0 \le y \le y_1\}$ is positively invariant.

(d) Prove that there is a periodic orbit in the first quadrant for $a > 0$ and $0 < b < 3a/5 - 25/a$.

Fiber Contractions (Stable Manifold Theory)

5.42. Let $\bar{B}^n(r) \subset \mathbb{R}^n$ be the closed ball in $\mathbb{R}^n$ of radius $r > 0$ about $\mathbf{0}$, and $\bar{B}^k(r') \subset \mathbb{R}^k$ be the closed ball in $\mathbb{R}^k$ of radius $r' > 0$ about $\mathbf{0}$. Assume $F : \bar{B}^n(r) \times \bar{B}^k(r') \to \mathbb{R}^n \times \bar{B}^k(r')$ is a C^1 map of the form $F(\mathbf{x}, \mathbf{y}) = (f(\mathbf{x}), g(\mathbf{x}, \mathbf{y}))$ such that (i) $f(\text{int}(\bar{B}^n(r))) \supset \bar{B}^n(r)$, and (ii) $f|\bar{B}^n(r)$ is a diffeomorphism from $\bar{B}^n(r)$ onto its image. Let $D_2 g_{(\mathbf{x},\mathbf{y})} : \mathbb{R}^k \to \mathbb{R}^k$ be the derivative with respect to the variables in $\mathbb{R}^k$. Let $\kappa_{\mathbf{x}} = \sup\{\|D_2 g_{(\mathbf{y},\mathbf{x})}\| : \mathbf{y} \in \mathbb{R}^k\}$. Assume that (iii) $\sup\{\kappa_{\mathbf{x}} : \mathbf{x} \in \bar{B}^n(r)\} < 1$.
   (a) Prove that for each $\mathbf{x} \in \bar{B}^n(r)$, there is a unique $\sigma(\mathbf{x}) \in \bar{B}^k(r')$ such that $(\mathbf{x}, \sigma(\mathbf{x}))$ has a backward orbit by F in : $\bar{B}^n(r) \times \bar{B}^k(r')$. Hint: Prove that the intersection $\bigcap_{n=0}^{\infty} F^n(\{f^{-n}(\mathbf{x}) \times \bar{B}^k(r'))$ is a single point.
   (b) Prove that $\sigma : \bar{B}^n(r) \to \bar{B}^k(r')$ is an invariant section, i.e.,

$$F(\mathbf{x}, \sigma^*(\mathbf{x})) = (f(\mathbf{x}), \sigma^* \circ f(\mathbf{x}))$$

   for all $\mathbf{x} \in f^{-1}(\bar{B}^n(r))$.
   (c) Prove that map $\sigma : \bar{B}^n(r) \to \bar{B}^k(r')$ is continuous.

5.43. Using the notation of the previous exercise, assume $F : \bar{B}^n(r) \times \mathbb{R}^k \to \mathbb{R}^n \times \mathbb{R}^k$ is a C^1 map of the form $F(\mathbf{x}, \mathbf{y}) = (f(\mathbf{x}), g(\mathbf{x}, \mathbf{y}))$ such that (i) $f(\text{int}(\bar{B}^n(r))) \supset \bar{B}^n(r)$, and (ii) $f|\bar{B}^n(r)$ is a diffeomorphism from $\bar{B}^n(r)$ onto its image, (iii) $\sup\{\kappa_{\mathbf{x}} : \mathbf{x} \in \bar{B}^n(r)\} < 1$, and (iv) $\sup\{\kappa_{\mathbf{x}} \lambda_{\mathbf{x}} : \mathbf{x} \in \bar{B}^n(r)\} < 1$, where $\lambda_{\mathbf{x}} = \|(Df_{\mathbf{x}})^{-1}\|$ and $\kappa_{\mathbf{x}} = \sup\{\|D_2 g_{(\mathbf{y},\mathbf{x})}\| : \mathbf{y} \in \mathbb{R}^k\}$. Let $\sigma^* : \bar{B}^n(r) \to \mathbb{R}^k$ be the unique continuous invariant section found in the previous exercise. Prove that σ^* is C^1. Hint: Construct a family of horizontal cones $C_{(\mathbf{x},\mathbf{y})}$ that are taken inside themselves by $DF_{(\mathbf{x},\mathbf{y})}$, $DF_{(\mathbf{x},\sigma(\mathbf{x}))} C_{(\mathbf{x},\sigma(\mathbf{x}))} \subset C_{F(\mathbf{x},\sigma(\mathbf{x}))}$.

Center Manifold

5.44. (A polynomial vector field without a C^∞ center manifold. This example is taken from Carr (1981) and is a modification of an example of van Strien (1979).) Consider the equations

$$\begin{aligned} \dot{x} &= xy + x^3 \\ \dot{y} &= 0 \\ \dot{z} &= z - x^2. \end{aligned}$$

   (a) Show that the center manifold of $(x, y, z) = \mathbf{0}$ can be written as a graph $z = h(z, y)$ for $|x| \le \delta$ and $|y| \le \delta$ for small $\delta > 0$.
   (b) Show that the fixed points $\{(0, y, 0) : |y| \le \delta\}$ all lie on $W^c(\mathbf{0})$.
   (c) Assume that $z = h(z, y)$ is C^{2n} for $|x| \le \delta$ and $|y| \le \delta$. Take the Taylor expansion of h in x about $x = 0$ (with coefficients which are functions of y):

$$h(x, y) = \sum_{j=1}^{2n} a_j(y) x^j + o(|x|^{2n}).$$

   Find a relationship between the coefficients by equating $\dot{z} = z - x^2 = h(x, y) - x^2$ and $\dot{z} = \dfrac{\partial h}{\partial x}\dot{x} + \dfrac{\partial h}{\partial y}\dot{y}$. In particular, show that (i) $a_1(y) = 0$, (ii) $a_2(y) \ne 0$, and (iii) $(1 - jy)a_j(y) = (j - 2)a_{j-2}(y)$ for $j > 2$.
   (d) Show that the point $(0, 1/(2n), 0)$ cannot lie in the domain where h is C^{2n}. Hint: Show that $a_{2i}(1/(2n)) \ne 0$ for $1 \le i < n$, and obtain a contradiction

for the relationship involving the coefficients $a_{2n}(1/(2n))$ and $a_{2n-2}(1/(2n))$. Remark: What makes this example work is the resonance between the eigenvalues at the fixed point $(0, 1/(2n), 0)$. The resonance forces the weak unstable manifold for the eigenvalue $1/(2n)$ to be C^{2n-1}, but not C^{2n}. In turn, this manifold is contained in the center manifold of $\mathbf{0}$, so it cannot be C^{2n} either.

(e) Show that there is no neighborhood of $\mathbf{0}$ on which the center manifold $W^c(\mathbf{0})$ is C^∞.

5.45. Assume $f : \mathbb{R}^n \times \mathbb{R} \to \mathbb{R}^n$ is a C^r function for $r \geq 1$. Write $f_\lambda(\mathbf{x})$ for $f(\mathbf{x}, \lambda)$. The map $F : \mathbb{R}^{n+1} \to \mathbb{R}^{n+1}$ be defined by $F(\mathbf{x}, \lambda) = (f(\mathbf{x}, \lambda), \lambda)$ is also C^r. Assume that $\mathbf{x}_0$ is a hyperbolic fixed point for f_0. Let $\mathbf{x}_\lambda$ be the corresponding hyperbolic fixed point for λ near 0. Using the C^r Center Manifold Theorem for F, prove that the stable manifold of $\mathbf{x}_\lambda$ for f_λ, $W^s(\mathbf{x}_\lambda, f_\lambda)$, depends C^r jointly on position and parameter λ, i.e., prove that $W^s(\mathbf{x}_\lambda, f_\lambda)$ for $|\lambda| < \epsilon$ and $\epsilon > 0$ small can be represented as the graph of a C^r function

$$\sigma : \mathbb{E}^s_{\mathbf{x}_0,0} \times (-\epsilon, \epsilon) \to \mathbb{E}^u_{\mathbf{x}_0,0}.$$

Inclination Lemma

5.46. Let $A = \begin{pmatrix} 0.5 & 0 \\ 0 & 2 \end{pmatrix}$, and consider the linear map $A\mathbf{x}$. Let $S = [-r, r] \times [-r, r]$ be the square. Let D be the line segment be the part of the line through the point $(1, 0)$ with slope s that lies within S. Assume s and r are chosen so that D intersects the top and the bottom of S and not the sides. Prove by a direct calculation that the nth iterate of the line segment, $A^n(D) \cap S$, converges to the part of the y-axis given by $\{0\} \times [-r, r]$. Prove the convergence both in terms of the points and the slope.

CHAPTER VI
Hamiltonian Systems

In this chapter we give a very brief introduction to Hamiltonian differential equations. These equations are one way to describing mechanical systems without any friction. We mainly give the local treatment near fixed points or periodic orbits. For a more complete treatment, see one of the several books devoted entirely to Hamiltonian systems: Meyer and Hall (1992) is a good recent book of approximately the level of this one; other books include Siegel and Moser (1971), Abraham and Marsden (1978), Arnold (1978), Arnold, Kozlov, and Neishtadt (1993), and Dankowicz (1997).

6.1 Hamiltonian Differential Equations

Newton's equations of motion of a particle moving in a configuration space $S = \{\mathbf{q} \in \mathbb{R}^n\}$ with a potential function $V : S \to \mathbb{R}$ and momentum $\mathbf{p} = m\,\mathbf{v}$ are given by

$$m\ddot{\mathbf{q}} = -\nabla V_{\mathbf{q}},$$

or

$$\begin{aligned}\dot{q}_j &= p_j/m\\ \dot{p}_j &= -\frac{\partial V}{\partial q_j}\end{aligned}$$

for $1 \le j \le n$. The full phase space of both positions and momenta is $S \times \mathbb{R}^n$. By forming the energy function

$$H(\mathbf{q},\mathbf{p}) = V(\mathbf{q}) + \frac{\mathbf{p}\cdot\mathbf{p}}{2m},$$

which is the sum of the kinetic and potential energy, the equations can be written in Hamiltonian form

$$\begin{aligned}\dot{q}_j &= \frac{\partial H}{\partial p_j}\\ \dot{p}_j &= -\frac{\partial H}{\partial q_j}\end{aligned}$$

for $1 \le j \le n$.

More generally, consider a phase space $S \times \mathbb{R}^n$, where S equals either (i) $\mathbb{R}^n$, (ii) the n-torus $\mathbb{T}^n$, or (iii) some combination $\mathbb{T}^k \times \mathbb{R}^{n-k}$. Assume the time-dependent Hamiltonian $H : S \times \mathbb{R}^n \times \mathbb{R} \to \mathbb{R}$ is a real valued function of $2n+1$

variables, $H(q_1, \ldots, q_n, p_1, \ldots, p_n, t)$ with $\mathbf{q} \in S$, $\mathbf{p} \in \mathbb{R}^n$, and $t \in \mathbb{R}$ time. The time-dependent *Hamiltonian differential equations* generated by H are given by

$$\dot{q}_j = \frac{\partial H}{\partial p_j}(\mathbf{q}, \mathbf{p}, t),$$
$$\dot{p}_j = -\frac{\partial H}{\partial q_j}(\mathbf{q}, \mathbf{p}, t)$$

for $j = 1, \ldots, n$. If H does not depend on t, then the equations are called *time-independent.*

These equations can be expressed in terms of the gradient of H by introducing the $2n \times 2n$ skew symmetric matrix $J = J_n$ given by

$$J_n = \begin{pmatrix} 0_n & I_n \\ -I_n & 0_n \end{pmatrix}$$

where 0_n is the $n \times n$ zero matrix and I_n is the $n \times n$ identity matrix. Letting $\mathbf{z} = (z_1, \ldots z_{2n}) = (\mathbf{q}, \mathbf{p})$ the differential equations become

$$\dot{\mathbf{z}} = J\nabla H_{\mathbf{z},t}$$

where the gradient is with respect to the variable $\mathbf{z}$ only and not t. The vector field $X_H(\mathbf{z}) = J\nabla H_{\mathbf{z}}$ is called the *Hamiltonian vector field* for the Hamiltonian function H.

We give a brief idea of a more invariant way to consider Hamiltonian equations. A more complete treatment of this approach uses differential forms. See Abraham and Marsden (1978) or Arnold (1978). The matrix J induces a bilinear form ω on $\mathbb{R}^{2n}$ by

$$\omega(\mathbf{v}, \mathbf{w}) = \mathbf{v}^T J\mathbf{w}.$$

This form ω is (i) linear in each variable separately (bilinear), (ii) $\omega(\mathbf{v}, \mathbf{w}) = -\omega(\mathbf{w}, \mathbf{v})$ (alternating), and (iii) for each $\mathbf{v} \in \mathbb{R}^{2n}$, there is a $\mathbf{w} \in \mathbb{R}^{2n}$ such that $\omega(\mathbf{v}, \mathbf{w}) \neq 0$ (nondegenerate). Any bilinear from satisfying these three properties is called a *symplectic form.*

We only consider symplectic forms which are independent of the base point of vectors; but using the theory of differential form, it is possible to let the symplectic form ω vary with the base point. In the case when ω is a differential form which can vary with the base point, the phase space can be a manifold of states and not just a Euclidean space, and the differential form acts on tangent vectors to the manifold. See Abraham and Marsden (1978) or Arnold (1978).

The symplectic form ω for J induces a map $\bar{\omega}$ between $\mathbb{R}^{2n}$ and its dual space $\mathbb{R}^{2n*}$ by

$$\bar{\omega}(\mathbf{v})(\mathbf{w}) = \omega(\mathbf{v}, \mathbf{w}).$$

Considering elements of $\mathbb{R}^{2n*}$ as row vectors, this map is given by $\bar{\omega}(\mathbf{v}) = \mathbf{v}^T J$ which is an isomorphism, and

$$\bar{\omega}^{-1}(\mathbf{u}^*) = (\mathbf{u}^* J^{-1})^T = J(\mathbf{u}^*)^T.$$

Now, if $H : \mathbb{R}^{2n} \to \mathbb{R}$ is a real valued differentiable function, then for each point $\mathbf{x}$, the derivative $DH_{\mathbf{x}} \in \mathbb{R}^{2n*}$ (it acts on a vector to give a number). Thus, $\bar{\omega}^{-1}(DH_{\mathbf{x}}) = J(DH_{\mathbf{x}})^T = X_H(\mathbf{x})$ is the Hamiltonian vector field for the Hamiltonian H.

Proposition 1.1. *Let H be a time-independent Hamiltonian for the Hamiltonian vector field X_H. Then, the value of H is constant along solutions of X_H.*

PROOF. The time derivative of H along a trajectory is calculated as follows.

$$\begin{aligned}\dot{H}(\mathbf{q}(t),\mathbf{p}(t)) &= \sum_j \Big(\frac{\partial H}{\partial q_j}\dot{q}_j + \frac{\partial H}{\partial p_j}\dot{p}_j\Big) \\ &= \sum_j \frac{\partial H}{\partial q_j}\frac{\partial H}{\partial p_j} + \frac{\partial H}{\partial p_j}\Big(-\frac{\partial H}{\partial q_j}\Big) \\ &\equiv 0.\end{aligned}$$

Since the time derivative is zero, H is constant along solutions. Therefore, the Hamiltonian function is a "weak Liapunov function," at least in terms of the time derivative. □

Example 1.1. The linear harmonic oscillator is the second-order differential equation in one variable, $n = 1$, given by

$$\ddot{q} + \omega^2 q = 0.$$

(We take $\omega > 0$.) The corresponding potential function is $V(q) = \omega^2 q^2$, so $H(q,p) = \dfrac{p^2 + \omega^2 q^2}{2}$. Therefore, solutions lie on the ellipses formed by setting the Hamiltonian equal to positive constants. The solutions can be found explicitly as $q(t) = A\cos(\omega(t-\delta))$ and $p(t) = -\omega A\sin(\omega(t-\delta))$ for arbitrary constants A and δ. See Figure IV.4.4 for a closely related phase portrait.

The fixed points of a Hamiltonian system are the critical points of the Hamiltonian function, i.e., points $\mathbf{x}_0$ where the derivative $DH_{\mathbf{x}_0} = \mathbf{0}$. For a system which has a Hamiltonian which is the sum of kinetic plus potential energy V, the fixed points are the points $\mathbf{x}_0 = (\mathbf{q}_0, \mathbf{0})$ where $\nabla V_{\mathbf{q}_0} = \mathbf{0}$.

Example 1.2. The equations

$$\begin{aligned}\dot{q} &= p, \\ \dot{p} &= q - q^3 = f(q)\end{aligned}$$

are Hamiltonian with potential energy

$$V(q) = -\int f(q)\,dq = -\frac{q^2}{2} + \frac{q^4}{4}$$

and Hamiltonian function $H(q,p) = p^2/2 + V(q)$. The potential function $V(q)$ is a quartic polynomial with two minima and one local maximum. See Figure 1.1. The fixed points correspond to these extrema of V and are $(q,p) = (0,0)$, $(\pm 1,\, 0)$. Exercise 6.4 proves that the minima correspond to elliptic fixed points and the maximum to a saddle fixed point. This can be checked explicitly

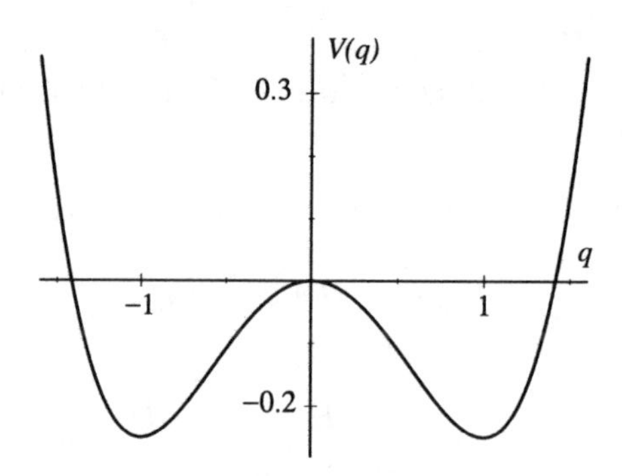

FIGURE 1.1. Graph of the Potential Function for Example 1.2

for this example by forming the matrix of partial derivatives at the various fixed points.

Consider the energy level $h_1 > 0$, i.e., $H^{-1}(h_1)$. Let $\pm q_1$ be the two points with $V(\pm q_1) = h_1$, so that $H(\pm q_1, 0) = h_1$ and the entire level curve lie in the vertical strip $-q_1 < q < q_1$. (For $|q| > q_1$, $V(q) > h_1$, so there are no points on the level curve $H^{-1}(h_1)$.) Because $h_1 - V(q) > 0$ for any $q \in (-q_1, q_1)$, for each such q, there are exactly two values of p with $H(q,p) = h_1$. Therefore, the level curve $H^{-1}(h_1)$ is a single closed curve surrounding the origin. An example of such a level curve is the outermost closed curve in Figure 1.2. Because the differential equation has no fixed point on this closed curve, we conclude that solutions of the differential equations corresponding to positive energy levels are periodic orbits which surround the origin.

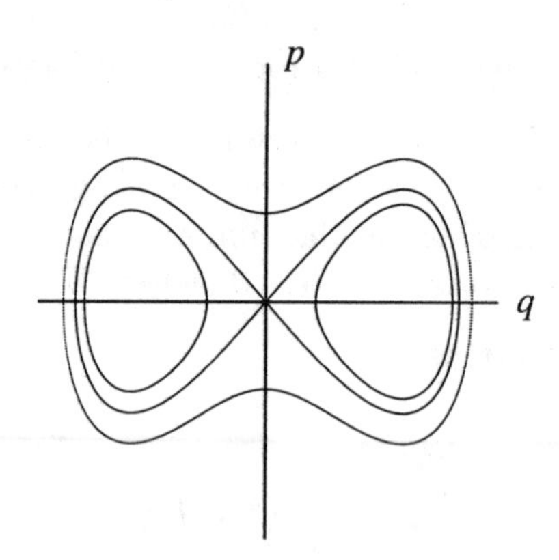

FIGURE 1.2. Phase Portrait for Example 1.2

Now consider the energy level $h_0 = 0$. Note that $V(0) = V(\pm\sqrt{2}) = 0$, so for $q = 0$ or $\pm\sqrt{2}$, p muxt be 0 for (q,p) to lie on the level curve $H^{-1}(0)$. For $0 < |q| < \sqrt{2}$, there are two points $\pm p$ on $H^{-1}(0)$. Therefore, the level curve of the energy is the two lobed curve with a crossing at $(0,0)$. This level set is the second level curve in Figure 1.2 which looks like a "figure 8" on its side. These solutions of the differential equations give orbits which have both α-limits and ω-limits at the fixed point $(0,0)$. Such orbits are called *homoclinic*.

For the energy level $-0.25 < h_{-1} < 0$, there are two ranges of q allowed where $V(q) \leq h_{-1}$. The level set of the Hamiltonian function has two separate closed curves which correspond to two separate periodic orbits. These are the two closed curves in Figure 1.2 which do not surround the origin.

Example 1.3. The equation of a pendulum of mass m and length L are given by

$$mL\ddot{\theta} = -mg\sin(\theta)$$

and has Hamiltonian $H(\theta, p) = \frac{p^2}{2} + \frac{g}{L}(1 - \cos(\theta))$. The variable θ can be considered as a variable on a circle (angular variable) or a real variable. We leave to the reader to determine the phase portrait.

Proposition 1.2. *A Hamiltonian vector field has zero divergence, so it preserves volume.*

PROOF. The divergence is

$$\begin{aligned} \operatorname{div}(X_H)(\mathbf{z}) &= \sum_{j=1}^{n} \frac{\partial}{\partial q_j}(\frac{\partial H}{\partial p_j})(\mathbf{z}) + \frac{\partial}{\partial p_j}(-\frac{\partial H}{\partial q_j})(\mathbf{z}) \\ &= 0. \end{aligned}$$

□

There is another way to write the equations of motion using the Poisson bracket. The *Poisson bracket* is an alternating bilinear map on real valued functions

$$\{\cdot, \cdot\} : C^r(\mathbb{R}^{2n}, \mathbb{R}) \times C^r(\mathbb{R}^{2n}, \mathbb{R}) \to C^{r-1}(\mathbb{R}^{2n}, \mathbb{R})$$

defined by

$$\{F, G\} = \sum_{j=1}^{n} \frac{\partial F}{\partial q_j} \cdot \frac{\partial G}{\partial p_j} - \frac{\partial F}{\partial p_j} \cdot \frac{\partial G}{\partial q_j}.$$

Notice that

$$\begin{aligned} \{F, G\}(\mathbf{z}) &= DF_{\mathbf{z}} \cdot X_G(\mathbf{z}) \\ &= -DG_{\mathbf{z}} \cdot X_F(\mathbf{z}) \\ &= \nabla F_{\mathbf{z}}^T J \nabla G_{\mathbf{z}}. \end{aligned}$$

From the definition, it follows that

$$\begin{aligned} \{q_j, H\} &= \frac{\partial H}{\partial p_j} = \dot{q}_j \\ \{p_j, H\} &= -\frac{\partial H}{\partial q_j} = \dot{p}_j, \end{aligned}$$

which expresses the differential equation in terms of the Poisson bracket.

6.2 Linear Hamiltonian Systems

In this section we study the properties of linear systems. In this section it is often important that J is orthogonal and skew symmetric, i.e.,

$$J^{-1} = J^T = -J.$$

Notice that at a fixed point $\mathbf{x}_0$ for a Hamiltonian H, the linearized equations are given by

$$\begin{aligned}\dot{\mathbf{x}} &= JD^2H_{\mathbf{x}_0}(\mathbf{x}-\mathbf{x}_0)\\ &= \begin{pmatrix} \dfrac{\partial^2 H}{\partial q_j \partial p_i} & \dfrac{\partial^2 H}{\partial p_j \partial p_i} \\ -\dfrac{\partial^2 H}{\partial q_j \partial q_i} & -\dfrac{\partial^2 H}{\partial p_j \partial q_i} \end{pmatrix}(\mathbf{x}-\mathbf{x}_0).\end{aligned}$$

The matrix of the linearization is equal to J times a symmetric matrix. A matrix A can be written as JS where S is symmetric if and only if A satisfies $A^TJ + JA = 0$.

Definition. Therefore, we define a $2n \times 2n$ matrix A to be *Hamiltonian* (or *infinitesimally symplectic*) provided $A^TJ + JA = 0$. The set of all $2n \times 2n$ Hamiltonian matrices with real entries is denoted by $sp(n, \mathbb{R})$.

Theorem 2.1. *(a) Consider the the fundamental matrix solution $M(t)$ of a time-dependent linear Hamiltonian system $\dot{\mathbf{z}} = A(t)\mathbf{z}$, where $A(t)$ is Hamiltonian for each t. Then, $M(t)$ satisfies $M(t)^TJM(t) = J$.*

(b) Conversely, if $M(t)$ is a differentiable curve of $2n \times 2n$ matrices which satisfy $M(t)^TJM(t) = J$, then M is a matrix solution of a time-dependent linear Hamiltonian system.

PROOF. (a) Let $P(t) = M(t)^TJM(t)$. For $t = 0$, $P(0) = IJI = J$. The time derivative

$$\begin{aligned}\dot{P}(t) &= \dot{M}(t)^TJM(t) + M(t)^TJ\dot{M}(t)\\ &= M(t)^TA(t)^TJM(t) + M(t)^TJA(t)M(t)\\ &= M(t)^T[A(t)^TJ + JA(t)]M(t)\\ &= 0,\end{aligned}$$

so $P(t) \equiv J$.

(b) Conversely, if $M(t)^TJM(t) = J$, then $0 = \dot{M}(t)^TJM(t) + M(t)^TJ\dot{M}(t)$. Thus, $(\dot{M}(t)M(t)^{-1})^TJ + J(\dot{M}(t)M(t)^{-1}) = 0$, $A(t) \equiv (\dot{M}(t)M(t)^{-1})$ is Hamiltonian, and $M(t)$ satisfies $\dot{M}(t) = A(t)M(t)$. □

Definition. Because of the above theorem, we define a $2n \times 2n$ matrix M to be *symplectic* provided $M^TJM = J$. This is equivalent to saying that the matrix M preserves the symplectic form ω associated with J. The set of all symplectic matrices with real entries is denoted by $Sp(n, \mathbb{R})$.

Corollary 2.2. *Let H be a Hamiltonian function and φ_H^t the flow of the corresponding Hamiltonian vector field X_H. Then, $D(\varphi_H^t)_{\mathbf{x}}$ is a symplectic matrix for all t.*

PROOF. The derivative of the flow satisfies the first variation equation

$$\frac{d}{dt}D(\varphi_H^t)_{\mathbf{x}} = D(X_H)_{\varphi_H^t(\mathbf{x})}D(\varphi_H^t)_{\mathbf{x}},$$

and $D(X_H)_{\varphi_H^t(\mathbf{x})}$ is Hamiltonian. By the previous result $D(\varphi_H^t)_{\mathbf{x}}$ is symplectic. □

For both Hamiltonian and symplectic matrices, there are restrictions on the eigenvalues as indicated in the following result.

Proposition 2.3. *(a) Let A be a Hamiltonian matrix. If λ is an eigenvalue of A, then so are $-\lambda$, $\bar{\lambda}$, and $-\bar{\lambda}$. In fact, the Jordan block structures of λ and $-\lambda$ are the same. Therefore, the direct sum of the eigenspaces for λ, $-\lambda$, $\bar{\lambda}$, and $-\bar{\lambda}$ has even dimension. Also, the eigenvalue 0 has even multiplicity. Finally, the trace $\mathrm{tr}(A) = 0$.*

(b) Let M be a symplectic matrix. If λ is an eigenvalue of M, then so are λ^{-1}, $\bar{\lambda}$, and $\bar{\lambda}^{-1}$. In fact, the Jordan block structures of λ and λ^{-1} are the same. Therefore, the direct sum of the eigenspaces for λ, λ^{-1}, $\bar{\lambda}$, and $\bar{\lambda}^{-1}$ has even dimension. The multiplicities of 1 and -1 as eigenvalues are even (possibly zero). Also, $\det(M) = 1$.

PROOF. Since A is real, the eigenvalues λ and $\bar{\lambda}$ have the same multiplicities.

(a) Since $A^T J + JA = 0$, $A^T = -JAJ^{-1}$, and so A^T is similar to $-A$. Since the eigenvalues of A and A^T are the same, the result about the eigenvalues and Jordan block structures follows. The total dimension is even and the sum of the multiplicities of λ and $-\lambda$ is even for $\lambda \neq 0$, therefore the multiplicity of 0 is even. The trace is the sum of the eigenvalues, and since the multiplicities of λ and $-\lambda$ are equal, the trace of A is zero.

(b) Since $M^T JM = J$, $M^T = JM^{-1}J^{-1}$, and so M^T is similar to M^{-1}. Again, the result about the eigenvalues and Jordan block structures follows. The sum of the multiplicities of 1 and -1 is even, because the total dimension is even and the sum of the multiplicities of λ and λ^{-1} is even. The fact that multiplicities of 1 and -1 are each separately even follows from the fact that the determinant is 1.

By taking the determinant of $M^T JM = J$, we get that $\det(M)^2 = 1$ so $\det(M) = \pm 1$. The fact that the determinant is actually $+1$ follows from the following two lemmas. □

REMARK 2.1. There are actually several ways to prove that the determinant of a symplectic matrix is $+1$. (i) It follows because the two-form induced by J is nondegenerate on the generalized eigenspace for the eigenvalue -1. (Also, J is nondegenerate on the generalized eigenspace for the eigenvalue -1). In order for the two-form to be nondegenerate, the dimension needs to be even. The determinant is the product of the eigenvalues and the muplicities of λ and λ^{-1} are equal for $\lambda \neq \pm 1$. Combining, $\det(M) = 1$. For a proof like this, see

Section D of Chapter II of Meyer and Hall (1992). This approach is part of the construction of a normal form for symplectic matrices. (ii) The two-form ω for J induces a volume which must be preserved by M since it preserves J. A volume preserving matrix has determinant 1. See Chapter III of Meyer and Hall (1992). (iii) The following two lemmas (following Section A of Chapter II of Meyer and Hall (1992)) give a proof using the polar decomposition of a matrix.

Lemma 2.4. *Let M be a symplectic matrix. Write it in its polar decomposition as the product of a positive definite and an orthogonal matrix, $M = PQ$. Then, both P and Q are symplectic.*

PROOF. The matrix M satisfies $M^T JM = J$ so $M^{-1} = J^{-1}M^T J$. Substituting in the polar decomposition gives

$$Q^{-1}P^{-1} = (J^T Q^T J)(J^T P^T J).$$

Since both J and Q^T are orthogonal, $J^T Q^T J$ is orthogonal. Since P^T is positive definite, $J^T P^T J$ is positive definite. By the uniqueness of the polar decomposition, $Q^{-1} = J^T Q^T J$ and $P^{-1} = J^T P^T J$, which implies that both Q and P are symplectic. □

Lemma 2.5. *Let M be a symplectic matrix. Then,* $\det(M) = 1$.

PROOF. We noted above that $\det(M) = \pm 1$, so we only need to show that $\det(M) > 0$. Writing M in its polar decomposition, $M = PQ$, we have that $\det(P) > 0$ so we only need to show that $\det(Q) > 0$, where Q is both symplectic and orthogonal.

We can write Q in block form with $n \times n$ submatrices:

$$Q = \begin{pmatrix} a & b \\ c & d \end{pmatrix}.$$

Then,

$$Q^T = \begin{pmatrix} a^T & c^T \\ b^T & d^T \end{pmatrix} = Q^{-1} = -JQ^T J = \begin{pmatrix} d^T & -b^T \\ -c^T & a^T \end{pmatrix},$$

so $a = d$ and $c = -b$, or

$$Q = \begin{pmatrix} a & b \\ -b & a \end{pmatrix}.$$

To see that a matrix with this block form has positive determinant, we conjugate the matrix to a complex block diagonal matrix. Let

$$S = \frac{1}{\sqrt{2}} \begin{pmatrix} I & -iI \\ I & iI \end{pmatrix}, \qquad S^{-1} = \frac{1}{\sqrt{2}} \begin{pmatrix} I & I \\ iI & -iI \end{pmatrix},$$

Then,

$$\begin{aligned} \det(Q) &= \det(SQS^{-1}) \\ &= \det \begin{pmatrix} a + ib & 0 \\ 0 & a - ib \end{pmatrix} \\ &= \det(a + ib)\det(a - ib) \\ &= |\det(a + ib)|^2 > 0, \end{aligned}$$

where the last equality holds because it is the product of complex conjugate numbers. This completes the proof of the lemma and the proposition. □

REMARK 2.2. For a Hamiltonian matrix in two dimensions (e.g., at a fixed point of a Hamiltonian vector field), there are quite explicit restrictions on the eigenvalues. There are only three cases: 1) (elliptic, center) both eigenvalues are purely imaginary; 2) (saddle) both eigenvalues are real with one negative and the other positive; or 3) (zero eigenvalue) both eigenvalues are zero. Because of these restrictions and the fact that the eigenvalues vary continuously with variation in the matrix, if a Hamiltonian matrix is elliptic, then a small Hamiltonian perturbation has to stay elliptic. This persistence carries over to Hamiltonian differential equations; a small perturbation of a Hamiltonian differential equation with an elliptic fixed point also has an elliptic fixed point.

REMARK 2.3. For a closed orbit of a Hamiltonian vector field, the multiplicity of the eigenvalue 1 of $D(\varphi_H^T)_{\mathbf{x}}$ must be at least two (and even). One multiplicity comes from being a closed orbit and preserving $X_H(\mathbf{x})$, and a second from the fact that φ_H^t preserves the energy H.

REMARK 2.4. The fact that the trace of a Hamiltonian matrix is zero is directly related to the fact that Hamiltonian vector fields are divergence free, Proposition 1.2. The fact that a symplectic matrix has determinant 1 means that it preserves the volume. Any symplectic form for ω induces a volume element Ω by taking its wedge product n times (in the language of differential forms). The fact that M preserves ω implies that it preserves Ω and hence is volume preserving.

6.3 Symplectic Diffeomorphisms

We have considered vector fields which preserve the symplectic structure. In this section we turn to maps which do so. We also consider ways of changing coordinates which preserve the symplectic structure.

Definition. A map $f : \mathbb{R}^{2n} \to \mathbb{R}^{2n}$ is called *symplectic* provided that the derivative at any point $\mathbf{x} \in \mathbb{R}^{2n}$ is a symplectic linear map, $(Df_{\mathbf{x}})^T J Df_{\mathbf{x}} = J$, i.e., $Df_{\mathbf{x}}$ preserves the associated symplectic form ω,

$$\omega(Df_{\mathbf{x}}\mathbf{v}, Df_{\mathbf{x}}\mathbf{w}) = \mathbf{v}^T (Df_{\mathbf{x}})^T J Df_{\mathbf{x}}\mathbf{w} = \mathbf{v}^T J \mathbf{w} = \omega(\mathbf{v}, \mathbf{w}).$$

A *symplectic diffeomorphism* is a symplectic map which is also a diffeomorphism.

REMARK 3.1. Symplectic diffeomorphisms are sometimes used as change of coordinates to put a Hamiltonian differential equation into a better form. In this context, they are often called *canonical transformations*. In the next section, we use such change of coordinates to put a Hamiltonian vector field into a normal form near a fixed point.

REMARK 3.2. Because a symplectic matrix has determinant 1, a symplectic diffeomorphism preserves volume.

The first result shows that a symplectic diffeomorphism preserves the form of a Hamiltonian vector field. Because change of coordinates are often only defined in an open set, we only require that the symplectic diffeomorphism be local, i.e., defined in such an open set.

Theorem 3.1. *Let $H : \mathbb{R}^{2n} \to \mathbb{R}$ a Hamiltonian function, and $g : \mathbb{R}^{2n} \to \mathbb{R}^{2n}$ be a (local) symplectic diffeomorphism. Then, the differential equation in the new coordinates $\mathbf{y} = g(\mathbf{x})$ given by $\dot{\mathbf{y}} = Y(\mathbf{y}) = Dg_{g^{-1}(\mathbf{y})} X_H(g^{-1}(\mathbf{y}))$ is Hamiltonian with Hamiltonian function $K(\mathbf{y}) = H \circ g^{-1}(\mathbf{y})$.*

PROOF. We consider the change of coordinates $\mathbf{y} = g(\mathbf{x})$ which can be defined on only an open set of $\mathbb{R}^{2n}$. The differential equation in the new variables is given by

$$\begin{aligned}\dot{\mathbf{y}} &= Dg_{\mathbf{x}}\dot{\mathbf{x}} \\ &= Dg_{\mathbf{x}} J (DH_{\mathbf{x}})^T \\ &= Dg_{\mathbf{x}} J (D(K \circ g)_{\mathbf{x}})^T \\ &= Dg_{\mathbf{x}} J Dg_{\mathbf{x}}^T DK_{\mathbf{y}}^T \\ &= J DK_{\mathbf{y}}^T \\ &= X_K(\mathbf{y}),\end{aligned}$$

where $\mathbf{x} = g^{-1}(\mathbf{y})$. □

The following theorem gives one of the principal ways in which symplectic diffeomorphisms arise as the time t map of a Hamiltonian vector field.

Theorem 3.2. *Let $H : \mathbb{R}^{2n} \times \mathbb{R} \to \mathbb{R}$ be a (possibly time-dependent) Hamiltonian function. Let $\varphi_H^t(\mathbf{x}, t_0)$ be the flow where $\varphi_H^{t_0}(\mathbf{x}, t_0) = \mathbf{x}$. Then, for each fixed t, $\varphi_H^t(\cdot, t_0) : \mathbb{R}^{2n} \to \mathbb{R}^{2n}$ is a symplectic diffeomorphism on the domain of its definition.*

PROOF. The map $\varphi_H^t(\cdot, t_0)$ is not necessarily defined on all $\mathbb{R}^{2n}$, but otherwise it is a diffeomorphism as usual. The derivative $D(\varphi_H^t)_{(\mathbf{x},t_0)}$ (with respect to initial conditions $\mathbf{x}$) satisfies the first variation equation

$$\frac{d}{dt} D(\varphi_H^t)_{(\mathbf{x},t_0)} = D(X_H)_{\varphi_H^t(\mathbf{x},t_0),t} D(\varphi_H^t)_{(\mathbf{x},t_0)},$$

and $D(X_H)_{\mathbf{y}}$ is a Hamiltonian matrix for any point $\mathbf{y}$. Then Theorem 2.1 proves that $D(\varphi_H^t)_{(\mathbf{x},t_0)}$ is a symplectic matrix. □

A second manner in which local symplectic diffeomorphisms can be constructed is by means of a generating function. These are useful in creating local coordinates in which to simplify the local form of a Hamiltonian system, as we do in the next section. The motivation and derivation of this method are usually given in terms of differential forms, which we have not introduced. However, in two dimensions, the explanation reduces to material usually covered in calculus. Therefore, we consider a map $(Q, P) = F(q, p)$ in $\mathbb{R}^2$. The fact that F preserves area can be express by the fact that for any closed curve γ

$$\oint_{F(\gamma)} P\, dQ = \oint_{\gamma} p\, dq = \int_R dq\, dp,$$

where R is the region inside γ. As we shall show below, sometimes it is possible to determine p and P from the variables of q and Q. Considering all the variables to be functions of q and Q,

$$\oint_{\gamma} [p(q,Q)\, dq - P(q,Q)\, dQ] = 0$$

for all closed paths γ. This says that the line integral of the integrand is independent of path and is an exact differential $dS(q,Q)$ (or the coefficients are the gradient of a function S). Therefore,

$$p = \frac{\partial S}{\partial q}(q,Q) \quad \text{and} \quad P = -\frac{\partial S}{\partial Q}(q,Q).$$

This pair of equations implicitly defines the map F. If $\dfrac{\partial^2 S}{\partial q \partial Q}(q,Q) \neq 0$, then it is possible to locally solve the equation $p = \dfrac{\partial S}{\partial q}(q,Q)$ for Q in terms of p and q, and so for P in terms of p and q. Therefore, starting with a real valued function S whose second derivative $\dfrac{\partial^2 S}{\partial q \partial Q}(q,Q) \neq 0$, it is possible to define a local map of (Q,P) defined in terms of (q,p), which is in fact a local diffeomorphism. By reversing the above discussion of line integrals, the diffeomorphism preserves area and so is symplectic. The function $S(q,Q)$ is called the *generating function* for the symplectic map. What seems strange at first is the fact that the generating function depends on some variables in the domain and other variables in the range.

The way the above argument works in higher dimensions is basically the same. The fact that F is symplectic for the matrix J can be expressed by saying that it preserves the associated two form $\sum_j dq_j \wedge dp_j$, or $\sum_j [dQ_j \wedge dP_j - dq_j \wedge dp_j] = 0$. Therefore, the exterior derivative

$$d[\sum_j -P_j\, dQ_j + p_j\, dq_j] = 0.$$

Because the form is closed, it is locally exact and there exists a real valued function $S(\mathbf{q}, \mathbf{Q})$ such that

$$p_j = \frac{\partial S}{\partial q_j}(\mathbf{q}, \mathbf{Q}) \quad \text{and} \quad P_j = -\frac{\partial S}{\partial Q_j}(\mathbf{q}, \mathbf{Q}).$$

(This is basically the same result that says a vector field whose curl is zero equals the gradient of a function.) Again, if $\det\left(\dfrac{\partial^2 S}{\partial q_j \partial Q_i}(\mathbf{q}, \mathbf{Q})\right) \neq 0$, then it is possible to locally solve the equation $\mathbf{p} = \left(\dfrac{\partial S}{\partial q_j}(\mathbf{q}, \mathbf{Q})\right)$ for $\mathbf{Q}$ in terms of $\mathbf{p}$ and $\mathbf{q}$, and so for $\mathbf{P}$ in terms of $\mathbf{p}$ and $\mathbf{q}$.

It is possible to use different combinations of variables in the above considerations. Therefore, we get the following theorem.

Theorem 3.3. *A real valued function* $S_1(\mathbf{q},\mathbf{Q})$, $S_2(\mathbf{q},\mathbf{P})$, $S_3(\mathbf{p},\mathbf{P})$, *or* $S_4(\mathbf{p},\mathbf{Q})$ *on* $\mathbb{R}^{2n}$ *locally defines a symplectic map by*

$$
\begin{aligned}
p_j &= \frac{\partial S_1}{\partial q_j}(\mathbf{q},\mathbf{Q}) \quad \text{and} \quad P_j = -\frac{\partial S_1}{\partial Q_j}(\mathbf{q},\mathbf{Q}),\\
p_j &= \frac{\partial S_2}{\partial q_j}(\mathbf{q},\mathbf{P}) \quad \text{and} \quad Q_j = \frac{\partial S_2}{\partial P_j}(\mathbf{q},\mathbf{P}),\\
q_j &= \frac{\partial S_3}{\partial p_j}(\mathbf{p},\mathbf{P}) \quad \text{and} \quad Q_j = -\frac{\partial S_3}{\partial P_j}(\mathbf{p},\mathbf{P}), \quad \text{or}\\
q_j &= \frac{\partial S_4}{\partial p_j}(\mathbf{p},\mathbf{Q}) \quad \text{and} \quad P_j = \frac{\partial S_4}{\partial Q_j}(\mathbf{p},\mathbf{Q}),
\end{aligned}
$$

provided

$$
\begin{aligned}
&\det\left(\frac{\partial^2 S_1}{\partial q_i \partial Q_j}\right) \neq 0,\\
&\det\left(\frac{\partial^2 S_2}{\partial q_i \partial P_j}\right) \neq 0,\\
&\det\left(\frac{\partial^2 S_3}{\partial p_i \partial P_j}\right) \neq 0, \quad \text{or}\\
&\det\left(\frac{\partial^2 S_4}{\partial p_i \partial Q_j}\right) \neq 0,
\end{aligned}
$$

respectively.

Any of the real valued functions S_i in the above theorem is called a *generating function.*

To obtain many aspects of the theory for area preserving maps on an annulus $A = S^1 \times \mathbb{R}$, we need to know that the map does not "shift" A along the $\mathbb{R}$-variable. Using the variable q as the angular variable in S^1 and p in $\mathbb{R}$, we say that $f : A \to A$ is called *exact symplectic* provided for any closed curve γ (including ones which go around S^1)

$$\oint_\gamma p\, dq = \oint_{f(\gamma)} P\, dQ.$$

The map f being exact symplectic implies that the signed area between $f(\gamma)$ and γ is zero, for any nontrivial closed curve which goes around the S^1 direction. In particular, for such γ, $f(\gamma) \cap \gamma \neq \emptyset$. This condition is used in studying so-called twist maps on annuli.

Example 3.1. One simple example which often considered in the *standard map* given by

$$
\begin{aligned}
Q &= q + p - \frac{k}{2\pi}\sin(2\pi q) \mod 1\\
P &= p - \frac{k}{2\pi}\sin(2\pi q),
\end{aligned}
$$

where p is a real variable and q is taken as modulo 1. This map is exact symplectic and has complicated dynamics for $0 < k < 1$. See Meiss (1992) for a discussion of the dynamics of the standard map.

6.4 Normal Form at Fixed Point

In order to understand the dynamics near a fixed point, we often want to put the equations in simple form. When the fixed point is hyperbolic, then the Hartman-Grobman Theorem or Sternberg Linearization Theorem proves that the system is conjugate to the linear terms by continuous or differentiable change of coordinates. For Hamiltonian systems, there are often elliptic points which cannot be perturbed away; and near such elliptic points, it is usually not possible to find a neighborhood in which the equations have only linear terms. In this section, we discuss the Birkhoff normal form, which states that there is a change of coordinates up to terms of a given order putting the formal Taylor expansion at the fixed point in a relatively simple form with a remainder term. If we try to let the order of the change of coordinates go to infinity, then it usually does not converge, and so only gives a formal change of coordinates to the normal form with no remainder term. In the next section, we use the low order normal form to state the KAM Theorem about the existence of invariant tori near an elliptic fixed point.

Chapter VII of Meyer and Hall (1992) gives a general treatment of normal forms using Lie transforms, which is applicable to both Hamiltonian and non-Hamiltonian systems. We restrict ourselves to Hamiltonian systems and use generating functions to make the symplectic change of coordinates.

We start by writing the Taylor expansion of the Hamiltonian at the fixed point. For simplicity, we take the fixed point to be the origin and the value $H(\mathbf{0}) = 0$. Then,

$$H(\mathbf{x}) = \sum_{k=2}^{\infty} H_k(\mathbf{x}),$$

where H_k is a homogeneous polynomial of degree k. The linear terms of the equations at the fixed point are

$$A\mathbf{x} = J\nabla(H_2)_{\mathbf{x}}.$$

We consider the case when all the eigenvalues of A are purely imaginary, $\pm\, i\,\omega_j$ for $j = 1, \ldots n$. If each eigenvalue has multiplicity one, then it is possible to take symplectic coordinates such that

$$\begin{aligned} H_2(\mathbf{x}) &= \sum_{j=1}^{n} \frac{\omega_j}{2}(x_j^2 + x_{j+n}^2) \\ &= \sum_{j=1}^{n} \omega_j \rho_j \end{aligned}$$

where $\rho_j = (x_j^2 + x_{j+n}^2)/2$. This form forces the choice of the sign of the ω_j: some may be positive and others negative. We allow multiplicities or resonances of the eigenvalues but assume that the quadratic terms can be put in the above form. The generalized Birkhoff normal form which allows for resonance of the eigenvalues is given in the following theorem. We only give the result to finite order. If the process continues for all degrees, then there is no assurance that the change of coordinates and the normal form converge.

Theorem 4.1. *Let H be a Hamiltonian with a fixed point at the origin with all the eigenvalues purely imaginary and the quadratic terms H_2 are the sums of squares as given above. Then, there is a symplectic change of coordinates up to degree d*

$$\mathbf{y} = h(\mathbf{x}) = \mathbf{x} + h_2(\mathbf{x}) + \cdots + h_d(\mathbf{x})$$

where each h_k is a homogeneous polynomial of degree k, which transforms the Hamiltonian to

$$K(\mathbf{y}) = H_2(\mathbf{y}) + \sum_{k=3}^{d} K_k(\mathbf{y}) + K_{d+1}(\mathbf{y}),$$

where each K_k for $3 \le k \le d$ is a homogeneous polynomial of degree k such that the Poisson bracket

$$\{K_k, H_2\} \equiv 0,$$

and K_{d+1} is the remainder of terms of degrees greater than or equal to $d+1$. (All partial derivatives of K_{d+1} up to order d are equal to zero at the origin.)

REMARK 4.1. Since the change of coordinates is the identity plus higher-order terms, $K_2(\mathbf{y}) = H_2(\mathbf{y})$. The condition on the Poisson bracket implies that $K_k(e^{At}\mathbf{y}) = K_k(\mathbf{y})$ for all $2 \le k \le d$.

REMARK 4.2. The eigenvalues are said to be *linearly independent over the integers up to order d* provided that

$$\sum_{j=1}^{n} m_j \omega_j \neq 0$$

for any $m_j \in \mathbb{Z}$ with $0 < \sum_j |m_j| \le d$. They are said to be *linearly independent over the integers* provided they are linearly independent for all orders. In this case, the proof shows that for the normal form, all the odd degree terms are zero and the even degree terms can be written as functions of the $\rho_j = (x_j^2 + x_{j+n}^2)/2$, i.e., they are independent of the angles $\theta_j = \tan^{-1}(x_{j+n}/x_j)$. If the change of coordinates converged, this would imply the stability of the fixed point.

REMARK 4.3. The following proof follows Moser (1968), which is based on the work of Gustavson. It possible to allow more general forms of the quadratic terms, but then the operator below does not have a basis of eigenfunctions. We restrict ourselves to the case where the proof is simpler. See Guckenheimer and Holmes (1983) for a more general treatment.

PROOF.

We eliminate terms in the expansion by induction on the degree using symplectic change of coordinates given by means of a generating function

$$G(x_1, \ldots, x_n, y_{n+1}, \ldots, y_{2n}).$$

We seek the function G which determines K by the equation

$$H(x_1, \ldots, x_n, \frac{\partial G}{\partial x_1}, \ldots, \frac{\partial G}{\partial x_n}) = K(\frac{\partial G}{\partial y_{n+1}}, \cdots \frac{\partial G}{\partial y_{2n}}, y_{n+1}, \ldots, y_{2n}).$$

We represent G as the sum of homogeneous polynomials G_k of degree k,

$$G = \sum_{j=1}^{n} x_j y_{n+j} + \sum_{k \geq 3} G_k.$$

The explicit quadratic terms given imply that the symplectic map starts out with the identity for the linear term. This is possible because H_2 is already in normal form.

By induction, assume that G_2 through G_{k-1} have been chosen so that K_2 through K_{k-1} are already in normal form. We write out the calculation as if H started in normal form for these terms, and we take $G_\ell = 0$ for $3 \leq \ell < k$. Substituting

$$\begin{aligned}
H(x_1, &\ldots, x_n, \frac{\partial G}{\partial x_1}, \ldots, \frac{\partial G}{\partial x_n}) \\
&= \frac{1}{2} \sum_j \omega_j \left[x_j^2 + (y_{n+j} + \frac{\partial G_k}{\partial x_j})^2\right] \\
&\quad + H_3(x_1, \ldots, x_n, y_{n+1}, \ldots, y_{2n}) + \cdots \\
&\quad + H_k(x_1, \ldots, x_n, y_{n+1}, \ldots, y_{2n}) + \cdots \\
&= \frac{1}{2} \sum_j \omega_j \left[x_j^2 + y_{n+j}^2\right] + H_3(x_1, \ldots, x_n, y_{n+1}, \ldots, y_{2n}) + \cdots \\
&\quad + H_k(x_1, \ldots, x_n, y_{n+1}, \ldots, y_{2n}) + \sum_j \omega_j y_{n+j} \frac{\partial G_k}{\partial x_j} + \cdots
\end{aligned}$$

and

$$\begin{aligned}
K(&\frac{\partial G}{\partial y_{n+1}}, \ldots, \frac{\partial G}{\partial y_{2n}}, y_{n+1}, \ldots, y_{2n}) \\
&= K_2(x_1, \ldots, x_n, y_{n+1}, \ldots, y_{2n}) + \cdots \\
&\quad + K_k(x_1, \ldots, x_n, y_{n+1}, \ldots, y_{2n}) + \sum_j \omega_j x_j \frac{\partial G_k}{\partial y_{n+j}} + \cdots
\end{aligned}$$

Equating the terms of order k in both expansions, we get

$$\begin{aligned}
K_k(x_1, \ldots, x_n, y_{n+1}, \ldots, y_{2n}) + \sum_j \omega_j \left[x_j \frac{\partial G_k}{\partial y_{n+j}} - y_{n+j} \frac{\partial G_k}{\partial x_j}\right] \\
= P_k(x_1, \ldots, y_{2n})
\end{aligned}$$

where P_k is a homogeneous polynomial determined by the previous steps.

To determine which terms can be eliminated and which need to be left, we need to determine the kernel and the cokernel (complement of the image) for the operator D defined by

$$DG_k = \sum_j \omega_j \big[x_j \frac{\partial G_k}{\partial y_{n+j}} - y_{n+j} \frac{\partial G_k}{\partial x_j}\big] = \{H_2, G_k\}.$$

Let $\omega = (\omega_1, \ldots, \omega_n)$. Let M be the set of all vectors $\mathbf{m} = (m_1, \ldots, m_n)$ with integer entries such that

$$\langle \mathbf{m}, \omega \rangle = \sum_{j=1}^{n} m_j \omega_j = 0.$$

If the eigenvalues are independent over the integers, then $M = \{\mathbf{0}\}$. In general, M is a subgroup of the integer lattice.

It is possible to give the proof using only real coordinates, but it is easier to introduce complex variables and check that the change of coordinates preserves the real subspace. Let $z_j = x_j + i\, y_{j+n}$ and $\bar{z}_j = x_j - i\, y_{j+n}$. Then, the Poisson bracket

$$\begin{aligned} \{H_2, G\} &= \sum_j \omega_j \big[x_j \frac{\partial G}{\partial y_{j+n}} - y_{j+n} \frac{\partial G}{\partial x_j}\big] \\ &= i \sum_j \omega_j \big[z_j \frac{\partial G}{\partial z_j} - \bar{z}_j \frac{\partial G}{\partial \bar{z}_j}\big]. \end{aligned}$$

We expand G in terms of z_j and $\bar{z}_j$, and a monomial term is of the form

$$\mathbf{z}^{\mathbf{k}} \bar{\mathbf{z}}^{\mathbf{m}} = \prod_{j=1}^{n} z_j^{k_j} \bar{z}_j^{m_j}.$$

Its Poisson bracket with $H_2(x_1, \ldots, y_{2n}) = \sum_j (\omega_j/2) z_j \bar{z}_j$ is

$$\{H_2, \mathbf{z}^{\mathbf{k}} \bar{\mathbf{z}}^{\mathbf{m}}\} = i \sum_{j=1}^{n} \omega_j (k_j - m_j)(\mathbf{z}^{\mathbf{k}} \bar{\mathbf{z}}^{\mathbf{m}}).$$

Therefore, all these monomials are eigenfunctions of the operator D. Those monomials which have $\mathbf{k} - \mathbf{m} \in M$ are in the kernel of D: these are terms which must be left in the normal form. We can decompose the homogeneous term $P_k = P_{k,N} + P_{k,I}$ where $D(P_{k,N}) = \{H_2, P_{k,N}\} = 0$ is in the kernel of the operator and $P_{k,I}$ is in the image of the operator. We can solve

$$\begin{aligned} D(G_k) &= P_{k,I} \\ K_k &= P_{k,N} \end{aligned}$$

for G_k and K_k. The terms for $\mathbf{z}^{\mathbf{k}}\bar{\mathbf{z}}^{\mathbf{m}}$ and $\mathbf{z}^{\mathbf{m}}\bar{\mathbf{z}}^{\mathbf{k}}$ will appear with complex conjugate coefficients, so the complex generating function will give a real generating function which preserves real variables. This completes the induction step and the proof. □

There is a corresponding result for symplectic maps. We give the statement for multiplicatively independent eigenvalues. Let $\lambda_1, \ldots, \lambda_n, \lambda_1^{-1}, \ldots, \lambda_n^{-1}$ be the eigenvalues. These eigenvalues are *multiplicatively independent up to order* d provided $\prod_{j=1}^{n} \lambda_j^{m_j} \neq 1$ for $m_j \in \mathbb{Z}$ with $0 < \sum_j |m_j| \leq d$. If the eigenvalues are multiplicatively independent, then the matrix of the derivative can be diagonalized into 2×2 blocks on which it is a rotation. The proof of the theorem is similar to that given above and is given in Easton (1998).

Theorem 4.2. *Let f be a symplectic diffeomorphism with a fixed point at the origin for which all the eigenvalues have absolute value equal to 1. Assume the eigenvalues are multiplicatively independent up to order at least d. Then there is a symplectic change of variables $\mathbf{x} = h(\mathbf{y})$ such that in terms of these coordinates, $F = h^{-1} \circ f \circ h$ has the form $F(\mathbf{q}, \mathbf{p}) = (\mathbf{Q}, \mathbf{P})$ with*

$$\begin{aligned} Q_j &= q_j \cos(\varphi_j) - p_j \sin(\varphi_j) + R_{d+1,j}, \\ P_j &= q_j \sin(\varphi_j) + p_j \cos(\varphi_j) + R_{d+1,n+j}, \quad \textit{where} \\ \varphi_j &= \omega_j + \beta_j(\rho_1, \ldots, \rho_n), \\ \rho_j &= (q_j^2 + p_j^2)/2, \qquad \textit{and} \\ \lambda_j &= e^{i\omega_j} \end{aligned}$$

and all the remainder terms $R_{d+1,j}$ have degree greater than or equal to $d+1$.

REMARK 4.4. The functions β_j give the dependence of the amount of rotation on the radii. For the KAM Theorem in the next section, we assume that the matrix of partial derivatives

$$\det\left(\frac{\partial^2 \beta_i}{\partial \rho_j}\right)(\mathbf{0}) \neq 0.$$

This says that the rotation in the different planes varies as the radii are varied.

6.5 KAM Theorem

There are several related theorems called KAM Theorems. The initials KAM stand for Kolmogorov, Arnold, and Moser. The normal forms at fixed points discussed in the last section do not necessarily converge. If they did converge, then the fixed points would be Liapunov stable and all solutions would lie on invariant tori given by setting the ρ_j equal to constants. What the KAM theorems say is that all the tori do not necessarily exist for the full nonlinear system, but some do exist: those which have very irrational rotations with the rotation numbers in the different factors irrationally related. The idea for the proof was originally given by Kolmogorov. For an English translation of his 1954 address see the appendix of Abraham and Marsden (1978). Later, Arnold supplied detailed proofs of some of the results for analytic systems. Still later, Moser showed that the theorems were valid for C^∞ systems using a smoothing and approximation method. Herman and others have refined many of the results. See Herman (1983, 1986). A good introduction to the results is given in Moser (1968, 1973) and Arnold and Avez (1968). Also see Pöschel (1982).

We first consider a Hamiltonian differential equation near a fixed point. The tori which persist are those which have highly irrational rotation numbers. The unperturbed system obtained from the first terms in the normal form is given by

$$\begin{aligned}\dot{\rho}_j &= 0\\ \dot{\theta}_j &= \Omega_j(\rho),\end{aligned}$$

where $\Omega_j(\rho) = \omega_j + \dfrac{\partial H_4}{\partial \rho_j}(\rho)$. If the frequencies $\Omega_j(\rho^0)$ are irrationally related, then the motion on the torus $\{\rho_j = \rho_j^0\}$ is topologically transitive. For the torus to persist for the full system, we need not only for the frequencies to be irrationally related but also that rational combinations are bounded away from zero. A vector of frequencies $\Omega = (\Omega_1, \ldots, \Omega_n)$ is said to satisfy the *Diophantine condition* $D(C, \mu)$ for positive constants C and μ, provided that

$$|k_1\Omega_1 + \cdots + k_n\Omega_n| > \frac{C}{|\mathbf{k}|^\mu}$$

for all choices of integers k_j where $|\mathbf{k}| = |k_1| + \cdots + |k_n|$. The tori which persist are those for which $(\Omega_1(\rho^0), \ldots, \Omega_n(\rho^0))$ satisfies the Diophantine condition $D(C, \mu)$ for some positive C and μ. For these frequencies, the flow densely fills the n-torus in a fairly uniform manner.

Theorem 5.1. *Assume a Hamiltonian system has a fixed point at $\mathbf{x}^0$ and all the eigenvalues are purely imaginary with quadratic terms $H_2(\mathbf{x}) = \sum_j \omega_j \rho_j$ where $\rho_j = [(x_j - x_j^0)^2 + (x_{n+j} - x_{n+j}^0)^2)/2$. The eigenvalues are assumed to be linearly independent up to order 4, so the normal form can be given up to terms of order 4 with H_4 a function of only the ρ_j. The nondegeneracy condition on the rotation is that*

$$\det\left(\frac{\partial^2 H_4}{\partial \rho_i \partial \rho_j}(\mathbf{0})\right) \neq 0.$$

Then, near the origin, there is a set of positive measure filled with tori of dimension n which are very near the tori $\{\rho_j = \rho_j^0 : 1 \le j \le n\}$. The tori which persist are those for which $(\Omega_1(\rho^0), \ldots, \Omega_n(\rho^0))$ satisfies the Diophantine condition $D(C, \mu)$ for some positive C and μ. The motion on one of the persistent tori is differentiably conjugate to the unperturbed constant flow for the equations $\dot{\theta}_j = \Omega_j(\rho^0)$, $1 \le j \le n$.

This theorem is often stated without reference to a fixed point by assuming the Hamiltonian function has the form

$$\sum_{j=1}^{n} \omega_j \rho_j + \sum_{i=1}^{n} \sum_{j=1}^{n} B_{ij} \rho_i \rho_j$$

plus small perturbing terms. If the perturbation is small enough, then the tori persist which satisfy the Diophantine condition.

A similar theorem is true for symplectic diffeomorphisms.

Theorem 5.2. *Assume a symplectic diffeomorphism f has a period m point $\mathbf{x}^0$ and all the eigenvalues of $Df^m_{\mathbf{x}_0}$ have absolute value of 1 and multiplicative independence up to order 4. Let the normal form for f^m be determined as in Theorem 4.2 and assume*

$$\det\left(\frac{\partial^2 \beta_i}{\partial \rho_j}\right)(\mathbf{0}) \neq 0.$$

where $\rho_j = [(y_j - x_j^0)^2 + (y_{n+j} - x_{n+j}^0)^2]/2$. Then, near $\mathbf{x}^0$, there is a set of positive measure, filled with tori of dimension n which are invariant by f^m and are very near the tori $\{\rho_j = \rho_j^0 : 1 \leq j \leq n\}$. The tori which persist are those for which their rotation vectors $(\Omega_1(\rho^0), \ldots, \Omega_n(\rho^0), 2\pi)$ with $\Omega_j(\rho^0) = \omega_j + \beta_j(\rho^0)$ satisfy the Diophantine condition $D(C, \mu)$ for some positive C and μ. The map on one of the persistent tori is differentiably conjugate to the constant rotation $(\theta_1, \ldots, \theta_n) \mapsto (\theta_1 + \Omega_1(\rho^0), \ldots, \theta_n + \Omega_n(\rho^0))$.

If the dimension is two ($n = 1$), then the periodic point $\mathbf{x}^0$ is Liapunov stable.

The motion on one of the tori which persist is called *quasi-periodic.* Since the rotation vector (which includes the number 2π) satisfies a Diophantine condition, all the rotations numbers of the maps $\theta_j \mapsto \theta_j + \Omega_j(\rho^0)$ are irrational. Also, the rotations numbers are irrationally related. For these frequencies, an orbit densely fills the n-torus in a fairly uniform manner.

Between the tori which persist, there are complicated dynamics with many periodic points (often hyperbolic). These regions between the tori are called the zones of instability. There are pictures generated by computer simulation in Moser (1968), Arrowsmith and Place (1990), and others. For dimensions greater than one, it is possible to an orbit to make its way between the tori and escape, a process called Arnold diffusion.

Even though it is not near an equilibrium, the KAM applies to the standard map given in Example 3.1. Therefore, it has invariant circles for small $k > 0$ which go around the angular q variable.

6.6 Exercises

Hamiltonian Differential Equations

6.1. Consider the differential equations for the pendulum given in Example 1.3.
  (a) What are the fixed points?
  (b) Draw the phase portrait. For the phase portrait, consider θ as a real variable.
  (c) What solutions are periodic? What solutions have both α-limits and ω-limits at fixed points? (For these equations, the fixed points are different points if θ is a real variable, so they are called heteroclinic orbits.)

6.2. Consider the differential equations for a pendulum with torque A given by $\ddot{\theta} = -(g/L)\sin(\theta) - A$.

(a) What is the potential energy function? (In order for the potential energy function to be well defined, consider θ as a real variable.)

(b) Draw the phase portrait.

(c) Explain the different types of orbits.

6.3. Consider the differential equations given by $\ddot{q} = -q + q^3$.

(a) What is the potential energy function?

(b) What are the fixed points?

(c) Draw the phase portrait.

(d) Explain the different types of orbits.

6.4. Consider the Hamiltonian function $H(q, p) = p^2 + V(q)$ where q and p are real variables.

(a) Prove that a nondegenerate minimum of V, $V'(q_0) = 0$ and $V''(q_0) > 0$, corresponds to an elliptic fixed point $(q_0, 0)$ for the corresponding Hamiltonian differential equations.

(b) Prove that a nondegenerate maximum of V, $V'(q_0) = 0$ and $V''(q_0) < 0$, corresponds to a saddle fixed point $(q_0, 0)$ for the corresponding Hamiltonian differential equations.

Linear Hamiltonian Systems

6.5. Let $\mathbf{v}^1, \ldots, \mathbf{v}^{2n}$ be a symplectic basis for the symplectic form with matrix J, and $S = (\mathbf{v}^1 \ \ldots \ \mathbf{v}^{2n})$ be the matrix with the vectors as columns. Prove that S is a symplectic matrix, $S^T J S = J$.

6.6. Let M be a 2×2 symplectic matrix with eigenvalue $\alpha \pm i\,\beta$ where $\alpha^2 + \beta^2 = 1$ and $\beta > 0$. Show that there is a symplectic basis $\{\mathbf{v}^1, \mathbf{v}^2\}$ in terms of which has the form

$$\begin{pmatrix} \alpha & \beta \\ -\beta & \alpha \end{pmatrix} \quad \text{or} \quad \begin{pmatrix} \alpha & -\beta \\ \beta & \alpha \end{pmatrix},$$

i.e., in the first case $M\mathbf{v}^1 = \alpha\mathbf{v}^1 + \beta\mathbf{v}^2$ and $M\mathbf{v}^2 = -\beta\mathbf{v}^1 + \alpha\mathbf{v}^2$. Note that the matrix for the basis is given by $S^{-1}MS$ where S is the symplectic matrix with the basis vectors as columns. (See the previous exercise.)

6.7. (a) Let M be a 2×2 matrix. Prove that if $\det(M) = 1$, then M is symplectic.

(b) Let A be a 2×2 matrix. Prove that if $\operatorname{tr}(A) = 0$, then A is Hamiltonian.

Symplectic Diffeomorphisms

6.8. Let $S(q,P) = qP + S_n(q,P)$ be a generating function where S_n is homogeneous of degree n. Prove that it induces a symplectic diffeomorphism

$$Q = F_1(q,p) = q + \frac{\partial S_n}{\partial P}(q,p) + O(|(q,p)|^n)$$
$$P = F_2(q,p) = p - \frac{\partial S_n}{\partial q}(q,p) + O(|(q,p)|^n)$$

where $O(|(q,p)|^n)$ are terms of order at least n.

6.9. Show that the standard map given in Example 3.1 is exact symplectic.

Normal Forms

6.10. Write the normal form for a symplectic map in $\mathbb{R}^2$ given in Theorem 4.2 using polar coordinates.

6.11. Consider the Duffing's equation with Hamiltonian

$$H(q,p) = \frac{1}{2}(q^2 + p^2) + \frac{\gamma}{4}q^4.$$

Show that the normal form through fourth-order terms is

$$K(q,p) = \frac{1}{2}(q^2 + p^2) + \frac{3\gamma}{8}(q^2 + p^2)^2 + \cdots .$$

CHAPTER VII
Bifurcation of Periodic Points

Throughout this chapter we consider a map with one parameter. These results also apply to differential equations near a periodic orbit by considering the Poincaré map. We write $f_\mu(\mathbf{x}) = f(\mathbf{x}, \mu)$, where $\mu \in \mathbb{R}$ and $\mathbf{x} \in \mathbb{R}^n$. We proved in Theorem V.6.4 that if $f_{\mu_0}(\mathbf{x}_0) = \mathbf{x}_0$ is a fixed point and 1 is not an eigenvalue of $D(f_{\mu_0})_{\mathbf{x}_0}$, then the fixed point can be continued for values of the parameter μ near μ_0. This is a non-bifurcation result. Notice that the tool we used to show that the fixed point could be continued in this case is the Implicit Function Theorem. We repeatedly use this theorem to study the bifurcations considered in this chapter.

The first type of bifurcation, called the *saddle-node bifurcation*, occurs where the above assumption on the eigenvalues is violated and 1 is an eigenvalue. With further assumptions on the higher derivatives of $f_\mu(\mathbf{x})$, it follows that (i) for μ on one side of μ_0, there are no fixed points near $\mathbf{x}_0$, and (ii) for μ on the other side of μ_0, there are two fixed points. Of the two fixed points, one is attracting and the other is repelling (at least in one dimension, $n = 1$).

The other types of bifurcations that we consider are ones where the fixed point persists (1 is not an eigenvalue), but the stability type of the fixed point changes as μ passes through μ_0 (one eigenvalue has absolute value equal to 1). Among this type of bifurcation, the second type we consider, called a *period doubling bifurcation* or *flip bifurcation*, occurs when one eigenvalue is -1. In one spatial dimension, $n = 1$, an attracting point of period 1 becomes repelling at μ_0 and a stable orbit of period 2 branches off from the fixed point.

The last type of bifurcation we consider is called the *Andronov – Hopf bifurcation.* For a family of diffeomorphisms, it occurs when a pair of complex eigenvalues have absolute value 1 when $\mu = \mu_0$, but are not equal to ± 1. As μ passes through μ_0, the absolute value of these eigenvalues change from less than 1 to greater than 1. For a pair of complex eigenvalues to occur, it is necessary to be considering a map in at least two dimensions. With further assumptions on the derivatives, for μ slightly bigger than μ_0, there is an invariant closed curve near x_0. This bifurcation is simpler for differential equations. In this case, for a differential equation in two dimensions, a fixed point changes from attracting to repelling as a pair of eigenvalues crosses the imaginary axis, and a stable periodic orbit branches off from the family of fixed points.

For a more complete treatment of bifurcation theory, see Guckenheimer and Holmes (1983), Chow and Hale (1982), Wiggins (1988, 1990), or Hale and Koçak (1991).

7.1 Saddle-Node Bifurcation

As stated in the introduction, the first bifurcation we consider is the one that occurs when the map fails to be hyperbolic because 1 is an eigenvalue of the derivative.

We first consider the case where x is a real variable. We want $f_{\mu_0}(x_0) = x_0$ and $f'_{\mu_0}(x_0) = 1$. We also want the tangency of the graph of f_{μ_0} to the diagonal $\{(x, y) : y = x\}$ to occur in the simplest possible fashion, so we assume that $f''_{\mu_0}(x_0) \neq 0$. Finally, we need that the graph of f_μ is moving upward or downward as the parameter varies,

$\frac{\partial f}{\partial \mu}(x_0, \mu_0) \neq 0$. With these assumptions, the fixed point disappears on one side of μ_0 and two fixed points branch off on the other side. Before stating the general theorem, we give an example.

Example 1.1. Let $f_\mu(x) = \mu + x - ax^2$ with $a > 0$. For $\mu < 0$, $f_\mu(x) - x = \mu - ax^2 < 0$ for all x. so there are no fixed points. For $\mu = 0$, $0 = f_0(x) - x = -ax^2$ has one root at $x = 0$, so f_0 has one fixed point at $x = 0$. For $\mu > 0$, $0 = f_\mu(x) - x = \mu - ax^2$ has two roots, $x_\pm = \pm\,(\mu/a)^{1/2}$. Thus, f_μ has two fixed points. See Figure 1.1.

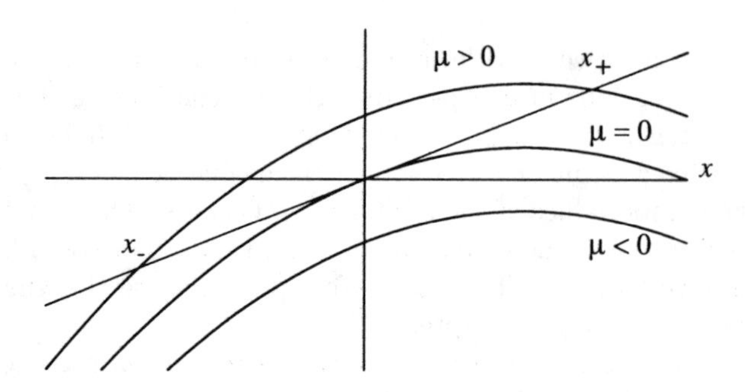

FIGURE 1.1. Creation of Two Fixed Points

Theorem 1.1 (Saddle-Node Bifurcation). *Assume that $f : \mathbb{R}^2 \to \mathbb{R}$ is a C^r function jointly in both variables with $r \geq 2$. We also write $f_\mu(x) = f(x, \mu)$. Make the following further assumptions:*

1. $f(x_0, \mu_0) = x_0$,
2. $f'_{\mu_0}(x_0) = 1$,
3. $f''_{\mu_0}(x_0) \neq 0$, *and*
4. $\frac{\partial f}{\partial \mu}(x_0, \mu_0) \neq 0$.

Then, there exist intervals I about x_0 and N about μ_0 and a C^r function $m : I \to N$ such that (i) $f_{m(x)}(x) = x$, (ii) $m(x_0) = \mu_0$, and (iii) the graph of m gives all the fixed points in $I \times N$. Moreover, $m'(x_0) = 0$ and

$$m''(x_0) = \frac{-\frac{\partial^2 f}{\partial x^2}(x_0, \mu_0)}{\frac{\partial f}{\partial \mu}(x_0, \mu_0)} \neq 0.$$

These fixed points are attracting on one side of x_0 and repelling on the other. See Figure 1.2.

PROOF. To find the fixed points of f_μ, we consider the new function

$$G(x, \mu) = f(x, \mu) - x.$$

Then, fixed points of f_μ exactly correspond to zeroes of G. First, we have $G(x_0, \mu_0) = 0$. We want to use the Implicit Function Theorem to solve for nearby zeroes of G. Note that $(\partial G/\partial x)(x_0, \mu_0) = f'_{\mu_0}(x_0) - 1 = 0$, so we cannot solve for x in terms of μ. However,

$$\frac{\partial G}{\partial \mu}(x_0, \mu_0) = \frac{\partial f}{\partial \mu}(x_0, \mu_0) \neq 0,$$

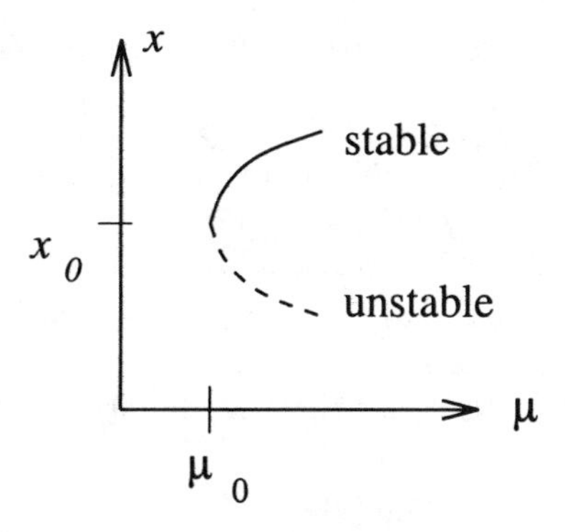

FIGURE 1.2. Bifurcation Diagram for Saddle-Node Bifurcation

so we can solve for μ in terms of x. In fact, there are intervals I about x_0 and N about μ_0 and a C^r function $m : I \to N$ such that $G(x, m(x)) \equiv 0$, and these give all the zeroes of G in $I \times N$. This construction proves the first facts about the fixed points.

To calculate the derivatives of $m(x)$, we use implicit differentiation. We use subscripts to designate partial derivatives. Thus, $G_x = \frac{\partial G}{\partial x}$. Differentiating $0 = G(x, m(x))$ with respect to x, we get $0 = G_x + G_\mu m'$. Evaluating this at x_0 and using the fact that $G_x(x_0, \mu_0) = 0$ and $G_\mu(x_0, \mu_0) \neq 0$, we get that $m'(x_0) = 0$. (Notice that this much of the proof only uses the fact that $f(x, \mu)$ is C^1 and does not use the second derivative.)

To get the second derivative of m, we differentiate the equation $0 = G_x(x, m(x)) + G_\mu(x, m(x))m'(x)$ a second time with respect to x and get

$$0 = G_{xx} + 2G_{\mu x}m' + G_{\mu\mu}(m')^2 + G_\mu m''.$$

Evaluating this expression at x_0 and using the fact that $m'(x_0) = 0$, we get that

$$m''(x_0) = \frac{-G_{xx}(x_0, \mu_0)}{G_\mu(x_0, \mu_0)} = \frac{-f_{xx}(x_0, \mu_0)}{f_\mu(x_0, \mu_0)}$$

as claimed. (In this notation, $f_\mu(x_0, \mu_0)$ is the μ-partial derivative evaluated at the point indicated.)

To find the stability of the fixed points, we use the Taylor expansion of f_x about (x_0, μ_0):

$$\begin{aligned}\frac{\partial f}{\partial x}(x, \mu) =& 1 + \frac{\partial^2 f}{\partial x^2}(x - x_0) + \frac{\partial^2 f}{\partial x \partial \mu}(\mu - \mu_0) \\ &+ O(|x - x_0|^2) + O(|x - x_0| \cdot |\mu - \mu_0|) + O(|\mu - \mu_0|^2).\end{aligned}$$

Because $m'(x_0) = 0$, it follows that $m(x) - \mu_0 = O(|x - x_0|^2)$. Therefore,

$$\frac{\partial f}{\partial x}(x, m(x)) = 1 + \frac{\partial^2 f}{\partial x^2}_{(x_0, \mu_0)}(x - x_0) + O(|x - x_0|^2).$$

Because $\frac{\partial^2 f}{\partial x^2}(x_0, \mu_0) \neq 0$, $\frac{\partial f}{\partial x}(x, m(x)) - 1$ has opposite signs on the two sides of x_0, $\frac{\partial f}{\partial x}(x, m(x))$ is less than 1 on one side and greater than 1 on the other, and so one side has an attracting fixed point and the other side a repelling fixed point. □

7.2 Saddle-Node Bifurcation in Higher Dimensions

In the last section, we discuss the saddle-node bifurcation in one spatial dimension. This section considers this same bifurcation in higher dimensions. Before stating the theorem, we consider the example of the Hénon map.

Example 2.1. Let $F_{AB}(x,y) = (A - By - x^2, x)$ be the Hénon family of maps. The fixed points have x-coordinate given by $x = -\frac{B+1}{2} \pm \left[(\frac{B+1}{2})^2 + A\right]^{1/2}$, so there are no fixed points when $A < -[(B+1)/2]^2$ and two fixed points when $A > -[(B+1)/2]^2$. The eigenvalues of the derivative $D(F_{AB})_{(x,y)}$ are $\lambda_\pm = -x \pm [x^2 - B]^{1/2}$. When $A = -[(B+1)/2]^2$ and $x = -(B+1)/2$, the eigenvalues are $\lambda_- = B$ and $\lambda_+ = 1$. Thus, there is a bifurcation from no fixed points to two fixed which occurs when $A = -[(B+1)/2]^2$, and one of the eigenvalues of this fixed point is 1 at this bifurcation value. We leave to the exercises for the reader to verify that this family satisfies the conditions of the theorem below for a saddle-node bifurcation. See Exercise 7.2.

To state the theorem in higher dimensions, it is necessary to specify the derivative of the coordinate function along the direction of the eigenvector for the eigenvalue 1. We assume that $\lambda_1 = 1$ is an eigenvalue of $D(f_{\mu_0})_{\mathbf{x}_0}$ of multiplicity one. We first state the theorem using a change of basis which puts the derivative in the form

$$D(f_{\mu_0})_{\mathbf{x}_0} = \begin{pmatrix} 1 & 0 & \cdots & 0 \\ 0 & * & \cdots & * \\ \cdot & \cdot & \cdots & \cdot \\ 0 & * & \cdots & * \end{pmatrix}, \tag{*}$$

so the eigenvector for $\lambda_1 = 1$ is $\mathbf{v}^1 = (1, 0, \ldots, 0)^T$. We write the first coordinate function of f as f_1.

Theorem 2.1. *Let $f : \mathbb{R}^n \times \mathbb{R} \to \mathbb{R}^n$ be C^2 jointly in all the variables, and write $f_\mu(\mathbf{x}) = f(\mathbf{x}, \mu)$. Assume that f_μ satisfies the following conditions.*

(1) *There is a parameter value μ_0 such that f_{μ_0} has a fixed point $\mathbf{x}_0$, $f(\mathbf{x}_0, \mu_0) = \mathbf{x}_0$. Let a be the first coordinate of $\mathbf{x}_0$, $a = (\mathbf{x}_0)_1$.*
(2) *The eigenvalues of $D(f_{\mu_0})_{\mathbf{x}_0}$ are $\{\lambda_j\}_{j=1}^n$ with $\lambda_1 = 1$ and $|\lambda_j| \neq 1$ for $2 \leq j \leq n$.*
(3) *For conditions (3) and (4), choose a basis so that $D(f_{\mu_0})_{\mathbf{x}_0}$ has the form (*). Assume the first coordinate function satisfies $(\partial^2 f_1/\partial x_1^2)(\mathbf{x}_0, \mu_0) \neq 0$.*
(4) *Finally, assume the derivative $(\partial f_1/\partial \mu)(\mathbf{x}_0, \mu_0) \neq 0$.*

Then it is possible to parameterize μ and $\mathbf{x}$ in terms of x_1, $\mu = m(x_1)$ and $\mathbf{x} = \mathbf{q}(x_1)$, such that $f(\mathbf{q}(x_1), m(x_1)) \equiv \mathbf{q}(x_1)$, $m(a) = \mu_0$, and $\mathbf{q}(a) = \mathbf{x}_0$. (The first coordinate of $\mathbf{x} = \mathbf{q}(x_1)$ is x_1.)

REMARK 2.1. If $n = 2$, $|\lambda_2| < 1$, then for each μ with a fixed point, one of the fixed points is a stable node (attracting fixed point with two unequal eigenvalues) and the other is a saddle. This is the reason for the name of the bifurcation.

In order to be able to apply the theorem more easily (e.g., to the Hénon map), we restate the theorem in the original coordinates using the the eigenvectors and a dual basis of the dual space. For $\mathbb{R}^n$ with the usual inner product, the dual space can be thought of as the space of row vectors, where the pairing of a dual vector $\mathbf{w}$ (row) with a vector $\mathbf{v}$ (column) is given by matrix multiplication, $\mathbf{wv}$. Let $\{\mathbf{v}^j\}_{j=1}^n$ be the basis of generalized eigenvectors for $D(f_{\mu_0})_{\mathbf{x}_0}$, where $\mathbf{v}^1$ corresponds to $\lambda_1 = 1$. Let $\{\mathbf{w}^i\}_{i=1}^n$ be the dual basis, so $\mathbf{w}^i\mathbf{v}^j = \delta_{ij}$, i.e., 1 if $i = j$ and is 0 if $i \neq j$. (Note, since $D(f_{\mu_0})_{\mathbf{x}_0}$ is usually not symmetric, the vectors $\mathbf{w}^j$ are usually not the transposes of the vectors $\mathbf{v}^j$.) The $\mathbf{w}^i$ turn out to be the generalized left (row) eigenvectors of $D(f_{\mu_0})_{\mathbf{x}_0}$. We only need to use $\mathbf{w}^1$, so we check that $\mathbf{w}^1 D(f_{\mu_0})_{\mathbf{x}_0} = \mathbf{w}^1$. By associativity of matrix multiplication,

$$(\mathbf{w}^1 D(f_{\mu_0})_{\mathbf{x}_0})\mathbf{v}^1 = \mathbf{w}^1(D(f_{\mu_0})_{\mathbf{x}_0}\mathbf{v}^1) = \mathbf{w}^1\mathbf{v}^1 = 1, \quad \text{and}$$

$$(\mathbf{w}^1 D(f_{\mu_0})_{\mathbf{x}_0})\mathbf{v}^j = \mathbf{w}^1(D(f_{\mu_0})_{\mathbf{x}_0}\mathbf{v}^j) = \mathbf{w}^1(\sum_{k=2}^n a_{jk}\mathbf{v}^k) = 0 \quad \text{for } 2 \leq j \leq n.$$

Since the condition $(\mathbf{w}^1 D(f_{\mu_0})_{\mathbf{x}_0})\mathbf{v}^j = \delta_{1j}$ for $1 \leq j \leq n$ characterizes $\mathbf{w}^1$, $\mathbf{w}^1 = \mathbf{w}^1 D(f_{\mu_0})_{\mathbf{x}_0}$ and $\mathbf{w}^1$ is the left (row) eigenvector for $\lambda_1 = 1$. We can now use $(\mathbf{w}^1 f_\mu(\mathbf{x}))$ as the component of $f_\mu(\mathbf{x})$ along the direction of $\mathbf{v}^1$ instead of the coordinate function f_1 of Theorem 2.1.

Theorem 2.2. *Let $f : \mathbb{R}^n \times \mathbb{R} \to \mathbb{R}^n$ be C^2 jointly in all the variables, and write $f_\mu(\mathbf{x}) = f(\mathbf{x}, \mu)$. Assume that f_μ satisfies the following conditions.*

(1) *There is a parameter value μ_0 such that f_{μ_0} has a fixed point $\mathbf{x}_0$, $f(\mathbf{x}_0, \mu_0) = \mathbf{x}_0$.*
(2) *The eigenvalues of $D(f_{\mu_0})_{\mathbf{x}_0}$ are $\{\lambda_j\}_{j=1}^n$ with $\lambda_1 = 1$ and $|\lambda_j| \neq 1$ for $2 \leq j \leq n$. Let $\mathbf{v}^1$ be the right (column) eigenvector and $\mathbf{w}^1$ be the left (row) eigenvector for $\lambda_1 = 1$.*
(3) *The second derivative of $\mathbf{w}^1 f_{\mu_0}$ in the direction of $\mathbf{v}^1$ is nonzero,*

$$\mathbf{w}^1 \left[D^2(f_{\mu_0})_{\mathbf{x}_0}(\mathbf{v}^1, \mathbf{v}^1)\right] \neq 0.$$

(4) *The partial derivative with respect to the parameter is nonzero,*

$$\mathbf{w}^1(\partial f/\partial \mu)(\mathbf{x}_0, \mu_0) \neq 0.$$

Then, it is possible to parameterize $\mu = m(s)$ and $\mathbf{x} = \mathbf{q}(s)$ such that $m(0) = \mu_0$, $\mathbf{q}(0) = \mathbf{x}_0$, $\mathbf{q}'(0) = \mathbf{v}^1$, and $f(\mathbf{q}(s), m(s)) \equiv \mathbf{q}(s)$.

We leave it to the reader to see how to adapt the proof of Theorem 2.1 to Theorem 2.2.

REMARK 2.2. There are two approaches to the proof. One uses the center manifold associated to $F(\mathbf{x}, \mu) = (f_\mu(\mathbf{x}), \mu)$ which is two-dimensional: one spatial dimension and one parameter direction. The restriction of F to this invariant manifold satisfies the assumptions of the earlier theorem.

A second approach uses the Implicit Function Theorem to do the reduction in dimensions. This method does not produce an invariant two-dimensional set, but it does produce a two-dimensional set on which all the fixed points must be found. This second method is often called Liapunov–Schmidt reduction.

We give proofs using both approaches below.

PROOF USING THE CENTER MANIFOLD THEOREM. Take coordinates so the derivative satisfies (*). Define the map on $\mathbb{R}^{n+1}$ given by $F(\mathbf{x}, \mu) = (f(\mathbf{x}, \mu), \mu)$. and its derivative is given in the following block form:

$$DF_{(\mathbf{x},\mu)} = \begin{pmatrix} D(f_\mu)_{\mathbf{x}} & \dfrac{\partial f}{\partial \mu}(\mathbf{x}, \mu) \\ 0 & 1 \end{pmatrix}.$$

The center subspace at $(\mathbf{x}_0, \mu_0)$ for F is spanned by the eigenvector $(1, 0, \ldots, 0)^T$ for λ_1 and the vector $(0, \ldots, 0, 1)^T$ in the μ-direction. By the Center Manifold Theorem, we can solve for an invariant manifold which is a graph over x_1 and μ, $(\mathbf{x}, \mu) = (x_1, \varphi(x_1, \mu), \mu)$. The surface given by the image of φ is tangent to center subspace

$$\frac{\partial \varphi}{\partial x_1}(a, \mu_0) = 0, \qquad \text{and}$$
$$\frac{\partial \varphi}{\partial \mu}(a, \mu_0) = 0$$

where $a = (\mathbf{x}_0)_1$ is the first coordinate of the fixed point. We use the notation $\Phi(x_1, \mu) = (x_1, \varphi(x_1, \mu))$ to simplify notation.

With these preliminaries, define

$$g(x_1,\mu) = f_1(x_1,\varphi(x_1,\mu),\mu) = f_1(\Phi(x_1,\mu),\mu),$$

which is the coordinate function in the direction of the eigenvector. To apply the Saddle-Node Bifurcation Theorem in one spatial dimension, we check the conditions on g:

$$\begin{aligned} g(a,\mu_0) &= f_1(\mathbf{x}_0,\mu_0) = a, \\ \frac{\partial g}{\partial x_1}(x_1,\mu) &= \frac{\partial f_1}{\partial x_1}(\Phi(x_1,\mu),\mu) + D(f_{1,\mu})_{\Phi(x_1,\mu)}\frac{\partial \varphi}{\partial x_1}(x_1,\mu), \quad \text{so} \\ \frac{\partial g}{\partial x_1}(a,\mu_0) &= \frac{\partial f_1}{\partial x_1}(\mathbf{x}_0,\mu_0) = 1. \end{aligned}$$

These are the first two conditions for a saddle-node bifurcation for g. Taking the derivative again,

$$\begin{aligned} \frac{\partial^2 g}{\partial x_1^2}(a,\mu_0) =& \frac{\partial^2 f_1}{\partial x_1^2}(\Phi(a,\mu_0),\mu_0) + 2\,D\Big(\frac{\partial f_{1,\mu_0}}{\partial x_1}\Big)_{\mathbf{x}_0}\frac{\partial \varphi}{\partial x_1}(a,\mu_0) \\ &+ D^2(f_{1,\mu_0})_{\mathbf{x}_0}\Big(\frac{\partial \varphi}{\partial x_1}(a,\mu_0), \frac{\partial \varphi}{\partial x_1}(a,\mu_0)\Big) + D(f_{1,\mu_0})_{\mathbf{x}_0}\frac{\partial^2 \varphi}{\partial x_1^2}(a,\mu_0). \end{aligned}$$

Since φ takes its values in the 2nd through the nth variables and f satisfies (*),

$$D(f_{1,\mu_0})_{\mathbf{x}_0}\frac{\partial^2 \varphi}{\partial x_1^2}(a,\mu_0) = (1,0,\ldots,0)\,\frac{\partial^2 \varphi}{\partial x_1^2}(a,\mu_0) = 0.$$

Combining with the fact that $\dfrac{\partial \varphi}{\partial x_1}(a,\mu_0) = 0$, it follows that

$$\frac{\partial^2 g}{\partial x_1^2}(a,\mu_0) = \frac{\partial^2 f_1}{\partial x_1^2}(\mathbf{x}_0,\mu_0),$$

which is nonzero by an assumption of the theorem. Finally,

$$\begin{aligned} \frac{\partial g}{\partial \mu}(a,\mu_0) &= \frac{\partial f_1}{\partial \mu}(\mathbf{x}_0,\mu_0) + D(f_{1,\mu_0})_{\mathbf{x}_0}\frac{\partial \varphi}{\partial \mu}(a,\mu_0) \\ &= \frac{\partial f_1}{\partial \mu}(\mathbf{x}_0,\mu_0) \neq 0 \end{aligned}$$

by the last assumption of the theorem. (Again, $D(f_{1,\mu_0})_{\mathbf{x}_0}\dfrac{\partial \varphi}{\partial \mu}(a,\mu_0) = 0$ because φ has no first variable.)

We have checked the assumptions of the saddle-node bifurcation in one spatial dimension applied to g, so we can solve for $\mu = m(x_1)$ such that $g(x_1, m(x_1)) = x_1$. Also, $m'(a) = 0$ and

$$\begin{aligned} m''(a) &= -\frac{\dfrac{\partial^2 g}{\partial x_1^2}(a,\mu_0)}{\dfrac{\partial g}{\partial \mu}(a,\mu_0)} \\ &= -\frac{\dfrac{\partial^2 f_1}{\partial x_1^2}(\mathbf{x}_0,\mu_0)}{\dfrac{\partial f_1}{\partial \mu}(\mathbf{x}_0,\mu_0)} \\ &\neq 0. \end{aligned}$$

Next we check that this function gives us the set of fixed points that we want. First,

$$f_1(\Phi(x_1, m(x_1)), m(x_1)) = g(\Phi(x_1, m(x_1)), m(x_1)) = x_1.$$

Both $(f(\Phi(x_1, m(x_1)), m(x_1)), m(x_1))$ and $(\Phi(x_1, m(x_1)), m(x_1))$ are in the center manifold and equal in the x_1 and μ components; therefore, they are equal,

$$f(\Phi(x_1, m(x_1)), m(x_1)) = \Phi(x_1, m(x_1)).$$

Letting $\mathbf{q}(x_1) = \Phi(x_1, m(x_1))$, we have the conclusion of the theorem. □

PROOF USING THE IMPLICIT FUNCTION THEOREM. This method uses only the fact that $\lambda_j \neq 1$ for $2 \leq j \leq n$ and not that $|\lambda_j| \neq 1$. Rather than using a center manifold, we reduce the problem to one spatial dimension by means of the Implicit Function Theorem, a method called Liapunov–Schmidt reduction.

Again, we take a basis so that the derivative of f satisfies (*). In terms of partial derivatives, $\frac{\partial f_1}{\partial x_j}(\mathbf{x}_0, \mu_0) = 0$ for $j \geq 2$. Define $\psi : \mathbb{R}^{n+1} \to \mathbb{R}^{n-1}$ by $\psi_j(\mathbf{x}, \mu) = f_j(\mathbf{x}, \mu) - x_j$ for $2 \leq j \leq n$, where the f_j's are the coordinate functions. Then, $\psi(\mathbf{x}_0, \mu_0) = 0$ and

$$\left(\frac{\partial \psi_i}{\partial x_j}\right)_{2 \leq i,j \leq n} = \left(\frac{\partial f_i}{\partial x_j}\right)_{2 \leq i,j \leq n} - id$$

is invertible at $(\mathbf{x}_0, \mu_0)$. Therefore, we can solve for $(x_2, \ldots, x_n)$ in terms of (x_1, μ),

$$(x_2, \ldots, x_n) = \varphi(x_1, \mu) = (\varphi_2(x_1, \mu), \ldots, \varphi_n(x_1, \mu)),$$

where $\psi(x_1, \varphi(x_1, \mu), \mu) \equiv 0$. Note that

$$\frac{\partial \psi}{\partial x_1} + \left(\frac{\partial \psi}{\partial x_j}\right)\left(\frac{\partial \varphi}{\partial x_1}\right) = 0$$

and $\frac{\partial \psi}{\partial x_1}(\mathbf{x}_0, \mu_0) = 0$, so $\frac{\partial \varphi}{\partial x_1}(a, \mu_0) = 0$ where $a = (\mathbf{x}_0)_1$. Also, note that $\psi^{-1}(0)$ is not an invariant manifold, but we can use it in the same manner as the invariant center manifold in the previous proof.

Writing a for $(\mathbf{x}_0)_1$, define

$$g(s, \mu) = f_1(s + a, \varphi(a + s, \mu), \mu) - a.$$

Then, $g(0, \mu_0) = 0$ is a fixed point, and

$$\frac{\partial g}{\partial s}(s, \mu_0) = \frac{\partial f_1}{\partial x_1}(a + s, \varphi(a + s, \mu_0), \mu_0) + \left(\frac{\partial f_1}{\partial x_j}(a + s, \varphi(a + s, \mu_0), \mu_0)\right)\left(\frac{\partial \varphi}{\partial x_1}(a + s, \mu_0)\right),$$

and at $(0, \mu_0)$,

$$\frac{\partial g}{\partial s}(0, \mu_0) = \frac{\partial f_1}{\partial x_1}(\mathbf{x}_0, \mu_0) = 1,$$

because $\frac{\partial f_1}{\partial x_j}(\mathbf{x}_0, \mu_0) = 0$ for $j \geq 2$. Thus, g satisfies the first two conditions for a saddle-node bifurcation.

Taking the second derivative with respect to s,

$$\begin{aligned}\frac{\partial^2 g}{\partial s^2}(0,\mu_0) &= \frac{\partial^2 f_1}{\partial x_1^2}(\mathbf{x}_0,\mu_0) + 2\Big(\frac{\partial^2 f_1}{\partial x_1 \partial x_j}(\mathbf{x}_0,\mu_0)\Big)\Big(\frac{\partial \varphi}{\partial x_1}(a,\mu_0)\Big) \\ &\quad + \Big(\frac{\partial f_1}{\partial x_j}(\mathbf{x}_0,\mu_0)\Big)\Big(\frac{\partial^2 \varphi}{\partial x_1}(a,\mu_0)\Big) \\ &= \frac{\partial^2 f_1}{\partial x_1^2}(\mathbf{x}_0,\mu_0) \neq 0\end{aligned}$$

because $\frac{\partial \varphi}{\partial x_1}(a,\mu_0) = 0$ and $\frac{\partial f_1}{\partial x_j}(\mathbf{x}_0,\mu_0) = 0$ for $j \geq 2$.

The final assumption on g for a saddle-node bifurcation involves the derivative with respect to μ. But

$$\begin{aligned}\frac{\partial g}{\partial \mu}(0,\mu_0) &= \frac{\partial f_1}{\partial \mu}(\mathbf{x}_0,\mu_0) + \Big(\frac{\partial f_1}{\partial x_j}(\mathbf{x}_0,\mu_0)\Big)\frac{\partial \varphi}{\partial \mu}(a,\mu_0) \\ &= \frac{\partial f_1}{\partial \mu}(\mathbf{x}_0,\mu_0) \neq 0\end{aligned}$$

because $\frac{\partial f_1}{\partial x_j}(\mathbf{x}_0,\mu_0) = 0$ for $j \geq 2$.

Thus, g has a saddle node bifurcation, and we can solve for μ, $\mu = m(s)$ such that $g(s, m(s)) \equiv s$. By the previous results, $m'(0) = 0$ and

$$\begin{aligned}m''(0) &= -\frac{\frac{\partial^2 g}{\partial s^2}(0,\mu_0)}{\frac{\partial g}{\partial \mu}(0,\mu_0)} \\ &= -\frac{\frac{\partial^2 f_1}{\partial x_1^2}(\mathbf{x}_0,\mu_0)}{\frac{\partial f_1}{\partial \mu}(\mathbf{x}_0,\mu_0)} \\ &\neq 0.\end{aligned}$$

This completes the proof using the Implicit Function Theorem. □

7.3 Period Doubling Bifurcation

Consider the quadratic family of maps, $F_\mu(x) = \mu x(1-x)$. In Chapter II we showed that the fixed points are 0 and $p_\mu = 1 - 1/\mu$. We also showed that p_μ is attracting for $1 < \mu < 3$ and repelling for $3 < \mu$. The reason the stability can change at $\mu = 3$ is that $F_3'(p_3) = -1$, so the fixed point is not hyperbolic. We further showed that for $1 < \mu < 3$, all points $x \in (0,1)$ have their forward orbit $F_\mu^j(x)$ converge to p_μ as j goes to infinity. Thus, there are no points of period 2 for $1 < \mu < 3$. A further calculation shows that for $\mu > 3$, there is an orbit of period 2, $q_\mu^\pm$, which bifurcates off from the fixed point. It can be shown that this orbit is attracting for $3 < \mu < 1 + \sqrt{6}$, and then repelling for $\mu > 1 + \sqrt{6}$. In Section 3.4, we discuss the repeated period doubling bifurcations which take place as the stable orbit increased in period through 1, 2, 4, 8, ..., 2^n, ... In this section we concentrate on one of these bifurcations, e.g., the one which occurs at $\mu = 3$ where a fixed point becomes repelling when its derivative equals -1 and a period 2 orbit is created. The following example is a model problem where the fixed point is always at $x = 0$.

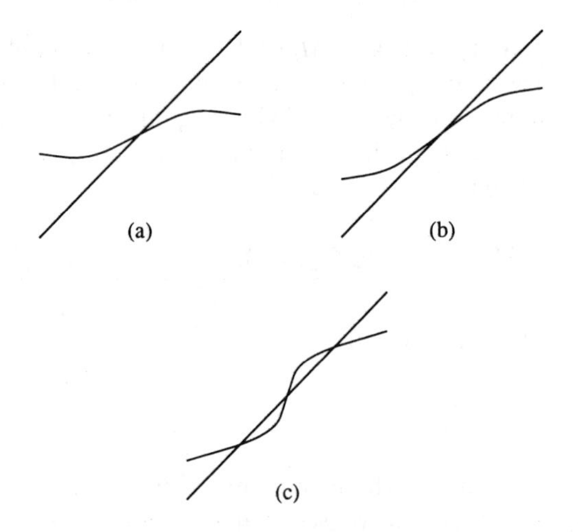

FIGURE 3.1. The Graph of f_μ^2 for (a) $\mu < 1$, (b) $\mu = 1$, and (c) $\mu > 1$

Example 3.1. Let $f_\mu(x) = -\mu x + ax^2 + bx^3$. Notice that $f_1'(0) = -1$. We want to find the points of period two for μ near 1. See Figure 3.1.

A calculation shows that

$$f_\mu^2(x) = \mu^2 x + x^2(-a\mu + a\mu^2) + x^3(-b\mu - 2a^2\mu - b\mu^3) + O(x^4).$$

We want to find zeroes of $f_\mu^2(x) - x = 0$. We know that $f_\mu^2(0) - 0 = 0$, since $f_\mu(0) = 0$. This is reflected in the fact that x is a factor of $f_\mu^2(x) - x$. Since we want to find the zeroes of $f_\mu^2(x) - x = 0$ other than 0, we define

$$\begin{aligned} M(x,\mu) &= \frac{f_\mu^2(x) - x}{x} \\ &= \mu^2 - 1 + x(-a\mu + a\mu^2) + x^2(-b\mu - 2a^2\mu - b\mu^3) + O(x^3). \end{aligned}$$

This function vanishes at $(x,\mu) = (0,1)$, $M(0,1) = 0$. Also, both the constant term, $\mu^2 - 1$, and the coefficient of x, $-a\mu + a\mu^2$, vanish at $(x,\mu) = (0,1)$. Using the fact that $\mu^2 - 1 = (\mu - 1)(\mu + 1) \approx 2(\mu - 1)$, to lowest-terms the zeroes of $M(x,\mu)$ are approximately equal to the zeroes of $0 = 2(\mu - 1) - 2(b + a^2)x^2$, which are

$$\begin{aligned} &\mu = 1 + (b + a^2)x^2, && \text{or} \\ &x = \pm \left(\frac{\mu - 1}{b + a^2}\right)^{1/2} && \text{for } \frac{\mu - 1}{b + a^2} > 0. \end{aligned}$$

In the proof of the general theorem, this is justified by applying the Implicit Function Theorem. The partial derivatives of M at $(0,1)$ are $M_x(0,1) = -a\mu + a\mu^2|_{\mu=1} = 0$ and $M_\mu(0,1) = 2\mu|_{\mu=1} = 2 \neq 0$. The Implicit Function Theorem says that μ can be solved for in terms of x to give zeroes of M, $\mu = m(x)$ with $M(x, m(x)) \equiv 0$ so $f_{m(x)}^2(x) = x$. To justify the approximation made above, we can calculate the derivatives of $m(x)$ by implicit differentiation and show that this gives the lowest order terms given above. Differentiating $0 = M(x, m(x))$ twice with respect to x gives

$$\begin{aligned} &0 = M_x(x, m(x)) + M_\mu(x, m(x))m'(x) \qquad \text{and} \\ &0 = M_{xx} + 2M_{x\mu}m' + M_{\mu\mu}(m')^2 + M_\mu m''. \end{aligned}$$

Using the fact that $M_x(0,1) = 0$ and $M_\mu(0,1) \neq 0$ in the first equation gives that $m'(0) = 0$. We use the second equation to determine $m''(0)$. Because M_{xx} is the only second derivative not multiplied by $m'(0)$ (which equals zero), this is the only one we need to calculate. Using the explicit expression for M, $M_{xx}(0,1) = 2(-b\mu - 2a^2\mu - b\mu^3)|_{\mu=1} = -4b - 4a^2$. Then,

$$\begin{aligned} m''(0) &= \frac{-M_{xx}(0,1) - 2M_{x\mu}(0,1)m'(0) - M_{\mu\mu}(0,1)(m'(0))^2}{M_\mu(0,1)} \\ &= \frac{4b + 4a^2 - 0 - 0}{2} \\ &= 2(b + a^2). \end{aligned}$$

Thus, to get a quadratic shape to the new points of period 2, we need to assume that $b + a^2 \neq 0$. In the general theorem, we see that the sign of $b + a^2$ also determines the stability of the period 2 orbit. Note that $-2(a^2 + b)$ is the coefficient of x^3 in f_1^2 where 1 is the bifurcation parameter value.

The above example is fairly general, but it does assume that the fixed point does not vary with the parameter, so $\frac{\partial f}{\partial \mu}(0,1) = 0$. In the theorem, we allow $\frac{\partial f}{\partial \mu}(x_0, \mu_0) \neq 0$ and show the effect of this term. We also give a condition for the spatial derivative of f to vary along the curve of fixed points as the parameter varies. This condition is given in terms of derivatives of f, so it is not necessary to calculate $f_\mu^2(x)$ to apply the theorem. The bifurcation described in the following theorem is called the *period doubling* or *flip bifurcation.*

Theorem 3.1 (Period Doubling Bifurcation). *Assume that $f : \mathbb{R}^2 \to \mathbb{R}$ is a C^r function jointly in both variables with $r \geq 3$, and that f satisfies the following conditions.*

1. *The point x_0 is a fixed point for $\mu = \mu_0$: $f(x_0, \mu_0) = x_0$.*
2. *The derivative of f_{μ_0} at x_0 is -1: $f'_{\mu_0}(x_0) = -1$. Since this derivative is not equal to 1, there is a curve of fixed points $x(\mu)$ for μ near μ_0.*
3. *The derivative of $f'_\mu(x(\mu))$ with respect to μ is nonzero (the derivative is varying along the family of fixed points):*

$$\alpha = \left[\frac{\partial^2 f}{\partial \mu \partial x} + \left(\frac{1}{2}\right)\left(\frac{\partial f}{\partial \mu}\right)\left(\frac{\partial^2 f}{\partial x^2}\right)\right]\Big|_{(x_0,\mu_0)} \neq 0.$$

4. *The graph of $f^2_{\mu_0}$ has nonzero cubic term in its tangency with the diagonal (the quadratic term is zero):*

$$\beta = \left(\frac{1}{3!}\frac{\partial^3 f}{\partial x^3}(x_0,\mu_0)\right) + \left(\frac{1}{2!}\frac{\partial^2 f}{\partial x^2}(x_0,\mu_0)\right)^2 \neq 0.$$

Then, there is a period doubling bifurcation at (x_0, μ_0). More specifically, there is a differentiable curve of fixed points, $x(\mu)$, passing through x_0 at μ_0, and the stability of the fixed point changes at μ_0. (Which side of μ_0 is attracting depends on the sign of α.) There is also a differentiable curve γ passing through (x_0, μ_0) so that $\gamma \setminus \{(x_0, \mu_0)\}$ is the union of hyperbolic period 2 orbits. The curve γ is tangent to the line $\mathbb{R} \times \{\mu_0\}$ at (x_0, μ_0), so γ is the graph of a function of x, $\mu = m(x)$ with $m'(x_0) = 0$ and $m''(x_0) = -2\beta/\alpha \neq 0$. The stability type of the period 2 orbit depends on the sign of

β: if $\beta > 0$, then the period 2 orbit is attracting; and if $\beta < 0$, then the period 2 orbit is repelling.

PROOF. By the assumptions, $f'_{\mu_0}(x_0) \neq 1$, so there is a curve of fixed points parameterized by μ, $x(\mu)$. Moreover,

$$\begin{aligned} x'(\mu_0) &= -\big(\frac{\partial f}{\partial x}(x_0,\mu_0) - 1\big)^{-1}\frac{\partial f}{\partial \mu}(x_0,\mu_0) \\ &= \frac{1}{2}\frac{\partial f}{\partial \mu}(x_0,\mu_0). \end{aligned}$$

We want to translate the coordinates so that 0 is a fixed point for all nearby μ, so we define

$$g(y,\mu) = f_\mu(y + x(\mu)) - x(\mu).$$

Then, $g(0,\mu) \equiv 0$. We write $g^2(y,\mu)$ when we mean $g(\cdot,\mu) \circ g(y,\mu)$. The partial derivatives of g with respect to y are the same as those of f with respect to x at the corresponding points,

$$\frac{\partial^j g}{\partial y^j}(0,\mu) = \frac{\partial^j f}{\partial x^j}(x(\mu),\mu).$$

The value of the partial derivative with respect to the position, $\dfrac{\partial g}{\partial y}(0,\mu)$, determines the stability of the fixed point, so $\dfrac{\partial^2 g}{\partial \mu \partial y}(0,\mu)$ measures the change along the curve of fixed points, and

$$\begin{aligned} \frac{\partial^2 g}{\partial \mu \partial y}(0,\mu_0) &= \frac{\partial}{\partial \mu}\frac{\partial f}{\partial x}(x(\mu),\mu)|_{\mu=\mu_0} \\ &= \frac{\partial^2 f}{\partial \mu \partial x}(x_0,\mu_0) + \frac{\partial^2 f}{\partial x^2}(x_0,\mu_0)x'(\mu_0) \\ &= \frac{\partial^2 f}{\partial \mu \partial x}(x_0,\mu_0) + \frac{1}{2}\frac{\partial^2 f}{\partial x^2}(x_0,\mu_0)\frac{\partial f}{\partial \mu}(x_0,\mu_0) \\ &= \alpha \neq 0. \end{aligned}$$

This calculation shows that α measures the quantity described in the statement of the theorem.

Let $a_j(\mu)$ be the coefficient of y^j in the Taylor expansion of g about $y = 0$, so

$$g(y,\mu) = a_1(\mu)y + a_2(\mu)y^2 + a_3(\mu)y^3 + O(y^4).$$

A direct calculation then shows that

$$g^2(y,\mu) = a_1^2 y + (a_1a_2 + a_2a_1^2)y^2 + (a_1a_3 + 2a_1a_2^2 + a_3a_1^3)y^3 + O(y^4)$$

where we do not exhibit the dependence of the coefficients a_j on μ. As in the example, we want to find points where $g^2(y,\mu) - y = 0$ that are different from 0. Since $y = 0$ is always a solution, we divide by y when $y \neq 0$. So we define

$$M(y,\mu) = \begin{cases} \dfrac{g^2(y,\mu) - y}{y} & \text{for } y \neq 0 \\ \dfrac{\partial}{\partial y}(g^2(y,\mu))|_{y=0} - 1 & \text{for } y = 0. \end{cases}$$

Notice that the definition for $y = 0$ is the natural extension of the definition for $y \neq 0$. Using the expansion of g^2,

$$M(y,\mu) = (a_1^2 - 1) + (a_1a_2 + a_2a_1^2)y + (a_1a_3 + 2a_1a_2^2 + a_3a_1^3)y^2 + O(y^3).$$

In order to show that the Implicit Function Theorem applies, we note that $M(0,\mu_0) = 0$, and the partial derivatives are

$$\begin{aligned} M_y(0,\mu_0) &= a_1a_2 + a_2a_1^2|_{\mu_0} = -a_2(\mu_0) + a_2(\mu_0) = 0 \quad \text{and} \\ M_\mu(0,\mu_0) &= \frac{\partial}{\partial\mu}\frac{\partial g^2}{\partial y}(0,\mu)|_{\mu_0} \\ &= \frac{\partial}{\partial\mu}\Big(\frac{\partial g}{\partial y}(0,\mu)\Big)^2|_{\mu_0} \\ &= 2\frac{\partial g}{\partial y}(0,\mu_0)\frac{\partial^2 g}{\partial\mu\partial y}(0,\mu_0) \\ &= -2\alpha \neq 0. \end{aligned}$$

Because $M_\mu(0,\mu_0) \neq 0$, the Implicit Function Theorem applies and there is a differentiable function $m(y)$ such that $M(y,m(y)) \equiv 0$.

By implicit differentiation, $0 = M_y(0,\mu_0) + M_\mu(0,\mu_0)m'(0)$, so

$$m'(0) = \frac{-M_y(0,\mu_0)}{M_\mu(0,\mu_0)} = 0.$$

To calculate the second derivative of $m(y)$, differentiate the equation $0 = M_y(y,m(y)) + M_\mu(y,m(y))m'(y)$ again:

$$0 = M_{yy} + 2M_{y\mu}m'(y) + M_{\mu\mu}(m')^2 + M_\mu m''.$$

Evaluating these at $y = 0$ where $m'(0) = 0$ gives

$$m''(0) = \frac{-M_{yy}(0,\mu_0)}{M_\mu(0,\mu_0)}.$$

Thus, we need to calculate the numerator,

$$\begin{aligned} M_{yy}(0,\mu_0) &= 2(a_1a_3 + 2a_1a_2^2 + a_3a_1^3)|_{\mu_0} \\ &= 2(-a_3 - 2a_2^2 - a_3)|_{\mu_0} = -4(a_3 + a_2^2)|_{\mu_0} \\ &= -4\beta \neq 0. \end{aligned}$$

Therefore,

$$m''(0) = \frac{4\beta}{-2\alpha} = -\frac{2\beta}{\alpha} \neq 0.$$

It can also be checked that $\dfrac{\partial^3 g^2}{\partial y^3}(0,\mu_0) = 3M_{yy}(0,\mu_0) = -12\beta \neq 0.$

This leaves only the stability of the period 2 orbit to check. For this we use the Taylor expansion for $\dfrac{\partial(g^2)}{\partial y}(y,m(y))$ about $y = 0$ and $\mu = \mu_0$:

$$\begin{aligned} \frac{\partial(g^2)}{\partial y}(y,\mu) &= \frac{\partial(g^2)}{\partial y}(0,\mu_0) + \frac{\partial^2(g^2)}{\partial y^2}(0,\mu_0)y \\ &\quad + \frac{\partial^2(g^2)}{\partial\mu\partial y}(0,\mu_0)(\mu-\mu_0) + \frac{1}{2}\frac{\partial^3(g^2)}{\partial y^3}(0,\mu_0)y^2 + \cdots \end{aligned}$$

We have already calculated the value of several of these coefficients:

$$\frac{\partial(g^2)}{\partial y}(0, \mu_0) = (-1)^2 = 1$$

(by the chain rule),

$$\frac{\partial^2(g^2)}{\partial y^2}(0, \mu_0) = 0$$

as is noted in the calculation of $M_y(0, \mu_0)$, and

$$\frac{\partial^2(g^2)}{\partial\mu\partial y}(0, \mu_0) = M_\mu(0, \mu_0) = -2\alpha,$$

so

$$\begin{aligned}\frac{\partial^2(g^2)}{\partial\mu\partial y}(0, \mu_0)(m(y) - \mu_0) &= M_\mu \frac{1}{2} m''(0) y^2 + O(y^3) \\ &= M_\mu \frac{1}{2}\left(\frac{-M_{yy}}{M_\mu}\right) y^2 \\ &= 2\beta y^2.\end{aligned}$$

Finally,

$$\frac{1}{2}\frac{\partial^3(g^2)}{\partial y^3}(0, \mu_0) = \left(\frac{1}{2}\right) 6(a_1 a_3 + 2a_1 a_2^2 + a_3 a_1^3) = -6\beta.$$

Combining these terms,

$$\begin{aligned}\frac{\partial(g^2)}{\partial y}(y, m(y)) &= 1 + 2\beta y^2 - 6\beta y^2 + O(y^3) \\ &= 1 - 4\beta y^2 + O(y^3).\end{aligned}$$

Thus, if $\beta > 0$ the period 2 orbit is attracting, and if $\beta < 0$, then it is repelling. This finishes checking all the conditions of the theorem. □

7.4 Andronov-Hopf Bifurcation for Differential Equations

The theorem for a Andronov-Hopf bifurcation is much simpler for differential equations than for diffeomorphisms, so we consider this case first. For a family of differential equations, the Andronov-Hopf Bifurcation a family of periodic orbits bifurcates from a fixed point; while for a family of diffeomorphisms, there is just a family of invariant closed curves which bifurcate (on which the dynamics is not exactly specified). This difference makes the analysis much simpler in differential equations case.

As stated in the introduction to the chapter, the Andronov-Hopf bifurcation for a system of differential equations occurs when a pair of eigenvalues for a fixed point changes from negative real part to positive real part, i.e., the fixed point changes from stable to unstable by a pair of eigenvalues crossing the imaginary axis. With further conditions on derivatives, it follows that a periodic orbit bifurcates off from the fixed point. We proceed to make some assumptions and constructions before we state the main bifurcation theorem of the section.

We consider a one parameter family of differential equations

$$\dot{\mathbf{x}} = f(\mathbf{x}, \mu) = f_\mu(\mathbf{x}) \tag{$*$}$$

with $\mathbf{x} \in \mathbb{R}^2$ that satisfies the following assumptions.

(1) The origin is a fixed point for all values of μ near 0: $f(\mathbf{0}, \mu) = \mathbf{0}$.

(2) The eigenvalues of $D(f_\mu)_\mathbf{0}$ are $\alpha(\mu) \pm i\,\beta(\mu)$ with $\alpha(0) = 0$, $\beta(0) = \beta_0 \neq 0$, and $\alpha'(0) \neq 0$, so the eigenvalues are crossing the imaginary axis.

The last assumption of the theorem involves the Taylor expansion of the differential equations, with a condition given on a combination of the coefficients when they are expressed in polar coordinates. Therefore, after we make some preliminary change of coordinates, we indicate in a lemma the form of the equations when transformed into polar coordinates.

By the Implicit Function Theorem, since $\alpha'(0) \neq 0$, the parameter can be changed so that $\alpha(\mu) = \mu$. We use this new parameter. Then, there is a change of basis on $\mathbb{R}^2$ such that

$$\dot{\mathbf{x}} = A(\mu)\mathbf{x} + F(\mathbf{x}, \mu)$$

with

$$A(\mu) = \begin{pmatrix} \mu & -\beta(\mu) \\ \beta(\mu) & \mu \end{pmatrix}$$

and

$$F(\mathbf{x}, \mu) = \begin{pmatrix} B_2^1(x_1, x_2, \mu) + B_3^1(x_1, x_2, \mu) + O(|\mathbf{x}|^4) \\ B_2^2(x_1, x_2, \mu) + B_3^2(x_1, x_2, \mu) + O(|\mathbf{x}|^4) \end{pmatrix}$$

where $B_j^k(x_1, x_2, \mu)$ is a homogeneous polynomial of degree j in x_1 and x_2. Next, we transform the equations to polar coordinates and obtain the form stated in the following lemma.

Lemma 4.1. *Consider the differential equations* $(*)$ *when expressed in polar coordinates,* $x_1 = r\cos(\theta)$ *and* $x_2 = r\sin(\theta)$. *Then,*

$$\begin{aligned} \dot{r} &= \mu r + r^2 C_3(\theta, \mu) + r^3 C_4(\theta, \mu) + O(r^4) \\ \dot{\theta} &= \beta(\mu) + r D_3(\theta, \mu) + r^2 D_4(\theta, \mu) + O(r^3) \end{aligned}$$

where $C_j(\cdot, \mu)$ *and* $D_j(\cdot, \mu)$ *are homogeneous polynomials of degree* j *in* $\sin(\theta)$ *and* $\cos(\theta)$. *In fact,*

$$\begin{aligned} C_3(\theta, \mu) &= \cos(\theta) B_2^1(\cos(\theta), \sin(\theta), \mu) + \sin(\theta) B_2^2(\cos(\theta), \sin(\theta), \mu) \\ D_3(\theta, \mu) &= -\sin(\theta) B_2^1(\cos(\theta), \sin(\theta), \mu) + \cos(\theta) B_2^2(\cos(\theta), \sin(\theta), \mu), \end{aligned}$$

where $B_j^k(\cdot, \cdot, \mu)$ *is the homogeneous term of degree* j *in terms of* x_1 *and* x_2 *of* $\dot{x}_k$. *Moreover,*

$$\int_0^{2\pi} C_3(\theta, \mu)\, d\theta = 0.$$

PROOF. Taking the time derivatives of the equations which define polar coordinates, we get

$$\begin{pmatrix} \dot{x}_1 \\ \dot{x}_2 \end{pmatrix} = \begin{pmatrix} \cos(\theta) & -r\sin(\theta) \\ \sin(\theta) & r\cos(\theta) \end{pmatrix} \begin{pmatrix} \dot{r} \\ \dot{\theta} \end{pmatrix}$$

$$\begin{aligned} \begin{pmatrix} \dot{r} \\ \dot{\theta} \end{pmatrix} &= \frac{1}{r} \begin{pmatrix} r\cos(\theta) & r\sin(\theta) \\ -\sin(\theta) & \cos(\theta) \end{pmatrix} \begin{pmatrix} \dot{x}_1 \\ \dot{x}_2 \end{pmatrix} \\ &= \begin{pmatrix} \cos(\theta)\dot{x}_1 + \sin(\theta)\dot{x}_2 \\ -r^{-1}\sin(\theta)\dot{x}_1 + r^{-1}\cos(\theta)\dot{x}_2 \end{pmatrix} \\ &= \begin{pmatrix} \mu r \\ \beta(\mu) \end{pmatrix} \\ &\quad + \begin{pmatrix} \cos(\theta)[B_2^1 + B_3^1] + \sin(\theta)[B_2^2 + B_3^2] + O(r^4) \\ -r^{-1}\sin(\theta)[B_2^1 + B_3^1] + r^{-1}\cos(\theta)[B_2^2 + B_3^2] + \frac{1}{r}O(r^4) \end{pmatrix}, \end{aligned}$$

where B_j^k are functions of $r\cos(\theta)$, $r\sin(\theta)$, and μ, $B_j^k(r\cos(\theta), r\sin(\theta), \mu)$. Factoring out r from the B_j^k (and remembering that B_j^k is homogeneous of degree j), we get the form of the statement of the lemma. To check the integral of $C_3(\theta,\mu)$, notice that $C_3(\theta,\mu)$ is a homogeneous cubic polynomial in $\sin(\theta)$ and $\cos(\theta)$ so that

$$\int_0^{2\pi} C_3(\theta,\mu)\,d\theta = 0.$$

□

(3) Using the coefficients defined in Lemma 4.1, we define

$$K \equiv \frac{1}{2\pi}\int_0^{2\pi} C_4(\theta,0) - \frac{1}{\beta_0}C_3(\theta,0)D_3(\theta,0)\,d\theta$$

and make the assumption that $K \neq 0$.

The significance of assumption (3) is best understood in terms of a "normal form." With assumptions (1) and (2), there is a change of coordinates $R = r + u_1(r,\theta,\mu)$ in terms of which the differential equations become the following:

$$\begin{aligned} \dot{R} &= \mu R + KR^3 + O(R^4) \\ \dot{\theta} &= \beta(\mu) + O(R). \end{aligned}$$

See Section 3.2 in Carr (1981). Thus, the fact that $K \neq 0$ means that the $\dot{R}$ equation has a nonzero cubic term and so has an invariant closed curve of approximate radius $(-\mu/K)^{1/2}$ (almost a circle of this radius). This situation is similar to the discussion we gave for the diffeomorphism case. In the following proof we avoid the use of the normal form (so we do not need to verify it), but merely build the necessary construction into the proof.

In the following theorem we use the radius of the solution at $t = 0$ as the parameter and denote it by ϵ. Thus, we find $T(\epsilon)$-periodic solutions in time, $\mathbf{x}^*(t,\mu(\epsilon))$, such that their initial conditions in polar coordinates are given by $r^*(0,\mu(\epsilon)) = \epsilon$ and $\theta^*(0,\mu(\epsilon)) = 0$, for some parameter $\mu(\epsilon)$ which is a function of ϵ, and where the period $T(\epsilon)$ is a function of ϵ. Thus, the period and the parameter value for which the periodic orbit occurs are functions of the approximate radius of the periodic solution. The bifurcation described in the following theorem is called the *Andronov-Hopf bifurcation* for flows.

Theorem 4.2 (Andronov-Hopf Bifurcation). *Make assumptions (1) and (2) on the differential equation,*

$$\dot{\mathbf{x}} = f(\mathbf{x}, \mu). \tag{$*$}$$

(a) Then, there exists an $\epsilon_0 > 0$ *such that for* $0 \le \epsilon \le \epsilon_0$, *there are (i) differentiable functions* $\mu(\epsilon)$ *and* $T(\epsilon)$ *with* $T(0) = 2\pi/\beta_0$, $\mu(0) = 0$, *and* $\mu'(0) = 0$ *and (ii) a* $T(\epsilon)$*-periodic function of* t, $\mathbf{x}^*(t, \epsilon)$, *that is a solution of* $(*)$ *for the parameter value* $\mu = \mu(\epsilon)$ *and with initial conditions in polar coordinates given by* $r^*(0, \epsilon) = \epsilon$ *and* $\theta^*(0, \epsilon) = 0$. *In fact, for all* t, $r^*(t, \epsilon) = \epsilon + o(\epsilon)$. *(Uniqueness) Further, there are* $\mu_0 > 0$ *and* $\delta_0 > 0$ *such that any* T*-periodic solution* $\mathbf{x}(t)$ *of* $(*)$ *with* $|\mu| \le \mu_0$, $|T - 2\pi/\beta_0| \le \delta_0$, *and* $|\mathbf{x}(t)| \le \delta_0$, *must be* $\mathbf{x}^*(t, \mu)$ *up to a phase shift, i.e.,* $\mathbf{x}(t + t_0) = \mathbf{x}^*(t, \mu)$ *where* $\mu = \mu(|\mathbf{x}(t_0)|)$ *and* t_0 *is chosen so that the polar angle* θ *is zero for* $x(t_0)$, $\theta(t_0) = 0$.

(b) If we also make assumption (3), then not only is $\mu'(0) = 0$ *but also* $\mu''(0) = -2K \neq 0$. *(This means the periodic solutions occur for* μ *on one side of* 0 *with the side determined by the sign of* K.*) Further, the periodic solution is attracting if* $K < 0$ *and is repelling if* $K > 0$. *See Figure 4.1.*

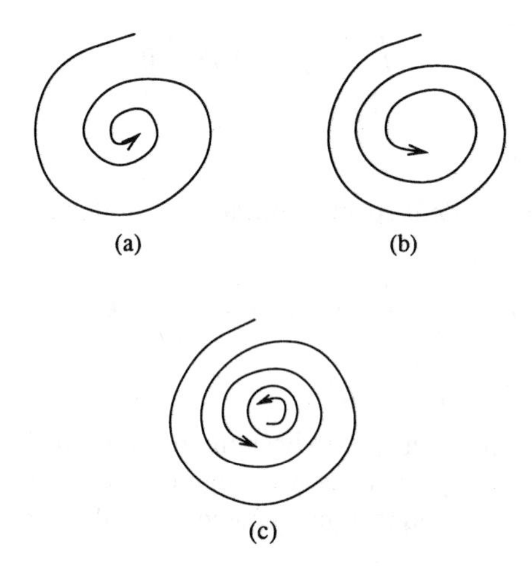

FIGURE 4.1. The Phase Portraits of the Andronov-Hopf Bifurcation with $K < 0$ for (a) $\mu < 0$, (b) $\mu = 0$, and (c) $\mu > 0$

REMARK 4.1. Examples of this bifurcation are found in the work of Poincaré. This theorem was explicitly stated and proved by Andronov (1929). Also see Andronov and Leontovich-Andronova (1939). Later, Hopf gave an independent proof of this theorem and extended it to higher dimensions where two eigenvalues cross the imaginary axis (Hopf, 1942). See Arnold (1983) and Chow and Hale (1982) for further discussion and references. See Marsden and McCracken (1976) for applications.

REMARK 4.2. The proof given below is a combination of those found in Carr (1981) and Chow and Hale (1982).

PROOF OF THEOREM 4.2(a). We first determine the rate of change of r with respect to θ to determine the Poincaré map. By using the equations in polar coordinates and dividing $\dot{r}$ by $\dot{\theta}$, we obtain

$$\begin{aligned}\frac{dr}{d\theta} = \frac{\mu}{\beta} r &+ r^2 \big[\frac{1}{\beta} C_3(\theta,\mu) - \frac{\mu}{\beta^2} D_3(\theta,\mu)\big] \\ &+ r^3 \big[\frac{1}{\beta} C_4(\theta,\mu) - \frac{1}{\beta^2} C_3(\theta,\mu) D_3(\theta,\mu) \\ &\qquad - \frac{\mu}{\beta^2} D_4(\theta,\mu) + \frac{\mu}{\beta^3} D_3(\theta,\mu)^2\big] \\ &+ O(r^4).\end{aligned}$$

(In this and subsequent equations, $O(r^4)$ is really a remainder term in an expansion, so it can be differentiated to give a term of the next lower order: $(\partial^j/\partial r^j)O(r^4) = O(r^{4-j})$ for $1 \le j \le 4$.) Let $r(\theta,\epsilon,\mu)$ be the solution for the parameter value μ with $r(0,\epsilon,\mu) = \epsilon$. If $\epsilon = 0$, then $r(\theta, 0, \mu) \equiv 0$, so this gives $\epsilon = 0$ as a fixed point for $r(2\pi,\epsilon,\mu)$. Since we want to find nonzero fixed points of the Poincaré map, we define

$$F(\epsilon,\mu) \equiv \frac{r(2\pi,\epsilon,\mu) - \epsilon}{\epsilon}.$$

We want to apply the Implicit Function Theorem to continue a zero of F. In fact, we show that $F(0,0) = 0$, $F_\epsilon(0,0) = 0$, and $F_\mu(0,0) \neq 0$. To verify these conditions, we need to show that $r(2\pi,\epsilon,\mu) - \epsilon = O(\epsilon^2)$, and $r(2\pi,\epsilon,0) - \epsilon = O(\epsilon^3)$. Such estimates can probably be proved using Gronwall's Inequality applied to the equation $(d/d\theta)e^{-\mu\theta/\beta} r = O(r^2)$. Instead of using this approach, we scale the variable r by ϵ, defining $r = \epsilon\rho$, and derive estimates on ρ.

Thus, we define $r = \epsilon\rho$, and let $\rho(\theta,\epsilon,\mu)$ be the solution with $\rho(0,\epsilon,\mu) = 1$ (which corresponds to $r(0,\epsilon,\mu) = \epsilon$). We need to verify that $\rho(2\pi,\epsilon,\mu) - 1 = O(\epsilon)$. The differential equation, $d\rho/d\theta$, is given by

$$\begin{aligned}\frac{d\rho}{d\theta} = \frac{\mu}{\beta} \rho &+ \epsilon\rho^2 \big[\frac{1}{\beta} C_3(\theta,\mu) - \frac{\mu}{\beta^2} D_3(\theta,\mu)\big] \\ &+ \epsilon^2\rho^3 \big[\frac{1}{\beta} C_4(\theta,\mu) - \frac{1}{\beta^2} C_3(\theta,\mu) D_3(\theta,\mu) \\ &\qquad - \frac{\mu}{\beta^2} D_4(\theta,\mu) + \frac{\mu}{\beta^3} D_3(\theta,\mu)^2\big] \\ &+ O(\epsilon^3\rho^4).\end{aligned}$$

We treat these equations as a perturbation of the linear terms. Using the integrating factor $e^{-\mu\theta/\beta}$, the derivative of $e^{-\mu\theta/\beta}\rho$ is as follows:

$$\begin{aligned}\frac{d}{d\theta}(e^{-\mu\theta/\beta}\rho) = \epsilon e^{-\mu\theta/\beta} \rho^2 &\big[\frac{1}{\beta} C_3 - \frac{\mu}{\beta^2} D_3\big] \\ &+ \epsilon^2 e^{-\mu\theta/\beta} \rho^3 \big[\frac{1}{\beta} C_4 - \frac{1}{\beta^2} C_3 D_3 - \frac{\mu}{\beta^2} D_4 + \frac{\mu}{\beta^3} D_3^2\big] \\ &+ O(\epsilon^3\rho^4).\end{aligned}$$

Integrating from 0 to 2π, we obtain

$$\begin{aligned}&e^{-2\pi\mu/\beta}\rho(2\pi,\epsilon,\mu) - 1 \\ &\quad = \epsilon \int_0^{2\pi} e^{-\mu\theta/\beta} \rho(\theta,\epsilon,\mu)^2 \big[\frac{1}{\beta} C_3(\theta,\mu) - \frac{\mu}{\beta^2} D_3(\theta,\mu)\big]\, d\theta \\ &\quad + \epsilon^2 \int_0^{2\pi} e^{-\mu\theta/\beta} \rho(\theta,\epsilon,\mu)^3 \big[\frac{1}{\beta} C_4(\theta,\mu) - \frac{1}{\beta^2} C_3(\theta,\mu) D_3(\theta,\mu) \\ &\qquad\qquad - \frac{\mu}{\beta^2} D_4(\theta,\mu) + \frac{\mu}{\beta^3} D_3(\theta,\mu)^2\big] + O(\epsilon^3\rho^4(\theta,\epsilon,\mu))\, d\theta \\ &\quad \equiv \epsilon\, h(\epsilon,\mu).\end{aligned}$$

Thus,

$$\begin{aligned}F(\epsilon,\mu) &= \rho(2\pi,\epsilon,\mu) - 1\\ &= [e^{2\pi\mu/\beta} - 1] + \epsilon e^{2\pi\mu/\beta} h(\epsilon,\mu).\end{aligned}$$

At $(\epsilon,\mu) = (0,0)$, $F(0,0) = 0$ and

$$\begin{aligned}F_\mu(0,0) &= (\frac{2\pi}{\beta})e^{2\pi\mu/\beta} - (\frac{2\pi\mu}{\beta^2})(\frac{\partial\beta}{\partial\mu})e^{2\pi\mu/\beta}\\ &\quad + \epsilon\frac{\partial}{\partial\mu}[e^{2\pi\mu/\beta}h(\epsilon,\mu)]\big|_{\epsilon=0,\mu=0}\\ &= \frac{2\pi}{\beta_0} \neq 0.\end{aligned}$$

Therefore, we can solve for μ as a function of ϵ, $\mu(\epsilon)$, such that $F(\epsilon,\mu(\epsilon)) \equiv 0$. These points are periodic orbits with $\theta_0 = 0$, $\rho(0,\epsilon,\mu(\epsilon)) = 1$ so $r(0,\epsilon,\mu(\epsilon)) = \epsilon$, and $\rho(2\pi,\epsilon,\mu(\epsilon)) = 1$ so $r(2\pi,\epsilon,\mu(\epsilon)) = \epsilon$.

We find the derivative of $\mu(\epsilon)$ by implicit differentiation:

$$F_\epsilon(0,0) + F_\mu(0,0)\mu'(0) = 0.$$

Thus, we need to calculate $F_\epsilon(0,0)$. Note that

$$\begin{aligned}F(\epsilon,0) &= 0 + \epsilon h(\epsilon,0)\\ &= \epsilon\int_0^{2\pi} \rho(\theta,\epsilon,0)^2(\frac{1}{\beta_0})C_3(\theta,0)\,d\theta\\ &\quad + \epsilon^2\int_0^{2\pi} \{\rho(\theta,\epsilon,0)^3[(\frac{1}{\beta_0})C_4(\theta,0) - (\frac{1}{\beta_0^2})C_3(\theta,0)D_3(\theta,0)]\\ &\qquad + O(\epsilon^3\rho^4(\theta,\epsilon,0))\}\,d\theta.\end{aligned}$$

Using the fact that $\rho(\theta,0,0) \equiv 1$,

$$\begin{aligned}F_\epsilon(0,0) &= \frac{1}{\beta_0}\int_0^{2\pi} \rho(\theta,0,0)^2 C_3(\theta,0)\,d\theta + 0\\ &= \frac{1}{\beta_0}\int_0^{2\pi} C_3(\theta,0)\,d\theta\\ &= 0.\end{aligned}$$

The integral is 0 because $C_3(\theta,0)$ is a homogeneous polynomial of odd degree in $\sin(\theta)$ and $\cos(\theta)$. Using that $\mu'(0) = -F_\epsilon(0,0)F_\mu(0,0)^{-1}$, $F_\epsilon(0,0) = 0$, and $F_\mu(0,0) \neq 0$, we get that $\mu'(0) = 0$. This completes the proof of part (a) of the theorem. □

PROOF OF THEOREM 4.2(b). By part (a), there is a function $\mu(\epsilon)$ such that $F(\epsilon,\mu(\epsilon)) \equiv 0$, $\mu'(0) = 0$, and $0 = F_\epsilon(\epsilon,\mu(\epsilon)) + F_\mu(\epsilon,\mu(\epsilon))\mu'(\epsilon)$. Differentiating this last equation again with respect to ϵ and evaluating at $(\epsilon,\mu) = (0,0)$ gives

$$0 = F_{\epsilon\epsilon}(0,0) + 2F_{\epsilon\mu}(0,0)\mu'(0) + F_{\mu\mu}(0,0)\mu'(0)^2 + F_\mu(0,0)\mu''(0),$$

so

$$\begin{aligned}\mu''(0) &= -\frac{F_{\epsilon\epsilon}(0,0)}{F_\mu(0,0)}\\ &= -(\frac{\beta_0}{2\pi})F_{\epsilon\epsilon}(0,0).\end{aligned}$$

Thus, we need only show that $F_{\epsilon\epsilon}(0,0) = (4\pi/\beta_0)K$ in order to verify the derivative given in the theorem.

In the calculation of $F_{\epsilon\epsilon}(0,0)$, we can fix $\mu = 0$ and differentiate $F(\epsilon, 0)$ twice. Using the expression for $F(\epsilon, 0)$ given in part (a),

$$
\begin{aligned}
F_\epsilon(\epsilon,0) &= (\frac{1}{\beta_0}) \int_0^{2\pi} \rho(\theta,\epsilon,0)^2 C_3(\theta,0)\,d\theta \\
&+ (\frac{\epsilon}{\beta_0}) \int_0^{2\pi} 2\rho(\theta,\epsilon,0)\frac{\partial\rho}{\partial\epsilon}(\theta,\epsilon,0)C_3(\theta,0)\,d\theta \\
&+ 2\epsilon \int_0^{2\pi} \rho(\theta,\epsilon,0)^3 \big[(\frac{1}{\beta_0})C_4(\theta,0) - (\frac{1}{\beta_0^2})C_3(\theta,0)D_3(\theta,0)\big]\,d\theta \\
&+ O(\epsilon^2),
\end{aligned}
$$

and

$$
\begin{aligned}
F_{\epsilon\epsilon}(0,0) &= 2\int_0^{2\pi} \rho(\theta,0,0)^3 \big[(\frac{1}{\beta_0})C_4(\theta,0) - (\frac{1}{\beta_0^2})C_3(\theta,0)D_3(\theta,0)\big]\,d\theta \\
&+ (\frac{4}{\beta_0}) \int_0^{2\pi} \rho(\theta,0,0)\frac{\partial\rho}{\partial\epsilon}(\theta,0,0)C_3(\theta,0)\,d\theta \\
&= \frac{4\pi}{\beta_0}K + (\frac{4}{\beta_0}) \int_0^{2\pi} \frac{\partial\rho}{\partial\epsilon}(\theta,0,0)C_3(\theta,0)\,d\theta,
\end{aligned}
$$

because $\rho(\theta,0,0) \equiv 1$. Thus, it is enough to show the last integral, which we denote by I, is 0. Since for $\mu = 0$,

$$
\frac{d\rho}{d\theta}(\theta,\epsilon,0) = (\frac{\epsilon}{\beta_0})\rho(\theta,\epsilon,0)^2 C_3(\theta,0) + O(\epsilon^2),
$$

the derivative with respect to ϵ gives (the first variation equation)

$$
\begin{aligned}
\frac{d}{d\theta}(\frac{\partial\rho}{\partial\epsilon})(\theta,\epsilon,0)\big|_{\epsilon=0} &= (\frac{1}{\beta_0})\rho(\theta,0,0)^2 C_3(\theta,0) \\
&= (\frac{1}{\beta_0})C_3(\theta,0).
\end{aligned}
$$

Integrating from 0 to θ gives

$$
\begin{aligned}
(\frac{\partial\rho}{\partial\epsilon})(\theta,0,0) &= (\frac{1}{\beta_0}) \int_0^{\theta} C_3(s,0)\,ds \\
&\equiv (\frac{\hat{C}_3(\theta,0)}{\beta_0}).
\end{aligned}
$$

Thus, the integral I is given as follows:

$$
\begin{aligned}
I &= (\frac{2}{\beta_0^2}) \int_0^{2\pi} 2\hat{C}_3(\theta,0)C_3(\theta,0)\,d\theta \\
&= (\frac{2}{\beta_0^2})\big[\hat{C}_3(2\pi,0)^2 - \hat{C}_3(0,0)^2\big] \\
&= 0
\end{aligned}
$$

because $C_3(\theta, 0)$ is a homogeneous polynomial of odd degree in $\sin(\theta)$ and $\cos(\theta)$.

This only leaves the stability of the periodic orbit to be determined. For $\beta_0 > 0$, the Poincaré map, $P_\mu(\epsilon)$, is given by $P(\epsilon, \mu) = r(2\pi, \epsilon, \mu) = \epsilon\rho(2\pi, \epsilon, \mu)$. (For $\beta_0 < 0$, the Poincaré map is given by $P(\epsilon, \mu) = r(-2\pi, \epsilon, \mu)$, and $P(\epsilon, \mu)^{-1} = r(2\pi, \epsilon, \mu)$. We do not indicate the changes for this case.) The derivative with respect to ϵ, which is used to determine the stability, is given by

$$P'_\mu(\epsilon) = \rho(2\pi, \epsilon, \mu) + \epsilon \frac{\partial \rho}{\partial \epsilon}(2\pi, \epsilon, \mu).$$

The Taylor expansion in ϵ and μ is given by

$$\begin{aligned}\rho(2\pi, \epsilon, \mu) &= F(\epsilon, \mu) + 1 \\ &= 1 + F_\epsilon(0,0)\epsilon + F_\mu(0,0)\mu \\ &\quad + \frac{1}{2}F_{\epsilon\epsilon}(0,0)\epsilon^2 + F_{\epsilon\mu}(0,0)\epsilon\mu + \frac{1}{2}F_{\mu\mu}(0,0)\mu^2 + O(|(\epsilon,\mu)|^3)\end{aligned}$$

and

$$\begin{aligned}\rho(2\pi, \epsilon, \mu(\epsilon)) &= 1 + 0 \cdot \epsilon + F_\mu(0,0)[\frac{\mu''(0)}{2}\epsilon^2 + O(\epsilon^3)] + \frac{1}{2}F_{\epsilon\epsilon}(0,0)\epsilon^2 + O(\epsilon^3) \\ &= 1 + F_\mu(0,0)(\frac{-F_{\epsilon\epsilon}(0,0)}{2\,F_\mu(0,0)})\epsilon^2 + \frac{1}{2}F_{\epsilon\epsilon}(0,0)\epsilon^2 + O(\epsilon^3) \\ &= 1 + O(\epsilon^3).\end{aligned}$$

Next,

$$\frac{\partial \rho}{\partial \epsilon}(2\pi, \epsilon, \mu) = F_\epsilon(0,0) + F_{\epsilon\epsilon}(0,0)\epsilon + F_{\epsilon\mu}(0,0)\epsilon\mu + O(|(\epsilon,\mu)|^2)$$

and

$$\frac{\partial \rho}{\partial \epsilon}(2\pi, \epsilon, \mu(\epsilon)) = F_{\epsilon\epsilon}(0,0)\epsilon + O(\epsilon^2).$$

Combining,

$$\begin{aligned}P'_\mu(\epsilon) &= 1 + F_{\epsilon\epsilon}(0,0)\epsilon^2 + O(\epsilon^3) \\ &= 1 + (\frac{4\pi}{\beta_0})K\epsilon^2 + O(\epsilon^3).\end{aligned}$$

Therefore, if $K < 0$, then $0 < P'_\mu(\epsilon) < 1$ for small ϵ and the orbit is attracting. Similarly, if $K > 0$, the orbit is repelling. (When $\beta_0 < 0$, it is still the case that the orbit is attracting for $K < 0$ and repelling for $K > 0$. We leave this verification to the reader.) This completes the proof of the theorem. □

7.5 Andronov-Hopf Bifurcation for Diffeomorphisms

As stated in the introduction to the chapter, the Andronov-Hopf bifurcation for a diffeomorphism occurs when a pair of eigenvalues for a fixed point changes from absolute value less than 1 to absolute value greater than 1, i.e., the fixed point changes from stable to unstable by a pair of eigenvalues crossing the unit circle. With further conditions on derivatives, it follows that an invariant closed curve bifurcates off from the fixed point. The motion on this invariant curve is a rotation, whose rotation number starts near the value determined by the eigenvalue. The formal statement of the theorem requires some notation and preliminary change of variables. We do this before we state the theorem. After stating the theorem, we give a sketch of the proof but leave the details to the references.

Assume that $F : \mathbb{R}^2 \times \mathbb{R} \to \mathbb{R}^2$ is a one-parameter family of C^r diffeomorphisms which satisfies the following conditions. (We do not state a version which allows the fixed point to vary with the parameter.)

(a) Assume $r \geq 5$.

(b) Assume that the origin is a fixed point of F_μ for μ near 0: $F_\mu(0,0) = (0,0)$.

(c) Assume that $D(F\mu)_{(0,0)}$ has two non-real eigenvalues, $\lambda(\mu)$ and $\bar{\lambda}(\mu)$, such that $|\lambda(0)| = 1$ and $\frac{d}{d\mu}|\lambda(\mu)| \neq 0$. By a change of parameter, we can assume that $|\lambda(\mu)| = 1 + \mu$.

(d) By a change of basis on $\mathbb{R}^2$ (which depends on μ), we can assume that

$$D(F\mu)_{(0,0)} = (1+\mu)\begin{pmatrix} \cos(\beta(\mu)) & -\sin(\beta(\mu)) \\ \sin(\beta(\mu)) & \cos(\beta(\mu)) \end{pmatrix}.$$

(e) We further assume that $\lambda(0)^m = e^{im\beta(0)} \neq 1$ for $m = 1, 2, \ldots, 5$. This means that $\lambda(0)$ is not a low root of unity (in addition to not being equal to ± 1).

(f) Because $\lambda(0)$ is not a low root of unity, there exists a change of coordinates that bring F_μ into the form

$$F_\mu(x,y) = N_\mu(x,y) + O(|(x,y)|^5),$$

where $N_\mu(x,y)$ is given in polar coordinates by

$$N_\mu(r,\theta) = ((1+\mu)r - f_1(\mu)r^3, \theta + \beta(\mu) + f_3(\mu)r^2).$$

We make the assumption that $f_1(0) \neq 0$. Thus, the radial component of the map is a nonlinear function of r. (Notice that $(1+\mu)r - f_1(\mu)r^3 = r$ has a solution $r^2 = \mu f_1(\mu)^{-1}$ for those μ with $\mu f_1(\mu)^{-1} > 0$. Thus, the normal form terms have an invariant circle for μ on one side of $\mu = 0$.) The bifurcation described in the following theorem is called the *Andronov-Hopf Bifurcation* for diffeomorphisms.

Theorem 5.1 (Andronov-Hopf Bifurcation). *Assume $F_\mu(x,y)$ satisfies assumptions (a)–(f). Then, for all sufficiently small μ with $\mu f_1(\mu)^{-1} > 0$, F_μ has an invariant closed curve surrounding the fixed point $(0,0)$ of radius approximately equal to $[\mu/f_1(\mu)]^{1/2}$. Further, if $f_1(0) > 0$, then the closed curve is attracting; and if $f_1(0) < 0$, then it is repelling.*

REMARK 5.1. This theorem was proved by Naimark (1967), Sacker (1964), and Ruelle and Takens (1971). Also see Marsden and McCracken (1976) and Iooss (1979). The name of Andronov-Hopf Bifurcation is commonly used because of the connection with the Andronov-Hopf bifurcation for differential equations given in the last section.

REMARK 5.2. The map on the invariant closed curve can vary. This map is most likely not conjugate to a rigid rotation for most parameter values. As μ approaches 0, the rotation number is approximately $\beta(\mu)/(2\pi)$.

REMARK 5.3. The proof of this theorem is more difficult than those for the other bifurcations which we treat. The reason that it is harder is that we need to find a whole closed curve of points and not just a single point. For this reason, we cannot use the Implicit Function Theorem to prove this theorem, but must apply a contraction mapping argument to a set of potential invariant curves, and the construction is much more involved and delicate. Because of these complications, we only give a sketch of the proof and refer the reader to the references for the details. In the last section we proved the simpler Hopf-Andronov bifurcation for differential equations where we can use the Poincaré map in polar coordinates from $\theta = 0$ to itself and reduce the problem to finding a fixed point of this map.

SKETCH OF THE PROOF. The first steps in the proof involve showing that there is a change of coordinates to put the map in normal form given above in polar coordinates $F_\mu(r,\theta) = N_\mu(r,\theta) + O(r^5)$. We consider the case when $f_1(0) > 0$, so the origin is weakly attracting at $\mu = 0$.

The ideas of the proof is that for small $\epsilon > 0$ and $0 < \mu \leq \epsilon$, the set

$$\Sigma_\mu = \{(r,\theta,\mu) : \ \mu = f_1(\mu)r^2\}$$

is normally attracting for N_μ. (The technical definition of normally attracting is given in Chapter XII, but here it just means that the vectors in the increasing r directions are contracted.) The rate of contraction toward Σ_μ goes to 0 as μ goes to 0, but the map F_μ also gets very near to N_μ. In fact, the normal form and the total map get near enough so that F_μ also contracts on an annulus $U_{\mu,\delta}$ about Σ_μ and has an invariant closed curve within $U_{\mu,\delta}$.

For $0 < \mu \leq \epsilon$, define the annulus

$$U_{\mu,\delta} = \{(r,\theta,\mu) : r > 0, \mu \in r^2[f_1(\mu) - \delta, f_1(\mu) + \delta]\ \}$$

for $0 < \delta << f_1(0)$. Then, for a point on the inner boundary of $U_{\mu,\delta}$ where $\mu = r^2 f_1(\mu) + r^2\delta$, the r coordinate of $N_\mu(r,\theta,\mu)$ is

$$(1 + r^2 f_1(\mu) + r^2\delta)r - f_1(\mu)r^3 = r + \delta r^3$$

and is larger than r, and so it is inside $U_{\mu,\delta}$. On the outer boundary, the r value of the image is $r - \delta r^3$ is smaller than r, and so $N_\mu(r,\theta,\mu)$ is inside $U_{\mu,\delta}$. In fact, $N_\mu(U_{\mu,\delta}) \subset U_{\mu,\delta}$. Since F_μ only differs from N_μ in terms of terms of order r^5, $F_\mu(\Sigma_\mu) \subset U_{\mu,\delta}$ and is the graph of a function giving r in terms of θ with slope less than μ. By iteration, if ϵ is small enough, then $F_\mu^n(\Sigma_\mu) \subset U_{\mu,\delta}$ and $F_\mu^n(\Sigma_\mu)$ is the graph of a function giving r in terms of θ with slope less than μ; i.e., there is a smooth function $g_{n,\mu}$ for $0 < \mu \leq \epsilon$ such that

$$F_\mu^n(\Sigma_\mu) = \{(g_{n,\mu_0}(\theta), \theta, \mu_0)\},$$

and

$$\frac{d}{d\theta}(f_{n,\mu}(\theta)) \leq \mu.$$

The curves $F_\mu^n(\Sigma_\mu)$ then have to converge to a curve $\Sigma_{\mu.F}$, which is the graph of a Lipschitz function with Lipschitz constant less than μ. Because it is the limit of iterates by F_μ, $\Sigma_{\mu.F}$ is invariant by F_μ. Once the existence of a Lipschitz curve is proven, it can be shown that that the curves are in fact as differentiable as the original map F.

An alternative way to generate the invariant curve is to show that

$$\bigcap_{n=0}^{\infty} F_\mu^n(U_{\mu,\delta} = \Sigma_{\mu.F}$$

is the graph of a function giving r in terms of θ, which has Lipschitz constant less than μ. This method would be more in the spirit of the proof given for the stable manifold theorem in this book. □

7.6 Exercises

Bifurcations for Maps

7.1. Let $F_\mu(x) = \mu + x^2$ for $x \in \mathbb{R}$.
   (a) Find the point and parameter value where there is a saddle-node bifurcation of fixed points. Verify the assumptions of the theorem.
   (b) Find the point and parameter value where there is a period doubling bifurcation from a fixed point to an orbit of period two. Verify the assumptions of the theorem.

7.2. Let $F_{AB}(x,y) = (A - By - x^2, x)$ be the Hénon family of maps. Prove that F_{AB} undergoes a saddle-node bifurcation when $A = -[(B+1)/2]^2$.

7.3. Let $f_\mu(x) = \mu x - x^3$ for $x \in \mathbb{R}$.
   (a) Find the fixed points. Note that the bifurcation at $\mu = 0$ is not one that we have studied. It is called the *pitchfork bifurcation*, and takes place naturally in systems with a symmetry $f_\mu(-x) = -f_\mu(x)$ for all μ.
   (b) Show that there is a period doubling bifurcation at $\mu = 2$. (Verify the conditions of the theorem.)

7.4. Assume that $f_\mu(x) = -f_\mu(-x)$ for all μ where $x \in \mathbb{R}$.
   (a) Prove that $f_\mu(0) \equiv 0$ and $f''_\mu(0) \equiv 0$.
   (b) Assume $f'_{\mu_0}(0) = 1$, $f'''_{\mu_0}(0) \neq 0$, and $\frac{\partial}{\partial\mu} f'_\mu(0)|_{\mu=\mu_0} \neq 0$. Prove that $f_\mu(x)$ undergoes a pitchfork bifurcation like the example in the previous exercise.

7.5. Assume $f : \mathbb{R}^n \times \mathbb{R} \to \mathbb{R}^n$ is C^3 and satisfies the following conditions.
   (1) There is an $\mathbf{x}_0 \in \mathbb{R}^n$ and $\mu_0 \in \mathbb{R}$ such that $f(\mathbf{x}_0, \mu_0) = \mathbf{x}_0$.
   (2) The derivative of f_{μ_0} at $\mathbf{x}_0$ has eigenvalues $\lambda_1(\mu_0) = -1$ and $\lambda_j(\mu_0)$ for $2 \le j \le n$ with $|\lambda_j(\mu_0)| \neq 1$. Let $\mathbf{v}^1$ be the right eigenvector for the eigenvalue $\lambda_1(\mu_0) = -1$ of $D(f_{\mu_0})_{\mathbf{x}_0}$.
   (3) Let $\mathbf{x}(\mu)$ be the curve of fixed points of f_μ. Let $\lambda_j(\mu)$ be the eigenvalues of $D(f_\mu)_{\mathbf{x}(\mu)}$. Assume

$$\frac{d}{d\mu}\lambda_1(\mu)|_{\mu_0} \neq 0.$$

   (a) Prove that there is a curve of points of period 2 bifurcating off from $(\mathbf{x}_0, \mu_0)$ in $\mathbb{R}^n \times \mathbb{R}$, i.e., there is a differentiable curve γ passing through (x_0, μ_0) so that $\gamma \setminus \{(x_0, \mu_0)\}$ is the union of period 2 orbits. The curve γ is tangent to the line $< \mathbf{v}^1 > \times\{\mu_0\}$ at $(\mathbf{x}_0, \mu_0)$.

For parts (b) through (d), assume $\mathbf{x}_0 = \mathbf{0}$. Let $\mathbf{w}$ be a left eigenvector for the eigenvalue -1 of $D(f_{\mu_0})_{\mathbf{0}}$, and let $\pi : \mathbb{R}^n \to \mathbb{R}^{n-1}$ be the projection along $\mathbf{v}^1$ onto $\langle \mathbf{v}^2, \ldots, \mathbf{v}^n \rangle$. Using coordinates with $\mathbf{v}^1$ along the x_1-axis, and $\psi(\mathbf{x}, \mu) = \pi[f_\mu^2(\mathbf{x}) - \mathbf{x}]$, construct $\varphi(\mathbf{x}_1, \mu)$ such that $\psi(x_1, \varphi(x_1, \mu), \mu) \equiv 0$ and let $g(x_1, \mu)$ be defined as in the proof of Theorem 2.1.

   (b) Prove that

$$\frac{\partial^2 g}{\partial s^2}(0, \mu_0) = \mathbf{w}D^2(f_{\mu_0})(\mathbf{v}^1, \mathbf{v}^1) \qquad \text{and}$$

$$\frac{\partial^3 g}{\partial s^3}(0, \mu_0) = \mathbf{w}D^3(f_{\mu_0})(\mathbf{v}^1, \mathbf{v}^1, \mathbf{v}^1) + 3\,\mathbf{w}D^2(f_{\mu_0})(\mathbf{v}^1, \frac{\partial^2 \varphi}{\partial s^2}(0, \mu_0)).$$

   (c) Prove that

$$\frac{\partial^2 \varphi}{\partial s^2}(0, \mu_0) = -[\pi\, D^2(f_{\mu_0}) - I]^{-1} \frac{\partial^2}{\partial x_1^2}(f_{\mu_0}^2)(\mathbf{0}).$$

(d) Use parts (b) and (c) to write the conditions on the derivatives of f_{μ_0} and $f^2_{\mu_0}$ which insure that the orbits of period 2 are on one side of $\mu = \mu_0$.

7.6. Let $F_{AB}(x, y) = (A - By - x^2, x)$ be the Hénon map. Fix $B = B_0$.

(a) Show that F_{AB_0} has a fixed point with one eigenvalue equal to -1 (and the other eigenvalue equal to $-B_0$) for $A = 3(1 + B_0)^2/4$.

(b) Prove that F_{AB_0} undergoes a period doubling bifurcation at $A = 3(1 + B_0)^2/4$ as A varies using the conditions derived in the last exercise. (Verify at least the conditions of part (a).)

Bifurcations for Differential Equations

7.7. Let

$$\dot{x} = y$$
$$\dot{y} = -x + \mu y + a y^3.$$

(a) Show that this system satisfies the eigenvalue conditions for a Andronov-Hopf bifurcation at the origin for $\mu = 0$.

(b) Find a (right handed) basis (of eigenvectors) which changes the linear terms at the origin to the system

$$\dot{u} = \nu u - \gamma v$$
$$\dot{v} = \gamma u + \nu v$$

for the appropriate choice of ν and γ. Also, calculate the nonlinear equations in terms of these variables.

(c) Express in polar coordinates the nonlinear equations found in part (b), where $r^2 = u^2 + v^2$ and $\tan\theta = v/u$.

(d) Find the constant K (used in the statement of the Andronov-Hopf bifurcation theorem), and show that it is nonzero. Here $a \neq 0$, but it can either be positive or negative. Hint: $C_3(0, \theta) = D_3(0, \theta) = 0$.

7.8. Consider the system of differential equations given by

$$\dot{x} = a - (1 + b)x + x^2 y$$
$$\dot{y} = bx - x^2 y.$$

This system of equations is a model of a certain chemical reaction and is called the "Brusselator."

(a) Prove that $(a, b/a)$ is the unique fixed point.

(b) Prove that the fixed point is stable for $b < 1 + a^2$ and unstable for $b > 1 + a^2$ with a pair of complex eigenvalues crossing the imaginary axis at $b = 1 + a^2$. (It can be further shown that a Andronov-Hopf bifurcation to a stable periodic orbit occurs at $b = 1 + a^2$. See Prigonine and Lefever (1968) and Lefever and Nicholis (1971).)

7.9. Consider the following systems in polar coordinates, where in each case $\dot{\theta} = 1$. In each case, find the periodic orbits and draw the phase portrait (in Cartesian coordinates) for $\mu < 0$, $\mu = 0$, and $\mu > 0$. Indicate which conditions of the full Andronov-Hopf Theorem are not true (if any).

(a) $\dot{r} = r(\mu^2 - r^2)$.

(b) $\dot{r} = \mu r(1 + r^2)$.

(c) $\dot{r} = r(\mu - r^2)(4\mu - r^2)$.

(d) $\dot{r} = r(\mu - r^4)$.

7.10. Consider the Lorenz system of differential equations:

$$\begin{aligned}
\dot{x} &= -10x + 10y, \\
\dot{y} &= \rho x - y - xz, \\
\dot{z} &= \frac{8}{3}z + xy.
\end{aligned}$$

(a) Find the fixed points, $\mathbf{0}$, $\mathbf{a}^{\pm}$.
(b) Find the eigenvalues at the fixed point $\mathbf{0}$.
(c) The rest of the problem deals with the eigenvalues at the fixed points $\mathbf{a}^{\pm}$. (The eigenvalues are the same at $\pm\mathbf{a}$.) In particular, part (h) asks the reader to verify that a pair of complex eigenvalues cross the imaginary axis (some of the conditions for a Hopf bifurcation) at a parameter value ρ_1 which is found in part (f). To start this process, we ask the reader to find the characteristic polynomial at $\mathbf{a}^{\pm}$. Show that the characteristic polynomial at $\mathbf{a}^{\pm}$ is given by

$$p(\lambda) = \lambda^3 + \left(\frac{41}{3}\right)\lambda^2 + \frac{8}{3}(10 + \rho)\lambda + \frac{160}{3}(\rho - 1).$$

(d) Show that $p(\lambda)$ has a negative real root. (Note that $p(0) > 0$ and all the coefficients are positive.)
(e) For $\rho \geq 14$, show that $p(\lambda)$ is a monotonically increasing function of λ, so it has exactly one real root, λ_0. Thus, the eigenvalues are λ_0 and $\alpha \pm i\beta$ with $\beta \neq 0$. (There is a pair of complex eigenvalues for smaller ρ as well, but it is not as easy to show.)
(f) Find a parameter value ρ_1 for which $\alpha = 0$. Hint: Note that $\lambda_0 + 2\alpha = -41/3$, where $41/3$ is the coefficient of λ^2. So, find $\rho = \rho_1$ such that $\lambda_0 = -41/3$ is a real root.
(g) Show that $\lambda_0 > -41/3$ for $\rho < \rho_1$, and $\lambda_0 < -41/3$ for $\rho > \rho_1$.
(h) Show that the complex eigenvalue $\alpha + i\beta$ crosses the imaginary axis as ρ increases through ρ_1. Note that this shows that the Lorenz equations satisfy some of the conditions for a Hopf bifurcation at the fixed point $\mathbf{a}$ for $\rho = \rho_1$.

CHAPTER VIII
Examples of Hyperbolic Sets and Attractors

In this chapter, we return to consider examples with complicated invariant sets. We introduce the idea of a hyperbolic invariant set and show that not only periodic orbits can have stable and unstable manifolds, but that a hyperbolic invariant set also has a family of stable manifolds. We give a number of different types of examples which are hyperbolic and give a method to show that the map is topologically transitive on the invariant set. One very important type of example arises from the intersection of stable and unstable manifolds of a saddle periodic orbit. This gives rise to an invariant set called a Smale horseshoe. It is very similar to the invariant set which we found for the quadratic map on the real line. Another important type of hyperbolic invariant set occurs where all nearby orbits tend toward the invariant set. Such an invariant set is called an attractor (with further conditions added). Thus, an attractor is like a periodic sink where the invariant set itself is more complicated topologically. The final class of examples we consider are those with only a finite number of periodic points, called Morse–Smale systems. For these systems we make a connection with the Lefschetz theory through the Morse–Smale inequalities.

8.1 Definition of a Manifold

In this chapter, we consider diffeomorphisms (or flows) on a manifold M having a dimension greater than or equal to 2. A manifold is merely a set on which there are local coordinates that make it a Euclidean space. We have already seen examples of manifolds: (i) the circle which is represented by $\pi : \mathbb{R}^1 \to S^1$, $\pi(t) = t \bmod 1$, and (ii) local stable and unstable manifolds which are represented as graphs, $\sigma^s : \mathbb{E}^s(r) \to \mathbb{E}^u(r)$ and $W_r^s(\mathbf{0}) = \{(\mathbf{x}, \sigma^s(\mathbf{x})) : \mathbf{x} \in \mathbb{E}^s(r)\}$. We now give a more formal definition of a manifold.

Definition. A C^r *n-dimensional manifold* M is a second countable metric space together with a collection of homeomorphisms $\varphi_\alpha : V_\alpha \subset \mathbb{R}^n \to U_\alpha \subset M$ for α in some index set A such that (i) $\varphi(V_\alpha) = U_\alpha$, (ii) $\{U_\alpha\}_{\alpha \in A}$ is an open cover of M, and (iii) if $U_\alpha \cap U_\beta \neq \emptyset$, then

$$\varphi_{\alpha,\beta} = \varphi_\beta^{-1} \circ \varphi_\alpha : \varphi_\alpha^{-1}(U_\alpha \cap U_\beta) \subset V_\alpha \to \varphi_\beta^{-1}(U_\alpha \cap U_\beta) \subset V_\beta$$

is a C^r diffeomorphism between open subsets of $\mathbb{R}^n$. One of the allowable maps $\varphi_\alpha : V_\alpha \subset \mathbb{R}^n \to U_\alpha \subset M$ is called a *coordinate chart* on M.

Example 1.1. For the circle, we can use the map $\pi : \mathbb{R} \to S^1$ to induce homeomorphisms on open intervals I_α of length less than one. If I_α and I_β are two such open intervals with $\pi(I_\alpha) \cap \pi(I_\beta) \neq \emptyset$, $U_\alpha = \pi(I_\alpha)$, and $U_\beta = \pi(I_\beta)$, then

$$\pi_{\alpha,\beta} = (\pi|I_\beta)^{-1} \circ \pi : (\pi|I_\alpha)^{-1}(U_\alpha \cap U_\beta) \to I_\beta$$

is given by $\pi_{\alpha,\beta}(t) = t + j$ for some integer j. Therefore, $\pi_{\alpha,\beta}$ is C^∞ and the collection of these maps clearly satisfies the conditions of the definition of a manifold.

REMARK 1.1. In the case of a local stable manifold, we allowed it to be the graph of a function between Banach spaces. We do not make the formal definition of a Banach manifold, but it is much the same as that given above for finite-dimensional manifolds. See Lang (1967).

REMARK 1.2. The local stable or unstable manifolds of a hyperbolic fixed point of a diffeomorphism can be represented as a graph, so it is clearly a manifold. The global stable and unstable manifolds do not always satisfy the full conditions stated above for a manifold. The problem is that the global stable manifold can accumulate on itself. (See the examples of the Smale horseshoe and the toral Anosov automorphism given in Sections 8.4 and 8.5.) More specifically, let $f : M \to M$ be a C^r diffeomorphism with a hyperbolic saddle fixed point $\mathbf{p}$. It is always possible to define a C^r map $\varphi : \mathbb{E}^s_{\mathbf{p}} \to M$ such that (i) φ is one to one, (ii) φ is onto $W^s(\mathbf{p})$, and (iii) the derivative of φ at each point is an isomorphism. (See the discussion of the derivative of maps into a manifold below.) However, the map φ does not always have a continuous inverse (i.e., φ is not a homeomorphism), so $W^s(\mathbf{p})$ is not a manifold in the full sense defined above.

A C^r map $\varphi : N \to M$ from one manifold into another is called an *immersion* provided the derivative of φ at each point is an isomorphism. The image of a one to one immersion is called an *immersed submanifold.* If an immersion is a homeomorphism, then it is called an *embedding* and its image is called an *embedded submanifold.* (Some people require that an embedding is also proper, i.e., the inverse image of a compact set is compact.) See Hirsch (1976) for discussion of these concepts.

REMARK 1.3. A one-dimensional manifold is usually called a *curve*; a two-dimensional manifold is usually called a *surface.*

Example 1.2. The *n-torus* $\mathbb{T}^n$ is the product of n copies of the circle, $\mathbb{T}^n = S^1 \times \cdots \times S^1$. To define a map from $\mathbb{R}^n$ onto $\mathbb{T}^n$, we first define an equivalence on $\mathbb{R}^n$ by $(x_1, \ldots, x_n) \sim (y_1, \ldots, y_n)$ if $x_j = y_j \bmod 1$ for all j. Define $\pi : \mathbb{R}^n \to \mathbb{T}^n$ by letting $\pi(\mathbf{x})$ be the equivalence class of $\mathbf{x}$ under $\sim$. The reader can check that this gives $\mathbb{T}^n$ the structure of a C^∞ manifold.

Example 1.3. Let $S^n = \{\mathbf{x} \in \mathbb{R}^{n+1} : |\mathbf{x}| = 1\}$ be the *n-sphere.* To show that S^n is a manifold, we represent pieces of S^n as graphs. Let D^n be the open unit ball in $\mathbb{R}^n$. For $1 \le j \le n$, define $\varphi_j^{\pm} : D^n \to S^n$ by

$$\varphi_j^{\pm}(y_1, \ldots, y_n) = (y_1, \ldots, y_{j-1}, \pm(1 - y_1^2 - \cdots - y_n^2)^{1/2}, y_j, \ldots, y_n).$$

Each of these maps $\varphi_j^{\pm}$ is a homeomorphism onto a "hemisphere" of S^n. We leave to the exercises the verification that these maps gives S^n the structure of a C^∞ manifold. See Exercise 8.1.

Example 1.4. A common method to specify a manifold is as the level set of a function. Let $F : \mathbb{R}^{n+1} \to \mathbb{R}$ be a C^r function for some $r \ge 1$. Assume $c \in \mathbb{R}$ is a value such that for each $\mathbf{p} \in F^{-1}(c)$, $DF_{\mathbf{p}} \ne \mathbf{0}$, i.e., $DF_{\mathbf{p}}$ has rank one, or some partial derivative of F is nonzero at $\mathbf{p}$. Let $M = F^{-1}(c)$. For each $\mathbf{p} \in M$, the Implicit Function Theorem proves that there is a neighborhood $U_{\mathbf{p}}$ of $\mathbf{p}$ and a C^r function $\sigma_{\mathbf{p}} : V_{\mathbf{p}} \subset \mathbb{R}^n \to \mathbb{R}$ such that the graph $\sigma_{\mathbf{p}}$ is onto $U_{\mathbf{p}}$. More specifically, if $\frac{\partial F}{\partial x_j}(\mathbf{p}) \ne 0$, there there is an open set $V_{\mathbf{p}}$ in $\mathbb{R}^n$ and a C^r function $\sigma_{\mathbf{p}} : V_{\mathbf{p}} \to \mathbb{R}$ such that

$$U_{\mathbf{p}} = \{(y_1, \ldots, y_{j-1}, \sigma_{\mathbf{p}}(y_1, \ldots, y_n), y_j, \ldots, y_n) : (y_1, \ldots, y_n) \in V_{\mathbf{p}}\}$$

is a neighborhood of $\mathbf{p}$ in M. These graphs give M the structure of a C^r manifold. This example is a generalization of the n-sphere of the last example.

See Guillemin and Pollack (1974), Hirsch (1976), Chillingworth (1976), or Lang (1967) for more details and examples of manifolds. The only explicit examples of manifolds that we use are Euclidean spaces (a trivial example), tori, and spheres.

Given the definition of a differentiable manifold, we can define a differentiable map between two manifolds.

Definition. Let M and N be two C^r manifolds for some $r \geq 1$. Assume $f : M \to N$ is a continuous map. We say that f is *r times continuously differentiable*, or C^r, provided for each point $\mathbf{p} \in M$ and coordinate charts $\varphi_\alpha : V_\alpha \to U_\alpha \subset M$ and $\varphi_\beta : V_\beta \to U_\beta \subset N$ at $\mathbf{p}$ and $f(\mathbf{p})$, respectively (i.e., $\mathbf{p} \in U_\alpha$ and $f(\mathbf{p}) \in U_\beta$), $\varphi_\beta^{-1} \circ f \circ \varphi_\alpha$ is differentiable at $\varphi_\alpha^{-1}(\mathbf{p})$. Note that if $\varphi_\beta^{-1} \circ f \circ \varphi_\alpha$ is C^r at $\varphi_\alpha^{-1}(\mathbf{p})$ for one pair of coordinate charts, and $\varphi_{\alpha'} : V_{\alpha'} \to U_{\alpha'} \subset M$ and $\varphi_{\beta'} : V_{\beta'} \to U_{\beta'} \subset N$ is another pair of coordinate charts at $\mathbf{p}$ and $f(\mathbf{p})$, then $\varphi_{\beta'}^{-1} \circ f \circ \varphi_{\alpha'}$ is C^r at $\varphi_{\alpha'}^{-1}(\mathbf{p})$ because both $\varphi_{\alpha,\alpha'}$ and $\varphi_{\beta,\beta'}$ are C^r. The set of all C^r maps from M to N is denoted by $C^r(M, N)$. A C^r maps from M to M is a *diffeomorphism* provided it is one to one, onto, and the derivative at each point (in local coordinates) is nonsingular. The set of all C^r diffeomorphisms on M is denoted by $\mathrm{Diff}^r(M)$.

8.1.1 Topology on Space of Differentiable Functions

In this chapter and the next, we consider the structural stability of diffeomorphisms or differential equations on a compact manifold. The definition of structural stability uses the notion that two functions are close in the C^1 or C^r topology. In this subsection, we give these definitions which are used later.

Definition. The definition of the C^r distance between functions is easier on the torus, so we start with this case. If $f : \mathbb{T}^n \to \mathbb{T}^n$, then there is a lift $F : \mathbb{R}^n \to \mathbb{R}^n$ such that (i) $\pi \circ F = f \circ \pi$, where π is the projection $\pi : \mathbb{R}^n \to \mathbb{T}^n$ defined in Example 1.2, and (ii) $F(\mathbf{x} + \mathbf{j}) = F(\mathbf{x}) + \mathbf{j}$ for all $\mathbf{x} \in \mathbb{R}^n$ and $\mathbf{j} \in \mathbb{Z}^n$. If $f, g : \mathbb{T}^n \to \mathbb{T}^n$ are two C^r maps with lifts $F, G : \mathbb{R}^n \to \mathbb{R}^n$, then the C^r *distance from f to g* is defined to be

$$\begin{aligned} d_r(f, g) = \sup\{d(f \circ \pi(\mathbf{x}), g \circ \pi(\mathbf{x})),\ &\|D^i F_{\mathbf{x}} - D^i G_{\mathbf{x}}\| : \\ &\mathbf{x} = (x_1, \ldots, x_n) \in \mathbb{R}^n \text{ satisfies } 0 \leq x_j \leq 1 \\ &\text{for } 1 \leq j \leq n,\ 1 \leq i \leq r\}. \end{aligned}$$

In this definition, d is the distance between points on $\mathbb{T}^n$.

The next easiest case to treat is functions from a compact manifold M to a Euclidean space $\mathbb{R}^N$. Let

$$\{\varphi_j : V_j \subset \mathbb{R}^n \to U_j \subset M\}_{j=1}^J$$

be a finite number of coordinate charts with $\bigcup_{j=1}^J U_j = M$. Let $C_j \subset U_j$ be compact subsets with $\bigcup_{j=1}^J C_j = M$. If $f, g : M \to \mathbb{R}^N$ are two C^r functions, then the C^r *distance from f to g* is defined to be

$$\begin{aligned} d_r(f, g) = \sup\{|f(\mathbf{x}) - g(\mathbf{x})|,\ &\|D^i(f \circ \varphi_j)_{\varphi_j^{-1}(\mathbf{x})} - D^i(g \circ \varphi_j)_{\varphi_j^{-1}(\mathbf{x})}\| : \\ &\mathbf{x} \in C_j,\ 1 \leq j \leq J, \text{ and } 1 \leq i \leq r\}. \end{aligned}$$

The case of maps between two manifolds is slightly more complicated. Rather than define a distance between two functions, we define a base of neighborhoods for the topology. Assume M and N are C^r compact manifolds. Let $\{\varphi_j : V_j \subset \mathbb{R}^n \to U_j \subset M\}_{j=1}^J$ be a finite number of coordinate charts on M, and $C_j \subset U_j$ be compact subsets as above. Let $\{\psi_k : V'_k \subset \mathbb{R}^n \to U'_k \subset N\}_{k=1}^{K}$ be a finite number of coordinate charts on N. Let $f \in C^r(M,N)$. Each C_j can be broken into finite number of compact subpieces $C_j = \bigcup_{\ell=1}^{L_j} C_{j,\ell}$, and an index $k(j,\ell)$ chosen so that $f(C_{j,\ell}) \subset U'_{k(j,\ell)}$ for $1 \leq \ell \leq L_j$ and $1 \leq j \leq J$. For $\epsilon > 0$, let $\mathcal{N}$ be the neighborhood of f given by

$$\begin{aligned}\mathcal{N} = \{g \in & C^r(M,N) : g(C_{j,\ell}) \subset U'_{k(j,\ell)}, \\ & |\psi^{-1}_{k(j,\ell)} \circ f(\mathbf{x}) - \psi^{-1}_{k(j,\ell)} \circ g(\mathbf{x})| < \epsilon, \\ & \|D^i(\psi^{-1}_{k(j,\ell)} \circ f \circ \varphi_j)_{\varphi_j^{-1}(\mathbf{x})} - D^i(\psi^{-1}_{k(j,\ell)} \circ g \circ \varphi_j)_{\varphi_j^{-1}(\mathbf{x})}\| < \epsilon \\ & \text{for all } \mathbf{x} \in C_{j,\ell},\ 1 \leq \ell \leq L_j,\ 1 \leq j \leq J,\ 1 \leq i \leq r\}.\end{aligned}$$

The C^0 estimate $|\psi^{-1}_{k(j,\ell)} \circ f(\mathbf{x}) - \psi^{-1}_{k(j,\ell)} \circ g(\mathbf{x})| < \epsilon$ can be replaced by the estimate $d(f(\mathbf{x}), g(\mathbf{x})) < \epsilon$ using distances between points on the manifold. These base neighborhoods $\mathcal{N}$ generate the topology on $C^r(M,N)$.

REMARK 1.4. Let M and N be manifolds with M compact. The manifold N can be embedded into some (large dimensional) Euclidean space $\mathbb{R}^L$. Using this embedding, it is possible to consider $C^r(M,N) \subset C^r(M,\mathbb{R}^L)$, i.e., $f \in C^r(M,N)$ if $f \in C^r(M,\mathbb{R}^L)$ and f takes its values in the subset $N \subset \mathbb{R}^L$. Using this, the topology on $C^r(M,N)$ can be inherited from $C^r(M,\mathbb{R}^L)$. Franks (1979) uses this approach.

Definition. A C^r diffeomorphism $f : M \to M$ is called C^r *structurally stable* provided there is a neighborhood $\mathcal{N}$ in $C^r(M,M)$ such that every $g \in \mathcal{N}$ is topologically conjugate to f. A C^1 structurally stable diffeomorphism f is also called just *structurally stable.*

Definition. For flows $\varphi^t, \psi^t : M \to M$, the C^1 topology measures their derivatives as functions from $[0,1] \times M$ to M. A C^1 flow $\varphi^t : M \to M$ is called *structurally stable* provided there is a neighborhood $\mathcal{N}$ of φ^t in the C^1 topology such that any flow $\psi^t \in \mathcal{N}$ is flow equivalent (topologically equivalent) to φ^t.

We give the definition of vector fields and the topology on the set of vector fields in the next subsection.

For more detail on the C^r topology, including the noncompact case, see Hirsch (1976).

8.1.2 Tangent Space

We are interested in invariant sets which are more complicated than a single fixed point. The first example is a horseshoe. In Chapter II we saw that the quadratic map had an invariant Cantor set. In two dimensions, the horseshoe is homeomorphic to the product of two Cantor sets. It has some directions which are expanding and some which are contracting. The directions which are expanding and contracting can vary from point to point. These directions of expansion and contraction at a point $\mathbf{p}$ can be thought of as infinitesimal displacements at $\mathbf{p}$ and are called tangent vectors. A tangent vector at $\mathbf{p}$ is also thought of as the derivative of a curve through $\mathbf{p}$: if $\gamma : (-\delta, \delta) \to M$ is a differentiable curve with $\gamma(0) = \mathbf{p}$, then $\mathbf{v} = \gamma'(0)$ is a tangent vector at $\mathbf{p}$. In order to organize these ideas, we need to distinguish vectors at different points. The following definition makes these ideas precise, even on a manifold.

Definition. We start by giving the definition of tangent vectors for Euclidean space. Fix a point $\mathbf{p} \in \mathbb{R}^n$. A *tangent vector at* $\mathbf{p}$ is a pair $(\mathbf{p}, \mathbf{v})$ where $\mathbf{v} \in \mathbb{R}^n$. This pair $(\mathbf{p}, \mathbf{v})$ is often written $\mathbf{v}_{\mathbf{p}}$. The set of all possible tangent vectors at $\mathbf{p}$ is denoted by $T_{\mathbf{p}}\mathbb{R}^n$ is called the *tangent space at* $\mathbf{p}$. The tangent space at $\mathbf{p}$ is a vector space where

$(\mathbf{p}, \mathbf{v}) + (\mathbf{p}, \mathbf{w}) = (\mathbf{p}, \mathbf{v} + \mathbf{w})$. The disjoint union of the tangent vectors at different points is called the *tangent bundle* or the *tangent space of* $\mathbb{R}^n$ and is denoted by $T\mathbb{R}^n$:

$$\begin{aligned} T\mathbb{R}^n &= \{(\mathbf{p}, \mathbf{v}) : \mathbf{p} \in \mathbb{R}^n \text{ and } \mathbf{v} \text{ is a tangent vector at } \mathbf{p}\} \\ &= \mathbb{R}^n \times \mathbb{R}^n. \end{aligned}$$

Thus, the tangent space of a Euclidean space $\mathbb{R}^n$ is isomorphic to the cross product of $\mathbb{R}^n$ with itself: the first copy of $\mathbb{R}^n$ is the set of the base points at which the vector is situated and the second copy is the set of vectors at a given point, or $(\mathbf{p}, \mathbf{v})$ is a vector at $\mathbf{p}$ in the direction of $\mathbf{v}$.

For a manifold M, we need to specify what we mean by a tangent vector at a point $\mathbf{p}$. Assume $\gamma : (-\delta, \delta) \subset \mathbb{R} \to M$ is a C^1 curve with $\gamma(0) = \mathbf{p}$. Assume $\varphi_\alpha : V_\alpha \to U_\alpha$ is a coordinate chart at $\mathbf{p}$. By the above definition of differentiability, $\varphi_\alpha^{-1} \circ \gamma(t)$ is C^1. In the coordinate chart, the tangent vector determined by γ is given by $(\varphi_\alpha^{-1} \circ \gamma)'(0) = \mathbf{v}_{\mathbf{p}}^\alpha$. If $\varphi_\beta : V_\beta \to U_\beta$ is another coordinate chart at $\mathbf{p}$, then the tangent vector determined by γ in this coordinate chart is given by $(\varphi_\beta^{-1} \circ \gamma)'(0) = \mathbf{v}_{\mathbf{p}}^\beta$. Note that

$$\mathbf{v}_{\mathbf{p}}^\beta = D(\varphi_\beta^{-1} \circ \varphi_\alpha)_{\mathbf{p}^\alpha} \mathbf{v}_{\mathbf{p}}^\alpha$$

where $\mathbf{p}^\alpha = \varphi_\alpha^{-1}(\mathbf{p})$. These two vectors, $\mathbf{v}_{\mathbf{p}}^\alpha$ and $\mathbf{v}_{\mathbf{p}}^\beta$, should be considered as representatives of the same vector in different coordinate charts because they are the derivative of coordinate representatives of the same curve. The *derivative of the curve on a manifold* (or *tangent vector to a curve*) is the equivalence class of representatives in different coordinate charts, where $\mathbf{v}_{\mathbf{p}}^\alpha \sim \mathbf{v}_{\mathbf{p}}^\beta$ provided $\mathbf{v}_{\mathbf{p}}^\beta = D(\varphi_\beta^{-1} \circ \varphi_\alpha)_{\mathbf{p}^\alpha} \mathbf{v}_{\mathbf{p}}^\alpha$. A *tangent vector at* $\mathbf{p}$ is a derivative of a differentiable curve through $\mathbf{p}$. The set of all tangent vectors at $\mathbf{p}$ is written $T_{\mathbf{p}}M$,

$$T_{\mathbf{p}}M = \{\mathbf{v}_{\mathbf{p}} : \mathbf{v}_{\mathbf{p}} \text{ is a tangent vector of a differentiable curve through } \mathbf{p}\}.$$

For the point $\mathbf{p}$ fixed, the set of vectors at $\mathbf{p}$, $T_{\mathbf{p}}M$, forms a vector space (using the addition in any one of the coordinate charts at $\mathbf{p}$). The disjoint union of the tangent vectors at different points gives the *tangent bundle* or the *tangent space of* M which is denoted by TM:

$$\begin{aligned} TM &= \{(\mathbf{p}, \mathbf{v}) : \mathbf{p} \in M \text{ and } \mathbf{v} \text{ is a tangent vector at } \mathbf{p}\} \\ &= \bigcup_{\mathbf{p} \in M} \{\mathbf{p}\} \times T_{\mathbf{p}}M. \end{aligned}$$

If M is a manifold and S is a subset of M, we denote the *tangent vectors to* M *at points of* S by

$$T_S M = \bigcup_{\mathbf{p} \in S} \{\mathbf{p}\} \times T_{\mathbf{p}}M.$$

REMARK 1.5. If $M \subset \mathbb{R}^{n+k}$ is a C^r manifold, then the tangent space of M at $\mathbf{p}$ can be thought of as a subspace of the tangent vectors of $\mathbb{R}^{n+k}$ at $\mathbf{p}$, $T_{\mathbf{p}}M \subset T_{\mathbf{p}}\mathbb{R}^{n+k}$.

Definition. With this idea of tangent vectors, if $f : M \to N$ is a C^1 map between manifolds, we consider the *derivative of* f *at* $\mathbf{p}$ to be a linear map from $T_{\mathbf{p}}M$ to $T_{f(\mathbf{p})}N$, $Df_{\mathbf{p}} : T_{\mathbf{p}}M \to T_{f(\mathbf{p})}N$. If $\varphi_\alpha : V_\alpha \to U_\alpha \subset M$ and $\varphi_\beta : V_\beta \to U_\beta \subset N$ are coordinate charts at $\mathbf{p}$ and $f(\mathbf{p})$, respectively, then

$$D(\varphi_\beta^{-1} \circ f \circ \varphi_\alpha)_{\mathbf{p}^\alpha} \mathbf{v}_{\mathbf{p}}^\alpha = \mathbf{w}_{f(\mathbf{p})}^\beta$$

takes the representative of a vector at $\mathbf{p}$ in the coordinate chart $(\varphi_\alpha, V_\alpha, U_\alpha)$ into the representative of a vector at $f(\mathbf{p})$ in the coordinate chart $(\varphi_\beta, V_\beta, U_\beta)$.

In fact, when we used cones of vectors to prove the stable manifold theorem, we used this idea of the derivative acting on tangent vectors at points implicitly in our description.

Definition. Below, we define a hyperbolic structure in terms of expanding and contracting directions. To do that, we need to be able to determine the length of vectors. The length can be determined from an inner product on each tangent space $T_{\mathbf{p}}M$. A *Riemannian metric* is an inner product on each tangent space,

$$\langle \cdot, \cdot \rangle_{\mathbf{p}} : T_{\mathbf{p}}M \times T_{\mathbf{p}}M \to \mathbb{R}$$

which is a symmetric, positive definite bilinear form. Such an inner product can be obtained by embedding M in some large Euclidean space $\mathbb{R}^N$ and inheriting the inner product from $\mathbb{R}^N$. For example, the two sphere inherits a Riemannian metric from $\mathbb{R}^3$ because it is a subset, $S^2 \subset \mathbb{R}^3$.

Once the inner product is known, then

$$|\mathbf{v_p}|_{\mathbf{p}} = \langle \mathbf{v_p}, \mathbf{v_p} \rangle_{\mathbf{p}}^{1/2}$$

defines a *Riemannian norm* on each tangent space. We are most often interested in the norm rather than the inner product.

Definition. A differential equation on a manifold M is given in terms of a *vector field*, i.e., a function $X : M \to TM$ such that $X(\mathbf{x}) \in T_{\mathbf{x}}M$ for all $\mathbf{x} \in M$. The vector field is C^r if the representatives of X are C^r in local coordinates. The set of C^r vector fields on M is denoted by $\mathcal{X}^r(M)$.

Let $1 \leq r < \infty$. The C^r topology on $\mathcal{X}^r(M)$ is determined by local coordinate charts. Assume M is compact. Let

$$\{\varphi_j : V_j \subset \mathbb{R}^n \to U_j \subset M\}_{j=1}^J$$

be a finite number of coordinate charts on M and $C_j \subset U_j$ be compact subsets with $\bigcup_{j=1}^J C_j = M$. If $X, Y \in \mathcal{X}^r(M)$, then the C^r *distance from X to Y* is defined to be

$$\begin{aligned} d_r(X,Y) = \sup\{&|X^j(\mathbf{x}) - Y^j(\mathbf{x})|,\ \|D^i(X^j)_{\varphi_j^{-1}(\mathbf{x})} - D^i(Y^j)_{\varphi_j^{-1}(\mathbf{x})}\| : \\ &\mathbf{x} \in C_j,\ 1 \leq j \leq J, \text{ and } 1 \leq i \leq r\}, \end{aligned}$$

where $X^j, Y^j : V_j \to \mathbb{R}^n$ are the representatives of X and Y in the local coordinate chart.

Given such C^r a vector field X on M, then

$$\dot{\mathbf{p}} = X(\mathbf{p})$$

is a *differential equation on M*. The *flow* φ^t of the vector field X satisfies

$$\frac{d}{dt}\varphi^t(\mathbf{p}) = X \circ \varphi^t(\mathbf{p}).$$

As before, a flow satisfies the group property.

For more details on the ideas of the tangent space, the derivative between tangent spaces, and Riemannian metrics see Guillemin and Pollack (1974), Hirsch (1976), Chillingworth (1976), or Lang (1967).

8.1.3 Hyperbolic Invariant Sets

We want to make more precise the concept of expanding and contracting directions at different points for a map. Let Λ be an invariant set for a diffeomorphism f. If a periodic point $\mathbf{p}$ for f is hyperbolic, there are subspaces $\mathbb{E}^u_{\mathbf{p}}$ and $\mathbb{E}^s_{\mathbf{p}}$ such that (i) $T_{\mathbf{p}}M$ is the direct sum of $\mathbb{E}^u_{\mathbf{p}}$ and $\mathbb{E}^s_{\mathbf{p}}$ (as vector spaces), (ii) $T_{\mathbf{p}}M = \mathbb{E}^u_{\mathbf{p}} \oplus \mathbb{E}^s_{\mathbf{p}}$, and (iii) $\mathbb{E}^u_{\mathbf{p}}$ is expanding and $\mathbb{E}^s_{\mathbf{p}}$ is contracting under $Df^n_{\mathbf{p}}$. An invariant set Λ is said to be hyperbolic provided that at each point $\mathbf{p}$ in Λ this same type of splitting into subspaces exists and the subspaces vary continuously as the point $\mathbf{p}$ varies. We give the formal definition of a hyperbolic invariant set in this subsection. We give examples of hyperbolic invariant sets in the rest of the chapter which should help to make the definition more understandable.

Definition. An invariant set Λ has a *hyperbolic structure* for a diffeomorphism f on M provided (i) at each point $\mathbf{p}$ in Λ the tangent space to M splits as the direct sum of $\mathbb{E}^u_{\mathbf{p}}$ and $\mathbb{E}^s_{\mathbf{p}}$, $T_{\mathbf{p}}M = \mathbb{E}^u_{\mathbf{p}} \oplus \mathbb{E}^s_{\mathbf{p}}$, (ii) the splitting is invariant under the action of the derivative map in the sense that $Df_{\mathbf{p}}(\mathbb{E}^u_{\mathbf{p}}) = \mathbb{E}^u_{f(\mathbf{p})}$ and $Df_{\mathbf{p}}(\mathbb{E}^s_{\mathbf{p}}) = \mathbb{E}^s_{f(\mathbf{p})}$, and (iii) there exist $0 < \lambda < 1$ and $C \geq 1$ independent of $\mathbf{p}$ such that for all $n \geq 0$,

$$|Df^n_{\mathbf{p}}\mathbf{v}^s| \leq C\lambda^n|\mathbf{v}^s| \quad \text{for } \mathbf{v}^s \in \mathbb{E}^s_{\mathbf{p}}, \qquad \text{and}$$
$$|Df^{-n}_{\mathbf{p}}\mathbf{v}^u| \leq C\lambda^n|\mathbf{v}^u| \quad \text{for } \mathbf{v}^u \in \mathbb{E}^u_{\mathbf{p}}.$$

If an invariant set Λ has a hyperbolic structure for f, we also say that Λ is a *hyperbolic invariant set.*

REMARK 1.6. With the assumptions above for a hyperbolic invariant set, it follows that $\mathbb{E}^u_{\mathbf{p}}$ and $\mathbb{E}^s_{\mathbf{p}}$ vary continuously with $\mathbf{p}$. In terms of cones, the continuity of the subspace $\mathbb{E}^u_{\mathbf{p}}$ can be expressed by saying that given $\epsilon > 0$, there exists a $\delta > 0$ such that if $d(\mathbf{q}, \mathbf{p}) < \delta$, then

$$\mathbb{E}^u_{\mathbf{q}} \subset C^u_{\mathbf{p}}(\epsilon) = \{\mathbf{v}^u + \mathbf{v}^s : \mathbf{v}^u \in \mathbb{E}^u_{\mathbf{p}},\ \mathbf{v}^s \in \mathbb{E}^s_{\mathbf{p}},\ \|\mathbf{v}^s\| \leq \epsilon\|\mathbf{v}^u\|\}.$$

(We have identified $\mathbb{E}^u_{\mathbf{q}}$ with the "parallel" plane in $T_{\mathbf{p}}M$.) The reader should add this condition to the definition when trying to understand its intuitive meaning. See Exercise 8.17.

REMARK 1.7. Notice that if m is a positive integer such that $\rho = C\lambda^m < 1$, then

$$|Df^m_{\mathbf{p}}\mathbf{v}^s| \leq \rho|\mathbf{v}^s| \quad \text{for } \mathbf{v}^s \in \mathbb{E}^s_{\mathbf{p}}, \qquad \text{and}$$
$$|Df^{-m}_{\mathbf{p}}\mathbf{v}^u| \leq \rho|\mathbf{v}^u| \quad \text{for } \mathbf{v}^u \in \mathbb{E}^u_{\mathbf{p}},$$

so $Df^m_{\mathbf{p}}|\mathbb{E}^s_{\mathbf{p}}$ is a contraction and $Df^m_{\mathbf{p}}|\mathbb{E}^u_{\mathbf{p}}$ is an expansion. Thus, the constant C determines the number of iterates of f which are necessary before the vectors get contracted (respectively, expanded) in the subbundle $\mathbb{E}^s_{\mathbf{p}}$ (respectively, $\mathbb{E}^u_{\mathbf{p}}$). Sometimes, an invariant set satisfying the above conditions is said to have a *uniform hyperbolic structure* because the constants λ and C, and so the number of iterates m, are independent of the point $\mathbf{p}$. By contrast, a diffeomorphism is sometimes said to be *nonuniformly hyperbolic on some set* provided it has nonzero Liapunov exponents for almost all points on this set.

REMARK 1.8. We use backward iterates to specify the unstable bundle because any vector with a nonzero component in the unstable subbundle expands exponentially under forward iteration. Therefore, forward iteration does not characterize the vectors in $\mathbb{E}^u_{\mathbf{p}}$. However, the above conditions characterize these subbundles (i.e., make them unique).

Just as for a fixed point, it is useful to change the norm so that the constant $C = 1$. In this case, the new norm $|\cdot|_{\mathbf{p}}^*$ must vary from point to point, i.e., the norm of a vector depends on the point which is the base point of the vector. In terms of this new norm, vectors are stretched or contracted by $Df_{\mathbf{p}}$, and not just by the derivative of some high iterate of f, $Df_{\mathbf{p}}^n$. This type of norm is called an *adapted norm.* The existence of an adapted norm for a hyperbolic invariant set cannot be proved using the Jordan Canonical Form, but must be proved using the averaging method given for a fixed point. See the appendix of Smale (1967) by Mather for one of the original proofs of the existence of an adapted norm.

Theorem 1.1. *Assume Λ is a hyperbolic invariant set for f with constants $0 < \lambda < 1$ and $C \geq 1$ giving the hyperbolic structure. Let λ' be any number with $0 < \lambda < \lambda' < 1$. Then there is a continuous change of norm $|\cdot|_{\mathbf{p}}^*$ for points $\mathbf{p} \in \Lambda$ such that, in terms of this norm,*

$$|Df_{\mathbf{p}}\mathbf{v}^s|_{f(\mathbf{p})}^* \leq \lambda' |\mathbf{v}^s|_{\mathbf{p}}^* \quad \text{for } \mathbf{v}^s \in \mathbb{E}_{\mathbf{p}}^s, \quad \text{and}$$
$$|Df_{\mathbf{p}}^{-1}\mathbf{v}^u|_{f^{-1}(\mathbf{p})}^* \leq \lambda' |\mathbf{v}^u|_{\mathbf{p}}^* \quad \text{for } \mathbf{v}^u \in \mathbb{E}_{\mathbf{p}}^u.$$

PROOF. As stated above, we must average the norm and cannot use some kind of Jordan canonical form.

For $\mathbf{p} \in \Lambda$, any vector $\mathbf{v} \in T_{\mathbf{p}}M$ can be decomposed into components $\mathbf{v}^s \in \mathbb{E}_{\mathbf{p}}^s$ and $\mathbf{v}^u \in \mathbb{E}_{\mathbf{p}}^u$, $\mathbf{v} = \mathbf{v}^s + \mathbf{v}^u$. Then, we can let

$$|\mathbf{v}|_{\mathbf{p}}^* = \max\{|\mathbf{v}^s|_{\mathbf{p}}^*, |\mathbf{v}^u|_{\mathbf{p}}^*\}.$$

Thus, it is enough to define the norm on each of the subspaces. (The maximum of two norms on subspaces defines a norm.) Let n_0 be a positive integer such that $C(\lambda/\lambda')^{n_0} < 1$, i.e., $C\lambda^{n_0} < (\lambda')^{n_0}$. For $\mathbf{v}^s \in \mathbb{E}_{\mathbf{p}}^s$, define

$$|\mathbf{v}^s|_{\mathbf{p}}^* = \sum_{j=0}^{n_0-1} (\lambda')^{-j} |D(f^j)_{\mathbf{p}}\mathbf{v}^s|.$$

We show this norm works on $\mathbb{E}_{\mathbf{p}}^s$:

$$\begin{aligned}
|Df_{\mathbf{p}}\mathbf{v}^s|_{f(\mathbf{p})}^* &= \sum_{j=0}^{n_0-1} (\lambda')^{-j} |D(f^{j+1})_{\mathbf{p}}\mathbf{v}^s| \\
&= \Big(\sum_{j=0}^{n_0-2} (\lambda')^{-j} |D(f^{j+1})_{\mathbf{p}}\mathbf{v}^s|\Big) + (\lambda')^{-n_0+1} |D(f^{n_0})_{\mathbf{p}}\mathbf{v}^s| \\
&\leq \lambda' \Big(\sum_{j=1}^{n_0-1} (\lambda')^{-j} |D(f^j)_{\mathbf{p}}\mathbf{v}^s|\Big) + (\lambda')^{-n_0+1} C\lambda^{n_0} |\mathbf{v}^s| \\
&\leq \lambda' \sum_{j=0}^{n_0-1} (\lambda')^{-j} |D(f^j)_{\mathbf{p}}\mathbf{v}^s|,
\end{aligned}$$

since $C(\lambda/\lambda')^{n_0} < 1$.

In the same way, for $\mathbf{v}^u \in \mathbb{E}_{\mathbf{p}}^u$, define

$$|\mathbf{v}^u|_{\mathbf{p}}^* = \sum_{j=0}^{n_0-1} (\lambda')^{-j} |D(f^{-j})_{\mathbf{p}}\mathbf{v}^u|.$$

The reader can check that a similar calculation shows that

$$|D(f^{-1})_{\mathbf{p}}\mathbf{v}^u|^*_{f^{-1}(\mathbf{p})} \le \lambda' |\mathbf{v}^u|^*_{\mathbf{p}}.$$

□

If a set Λ is a hyperbolic invariant set, then it can be proved that there are stable and unstable manifolds through each point in Λ. The following theorem states this result more precisely.

Theorem 1.2 (Stable Manifold Theorem for a Hyperbolic Set). *Let $f : M \to M$ be a C^k diffeomorphism. Let Λ be a hyperbolic invariant set for f with hyperbolic constants $0 < \lambda < 1$ and $C \ge 1$. Then there is an $\epsilon > 0$ such that for each $\mathbf{p} \in \Lambda$, there are two C^k embedded disks $W^s_\epsilon(\mathbf{p}, f)$ and $W^u_\epsilon(\mathbf{p}, f)$ which are tangent to $\mathbb{E}^s_{\mathbf{p}}$ and $\mathbb{E}^u_{\mathbf{p}}$, respectively. In order to consider these disks as graphs of functions, we identify a neighborhood of each point $\mathbf{p}$ with $\mathbb{E}^u_{\mathbf{p}}(\epsilon) \times \mathbb{E}^s_{\mathbf{p}}(\epsilon)$. (On a manifold, this identification can be realized by either local coordinates or the exponential map.) Using this identification, $W^s_\epsilon(\mathbf{p}, f)$ is the graph of a C^k function $\sigma^s_{\mathbf{p}} : \mathbb{E}^s_{\mathbf{p}}(\epsilon) \to \mathbb{E}^u_{\mathbf{p}}(\epsilon)$ with $\sigma^s_{\mathbf{p}}(\mathbf{0}_{\mathbf{p}}) = \mathbf{0}_{\mathbf{p}}$ and $D(\sigma^s_{\mathbf{p}})_{\mathbf{0}} = \mathbf{0}$:*

$$W^s_\epsilon(\mathbf{p}, f) = \{(\sigma^s_{\mathbf{p}}(\mathbf{y}), \mathbf{y}) : \mathbf{y} \in \mathbb{E}^s_{\mathbf{p}}(\epsilon)\}.$$

Also, the function $\sigma^s_{\mathbf{p}}$ and its first k derivatives vary continuously as $\mathbf{p}$ varies. Similarly, there is a C^k function $\sigma^u_{\mathbf{p}} : \mathbb{E}^u_{\mathbf{p}}(\epsilon) \to \mathbb{E}^s_{\mathbf{p}}(\epsilon)$ with $\sigma^u_{\mathbf{p}}(\mathbf{0}_{\mathbf{p}}) = \mathbf{0}_{\mathbf{p}}$ and $D(\sigma^u_{\mathbf{p}})_{\mathbf{0}} = \mathbf{0}$ and with the function $\sigma^u_{\mathbf{p}}$ and its first k derivatives varying continuously as $\mathbf{p}$ varies such that

$$W^u_\epsilon(\mathbf{p}, f) = \{(\mathbf{x}, \sigma^u_{\mathbf{p}}(\mathbf{x})) : \mathbf{x} \in \mathbb{E}^u_{\mathbf{p}}(\epsilon)\}.$$

Moreover, identifying a neighborhood of $\mathbf{p}$ with $\mathbb{E}^u_{\mathbf{p}}(\epsilon) \times \mathbb{E}^s_{\mathbf{p}}(\epsilon)$ and taking $\lambda < \lambda' < 1$, for $\epsilon > 0$ small enough

$$\begin{aligned} W^s_\epsilon(\mathbf{p}, f) &= \{\mathbf{q} \in \mathbb{E}^u_{\mathbf{p}}(\epsilon) \times \mathbb{E}^s_{\mathbf{p}}(\epsilon) : f^j(\mathbf{q}) \in \mathbb{E}^u_{f^j(\mathbf{p})}(\epsilon) \times \mathbb{E}^s_{f^j(\mathbf{p})}(\epsilon) \text{ for } j \ge 0\} \\ &= \{\mathbf{q} \in \mathbb{E}^u_{\mathbf{p}}(\epsilon) \times \mathbb{E}^s_{\mathbf{p}}(\epsilon) : f^j(\mathbf{q}) \in \mathbb{E}^u_{f^j(\mathbf{p})}(\epsilon) \times \mathbb{E}^s_{f^j(\mathbf{p})}(\epsilon) \text{ for } j \ge 0 \text{ and} \\ &\qquad d(f^j(\mathbf{q}), f^j(\mathbf{p})) \le C(\lambda')^j d(\mathbf{q}, \mathbf{p}) \text{ for all } j \ge 0\}. \end{aligned}$$

Similarly,

$$\begin{aligned} W^u_\epsilon(\mathbf{p}, f) &= \{\mathbf{q} \in \mathbb{E}^u_{\mathbf{p}}(\epsilon) \times \mathbb{E}^s_{\mathbf{p}}(\epsilon) : \\ &\qquad f^{-j}(\mathbf{q}) \in \mathbb{E}^u_{f^{-j}(\mathbf{p})}(\epsilon) \times \mathbb{E}^s_{f^{-j}(\mathbf{p})}(\epsilon) \text{ for } j \ge 0\} \\ &= \{\mathbf{q} \in \mathbb{E}^u_{\mathbf{p}}(\epsilon) \times \mathbb{E}^s_{\mathbf{p}}(\epsilon) : \\ &\qquad f^{-j}(\mathbf{q}) \in \mathbb{E}^u_{f^{-j}(\mathbf{p})}(\epsilon) \times \mathbb{E}^s_{f^{-j}(\mathbf{p})}(\epsilon) \text{ for } j \ge 0 \text{ and} \\ &\qquad d(f^{-j}(\mathbf{q}), f^{-j}(\mathbf{p})) \le C(\lambda')^j d(\mathbf{q}, \mathbf{p}) \text{ for all } j \ge 0\}. \end{aligned}$$

The proof is very similar to that for a hyperbolic fixed point. We delay the proof until Section 10.2.

After the local stable and unstable manifolds are obtained by the above theorem, the global manifolds are determined as follows:

$$\begin{aligned} W^s(\mathbf{p}, f) &\equiv \{\mathbf{q} \in M : d(f^j(\mathbf{q}), f^j(\mathbf{p})) \to 0 \text{ as } j \to \infty\} \\ &= \bigcup_{n \ge 0} f^{-n}(W^s_\epsilon(f^n(\mathbf{p}), f)) \qquad \text{and} \\ W^u(\mathbf{p}, f) &\equiv \{\mathbf{q} \in M : d(f^{-j}(\mathbf{q}), f^{-j}(\mathbf{p})) \to 0 \text{ as } j \to \infty\} \\ &= \bigcup_{n \ge 0} f^{n}(W^u_\epsilon(f^{-n}(\mathbf{p}), f)). \end{aligned}$$

An important fact about the stable (respectively, unstable) manifold of each point is that it is an immersed copy of the linear spaces $\mathbb{E}^s_{\mathbf{p}}$ (respectively, $\mathbb{E}^u_{\mathbf{p}}$). This means that the map $\sigma^s : \mathbb{E}^s_{\mathbf{p}} \to M$ (respectively, $\sigma^u : \mathbb{E}^u_{\mathbf{p}} \to M$) whose image equals $W^s(\mathbf{p}, f)$ (respectively, $W^u(\mathbf{p}, f)$) is one to one but need not have a continuous inverse. The fact that the stable and unstable manifolds of points are immersed copy of the linear spaces implies that they cannot be circles or cylinders.

Proposition 1.3. *Let Λ be a hyperbolic invariant set for a diffeomorphism f. Then, for each $\mathbf{p} \in \Lambda$, $W^s(\mathbf{p}, f)$ is an immersed copy of $\mathbb{E}^s_{\mathbf{p}}$, and $W^u(\mathbf{p}, f)$ is an immersed copy of $\mathbb{E}^s_{\mathbf{p}}$.*

PROOF. By the Stable Manifold Theorem, each $W^s_\epsilon(\mathbf{p}, f)$ is the embedded image of the closed disk $\mathbb{E}^s_{\mathbf{p}}(\epsilon)$ by a map $\sigma^s_{\mathbf{p}}$. Because f is a diffeomorphism, $f^j(W^s_\epsilon(\mathbf{p}, f))$ is thus the embedded image of the closed disk $\mathbb{E}^s_{\mathbf{p}}(\epsilon)$ by the map $\sigma^s_{\mathbf{p},j} = f^j \circ \sigma^s_{\mathbf{p}}$. Thus, $W^s(\mathbf{p}, f)$ is the union of these sets, each of which is an embedded copy of a disk, and so $W^s(\mathbf{p}, f)$ is the immersed copy of the linear subspace. The proof for the unstable manifold is similar. □

In the following sections, we give examples of hyperbolic invariant sets and determine their stable and unstable manifolds. These examples should make both the definition and the meaning of the theorem more understandable.

Hyperbolic Structure for Flows

We end the section by mentioning the changes needed in the definitions for a flow or differential equation.

Definition. An invariant set Λ for the flow φ^t has a *hyperbolic structure*, or Λ is a *hyperbolic invariant set*, provided (i) at each point $\mathbf{p}$ in Λ, the tangent space to M splits as the direct sum of $\mathbb{E}^u_{\mathbf{p}}$, $\mathbb{E}^s_{\mathbf{p}}$, and span$(X(\mathbf{p}))$,

$$T_{\mathbf{p}}M = \mathbb{E}^u_{\mathbf{p}} \oplus \mathbb{E}^s_{\mathbf{p}} \oplus \operatorname{span}(X(\mathbf{p})),$$

(ii) the splitting is invariant under the action of the derivative in the sense that

$$\begin{aligned} D(\varphi^t)_{\mathbf{p}}\mathbb{E}^u_{\mathbf{p}} &= \mathbb{E}^u_{\varphi^t(\mathbf{p})}, \\ D(\varphi^t)_{\mathbf{p}}\mathbb{E}^s_{\mathbf{p}} &= \mathbb{E}^s_{\varphi^t(\mathbf{p})}, \qquad \text{and} \\ D(\varphi^t)_{\mathbf{p}}X(\mathbf{p}) &= X(\varphi^t(\mathbf{p})), \end{aligned}$$

(iii) $\mathbb{E}^u_{\mathbf{p}}$ and $\mathbb{E}^s_{\mathbf{p}}$ vary continuously with $\mathbf{p}$, and (iv) there exist $\mu > 0$ and $C \geq 1$ such that for $t \geq 0$,

$$\begin{aligned} |D\varphi^t_{\mathbf{p}}\mathbf{v}^s| &\leq Ce^{-\mu t}|\mathbf{v}^s| \quad \text{for } \mathbf{v}^s \in \mathbb{E}^s_{\mathbf{p}}, \quad \text{and} \\ |D\varphi^{-t}_{\mathbf{p}}\mathbf{v}^u| &\leq Ce^{-\mu t}|\mathbf{v}^u| \quad \text{for } \mathbf{v}^u \in \mathbb{E}^u_{\mathbf{p}}. \end{aligned}$$

REMARK 1.9. If Λ is a hyperbolic invariant set for φ^t and is a single chain component for φ^t, then either Λ is a single fixed point or Λ does not contain any fixed points (because of the assumption that the splitting varies continuously). In Section 8.11, we consider the Lorenz equations which have an invariant set which contains a fixed point together with other nonfixed points. This set cannot have a standard hyperbolic structure. In that situation, we discuss the possibility of a "generalized hyperbolic structure."

8.2 Transitivity Theorems

In the following sections, we often want to prove that a map is ***(topologically) transitive*** on a set X, i.e., the (forward) orbit of some point $\mathbf{p}$ is dense in X. When considering the horseshoe for the quadratic map on the line, we were able to verify this condition by means of the conjugacy with the shift map and then constructing an explicit point with a dense orbit for the (one-sided) shift map. We will also be able to do this in the first set of examples below which are horseshoes for a diffeomorphism in two dimensions using the conjugacy to the two-sided shift map. In other cases, this type of approach is not possible. The following theorem of Birkhoff shows that it is enough to show that the orbit of any open set intersects every other open set. In fact, it follows from this hypothesis that there are many points (a residual set in the sense of Baire category) with dense orbits. Remember that a set A is residual in a space X if there is a countable number of dense open sets $\{U_j\}_{j=1}^{\infty}$ in X such that $A = \bigcap_{j=1}^{\infty} U_j$. The Baire category theorem says that any residual subset of a complete metric space is dense.

Theorem 2.1 (Birkhoff Transitivity Theorem). *Let X be a complete metric space with countable basis and $f : X \to X$ a continuous map.*

(a) Assume that for every open set U of X, $\mathcal{O}^-(U) = \bigcup_{n\le 0} f^n(U)$ is dense in X. Then, there is a residual set (large in the sense of Baire category) $\mathcal{R}^+$ such that for every $\mathbf{p} \in \mathcal{R}^+$, the forward orbit of $\mathbf{p}$, $\mathcal{O}^+(\mathbf{p})$, is dense in X.

(b) Assume that f is a homeomorphism, and that for every open set U of X, $\mathcal{O}^+(U) = \bigcup_{n\ge 0} f^n(U)$ is dense in X. Then, there is a residual set $\mathcal{R}^-$ such that for every $\mathbf{p} \in \mathcal{R}^-$, the backward orbit of $\mathbf{p}$, $\mathcal{O}^-(\mathbf{p})$, is dense in X.

(c) Combining the first two parts, if f is a homeomorphism and every open set U has both $\mathcal{O}^+(U)$ and $\mathcal{O}^-(U)$ dense in X, then there is a residual subset $\mathcal{R} \subset X$ such that any $\mathbf{p} \in \mathcal{R}$ has both $\mathcal{O}^+(\mathbf{p})$ and $\mathcal{O}^-(\mathbf{p})$ dense in X.

PROOF. Let $\{V_j : j \in \mathbb{Z}\}$ be a countable basis of X. Then,

$$\mathcal{R}^+ = \bigcap\{\mathcal{O}^-(V_j) : j \in \mathbb{Z}\}$$

is the intersection of dense open sets and so is residual. Let $\mathbf{p} \in \mathcal{R}^+$. Then, $\mathbf{p} \in \mathcal{O}^-(V_j)$ for all j, and so $\mathcal{O}^+(\mathbf{p}) \cap V_j \neq \emptyset$. This proves that $\mathcal{O}^+(\mathbf{p})$ is dense in X, and so part (a). In the proof of (b), f is assumed to be a homeomorphism so the backward orbit of a point is well defined. The rest of the proof is similar. Part (c) is a combination of parts (a) and (b). □

Ergodicity and Birkhoff Ergodic Theorem

In connection with a few concepts we refer to invariant measures: Liapunov exponents (Sections 3.6 and 9.2), measure theoretic entropy (Section 9.1), Sinai-Ruelle-Bowen measure (Section 9.3), and Hausdorff dimension (Section 9.4). In some of these situations, we refer to a system being ergodic. Because the concept of ergodicity is measure theoretic type of transitivity, and so relates to the Birkhoff Transitivity Theorem, we introduce it in this section.

A measure μ on a space X is called a *Borel measure* provided it is a measure for the sigma algebra of Borel sets (generated by open sets). A measurable set A in X is said to be of *full measure in* X provided $\mu(X \setminus A) = 0$. If the measure of X is finite, a set A is of full measure in X provided $\mu(A) = \mu(X)$. For a measure μ on X, the *support of the measure*, $\text{supp}(\mu)$, is the smallest closed set with full measure, or

$$\text{supp}(\mu) = \bigcap\{C : C \text{ is closed, and } \mu(X \setminus C) = 0\}.$$

A measure μ is *invariant* for a map $f : X \to X$ provided $\mu(f^{-1}(A)) = \mu(A)$ for all measurable sets A. If μ is an invariant measure for f, f is also said to be a *measure preserving transformation for* μ. The flow of a vector field with zero divergence preserves Lebesgue measure, so this gives one example of a measure preserving system. (Geodesic flows are such systems. See Katok and Hasselblatt (1994).)

Finally, a map $f : X \to X$ is called *ergodic* with respect to invariant measure μ provided $\mu(X \setminus A) = 0$ for any measurable invariant set A for f with $\mu(A) > 0$. Thus, for an ergodic map, all invariant measurable sets either have zero measure or full measure in X. If f is ergodic with respect to the measure μ, we also say that μ is an *ergodic measure for* f.

Exercises 8.9–11 make some connections between a measure preserving map and recurrence. In particular, Exercise 8.10 states that if f is a measure preserving homeomorphism for an invariant measure μ, then the set $\{\mathbf{x} \in X : \mathbf{x} \in \alpha(\mathbf{x}) \text{ and } \mathbf{x} \in \omega(\mathbf{x})\}$ has full measure. Thus, for such a homeomorphism, μ-almost every point is recurrent.

In the discussion in Section 3.6, the Birkhoff Ergodic Theorem is used to show that if $f : \mathbb{R} \to \mathbb{R}$ has an invariant measure μ with compact support, then the Liapunov exponents exists μ-almost all points. We state this theorem next but leave the proof to the reference.

Theorem 2.2 (Birkhoff Ergodic Theorem). *Assume $f : X \to X$ is a measure preserving transformation for the measure μ. Assume $g : X \to \mathbb{R}$ is a μ-integrable function. Then, $\lim_{n\to\infty}(1/n)\sum_{j=0}^{n-1} g \circ f^j(\mathbf{x})$ converges μ-almost everywhere to an integrable function g^*. Also, g^* is f invariant wherever it is defined, i.e., $g^* \circ f(\mathbf{x}) = g^*(\mathbf{x})$ for μ-almost all $\mathbf{x}$. Also, (i) if $\mu(X) < \infty$, then $\int_X g^*(\mathbf{x})\, d\mu(\mathbf{x}) = \int_X g(\mathbf{x})\, d\mu(\mathbf{x})$, and (ii) if μ is an ergodic measure for f, then g^* is a constant μ-almost everywhere.*

REMARK 2.1. Consider the case when μ is an ergodic measure for f. The value $\int_X g(\mathbf{x})\, d\mu(\mathbf{x})$ is the space average of the function g; the value $g^*(\mathbf{p})$ is the time average of g along the orbit of $\mathbf{p}$. Thus, for any integrable function (e.g., the characteristic function of an open set), the time average along almost all orbits equals the space average of the function. This fact indicates that almost all orbits for an ergodic measure are dense in the support of the measure.

The following theorem gives a result for ergodic maps which is similar to the Birkhoff Transitivity Theorem.

Theorem 2.3. *Let X be a complete metric space with countable basis. Assume $f : X \to X$ is an ergodic map with respect to the measure μ. Assume that μ is positive on open sets and $\mu(X) < \infty$. Then, there is a set $\mathcal{R}$ of full measure such that the forward orbit of every point in $\mathcal{R}$ is dense in X.*

PROOF. Let $\{V_j : j \in \mathbb{Z}\}$ be a countable basis of X. For each V_j, $\mathcal{O}^-(V_j)$ is a measurable invariant set of positive measure, so it has full measure, i.e., $X \setminus \mathcal{O}^-(V_j)$ has zero measure. Therefore,

$$\bigcup \{X \setminus \mathcal{O}^-(V_j) : j \in \mathbb{Z}\}$$

has zero measure and

$$\mathcal{R}^+ = \bigcap \{\mathcal{O}^-(V_j) : j \in \mathbb{Z}\}$$

has full measure. For any $\mathbf{p} \in \mathcal{R}^+$, $\mathcal{O}^+(\mathbf{p})$ is dense in X, just as in the proof of the Birkhoff Transitivity Theorem, □

For details and an introduction to ergodic theory, see Walters (1982). For further discussion of ergodic theory in the context of smooth dynamical systems, see Katok and Hasselblatt (1994).

8.3 Two-Sided Shift Spaces

In Chapters II and III, symbolic dynamics is used to specify the forward itinerary for the orbit of a noninvertible map, and so one-sided shifts and subshifts are used. In this chapter, symbolic dynamics is used to specify both the forward and backward itinerary for the orbit of an invertible map, and so two-sided shift spaces are introduced. A point in a two-sided shift space is given by $\mathbf{s} = (\ldots, s_{-1}, s_0, s_1, \ldots)$, which includes a symbol s_j for all $j \in \mathbb{Z}$. If Section 3.2 has not been read previously, it should be read at this time.

Definition. Let n be an integer that is greater than 1. Let

$$\begin{aligned}\Sigma_n &= \{1, \ldots, n\}^{\mathbb{Z}} \\ &= \{\mathbf{s} = (\ldots s_{-1}, s_0, s_1, \ldots) : s_j \in \{1, \ldots, n\} \text{ for all } j \in \mathbb{Z}\}.\end{aligned}$$

We define the distance d on Σ_n:

$$d(\mathbf{s}, \mathbf{t}) = \sum_{j=-\infty}^{\infty} \frac{\delta(s_j, t_j)}{4^{|j|}}.$$

where

$$\delta(a, b) = \begin{cases} 0 & \text{if } a = b \\ 1 & \text{if } a \neq b. \end{cases}$$

As in the case of the one-sided shift, this makes Σ_n into a complete metric space. Exercise 8.12 proves that making the factor in the denominator greater than 3 makes the cylinder sets into open balls in terms of the metric. There is also a shift map σ defined on Σ_n by $\sigma(\mathbf{s}) = \mathbf{t}$ where $t_j = s_{j+1}$. The space Σ_n together with the map σ is called the *full two-sided n-shift*, or sometimes simply the *n-shift*.

The shift space is called two-sided because the index is both positive and negative. It is called the *full* n-shift because all possible sequences are allowed on n symbols. Notice that this σ is one to one on this Σ_n and so is invertible, while the one-sided shift map is n to one.

As for one-sided shifts, we can also define subshifts of finite type. If $A = (a_{ij})$ is an $n \times n$ matrix with each $a_{ij} \in \{0, 1\}$, then $\Sigma_A \subset \Sigma_n$ is the subspace of all points $\mathbf{s} = (s_j)$ for which $a_{s_j, s_{j+1}} = 1$ for all j. The shift map restricted to Σ_A is designated by $\sigma_A = \sigma|\Sigma_A$. The map σ_A on Σ_A is called a *subshift of finite type.* As for one-sided shifts, the number of fixed points of σ_A^k is given by the trace of A^k, $\#\,\mathrm{Fix}(\sigma_A^k) = \mathrm{tr}(A^k)$. Also, σ_A has a dense forward orbit in Σ_A if and only if A is irreducible (where irreducible is defined as before).

8.3.1 Subshifts for Nonnegative Matrices

In this subsection we extend the definition of a subshift of finite type by associating a shift space with a matrix with nonnegative integer entries. For $F_\mu = \mu x(1-x)$ with $\mu > 4$, there are two subintervals $I_1, I_2 \subset [0, 1]$ such that $F_\mu(I_\sigma) \supset I_1 \cup I_2$. Using the itinerary map for these intervals, we get sequences in Σ_2, i.e., for the transition matrix $\begin{pmatrix} 1 & 1 \\ 1 & 1 \end{pmatrix}$. However, the one interval $I = [0, 1]$ crosses itself twice, so we could asssocate the matrix (2). There are several advantages to this extension. First, the matrix which gives the symbolic dynamics can be made to have smaller size. In the above example, we get the 1×1 matrix (2) rather than the 2×2 matrix $\begin{pmatrix} 1 & 1 \\ 1 & 1 \end{pmatrix}$. Second, if A is a transition matrix, then we can from the subshift of finite type for A^k for any positive power. For example, if $A = \begin{pmatrix} 1 & 1 \\ 1 & 1 \end{pmatrix}$, then $A^2 = \begin{pmatrix} 2 & 2 \\ 2 & 2 \end{pmatrix}$ also has a subshift of finite type associated to it. Finally, when we consider Markov partitions for hyperbolic toral automorphisms later in the chapter, the matrix for the subshift of finite type associated to a hyperbolic toral automorphism can be taken to be the same as the matrix which induces the hyperbolic toral automorphism, even when it has entries which are greater than 1.

We call A an *adjacency matrix* provided $A = (a_{ij})$ is an $n \times n$ matrix with entries $a_{ij} \in \mathbb{N}$. For example, (2) and $\begin{pmatrix} 2 & 1 \\ 1 & 1 \end{pmatrix}$ are adjacency matrices. We call A a *transition matrix* provided $A = (a_{ij})$ is an $n \times n$ matrix with entries $a_{ij} \in \{0, 1\}$.

Now, we turn to understanding how an adjacency matrix corresponds to a subshift. The previous subshift constructed for a matrix with entries of 0's and 1's is called the *vertex subshift*. For an adjacency matrix, we construct an edge subshift. Let $S = \{V_1, V_2, \dots, V_n\}$ be a set with n elements. We use the adjacency matrix A to form an oriented transition graph G_A with vertices the elements of S. If the entry $a_{ij} > 0$, then we draw a_{ij} directed edges from vertex V_i to vertex V_j. For example, if $a_{ij} = 2$, then there are two edges from V_i to V_j. Figure 3.1 gives the transition graphs for the matrices (2) and $\begin{pmatrix} 2 & 1 \\ 1 & 1 \end{pmatrix}$. Thus, there are $N = \sum_{i,j} a_{ij}$ edges in the transition graph G_A. Notice that the adjacency matrix tells which vertices are adjacent to each other in the sense that there is a path in the transition graph directly from the first vertex to the second. Let $\mathcal{E} = \{E_1, E_2, \dots, E_N\}$ be the set of all edges. If A is a $\{0,1\}$-matrix, then we get the same transition graph which we have considered before with either no edge or one edge from vertex V_i to vertex V_j.

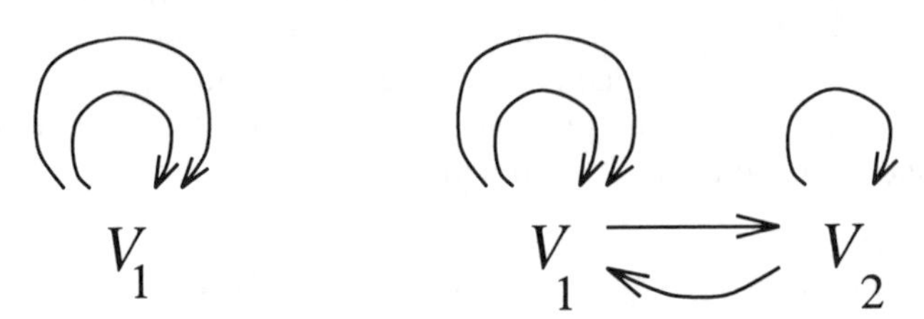

FIGURE 3.1. Graphs for the Matrices (2) and $\begin{pmatrix} 2 & 1 \\ 1 & 1 \end{pmatrix}$.

Next, we form the transition matrix $T = (t_{ij})$ on $\mathcal{E}$. This is an $N \times N$ matrix whose entries are 0's or 1's. For the edge $E_j \in \mathcal{E}$, let $b(E_j) \in S$ be the beginning vertex of edge E_j and $e(E_j) \in S$ be the end vertex, i.e., E_j is an edge from $b(E_j)$ to $e(E_j)$. The entries of T are defined as follows:

$$t_{ij} = \begin{cases} 1 & \text{if } e(E_i) = b(E_j) \\ 0 & \text{if } e(E_i) \neq b(E_j). \end{cases}$$

Thus, the transition on $\mathcal{E}$ from E_i to E_j is allowable provided the end of the edge E_i is at the vertex which is the beginning of the edge E_j. Using the transition matrix T, we can form the vertex subshift Σ_T. This process takes an adjacency matrix A and constructs a subshift $\sigma_T : \Sigma_T \to \Sigma_T$, called the *edge subshift for A*. (This induced subshift on N symbols is often labeled as $\sigma_A : \Sigma_A \to \Sigma_A$, but we do not do this. We always label the subshift by the transition matrix which induces it.)

An allowable sequence $\mathbf{s} \in \Sigma_T$ can be visualized as a curve in the transition graph G_A. An *allowable curve* in the transition graph G_A is a continuous function $\gamma : \mathbb{R} \to G_A$ such that for each i, $\gamma(i)$ is a vertex and $\gamma([i, i+1])$ is an edge from $\gamma(i)$ to $\gamma(i+1)$. Thus, an allowable sequence $\mathbf{s} \in \Sigma_T$ can be visualized as the allowable curve γ in G_A with $\gamma(i) = s_i \in \mathcal{E}$.

Notice for the adjacency matrix $A = (2)$, its transition matrix is $T = \begin{pmatrix} 1 & 1 \\ 1 & 1 \end{pmatrix}$. (Either of the two edges can follow each other.) Therefore, the 1×1 matrix (2) corresponds to the full two-shift. In the same way, the adjacency matrix $A = (n)$ corresponds to the full n-shift.

One case that is interesting is when A is already has all its entries $a_{ij} \in \{0, 1\}$. Let T be the transition matrix formed above. (The matrix T is usually a different size than A.) We leave to Exercise 8.16 to show that vertex subshift $\sigma_A : \Sigma_A \to \Sigma_A$ is topologically conjugate to the edge subshift formed from A, $\sigma_T : \Sigma_T \to \Sigma_T$. This result shows that the edge subshift is a generalization of the vertex subshift for matrices with nonnegative integer entries.

The adjacency matrix A is called *irreducible* provided that for each pair $1 \leq i, j \leq n$, there is a positive integer $k = k(i, j)$ such that $(A^k)_{ij} > 0$. (This is really the same definition as for a $\{0, 1\}$-matrix.) We leave to Exercise 8.15 to verify that if A is an irreducible adjacency matrix and T is its induced transition matrix, then T is irreducible. Thus, if A is irreducible, then σ_T is topologically transitive.

For further development subshifts of finite type, see Boyle (1993), Lind and Marcus (1995), and Franks (1982).

8.4 Geometric Horseshoe

In this section, we give an example of a diffeomorphism $f : S^2 \to S^2$ (or from $\mathbb{R}^2$ to itself) that has an invariant set which is a Cantor set. This example is closely related to the map $f_\mu(x) = \mu x(1 - x)$ on $\mathbb{R}$ for $\mu > 4$ discussed in Chapter II. It was introduced by Smale and is one of the important early examples with complicated (chaotic) dynamics on an invariant set. It is called the *Smale horseshoe* or *geometric model horseshoe.* It has been at the heart of much of modern Dynamical Systems. See Smale (1965, 1967). In the next subsection, we show how a horseshoe arises for a polynomial map, the Hénon map. The horseshoe also leads to a better understanding of the dynamics of points near a transverse homoclinic point, as we discuss in the second subsection below. Finally, we show how transverse homoclinic points, and so horseshoes, can be proven to exist for small parameter, time periodic perturbations of differential equations (Melnikov method).

Let $S = [0, 1] \times [0, 1]$ be the unit square. Let H_j for $j = 1, 2$ be two disjoint horizontal substrips, $H_j = \{(x, y) : 0 \leq x \leq 1, y_1^j \leq y \leq y_2^j\}$ with $0 \leq y_1^1 < y_2^1 < y_1^2 < y_2^2 \leq 1$. Similarly, let V_j for $j = 1, 2$ be two disjoint vertical substrips, $V_j = \{(x, y) : x_1^j \leq x \leq x_2^j, 0 \leq y \leq 1\}$ with $0 \leq x_1^1 < x_2^1 < x_1^2 < x_2^2 \leq 1$. We assume that f is a diffeomorphism such that $f(H_j) = V_j$ for $j = 1, 2$, $S \cap F^{-1}(S) = H_1 \cup H_2$, and for $\mathbf{p} \in H_1 \cup H_2$

$$Df_{\mathbf{p}} = \begin{pmatrix} a_{\mathbf{p}} & 0 \\ 0 & b_{\mathbf{p}} \end{pmatrix}$$

with $|a_{\mathbf{p}}| = \mu < 1/2$ and $|b_{\mathbf{p}}| = \lambda > 2$. See Figure 4.1.

This map f can be extended to all of S^2 as follows. Let A be the semidisk of radius $\frac{1}{2}$ on the bottom of S; B be the semidisk of radius $\frac{1}{2}$ on the top of S; and $N = S \cup A \cup B$ be the topological disk which is the union of these three regions. Let G be the horizontal

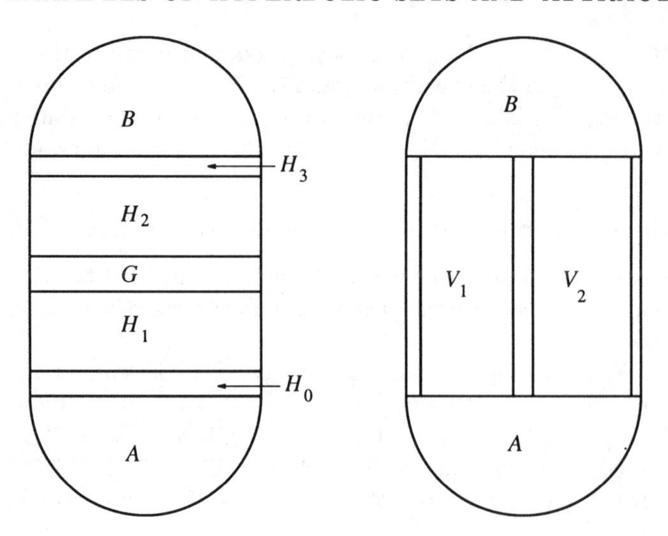

FIGURE 4.1. Horizontal and Vertical Strips in S

gap between H_1 and H_2, H_3 be the top horizontal strip in S above H_2, and H_0 be the bottom horizontal strip in S below H_1. We extend f so that $f(G) \subset B$ and the image arcs from the top of V_1 to the top of V_2. This part of the map can be taken so that $\left|\frac{\partial f}{\partial x}(\mathbf{q})\right| = \mu$ for $\mathbf{q} \in G$. See Figure 4.2. For the extension, the image of $H_0 \cup H_3$ is contained in A; the first coordinate function of f on $H_0 \cup H_3$ is a contraction by a factor of μ; the second coordinate function of f on $H_0 \cup H_3$ changes from an expansion by λ on the boundary with $H_1 \cup H_2$ to a contraction at the boundary with $A \cup B$. We take f so that $f(A) \subset A$, and that f is a contraction on A, so f has a unique fixed point $\mathbf{p}_0 \in A$. Also, $f(B) \subset A$. Thus, f maps the topological disk N into itself. The map can be extended to all of $\mathbb{R}^2$ so all other points enter this topological disk under forward iteration. Finally, f can be extended to S^2 so that it takes the point at infinity, p_∞, to itself, and p_∞ is a source for f on S^2.

This map f restricted to N can be thought of as the composition of two maps. The first map, L, takes the region and stretches it out to be over twice as tall and less than half as wide, $L(x,y) = (\mu x, \lambda y)$. The second map g takes this longer and thinner rectangle and bends it in the middle and puts it down so it crosses S twice. The map g must change the image near the ends so that $f = g \circ L$ is a contraction on A and $f(B) \subset A$. The image $f(N)$ is drawn in Figure 4.2 together with the region N itself. Since $f(N) \subset N$, it follows that $f^2(N) \subset f(N) \subset N$. Notice that $L \circ f(N)$ is a longer and thinner version of $f(N)$ with two vertical strips. Then, $f^2(N) = g \circ L \circ f(N)$ is bent again in the middle and has its image inside $f(N)$. Thus, the part of the image in S, $f^2(N) \cap S$, is four vertical strips of width μ^2. See Figure 4.2.

Next, we consider the chain recurrent set of f. If $\mathbf{q} \in A \cup B$, then $\omega(\mathbf{q}) = \{\mathbf{p}_0\}$, so $\mathcal{R}(f) \cap (A \cup B) = \{\mathbf{p}_0\}$. If $\mathbf{q} \in S^2 \setminus N$, then $\alpha(\mathbf{q}) = \{\mathbf{p}_\infty\}$, so $\mathcal{R}(f) \cap (S^2 \setminus N) = \{\mathbf{p}_\infty\}$. Finally, if $\mathbf{q} \in S \cap \mathcal{R}(f)$, then $f^j(\mathbf{q})$ must be in S for all $j \in \mathbb{Z}$ (or else it wanders forward to $\mathbf{p}_0$ or backward to $\mathbf{p}_\infty$). Combining,

$$\mathcal{R}(f) \subset \Lambda \cup \{\mathbf{p}_0\} \cup \{\mathbf{p}_\infty\}$$

where

$$\Lambda = \bigcap_{j \in \mathbb{Z}} f^j(S).$$

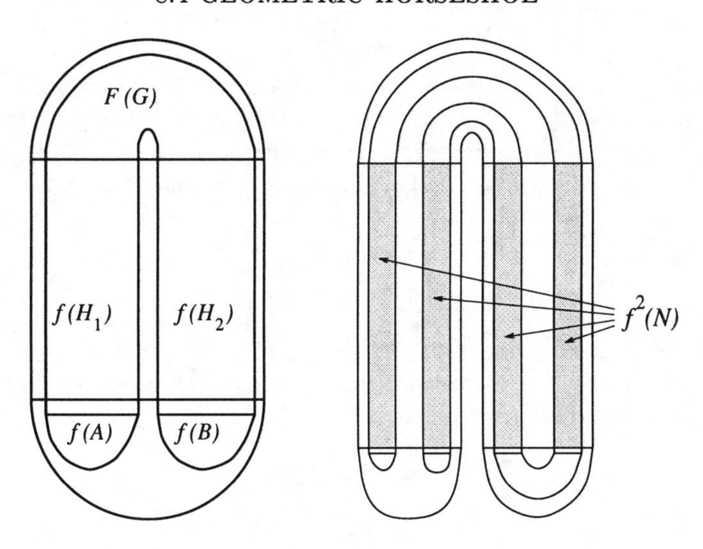

FIGURE 4.2. First and Second Image of the Neighborhood N for the Geometric Horseshoe

Below we prove that $f|\Lambda$ is conjugate to the shift map on a symbol space. As a corollary of this result, $\Lambda \subset \mathcal{R}(f)$ so $\mathcal{R}(f) = \Lambda \cup \{\mathbf{p}_0\} \cup \{\mathbf{p}_\infty\}$.

Now, we turn to the analysis of Λ. We define sets in a manner similar to the construction of the Cantor set for the quadratic map on the line:

$$\mathcal{S}_m^n = \bigcap_{j=m}^{n} f^j(S).$$

By the description above, $\mathcal{S}_0^1$ is the union of the two vertical strips V_1 and V_2 of width μ. As in the one-dimensional case,

$$\begin{aligned}\mathcal{S}_0^n &= f(\mathcal{S}_0^{n-1}) \cap S \\ &= [f(\mathcal{S}_0^{n-1}) \cap V_1] \cup [f(\mathcal{S}_0^{n-1}) \cap V_2] \\ &= f(\mathcal{S}_0^{n-1} \cap H_1) \cup f(\mathcal{S}_0^{n-1} \cap H_2).\end{aligned}$$

In particular for $n = 2$,

$$\begin{aligned}\mathcal{S}_0^2 &= [f(\mathcal{S}_0^1) \cap V_1] \cup [f(\mathcal{S}_0^1) \cap V_2] \\ &= f([V_1 \cup V_2] \cap H_1) \cup f([V_1 \cup V_2] \cap H_2).\end{aligned}$$

Then, for $k = 1$ or 2, $f(\mathcal{S}_0^1) \cap V_k = f([V_1 \cup V_2] \cap H_k)$ is the union of two vertical strips of width μ^2. Therefore, $\mathcal{S}_0^2$ is the union of 2^2 vertical strips of width μ^2. By induction

$$\mathcal{S}_0^n = f(\mathcal{S}_0^{n-1} \cap H_1) \cup f(\mathcal{S}_0^{n-1} \cap H_2)$$

is the union of 2^n vertical strips of width μ^n. Taking the infinite intersection,

$$\mathcal{S}_0^\infty = \bigcap_{n=0}^{\infty} \mathcal{S}_0^n = C_1 \times [0,1]$$

is a Cantor set of vertical line segments. If $\mathbf{q} \in \mathcal{S}_0^\infty$, then $\mathbf{q} \in f^j(S)$ and $f^{-j}(\mathbf{q}) \in S$ for all $j \geq 0$. Thus, $\mathcal{S}_0^\infty$ is the set of points whose backward iterates stay in S.

Considering the sets $\mathcal{S}_{-m}^0$, $\mathcal{S}_{-1}^0 = H_1 \cup H_2$ is the union of two horizontal strips of height λ^{-1}. Then, $\mathcal{S}_{-2}^0$ is the union of four horizontal strips of height λ^{-2}. Continuing by induction, $\mathcal{S}_{-m}^0$ is the union of 2^m horizontal strips of height λ^{-m}, and

$$\mathcal{S}_{-\infty}^0 = \bigcap_{m=0}^{\infty} \mathcal{S}_{-m}^0 = [0,1] \times C_2$$

is a Cantor set of horizontal line segments. If $\mathbf{q} \in \mathcal{S}_{-\infty}^0$, then for all $j \geq 0$, $\mathbf{q} \in f^{-j}(S)$ and $f^j(\mathbf{q}) \in S$. Thus, $\mathcal{S}_{-\infty}^0$ is the set of points whose forward iterates stay in S.

Intersecting these two sets , we get that

$$\begin{aligned}\Lambda &= \mathcal{S}_{-\infty}^\infty \\ &= \mathcal{S}_0^\infty \cap \mathcal{S}_{-\infty}^0 \\ &= C_1 \times C_2\end{aligned}$$

is the product of two Cantor sets, and Λ is the set of points such that both the forward and backward iterates stay in S. We want to show that Λ has the three properties of a Cantor set in the line: Λ is perfect and its connected components are points. It is perfect because both the sets C_j are perfect. From the fact that the set $\mathcal{S}_{-n}^n$ is the union of 2^{2n} rectangles of dimensions μ^n by λ^{-n}, it follows that the connected components of Λ are points. Thus, Λ has the three properties of a Cantor set in the line. There is, in fact, a theorem which says that such a set is homeomorphic to a Cantor set in the line. (See Hocking and Young (1961), page 97.)

We now turn to considering the stable and unstable manifolds of points. We first consider the set

$$\mathcal{S}_{-\infty}^0 = \bigcap_{m=0}^{\infty} \mathcal{S}_{-m}^0 = [0,1] \times C_2.$$

A point $\mathbf{q}$ is in $\mathcal{S}_{-\infty}^0$ if and only if $\mathbf{q} \in f^j(S)$ for all $j \leq 0$, if and only if $f^j(\mathbf{q}) \in S$ for all $j \geq 0$. Thus, such a point $\mathbf{q}$ is in the stable manifold of the whole set Λ. In fact, if $\mathbf{q} \in [0,1] \times C_2$ and $\mathbf{p} \in C_1 \times C_2 = \Lambda$ have the same y coordinate, then $|f^j(\mathbf{q}) - f^j(\mathbf{p})| \leq \mu^j$ and

$$\mathbf{q} \in \text{comp}_{\mathbf{p}}(W^s(\mathbf{p}) \cap S).$$

Thus, these horizontal line segments are the local stable manifolds of points in Λ. Similarly for $\mathbf{p} \in \Lambda$, $\text{comp}_{\mathbf{p}}(W^u(\mathbf{p}) \cap S)$ is the vertical line segment through $\mathbf{p}$.

The union of the local unstable manifolds for all points in Λ, $W^u_{loc}(\Lambda)$, is thus the set we labeled $\mathcal{S}_0^\infty$. The global unstable manifold of Λ, $W^u(\Lambda)$, is the forward orbit of $\mathcal{S}_0^\infty$. Because the forward images of the larger neighborhood N are nested, $f^j(N) \subset f^k(N)$ for $j > k$, it is not hard to see that $W^u(\Lambda)$ is the intersection of these images,

$$\begin{aligned}W^u(\Lambda) &= \bigcup_{j=0}^{\infty} f^j(\mathcal{S}_0^\infty) \\ &= \bigcap_{j=0}^{\infty} f^j(N).\end{aligned}$$

This set winds around and accumulates on itself. In fact, it is a continuum which cannot be written as the union of two proper subcontinua. For this reason, it is an example of an indecomposable continuum. This particular example is called the Knaster continuum. See Barge (1986) for more discussion of this property.

To introduce the *symbolic dynamics* of the map f on Λ, we need to consider the two-sided shift on two symbols because f involves both forward and backward iterates to determine Λ. The next theorem proves that $f|\Lambda$ is topologically conjugate to σ on Σ_2. The idea is to follow the orbit of a point $\mathbf{q} \in \Lambda$ and see in which of the horizontal boxes H_i it lies for each iterate. The *itinerary map* h is the map which assigns to the point $\mathbf{q} \in \Lambda$ the sequence $\mathbf{s} \in \Sigma_2$ which labels the box H_{s_j} in which the point $f^j(\mathbf{q})$ lies. In this construction, the alignment of the images $f(H_i)$ for $i = 1, 2$ with the boxes H_1 and H_2 is important: we say that the set of ambient boxes $\{H_1, H_2\}$ satisfy the *Markov property* because of the similarity to the condition given in Section 8.5.1.

Theorem 4.1. *Let Σ_2 be the full two-sided two-shift space with shift map σ. Define $h : \Lambda \to \Sigma_2$ by $h(\mathbf{q}) = \mathbf{s}$ where $f^j(\mathbf{q}) \in H_{s_j}$ for all j. Then, h is a topological conjugacy from $f|\Lambda$ to σ on Σ_2.*

PROOF. Let $h(\mathbf{q}) = \mathbf{s}$ and $h(f(\mathbf{q})) = \mathbf{t}$. Then $f^{j+1}(\mathbf{q}) \in H_{s_{j+1}}$, but also $f^{j+1}(\mathbf{q}) = f^j \circ f(\mathbf{q}) \in H_{t_j}$. Therefore, $s_{j+1} = t_j$, and $\sigma(\mathbf{s}) = \mathbf{t}$ or $\sigma(h(\mathbf{q})) = h(f(\mathbf{q}))$. This proves the first property of a conjugacy.

Next we show that h is continuous. Let $h(\mathbf{q}) = \mathbf{s}$. A neighborhood of $\mathbf{s}$ is given by

$$\mathcal{N} = \{\mathbf{t} : t_j = s_j \text{ for } -n_0 \le j \le n_0\}.$$

With n_0 fixed, the continuity of f insures that there is a $\delta > 0$ such that for $\mathbf{p} \in \Lambda$ and $|\mathbf{p} - \mathbf{q}| \le \delta$, $f^j(\mathbf{p}) \in H_{s_j}$ for $-n_0 \le j \le n_0$. Thus, if $\mathbf{t} = h(\mathbf{p})$ and $|\mathbf{p} - \mathbf{q}| \le \delta$, then $\mathbf{t} \in \mathcal{N}$. This proves the continuity of h.

To check that h is onto, we apply induction on n to show that $\bigcap_{j=1}^{n} f^j(H_{s_{-j}})$ is a vertical strip of width μ^n for all strings of symbols $\mathbf{s} \in \Sigma_2$. Let $\mathbf{s} \in \Sigma_2$. For $n = 1$, this set is just $f(H_{s_{-1}}) = V_{s_{-1}}$, which is a vertical strip of width μ. Then,

$$\bigcap_{j=1}^{n} f^j(H_{s_{-j}}) = f(\bigcap_{j=2}^{n} f^{j-1}(H_{s_{-j}})) \cap f(H_{s_{-1}})$$

is a strip of width μ^n since $\bigcap_{j=2}^{n} f^{j-1}(H_{s_{-j}})$ is a strip of width μ^{n-1}. Letting n go to infinity,

$$\bigcap_{j=1}^{\infty} f^j(H_{s_{-j}})$$

is a vertical line segment. Similarly, $\bigcap_{j=-\infty}^{0} f^j(H_{s_{-j}})$ is a horizontal line segment, and $\bigcap_{j=-\infty}^{\infty} f^j(H_{s_{-j}})$ is a (single) point, $\mathbf{q}$. In particular, the intersection is nonempty. For this $\mathbf{q}$, $h(\mathbf{q}) = \mathbf{s}$, and h is onto.

Finally, to show that h is one to one, assume that $h(\mathbf{p}) = h(\mathbf{q}) = \mathbf{s}$. Then, for all j, both $f^{-j}(\mathbf{p})$ and $f^{-j}(\mathbf{q})$ are in $H_{s_{-j}}$, so $\mathbf{p}, \mathbf{q} \in f^j(H_{s_{-j}})$. Letting j run from 1 to ∞, we see that $\mathbf{p}, \mathbf{q} \in \bigcap_{j=1}^{\infty} f^j(H_{s_{-j}})$, so they are in the same vertical line segment. Letting j run from $-\infty$ to 0, we see that $\mathbf{p}, \mathbf{q} \in \bigcap_{j=-\infty}^{0} f^j(H_{s_{-j}})$, so they are in the same horizontal line segment. Using all j from $-\infty$ to ∞, we see that $\mathbf{p} = \mathbf{q}$. This proves h is one to one. This completes the proof that h is a conjugacy. □

It is possible to have other horseshoes which are conjugate to subshifts of finite type.

Example 4.1. Let the region N in the plane be made up of three disks A, B, and C and two strips S_1 and S_2 connecting them as in Figure 4.3. The map f takes N inside

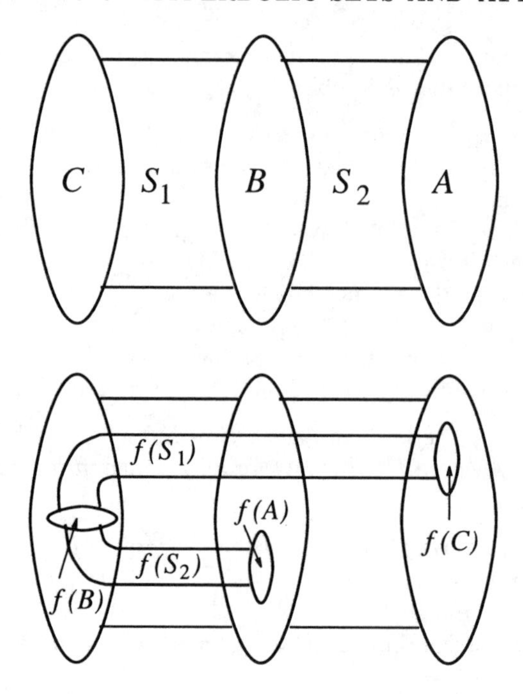

FIGURE 4.3. Horseshoe for a Subshift of Finite Type

itself. The disks A, B, and C are permuted with $f(A) \subset B$, $f(B) \subset C$, and $f(C) \subset A$. The map on these sets is a contraction and there is a unique attracting orbit of period 3, $\{\mathbf{p}_1, \mathbf{p}_2, \mathbf{p}_3\}$ with $\mathbf{p}_1 \in A$, $\mathbf{p}_2 \in B$, and $\mathbf{p}_3 \in C$. The strip S_1 is stretched across S_1, B, and S_2 with a contraction in the vertical direction as indicated. Finally, S_2 is stretched across S_1. This map can be extended to S^2 so it has a fixed point source at infinity, $\mathbf{p}_\infty$. As in the geometric horseshoe above, it can be shown that the chain recurrent set is given by

$$\mathcal{R}(f) = \{\mathbf{p}_1, \mathbf{p}_2, \mathbf{p}_3, \mathbf{p}_\infty\} \cup \Lambda$$

where

$$\Lambda = \bigcap_{j=-\infty}^{\infty} f^j(S_1 \cup S_2).$$

The map f can be taken so that it has a hyperbolic structure on Λ with one expanding direction and one contracting direction. Because of the manner in which f maps the strips, $f|\Lambda$ is not conjugate to the the full two-shift, but is conjugate to the two-sided subshift of finite type Σ_B for the transition matrix

$$B = \begin{pmatrix} 1 & 1 \\ 1 & 0 \end{pmatrix}.$$

There are restrictions on what combinations of periodic orbits and subshifts of finite type which can be realized on a manifold such as the two sphere. For more details, see Franks (1982).

8.4.1 Horseshoe for the Hénon Map

Let $F_{AB} : \mathbb{R}^2 \to \mathbb{R}^2$ be given by

$$F_{AB}(x,y) = (A - By - x^2, x), \qquad \text{so}$$
$$F_{AB}^{-1}(x,y) = (y, \frac{A - x - y^2}{B}).$$

We will often write F for F_{AB} in this section. This map is called the *Hénon map*. It was written down by Hénon to realize the Smale horseshoe for a specific function which could be iterated on the computer (Hénon, 1976). He also observed what appeared to be an attractor. We return to this aspect of the map in Section 8.10. Notice that

$$\det\,(DF_{AB})_{(x,y)} = \det \begin{pmatrix} -2x & -B \\ 1 & 0 \end{pmatrix} = B$$

is the amount that F changes area. If $B > 0$, then F preserves orientation; and if $B < 0$, then it reverses orientation.

The usual parameters discussed are $A = 1.4$ and $B = -0.3$ for which computer iteration indicates there is a "strange attractor." Notice that for these parameter values, F_{AB} decreases area and reverses orientation. It is still unproven that there is a "transitive" attractor for these parameter values. We will discuss this more fully when we come to examples of attractors.

The following theorem is the main result of the section. It proves that F_{AB} has a horseshoe for larger A, namely $A = 5$ and $B = \pm 0.3$ or, more generally, $A \geq (5 + 2\sqrt{5})(1 + |B|)^2/4$. Note that for $B = 0$, F_{A0} is conjugate to the quadratic map $F_\mu(x) = \mu x(1 - x)$. Therefore, the following theorem is analogous to Theorem II.4.1.

Theorem 4.2. *Let $B \neq 0$ and $A \geq (5 + 2\sqrt{5})(1 + |B|)^2/4$. In particular, $B = \pm 0.3$ and $A = 5$ are allowed. Let $R = \frac{1}{2}\{1 + |B| + [(1 + |B|)^2 + 4A]^{1/2}\}$ and let the square*

$$S = \{(x,y) \in \mathbb{R}^2 : |x| \leq R \text{ and } |y| \leq R\}.$$

Let $\Lambda = \bigcap_{j=-\infty}^{\infty} F_{AB}^j(S)$. Then, (a) all points which are nonwandering are contained inside S, (b) F_{AB} has a hyperbolic structure on Λ, (c) Λ is a Cantor set in the plane, and (d) $F_{AB}|\Lambda$ is topologically conjugate to the two-sided shift on two symbols. Thus, for these parameter values, the nonwandering set of F_{AB} is a horseshoe.

REMARK 4.1. The proof given below is based on Devaney and Nitecki (1979).

PROOF. To make the calculations somewhat easier, we take $A = 5$ and $B = 0.3$. Most of the calculations are very similar with $B = -0.3$. We also take $S = \{(x,y) : |x| \leq 3, |y| \leq 3\}$ which is a slight enlargement of the size given in the general definition in the statement of the theorem.

A direct analysis of the effect of the map F on points in different regions outside S shows that for $\mathbf{p} \notin S$, $\mathbf{p}$ is wandering with $|F^j(\mathbf{p})|$ going to infinity as j goes to either ∞ or $-\infty$. See Devaney and Nitecki (1979).

To find the image of S by F, we look at the image of the corners, the middle vertical line, and horizontal lines:

$$F(\pm 3,3) = \begin{pmatrix} 5-0.9-9 \\ \pm 3 \end{pmatrix} = \begin{pmatrix} -4.9 \\ \pm 3 \end{pmatrix},$$

$$F(\pm 3,-3) = \begin{pmatrix} 5+0.9-9 \\ \pm 3 \end{pmatrix} = \begin{pmatrix} -3.1 \\ \pm 3 \end{pmatrix},$$

$$F(0,y) = \begin{pmatrix} 5-0.3y \\ 0 \end{pmatrix}, \qquad \text{and}$$

$$F(x,y_0) = \begin{pmatrix} 5-0.3y_0-x^2 \\ x \end{pmatrix}.$$

These images show that the four corners go to points to the left of the box S. Since $5-0.3y > 3$ for $-3 \le y \le 3$, the middle vertical line is mapped to the right of the box. Finally, each horizontal line goes to a parabola. See Figure 4.4.

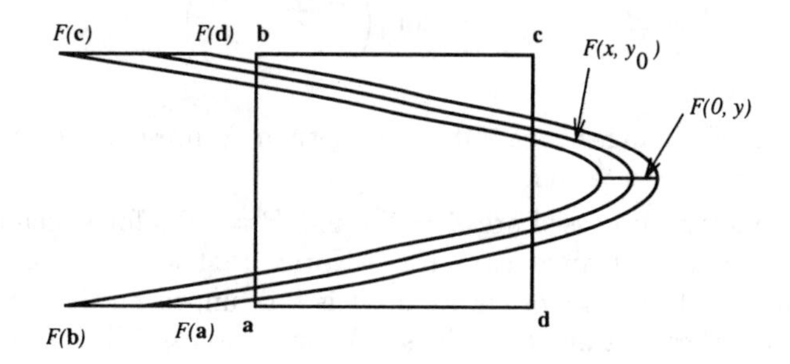

FIGURE 4.4. Image of the Square by the Hénon Map

Thus, $F(S) \cap S$ has two horizontal strips. Similarly, $F^{-1}(S) \cap S$ has two vertical strips, and $F^{-1}(S) \cap S \cap F(S)$ has four components. By induction, $\bigcap_{j=-n}^{n} F^j(S)$ has 4^n components. With this much information we can define a semi-conjugacy $h : \Lambda \to \{1,2\}^{\mathbb{Z}}$ which is onto.

The next step is to show that Λ has a hyperbolic structure, so it is possible to prove that the components of Λ are points. This will enable us to conclude that Λ is a Cantor set and that the semi-conjugacy h is one to one, and thus is a conjugacy.

To prove that the hyperbolic structure exists, we define cones (or sectors) at each point that are mapped into each other. This follows the ideas that we used for the proof of the stable manifold theorem. Also see the discussion in Moser (1973).

We define the cones $C^u(\mathbf{p})$ and $C^s(\mathbf{p})$ by

$$C^u(\mathbf{p}) = \{(\xi,\eta) \in T_{\mathbf{p}}\mathbb{R}^2 : |\eta| \le \lambda^{-1}|\xi|\},$$
$$C^s(\mathbf{p}) = \{(\xi,\eta) \in T_{\mathbf{p}}\mathbb{R}^2 : |\eta| \ge \lambda|\xi|\}.$$

For ease of estimation, we use the norm which measures the larger component of a vector: $|(\xi,\eta)|_* = \max\{|\xi|,|\eta|\}$. We find a $\lambda > 1$ that makes the cones invariant. (For $A = 5$ and $B = \pm 0.3$, λ can be taken to be 1.7.)

Lemma 4.3. *There is a $\lambda > 1$ (which can be taken to be 1.7 for $A = 5$ and $B = \pm 0.3$) for which the following two statements are true.*

(a) *For all $\mathbf{p} \in S \cap F^{-1}(S)$ and $\mathbf{v} \in C^u(\mathbf{p})$, $DF_{\mathbf{p}}\mathbf{v} \in C^u(F(\mathbf{p}))$ and $|DF_{\mathbf{p}}\mathbf{v}|_* \ge \lambda|\mathbf{v}|_*$.*
(b) *For all $\mathbf{p} \in S \cap F(S)$ and $\mathbf{v} \in C^s(\mathbf{p})$, $DF_{\mathbf{p}}^{-1}\mathbf{v} \in C^s(F^{-1}(\mathbf{p}))$ and $|DF_{\mathbf{p}}^{-1}\mathbf{v}|_* \ge \lambda|\mathbf{v}|_*$.*

PROOF. In the proof, we need some estimates which we prove first in a sublemma.

Sublemma 4.4. *(a) If $(x, y), F(x, y) \in S$, then $|x| > 1$ and $2|x| - |B| \geq 1.7 = \lambda > 1$.*
(b) If $(x, y), F^{-1}(x, y) \in S$, then $|y| > 1$.

PROOF. Let $(x_1, y_1) = F(x, y)$. If $|y| \leq 3$ and $|x_1| \leq 3$, then $5 - |B|y - x^2 \leq 3$, or $x^2 \geq 2 - |B|(3) = 1.1$. Thus, $|x| > 1$. The second estimate follows from the first: $2|x| - |B| \geq 2 - 0.3 = 1.7$.

Let $(x_{-1}, y_{-1}) = F^{-1}(x, y)$. Then, $(x, y) = F(x_{-1}, y_{-1})$, so $|y| = |x_{-1}| > 1$ by part (a). □

Now, take $\mathbf{p} = (x, y)$ with $\mathbf{p}, F(\mathbf{p}) \in S$ and $\begin{pmatrix} \xi \\ \eta \end{pmatrix} \in C^u(\mathbf{p})$. Because $|\xi| \geq \lambda|\eta| > |\eta|$, we have that $|(\xi, \eta)|_* = |\xi|$. The image of the vector by the derivative is given by

$$\begin{pmatrix} \xi_1 \\ \eta_1 \end{pmatrix} = \begin{pmatrix} -2x & -B \\ 1 & 0 \end{pmatrix} \begin{pmatrix} \xi \\ \eta \end{pmatrix}.$$

Then, $|\eta_1| = |\xi|$ and $|\xi_1| = |-2x\xi - B\eta| \geq |2x||\xi| - |B||\eta| \geq (2|x| - |B|)|\xi| \geq \lambda|\xi| \geq \lambda|\eta_1|$. Therefore, $\begin{pmatrix} \xi_1 \\ \eta_1 \end{pmatrix} \in C^u(F(\mathbf{p}))$, and

$$|(\xi_1, \eta_1)|_* = |\xi_1| \geq \lambda|\xi| = \lambda|(\xi, \eta)|_*.$$

This proves the first part of the lemma.

For the second part, take $\mathbf{p} = (x, y)$ with $\mathbf{p}, F^{-1}(\mathbf{p}) \in S$ and $\begin{pmatrix} \xi \\ \eta \end{pmatrix} \in C^s(\mathbf{p})$. Because $|\xi| < \lambda|\xi| \leq |\eta|$, we have that $|(\xi, \eta)|_* = |\eta|$. The image of the vector by the derivative of the inverse is given by

$$\begin{pmatrix} \xi_{-1} \\ \eta_{-1} \end{pmatrix} = \begin{pmatrix} 0 & 1 \\ -B^{-1} & -2yB^{-1} \end{pmatrix} \begin{pmatrix} \xi \\ \eta \end{pmatrix}.$$

Then, $|\eta_{-1}| \geq (|2y| - 1)|B|^{-1}|\eta| \geq (2 - 1)|B|^{-1}|\eta| \geq \lambda|\eta| = \lambda|\xi_{-1}|$. Therefore,

$$\begin{pmatrix} \xi_{-1} \\ \eta_{-1} \end{pmatrix} \in C^s(F^{-1}(\mathbf{p})),$$

and

$$|(\xi_{-1}, \eta_{-1})|_* = |\eta_{-1}| \geq \lambda|\eta| = \lambda|(\xi, \eta)|_*.$$

This completes the proof of the lemma. □

Now, to consider the hyperbolic structure on Λ, the lemma shows that for each $\mathbf{p} \in \Lambda$,

$$\bigcap_{j=0}^{n} DF^j_{F^{-j}(\mathbf{p})} C^u(F^{-j}(\mathbf{p}))$$

is a nested set of cones at $\mathbf{p}$, so the infinite intersection is a nonempty cone (scalar multiples of vectors in the intersection are still in the intersection):

$$\bigcap_{j=0}^{\infty} DF^j_{F^{-j}(\mathbf{p})} C^u(F^{-j}(\mathbf{p})) \neq \emptyset.$$

Similar statements are true for the stable cones:

$$\bigcap_{j=0}^{-\infty} DF^j_{F^{-j}(\mathbf{p})} C^s(F^{-j}(\mathbf{p})) \neq \emptyset.$$

Since Lemma 4.3 proves the expansion and contraction in these two sets of intersections, to complete the proof that the set is hyperbolic, all we need to show is that this intersection is a line for each $\mathbf{p} \in \Lambda$.

In fact, the maximal angle between two vectors in the intersection

$$\bigcap_{j=0}^{n} DF^j_{F^{-j}(\mathbf{p})} C^u(F^{-j}(\mathbf{p}))$$

goes to 0 as n goes to infinity as the following calculation shows. Take

$$\begin{pmatrix} \xi \\ \eta \end{pmatrix}, \begin{pmatrix} \xi' \\ \eta' \end{pmatrix} \in C^u(\mathbf{q})$$

with $\xi = \xi' = 1$, $|\eta|, |\eta'| \leq \lambda^{-1}$, and $\mathbf{q} = F^{-j}(\mathbf{p})$ for some $j > 0$. Then,

$$\eta_1 = \xi = 1 = \xi' = \eta_1',$$

and

$$\begin{aligned} \xi_1 &= -2x\xi - B\eta \\ &= -2x - B\eta \\ \xi_1' &= -2x\xi' - B\eta' \\ &= -2x - B\eta'. \end{aligned}$$

Thus,

$$\begin{aligned} \left|\frac{\eta_1}{\xi_1} - \frac{\eta_1'}{\xi_1'}\right| &= \left|\frac{1}{-2x - B\eta} - \frac{1}{-2x - B\eta'}\right| \\ &= \left|\frac{B(\eta - \eta')}{(-2x - B\eta)(-2x - B\eta')}\right| \\ &\leq \frac{|B||\eta - \eta'|}{\lambda^2} \\ &\leq \frac{|B|}{\lambda^2}\left|\frac{\eta}{\xi} - \frac{\eta'}{\xi'}\right| \end{aligned}$$

since $|-2x - B\eta|, |-2x - B\eta'| \geq \lambda$ and $|\xi| = |\xi'| = 1$. Therefore, the angle between two vectors is contracted by $|B|\lambda^{-2}$ and so the intersection of the cones goes to a single line in the tangent space at each point $\mathbf{p}$:

$$\bigcap_{j=0}^{\infty} DF^j_{F^{-j}(\mathbf{p})} C^u(F^{-j}(\mathbf{p})) = \mathbb{E}^u_{\mathbf{p}}.$$

Notice that the line $\mathbb{E}^u(\mathbf{p})$ can depend on $\mathbf{p}$ even though the original cones did not, because its definition involves the derivative of F along the backward orbit of $\mathbf{p}$. Similarly,

$$\bigcap_{j=0}^{-\infty} DF^j_{F^{-j}(\mathbf{p})} C^s(F^{-j}(\mathbf{p})) = \mathbb{E}^s_{\mathbf{p}}$$

is a line in the tangent space at $\mathbf{p}$. This completes the proof of the hyperbolic structure on Λ, or part (b) of the theorem.

For $\mathbf{p} \in \Lambda$, $W^s(\mathbf{p})$ has to start inside the cone $C^s(\mathbf{p}) + \{\mathbf{p}\}$. In fact, for $\mathbf{q} \in W^s(\mathbf{p})$, we have $T_{\mathbf{q}} W^s(\mathbf{p}) \subset C^s(\mathbf{q})$. Let H_1 and H_2 be the two (horizontal) components of $S \cap F(S)$. Then, by using the estimates of Lemma 4.3, we see that for each iterate by F, $\text{comp}_{\mathbf{p}}\,(W^s(\mathbf{p}) \cap H_j)$ is shrunk by a factor of λ^{-1}. In this way we see that

$$\sup_{\mathbf{p}\in\Lambda} \text{length}\,\{\,\text{comp}_{\mathbf{p}}\,[W^s(\mathbf{p}) \cap \bigcap_{j=-n}^{n} F^j(S)]\,\} \leq 6\lambda^{-n}$$

which goes to 0 as n goes to infinity. Similarly

$$\sup_{\mathbf{p}\in\Lambda} \text{length}\,\{\,\text{comp}_{\mathbf{p}}\,[W^u(\mathbf{p}) \cap \bigcap_{j=-n}^{n} F^j(S)]\,\} \leq 6\lambda^{-n}$$

which also goes to 0. Therefore, for $\mathbf{q} \in \Lambda$, $\text{comp}_{\mathbf{q}}(\bigcap_{j=-n}^{n} F^j(S))$ have diameters which go to 0 and so converge to the single point $\{\mathbf{q}\}$. Therefore, the connected components of Λ are points. This shows that Λ is a Cantor set. Also, by arguments as before, we can prove that the semiconjugacy of $F|\Lambda$ to Σ_2 is one to one, and so is a topological conjugacy. This completes the proof of the theorem. □

REMARK 4.2. Fix B. For $A < -(B+1)^2/4$, F_{AB} has (no periodic points and) empty nonwandering set. For $A > (5 + 2\sqrt{5})(1 + |B|)^2/4$, F_{AB} has a horseshoe. Therefore, as A varies, F_{AB} forms a horseshoe. There are many bifurcations which take place as this horseshoe is formed. Many people have studied this process but it is not yet completely understood. See Newhouse (1979), Mallet-Paret and Yorke (1982), Robinson (1983), Holmes and Whitley (1984), Holmes (1984), Yorke and Alligood (1985), Easton (1986, 1991), and Patterson and Robinson (1988).

8.4.2 Horseshoe from a Homoclinic Point

In the last two sections, we have analyzed the geometric model horseshoe and have shown how it arises in the Hénon map for certain parameter values. In this section, we show how it arises from a transverse intersection of the stable and unstable manifolds of a periodic point. Such an intersection is called a homoclinic point.

Definition. Let $\mathbf{p}$ be a hyperbolic periodic point of period n for a diffeomorphism f. Let

$$W^\sigma(\mathcal{O}(\mathbf{p})) = \bigcup_{j=0}^{n-1} W^\sigma(f^j(\mathbf{p})) \qquad \text{and}$$

$$\hat{W}^\sigma(\mathcal{O}(\mathbf{p})) = W^\sigma(\mathcal{O}(\mathbf{p})) \setminus \mathcal{O}(\mathbf{p})$$

for $\sigma = s, u$. A point $\mathbf{q} \in \hat{W}^s(\mathcal{O}(\mathbf{p})) \cap \hat{W}^u(\mathcal{O}(\mathbf{p}))$ is called a *homoclinic point for* $\mathbf{p}$. A point $\mathbf{q}$ is called a *transverse homoclinic point* provided the manifolds $\hat{W}^s(\mathcal{O}(\mathbf{p}))$ and $\hat{W}^u(\mathcal{O}(\mathbf{p}))$ have a nonempty transverse intersection at $\mathbf{q}$.

The following theorem proves that the existence of a transverse homoclinic point implies the existence of a Smale horseshoe. Figure 4.7 indicates why the invariant set of part (a) is called a horseshoe. By analogy, we also call any invariant set like the one constructed in part (b) a *horseshoe*.

Theorem 4.5. *Suppose that* $\mathbf{q}$ *is a transverse homoclinic point for a hyperbolic periodic point* $\mathbf{p}$ *for a diffeomorphism* f.

(a) For each neighborhood U *of* $\{\mathbf{p},\mathbf{q}\}$, *there is a positive integer* n *such that* f^n *has a hyperbolic invariant set* $\Lambda \subset U$ *with* $\mathbf{p},\mathbf{q} \in \Lambda$ *and on which* f^n *is topologically conjugate to the two-sided shift map on two symbols,* σ *on* Σ_2. *Thus,* $\Lambda \subset \text{cl}(\text{Per}(f)) \subset \Omega(f)$ *and* $\mathbf{q} \in \text{cl}(\text{Per}(f))$.

(b) Let V *be a neighborhood of* $\mathcal{O}(\mathbf{p}) \cup \mathcal{O}(\mathbf{q})$. *Then, there is a smaller neighborhood* $\mathcal{B} = \bigcup_{i=1}^{n} B_i$ *of* $\mathcal{O}(\mathbf{p}) \cup \mathcal{O}(\mathbf{q})$ *with* $n \geq 2$, $\mathcal{B} \subset V$, *such that* $\Lambda_{\mathcal{B}} = \bigcap_{j\in\mathbb{Z}} f^j(\mathcal{B}) \subset V$ *is a hyperbolic invariant set for* f *and* $f|\Lambda_{\mathcal{B}}$ *is topologically conjugate to the shift map* σ_A *on a transitive two-sided subshift of finite type on* n *symbols,* $\Sigma_A \subset \Sigma_n$. *The conjugacy* $h : \Lambda_{\mathcal{B}} \to \Sigma_A$ *is the itinerary function given by* $h(\mathbf{x}) = \mathbf{s}$, *where* $f^j(\mathbf{x}) \in B_{s_j}$ *for all* j. *Notice that* $\mathbf{q} \in \mathcal{O}(\mathbf{p}) \cup \mathcal{O}(\mathbf{q}) \subset \Lambda_{\mathcal{B}} \subset \text{cl}(\text{Per}(f)) \subset \Omega(f)$, *so* $\mathbf{q} \in \text{cl}(\text{Per}(f))$. *If* $\mathbf{p}$ *is a fixed point, then the transition matrix* $A = (a_{i,j})$ *for the subshift of finite type is given by*

$$a_{1,j} = \begin{cases} 1 & \text{for } j = 1,2 \\ 0 & \text{for } j \neq 1,2 \end{cases}$$
$$a_{i,j} = \begin{cases} 1 & \text{for } j = i+1, \ 2 \leq i < n \\ 0 & \text{for } j \neq i+1, \ 2 \leq i < n \end{cases}$$
$$a_{n,j} = \begin{cases} 1 & \text{for } j = 1 \\ 0 & \text{for } j \neq 1. \end{cases}$$

See Remark 4.8 for the transition matrix when $\mathbf{p}$ *is not a fixed point.*

REMARK 4.3. The theorem is most often stated as in part (a). See Smale (1965, 1967). There are many other references, including Guckenheimer and Holmes (1983), pages 252–3; Newhouse (1980), pages 13–24; Moser (1973); and Wiggins (1990), pages 470–483.

REMARK 4.4. Given a set $\mathcal{B}$, $\Lambda_{\mathcal{B}} = \bigcap_{j\in\mathbb{Z}} f^j(\mathcal{B})$ is the *maximal invariant set* in $\mathcal{B}$, i.e., the largest invariant set contained in $\mathcal{B}$. (See Section 10.3 for further discussion of the maximal invariant set.) In part (b), the set $\mathcal{B}$ needs to be chosen carefully so that this maximal invariant set is conjugate to a relatively simple subshift of finite type.

REMARK 4.5. One way to think of the subshift of finite type in part (b) is in terms of ϵ-chains. Consider the case when $\mathbf{p}$ is a fixed point. For big enough N_1 and N_2, $\Lambda'_{\mathbf{q}} = \{\mathbf{p}, f^{-N_1}(\mathbf{q}), f^{-N_1+1}(\mathbf{q}), \ldots, f^{N_2}(\mathbf{q})\}$ is a periodic ϵ-chain. Let $n = 2 + N_1 + N_2$, $\mathbf{q}_1 = \mathbf{p}$, and $\mathbf{q}_j = f^{-N_1-2+j}(\mathbf{q})$ for $2 \leq j \leq n$. The subshift of finite type given in the theorem corresponds to allowing the following transitions: (i) from $\mathbf{q}_1 = \mathbf{p}$ to $\mathbf{q}_1$ or $\mathbf{q}_2 = f^{-N_1}(\mathbf{q})$, (ii) from $\mathbf{q}_j$ to $\mathbf{q}_{j+1}$ for $2 \leq j < n$, and (iii) from $\mathbf{q}_n = f^{N_2}(\mathbf{q})$ to $\mathbf{q}_1 = \mathbf{p}$. The space $\Sigma_A \subset \Sigma_n$ for this subshift is a Cantor set with dense periodic points. (The proof is the same as for the section on one-sided subshifts since one of the n symbols has more than one option.)

The fact that any sequence of symbols in Σ_A can be realized by an actual orbit follows from shadowing. We prove below that the set $\Lambda_{\mathbf{q}} = \mathcal{O}(\mathbf{p}) \cup \mathcal{O}(\mathbf{q})$ has a hyperbolic structure. By the Shadowing Theorem X.3.1, any ϵ-chain in $\Lambda'_{\mathbf{q}} \subset \Lambda_{\mathbf{q}}$ can be shadowed by an orbit of f. The sets B_i are taken to be disjoint and neighborhoods of the points $\mathbf{q}_i \in \Lambda'_{\mathbf{q}}$; the orbit which shadows the ϵ-chain $\{\mathbf{x}_j\} \subset \Lambda'_{\mathbf{q}}$ can be taken with $f^j(\mathbf{x})$ in the same B_{s_j} as $\mathbf{x}_j$. In particular, the Shadowing Theorem proves that the homoclinic point $\mathbf{q}$ is in the closure of the periodic points. However, we want to prove that the maximal invariant set $\Lambda_{\mathcal{B}}$ in this neighborhood $\mathcal{B}$ of $\Lambda_{\mathbf{q}}$ is conjugate to a subshift of finite type Σ_A. (This is Smale's main contribution.) If we prove the theorem using the Shadowing

Theorem, we would need to prove that the set of all these orbits which shadow the ϵ-chains in $\Lambda_{\mathbf{q}}$ is the maximal invariant set in $\mathcal{B}$. This last step is not obvious from the statement of the Shadowing Theorem and involves shadowing all ϵ-chains in $\Lambda'_{\mathbf{q}}$ at once. For this reason, we give a proof below which is independent of the Shadowing Theorem and duplicates some of its proof.

REMARK 4.6. Notice that in addition to the transverse homoclinic point $\mathbf{q}$, the theorem allows that f can have nontransverse homoclinic points for $\mathbf{p}$ at points other than $\mathbf{q}$.

PROOF. First, we give an outline of the proof. Let $\Lambda_{\mathbf{q}} = \mathcal{O}(\mathbf{p}) \cup \mathcal{O}(\mathbf{q})$. This invariant set is closed because $\alpha(\mathbf{q}) = \omega(\mathbf{q}) = \mathcal{O}(\mathbf{p})$. The first step is to prove that $\Lambda_{\mathbf{q}}$ has a hyperbolic structure. The second step is to prove that if V is a small enough neighborhood of $\Lambda_{\mathbf{q}}$, then the maximal invariant set in V, $\Lambda_V = \bigcap_{i \in \mathbb{Z}} f^i(V)$, also has a hyperbolic structure. For the third step, the proofs of both parts (a) and (b) use boxes contained in the ambient space. These boxes are similar to those used for the geometric horseshoe and the Hénon map. For part (a), we find two disjoint boxes V_1 and V_2 contained in V and an $n > 0$ such that each of the images $f^n(V_i)$ crosses both V_1 and V_2. It follows that f^n restricted to $\Lambda = \bigcap_{m=-\infty}^{\infty} f^{mn}(V_1 \cup V_2)$ is conjugate to the full two-sided subshift on two symbols. For part (b), the construction uses disjoint closed boxes $B_i \subset V$ such that the union $\mathcal{B} = \bigcup_i B_i$ is a neighborhood of $\Lambda_{\mathbf{q}}$. This neighborhood $\mathcal{B}$ is contained inside V so $\Lambda_{\mathcal{B}} = \bigcap_{m \in \mathbb{Z}} f^m(\mathcal{B}) \subset \Lambda_V$ has a hyperbolic structure. One difference between this case and part (a) (or the geometric horseshoe) is that the image of each $f(B_i)$ does not intersect all the other B_j. However, each nonempty intersection $f(B_i) \cap B_j$ completely crosses B_j, so $\Lambda_{\mathcal{B}}$ can be shown to be conjugate to a subshift of finite type.

Throughout much of the proof we assume $\mathbf{p}$ is a fixed point. (The notation and labeling becomes a little simpler in this case.) The reader can make the necessary changes if it is not fixed. We start the proof as if we were proving part (b), although the first part of the proofs is the same for both parts. We indicate where the bifurcation in the proof of the two parts takes place and the reader can then choose which part to read first.

As indicated in the outline, the first step is to show that $\Lambda_{\mathbf{q}}$ has a hyperbolic structure. Let $\mathbf{q}_m = f^m(\mathbf{q})$. For each point $\mathbf{x} \in \Lambda_{\mathbf{q}}$, let $\mathbb{E}^\sigma_{\mathbf{x}} = T_{\mathbf{x}}(W^\sigma(\mathbf{p}))$ for $\sigma = s, u$. This is clearly a continuous splitting on $\mathcal{O}(\mathbf{q})$ and $\mathbf{p}$ separately. We must check that it is continuous as $\mathcal{O}(\mathbf{q})$ approaches $\mathbf{p}$. Consider $\mathbf{q}_m$ as m goes to ∞, so $\mathbf{q}_m$ approaches $\mathbf{p}$. The stable bundle $\mathbb{E}^s_{\mathbf{q}_m} = T_{\mathbf{q}_m}(W^s(\mathbf{p}))$ approaches $\mathbb{E}^s_{\mathbf{p}}$ because the stable manifold $W^s(\mathbf{p})$ is C^1. The unstable bundle $\mathbb{E}^u_{\mathbf{q}_m} = T_{\mathbf{q}_m}(W^u(\mathbf{p}))$ approaches $\mathbb{E}^u_{\mathbf{p}}$ by the linear estimates in the Inclination Lemma. The argument as m goes to $-\infty$ is similar. Therefore, we have a continuous splitting on $\Lambda_{\mathbf{q}}$.

Next we check that vectors in $\mathbb{E}^s|\Lambda_{\mathbf{q}}$ are uniformly contracted. We can take an adapted metric so that on $\mathbb{E}^s_{\mathbf{p}}$ the derivative is an immediate contraction, $\|Df_{\mathbf{p}}|\mathbb{E}^s_{\mathbf{p}}\| < \lambda < 1$. By continuity, there is a neighborhood W of $\mathbf{p}$ such that $\|Df_{\mathbf{q}_m}|\mathbb{E}^s_{\mathbf{q}_m}\| < \lambda$ for all $\mathbf{q}_m \in W$. Since there are only finitely many $\mathbf{q}_m \in \Lambda_{\mathbf{q}} \setminus W$, there is a $C \geq 1$ such that if $f^i(\mathbf{q}_m) \notin W$ for $0 \leq i < k$, then $\|Df^i_{\mathbf{q}_m}|\mathbb{E}^s_{\mathbf{q}_m}\| \leq C\lambda^i$ for $0 < i \leq k$. For any $\mathbf{q}_m \in \Lambda_{\mathbf{q}} \setminus \{\mathbf{p}\}$, there is at most one string of iterates for which $f^i(\mathbf{q}_m) \notin W$: there are i_1 and i_2 such that $f^i(\mathbf{q}_m) \notin W$ for $i_1 \leq i < i_2$ and $f^i(\mathbf{q}_m) \in W$ for $i < i_1$ or $i \geq i_2$. Combining the estimates on and off W, $\|Df^i_{\mathbf{q}_m}|\mathbb{E}^s_{\mathbf{q}_m}\| \leq C\lambda^i$ for any $\mathbf{q}_m \in \Lambda_{\mathbf{q}}$ and all $0 < i$. The estimates for $\mathbb{E}^u|\Lambda_{\mathbf{q}}$ are similar interchanging f and f^{-1}. This proves the hyperbolicity of $\Lambda_{\mathbf{q}}$.

Now, we turn to step 2 which proves that there is a small neighborhood V of $\Lambda_{\mathbf{q}}$ such that the maximal invariant set $\Lambda_V = \bigcap_{i \in \mathbb{Z}} f^i(V)$ has a hyperbolic structure. Note that there is no neighborhood V of $\Lambda_{\mathbf{q}}$ for which $\Lambda_{\mathbf{q}} = \Lambda_V$. (This might not be obvious but is true from the conclusion of this theorem or by the Shadowing Theorem X.3.1.) This means that $\Lambda_{\mathbf{q}}$ is not an isolated invariant set. (An invariant set Λ is called *isolated* provided there is a neighborhood V for which $\Lambda = \Lambda_V$ with Λ_V defined as above. See Section 10.3 for further discussion of isolated invariant sets.)

For simplicity below, we take an adapted metric on $\Lambda_{\mathbf{q}}$. (The adapted metric implies that for $\mathbf{x} \in \Lambda_{\mathbf{q}}$, $Df_{\mathbf{x}}$ is an immediate contraction on $\mathbb{E}^s_{\mathbf{x}}$ and an immediate expansion on $\mathbb{E}^u_{\mathbf{x}}$.) We extend the splitting $\mathbb{E}^s_{\mathbf{x}} \oplus \mathbb{E}^u_{\mathbf{x}}$ on Λ to a continuous (probably noninvariant) splitting $\hat{\mathbb{E}}^s_{\mathbf{x}} \oplus \hat{\mathbb{E}}^u_{\mathbf{x}}$ on a (perhaps smaller) neighborhood V of $\Lambda_{\mathbf{q}}$. We use cones to show there is an invariant splitting which approximates $\hat{\mathbb{E}}^s_{\mathbf{x}} \oplus \hat{\mathbb{E}}^u_{\mathbf{x}}$ and extends the splitting $\mathbb{E}^s_{\mathbf{x}} \oplus \mathbb{E}^u_{\mathbf{x}}$ on $\Lambda_{\mathbf{q}}$. For $\mathbf{x} \in V$, using the adapted metric, let

$$C^s(\mathbf{x}) = \{(\xi, \eta) \in \hat{\mathbb{E}}^s_{\mathbf{x}} \oplus \hat{\mathbb{E}}^u_{\mathbf{x}} : |\eta| \le \mu|\xi|\}$$

and

$$C^u(\mathbf{x}) = \{(\xi, \eta) \in \hat{\mathbb{E}}^s_{\mathbf{x}} \oplus \hat{\mathbb{E}}^u_{\mathbf{x}} : |\xi| \le \mu|\eta|\}$$

some $0 < \mu < 1$. By the hyperbolicity on $\Lambda_{\mathbf{q}}$ and continuity, there is a (perhaps smaller) neighborhood V of $\Lambda_{\mathbf{q}}$ such that

$$\begin{aligned}
Df_{\mathbf{x}} C^u(\mathbf{x}) &\subset C^u(f(\mathbf{x})) && \text{provided } \mathbf{x}, f(\mathbf{x}) \in V,\\
Df_{\mathbf{x}}^{-1} C^s(\mathbf{x}) &\subset C^s(f^{-1}(\mathbf{x})) && \text{provided } \mathbf{x}, f^{-1}(\mathbf{x}) \in V,\\
|Df^m_{\mathbf{x}} \mathbf{v}^u| &\ge \lambda^{-m}|\mathbf{v}^u| && \text{provided } f^i(\mathbf{x}) \in V \text{ for } 0 \le i < m, \text{ and}\\
|Df^{-m}_{\mathbf{x}} \mathbf{v}^s| &\ge \lambda^{-m}|\mathbf{v}^s| && \text{provided } f^{-i}(\mathbf{x}) \in V \text{ for } 0 \le i < m
\end{aligned}$$

for any $\mathbf{v}^u \in C^u(\mathbf{x})$ and $\mathbf{v}^s \in C^s(\mathbf{x})$. Let $\Lambda_V = \bigcap_{i \in \mathbb{Z}} f^i(V) \supset \Lambda_{\mathbf{q}}$ as before. For this neighborhood V and $\mathbf{x} \in \Lambda_V$, define

$$\begin{aligned}
\mathbb{E}^u_{\mathbf{x}} &= \bigcap_{i \ge 0} Df^i_{f^{-i}(\mathbf{x})} C^u(f^{-i}(\mathbf{x}))\\
\mathbb{E}^s_{\mathbf{x}} &= \bigcap_{i \ge 0} Df^{-i}_{f^i(\mathbf{x})} C^s(f^i(\mathbf{x})).
\end{aligned}$$

Because of the expansion esimates, these are subspaces for each $\mathbf{x} \in \Lambda_V$ which depend continuously on $\mathbf{x}$, have the usual invariance properties, and satisfy $\mathbb{E}^s_{\mathbf{x}} \oplus \mathbb{E}^u_{\mathbf{x}} = T_{\mathbf{x}}M$. By the inequalities for all vectors in the cones, vectors in $\mathbb{E}^u_{\mathbf{x}}$ and $\mathbb{E}^s_{\mathbf{x}}$ are expanded and contracted, respectively, so Λ_V is a hyperbolic invarinat set. This completes the proof of step 2.

The invariant set Λ_V defined above is probably not conjugate to a simple subshift of finite type. To get such an invariant set, we carefully construct a smaller neighborhood of $\Lambda_{\mathbf{q}}$, $\mathcal{B} \subset V$, which is a finite union of "boxes." Each box B_i corresponds to a symbol in the subshift. The transitions which are allowed in the subshift are exactly those for which $f(B_i) \cap B_j \neq \emptyset$. The images of the boxes B_i are correctly aligned so that a string of symbols is allowable if and only if there is a point whose orbits goes through this sequence of boxes. Since $\mathcal{B} \subset V$, $\Lambda_{\mathcal{B}} \subset \Lambda_V$, so $\Lambda_{\mathcal{B}}$ has a hyperbolic structure.

We take coordinates near $\mathbf{p}$ induced by the hyperbolic splitting. In fact, identifying $\mathbb{E}^\sigma_{\mathbf{p}}$ with a subspace for $\sigma = s, u$, we can take coordinates so a neighborhood can be considered as a subset of $\mathbb{E}^s_{\mathbf{p}} \times \mathbb{E}^u_{\mathbf{p}}$, and the local stable and unstable manifolds are disks in the subspaces given by the splitting, $W^s_r(\mathbf{p}) = \mathbb{E}^s_{\mathbf{p}}(r) \times \{\mathbf{0}\}$ and $W^u_r(\mathbf{p}) = \{\mathbf{0}\} \times \mathbb{E}^u_{\mathbf{p}}(r)$, and identify these two local stable and unstable manifolds with $\mathbb{E}^s_{\mathbf{p}}(r)$ and $\mathbb{E}^u_{\mathbf{p}}(r)$, respectively. For $\delta_s, \delta_u > 0$, we let $D^s = W^s_{\delta_s}(\mathbf{p})$ and $D^u = W^u_{\delta_u}(\mathbf{p})$. We also consider $D^s = \mathbb{E}^s_{\mathbf{p}}(\delta_s)$ and $D^u = \mathbb{E}^s_{\mathbf{p}}(\delta_u)$ and take their cross product in local coordinates near $\mathbf{p}$. We can take $\delta_s, \delta_u > 0$ and $k > 0$ such that

$$\mathbf{q} \in \operatorname{int}[f^{-k}(D^s) \setminus f^{-(k-1)}(D^s)]$$
$$\mathbf{q} \in \operatorname{int}[f^{k}(D^u) \setminus f^{(k-1)}(D^u)],$$

where the interiors are taken relative to $W^s(\mathbf{p})$ and $W^u(\mathbf{p})$, respectively. See Figure 4.5. We can also insure that $D^s \times D^u \subset V$ (resp. U) where V (resp. U) is the neighborhood of $\Lambda_{\mathbf{q}}$ (resp. $\{\mathbf{p}, \mathbf{q}\}$) given in the statement of part (b) (resp. of part (a)). For the same k to work both forward and backward, the relative sizes of δ_s and δ_u need to be adjusted.

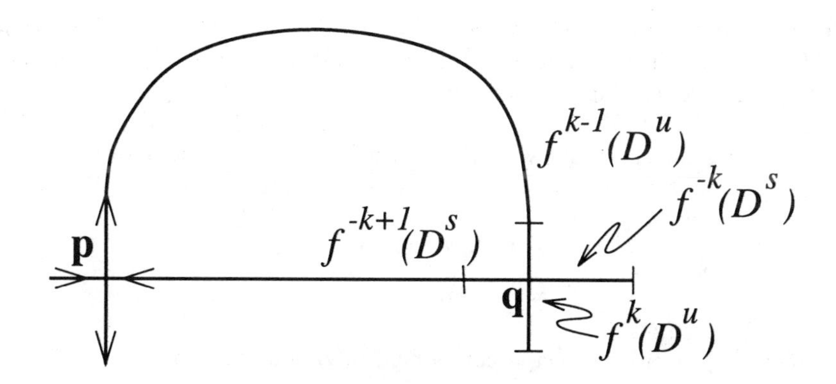

FIGURE 4.5. Images of the Disks D^s and D^u

Having fixed k, take $j_1 \geq 0$ such that for $j \geq j_1$, $f^k(D^u)$ crosses $f^{-k}(D^s \times f^{-j}(D^u))$ transversally in the components of the intersection containing $\mathbf{q}$ and $\mathbf{p}$. By transversally, we mean that it hits transversally each "horizontal fiber" $f^{-k}(D^s \times \{\mathbf{y}\})$ once and only once for each $\mathbf{y} \in f^{-j}(D^u)$. Thus, $f^k(D^u)$ is a "vertical disk" through $\mathbf{q}$. See Figure 4.6.

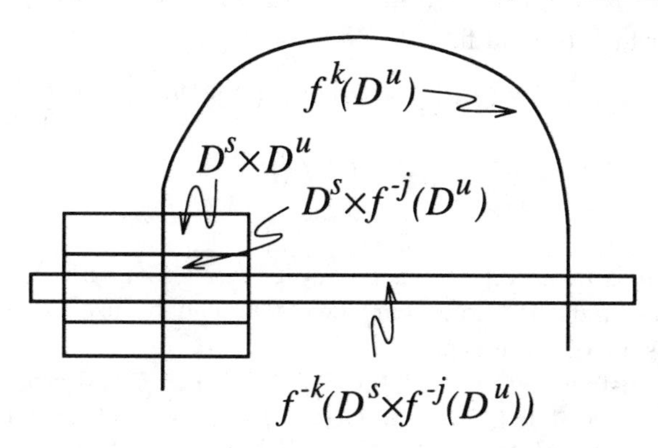

FIGURE 4.6. Choice of $f^{-k}(D^s \times f^{-j}(D^u))$

For $j \geq j_1$, the set

$$f^{2k+j} \circ f^{-k}(D^s \times f^{-j}(D^u)) = f^{k+j}(D^s \times f^{-j}(D^u))$$

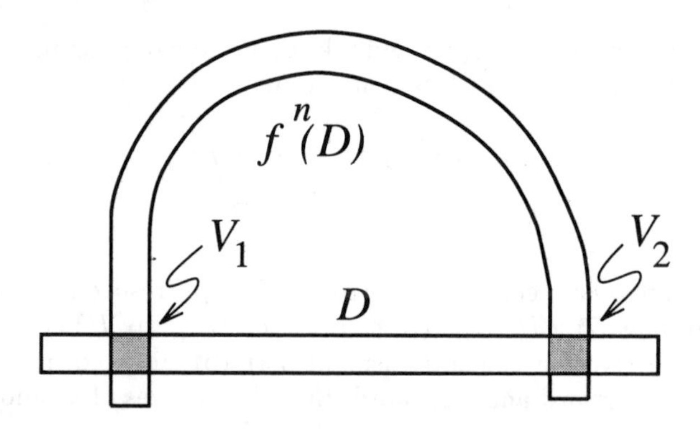

FIGURE 4.7. $\mathcal{D}$ and $f^n(\mathcal{D}) = f^{2k+j}(\mathcal{D})$ for $\mathcal{D} = f^{-k}(D^s \times f^{-j}(D^u))$

is a thin neighborhood of $f^k(D^u)$ by the Inclination Lemma. Therefore, for $j \geq j_1$ large enough, $f^{k+j}(D^s \times f^{-j}(D^u))$ crosses $f^{-k}(D^s \times f^{-j}(D^u))$ transversally in the components of the intersection containing $\mathbf{q}$ and $\mathbf{p}$. In particular, each "fiber" $f^{k+j}(\{\mathbf{x}\} \times f^{-j}(D^u))$ crosses each $f^{-k}(D^s \times \{\mathbf{y}\})$ once and only once for each $\mathbf{x} \in D^s$ and $\mathbf{y} \in f^{-j}(D^u)$. See Figure 4.7. Notice the similarity with the figure for the geometric horseshoe.

Fix an $j \geq j_1$ large enough to satisfy the above conditions. Let $n = 2k + j$,

$$\begin{aligned} B_1 &= D^s \times f^{-j}(D^u), \qquad \text{and} \\ \mathcal{D} &= f^{-k}(B_1). \end{aligned}$$

Letting $\text{comp}_{\mathbf{z}}(B)$ be the connected component of B containing $\mathbf{z}$, set

$$\begin{aligned} V_1 &= \text{comp}_{\mathbf{p}}(\mathcal{D} \cap f^n(\mathcal{D})) \subset B_1, \\ V_2 &= \text{comp}_{\mathbf{q}}(\mathcal{D} \cap f^n(\mathcal{D})), \qquad \text{and} \\ B_i &= f^{i-1-k-j}(V_2) \end{aligned}$$

for $2 \leq i \leq n$. See Figures 4.7 and 4.8. The set of boxes $\{V_1, V_2\}$ is used in the proof of part (a), and the set of boxes $\{B_i : 1 \leq i \leq n\}$ is used in the proof of part (b).

At this point we are in position to prove either part (a) or (b) of the theorem. The reader can choose which to read first.

PROOF OF PART (b). Let $\mathcal{B} = \bigcup_{1 \leq i \leq n} B_i$. For j large enough, $\mathcal{B} \subset V$. Finally, let

$$\Lambda_{\mathcal{B}} = \bigcap_{i \in \mathbb{Z}} f^i(\mathcal{B}),$$

so $\Lambda_{\mathcal{B}}$ is the maximal invariant set in $\mathcal{B}$, $\Lambda_{\mathcal{B}} \subset \Lambda_V \subset V$, and $\Lambda_{\mathcal{B}}$ has a hyperbolic structure. We show that the (nonlinear) boxes B_i can be used as symbols, so $\Lambda_{\mathcal{B}}$ is conjugate to a subshift of finite type.

Because of the construction, $f^{-k}(B_1)$ and $f^{k+j}(B_1) = f^n(\mathcal{D})$ cross V_2, but $f^\ell(B_1) \cap V_2 = \emptyset$ for $-k < \ell < k + j$. It follows that (i) $f(B_1)$ crosses B_1 and $f^{-k-j+1}(V_2) = B_2$ but not B_i for $i > 2$, (ii) $f^k(V_2) = f \circ f^{2k+j-1-k-j}(V_2) = f(B_n)$ crosses B_1, and (iii) $f^\ell(V_2) \cap B_1 = \emptyset$ for $-k - j < \ell < k$, so $B_i \cap B_1 = \emptyset$ for $2 \leq i \leq n$. Thus, we have constructed the desired disjoint boxes for the symbols. The first symbol, B_1, goes to either itself or B_2. The other symbols can only go to the next symbol: B_i goes to B_{i+1} for $2 \leq i < n$, and B_n goes to B_1. Therefore, the transition matrix for the subshift is given as in the statement of the theorem. (Note that we have assumed that $\mathbf{p}$ is a fixed

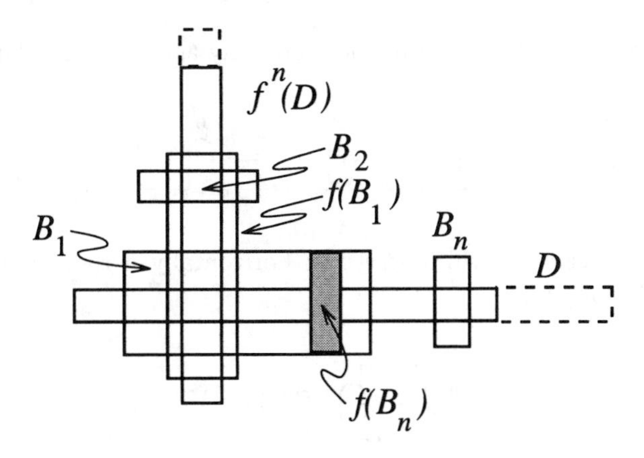

FIGURE 4.8. Choice of the Boxes $B_1, \ldots, B_n$

point. The form of the transition matrix when $\mathbf{p}$ is not fixed is given in Remark 4.8 below.)

In the construction of these boxes, they are correctly aligned: each of these boxes can be assigned coordinates so that the image of an unstable disk in B_i crosses B_{i+1} in the unstable direction, and the inverse images of a stable disk in B_i crosses B_{i-1} in the stable direction.

The conditions on the boxes are similar to the properties of a Markov partition for a hyperbolic invariant set which we define in Section 8.5.1. One difference is that the Markov partition is made up of "boxes" which are subsets of the hyperbolic invariant set while these boxes are diffeomorphic to "Euclidean boxes" and are neighborhoods in the ambient space. Because of this difference, we say the set of ambient boxes satisfies the *Markov property* rather than calling them a Markov partition.

Define $h : \Lambda_{\mathcal{B}} \to \Sigma_A$ as the itinerary function by $h(\mathbf{x}) = \mathbf{s}$ provided $f^i(\mathbf{x}) \in B_{s_i}$ for all $i \in \mathbb{Z}$. Because the boxes are disjoint, h is well defined. Fix any symbol $\mathbf{s} \in \Sigma_A$. For any $m \geq 0$, because the images of the boxes have the correct topological alignment, $\bigcap_{i=0}^{m} f^i(B_{s_{-i}})$ is a nonempty nonlinear sub-box of B_{s_0} which stretches all the way across the unstable direction. Similarly, $\bigcap_{i=-m}^{0} f^i(B_{s_{-i}})$ is a nonempty nonlinear sub-box of B_{s_0} which stretches all the way across the stable direction. Therefore, $\bigcap_{i=-m}^{m} f^i(B_{s_{-i}})$ and $\bigcap_{i\in\mathbb{Z}} f^i(B_{s_{-i}})$ are nonempty. Thus, h is onto Σ_A. By an argument like we used before, $h \circ f|\Lambda_{\mathcal{B}} = \sigma \circ h$ so h is a semiconjugacy. (This much of the argument does not use that $\Lambda_{\mathcal{B}}$ has a hyperbolic structure, but can be made to work if there is a "topologically transverse" intersection. See Burns and Weiss (1994).)

The fact that h is one to one follows from the fact that f has a hyperbolic structure on $\Lambda_{\mathcal{B}}$. The contraction and expansion implies that for any symbol sequence $\mathbf{s} \in \Sigma_A$, there is only one point in the intersection $\bigcap_{i\in\mathbb{Z}} f^i(B_{s_{-i}})$, i.e., there is only one point $\mathbf{x} \in \Lambda$ such that $f^i(\mathbf{x})$ is in the box V_{s_i} for all i, i.e., h is one to one. □

PROOF OF PART (a). Let U be an open set of $\mathbf{p}$ and $\mathbf{q}$ which is contained in the set V defined above. Now, fix k, j, $n = 2k + j$, δ_s, and δ_u as above. By the above choices, V_1 and V_2 are two correctly aligned sets, and $V_1 \cup V_2 \subset U$. See Figure 4.7. By the fact that the images of V_1 and V_2 by f^n stretch vertically across $\mathcal{D}$,

$$\mathcal{S}_0^{m-1} = \bigcap_{i=0}^{m-1} f^{in}(V_1 \cup V_2) = \bigcap_{i=0}^{m} f^{in}(\mathcal{D})$$

has 2^m components each of which stretches vertically across $\mathcal{D}$. Similarly,

$$\mathcal{S}_{-m}^{-1} = \bigcap_{i=-m}^{-1} f^{in}(V_1 \cup V_2) = \bigcap_{i=-m}^{0} f^{in}(\mathcal{D})$$

has 2^m components each of which stretches horizontally across $\mathcal{D}$ transverse to the vertical fibers. Combining,

$$\mathcal{S}_{-m}^{m-1} = \bigcap_{i=-m}^{m-1} f^{in}(V_1 \cup V_2)$$

has 2^{2m} components. (So far, we have not shown that the maximum of the diameters of these components goes to 0 as m goes to infinity.) By arguments like those for the geometric horseshoe, it follows that there is a semiconjugacy $h : \Lambda \to \Sigma_2$, where

$$\Lambda = \bigcap_{i=-\infty}^{\infty} f^{in}(V_1 \cup V_2),$$

which is onto and such that $h \circ f^n | \Lambda = \sigma \circ h$.

The fact that h is one to one follows from the fact that f^n has a hyperbolic structure on Λ: the contraction and expansion implies that for any one symbol sequence $\mathbf{s}$, there is only one point $\mathbf{x} \in \Lambda$ such that $f^{in}(\mathbf{x})$ is in the box V_{s_i} for all i. □

REMARK 4.7. It might seem that the invariant set for f, $\Lambda_{\mathcal{B}}$, should be the orbit of the invariant set for f^n, $\mathcal{O}(\Lambda, f)$. However, $\mathcal{O}(\Lambda, f)$ is not the largest natural invariant set in a neighborhood of $\Lambda_{\mathbf{q}}$, because $\mathcal{O}(\Lambda, f)$ only has periodic points which are multiples of n; $\Lambda_{\mathcal{B}}$ has points of all periods larger than n (if $\mathbf{p}$ is a fixed point). Another way to see the difference is that $f|\Lambda_{\mathcal{B}}$ is *topologically mixing* (if $\mathbf{p}$ is a fixed point) while $f|\mathcal{O}(\Lambda)$ is topologically transitive but not topologically mixing.

Still another way to characterize the difference is in terms of the topological entropy, which we define in Section 9.1, which is a measure of the complexity. The characteristic polynomial of the matrix A of Theorem 4.5(b) is $p(\lambda) = \lambda^n - \lambda^{n-1} - 1$. Since $p(2^{1/n}) = 2 - 2^{(n-1)/n} - 1 < 0$, the largest eigenvalue of A is larger than $2^{1/n}$. In Section 9.1 we show that the logarithm of this eigenvalue is equal the topological entropy of $f|\Lambda_{\mathcal{B}}$. On the other hand, $f^n|\Lambda$ has the same entropy as $\sigma_2|\Sigma_2$ which equals $\log(2)$. By further results in Section 9.1, $f|\mathcal{O}(\Lambda)$ has entropy $(1/n)\log(2)$. Therefore, $f|\Lambda_{\mathcal{B}}$ has more entropy than $f|\mathcal{O}(\Lambda)$ and so has more complex dynamics.

REMARK 4.8. If $\mathbf{p}$ is not a fixed point but has period p, then the subshift has a cycle of period p rather than a fixed point. Therefore, $a_{i,j} = 1$ in the following cases:

$$\begin{gathered} i = 1 \text{ and } j = 2, p+1, \\ 2 \le i < p \text{ and } j = i+1, \\ i = p \text{ and } j = 1, \\ p+1 \le i < n \text{ and } j = i+1, \text{ and} \\ i = n \text{ and } j = 1. \end{gathered}$$

For all other (i,j), $a_{i,j} = 0$. Thus, the transition matrix is

$$A = \begin{pmatrix} 0 & 1 & \dots & 0 & 1 & 0 & \dots & 0 & 0 \\ \vdots & & & & & & & & \vdots \\ 0 & 0 & \dots & 1 & 0 & 0 & \dots & 0 & 0 \\ 1 & 0 & \dots & 0 & 0 & 0 & \dots & 0 & 0 \\ 0 & 0 & \dots & 0 & 0 & 1 & \dots & 0 & 0 \\ \vdots & & & & & & & & \vdots \\ 0 & 0 & \dots & 0 & 0 & 0 & \dots & 1 & 0 \\ 0 & 0 & \dots & 0 & 0 & 0 & \dots & 0 & 1 \\ 1 & 0 & \dots & 0 & 0 & 0 & \dots & 0 & 0 \end{pmatrix}.$$

We leave the details to the reader.

In the next subsection, we show how a transverse homoclinic point arises from a time periodic perturbation of a differential equation with a nontransverse homoclinic connection. In the remainder of this subsection, we discuss a more geometric construction of a perturbation which changes a nontransverse homoclinic connection into a transverse homoclinic point.

Example 4.1. We start by giving the construction of a diffeomorphism in $\mathbb{R}^2$ with a nontransverse homoclinic point. In fact, our example has one branch of the stable manifold coinciding with one branch of the unstable manifold for a saddle fixed point. The simplest construction of such a diffeomorphism is by means of the flow of a system of differential equations. Let φ^t be the flow of the system of differential equations

$$\begin{aligned} \dot{x}_1 &= x_2 \\ \dot{x}_2 &= x_1 - x_1^2. \end{aligned}$$

The origin is a saddle fixed point for φ^t. The real valued function $H(\mathbf{x}) = x_2^2 - x_1^2/2 + x_1^3/3$ is an integral of motion, $\dot{H}(\mathbf{x}) \equiv 0$. (The analysis of the next subsection derives this function as the sum of the kinetic and potential energies.) Using the level sets of H, it can be seen that

$$\begin{aligned} W^s(\mathbf{0}, \varphi^t) \cap \{\mathbf{x} : x_1 > 0\} &= W^u(\mathbf{0}, \varphi^t) \cap \{\mathbf{x} : x_1 > 0\} \\ &= \{\mathbf{x} : x_2 = \pm\Big(\frac{x_1^2}{2} - \frac{x_1^3}{3}\Big)^{1/2}, x_1 > 0\}. \end{aligned}$$

See Figure 4.9. Let f be the time one flow of φ^t, $f(\mathbf{x}) = \varphi^1(\mathbf{x})$. Then,

$$\begin{aligned} W^s(\mathbf{0}, f) \cap \{\mathbf{x} : x_1 > 0\} &= W^u(\mathbf{0}, f) \cap \{\mathbf{x} : x_1 > 0\} \\ &= \{\mathbf{x} : x_2 = \pm\Big(\frac{x_1^2}{2} - \frac{x_1^3}{3}\Big)^{1/2}, x_1 > 0\}, \end{aligned}$$

so f has a nontransverse homoclinic connection for the fixed point $\mathbf{0}$.

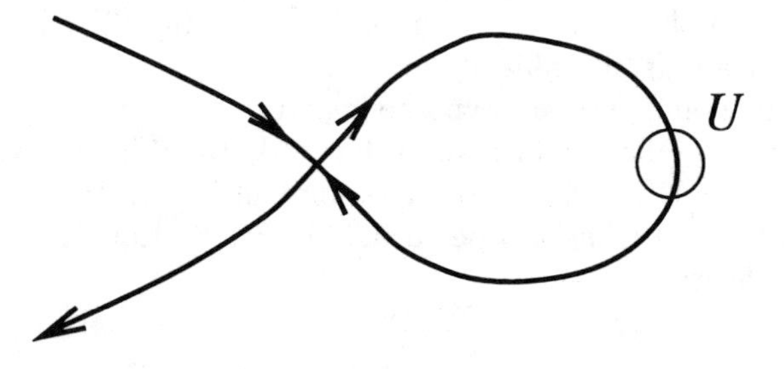

FIGURE 4.9. Homoclinic Connection for Example 4.1

The next step in the construction is to perturb f to a new diffeomorphism g which has a transverse homoclinic point. Let $\mathbf{x}^* = (1.5, 0)$ be the point where the homoclinic connection crosses the x_1-axis. Let U be a relatively small neighborhood of $\mathbf{x}^*$ which satisfies the following properties:

(i) $f^{-1}(U) \cap U = \emptyset$ and $f(U) \cap U = \emptyset$, and

(ii) letting $I = W^s(\mathbf{0}, f) \cap U$, $U \cap \bigcup_{j \neq 0} f^j(I) = \emptyset$.

Let $U' \subset U$ be a smaller neighborhood of $\mathbf{x}^*$. Let $\beta(\mathbf{x})$ be a nonnegative real valued bump function such that

$$\beta(\mathbf{x}) = \begin{cases} 0 & \text{for } \mathbf{x} \notin U \quad \text{and} \\ 1 & \text{for } \mathbf{x} \in U'. \end{cases}$$

We use the bump function to define the perturbation k_ϵ by

$$k_\epsilon(\mathbf{x}) = \mathbf{x} + \epsilon\beta(\mathbf{x}) \begin{pmatrix} x_2 \\ 0 \end{pmatrix},$$

and the new diffeomorphism g_ϵ by

$$g_\epsilon(\mathbf{x}) = k_\epsilon \circ f(\mathbf{x}).$$

(Notice that k_ϵ is a sheer near $\mathbf{x}^*$.)

We defer to Exercise 8.20 the verification of the following statements for small enough $\epsilon > 0$:

(a) g_ϵ is a diffeomorphism,

(b) $\mathbf{0}$ is a saddle fixed point for g_ϵ,

(c) letting $I = W^s(\mathbf{0}, f) \cap U$,

$$\bigcup_{j=0}^{\infty} f^j(I) \subset W^s(\mathbf{0}, g_\epsilon),$$

$$\bigcup_{j=-1}^{-\infty} f^j(I) \subset W^u(\mathbf{0}, g_\epsilon), \quad \text{and}$$

$$k_\epsilon(I) \subset W^u(\mathbf{0}, g_\epsilon),$$

and

(d) $\mathbf{x}^*$ is a transverse homoclinic point for g_ϵ.

Notice that the perturbation by composition with k_ϵ changes the unstable manifold in U but leaves the stable manifold unchanged in U. desired result.

REMARK 4.9. The Kupka-Smale Theorem states that any diffeomorphism f can be C^r-approximated by a diffeomorphism g for which

(i) all the periodic points of g are hyperbolic and

(ii) for any pair of periodic points $\mathbf{p}$ and $\mathbf{q}$ for g, $W^s(\mathbf{p}, g)$ is transverse to $W^u(\mathbf{q}, g)$.

The above example (and Exercise 8.20) gives an explicit example of the construction which makes condition (ii) true for a perturbation. See Section 11.1 for a discussion of the Kupka-Smale Theorem.

8.4.3 Nontransverse Homoclinic Point

In the last section, we analyzed the dynamics which are the result of a transverse homoclinic point. In this section, we consider the case when the stable and unstable manifolds are not transverse at the homoclinic point but they "cross" (are topologically transverse). The ideas of this section go back to work of Easton and McGehee on correctly aligned windows. The application to nontransverse homoclinic points was not made until the recent work of Burns and Weiss (1995) and Carbinatto, Kwapisz, and Mischaikow (1997). More recently, Kennedy and Yorke have given talks on related results.

We start by considering the case of a diffeomorphism in two dimensions where the idea of "topologically transverse" is simpler and reduces to the idea of the two sides of a curve. Two curves N_1 and N_2 in a two-dimensional ambient manifold M^2 which intersect at a point $\mathbf{q}$ are *topologically transverse* at $\mathbf{q}$ provided that locally near $\mathbf{q}$, N_1 crosses from one side of N_2 to the other side. We are assuming that the point of intersection $\mathbf{q}$ is isolated in N_1 and N_2. In the context of homoclinic points, we are assuming that the stable and unstable manifolds have a locally isolated intersection at the homoclinic point. This is not necessary, as is discussed in Burns and Weiss (1995).

If the the stable and unstable manifolds are tangent and do not even cross at a homoclinic point, then there is not necessarily an invariant horseshoe: it is possible for the map to have a finite or countable chain recurrent set. Therefore, to get complicated dynamics, we need to assume that the stable and unstable manifolds cross at the homoclinic point.

The following theorem states that if f has a topologically transverse homoclinic point $\mathbf{q}$ for a periodic point $\mathbf{p}$, then f has an invariant set $\Lambda_{\mathcal{B}}$ that is semi-conjugate to a subshift of finite type, i.e., f contains dynamics which are at least as complicated as a diffeomorphism with a transverse homoclinic point. The statement of the theorem is the same in higher dimensions, as long as we define the concept of a topologically transverse intersection, which we do later in the section.

Theorem 4.6. *Assume f is a diffeomorphism with a periodic point $\mathbf{p}$ and a topologically transverse homoclinic point for $\mathbf{q}$ for $\mathbf{p}$. Let V be a neighborhood of $\mathcal{O}(\mathbf{p}) \cup \mathcal{O}(\mathbf{q})$. Then, there is a smaller neighborhood $\mathcal{B} = \bigcup_{i=1}^{n} B_i$ of $\mathcal{O}(\mathbf{p}) \cup \mathcal{O}(\mathbf{q})$ with $n \geq 2$, $\mathcal{B} \subset V$, such that for the maximal invariant set in $\mathcal{B}$, $\Lambda_{\mathcal{B}} = \bigcap_{j \in \mathbb{Z}} f^j(\mathcal{B}) \subset V$, there is a semiconjugacy h of $f|\Lambda_{\mathcal{B}}$ to the shift map σ_A on a transitive two-sided subshift of finite type on n symbols, $\Sigma_A \subset \Sigma_n$. The semiconjugacy $h : \Lambda_{\mathcal{B}} \to \Sigma_A$ is the itinerary function given by $h(\mathbf{x}) = \mathbf{s}$ where $f^j(\mathbf{x}) \in B_{s_j}$ for all j. The subshift Σ_A is the same as in the case of a transverse homoclinic point considered in Theorem 4.5. Both $\mathbf{q}, \mathbf{p} \in \Lambda_{\mathcal{B}}$, so $\mathbf{q} \in \mathrm{cl}(\mathrm{Per}(f))$.*

PROOF. We only consider the two-dimensional case. We will not give the details of the proof but only indicate the main differences from the proof of Theorem 4.5. The main idea is to look at the intersection $\bigcap_{i \in \mathbb{Z}} f^i(\mathcal{B})$ without first shrinking the neighborhood to give hyperbolicity.

In the construction of the set $\mathcal{B}$, it is important that the image $f^{k+j}(D^s \times f^{-j}(D^u))$ extends from the top to the bottom of $f^{-k}(D^s \times f^{-j}(D^u))$ and does not intersect the sides. See Figure 4.10. We no longer claim that the "fibers" $f^{k+j}(\{\mathbf{x}\} \times f^{-j}(D^u))$ cross the horizontal disks $f^{-k}(D^s \times \{\mathbf{y}\})$ in unique points.

The boxes B_j which make up $\mathcal{B}$ are constructed so that (i) $f(B_1)$ crosses from top to bottom of both B_1 and B_2, but does not intersect B_i for $i > 2$; (ii) $f(B_n)$ crosses B_1 from top to bottom, but $f(B_n) \cap B_i = \emptyset$ for $i \geq 2$; and (iii) $f(B_j) = B_{j+1}$ for $2 \leq j \leq n-1$. In each case, these images do not come out the "sides" of the boxes that they intersect and extend all the way from the top to the bottom.

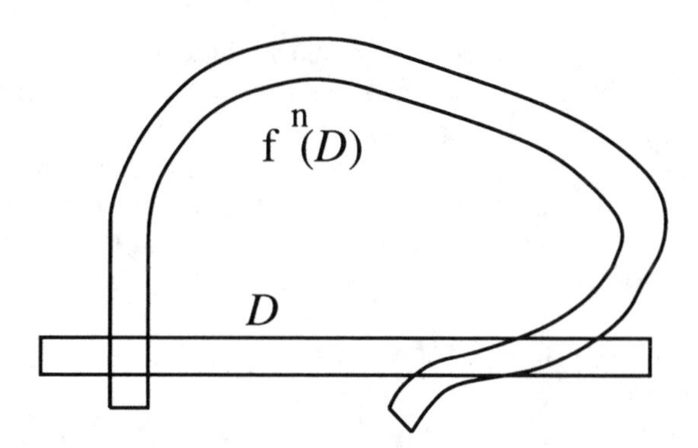

FIGURE 4.10. Nontransverse Intersection of $\mathcal{D}$ and $f^n(\mathcal{D}) = f^{2k+j}(\mathcal{D})$ for $\mathcal{D} = f^{-k}(D^s \times f^{-j}(D^u))$

Just as in the in proof of Theorem 4.5, for an allowable sequence $\mathbf{s}$, the intersection $\bigcap_{0\le i\le n} f^i(B_{s_{-i}})$ and $\bigcap_{0\le i} f^i(B_{s_{-i}})$ extends from the top to the bottom of B_{s_0} and separates the two sides of B_{s_0}. Similarly, $\bigcap_{0\ge i} f^i(B_{s_{-i}})$ extends between the two sides and separates the top from the bottom. Therefore, $\bigcap_{i\in\mathbb{Z}} f^i(B_{s_{-i}})$ is nonempty. Thus, the map $h : \Lambda_{\mathcal{B}} \to \Sigma_A$ is onto and a semiconjugacy. Since we do not have the hyperbolicity of f on $\Lambda_{\mathcal{B}}$, we cannot prove that $\bigcap_{i\in\mathbb{Z}} f^i(B_{s_{-i}})$ is a single point, and so do not know whether h is one to one or not. □

We end the section by discussing the definition of topologically transverse intersection of two manifolds. Consider the example of ambient space $\mathbb{R}^4$ which we write as $\mathbb{C}^2$. One manifold is $N_1 = \{(z,0) : z \in \mathbb{C}\}$ and the other is $N_2 = \{(z,z^2) : z \in \mathbb{C}\}$. They only intersect at $(0,0)$ in a nontransverse intersection. If we consider a small 1-sphere (circle) in the second manifold around $(0,0)$, it wraps twice around the first manifold. These two manifolds have an intersection number of 2 at $(0,0)$ and so are topologically transverse. Under any small perturbations of the two manifolds, they will continue to intersect and will have at least two points of intersection if the intersections are transverse.

For the general definition of intersection number and topologically transverse, consider two submanifolds N_1 and N_2 of an ambient manifold M such that the sum of the dimensions of N_1 and N_2 equals the dimension of M. Assume that $\mathbf{q}$ is an isolated point in the intersection $N_1 \cap N_2$. Let U be a small neighborhood of $\mathbf{q}$ such that $N_1 \cap U$ and $N_2 \cap U$ are small disks. (These are not necessarily flat disks but the diffeomorphic image of disks.) Let $S(\epsilon, \mathbf{q}, N_2) \subset U$ be the set of points in N_2 which are exactly distance ϵ from $\mathbf{q}$, for small $\epsilon > 0$. (This is the sphere of radius ϵ about $\mathbf{q}$ in N_2.) The *intersection number* of N_1 and N_2 at $\mathbf{q}$ is nonzero provided $S(\epsilon, \mathbf{q}, N_2)$ cannot be deformed to a point within $U \setminus N_1$. In terms of algebraic topology, this can be expressed by saying that the homology class of $S(\epsilon, \mathbf{q}, N_2)$ is nonzero in $H^{n_2-1}(U \setminus N_1)$ where n_2 is the dimension of N_2. The two submanifolds N_1 and N_2 are *topologically transverse* at $\mathbf{q}$ provided the intersection number of N_1 and N_2 at $\mathbf{q}$ is nonzero.

If N_1 and N_2 are one-dimensional, then $U \setminus N_1$ has two connected components and the curves are topologically transverse if the two points in the zero sphere $S(\epsilon, \mathbf{q}, N_2)$ are in different connected components. In the example above in $\mathbb{C}^2$, the one sphere $S(\epsilon, \mathbf{0}, N_2)$ wraps twice around the plane N_1, so the homology class of $S(\epsilon, \mathbf{0}, N_2)$ in $H^1(U \setminus N_1)$ is ± 2.

The reader can see Hirsch (1976) and Guillemin and Pollack (1974) for further discussion of intersection number. These books define it more globally and use the approximation of the manifolds by ones which are transverse.

8.4.4 Homoclinic Points and Horseshoes for Flows

In this subsection, we explain the structure of the natural invariant set which results from a transverse homoclinic point to a periodic orbit for a flow. The goal is to show that this invariant set is flow equivalent to the suspension of a subshift of finite type.

We start by reviewing (1) the Poincaré map, both local and global, and (2) the suspension of a homeomorphism. After extending those ideas from the previous treatments, we show how they can be applied to a transverse homoclinic point for a periodic orbit of a flow.

In Section 5.5.7, we discussed the local Poincaré map for a cross section (transversal) to a periodic orbit. In this context, the Poincaré map F is the first return map from a neighborhood U in a cross section Σ back to Σ. The same method can be applied between two cross sections. Assume φ^t is a differentiable, C^r $r \geq 1$, flow on a phase space M. Assume that $\frac{d}{dt}\varphi^t(\mathbf{p}) \neq 0$, $T > 0$, and $\mathbf{q} = \varphi^T(\mathbf{p})$. Let $\Sigma_{\mathbf{p}}$ be a local transversal at $\mathbf{p}$, and $\Sigma_{\mathbf{q}}$ be a local transversal at $\mathbf{q}$. The same proof as that of Theorem 5.8.1 using the Implicit Function Theorem proves that there exists a neighborhood $V_{\mathbf{p}}$ of $\mathbf{p}$ in $\Sigma_{\mathbf{p}}$ and a C^r function $\tau : V_{\mathbf{p}} \to \mathbb{R}$ such that $\tau(\mathbf{p}) = T$, and

$$F(\mathbf{x}) = \varphi^{\tau(\mathbf{x})}(\mathbf{x}) \in \Sigma_{\mathbf{q}}.$$

In particular, $F(\mathbf{p}) = \mathbf{q}$. The map $F : V_{\mathbf{p}} \subset \Sigma_{\mathbf{p}} \to \Sigma_{\mathbf{q}}$ is the local *Poincaré map.*

A periodic differential equation is one type of system which naturally results in a global Poincaré map. An example of a such an equation is given by

$$\begin{aligned}
\frac{d}{dt}q &= p \\
\frac{d}{dt}p &= q - q^3 + \epsilon\gamma\cos(\omega t) - \epsilon\delta p.
\end{aligned}$$

This system has period $2\pi/\omega$ in t. The general case is a differential equation of the form

$$\frac{d}{dt}\mathbf{x} = f(\mathbf{x}, t)$$

where there is a $T > 0$ such that $f(\mathbf{x}, t) = f(\mathbf{x}, t + T)$ for all $\mathbf{x}$ and t. Since these equations explicitly depend on the time variable t, they are call *non-autonomous.* It is easier to analyze these equations if we introduce a new variable s to make an extended phase space $\tilde{M}$ on which the equations are independent of time, i.e., *autonomous.* The above equations are equivalent to the autonomous equations that are given by

$$\begin{aligned}
\frac{d}{dt}\mathbf{x} &= f(\mathbf{x}, s) \\
\frac{d}{dt}s &= 1 \qquad \mod T.
\end{aligned}$$

Let M be the set of all allowable $\mathbf{x}$ in the original equations. This space M is often some $\mathbb{R}^n$ or $\mathbb{R}^{n-k} \times \mathbb{T}^k$ (k angles). The (extended) phase space of the autonomous differential equations and flow is

$$\tilde{M} = M \times \mathbb{R}/\sim$$

where $(\mathbf{x}, s + T) \sim (\mathbf{x}, s)$. The flow on $\tilde{M}$ is of the form $\tilde{\varphi}^t(\mathbf{x}, s) = (\varphi^t(\mathbf{x}, s), s + t)/\sim$, i.e., the time $s + t$ is taken modulo T. We identify a copy of M with $M \times \{0\}$ and get a global Poincaré map $F : M \to M$ by $F(\mathbf{x}) = \varphi^T(\mathbf{x}, 0)$. Notice that in this context, the return time is a constant T. The Poincaré map gives a stroboscopic view of the flow, looking at the location of the trajectories at times which are multiples of the period T. The reason why this is helpful is that the trajectories $\varphi^t(\mathbf{x}, 0)$ in M can cross each other in M if they get to the same point at different times in the cycle of the periodic variable s. By uniqueness of solutions, the trajectories of $\tilde{\varphi}^t$ cannot cross in $\tilde{M}$. The advantage of using the Poincaré map rather than the extended flow $\tilde{\varphi}^t$ is that it does not increase the dimension. Notice that if $F(\mathbf{x}_0) = \mathbf{x}_0$, then $\{\tilde{\varphi}^t(\mathbf{x}_0, 0)\}_{0 \le t \le T}$ is a periodic orbit of period T. If $F^k(\mathbf{x}_0) = \mathbf{x}_0$, then $\{\tilde{\varphi}^t(\mathbf{x}_0, 0)\}_{0 \le t \le kT}$ is a periodic orbit of period kT.

In Subsection 8.4.3 we show that the equations

$$\begin{aligned}
\frac{d}{dt}q &= p \\
\frac{d}{dt}p &= q - q^3 + \epsilon\gamma\cos(\omega s) - \epsilon\delta p \\
\frac{d}{dt}s &= 1 \qquad \text{mod } 2\pi/\omega
\end{aligned}$$

have a transverse homoclinic point to a periodic orbit. In particular, there is a periodic orbit of period $2\pi/\omega$ near $\{(0,0,s)\}_{0 \le s \le 2\pi/\omega}$ which is a saddle point for small ϵ and δ. Then, there is an $\epsilon_0(\delta, \gamma, \omega) > 0$ such that for δ, γ, and ω satisfying

$$0 < \frac{|\delta|}{\gamma} < \frac{3 \cdot 2^{\frac{1}{2}}}{4}\pi\omega\,\mathrm{sech}\,(\frac{\pi\omega}{2})$$

and $0 < |\epsilon| \le \epsilon_0(\delta, \gamma, \omega)$, there is a transverse homoclinic point of this saddle periodic orbit of the flow and also for a saddle fixed point of the Poincaré map. Below we explain the structure of the invariant set for the flow which corresponds to the horseshoe of the Poincaré map.

As indicated in Section 5.8.1, it is always possible to take the suspension of a homeomorphism. We sometimes consider maps which are not differentiable; in particular, we talk about the suspensions of a subshift of finite type for which the domain is a metric space but not a manifold. Given a homeomorphism $f : M \to M$ and $T > 0$, we consider the space $M \times \mathbb{R}$ with the equivalence relation, $(\mathbf{x}, s+T) \sim (f(\mathbf{x}), s)$. Then, we consider the quotient space

$$\tilde{M} = M \times \mathbb{R}/\sim$$

where we identify points which are equivalent under the above equivalence relation. To get all points in $\tilde{M}$, it is enough to consider $0 \le s \le T$, but the other points are included because they make it clear that the quotient space has a C^r structure if f is a C^r diffeomorphism. Now, we consider the differential equations on $M \times \mathbb{R}$ given by

$$\begin{aligned}
\dot{\mathbf{x}} &= 0 \\
\dot{s} &= 1.
\end{aligned}$$

This differential equation induces a flow φ^t on $M \times \mathbb{R}$ given by $\varphi^t(\mathbf{x}, s) = (\mathbf{x}, s+t)$ which passes to a flow $\tilde{\varphi}^t$ on the quotient space $\tilde{M}$ where $\tilde{\varphi}^t(\mathbf{x}, s) = (\mathbf{x}, s+t)/\sim$. In particular, $s + t$ for this flow is taken modulo T. Notice that the set $\Sigma = M \times \{0\}$ forms a global cross section, for which the Poincaré map $F : \Sigma \to \Sigma$ is given by $F(\mathbf{x}, 0) = \tilde{\varphi}^T(\mathbf{x}, 0) = (\mathbf{x}, T) \sim (f(\mathbf{x}), 0)$. Therefore, the flow on $\tilde{M}$ indeed has the original homeomorphism f as its (global) Poincaré map. This ends our discussion of the construction of the suspension for constant time T.

As preparation for the development below, it is helpful to introduce the idea of a variable time suspension. Let $\tau : M \to (0, \infty)$ be a Lipschitz function (often C^1). Consider

$$\hat{M} = \{(\mathbf{x}, s) : \mathbf{x} \in M, \ 0 \leq s \leq \tau(\mathbf{x})\}.$$

Introduce the equivalence $(\mathbf{x}, \tau(\mathbf{x})) \sim (f(\mathbf{x}), 0)$, and let $\tilde{M} = \hat{M}/\sim$. The flow $\varphi^t(\mathbf{x}, s) = (\mathbf{x}, s+t)$ on $\hat{M}$ induces a flow $\tilde{\varphi}^t$ on $\tilde{M}$ which has the Poincaré map $F : M \times \{0\} \to M \times \{0\}$ with return time $\tau(\mathbf{x})$ and satisfying $F(\mathbf{x}, 0) = (f(\mathbf{x}), 0)$.

To start our description of a horseshoe for a flow, we consider the suspension of a diffeomorphism with a transverse homoclinic point. Let $f : M \to M$ be such a diffeomorphism with $f(\mathbf{p}) = \mathbf{p}$ and $\mathbf{q}$ a transverse homoclinic point to $\mathbf{p}$. Let $\{B_j\}_{j=1}^N$ be the ambient boxes which satisfy the Markov property as given in Section 8.4.2, with $\mathbf{p} \in B_1$, $\mathcal{B} = \bigcup_{j=1}^N B_j$, and $\Lambda = \bigcap_{n \in \mathbb{Z}} f^n(\mathcal{B})$. By taking a different point on the orbit of $\mathbf{q}$, we can take $\mathbf{q} \in B_2$ (although this is for notation convenience and is not really necessary). Let (σ_A, Σ_A) be the subshift of finite type which is conjugate to $f|\Lambda$ by the homeomorphism $h : \Lambda \to \Sigma_A$. Take the suspension of f by the variable time $\tau : M \to \mathbb{R}$, which yields the flow $\tilde{\varphi}^t$ on $\tilde{M}$. Let $\tilde{\Lambda} = \Lambda \times \mathbb{R}/\sim$ be the suspension of Λ using τ, which has the flow $\tilde{\varphi}^t|\tilde{\Lambda}$. It can be checked that $\tilde{\Lambda}$ is an isolated invariant set for the flow $\tilde{\varphi}^t$ with isolating neighborhood

$$\{(\mathbf{x}, s) : \mathbf{x} \in \mathcal{B}, \ 0 \leq s \leq \tau(\mathbf{x})\}/\sim .$$

Let $\tilde{\Sigma}_A$ be the suspension of σ_A using the time $\tau \circ h^{-1}$ with flow $\tilde{\psi}^t$. Then, h induces a flow conjugacy $\tilde{h}$ of $\tilde{\varphi}^t|\tilde{\Lambda}$ to $\tilde{\psi}^t$ on $\tilde{\Sigma}_A$, i.e., the flow on the invariant set for the suspension of the diffeomorphism is flow conjugate to the flow induced on the suspension of a subshift of finite type. If we use the constant time suspension of σ_A, then it is only flow equivalent to $(\tilde{\varphi}^t|\tilde{\Lambda})$.

We next show that we can always use the suspension of a simpler subshift type. In the context of a flow, this is natural, while the corresponding construction for a diffeomorphism does not seem so natural. Let $\tilde{\tau} : \Lambda \cap (B_1 \cup B_2) \to \mathbb{R}$ be the return time for $\tilde{\varphi}^t$ from $[\Lambda \cap (B_1 \cup B_2)] \times \{0\}$ to $[\Lambda \cap (B_1 \cup B_2)] \times \{0\}$. This return time induces a Poincaré map $P : \Lambda \cap (B_1 \cup B_2) \to \Lambda \cap (B_1 \cup B_2)$. Since a point in $\Lambda \cap B_1$ can flow to either $\Lambda \cap B_1$ or $\Lambda \cap B_2$, while a point in $\Lambda \cap B_2$ can only flow to $\Lambda \cap B_1$, the Poincaré map P on $\Lambda \cap (B_1 \cup B_2)$ is conjugate by a homeomorphism k to the subshift of finite type (σ_B, Σ_B) with transition matrix

$$A' = \begin{pmatrix} 1 & 1 \\ 1 & 0 \end{pmatrix}.$$

The suspension of $P|[\Lambda \cap (B_1 \cup B_2)]$ by variable time $\tilde{\tau}$ is easily flow conjugate to (identified with) $\tilde{\varphi}^t|\tilde{\Lambda}$. Thus, the flow $(\tilde{\varphi}^t, \tilde{\Lambda})$ is flow conjugate to the suspension of the subshift of finite type $(\sigma_{A'}, \Sigma_{A'})$ with return time $\tilde{\varphi}^t \circ k^{-1}$. Thus, if we are willing to use longer return times, it is always possible to take a subshift of finite type with just two symbols.

There is not much difference if we could start with a flow $\tilde{\varphi}^t$ on a manifold $\tilde{M}$ with a global cross section M with Poincaré map f (e.g., from a periodically forced differential equation). If the flow $\tilde{\varphi}^t$ has a transverse homoclinic point $\mathbf{q}$ to a periodic orbit γ with $\mathbf{p} \in \gamma$, then we could repeat the above construction and get an isolated invariant set $\tilde{\Lambda}$ which is flow conjugate (or flow equivalent) to the suspension of a subshift of finite type.

Now, we turn to the general case of a transverse homoclinic point $\mathbf{q}$ to a periodic orbit γ for a flow φ^t. We take local cross sections $\Sigma_{\mathbf{q}}$ at $\mathbf{q}$ and $\Sigma_{\mathbf{p}}$ at $\mathbf{p} \in \gamma$. It is possible to carefully choose rectangles B_1 and B_2 with $\mathbf{p} \in B_1 \subset \Sigma_{\mathbf{p}}$ and $\mathbf{q} \in B_2 \subset \Sigma_{\mathbf{q}}$ and neighborhoods U_1 of B_1 in $\Sigma_{\mathbf{p}}$ and U_2 of B_2 in $\Sigma_{\mathbf{q}}$, and a (discontinuous) Poincaré map $P : U_1 \cup U_2 \to \Sigma_{\mathbf{p}} \cup \Sigma_{\mathbf{q}}$ such that (i) $P|U_1$ is the first return to $U_1 \cup \Sigma_{\mathbf{q}}$, (ii) $P|U_2$ is the first return to $\Sigma_{\mathbf{p}}$ (and does not intersect $\Sigma_{\mathbf{q}}$ in between), and (iii) P has the Markov property for $\{B_1, B_2\}$. Since P is continuous on $B_2 \cup \{B_1 \cap [P^{-1}(B_1) \cup P^{-1}(B_2)]\}$, it is continuous on the invariant set. The reason that P is discontinuous is that nearby to an orbit which returns to the boundary of U_1 is one which misses and first intersects $\Sigma_{\mathbf{q}}$. In terms of the transitions allowable, $P(B_1) \cap B_1 \neq \emptyset$, $P(B_1) \cap B_2 \neq \emptyset$, $P(B_2) \cap B_1 \neq \emptyset$, and $P(B_2) \cap B_2 = \emptyset$, so the invariant set for P has the matrix A' above as its transition matrix. Then, the corresponding isolated invariant set Λ for φ^t is flow conjugate to the suspension of the subshift of finite type $\Sigma_{A'}$. We leave the details to the reader.

Finally, we remark that Moser (1973) uses a different set of symbolic dynamics. He takes infinitely many ambient rectangles

$$R_j = \{\mathbf{x} \in B_2 : P^j(\mathbf{x}) \in B_2 \text{ and } P^i(\mathbf{x}) \notin B_2 \text{ for } 0 < i < j\}$$

for $j \geq N$, which satisfy the Markov property with infinitely many symbols. All of these rectangles are near the homoclinic point $\mathbf{q}$ and not $\mathbf{p}$. The invariant set which corresponds to these symbols is a subset of the invariant set Λ which we described above (it does not include the periodic orbit γ itself). In the case of a flow with a global cross section (e.g., periodically forced differential equation), the ambient rectangles R_j which Moser uses correspond to the number of iterates of the Poincaré map (periods for periodically force) that occur before the point returns to the transversal $\Sigma_{\mathbf{q}}$. Thus, the symbol has an intrinsic meaning. The Poinaré map from the cross section near the homoclinic point to itself is used to describe the dynamics rather than also using a transversal near the periodic orbit. (It should be mentioned that in the situation described in Moser's book, the periodic orbit is at "infinity" in an extended phase space outside of the physical positions which correspond to the usual phase space.)

Finally, we note why we have been talking about a transverse homoclinic point to a periodic orbit and not a fixed point. For a flow, it the stable and unstable manifolds intersect in a point $\mathbf{q}$, then they intersect along the whole orbit through $\mathbf{q}$. Thus, for the stable and unstable manifolds to be transverse, the sum of their dimensions must be at least one bigger than the dimension of the ambient space. For a fixed point $\mathbf{p}$,

$$\dim(W^s(\mathbf{p})) + \dim(W^u(\mathbf{p})) = \dim(M),$$

while for a periodic orbit γ,

$$\dim(W^s(\gamma)) + \dim(W^u(\gamma)) = \dim(M) + 1.$$

Therefore, it is possible for $W^s(\gamma)$ and $W^u(\gamma)$ to be transverse but not for $(W^s(\mathbf{p})$ and $W^u(\mathbf{p})$. Shilnikov has discovered how homoclinic points for fixed points can induce a "horseshoe" for a flow in certain situations, but it is more complicated than what we have described above.

8.4.5 Melnikov Method for Homoclinic Points

In the last section we showed how a horseshoe arises from a transverse homoclinic point for a hyperbolic periodic point. In this section, we give one way to verify that a system of Hamiltonian differential equations has a transverse homoclinic point. This approach goes back to Poincaré, but its recent use starts with Melnikov (1963). Some people call this the Poincaré-Melnikov-Arnold method. The method can be applied to time-independent Hamiltonian systems (see Robinson (1988)) or even non-Hamiltonian differential equations, Gruendler (1985)). For a more complete treatment of this type of result, see Wiggins (1988) or (1990).

The simplest and original context for the Melnikov method is for a time periodic perturbation of a Hamiltonian system. Such systems are introduced in Section 6.1. In particular, Example VI.1.2 discusses the phase portrait of the equations

$$\begin{aligned}\dot{q} &= p,\\ \dot{p} &= q - q^3 = f(q)\end{aligned}$$

The perturbation of these equations we analyze below using the Melnikov integral is obtained by adding a periodic forcing term and a "frictional" term:

$$\begin{aligned}\dot{q} &= p,\\ \dot{p} &= q - q^3 + \epsilon\gamma\cos(\omega t) - \epsilon\delta p.\end{aligned}$$

This system of equations is of the form

$$\begin{pmatrix}\dot{\mathbf{q}}\\ \dot{\mathbf{p}}\end{pmatrix} = X_H\begin{pmatrix}\mathbf{q}\\ \mathbf{p}\end{pmatrix} + \epsilon Y(\mathbf{q},\mathbf{p},t)$$

where Y has period T in t. We can add another variable τ to make the equations independent of time:

$$\begin{pmatrix}\dot{\mathbf{q}}\\ \dot{\mathbf{p}}\\ \dot{\tau}\end{pmatrix} = \begin{pmatrix}X_H\begin{pmatrix}\mathbf{q}\\ \mathbf{p}\end{pmatrix} + \epsilon Y(\mathbf{q},\mathbf{p},\tau)\\ 1\end{pmatrix} = \hat{X}_\epsilon\begin{pmatrix}\mathbf{q}\\ \mathbf{p}\\ \tau\end{pmatrix} \qquad (*)$$

where the variable τ is taken modulo T. (In the explicit example above, we could take $\dot{\tau} = \omega$ and $\tau = \omega t$ rather than $\tau = t$.)

For the rest of the section we assume that $n = 1$, so p and q are each real variables (or a single angle variable). We assume that X_H has a hyperbolic saddle fixed point (q_0, p_0), so equations $(*)$ have a closed orbit γ_0 for $\epsilon = 0$. Because the eigenvalues (characteristic multipliers) of γ_0 are not equal to 1, for $\epsilon \neq 0$ but small, there persists a closed orbit γ_ϵ.

Next, we assume that X_H has a homoclinic orbit for the fixed point, i.e., a point

$$(q,p) \in [W^s((q_0,p_0),X_H) \cap W^u((q_0,p_0),X_H)] \setminus \{(q_0,p_0)\}.$$

In the (q,p,τ)-space, the homoclinic orbit of X_H becomes a homoclinic surface for γ_0 for the equations $(*)$,

$$\Gamma = \{(q,p,\tau) : (q,p) \text{ is on a homoclinic orbit for } X_H\}.$$

For $\epsilon > 0$, the closed orbit γ_ϵ remains hyperbolic and its stable and unstable manifolds vary smoothly with ϵ on compact subsets. For each $\mathbf{z}_0 \in \Gamma$, let $\mathbf{z}^s(\mathbf{z}_0,\epsilon)$ be the point where $W^s(\gamma_\epsilon, \hat{X}_\epsilon)$ intersects the normal to Γ through $\mathbf{z}_0$. By the smooth dependence on ϵ, $\mathbf{z}^s(\mathbf{z}_0, 0) = \mathbf{z}_0$ and $\mathbf{z}^s(\mathbf{z}_0,\epsilon)$ is a smooth function of ϵ. Similarly define $\mathbf{z}^u(\mathbf{z}_0,\epsilon)$.

We want to measure the separation of the stable and unstable manifolds in the directions orthogonal to Γ, i.e., the separation $\mathbf{z}^u(\mathbf{z}_0,\epsilon)$ from $\mathbf{z}^s(\mathbf{z}_0,\epsilon)$. The function H is a good measure of a displacement in these directions (since the gradient of H is nonzero at points in Γ), so we want to measure $\hat{G}(\mathbf{z}_0, \epsilon) = H(\mathbf{z}^u(\mathbf{z}_0, \epsilon)) - H(\mathbf{z}^s(\mathbf{z}_0, \epsilon))$. Since $\hat{G}(\mathbf{z}_0, 0) \equiv 0$, it is possible to write

$$\hat{G}(\mathbf{z}_0, \epsilon) = H(\mathbf{z}^u(\mathbf{z}_0, \epsilon)) - H(\mathbf{z}^s(\mathbf{z}_0, \epsilon)) = \epsilon G(\mathbf{z}_0, \epsilon).$$

A zero of $G(\mathbf{z}_0, \epsilon)$ corresponds to a homoclinic point. Since we want to measure the rate of separation with respect to ϵ (the infinitesimal separation), we define

$$\begin{aligned} M(\mathbf{z}_0) &= \frac{\partial}{\partial \epsilon} H(\mathbf{z}^u(\mathbf{z}_0, \epsilon))|_{\epsilon=0} - \frac{\partial}{\partial \epsilon} H(\mathbf{z}^s(\mathbf{z}_0, \epsilon))|_{\epsilon=0} \\ &= G(\mathbf{z}_0, 0), \end{aligned}$$

which is called the *Melnikov function.* The function M is considered a function from Γ to the real numbers. Then, $G(\mathbf{z}_0, \epsilon)$ equals $M(\mathbf{z}_0)$ plus terms involving ϵ. A zero of M corresponds to a place where infinitesimally the stable and unstable manifold continue to intersect. In fact, the following theorem, which is a direct consequence of the implicit function theorem applied to G, gives a criterion that the manifolds actually intersect for $\epsilon \neq 0$. (See Melnikov (1963), Holmes (1980), or Marsden (1984) for a proof.) The use of this function to prove the existence of a transverse homoclinic point and a horseshoe is referred to as applying the *Melnikov method.*

Theorem 4.6. *Suppose $\mathbf{z}_0$ is a point on Γ with $M(\mathbf{z}_0) = 0$ and some directional derivative $\frac{\partial M}{\partial \mathbf{v}}(\mathbf{z}_0) \neq 0$ for $\mathbf{v}$ tangent to Γ. Then, for small enough $\epsilon \neq 0$, γ_ϵ has a transverse homoclinic intersection near $\mathbf{z}_0$. In fact, the point of transverse homoclinic intersection varies smoothly with ϵ.*

As a consequence of this theorem, $\hat{X}_\epsilon$ has a hyperbolic horseshoe near $\mathbf{z}_0$ of the type indicated. In order for this result to be useful, we need a method of calculating M. The next theorem gives just such a result.

Theorem 4.7. *The Melnikov function is given by the following improper integral:*

$$M(\mathbf{z}_0) = \int_{-\infty}^{\infty} DH_{\varphi_0(t,\mathbf{z}_0)} Y(\varphi_0(t, \mathbf{z}_0))\, dt$$

where φ_0 is the flow of X_H for $\epsilon = 0$ and $X_\epsilon = X_H + \epsilon Y$.

PROOF. We need to calculate $\frac{\partial}{\partial \epsilon} H(\mathbf{z}^\sigma(\mathbf{z}_0, \epsilon))$ for $\sigma = u, s$. To do this, we calculate

$$\frac{\partial}{\partial \epsilon} H \circ \varphi(t, \mathbf{z}^\sigma(\mathbf{z}_0, \epsilon), \epsilon)|_{\epsilon=0}$$

along the whole trajectory and, in fact,

$$\frac{d}{dt}\frac{\partial}{\partial \epsilon} H \circ \varphi(t, \mathbf{z}^\sigma(\mathbf{z}_0, \epsilon), \epsilon)|_{\epsilon=0}.$$

From now on, all the derivatives with respect to ϵ are evaluated at $\epsilon = 0$ even though this is not explicitly noted. Then, using several rules of differentiation (including the chain rule and a Leibniz rule),

$$\begin{aligned} \frac{d}{dt}\frac{\partial}{\partial \epsilon} H \circ \varphi(t, \mathbf{z}^\sigma(\mathbf{z}_0, \epsilon), \epsilon) &= \frac{\partial}{\partial \epsilon}\frac{d}{dt} H \circ \varphi(t, \mathbf{z}^\sigma(\mathbf{z}_0, \epsilon), \epsilon) \\ &= \frac{\partial}{\partial \epsilon}\left[DH \cdot (X_H + \epsilon Y)\right]_{\varphi(t,\mathbf{z}^\sigma(\mathbf{z}_0,\epsilon),\epsilon)} \\ &= (DH \cdot Y)_{\varphi(t,\mathbf{z}^\sigma(\mathbf{z}_0,0),0)} \\ &\quad + \frac{\partial}{\partial \epsilon}\left[DH \cdot X_H\right]_{\varphi(t,\mathbf{z}^\sigma(\mathbf{z}_0,\epsilon),\epsilon)} \\ &= (DH \cdot Y)_{\varphi_0(t,\mathbf{z}_0)} \end{aligned}$$

because $DH \cdot X_H \equiv 0$ at all points and $\mathbf{z}^\sigma(\mathbf{z}_0, 0) = \mathbf{z}_0$. Now, integrating between two times T_1 and T_2, we get

$$\frac{\partial}{\partial \epsilon} H \circ \varphi(T_2, \mathbf{z}^\sigma(\mathbf{z}_0, \epsilon), \epsilon) - \frac{\partial}{\partial \epsilon} H \circ \varphi(T_1, \mathbf{z}^\sigma(\mathbf{z}_0, \epsilon), \epsilon) = \int_{T_1}^{T_2} (DH \cdot Y)_{\varphi_0(t, \mathbf{z}_0)} \, dt.$$

For $\sigma = s$, we use $T_1 = 0$ and let $T_2 = T_2'$ go to ∞; for $\sigma = u$, we use $T_2 = 0$ and let $T_1 = T_1'$ go to $-\infty$. Substituting these into the definition of $M(\mathbf{z}_0)$, we get

$$\begin{aligned} M(\mathbf{z}_0) = &\int_{T_1'}^{T_2'} (DH \cdot Y)_{\varphi_0(t, \mathbf{z}_0)} \, dt \\ &+ \frac{\partial}{\partial \epsilon} H \circ \varphi(T_1', \mathbf{z}^u(\mathbf{z}_0, \epsilon), \epsilon) - \frac{\partial}{\partial \epsilon} H \circ \varphi(T_2', \mathbf{z}^s(\mathbf{z}_0, \epsilon), \epsilon). \end{aligned}$$

By the chain rule,

$$\frac{\partial}{\partial \epsilon} H \circ \varphi(T_2', \mathbf{z}^s(\mathbf{z}_0, \epsilon), \epsilon) = DH_{\varphi(T_2', \mathbf{z}_0, 0)} \frac{\partial}{\partial \epsilon} \varphi(T_2', \mathbf{z}^s(\mathbf{z}_0, \epsilon), \epsilon)$$

As T_2' goes to ∞, the point $\varphi(T_2', \mathbf{z}_0, 0)$ goes to the fixed point $\mathbf{P}$ of X_H in (q,p)-space where $DH_{\mathbf{P}} = \mathbf{0}$, so $DH_{\varphi(T_2', \mathbf{z}_0, 0)}$ converges to the zero row vector. This quantity acts on $\frac{\partial}{\partial \epsilon} \varphi(T_2', \mathbf{z}^s(\mathbf{z}_0, \epsilon), \epsilon)$. The point at which this is evaluated, $\varphi(T_2', \mathbf{z}^s(\mathbf{z}_0, \epsilon), \epsilon)$, goes to γ_ϵ, so the limit of $\frac{\partial}{\partial \epsilon} \varphi(T_2', \mathbf{z}^s(\mathbf{z}_0, \epsilon), \epsilon)$ goes to the infinitesimal movement of the closed orbit which is bounded. (This uses the smooth dependence of the stable manifold on the parameter.) Combining, we get that $\frac{\partial}{\partial \epsilon} H \circ \varphi(T_2', \mathbf{z}^s(\mathbf{z}_0, \epsilon), \epsilon)$ goes to 0 as T_2' goes to ∞. Similarly, $\frac{\partial}{\partial \epsilon} H \circ \varphi(T_2', \mathbf{z}^u(\mathbf{z}_0, \epsilon), \epsilon)$ goes to 0 as T_1' goes to $-\infty$. Thus, letting T_1' go to $-\infty$ and T_2' go to ∞, we have proven that the integral converges absolutely and the equality given in the theorem is true. □

We end the section by giving the calculation for the example given above.

Example 4.2. We consider the equations

$$\begin{aligned} \dot{q} &= p \\ \dot{p} &= q - q^3 + \epsilon \gamma \cos(\tau) - \epsilon \delta p \\ \dot{\tau} &= \omega. \end{aligned}$$

The Hamiltonian function is $H(q,p) = p^2/2 - q^2/2 + q^4/4$. The point $(q,p) = (0,0) = \mathbf{0}$ is a saddle fixed point with a homoclinic connection on two sides. A parameterized form of these two connections is given by

$$\begin{pmatrix} q_0^{\pm}(t) \\ p_0^{\pm}(t) \\ \tau(t) \end{pmatrix} = \begin{pmatrix} \pm \sqrt{2} \operatorname{sech}(t) \\ \mp \sqrt{2} \operatorname{sech}(t) \tanh(t) \\ \tau_0 + t\omega \end{pmatrix}$$

as can be verified by differentiation. To apply the theorem, we use one of the two homoclinic orbits. We use the positive side and drop the label '+'. Other parameterizations of homoclinic orbits are given by $(q_0(t - t_0), p_0(t - t_0), \tau(t))$, where t_0 is the time at which $p = 0$. The two variables (t_0, τ_0) parameterize the points on Γ. We actually only need to calculate M on some cross section. Either $t_0 = 0$ or $\tau_0 = 0$ is a reasonable cross section. For now, we keep both variables. Letting $s = t - t_0$ in the integral,

$$
\begin{aligned}
M(t_0, \tau_0) &= \int_{-\infty}^{\infty} \left(-q_0(t-t_0) + q_0^3(t-t_0), p_0(t-t_0) \right) \times \\
&\qquad \begin{pmatrix} 0 \\ \gamma \cos(\tau_0 + t\omega) - \delta p_0(t-t_0) \end{pmatrix} dt \\
&= -\delta \int_{-\infty}^{\infty} p_0(s)^2 \, ds + \gamma \int_{-\infty}^{\infty} p_0(s) \cos(\tau_0 + t_0\omega + \omega s) \, ds \\
&= -\delta \int_{-\infty}^{\infty} p_0(s)^2 \, ds + \gamma \cos(\tau_0 + t_0\omega) \int_{-\infty}^{\infty} p_0(s) \cos(\omega s) \, ds \\
&\qquad - \gamma \sin(\tau_0 + t_0\omega) \int_{-\infty}^{\infty} p_0(s) \sin(\omega s) \, ds.
\end{aligned}
$$

The integrand of the second integral is an odd function of s because $p_0(s)$ is an odd function and $\cos(\omega s)$ is an even function. Therefore, the second integral is 0. Substituting in for $p_0(s)$ in the first integral,

$$
-\delta \int_{-\infty}^{\infty} p_0(s)^2 \, ds = -2\delta \int_{-\infty}^{\infty} \operatorname{sech}^2(s) \tanh^2(s) \, ds = \frac{-4\delta}{3}
$$

as can be directly calculated by a substitution. The third integral is given as follows:

$$
\begin{aligned}
-\gamma \sin(\tau_0 + t_0\omega) &\int_{-\infty}^{\infty} p_0(s) \sin(\omega s) \, ds \\
&= \gamma \sin(\tau_0 + t_0\omega) 2^{\frac{1}{2}} \int_{-\infty}^{\infty} \operatorname{sech}(s) \tanh(s) \sin(\omega s) \, ds \\
&= \gamma \sin(\tau_0 + t_0\omega) 2^{\frac{1}{2}} \omega \int_{-\infty}^{\infty} \cosh(s) \cos(\omega s) \, ds,
\end{aligned}
$$

where the last equality follows by integration by parts. This last integral can be calculated by means of residues (and a line integral over regions with vertices at $\pm R$ and $\pm R + i\pi$) to give

$$
\gamma \sin(\tau_0 + t_0\omega) 2^{\frac{1}{2}} \pi\omega \operatorname{sech}\left(\frac{\pi\omega}{2}\right).
$$

Therefore,

$$
M(t_0, \tau_0) = \frac{-4\delta}{3} + \gamma 2^{\frac{1}{2}} \pi\omega \operatorname{sech}\left(\frac{\pi\omega}{2}\right) \sin(\tau_0 + t_0\omega).
$$

Now, we take the cross section $\tau_0 = 0$, and set

$$
\bar{M}(t_0) = M(t_0, 0) = \frac{-4\delta}{3} + \gamma 2^{\frac{1}{2}} \pi\omega \operatorname{sech}\left(\frac{\pi\omega}{2}\right) \sin(t_0\omega).
$$

This function has a nondegenerate zero as long as

$$
0 < \frac{|\delta|}{\gamma} < \frac{3 \cdot 2^{\frac{1}{2}}}{4} \pi\omega \operatorname{sech}\left(\frac{\pi\omega}{2}\right).
$$

The meaning of this inequality is that as long as there is not too much friction in comparison with the periodic forcing, there is a transverse homoclinic orbit.

8.4.6 Fractal Basin Boundaries

A certain perspective on Dynamical Systems places most of the emphasis on the attracting sets, i.e., sets A such that $\{\mathbf{q} : \omega(\mathbf{q}) \subset A\}$ is an open neighborhood of A. (See Section 8.6 for the actual definition.) If we are mainly interested in attracting sets, then in what way are horseshoes important? One answer to this question is that the stable manifolds of a horseshoe can form the separating set between two different basins of attraction. A good introduction to this type of result is contained in Alligood and Yorke (1989). Also see Grebogi and Yorke (1987). We start with an example.

Example 4.3. Instead of the horseshoe which gives the full two-shift, we consider the horseshoe which gives the full three shift. Figure 4.13 shows a neighborhood N and its image $f(N)$ by a diffeomorphism f on the two sphere. We assume that $f(A) \subset A$ and $f(B) \subset B$ and there is a fixed point sink in each of these regions, $\mathbf{q}_1 \in A$ and $\mathbf{q}_2 \in B$. The square middle region S contains a hyperbolic horseshoe, Λ, such that $f|\Lambda$ is conjugate to the two-sided shift map on three symbols. Considering the diffeomorphism on S^2, we add a source at infinity, $\mathbf{q}_\infty$. If we are careful in the construction, then $\mathcal{R}(f) = \Lambda \cup \{\mathbf{q}_1, \mathbf{q}_2, \mathbf{q}_\infty\}$.

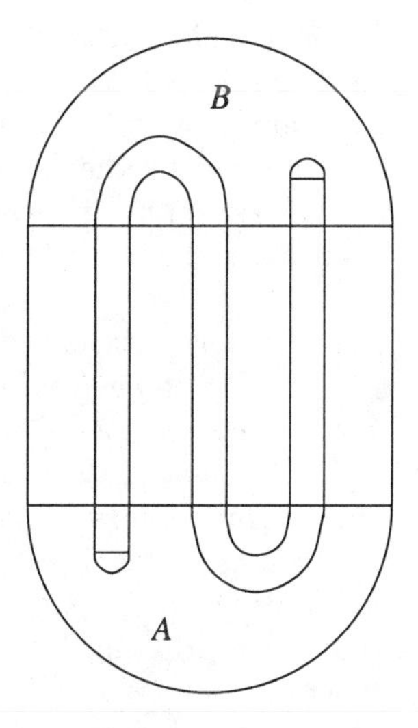

FIGURE 4.13

The sphere can be split up into the stable manifolds of the invariant sets,

$$S^2 = W^s(\mathbf{q}_1) \cup W^s(\mathbf{q}_2) \cup W^s(\Lambda) \cup \{\mathbf{q}_\infty\}.$$

The attracting sets for this example are merely the two fixed point sinks. Their basins are dense in the sphere. The stable manifold of the horseshoe separates them. The reader should determine which points in the gaps $S \setminus f^{-1}(S)$ are in $W^s(\mathbf{q}_1)$ and which are in $W^s(\mathbf{q}_2)$. In this case, the boundary of the basin $W^s(\mathbf{q}_j)$ for either fixed point sink is $W^s(\Lambda) \cup \{\mathbf{q}_\infty\}$. This set has the structure of a Cantor set across the stable manifolds and so is not a smooth manifold. Such sets are called fractal because they have fractional Hausdorff (or box) dimension. We discuss these concepts in Section 9.4.

If $\mathbf{q}$ is a fixed point sink for a diffeomorphism f, let B be the boundary of its basin of attraction. This is the boundary in the sense of point set topology, $B = \text{cl}(W^s(\mathbf{q})) \setminus W^s(\mathbf{q})$. This set B is called the *basin boundary.* In studying a basin boundary, an important feature is which points are accessible from within $W^s(\mathbf{q})$. A point $\mathbf{x} \in B$ is called *accessible* provided there is a curve γ starting in $W^s(\mathbf{q})$ and $\mathbf{x}$ is the first point on γ which is not in $W^s(\mathbf{q})$. In the above example, not all points of $W^s(\Lambda)$ are accessible. If $\mathbf{p}_1$, $\mathbf{p}_2$, and $\mathbf{p}_3$ are the three saddle fixed points in the corresponding vertical strips, then the accessible points in this example are $W^s(\mathbf{p}_1) \cup W^s(\mathbf{p}_3) \cup \{\mathbf{q}_\infty\}$. To understand this fact, let L be a vertical line segment in S from top to bottom. Then, $W^s(\Lambda) \cap L$ is a Cantor set. The only accessible points in $W^s(\Lambda) \cap L$ from within L are the end points of the Cantor set (at a finite stage in its formation). But these end points are exactly the points on the stable manifolds of $\mathbf{p}_1$ and $\mathbf{p}_3$ (and not the middle fixed point $\mathbf{p}_2$). Alligood and Yorke (1989) discusses the question of showing that the basin boundary contains a periodic point. Also see Barge and Gillette (1991).

There have also been a number of papers on understanding how the basin boundary changes from a smooth set to a fractal set under a deformation of the diffeomorphism. See Alligood and Yorke (1989), Grebogi, Ott, and Yorke (1987), Hammel and Jones (1989), and Alligood, Tedeschini-Lalli, and Yorke (1991).

8.5 Hyperbolic Toral Automorphisms

The horseshoe is an example of an invertible map with infinitely many periodic points. It is created on a piece of Euclidean space and so can occur on any manifold. In this section we consider another type of example with infinitely many periodic points which is more global in nature; the dynamics occur on the whole torus. As such, this type of dynamics is not as prevalent as the horseshoe, but it is an important construction of a more global example. We use it later in the chapter to construct an attractor called the DA attractor.

The hyperbolic toral automorphisms were introduced about the same time as the horseshoe examples. R. Thom is credited with suggesting them to Smale as examples with infinitely many periodic points. They are special cases of systems called Anosov diffeomorphisms or flows. Anosov (1967) showed that the geodesic flow on a manifold with negative curvature are examples of Anosov flows. In this section we define a general Anosov diffeomorphism but only analyze the hyperbolic toral automorphisms.

To prove that a horseshoe has infinitely many periodic points, we show that it is conjugate to a subshift of finite type. The proof in this section that a hyperbolic toral automorphism has infinitely many periodic points is more straightforward and algebraic. In Section 10.4, we prove that any Anosov diffeomorphism has infinitely many periodic points by a more analytic or geometric argument.

We start with the definitions.

Definition. A diffeomorphism $f : M \to M$ is called an *Anosov diffeomorphism* provided f has a hyperbolic structure on all of M. It is called a *toral Anosov diffeomorphism* provided in addition M is a torus, $\mathbb{T}^n$.

Construction of hyperbolic toral automorphisms. The examples of toral Anosov diffeomorphisms which we introduce are induced by a linear map on $\mathbb{R}^n$. Let $A = (a_{ij})$ be an n by n matrix with all the a_{ij} integers, $\det(A) = \pm 1$, and A hyperbolic (no eigenvalue of absolute value 1). Let $L_A : \mathbb{R}^n \to \mathbb{R}^n$ be the induced linear map on $\mathbb{R}^n$. Because A has all integer entries, L_A takes the integer lattice $\mathbb{Z}^n$ (points with all integer components) into itself. Because in addition the determinant of A is $\pm$ 1, A^{-1} has integer entries and $(L_A)^{-1} = L_{A^{-1}}$ also takes $\mathbb{Z}^n$ into itself, so $L_A(\mathbb{Z}^n) = \mathbb{Z}^n$. Let $\pi : \mathbb{R}^n \to \mathbb{T}^n$ be the projection which takes a point $\bar{\mathbf{x}} = (x_1, \ldots, x_n)$ in $\mathbb{R}^n$ to the point in the torus by taking each component modulo 1. Because $L_A(\mathbb{Z}^n) = \mathbb{Z}^n$, L_A induces a map f_A from the torus $\mathbb{T}^n$ to itself such that $f_A \circ \pi(\bar{\mathbf{x}}) = \pi \circ L_A(\bar{\mathbf{x}})$: if $\pi(\bar{\mathbf{x}}) = \pi(\bar{\mathbf{x}}')$, then $\bar{\mathbf{x}} = \bar{\mathbf{x}}' + \mathbf{m}$ with $\mathbf{m} \in \mathbb{Z}^n$, so $\pi \circ L_A(\bar{\mathbf{x}}) = \pi \circ L_A(\bar{\mathbf{x}}') + \pi \circ L_A(\mathbf{m}) = \pi \circ L_A(\bar{\mathbf{x}}')$

and $f_A \circ (\mathbf{x})$ is well defined. Because $\det(A) = \pm 1$, the inverse A^{-1} is again a matrix with integer entries, so f_A is a diffeomorphism with $f_A^{-1} = f_{A^{-1}}$. These maps are called *hyperbolic toral automorphisms.*

Example 5.1. Let

$$A = \begin{pmatrix} 1 & 1 \\ 1 & 0 \end{pmatrix} \text{ and } A^2 = \begin{pmatrix} 2 & 1 \\ 1 & 1 \end{pmatrix}.$$

These are both hyperbolic matrices. The eigenvalues of A are $\lambda^{\pm} = (1 \pm \sqrt{5})/2$, $\lambda^{+} \approx 1.618$ and $\lambda^{-} \approx -0.618$, so $|\lambda^{+}| > 1$ and $|\lambda^{-}| < 1$. The eigenvectors are

$$\mathbf{v}^s = \begin{pmatrix} 2 \\ -1 - \sqrt{5} \end{pmatrix} \quad \text{and} \quad \mathbf{v}^u = \begin{pmatrix} 2 \\ -1 + \sqrt{5} \end{pmatrix}$$

for λ^{-} and λ^{+}, respectively. Notice that both eigenvectors have irrational slope. It can be shown that this must be the case for hyperbolic integer matrices of determinant one. Notice that A reverses orientation and has one negative eigenvalue, while A^2 preserves orientation, with two positive eigenvalues, $(\lambda^{\pm})^2$. The eigenvectors for A^2 are the same as those for A.

With the above example as a model, we can now state the main theorem.

Theorem 5.1. *Let f_A be a hyperbolic toral automorphism. Then, the following are true.*
(a) The periodic points are dense in $\mathbb{T}^n$. In particular, there are an infinite number of periodic points. Also, the nonwandering set of f_A is all of $\mathbb{T}^n$.
(b) The toral automorphism f_A has a hyperbolic structure on all of $\mathbb{T}^n$:
(i) *$\mathbb{E}^s$ is the space of generalized stable eigenvectors of A,*
(ii) *$\mathbb{E}^u$ is the space of generalized unstable eigenvectors of A,*
(iii) *$\mathbb{E}^s_{\mathbf{p}}$ is the translation of $\mathbb{E}^s$ to $T_{\mathbf{p}}\mathbb{T}^n$,*
(iv) *$\mathbb{E}^u_{\mathbf{p}}$ is the translation of $\mathbb{E}^u$ to $T_{\mathbf{p}}\mathbb{T}^n$,*
(v) *f_A is an Anosov diffeomorphism, and*
(vi) *if $\pi(\bar{\mathbf{p}}) = \mathbf{p}$, then $W^s(\mathbf{p}) = \pi(\bar{\mathbf{p}} + \mathbb{E}^s)$ and $W^u(\mathbf{p}) = \pi(\bar{\mathbf{p}} + \mathbb{E}^u)$.*
(c) The toral automorphism f_A is topologically transitive.
(d) The toral automorphism f_A is expansive and so it has sensitive dependence on initial conditions.
(e) The toral automorphism f_A is structurally stable.

PROOF. (a) For fixed positive integer k, let $\text{Rat}(k)$ be the rational points in $\mathbb{T}^n$ with denominators k:

$$\text{Rat}(k) = \pi\{(i_1/k, \dots, i_n/k) : i_j \in \mathbb{Z}\} \subset \mathbb{T}^n.$$

Then $L_A(\{(i_1/k, \dots, i_n/k) : i_j \in \mathbb{Z}\}) \subset \{(i_1/k, \dots, i_n/k) : i_j \in \mathbb{Z}\}$, so $f_A(\text{Rat}(k)) \subset \text{Rat}(k)$. This set has a finite number of points (k^n points) and f_A is one to one, so $f_A|\text{Rat}(k)$ is a permutation of $\text{Rat}(k)$ and every point in this set is periodic. Finally, $\bigcup_k \text{Rat}(k)$ is dense in the torus, so the periodic points are dense.

The other two statememts in part (a) easily follow from the first.

(b) Because $\pi : \mathbb{R}^n \to \mathbb{T}^n$ is onto, these give global coordinates. If $U \subset \mathbb{T}^n$ is an open set and $\varphi_1, \varphi_2 : U \to \mathbb{R}^n$ are two sets of local coordinates on U which are inverses of π, then $\varphi_2 \circ \varphi_1^{-1}(\bar{\mathbf{x}}) = \bar{\mathbf{x}} + \mathbf{m}$, where $\mathbf{m} \in \mathbb{Z}^n$. Therefore, for $\mathbf{p} \in \mathbb{T}^n$, the tangent space at $\mathbf{p}$ can be thought of as $\{\mathbf{p}\} \times \mathbb{R}^n$.

In the local coordinates, the map f_A is given by L_A. As a map on $\mathbb{R}^n$, the derivative of L_A at a point $\bar{\mathbf{p}}$ is equal to the matrix A (or the linear map L_A), $D(L_A)_{\bar{\mathbf{p}}} = A$. Therefore, $D(f_A)_{\mathbf{p}} = A$ as a map from $T_{\mathbf{p}}\mathbb{T}^n = \{\mathbf{p}\} \times \mathbb{R}^n$ to $T_{f_A(\mathbf{p}}\mathbb{T}^n = \{f(\mathbf{p})\} \times \mathbb{R}^n$.

In dimension two, the span of the eigenvectors gives the stable and unstable directions for A, $\mathbb{E}^s = \{t\mathbf{v}^s : t \in \mathbb{R}\}$ and $\mathbb{E}^u = \{t\mathbf{v}^u : t \in \mathbb{R}\}$. In higher dimensions they are the generalized stable and unstable eigenspaces of A, respectively, as stated in the theorem. There is a $C \geq 1$, $0 < \mu < 1$, and $\lambda > 1$ such that $\|A^k|\mathbb{E}^s\| \leq C\mu^k$ and $\|A^{-k}|\mathbb{E}^u\| \leq C\lambda^{-k}$ for $k \geq 1$. For any point $\mathbf{p} \in \mathbb{T}^n$, define the subspaces at $\mathbf{p}$ to be the translates of the stable and unstable eigenspaces of A, $\mathbb{E}^s_{\mathbf{p}} = \{\mathbf{p}\} \times \mathbb{E}^s$ and $\mathbb{E}^u_{\mathbf{p}} = \{\mathbf{p}\} \times \mathbb{E}^u$.

For $\mathbf{p} \in \mathbb{T}^n$ and $\mathbf{v} \in \mathbb{E}^s_{\mathbf{p}}$, $|D(f_A^k)_{\mathbf{p}}\mathbf{v}| = |A^k\mathbf{v}| \leq C\mu^k|\mathbf{v}|$, where $0 < \mu < 1$ and $C \geq 1$ are the constants given above. The bound $C\mu^k|\mathbf{v}|$ goes to 0 as k goes to ∞, so $\mathbb{E}^s_{\mathbf{p}}$ is made up of stable vectors. A similar argument holds for $\mathbf{v} \in \mathbb{E}^u_{\mathbf{p}}$ as k goes to $-\infty$. This proves the hyperbolic structure.

Because f_A has a hyperbolic structure on all of $\mathbb{T}^n$ and all points are nonwandering, f_A is an Anosov diffeomorphism.

Next we turn to the stable manifold for a point $\mathbf{p} = \pi(\bar{\mathbf{p}})$. For $\epsilon > 0$ and $\mathbf{q} = \pi(\bar{\mathbf{q}})$, let $B(\bar{\mathbf{q}}, \epsilon) \subset \mathbb{R}^n$ be the ball of radius ϵ and $U(\mathbf{q}, \epsilon) = \pi(B(\bar{\mathbf{q}}, \epsilon)) \subset \mathbb{T}^n$. For ϵ small enough, $f_A^{-1}(U(f(\mathbf{q}), \epsilon))$ does not wrap around the handles of the torus, so

$$f_A^{-1}(U(f(\mathbf{q}), \epsilon)) \cap U(\mathbf{q}, \epsilon) = \pi[L_A^{-1}(B(L_A(\bar{\mathbf{q}}), \epsilon)) \cap B(\bar{\mathbf{q}}, \epsilon)].$$

Then, the local stable manifold is given as follows:

$$\begin{aligned} W^s_\epsilon(\mathbf{p}, f_A) &= \bigcap_{n \geq 0} f_A^{-n}(U(f_A^n(\mathbf{p}), \epsilon)) \\ &= \pi\Big[\bigcap_{n \geq 0} L_A^{-n}(B(L_A^n(\bar{\mathbf{p}}), \epsilon))\Big] \\ &= \pi\Big[\bar{\mathbf{p}} + \bigcap_{n \geq 0} A^{-n}(B(\mathbf{0}, \epsilon))\Big]. \end{aligned}$$

By the properties of the linear map, $\pi\big[\bar{\mathbf{p}} + \mathbb{E}^s(C^{-1}\epsilon)\big] \subset W^s_\epsilon(\mathbf{p}, f_A) \subset \pi\big[\bar{\mathbf{p}} + \mathbb{E}^s(\epsilon)\big]$. Therefore, the global stable manifold is given as stated:

$$\begin{aligned} W^s(\mathbf{p}, f_A) &= \bigcup_{n \geq 0} f_A^{-n}(W^s_\epsilon(f_A^n(\mathbf{p}), f_A)) \\ &= \pi\big[\bar{\mathbf{p}} + \mathbb{E}^s\big]. \end{aligned}$$

The result about the unstable manifold is proved similarly. This proves part (b) of the theorem.

Corollary 5.2. *For any point $\mathbf{p}$, $W^s(\mathbf{p})$ and $W^u(\mathbf{p})$ are each dense in $\mathbb{T}^n$ and so are their intersections, which are transverse homoclinic points.*

PROOF. For $n = 2$, the slopes of the lines $\mathbb{E}^s$ and $\mathbb{E}^u$ are irrational in $\mathbb{R}^2$, so the lines $W^s(\mathbf{p}) = \pi(\bar{\mathbf{p}} + \mathbb{E}^s)$ and $W^u(\mathbf{p}) = \pi(\bar{\mathbf{p}} + \mathbb{E}^u)$ are each dense in $\mathbb{T}^2$, and their intersections are dense.

For $n > 2$, we replace the lines with subspaces. The stable and unstable manifolds of L_A of a point $\bar{\mathbf{q}}$ are given by $W^s(\bar{\mathbf{q}}, L_A) = \bar{\mathbf{q}} + \mathbb{E}^s$ and $W^u(\bar{\mathbf{q}}, L_A) = \bar{\mathbf{q}} + \mathbb{E}^u$. Let $\mathbf{q}$ be a periodic point of period m with lift $\bar{\mathbf{q}}$. Then, $W^u(\mathbf{0}, L_A)$ intersects $W^s(\bar{\mathbf{q}}, L_A)$ in a point $\bar{\mathbf{z}}$ (because these two affine subspaces are complementary dimensions and not at all parallel). Letting $\mathbf{z} = \pi(\bar{\mathbf{z}})$, $\mathbf{z} \in W^u(\pi(\mathbf{0}), f_A) \cap W^s(\mathbf{q}, f_A)$ and $d(f_A^n(\mathbf{z}), f_A^n(\mathbf{q}))$ goes to 0 as n goes to ∞. Therefore, $W^u(\pi(\mathbf{0}), f_A)$ accumulates on $\mathbf{q}$ at the points $f^{mn}(\mathbf{z})$. Because the periodic points are dense in $\mathbb{T}^n$, $W^u(\pi(\mathbf{0}), f_A)$ is dense in $\mathbb{T}^n$.

Similarly, $W^s(\pi(\mathbf{0}), f_A)$ accumulates on $\mathbf{q}$ and so $W^s(\pi(\mathbf{0}), f_A)$ is dense in $\mathbb{T}^n$. Because $W^u(\pi(\mathbf{0}), f_A)$ and $W^s(\pi(\mathbf{0}), f_A)$ are projections of complementary subspaces, they intersect transversely arbitrarily near $\mathbf{q}$, so the transverse homoclinic points are dense in $\mathbb{T}^n$.

For an arbitrary point $\mathbf{p}$ in $\mathbb{T}^n$, $W^u(\mathbf{p}, f_A)$ and $W^s(\mathbf{p}, f_A)$ are translates of the manifolds of $\mathbf{0}$, so they are each dense in $\mathbb{T}^n$, and the homoclinic intersections for $\mathbf{p}$ are dense in $\mathbb{T}^n$. □

PROOF OF THEOREM 5.1 CONTINUED. (c) To prove that f_A is topologically transitive, we verify the hypothesis of the Birkhoff Transitivity Theorem. Let U and V be any two open sets in $\mathbb{T}^n$. The stable manifold of the origin, $W^s(\pi(\mathbf{0}))$, is dense in $\mathbb{T}^n$, so it intersects U in a point $\mathbf{q}$. Let $J = \pi(\bar{\mathbf{q}} + \mathbb{E}^u(r))$, where $r > 0$ is small enough so that $J \subset U$. If λ is a lower bound on the unstable eigenvalues, then the kth iterate of the disk, $f_A^k(J)$, contains a disk of radius at least least $C^{-1}\lambda^k r$ in $W^u(f_A^k(\mathbf{q}))$. As k increases, the "radius" of $f_A^k(J)$ becomes larger, so $f_A^k(J)$ accumulates on compact pieces of $W^u(\pi(\mathbf{0}))$, and it must intersect V. For this iterate,

$$\emptyset \neq f_A^k(J) \cap V \subset f_A^k(U) \cap V.$$

Thus, $\mathcal{O}^+(U) \cap V \neq \emptyset$. Similarly, $\mathcal{O}^-(U) \cap V \neq \emptyset$. By the Birkhoff Transitivity Theorem, f_A is topologically transitive.

(d) Fix a point $\mathbf{p}$. For $\mathbf{q} \in W^u(\mathbf{p}) \setminus \{\mathbf{p}\}$, the distance $d(f_A^k(\mathbf{p}), f_A^k(\mathbf{q})) \geq \lambda^k d(\mathbf{p}, \mathbf{q})$ as long as the distance stays less than one half. The quantity $\lambda^k d(\mathbf{p}, \mathbf{q})$ grows as $k > 0$ increases, so this distance gets bigger than $1/(2\lambda)$ for some $k > 0$. This proves that f_A has sensitive dependence. Similarly, for $\mathbf{q} \in W^s(\mathbf{p}) \setminus \{\mathbf{p}\}$, the distance $d(f_A^k(\mathbf{p}), f_A^k(\mathbf{q})) \geq \mu^k d(\mathbf{p}, \mathbf{q})$ grows as $k < 0$ becomes more negative, so it becomes bigger than $\mu/2$ for some $k < 0$.

Finally, there is a $\delta_0 > 0$ such that for any point $\mathbf{q} \neq \mathbf{p}$ with $d(\mathbf{p}, \mathbf{q}) < \delta_0$, the two points $\mathbf{q}_u \in W^u(\mathbf{p}) \cap W^s(\mathbf{q})$ and $\mathbf{q}_s \in W^s(\mathbf{p}) \cap W^u(\mathbf{q})$ are within $1/(2\lambda^2)$ and $\mu^2/2$, respectively, of $\mathbf{p}$. As least one of these two points is distinct from $\mathbf{p}$; let us take the case when $d(\mathbf{p}, \mathbf{q}_u) \geq d(\mathbf{p}, \mathbf{q}_s)$. Because the stable and unstable manifolds are parallel lines in the universal covering space, $d(\mathbf{p}, \mathbf{q}_u) = d(\mathbf{q}_s, \mathbf{q})$ and $d(\mathbf{p}, \mathbf{q}_s) = d(\mathbf{q}_u, \mathbf{q})$. There is a first $k_1 \geq 1$ such that $d(f^{k_1}(\mathbf{p}), f^{k_1}(\mathbf{q}_u)) \geq 1/(2\lambda)$. Then

$$\begin{aligned}
d(f^{k_1}(\mathbf{p}), f^{k_1}(\mathbf{q})) &\geq d(f^{k_1}(\mathbf{p}), f^{k_1}(\mathbf{q}_u)) - d(f^{k_1}(\mathbf{q}_u), f^{k_1}(\mathbf{q})) \\
&\geq \frac{1}{2\lambda} - \mu\, d(\mathbf{q}_u, \mathbf{q}) \\
&\geq \frac{1}{2\lambda} - \mu\, d(\mathbf{p}, \mathbf{q}_s) \\
&\geq \frac{1}{2\lambda} - \mu \left(\frac{1}{2\lambda^2}\right) \\
&= \left(1 - \frac{\mu}{\lambda}\right)\left(\frac{1}{2\lambda}\right).
\end{aligned}$$

Similarly, if $d(\mathbf{p}, \mathbf{q}_s) \geq d(\mathbf{p}, \mathbf{q}_u)$, then for some $k_1 < 0$,

$$d(f^{k_1}(\mathbf{p}), f^{k_1}(\mathbf{q})) \geq \left(1 - \frac{\mu}{\lambda}\right)\left(\frac{\mu}{2}\right).$$

Therefore, f_A is expansive with expansive constant

$$\min\left\{\delta_0,\ \left(1 - \frac{\mu}{\lambda}\right)\frac{1}{2\lambda},\ \left(1 - \frac{\mu}{\lambda}\right)\frac{\mu}{2}\right\}.$$

This proves part (d).

(e) The proof of structural stability uses the proof of the Hartman-Grobman Theorem, Section 5.7. After looking at the lifts of the diffeomorphisms to $\mathbb{R}^n$, the main change is that a little extra checking needs to be done in the proof that the conjugacy is one to one.

The map L_A is the lift of f_A to a map on $\mathbb{R}^n$. Let g be a C^1 perturbation of f_A. It is possible to choose the lift $G : \mathbb{R}^n \to \mathbb{R}^n$ for which $G(\mathbf{0})$ is near $\mathbf{0} = L_A(\mathbf{0})$. Let $\hat{G} = G - L_A$. Then, for any lattice point $\mathbf{w} \in \mathbb{Z}^n$, $L_A(\mathbf{x}+\mathbf{w}) - L_A(\mathbf{x}) = L_A(\mathbf{w})$, and $G(\mathbf{x}+\mathbf{w}) - G(\mathbf{x}) = L_A(\mathbf{w})$. (This latter equality can be proved by taking a homotopy g_t with $g_0 = f_A$ and $g_1 = g$. The lift G_t will have $G_t(\mathbf{x}+\mathbf{w}) - G_t(\mathbf{x})$ a lattice point for each $0 \le t \le 1$, so $G(\mathbf{x}+\mathbf{w}) - G(\mathbf{x}) = G_0(\mathbf{x}+\mathbf{w}) - G_0(\mathbf{x}) = L_A(\mathbf{w})$.) Then, $\hat{G}(\mathbf{x}+\mathbf{w}) = \hat{G}(\mathbf{x})$. Because of the periodicity of $\hat{G}$, it is C^1 small on all of $\mathbb{R}^n$. This shows that G is C^1 close to L_A on all of $\mathbb{R}^n$.

To conjugate L_A and G, we solve for $H = id + v$. The map v should be a bounded function on $\mathbb{R}^n$ and periodic. As before, let $C_b^0(\mathbb{R}^n)$ be the space of all bounded continuous maps from $\mathbb{R}^n$ to itself. Now, we let

$$C^0_{b,per}(\mathbb{R}^n) = \{v \in C^0_b(\mathbb{R}^n) : v(\mathbf{x}+\mathbf{w}) = v(\mathbf{x}) \quad \text{for all } \mathbf{w} \in \mathbb{Z}^n, \mathbf{x} \in \mathbb{R}^n\}$$

and

$$C^1_{b,per}(\mathbb{R}^n) = C^0_{b,per}(\mathbb{R}^n) \cap C^1(\mathbb{R}^n).$$

Then, if $\hat{G} \in C^1_{b,per}(\mathbb{R}^n)$. Because of the periodicity, there is a uniform bound on $\|D\hat{G}_{\mathbf{x}}\|$ for all $\mathbf{x} \in \mathbb{R}^n$.

For $\hat{G} \in C^1_{b,per}(\mathbb{R}^n)$ and $v \in C^0_{b,per}(\mathbb{R}^n)$, as in the proof of Hartman-Grobman, we let

$$\begin{aligned} \Theta(\hat{G}, v) &= \mathcal{L}^{-1}\{\hat{G} \circ (id + v) \circ L_A^{-1}\}, \qquad \text{where} \\ \mathcal{L}(v) &= (id - (L_A)_{\#})v = v - L_A \circ v \circ L_A^{-1}. \end{aligned}$$

A direct check shows that $\Theta(\hat{G}, \cdot)$ preserves $C^0_{b,per}(\mathbb{R}^n)$. (We leave this verification to the exercises. See Exercise 8.25.) Exactly as in the proof of the Hartman-Grobman Theorem, if $\mathrm{Lip}(\hat{G})$ is small relative to the distance of the contraction and expansion rates away from one, $\Theta(\hat{G}, \cdot)$ has a fixed point $v_{\hat{G}} \in C^0_{b,per}(\mathbb{R}^n)$. Letting $H_G = id + v_{\hat{G}}$, we get that

$$\begin{aligned} \mathcal{L}(v_{\hat{G}}) &= \hat{G} \circ (id + v_{\hat{G}}) \circ L_A^{-1} \\ H_G = id + v_{\hat{G}} &= L_A \circ L_A^{-1} + L_A \circ v_{\hat{G}} \circ L_A^{-1} + \mathcal{L}(v_{\hat{G}}) \\ &= L_A \circ H_G \circ L_A^{-1} + \hat{G} \circ H_G \circ L_A^{-1} \\ &= G \circ H_G \circ L_A^{-1} \end{aligned}$$

on $\mathbb{R}^n$. For $\mathbf{w} \in \mathbb{Z}^n$, $H_G(\mathbf{x}+\mathbf{w}) = H_G(\mathbf{x}) + \mathbf{w}$ so H_G induces a map h_g on $\mathbb{T}^n$ that satisfies $g \circ h_g \circ f_A^{-1} = h_g$.

Next we check that h_g is one to one. If $h_g(\mathbf{x}) = h_g(\mathbf{y})$, $\bar{\mathbf{x}}$ is a lift of $\mathbf{x}$, and $\bar{\mathbf{y}}$ is a lift of $\mathbf{y}$, then $H_G(\bar{\mathbf{x}}) = H_G(\bar{\mathbf{y}}) + \mathbf{w} = H_G(\bar{\mathbf{y}} + \mathbf{w})$ for some $\mathbf{w} \in \mathbb{Z}^n$. Replacing $\bar{\mathbf{y}}$ with $\bar{\mathbf{y}}' = \bar{\mathbf{y}} + \mathbf{w}$, we get another lift of $\mathbf{y}$ with $H_G(\bar{\mathbf{x}}) = H_G(\bar{\mathbf{y}}')$. Because H_G is one to one, $\bar{\mathbf{x}} = \bar{\mathbf{y}}'$ and $\mathbf{x} = \mathbf{y}$. (The proof that H_G is one to one uses the fact that L_A is expansive: $H_G \circ L_A^n(\bar{\mathbf{x}}) = H_G \circ L_A^n(\bar{\mathbf{y}}')$ for all n so $\bar{\mathbf{x}} = \bar{\mathbf{y}}'$.) Thus, h_g is one to one.

By invariance of domain, $h_g(\mathbb{T}^n)$ is open in $\mathbb{T}^n$. Since it is also close, $h_g(\mathbb{T}^n) = \mathbb{T}^n$, and h_g is onto. This completes the proof that h_g is a homeomorphism, that f_A is structurally stable, and the proof of the theorem. □

REMARK 5.1. In Section 10.7 we prove that all Anosov diffeomorphisms are structurally stable. Manning (1974) proved that any Anosov diffeomorphism on a torus is topologically conjugate to a hyperbolic toral automorphism. One conjecture which is still unknown is whether being Anosov implies that all points are nonwandering (or chain recurrent).

8.5.1 Markov Partitions for Hyperbolic Toral Automorphisms

We want to connect the dynamics of a hyperbolic toral automorphism, $f : \mathbb{T}^n \to \mathbb{T}^n$, with that of a subshift of finite type, i.e., to see how symbolic dynamics can be applied to a hyperbolic toral automorphism. We need to find (and define) the replacements for the geometric boxes of the horseshoe which are used to define the symbol sequences. The theory which we give is for all dimensions, but the examples are all in two dimensions where the situation is simpler.

Example 5.2. We introduce the ideas of rectangles, a Markov partition, and the semi-conjugacy using the toral automorphism f_A induced by the matrix $A = \begin{pmatrix} 1 & 1 \\ 1 & 0 \end{pmatrix}$. We want to subdivide the total space into *rectangles* (which can be taken to be actual parallelepipeds in two dimensions but not higher dimensions). The eigenvalues are $\lambda_u = (1+\sqrt{5})/2$ with eigenvector $\mathbf{v}^u = (2, \sqrt{5}-1)$ and $\lambda_s = (1-\sqrt{5})/2$ with eigenvector $\mathbf{v}^s = (2, -\sqrt{5}-1)$. Note that $\lambda_u + \lambda_s = 1 = \mathrm{tr}(A) > 0$. Thus, $\lambda_u = \mathrm{tr}(A) - \lambda_s$, and the fact that the trace of A is a positive integer insures that $\lambda_u > 0$. Then, $\lambda_u \lambda_s = \det(A) = -1 < 0$, so this insures that $\lambda_s < 0$. Also, $\mathbf{v}^u$ has positive slope and $\mathbf{v}^s$ has negative slope.

To form the rectangles for A, we look in the covering space, $\mathbb{R}^2$. From the origin and other lattice points, take the part of the unstable manifold of this point in $\mathbb{R}^2$ that crosses the fundamental domain above and to the right of the lattice point. See Figure 5.1. Next, extend the stable manifold from the lattice point downward to the point $\mathbf{a}$, where it hits the part of the unstable line segment drawn above. Similarly, extend the stable manifold upward from a lattice point to the point $\mathbf{b}$, where it hits the part of the unstable manifold drawn above. Finally, extend the unstable manifold to the point $\mathbf{c}$, where it hits the line segment $[\mathbf{a}, \mathbf{b}]_s$ in the stable manifold. These line segments, $[\mathbf{a}, \mathbf{b}]_s$ in $W^s(\mathbf{0})$ and $[\mathbf{0}, \mathbf{c}]_u$ in $W^u(\mathbf{0})$ (and their translates in $\mathbb{R}^2$), define two rectangles R_1 and R_2 in $\mathbb{T}^2$. See Figure 5.1.

To find the images of the rectangles, we first consider the images of the points $\mathbf{a}$, $\mathbf{b}$, and $\mathbf{c}$: $f_A(\mathbf{a}) = \mathbf{b}$, $f_A(\mathbf{b}) = \mathbf{c}$, and $f_A(\mathbf{c}) \in [\mathbf{0}, \mathbf{b}]_s$, where $[\mathbf{x}, \mathbf{y}]_s$ is a line segment in the stable manifold from $\mathbf{x}$ to $\mathbf{y}$. See Figure 5.1. Using these images, it follows that

$$f_A(R_1) \text{ crosses } R_1 \text{ and } R_2,$$
$$f_A(R_2) \text{ crosses } R_1.$$

See Figure 5.2. The pair of rectangles $\{R_1, R_2\}$ have the properties of a *Markov partition* for f_A: (i) the collection of rectangles covers $\mathbb{T}^2$, (ii) the interiors of R_1 and R_2 are disjoint, and (iii) if $f_A(\mathrm{int}(R_i)) \cap \mathrm{int}(R_j) \neq \emptyset$, then $f_A(R_i)$ reaches all the way across R_j in the unstable direction and does not cross the edges of R_j is the stable direction. (There is a fourth condition which we only discuss implicitly below in terms of the semi-conjugacy.) We give the general definition below.

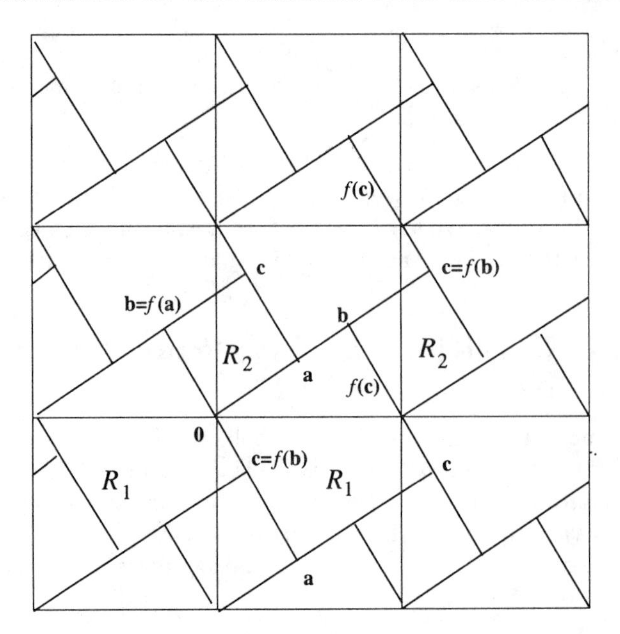

FIGURE 5.1. Rectangles for Example 5.2

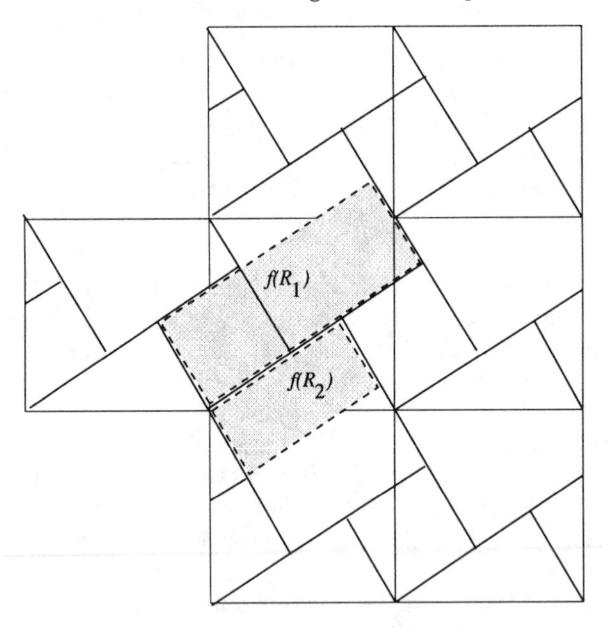

FIGURE 5.2. Images of Rectangles for Example 5.2

We define a transition matrix that indicates which itineraries for the orbit of a point are allowable: for a transition from rectangle R_i to R_j to be allowable, it must be possible for an orbit of a point to pass from the interior of R_i to the interior of R_j. (We disregard the fact that the image of the boundary of R_2 hits the boundary of R_2.) In this example the transition matrix is given by

$$B = \begin{pmatrix} 1 & 1 \\ 1 & 0 \end{pmatrix}.$$

Notice that this transition matrix B is the same matrix as the original matrix A which induced the toral automorphism. The shift space for B is the two-sided subshift of finite type

$$\Sigma_B = \{\mathbf{s} : \mathbb{Z} \to \{1,2\} : b_{s_i s_{i+1}} = 1\}$$

with shift map $\sigma_B = \sigma|\Sigma_B$.

To define the symbolic dynamics, we cannot get a continuous map (conjugacy or semiconjugacy) h from $\mathbb{T}^2$ to Σ_B because $\mathbb{T}^2$ is connected and Σ_B is a totally disconnected Cantor set. Also, for a point $\mathbf{p} \in \partial(R_i)$, there are at least two choices of rectangles to which $\mathbf{p}$ belongs. Therefore, there is no way to assign a unique symbol sequence to points on the boundary of a rectangle. Instead, we define a map going the other direction, $h : \Sigma_B \to \mathbb{T}^2$, which plays the role of the itinerary map. We want h to be a semiconjugacy (continuous, onto, and $f_A \circ h = h \circ \sigma_B$). To do this, we define $h : \Sigma_B \to \mathbb{T}^2$ by

$$h(\mathbf{s}) = \bigcap_{n=0}^{\infty} \mathrm{cl}\,\Big(\bigcap_{j=-n}^{n} f_A^{-j}(\mathrm{int}(R_{s_j}))\Big).$$

We take the images of the interiors because $R_1 \cap f_A^{-1}(R_2)$ does not always equal $\mathrm{cl}(\mathrm{int}(R_1) \cap f_A^{-1}(\mathrm{int}(R_2))$ but can have extra points whose images are on the boundary of R_2. (We must put up with the annoyance to be able to use fewer rectangles.) Using the general theory, Theorem 5.3 proves that this h is a semi-conjugacy. In fact, it proves that h is at most four to one.

In order to give the precise definitions of rectangle and Markov partition, it is necessary to indicate what we mean by the component of a stable or unstable manifold for a point in a rectangle. As we have done before, we use the notation of $\mathrm{comp}_{\mathbf{z}}(S)$ to be the connected component of the set S containing the point $\mathbf{z}$. We think of $W^\sigma(\mathbf{z}, R)$ as equal to $\mathrm{comp}_{\mathbf{z}}(R \cap W^\sigma(\mathbf{z}))$ for $\sigma = u, s$ and R one of the rectangles (if it is connected). However, this definition does not quite work, because even in the rectangles for Example 5.2 there is a difficulty: in $\mathbb{T}^2$, R_1 touches itself along the projection of the line segment from $\mathbf{0}$ to $\mathbf{c}$, $\pi([\mathbf{0}, \mathbf{c}]_s)$. When the total ambient manifold is a torus, a better definition of the stable and unstable manifolds in a rectangle uses the covering space $\mathbb{R}^2$ as follows. Let $\bar{R}$ be one lift of a rectangle R in $\mathbb{T}^2$ to a rectangle in $\mathbb{R}^2$, so $\pi : \bar{R} \to R$ is onto and one to one in the interior. Let $\bar{\mathbf{z}}$ be a lift of $\mathbf{z}$, $\bar{\mathbf{z}} \in \bar{R}$ and $\pi(\bar{\mathbf{z}}) = \mathbf{z}$. For $\sigma = u, s$, define

$$W^\sigma(\mathbf{z}, R) = \pi(W^\sigma(\bar{\mathbf{z}}) \cap \bar{R}).$$

Note in Example 5.2, for $\mathbf{z} = \mathbf{0}$ and rectangle R_1, there are two choices for the lift $\bar{R}_1$ which touches the origin $\bar{\mathbf{0}}$ in $\mathbb{R}^2$. (There is one choice above and to the right of $\bar{\mathbf{0}}$ and one below and to the left.) Making either of these choices, $W^\sigma(\mathbf{0}, R_1) = \pi(W^\sigma(\bar{\mathbf{0}}, \bar{R}_1))$ is a proper subset of $\mathrm{comp}_{\mathbf{0}}(R_1 \cap W^\sigma(\mathbf{0}))$. In fact, $\mathrm{comp}_{\mathbf{0}}(R_1 \cap W^\sigma(\mathbf{0}))$ is the union of the two choices for $W^\sigma(\mathbf{0}, R_1)$.

We now use the motivation of the rectangles defined above for the specific example to give a general definition of both a rectangle and a Markov partition.

Definition. For a hyperbolic toral automorphism on the n-torus, $\mathbb{T}^n$, we proceed as follows. Let R be a subset of $\mathbb{T}^n$ and $\mathbf{z} \in R$. Let $\bar{R}$ be a lift of R to $\mathbb{R}^n$ and $\bar{\mathbf{z}} \in \bar{R}$ be a lift of $\mathbf{z}$, i.e., $\pi : \bar{R} \subset \mathbb{R}^n \to R$ is a homeomorphism, and $\pi(\bar{\mathbf{z}}) = \mathbf{z}$. If R is connected, then $\bar{R}$ should be taken to be connected; if R is not connected, then care must be taken to choose the points in $(\pi)^{-1}(R)$ in a reasonable manner, e.g., $\bar{R}$ should be in one fundamental region of $\pi : \mathbb{R}^n \to \mathbb{T}^n$. For $\sigma = u, s$, let

$$W^\sigma(\mathbf{z}, R) = \pi(W^\sigma(\bar{\mathbf{z}}) \cap \bar{R}).$$

We do not give a completely precise definition for the general case of a hyperbolic invariant set Λ. An isolated hyperbolic invariant set has a property called a *local product structure* provided for $\epsilon > 0$ small enough, there is a $\delta > 0$ such that if $d(\mathbf{x}, \mathbf{y}) < \delta$ for $\mathbf{x}, \mathbf{y} \in \Lambda$, then $W^u_\epsilon(\mathbf{x}) \cap W^s_\epsilon(\mathbf{y})$ is a single point in Λ. Let Λ be a hyperbolic invariant set with a local product structure, let R be a subset of Λ that has diameter less than δ, and let $\mathbf{z} \in R$. Then,

$$W^\sigma(\mathbf{z}, R) \equiv R \cap W^\sigma_\epsilon(\mathbf{z})$$

using the local stable and unstable manifolds of size ϵ. This general case was considered by Bowen (1970a, 1975). Also see Section 10.6.

Definition. Let f be a diffeomorphism with a hyperbolic invariant set with a local product structure. (This includes the case where f is a hyperbolic toral automorphism.) A nonempty set R of $\mathbb{T}^n$ (or of Λ) is a *(proper) rectangle* provided

(i) $R = \mathrm{cl}(\mathrm{int}(R))$ (where the interior is relative to Λ) so that it is closed, and
(ii) $\mathbf{p}, \mathbf{q} \in R$ implies that $W^s(\mathbf{p}, R) \cap W^u(\mathbf{q}, R)$ is exactly one point, and this point is in R. If we are considering a hyperbolic toral automorphism, then the same lift must be used for R to determine both $W^s(\mathbf{p}, R)$ and $W^u(\mathbf{q}, R)$.

REMARK 5.2. In his general definition, Bowen defines $W^s_\epsilon(\mathbf{p}) \cap W^u_\epsilon(\mathbf{q}) \equiv [\mathbf{p}, \mathbf{q}]$. He then demands that for $\mathbf{p}, \mathbf{q} \in R$, $[\mathbf{p}, \mathbf{q}]$ is exactly one point, and that this point is in R. Note, if we use Bowen's definition, then R_1 is not a rectangle in Example 5.2 because there are points $\mathbf{p}$ and $\mathbf{q}$ in R_1 near $\mathbf{0}$, for which $W^s_\epsilon(\mathbf{p}) \cap W^u_\epsilon(\mathbf{q})$ is in R_2 and not in R_1. Using the fact that the manifold is a torus and our definition of the subsets of the stable and unstable manifolds using lifts, the sets R_1 and R_2 given in the above example are indeed rectangles.

Below, we define a collection of rectangles (a Markov partition) which have the properties needed to use them to define symbolic dynamics. The definitions use the notion of the interior and boundary of a rectangle. A point $\mathbf{p} \in R$ is a *boundary point of* R if arbitrarily near to $\mathbf{p}$ there is a point $\mathbf{q}$ in Λ such that $\mathbf{q} \notin R$. (This is the usual pointset boundary of a subset.) If $\mathbf{p}$ is a boundary point of R, it follows for such $\mathbf{q}$, that either $W^s(\mathbf{p}) \cap W^u(\mathbf{q})$ or $W^s(\mathbf{q}) \cap W^u(\mathbf{p})$ is not in R. Let $\partial(R)$ be the set of all boundary points of R, and the interior of R be the complement of $\partial(R)$ in R, $\mathrm{int}(R) = R \setminus \partial(R)$.

Definition. Assume that $f : M \to M$ is a diffeomorphism which has an isolated hyperbolic invariant set Λ with a local product structure. (This includes the case where f is a hyperbolic toral automorphism with $\Lambda = M$.) A *Markov partition* for f is a finite collection of rectangles, $\mathcal{R} = \{R_j\}_{j=1}^m$, that satisfies the following four conditions. (All interiors are taken relative to Λ.)

(i) The collection of rectangles cover Λ, $\Lambda = \bigcup_{j=1}^m R_j$.
(ii) If $i \neq j$, then $\mathrm{int}(R_i) \cap \mathrm{int}(R_j) = \emptyset$ (so $\mathrm{int}(R_i) \cap R_j = \emptyset$).
(iii) If $\mathbf{z} \in \mathrm{int}(R_i)$ and $f(\mathbf{z}) \in \mathrm{int}(R_j)$, then

$$f(W^u(\mathbf{z}, R_i)) \supset W^u(f(\mathbf{z}), R_j) \quad \text{and}$$
$$f(W^s(\mathbf{z}, R_i)) \subset W^s(f(\mathbf{z}), R_j).$$

(iv) (The rectangles are small enough.) If $\mathbf{z} \in \mathrm{int}(R_i) \cap f^{-1}(\mathrm{int}(R_j))$, then

$$\mathrm{int}(R_j) \cap f\big(W^u(\mathbf{z}, \mathrm{int}(R_i))\big) = W^u(f(\mathbf{z}), \mathrm{int}(R_j)) \quad \text{and}$$
$$\mathrm{int}(R_i) \cap f^{-1}\big(W^s(f(\mathbf{z}), \mathrm{int}(R_j))\big) = W^s(\mathbf{z}, \mathrm{int}(R_i)),$$

where $W^\sigma(\mathbf{z}', \mathrm{int}(R_k)) = W^\sigma(\mathbf{z}', \mathrm{int}(R_k)) \cap \mathrm{int}(R_k)$ for $\sigma = u, s$, any point $\mathbf{z}'$, and rectangle R_k.

Definition. Once we have a Markov partition, we want to set up the symbolic dynamics of the subshift of finite type by means of a transition matrix. Given a Markov partition $\mathcal{R} = \{R_j\}_{j=1}^m$, the *transition matrix* $B = (b_{ij})$ is defined by

$$b_{ij} = \begin{cases} 1 & \text{if } \operatorname{int}(f(R_i)) \cap \operatorname{int}(R_j) \neq \emptyset \\ 0 & \text{if } \operatorname{int}(f(R_i)) \cap \operatorname{int}(R_j) = \emptyset. \end{cases}$$

The *shift space for* B is defined as

$$\Sigma_B = \{\mathbf{s} : \mathbb{Z} \to \{1, \dots, m\} : b_{s_i s_{i+1}} = 1\}.$$

Letting σ be the shift map on the full m-shift, $\Sigma_m = \{1, \dots, m\}^{\mathbb{Z}}$, define $\sigma_B = \sigma|\Sigma_B : \Sigma_B \to \Sigma_B$.

Example 5.3. For the geometric horseshoe, let $R_i = H_i \cap \Lambda$ for $i = 1, 2$. These two rectangles form a Markov partition with transition matrix $\begin{pmatrix} 1 & 1 \\ 1 & 1 \end{pmatrix}$.

For the hyperbolic invariant set created for a homoclinic point, the sets $R_i = \Lambda \cap A_i$ form a Markov partition with transition matrix B given in the proof of Theorem 4.4(b).

REMARK 5.3. Notice that we do not demand that a rectangle be connected, although the examples we give for hyperbolic toral automorphisms are connected. There are examples where a rectangle has countably many components even for a Markov partition of a total space which is connected. In general, for a Markov partition of a hyperbolic invariant set, the total space is often not connected or even locally connected, so a rectangle certainly could not be connected in this case.

REMARK 5.4. Let $\partial(R_i)$ be the boundary of R_i relative to Λ. Conditions (i) and (ii) in the definition of a Markov partition imply that $\partial(R_i) = \{\mathbf{p} \in R_i : \mathbf{p} \in R_j \text{ for some } j \neq i\}$. This holds because clearly $\operatorname{int}(R_i) \cap \{\mathbf{p} \in R_i : \mathbf{p} \in R_j \text{ for some } j \neq i\} = \emptyset$, so $\partial(R_i) \supset \{\mathbf{p} \in R_i : \mathbf{p} \in R_j \text{ for some } j \neq i\}$. Next, if $\mathbf{p} \in \partial(R_i)$, then there are $\mathbf{q}_k \in R_{j_k}$ with $j_k \neq i$ and $\mathbf{q}_k$ converging to $\mathbf{p}$. Because there are a finite number of rectangles, by taking a subsequence we can take all the $j_k = j$ to be the same. Because R_j is closed, it follows that $\mathbf{p} \in R_j$. This proves that $\partial(R_i) \subset \{\mathbf{p} \in R_i : \mathbf{p} \in R_j \text{ for some } j \neq i\}$.

REMARK 5.5. Condition (iii) in the definition of a Markov partition insures that if the image of a rectangle hits the interior of another rectangle, then it goes all the way across in the unstable direction and is a subset in the stable direction (goes all the way across in the stable direction when looking at the inverse). Note that if a point $\mathbf{z}$ is on the boundary of a rectangle R_i, then the image of R_i can abut on another rectangle R_j without even going into the interior of rectangle R_j. (Thus, Condition (iii) does not necessarily hold for the points on the boundary.)

REMARK 5.6. Condition (iv) is not included in Bowen's definition because he only used small rectangles. It is added to our list to make the point determined by a sequence of rectangles allowed by the transition matrix well defined. This condition prohibits the image of a rectangle R_i from crossing a rectangle R_j twice. Note that it does allow the image to intersect the boundary a second time. (See $f(R_1)$ and R_2 in Figure 5.2.)

We could strengthen Condition (iv) to the following assumption:
(iv)′ for $\mathbf{z} \in \operatorname{int}(R_i) \cap f^{-1}(\operatorname{int}(R_j))$,

$$R_j \cap f\big(W^u(\mathbf{z}, R_i)\big) = W^u(f(\mathbf{z}), R_j) \qquad \text{and}$$
$$R_i \cap f^{-1}\big(W^s(f(\mathbf{z}), R_j)\big) = W^s(\mathbf{z}, R_i).$$

This condition does not allow the image of a rectangle R_i to cross the rectangle R_j once and then intersect the boundary a second time. Therefore, the partition constructed in Example 5.2 satisfies assumption (iv) but not assumption (iv)$'$. The advantage of assumption (iv)$'$ over (iv) is that the definition of the conjugacy in Theorem 5.3 without taking interiors and closures. See Remark 5.10.

For some purposes, people allow the image of a rectangle to cross more than one time. If multiple crossings are allowed, then (1) Condition (iv) is not included in the definition and (2) the transition matrix must be allowed to have integer entries which are larger than 1, i.e., we get an adjacency matrix as defined in Section 7.3.1 on subshifts for matrices with nonnegative integer entries. More precisely, assume there is a partition by rectangles $\{R_i\}_{i=1}^n$ which satisfies conditions (i – iii) for a Markov partition but not necessarily condition (iv). To such a partition, we can associate an adjacency matrix $A = (a_{ij})$, where the entry a_{ij} equals the number of times that the image $f(R_i)$ crosses the rectangle R_j. Thus, if $a_{ij} = 2$, then $f(R_i)$ crosses R_j twice. We do not pursue this connection. See Franks (1982).

REMARK 5.7. Adler and Weiss (1970) gave a method of constructing simple Markov partitions for hyperbolic toral automorphisms on $\mathbb{T}^2$. Assume A is a 2×2 adjacency matrix with all positive entries and which induces a hyperbolic toral automorphism on $\mathbb{T}^2$. Then, there is always has a partition by two rectangles, $\{R_1, R_2\}$, such that (1) the partition satisfies all the properties of a Markov partition except (iv), and (2) the image $f(R_i)$ has a_{ij} geometric crossings of R_j. The recent theses by Snavely (1990) and Rykken (1993) give more details on constructing such a Markov partition.

REMARK 5.8. It should be noted however that even for Markov partitions for hyperbolic toral automorphisms in $\mathbb{T}^n$ with $n \geq 3$, the boundaries of the rectangles are not smooth. Thus, the "rectangles" are much different than the simple two-dimensional example leads one to believe. See Bowen (1978b).

REMARK 5.9. Bowen (1970a) proved that any hyperbolic invariant set with a local product structure has a Markov partition. We prove this result in Section 10.6. In this chapter, we restrict ourselves to finding Markov partitions for hyperbolic toral automorphisms on $\mathbb{T}^2$ and the solenoid which is defined in Section 8.7.

We can now state the main result.

Theorem 5.3. *Let $\mathcal{R} = \{R_j\}_{j=1}^m$ be a Markov partition for a hyperbolic toral automorphism on $\mathbb{T}^2$ with transition matrix B. Let (Σ_B, σ_B) be the shift space and $h : \Sigma_B \to \mathbb{T}^2$ be defined by*

$$h(\mathbf{s}) = \bigcap_{n=0}^{\infty} \operatorname{cl}\Big(\bigcap_{j=-n}^{n} f^{-j}(\operatorname{int}(R_{s_j}))\Big).$$

Then, h is a finite to one semiconjugacy from σ_B to f. In fact, h is at most m^2 to one, where m is the number of rectangles in the partition.

REMARK 5.10. If we used assumption (iv)$'$ given in Remark 5.6 above, then we could just use the intersection of the images $f^{-j}(R_{s_j})$ to define h,

$$h(\mathbf{s}) = \bigcap_{j=-\infty}^{\infty} f^{-j}(R_{s_j}).$$

This latter intersection is usually used to define the conjugacy. The problem is that $f(\operatorname{int}(R_i)) \cap \operatorname{int}(R_j)$ can be nonempty and $f(R_i)$ abut on the boundary of R_j at points for which there are no nearby interior points, so

$$\operatorname{cl}\big(f(\operatorname{int}(R_i)) \cap \operatorname{int}(R_j)\big) \neq f(R_i) \cap R_j.$$

See Example 5.2. We allow such intersections on the boundary in order to find Markov partitions with fewer rectangles. This forces us to use this slightly more complicated definition of h given above.

REMARK 5.11. This theorem is used in Section IX.1.2 to prove that the topological entropy of F_A can be calculated by the largest eigenvalue of B.

PROOF. By condition (iv), $\mathrm{cl}(\mathrm{int}(R_{s_k}) \cap f^{-1}(\mathrm{int}(R_{s_{k+1}})))$ is a nonempty subrectangle that reaches all the way across in the stable direction. By induction,

$$\mathrm{cl}\,\Big(\bigcap_{j=k}^{k+i} f^{-j}(\mathrm{int}(R_{s_j}))\Big)$$

is a nonempty subrectangle that reaches all the way across in the stable direction for any $k \in \mathbb{Z}$ and $i \in \mathbb{N}$. The width of this set in the unstable direction decreases exponentially at the rate given by the inverse of the minimum expansion constant. Thus,

$$\bigcap_{n=0}^{\infty} \mathrm{cl}\,\Big(\bigcap_{j=0}^{n} f^{-j}(\mathrm{int}(R_{s_j}))\Big) = W^s(\mathbf{p}_s, R_{s_0})$$

for some $\mathbf{p}_s \in R_{s_0}$. Similarly,

$$\bigcap_{n=0}^{-\infty} \mathrm{cl}\,\Big(\bigcap_{j=n}^{0} f^{-j}(\mathrm{int}(R_{s_j}))\Big) = W^u(\mathbf{p}_u, R_{s_0})$$

for some $\mathbf{p}_u \in R_{s_0}$. Therefore,

$$\bigcap_{n=-\infty}^{\infty} \mathrm{cl}\,\Big(\bigcap_{j=-n}^{n} f^{-j}(\mathrm{int}(R_{s_j}))\Big) = W^s(\mathbf{p}_s, R_{s_0}) \cap W^u(\mathbf{p}_u, R_{s_0})$$

is a unique point $\mathbf{p} = h(\mathbf{s})$. This shows that h is a well defined map.

By arguments like those used for the horseshoe, h is continuous, onto, and a semi-conjugacy.

If $f^j(\mathbf{p}) \in \mathrm{int}(R_{s_j})$ for all j, then $h^{-1}(\mathbf{p})$ is a unique symbol sequence, $\mathbf{s}$, because $f^j(\mathbf{p}) \notin R_k$ for $k \neq s_j$. Thus, h is one to one on the residual subset (in the sense of Baire category)

$$\bigcap_j f^{-j}\Big(\bigcup_i \mathrm{int}(R_i)\Big).$$

Next we show that h is at most m^2 to one, where m is the number of partitions. Let $\mathbf{p} = h(\mathbf{s})$. As we showed above, we only have to worry if $f^n(\mathbf{p})$ is on the boundary of some rectangle R_j.

We want to distinguish the boundary points of a rectangle R which are on the edge of an unstable manifold in the rectangle, $W^u(\mathbf{z}, R)$, and those which are on the edge of a stable manifold, $W^s(\mathbf{z}, R)$. Let

$$\partial^s(R) = \{\mathbf{x} \in \partial(R) : \mathbf{x} \notin \mathrm{int}(W^u(\mathbf{x}, R))\} \qquad \text{and}$$
$$\partial^u(R) = \{\mathbf{x} \in \partial(R) : \mathbf{x} \notin \mathrm{int}(W^s(\mathbf{x}, R))\}.$$

Here, $\mathrm{int}(W^u(\mathbf{x}, R))$ is the interior relative to a compact part of the manifold $W^u(\mathbf{x}, R)$. Similarly for $\mathrm{int}(W^s(\mathbf{x}, R))$. Then, $\partial^s(R)$ is the union of stable manifolds $W^s(\mathbf{z}, R)$, and $\partial^u(R)$ is the union of such unstable manifolds.

If $f^n(\mathbf{p}) \in \partial^s(R_{s_n})$, then $f^j(\mathbf{p}) \in \partial^s(R_{s_j})$ for $j \geq n$. There are at most m choices for s_n. (The reader can check that for a hyperbolic toral automorphism on $\mathbb{T}^2$, there are at most four choices.) Since the transitions of interiors are unique, a choice for s_n determines the choices of s_j for $j \geq n$. Similarly, if $f^{n'}(\mathbf{p}) \in \partial^u(R_{s_{n'}})$, then a choice for $s_{n'}$ determines the choices of s_j for $j \leq n'$. Combining, there are at most m^2 choices as claimed. □

Example 5.4. As a second example of a hyperbolic toral automorphism, let $A_2 = \begin{pmatrix} 2 & 1 \\ 1 & 1 \end{pmatrix}$. As we noted above, if $A = \begin{pmatrix} 1 & 1 \\ 1 & 0 \end{pmatrix}$, then $A^2 = A_2$. The rectangles R_1 and R_2 from Example 5.2 are still rectangles for this matrix. However, the image of R_1 by f_{A_2} crosses R_1 twice. This partition satisfies conditions (i) – (iii) and has A_2 as an adjacency matrix.

If we want to get a transition matrix with only 0's and 1's, we must subdivide the rectangles (split symbols) by taking components of $R_1 \cap f_{A_2}(R_1)$: let the rectangle

$$\begin{aligned} R_{1a} &= \operatorname{comp}\left(\pi(\mathbf{0}), \operatorname{cl}(\operatorname{int}(R_1) \cap f_{A_2}(\operatorname{int}(R_1)))\right) \\ &= \pi(\bar{R}_1 \cap L_{A_2}(\bar{R}_1)), \end{aligned}$$

where L_{A_2} is the map on $\mathbb{R}^2$, and

$$R_{1b} = \operatorname{cl}(R_1 \setminus R_{1a}).$$

These rectangles can also be formed by extending the unstable manifold of the origin until it intersects the stable line segment $[\mathbf{0}, \mathbf{b}]_s$ at the point $\mathbf{e} = f(\mathbf{c})$. See Figures 5.3 and 5.1. The reader can check that

$$\begin{aligned} f_{A_2}(R_{1a}) &\quad \text{crosses} \quad R_{1a}, R_{1b} \text{ and } R_2, \\ f_{A_2}(R_{1b}) &\quad \text{crosses} \quad R_{1a}, R_{1b} \text{ and } R_2, \\ f_{A_2}(R_2) &\quad \text{crosses} \quad R_{1b} \text{ and } R_2. \end{aligned}$$

Thus, the transition matrix is

$$B = \begin{pmatrix} 1 & 1 & 1 \\ 1 & 1 & 1 \\ 0 & 1 & 1 \end{pmatrix}.$$

This transition matrix has characteristic polynomial $p(\lambda) = -\lambda(\lambda^2 - 3\lambda + 1)$, and eigenvalues 0, $(\lambda_u)^2$, and $(\lambda_s)^2$, where λ_u, and λ_s are the eigenvalues of A. Thus, the eigenvalues of B are those of A_2 together with 0. We do not prove it, but the eigenvalues of the transition matrix are always the eigenvalues of the original matrix A together with possibly 0 and/or roots of unity. See Snavely (1990).

8.5.2 Ergodicity of Hyperbolic Toral Automorphisms

In this subsection, we show that a hyperbolic toral automorphism on $\mathbb{T}^2$ is ergodic with respect to the normal Lebesgue measure μ on $\mathbb{T}^2$ induced from $\mathbb{R}^2$. Very little needs to be changed in the proof to show that the result is also true for hyperbolic toral automorphisms in higher dimensions. This property is also true for a general C^2 Anosov diffeomorphisms which preserves a measure equivalent to Lebesgue. The basic idea of the proof remains unchanged, but there are more technical details (including the absolute continuity of the set of stable and unstable manifolds).

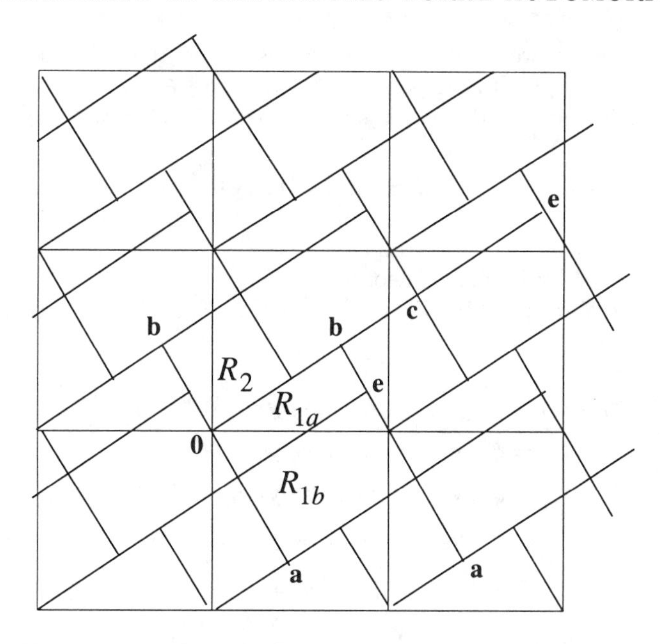

FIGURE 5.3. Markov Partition for Example 5.4

Theorem 5.8. *Let $f = f_A$ be a hyperbolic toral automorphism on $\mathbb{T}^2$. Let μ be Lebesgue measure on $\mathbb{T}^2$ (inherited from $\mathbb{R}^2$). Then, f is ergodic with respect to μ.*

PROOF. Since $\det(A) = \pm 1$, f preserves μ.

For any real valued function $g : \mathbb{T}^2 \to \mathbb{R}$, let

$$g^+(\mathbf{x}) = \lim_{n\to\infty} (1/n) \sum_{j=0}^{n-1} g \circ f^j(\mathbf{x})$$

be the forward time average, and

$$g^-(\mathbf{x}) = \lim_{n\to\infty} (1/n) \sum_{j=0}^{n-1} g \circ f^{-j}(\mathbf{x})$$

be the backward time average.

We start with a continuous real valued function $g : \mathbb{T}^2 \to \mathbb{R}$. We want to show that g^- is constant almost everywhere. By the Birkhoff Ergodic Theorem,

$$G = \{\mathbf{x} \in \mathbb{T}^2 : g^+(\mathbf{x}) \text{ and } g^-(\mathbf{x}) \text{ exist, and } g^+(\mathbf{x}) = g^-(\mathbf{x})\}$$

has full measure.

We remark that Fubini's Theorem implies that a set $X \subset \mathbb{T}^2$ has full measure if and only if X has full measure in almost every leaf $W^s(\mathbf{x})$. (Here we are using the fact that the stable and unstable directions form a coordinate system locally.) We use this fact several times below and do not distinguish between being a set of full measure in $\mathbb{T}^2$ and a set of full measure on almost every $W^s(\mathbf{x})$.

For $r > 0$ and $\mathbf{x}_0 \in \mathbb{T}^2$, let $I^s = W^s_r(\mathbf{x}_0)$ and

$$\mathcal{R} = \bigcup_{\mathbf{y} \in I^s} W^u_r(\mathbf{y}).$$

In the argument below, we want $r > 0$ small enough so that $\mathcal{R}$ is a rectangle for any $\mathbf{x}_0$, i.e., it is homeomorphic to $W^s_r(\mathbf{x}_0) \times W^u_r(\mathbf{y}_0)$ for $\mathbf{y}_0 \in I^s$.

Take an $\mathbf{x}_0$ such that $G \cap W^s_r(\mathbf{x}_0)$ has full measure in $I^s = W^s_r(\mathbf{x}_0)$. The set

$$H = \bigcup_{\mathbf{y} \in G \cap I^s} W^u_r(\mathbf{y})$$

has full measure in $\mathcal{R}$ by Fubini's Theorem.

Take $\mathbf{z}_1, \mathbf{z}_2 \in H$. Let $\mathbf{y}_i = I^s \cap W^u_r(\mathbf{z}_i)$ for $i = 1, 2$. Note that $\mathbf{y}_i \in G$ by the definition of H. Then,

$$\begin{aligned} g^-(\mathbf{z}_1) &= g^-(\mathbf{y}_1) && \text{because } \mathbf{y}_1 \in W^u_r(\mathbf{z}_1) \\ &= g^+(\mathbf{y}_1) && \text{because } \mathbf{y}_1 \in G \\ &= g^+(\mathbf{y}_2) && \text{because } \mathbf{y}_1, \mathbf{y}_2 \in I^s \subset W^s(\mathbf{x}_0) \\ &= g^-(\mathbf{y}_2) && \text{because } \mathbf{y}_2 \in G \\ &= g^-(\mathbf{z}_2) && \text{because } \mathbf{y}_2 \in W^u(\mathbf{z}_2). \end{aligned}$$

Therefore, $g^-(\mathbf{z}_1) = g^-(\mathbf{z}_2)$ for any two points $\mathbf{z}_1, \mathbf{z}_2 \in H$.

We have shown that for any rectangle $\mathcal{R}$ determined by a relatively small $r > 0$, g^- is constant on H, where H is full measure in $\mathcal{R}$. By varying these rectangles (for a fixed r), we get that g^- is constant on a set of full measure in $\mathbb{T}^2$. Thus, we have proved that for a continuous function g, g^- is constant almost everywhere.

Because (i) the projection $g \mapsto g^-$ is a bounded linear map on $L^1(\mathbb{T}^2, \mathbb{R})$ (norm one by the Birkhoff Theorem), (ii) $C^0(\mathbb{T}^2, \mathbb{R})$ is dense in $L^1(\mathbb{T}^2, \mathbb{R})$, and (iii) the set of functions which are constant almost everywhere is closed in $L^1(\mathbb{T}^2, \mathbb{R})$, g^- is constant almost everywhere for any $g \in L^1(\mathbb{T}^2, \mathbb{R})$.

Finally, let $A \subset \mathbb{T}^2$ be a measurable invariant set for f. Let χ be the characteristic function for A. By the above argument, χ^- is constant almost everywhere. Because A is invariant, $\chi = \chi^-$, and so χ is constant almost everywhere. Since χ takes on only the values 0 and 1, χ is either constantly equal to 0 almost everywhere or constantly equal to 1 almost everywhere, i.e., $\mu(A)$ equals either 0 or 1. Thus, we have proved that any invariant set has either measure zero or one, and so we have proved that f is ergodic. □

8.5.3 The Zeta Function for Hyperbolic Toral Automorphisms

As we mentioned in Section 3.3, the *zeta function* for a map f is defined by

$$\zeta_f(t) = \exp\Big(\sum_{j=1}^{\infty} \frac{t^j}{j} N_j\Big)$$

with $N_j = \#(\mathrm{Fix}(f^j))$. In that section, we prove that if σ_A is the subshift of finite type for the matrix A, then $\zeta_{\sigma_A}(t) = [\det(I - tA)]^{-1}$, which is a rational function of t. The zeta function has been proved to be a rational function for many more diffeomorphisms. In this subsection, we prove this is true for a toral Anosov diffeomorphism. The fact that the zeta function is rational means that the number of all the periodic points of all periods can be determined by the finite number of invariants given by the coefficients of the rational function.

Theorem 5.4. *(a) Assume that $f : \mathbb{T}^n \to \mathbb{T}^n$ is an Anosov diffeomorphism. Then, the zeta function of f is rational.*

(b) Assume that $f_A : \mathbb{T}^2 \to \mathbb{T}^2$ is a hyperbolic toral automorphism for the matrix A, and λ_u the the unstable eigenvalue. Then,

$$\zeta_f(t) = \begin{cases} \dfrac{(1-t)^2}{\det(I-tA)} & \text{if } \lambda_u > 0 \text{ and } \det(A) > 0, \\ \dfrac{(1+t)^2}{\det(I+tA)} & \text{if } \lambda_u < 0 \text{ and } \det(A) > 0, \\ \dfrac{(1-t^2)}{\det(I-tA)} & \text{if } \lambda_u > 0 \text{ and } \det(A) < 0, \text{ and} \\ \dfrac{(1-t^2)}{\det(I+tA)} & \text{if } \lambda_u < 0 \text{ and } \det(A) < 0. \end{cases}$$

REMARK 5.11. For the proof to be valid, f could be any Anosov diffeomorphism which induces the map A on the first homology group, $H_1(\mathbb{T}^2, \mathbb{R}) = \mathbb{R} \oplus \mathbb{R}$.

REMARK 5.12. There are two types of proof of the rationality of the zeta function. Manning (1971) gave a proof using Markov partitions. Also see Bowen (1978a). Manning proved that the zeta function is a rational function whenever the chain recurrent set has a hyperbolic structure. (Actually, he uses the slightly weaker hypothesis that the nonwandering set is hyperbolic and is equal to the closure of the periodic points.) We discuss this type of proof in the exercises. See Exercise 8.34.

In this section, we use the proof using the Lefschetz index and the Lefschetz Fixed Point Theorem. This proof goes back to Smale (1967) in the case of a toral Anosov diffeomorphism. The zeta function for a diffeomorphism on a compact manifold was later proved to be a rational function using this type of proof by Williams (1968) for an attractor and by Guckenheimer (1970) whenever the chain recurrent set has a hyperbolic structure. Finally, Fried (1987) proved that the zeta function is rational whenever the diffeomorphism is expansive on the chain recurrent set. For more discussion of zeta functions, see Franks (1982).

Before starting the proof, we need to define the Lefschetz number of a fixed point and state the Lefschetz Fixed Point Theorem. We only give these definitions in the case when no periodic point has an eigenvalue which is a root of unity.

Definition. Let $f : M \to M$ be a C^1 diffeomorphism on a compact manifold M of dimension n. If $\mathbf{p}$ is a fixed point, the *Lefschetz index of* $\mathbf{p}$, $I_{\mathbf{p}}(f)$, is defined to be the $\text{sign}(\det(I - Df_{\mathbf{p}}))$ (where the derivative and determinant are calculated using some local coordinates). If $\mathbf{p}$ is a hyperbolic fixed point, then an easy analysis shows that

$$I_{\mathbf{p}}(f) = (-1)^u \Delta$$

where $u = \dim(\mathbb{E}^u_{\mathbf{p}})$, and $\Delta = 1$ provided $Df_{\mathbf{p}}|\mathbb{E}^u_{\mathbf{p}} \to \mathbb{E}^u_{\mathbf{p}}$ preserves orientation and $\Delta = -1$ provided this linear map reverses orientation. (See Exercise 8.33.)

Let f_{*k} be the induced map on the kth homology group, $H_k(M, \mathbb{R})$. The *Lefschetz number of* f is defined as follows:

$$L(f) = \sum_{k=0}^{n} (-1)^k \operatorname{tr}(f_{*k})$$

where n is the total dimension of M.

Theorem 5.5 (Lefschetz). *Let $f : M \to M$ be a C^1 diffeomorphism on a compact manifold M. Then,*

$$L(f) = \sum_{\mathbf{p} \in \mathrm{Fix}(f)} I_{\mathbf{p}}(f).$$

See Dold (1972).

To prove the main theorem using the Lefschetz number, we need make a connection between the Lefschetz numbers of the iterates of f and the induced map on homology. We start by defining a zeta function using the Lefschetz numbers of the iterates of f which can be related to the induced map on homology by means of the Lefschetz Fixed Point Theorem.

Definition. Let $f : M \to M$ be a map for which none of the periodic points have eigenvalues which are roots of unity. The *homology zeta function*, $Z_f(t)$, is defined as follows:

$$Z_f(t) = \exp\Big(\sum_{j=1}^{\infty} \frac{t^j}{j} K_j\Big)$$

where $K_j = L(f^j)$.

We first relate the homology zeta function with the regular zeta function.

Proposition 5.6. *Let $f : \mathbb{T}^n \to \mathbb{T}^n$ be a C^1 Anosov toral diffeomorphism and $u = \dim(\mathbb{E}^u_{\mathbf{p}})$. If $Df|\mathbb{E}^u$ preserves orientation, then*

$$\zeta_f(t) = Z_f(t)^{(-1)^u}.$$

If $Df|\mathbb{E}^u$ reverses orientation, then

$$\zeta_f(t) = Z_f(-t)^{(-1)^u}.$$

REMARK 5.13. For a Anosov diffeomorphism on $\mathbb{T}^n$, the bundle

$$\bigcup_{\mathbf{p} \in \mathbb{T}^n} \{\mathbf{p}\} \times \mathbb{E}^u_{\mathbf{p}}$$

is oriented, and the derivative

$$Df_{\mathbf{p}} : \mathbb{E}^u_{\mathbf{p}} \to \mathbb{E}^u_{f(\mathbf{p})}$$

either preserves orientation for all points $\mathbf{p}$ or reverses orientation for all points $\mathbf{p}$. Therefore, the statement of the proposition takes care of all cases. This is the only use we make of the manifold being a torus except for the case of the two-dimensional hyperbolic toral automorphism.

PROOF. Remember that $K_j = L(f^j)$ and $N_j = \#\mathrm{Fix}(f^j)$. If $Df|\mathbb{E}^u$ preserves orientation, then the indices of all the periodic points are $(-1)^u$, and $K_j = (-1)^u N_j$ or $N_j = (-1)^u K_j$. Therefore,

$$\begin{aligned}
\zeta_f(t) &= \exp\Big(\sum_{j=1}^{\infty} \frac{t^j}{j} N_j\Big) \\
&= \exp\Big(\sum_{j=1}^{\infty} \frac{t^j}{j} (-1)^u K_i\Big) \\
&= \exp\Big(\sum_{j=1}^{\infty} \frac{t^j}{j} K_i\Big)^{(-1)^u} \\
&= Z_f(t)^{(-1)^u}.
\end{aligned}$$

Similarly, if $Df|\mathbb{E}^u$ reverses the orientation, then $Df^j|\mathbb{E}^u$ reverses the orientation for j odd and preserves the orientation for j even. Therefore, the index of any fixed point of f^j is $(-1)^u(-1)^j$, $N_j = (-1)^u(-1)^j K_j$, and

$$\begin{aligned}\zeta_f(t) &= \exp\big(\sum_{j=1}^{\infty} \frac{-t^j}{j}(-1)^u K_i\big)\\ &= Z_f(-t)^{(-1)^u}.\end{aligned}$$

□

By the above proposition, to prove the theorem it is enough to prove that the homology zeta function is rational. The next proposition relates the homology zeta function with the linear maps on the homology groups which are induced by f.

Proposition 5.7. *Let $f : M \to M$ be a C^1 diffeomorphism on a n-dimensional manifold M, for which none of the periodic points have eigenvalues which are roots of unity. Then,*

$$Z_f(t) = \prod_{k=0}^{n} \det(I - tf_{*k})^{(-1)^{k+1}}.$$

The proof of the proposition uses the following lemma.

Lemma 5.8. *For an arbitrary matrix B (which we take as some f_{*k} below),*

$$\exp\big(\operatorname{tr}\big(\sum_{j=1}^{\infty} \frac{t^j}{j} B^j\big)\big) = \big(\det(I - tB)\big)^{-1}.$$

PROOF. The infinite series $\sum_{j=1}^{\infty} B^j t^j / j$ is the formal power series for the function $-\log(I - tB)$, so

$$\exp\big(\sum_{j=1}^{\infty} \frac{t^j}{j} B^j\big) = (I - tB)^{-1}.$$

Applying Liouville's Formula, we get

$$\begin{aligned}\exp\big(\operatorname{tr}\big(\sum_{j=1}^{\infty} \frac{t^j}{j} B^j\big)\big) &= \det\big(\exp\big(\sum_{j=1}^{\infty} \frac{t^j}{j} B^j\big)\big)\\ &= \det\big((I - tB)^{-1}\big)\\ &= \big(\det(I - tB)\big)^{-1}.\end{aligned}$$

□

PROOF OF PROPOSITION 5.7. By the Lefschetz Fixed Point Theorem,

$$K_j = L(f^j) = \sum_{k=0}^{n} (-1)^k \operatorname{tr}(f_{*k}^j).$$

Substituting this expression for K_j in the definition of the homology zeta function $Z_f(t)$ as follows:

$$\begin{aligned} Z_f(t) &= \exp\big(\sum_{j=1}^{\infty} \frac{t^j}{j} K_j\big) \\ &= \exp\big(\sum_{j=1}^{\infty}\sum_{k=0}^{n} \frac{t^j}{j}(-1)^k \operatorname{tr}(f_{*k}^j)\big) \\ &= \prod_{k=0}^{n} \big(\exp\big(\sum_{j=1}^{\infty} \operatorname{tr}\big(\frac{t^j}{j} f_{*k}^j\big)\big)\big)^{(-1)^k} \\ &= \prod_{k=0}^{n} \det(I - t f_{*k})^{(-1)^{k+1}}, \end{aligned}$$

where the last equality uses Lemma 5.8. □

PROOF OF THEOREM 5.4. Since the homology zeta function is rational by Proposition 5.7, the usual zeta function is rational by Proposition 5.6.

For a hyperbolic toral automorphism induced by A on $\mathbb{T}^2$, $f_{*0} = 1$, $f_{*1} = A$, and $f_{*2} = \operatorname{sign}(\det(A)) \cdot 1$. Therefore, if $\det(A) > 0$, then

$$Z_f(t) = \frac{\det(I - tA)}{(1-t)^2}$$

by Proposition 5.7. Similarly if $\det(A) < 0$, then

$$\begin{aligned} Z_f(t) &= \frac{\det(I - tA)}{(1-t)(1+t)} \\ &= \frac{\det(I - tA)}{1 - t^2}. \end{aligned}$$

Since $u = \dim(\mathbb{E}^u_{\mathbf{p}}) = 1$, using the relationship between the homology zeta function and the usual zeta function given in Proposition 5.6, we get the four cases as stated in the theorem. □

8.6 Attractors

There are various definitions of an attractor. The main difference involves which points in a neighborhood of the attractor have to approach the set. We demand that this holds for a whole neighborhood. There is also the question of whether the dynamics are "indecomposable" on the attractor itself. We demand that the map (or flow) is chain transitive on the attractor but this requirement changes from author to author quite drastically.

We write the definitions for a diffeomorphism $f : M \to M$ but only slight changes are needed to apply for a flow.

Definitions. A compact region $N \subset M$ is called a *trapping region* for f provided $f(N) \subset \operatorname{int}(N)$. A set Λ is called an *attracting set* (or an attractor by the terminology of Conley) provided there is a trapping region N such that $\Lambda = \bigcap_{k \geq 0} f^k(N)$. A set Λ is called an *attractor* provided it is an attracting set and $f|\Lambda$ is chain transitive, so $\Lambda \subset \mathcal{R}(f)$. (Sometimes, we might want to assume that $f|\Lambda$ is topologically transitive.) An invariant set Λ is called a *chaotic attractor* provided it is an attractor and f has sensitive dependence on initial conditions on Λ. (Sometimes people require f to have a positive Liapunov exponent on Λ instead of sensitive dependence. See Section 9.2 for the definition of Liapunov exponents for a map in several dimensions. Compare with the discussion on chaos in Chapter III.) Finally, an attractor with a hyperbolic structure is called a *hyperbolic attractor*.

REMARK 6.1. Some authors do not require Λ to attract a whole neighborhood but only to attract a set of positive measure in order to call it an attractor; however, this leads to a very different concept which I might call the *core of the limit set*. See Milnor (1985) for further discussion along these lines. Also, see Section 10.1 for a discussion of Conley's theory.

It can easily be checked that an attracting set is both negatively and positively invariant and closed. (These properties hold because it is the intersection of the nested sets $f^k(N)$.)

It is useful to give a definition of an attracting set which is given more in terms of properties of the invariant set A, and not determined by intersections of a trapping set. The following example shows that for A to be an attracting set, it is not enough that there is one neighborhood V of A such that $\omega(\mathbf{p}) \subset A$ for all $\mathbf{p} \in V$. Also see the example of Vinograd given in Example V.5.3.

Example 6.1. Consider the equations

$$\dot{\theta} = \theta^2(2\pi - \theta)^2$$
$$\dot{r} = r(1 - r^2),$$

where θ is an angular variable modulo 2π. See Figure 6.1 for the phase portrait. In these polar coordinates, let $B = \{(0,1)\}$ and $V = \{(\theta, r) : |r - 1| < \epsilon\}$. Then, V is a trapping set which is a neighborhood of B and $\omega(\mathbf{p}) = B$ for all $\mathbf{p} \in V$. However, the attracting set and attractor for V is the circle $A = \{(\theta, 1) : 0 \leq \theta \leq 2\pi\}$. (This is an example where the chain recurrent set is larger than the limit set.) (The set B is an attractor in the sense of Milnor.)

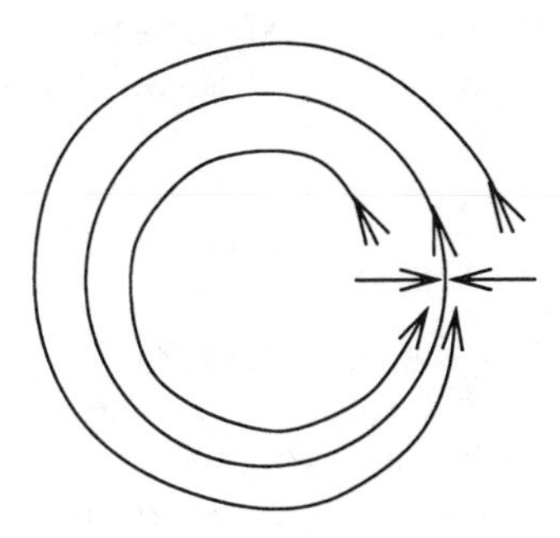

FIGURE 6.1. Phase Portrait of Example 6.1

Taking into consideration the above example, a compact attracting set can be characterized by ω-limit sets of points in arbitrarily small neighborhoods as given in the following proposition.

Proposition 6.1. *Let Λ be a compact invariant set in a finite-dimensional manifold. Then, Λ is an attracting set if and only if there are arbitrarily small neighborhoods V of Λ such that (i) V is positively invariant and (ii) $\omega(\mathbf{p}) \subset \Lambda$ for all $\mathbf{p} \in V$.*

We leave the proof of this result to the exercises. (See Exercise 8.35.) We also have the following result about the unstable manifolds for points in an attracting set.

Theorem 6.2. *Let Λ be an attracting set for f. Assume either that (i) $\mathbf{p} \in \Lambda$ is a hyperbolic periodic point or (ii) Λ has a hyperbolic structure and $\mathbf{p} \in \Lambda$. Then, $W^u(\mathbf{p}, f) \subset \Lambda$.*

PROOF. The set Λ is contained in the interior of a trapping region N, so there is an $\epsilon > 0$ such that $W^u_\epsilon(f^k(\mathbf{p})) \subset N$ for all $k \in \mathbb{Z}$. Therefore, for any $k \geq 0$,

$$W^u(f^{-k}(\mathbf{p})) = \bigcup_{j \geq 0} f^j W^u_\epsilon(f^{-j-k}(\mathbf{p})) \subset N \qquad \text{and}$$
$$W^u(\mathbf{p}) = f^k W^u(f^{-k}(\mathbf{p})) \subset f^k(N).$$

Taking the intersection for $k \geq 0$ of $f^k(N)$, we get that $W^u(\mathbf{p}) \subset \Lambda$. □

In the next few sections, we give examples of attractors which are locally the cross product of the unstable manifold by a Cantor set.

It is often useful to talk about the dimension of an attractor or other hyperbolic invariant set. Since such sets are often not manifolds, we need another concept. What we use is the topological dimension, which is always an integer. In other contexts, the fractal or Hausdorff dimension is useful, which is a non-negative real number. We postpone discussion of this concept until Section 9.4.

Definition. The definition of *topological dimension* is given inductively. A set Λ has topological dimension zero provided for each point $\mathbf{p} \in \Lambda$, there is an arbitrarily small neighborhood U of $\mathbf{p}$ such that $\partial(U) \cap \Lambda = \emptyset$. (It is not always possible to take U as a ball, as the example of Antoine's necklace shows.) Then, inductively, a set Λ is said to have dimension $n > 0$ provided for each point $\mathbf{p} \in \Lambda$, there is an arbitrarily small neighborhood U of $\mathbf{p}$ such that $\partial(U) \cap \Lambda$ has dimension $n - 1$. See Hurewicz and Wallman (1941) or Edgar (1990) for a more complete discussion of topological dimension.

Definition. Using the concept of topological dimension, we say that a hyperbolic attractor Λ is an *expanding attractor* provided the topological dimension of Λ is equal to the dimension of the unstable splitting. (Since $W^u(\mathbf{p}) \subset \Lambda$, we always have that the topological dimension is greater than or equal to the dimension of the unstable splitting.) See Williams (1967, 1974).

8.7 The Solenoid Attractor

In this section, we introduce an example of a hyperbolic attractor which can be given by a specific map, *the solenoid.* The solenoid was known to people studying topology; see Hocking and Young (1961), but was introduced as an example in Dynamical Systems by Smale (1967). Using this example as a model, R. Williams developed the theory of one-dimensional attractors, of which we see further examples in the sections on the Plykin attractors and the DA-attractor. See Williams (1967, 1974).

Let $D^2 = \{z \in \mathbb{C} : |z| \leq 1\}$. We think of D^2 as a subset of $\mathbb{R}^2$ even though we use complex notation for a point (and complex multiplication). Let $S^1 = \{t \in \mathbb{R} \text{ modulo } 1\}$ be the circle. The neighborhood of the attractor is the solid torus given by $N = S^1 \times D^2$. We define the embedding f of N into itself by means of a map g on S^1, $g : S^1 \to S^1$, given by $g(t) = 2t \bmod 1$. (The circle can also be thought of as a subset of $\mathbb{C}$. Using complex notation, $g(z) = z^2$ for $|z| = 1$.) The map g is called the doubling map (or squaring map). Using g, the embedding $f : N \to N$ (into N, not onto) is defined by

$$f(t, z) = (g(t), \frac{1}{4}z + \frac{1}{2}e^{2\pi t i}).$$

(Below, we call f a diffeomorphism even though it is not onto N.) The constants $1/4$ and $1/2$ in the definition are somewhat arbitrary as long as the first is small enough to make f one to one and the combination of the two insures that $f(N) \subset N$: $1/2 - 1/4 > 0$ and $1/2 + 1/4 < 1$. Geometrically, the map can be described as stretching the solid torus out to be twice as long in the S^1 direction and wrapping it twice around the S^1 direction. The image is thinner across in the D^2 direction by a factor of $1/4$. See Figure 7.1.

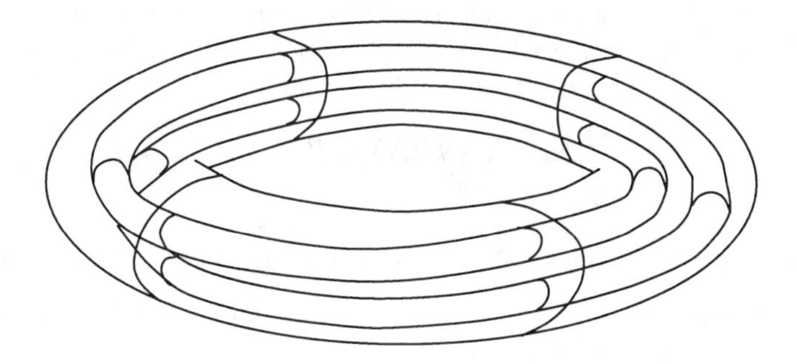

FIGURE 7.1. Image of Neighborhood N inside of N

Let

$$D(t) = \{t\} \times D^2$$

be the fiber with fixed "angle" t. The map f is a bundle map which takes a fiber $D(t)$ into the fiber $D(2t)$ and is a contraction by a factor of $1/4$ on each such fiber. We also use the notation

$$D([t_1, t_2]) = \bigcup \{D(t) : t \in [t_1, t_2]\}.$$

Theorem 7.1. *Let $\Lambda = \bigcap_{k=0}^{\infty} f^k(N)$. Then, Λ is a hyperbolic expanding attractor for f of topological dimension one, called the solenoid.*

We give various other properties of the solenoid throughout the section as we prove the theorem and further analyze this example. First note that N is a trapping region for f, so Λ is an attracting set. We start with the following proposition.

Proposition 7.2. *For each fixed t_0, $\Lambda \cap D(t_0)$ is a Cantor set.*

PROOF. If $f(t, z) \in D(t_0)$, then $g(t) = t_0 \mod 1$, so t is $t_0/2$ or $t_0/2 + 1/2$. The image $f(N) \cap D(t_0)$ is shown in Figure 7.2 for $t_0 = 0$ and $0 < t_0 < 1/2$.

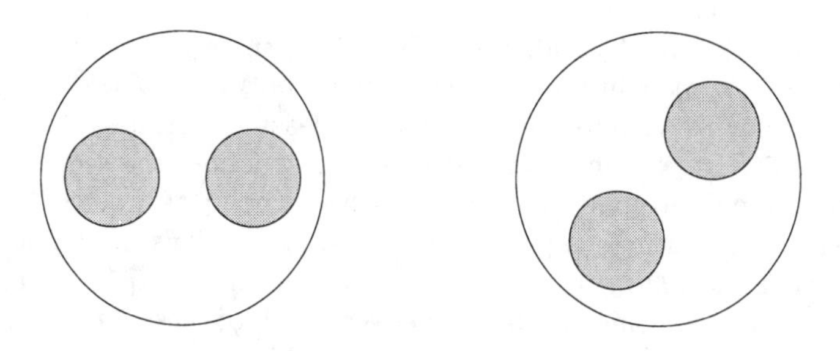

FIGURE 7.2. $f(N) \cap D(t_0)$ for $t_0 = 0$ and $0 < t_0 < 1/2$

Notice that

$$f(D(\frac{t_0}{2})) = (t_0, \frac{1}{4}D^2 + \{\frac{1}{2}e^{\pi t_0 i}\}) \quad \text{and}$$
$$f(D(\frac{t_0+1}{2})) = (t_0, \frac{1}{4}D^2 - \{\frac{1}{2}e^{\pi t_0 i}\})$$

since $e^{\pi t_0 i + \pi i} = -e^{\pi t_0 i}$. Thus, the two images are in the same fiber, but are reflections of each other through the origin in $D(t_0)$. They are disjoint because $1/2 - 1/4 > 0$. Both images are in $D(t_0)$ because $1/2 + 1/4 < 1$, so $f(N) \subset N$. Now, let

$$\mathcal{N}_k \equiv \bigcap_{j=0}^{k} f^j(N) = f^k(N).$$

Claim 7.3. *For all $t \in S^1$, the set $\mathcal{N}_k \cap D(t)$ is the union of 2^k disks of radius $(1/4)^k$.*

PROOF. The claim is trivially true for $k = 0$, and for $k = 1$, it follows from the image of N as discussed above. See Figures 7.1 and 7.2. Next,

$$\mathcal{N}_k \cap D(t) = f(\mathcal{N}_{k-1} \cap D(t/2)) \cup f(\mathcal{N}_{k-1} \cap D(t/2 + 1/2)).$$

By induction, $\mathcal{N}_{k-1} \cap D(t/2)$ and $\mathcal{N}_{k-1} \cap D(t/2+1/2)$ are each the union of 2^{k-1} disks of radius $(1/4)^{k-1}$. It follows from the fact that f is a contraction on fibers by a factor of $1/4$ that $f(\mathcal{N}_{k-1} \cap D(t/2))$ and $f(\mathcal{N}_{k-1} \cap D(t/2 + 1/2))$ are each the union of 2^{k-1} disks of radius $(1/4)^k$. Together, they are the union of 2^k disks of the stated radius. □

Now, $\Lambda = \bigcap_{j=0}^{\infty} f^j(N) = \bigcap_{j=0}^{\infty} \mathcal{N}_j$, so $\Lambda \cap D(t_0)$ is a Cantor set, as in the earlier example of a horseshoe. This proves Proposition 7.2. □

Next, we give several other topological properties of Λ.

Proposition 7.4. *The set Λ has the following properties:*

(a) *Λ is connected,*
(b) *Λ is not locally connected,*
(c) *Λ is not path connected, and*
(d) *the topological dimension of Λ is one.*

PROOF. (a) The sets $\mathcal{N}_k$ are compact, connected, and nested so their intersection Λ is connected. (See Exercise 5.11.)

(b) For $0 < t_2 - t_1 < 1$, $D([t_1, t_2]) \cap \mathcal{N}_k$ is the union of 2^k twisted tubes. For any neighborhood U of a point $\mathbf{p}$ in Λ, there is a choice of t_1 and t_2 and large k such that U contains two of these tubes. Since each tube contains some points of Λ, this shows that Λ is not locally connected.

(c) Fix $\mathbf{p} = (t_0, z_0) \in \Lambda$. By induction, for $k \geq 1$, there is a point $\mathbf{q}_k \in \Lambda \cap D(t_0)$ such that (i) for $k \geq 2$, $\mathbf{q}_k$ is in the same component of $\mathcal{N}_{k-1} \cap D(t_0)$ as $\mathbf{q}_{k-1}$, and (ii) any path from $\mathbf{p}$ to $\mathbf{q}_k$ in $\mathcal{N}_k$ must go around S^1 at least 2^{k-1} times. (This specifies the component of $\mathcal{N}_k \cap D(t_0)$ in which $\mathbf{q}_k$ lies.) By construction, the sequence $\{\mathbf{q}_k\}_{k=1}^{\infty}$ is Cauchy. Let $\mathbf{q}$ be the limit point of the $\mathbf{q}_k$, which is a point of Λ since Λ is closed. We claim that there is no continuous path in Λ from $\mathbf{p}$ to $\mathbf{q}$. This limit point $\mathbf{q}$ is in the same component of $\mathcal{N}_k \cap D(t_0)$ as $\mathbf{q}_k$, and any path from $\mathbf{p}$ to $\mathbf{q}$ in $\mathcal{N}_k$ must intersect $D(t_0)$ at least 2^{k-1} times, going around S^1 between each of these intersections. Thus, if there were a continuous path in Λ from $\mathbf{p}$ to $\mathbf{q}$, it would have to interest $D(t_0)$ infinitely many times (at least 2^{k-1} times for arbitrarily large k), going around S^1 between each of these intersections. A continuous path cannot go around S^1 an infinite number of times, so this contradicts the assumption that there is a continuous path in Λ from $\mathbf{p}$ to $\mathbf{q}$, and proves that Λ is not path connected.

(d) In each fiber, $\Lambda \cap D(t_0)$ is totally disconnected, and thus has topological dimension zero. In a segment of N, $\Lambda \cap D([t_1, t_2])$ is homeomorphic to the product of $\Lambda \cap D(t_1)$ and the interval $[t_1, t_2]$, and so has topological dimension one. This completes the proof of the proposition. □

Next, we state several properties of f on Λ which together imply Theorem 7.1.

Proposition 7.5. *The map f restricted to Λ has the following properties:*

(a) *the periodic points of f are dense in Λ,*
(b) *f is topologically transitive on Λ, and*
(c) *f has a hyperbolic structure on Λ.*

PROOF. As a first step in the proof of part (a), we give the following lemma about g.

Lemma 7.6. *The periodic points of g are dense in S^1.*

PROOF. The point t_0 is fixed by g^k, $g^k(t_0) = t_0$, if and only if $2^k t_0 = t_0 + j$ for some integer j. Thus, $t_0 = j/(2^k - 1)$. For k fixed, let $t_{k,j} = j/(2^k - 1)$ for $0 \leq j \leq 2^k - 2$. These points are evenly spaced on the circle with separation $1/(2^k - 1)$. As k goes to infinity, it follows that the set of all periodic points is dense. □

Now, if $g^k(t_0) = t_0$, $f^k(D(t_0)) \subset D(t_0)$. The set $D(t_0)$ is a disk, and f^k takes it into itself with a contraction factor of 4^{-k}, so f^k has a fixed point in $D(t_0)$. By the lemma, it follows that the fibers with a periodic point for f are dense in the set of all fibers.

We want to show the periodic points are actually dense in Λ. Take a point $\mathbf{p} \in \Lambda$ and a neighborhood U of $\mathbf{p}$. There is a choice of k and t_1, t_2 such that the tube, $f^k(D([t_1, t_2]) \subset U$. We showed above that f has a periodic point in $D([t_1, t_2])$ and so in $f^k(D([t_1, t_2]) \subset U$. This proves the first part of the proposition.

(b) We need to verify on Λ the hypothesis of the Birkhoff Transitivity Theorem, Theorem 2.1. Let U and V be two open subsets of Λ. Thus, there are open set in $S^1 \times D^2$, U' and V', such that $U' \cap \Lambda = U$ and $V' \cap \Lambda = V$. There exist $k \in \mathbb{N}$, $0 < t_2 - t_1 < 1$, and $0 < t_2' - t_1' < 1$ such that $f^k(D([t_1, t_2])) \subset U'$ and $f^k(D([t_1', t_2'])) \subset V'$. Then, there is a $j > 0$ such that $f^j(D([t_1, t_2])) \cap D([t_1', t_2']) \neq \emptyset$. In fact, since we can take j such that $f^j(D([t_1, t_2]))$ goes all the way across $D([t_1', t_2'])$, we can require that

$$f^j(D([t_1, t_2]) \cap \Lambda) \cap D([t_1', t_2']) \cap \Lambda \neq \emptyset.$$

Thus,

$$\begin{aligned} &f^j(f^k(D([t_1, t_2])) \cap \Lambda) \cap f^k(D([t_1', t_2'])) \cap \Lambda \neq \emptyset \qquad \text{and} \\ &f^j(U) \cap V = f^j(U' \cap \Lambda) \cap [V' \cap \Lambda] \neq \emptyset. \end{aligned}$$

By the Birkhoff Transitivity Theorem, Theorem 2.1, $f|\Lambda$ is transitive, and we have proved part (b) of the proposition.

(c) In terms of the coordinates on $S^1 \times D^2$,

$$Df_{(t,z)} = \begin{pmatrix} 2 & 0 \\ \pi i e^{2\pi t i} & \frac{1}{4} I_2 \end{pmatrix},$$

where I_2 is the identity matrix on $\mathbb{C}$ or $\mathbb{R}^2$. Let $\mathbb{E}^s_{\mathbf{p}} = \{0\} \times \mathbb{R}^2$. Then, for $(0, \mathbf{v}) \in \mathbb{E}^s_{\mathbf{p}}$,

$$Df_{\mathbf{p}} \begin{pmatrix} 0 \\ \mathbf{v} \end{pmatrix} = \begin{pmatrix} 0 \\ \frac{1}{4}\mathbf{v} \end{pmatrix}, \quad \text{and}$$

$$Df^k_{\mathbf{p}} \begin{pmatrix} 0 \\ \mathbf{v} \end{pmatrix} = \begin{pmatrix} 0 \\ \frac{1}{4^k}\mathbf{v} \end{pmatrix}$$

which goes to 0 as k goes to ∞. Therefore, this is indeed the stable bundle at each point $\mathbf{p} \in \Lambda$.

To find $\mathbb{E}^u_{\mathbf{p}}$, it is necessary to use cones. Let

$$C^u_{\mathbf{p}} = \{\begin{pmatrix} \mathbf{v}_1 \\ \mathbf{v}_2 \end{pmatrix} : \mathbf{v}_1 \in TS^1, \mathbf{v}_2 \in \mathbb{R}^2, \text{ and } |\mathbf{v}_1| \geq \frac{1}{2}|\mathbf{v}_2|\}.$$

Step 1. $Df_{\mathbf{p}}\, C^u_{\mathbf{p}} \subset C^u_{f(\mathbf{p})}$.

PROOF. Let $\begin{pmatrix} \mathbf{v}_1 \\ \mathbf{v}_2 \end{pmatrix} \in C^u_{\mathbf{p}}$, and

$$Df_{\mathbf{p}} \begin{pmatrix} \mathbf{v}_1 \\ \mathbf{v}_2 \end{pmatrix} = \begin{pmatrix} 2\mathbf{v}_1 \\ \pi i e^{2\pi t i}\mathbf{v}_1 + \frac{1}{4}\mathbf{v}_2 \end{pmatrix} \equiv \begin{pmatrix} \mathbf{v}'_1 \\ \mathbf{v}'_2 \end{pmatrix}.$$

Then,

$$\begin{aligned} |\mathbf{v}'_1| = 2|\mathbf{v}_1| &= \frac{1}{2}(4|\mathbf{v}_1|) \\ &> \frac{1}{2}(\pi|\mathbf{v}_1| + \frac{1}{2}|\mathbf{v}_1|) \\ &\geq \frac{1}{2}(\pi|\mathbf{v}_1| + \frac{1}{4}|\mathbf{v}_2|) \\ &\geq \frac{1}{2}|\mathbf{v}'_2|. \end{aligned}$$

This last inequality shows that $\begin{pmatrix} \mathbf{v}'_1 \\ \mathbf{v}'_2 \end{pmatrix} \in C^u_{f(\mathbf{p})}$, completing the proof of the first step. □

Step 2. *The intersection* $\bigcap_{k=0}^{\infty} Df^k_{f^{-k}(\mathbf{p})}\, C^u_{f^{-k}(\mathbf{p})} = \mathbb{E}^u_{\mathbf{p}}$ *is a line in the tangent space.*

PROOF. By Step 1, the finite intersections

$$\bigcap_{j=0}^{k} Df^j_{f^{-j}(\mathbf{p})}\, C^u_{f^{-j}(\mathbf{p})} = Df^k_{f^{-k}(\mathbf{p})}\, C^u_{f^{-k}(\mathbf{p})}$$

are nested. To prove that the intersection is a line, we prove that the angle between two vectors in these finite intersections goes to 0 as k goes to ∞. Let

$$\begin{pmatrix} \mathbf{v}_1 \\ \mathbf{v}_2 \end{pmatrix}, \begin{pmatrix} \mathbf{w}_1 \\ \mathbf{w}_2 \end{pmatrix} \in C^u_{f^{-k}(\mathbf{p})}$$

with $\mathbf{v}_1, \mathbf{w}_1 > 0$,

$$\begin{pmatrix} \mathbf{v}_1^k \\ \mathbf{v}_2^k \end{pmatrix} = Df^k_{f^{-k}(\mathbf{p})} \begin{pmatrix} \mathbf{v}_1 \\ \mathbf{v}_2 \end{pmatrix}, \quad \text{and} \quad \begin{pmatrix} \mathbf{w}_1^k \\ \mathbf{w}_2^k \end{pmatrix} = Df^k_{f^{-k}(\mathbf{p})} \begin{pmatrix} \mathbf{w}_1 \\ \mathbf{w}_2 \end{pmatrix}.$$

Then,

$$\begin{aligned} \left| \frac{\mathbf{v}_2^1}{\mathbf{v}_1^1} - \frac{\mathbf{w}_2^1}{\mathbf{w}_1^1} \right| &= \left| \frac{\pi i e^{2\pi t i}\mathbf{v}_1 + \frac{1}{4}\mathbf{v}_2}{2\mathbf{v}_1} - \frac{\pi i e^{2\pi t i}\mathbf{w}_1 + \frac{1}{4}\mathbf{w}_2}{2\mathbf{w}_1} \right| \\ &= \frac{1}{8} \left| \frac{\mathbf{v}_2}{\mathbf{v}_1} - \frac{\mathbf{w}_2}{\mathbf{w}_1} \right|, \end{aligned}$$

which shows there is a contraction on the difference of the slopes. By induction on k,

$$\left| \frac{\mathbf{v}_2^k}{\mathbf{v}_1^k} - \frac{\mathbf{w}_2^k}{\mathbf{w}_1^k} \right| = (\frac{1}{8})^k \left| \frac{\mathbf{v}_2}{\mathbf{v}_1} - \frac{\mathbf{w}_2}{\mathbf{w}_1} \right|,$$

which goes to 0 as k goes to ∞. Since the difference of slopes goes to 0, the cones converge to a line. □

Step 3. *The derivative of f restricted to $\mathbb{E}^u_{\mathbf{p}}$, $Df_{\mathbf{p}}|\mathbb{E}^u_{\mathbf{p}}$, is an expansion.*

PROOF. Since $\mathbb{E}^u_{\mathbf{p}}$ is a graph over $T_tS^1 \times \{0\}$, let $\left| \begin{pmatrix} \mathbf{v}_1 \\ \mathbf{v}_2 \end{pmatrix} \right|_* = |\mathbf{v}_1|$. This is a norm on the cone. Then,

$$\begin{aligned} \left| Df_{\mathbf{p}} \begin{pmatrix} \mathbf{v}_1 \\ \mathbf{v}_2 \end{pmatrix} \right|_* &= \left| \begin{pmatrix} 2\mathbf{v}_1 \\ \pi i e^{2\pi t i}\mathbf{v}_1 + \frac{1}{4}\mathbf{v}_2 \end{pmatrix} \right|_* \\ &= |2\mathbf{v}_1| \\ &= 2\left| \begin{pmatrix} \mathbf{v}_1 \\ \mathbf{v}_2 \end{pmatrix} \right|_*. \end{aligned}$$

Thus, $Df_{\mathbf{p}}$ is an expansion on vectors in this bundle in terms of this norm, and hence the standard norm. This completes the proof of Step 3, part (c) of the proposition, and the proposition. □

REMARK 7.1. Given the bundle of vectors which expand and contract, it is easy to see that $W^s(\mathbf{p}) \supset \{t\} \times D^2$ if $\mathbf{p} = (t, z)$. The unstable manifold, $W^u(\mathbf{p})$, winds around through Λ. Each $W^u(\mathbf{p})$ is an immersed line. (Since f is one to one and an expansion on $W^u(\mathbf{p})$, it cannot be a circle, which is the only other one-dimensional possibility.) The unstable manifold $W^u(\mathbf{p})$ hits $D(t)$ in a countable number of points. Since $\Lambda \cap D(t)$ is uncountable, there are many other points in $\Lambda \cap D(t)$ which are not in $W^u(\mathbf{p})$. These are points $\mathbf{q}$ for which there is no curve in Λ from $\mathbf{p}$ to $\mathbf{q}$.

REMARK 7.2. As an exercise, we ask the reader to construct Markov partitions for f and the doubling map g. See Exercises 8.37 and 8.38.

8.7.1 Conjugacy of the Solenoid to an Inverse Limit

Williams introduced the idea of representing certain attractors (expanding attractors) as inverse limits. See Williams (1967, 1974).

As before, $\mathbb{N}$ is the natural numbers, $\{0, 1, 2, 3, \ldots\}$. Let $g(t) = 2t \bmod 1$ as before. Let

$$\Sigma^- = \{\mathbf{s} \in (S^1)^{\mathbb{N}} : g(s_{j+1}) = s_j\}.$$

Define the shift map, σ, on Σ^- by $\sigma(\mathbf{s}) = \mathbf{t}$ if

$$t_j = \begin{cases} s_{j-1} & \text{if } j \geq 1 \\ g(s_0) & \text{if } j = 0. \end{cases}$$

If $\mathbf{s} \in \Sigma^-$, then $g(s_{j+1}) = s_j$ so $s_{j+1} \in g^{-1}(s_j)$ is one of the two preimages of s_j. The pair (Σ^-, σ) is called the *inverse limit of* g.

A point $\mathbf{p} = (t, z) \in \Lambda$ is determined by a sequence of descending disks in $D(t)$ as we saw above. These disks are in turn determined by the preimages of t by the map g. When a map is expanding (like the one-dimensional horseshoe), we use forward images of a point p to determine p. Because f contracts on fibers, we use backward images. Define $h : \Lambda \to (S^1)^{\mathbb{N}}$ by $h(\mathbf{p}) = \mathbf{s}$ where $f^{-j}(\mathbf{p}) \in D(s_j)$ with $s_j \in S^1$ for $j = 0, 1, \ldots$.

Theorem 7.7. *The map h defined above is a conjugacy from f on Λ to the inverse limit of g, σ on Σ^-.*

PROOF.

Step 1. $h(\Lambda) \subset \Sigma^-$.

PROOF. Let $h(\mathbf{p}) = \mathbf{s}$. Then, $f^{-j}(\mathbf{p}) \in D(s_j)$ and $f^{-j-1}(\mathbf{p}) \in D(s_{j+1})$. Therefore, the intersection $f(D(s_{j+1})) \cap D(s_j) \neq \emptyset$, so $f(D(s_{j+1})) \subset D(s_j)$. Thus, $g(s_{j+1}) = s_j$ for all j and $\mathbf{s} \in \Sigma^-$. □

Step 2. $h \circ f = \sigma \circ h$.

PROOF. For $\mathbf{p} \in \Lambda$, let $h(\mathbf{p}) = \mathbf{s}$ and $h(f(\mathbf{p})) = \mathbf{t}$. $f^{-(j+1)}(f(\mathbf{p})) = f^{-j}(\mathbf{p})$ and so is in both $D(t_{j+1})$ and $D(s_j)$. Therefore, $t_{j+1} = s_j$ for all $j \geq 0$. Similarly, $f(\mathbf{p})$ is in both $D(t_0)$ and $f(D(s_0))$, so $t_0 = g(s_0)$. This proves that $\sigma(\mathbf{s}) = \mathbf{t}$, as required. □

Step 3. *The map h is one to one.*

PROOF. If $h(\mathbf{p}) = h(\mathbf{q}) = \mathbf{s}$, then $\mathbf{p}, \mathbf{q} \in \bigcap_{j=0}^k f^j(D(s_j))$. This is a nested sequence of disks whose radii go to 0. Therefore, there is only one point in the intersection and $\mathbf{p} = \mathbf{q}$. □

Step 4. *The map h is onto Σ^-.*

PROOF. Take $\mathbf{s} \in \Sigma^-$. Then, $g(s_{j+1}) = s_j$ so $f(D(s_{j+1})) \subset D(s_j)$, and

$$f^j(D(s_j)) \subset f^{j-1}(D(s_{j-1})) \subset \cdots \subset D(s_0).$$

Therefore, $\bigcap_{j=0}^k f^j(D(s_j))$ is a nested sequence of disks with nonempty intersection; hence,

$$\bigcap_{j=0}^{\infty} f^j(D(s_j)) \neq \emptyset.$$

If $\mathbf{p}$ is a point in this intersection, then $h(\mathbf{p}) = \mathbf{s}$. This completes the proof of the fourth step and the theorem. □

8.8 The DA Attractor

The next example of an attractor we consider is constructed by modifying a toral Anosov diffeomorphism on the two-dimensional torus. For this reason, it is called the *Derived-from-Anosov-diffeomorphism* or the *DA-diffeomorphism*. It was first introduced by Smale (1967).

Let $g : \mathbb{T}^2 \to \mathbb{T}^2$ be the Anosov diffeomorphism induced by the linear matrix $\begin{pmatrix} 2 & 1 \\ 1 & 1 \end{pmatrix}$. (Any other example with a fixed point would work as well.) Let $\mathbf{p}_0$ be a fixed point of g corresponding to $\mathbf{0}$ in $\mathbb{R}^2$. Let $\mathbf{v}^u$ and $\mathbf{v}^s$ be the unstable and stable eigenvectors of the matrix and use coordinates $u_1\mathbf{v}^u + u_2\mathbf{v}^s$ in a (relatively small) neighborhood, U, of $\mathbf{p}_0$. Let $r_0 > 0$ be small enough so the ball of radius r_0 about $\mathbf{p}_0$ is contained in U. Let $\delta(x)$ be a bump function of a single variable such that $0 \le \delta(x) \le 1$ for all x, and

$$\delta(x) = \begin{cases} 0 & \text{for } x \ge r_0 \\ 1 & \text{for } x \le r_0/2. \end{cases}$$

Consider the differential equations

$$\begin{aligned} \dot{u}_1 &= 0 \\ \dot{u}_2 &= u_2\delta(|(u_1, u_2)|). \end{aligned}$$

Let φ^t be the flow of these differential equations, $\varphi^t(u_1, u_2) = (u_1, \varphi_2^t(u_1, u_2))$. Then, the support, $\text{supp}(\varphi^t - id) \subset U$. Also, the derivative of the flow at $\mathbf{p}_0$ in terms of the (u_1, u_2)-coordinates is

$$D\varphi^t_{\mathbf{p}_0} = \begin{pmatrix} 1 & 0 \\ 0 & e^t \end{pmatrix}.$$

Define $f = \varphi^\tau \circ g$ for a fixed $\tau > 0$ such that $e^\tau \lambda_s > 1$, where λ_s is the stable eigenvalue. The map f is called the *DA-diffeomorphism.* Note that in the (u_1, u_2)-coordinates, the derivative of f at $\mathbf{p}_0$ is

$$Df_{\mathbf{p}_0} = D\varphi^\tau_{\mathbf{p}_0}\, Dg_{\mathbf{p}_0} = \begin{pmatrix} \lambda_u & 0 \\ 0 & e^\tau \lambda_s \end{pmatrix},$$

so $\mathbf{p}_0$ is a source.

Theorem 8.1. *The DA-diffeomorphism f described above has $\Omega(f) = \{\mathbf{p}_0\} \cup \Lambda$, where $\mathbf{p}_0$ is a fixed point source and Λ is an expanding attractor of topological dimension one. The map f is transitive on Λ and the periodic points are dense in Λ.*

PROOF. Because the neighborhood U can be taken arbitrarily small, f can be made arbitrarily C^0 near g, but not C^1 near g since e^τ cannot be arbitrarily small. Also note that the flow φ^t preserves each stable manifold of a point for g, $W^s(\mathbf{q}, g)$, because of the form of the differential equations. Therefore, f preserves each $W^s(\mathbf{q}, g)$.

The new map f has three fixed points on $W^s(\mathbf{p}_0, g)$, $\mathbf{p}_0$ and two new fixed points $\mathbf{p}_1$ and $\mathbf{p}_2$. This fact can be seen to be true because $f(\mathbf{p}_0) = \mathbf{p}_0$ is a source and outside U the slope of the graph of f on $W^s(\mathbf{q}, g)$ is still less than 1. Therefore, there must be a fixed point on each side of $\mathbf{p}_0$ along $W^s(\mathbf{q}, g)$. See Figure 8.1. We claim that both $\mathbf{p}_1$ and $\mathbf{p}_2$ are saddles. To see this, note that in U, $Df_{\mathbf{q}} = \begin{pmatrix} a_{11} & 0 \\ a_{21} & a_{22} \end{pmatrix}$, with $a_{11} = \lambda_u$ for all $\mathbf{q}$, and with $0 < a_{22} < 1$ at $\mathbf{p}_1$ and $\mathbf{p}_2$ because of the nature of the graph of $f|W^s(\mathbf{p}_0, g)$ indicated in Figure 8.1.

Let V be a neighborhood of $\mathbf{p}_0$ (not containing $\mathbf{p}_1$ and $\mathbf{p}_2$) contained in U such that (i) $a_{22} > 1$ for $\mathbf{q} \in V$ (f is an expansion along $\mathbb{E}^s$ in V), (ii) $0 < a_{22} < 1$ for $\mathbf{q} \notin f(V)$ (f is a contraction along $\mathbb{E}^s$ outside of V), and (iii) $f(V) \supset V$. See Figure 8.2. (We leave as an exercise the existence of such a neighborhood V. See Exercise 8.41.) Clearly, $V \subset W^u(\mathbf{p}_0, f)$, so it is the local unstable manifold of $\mathbf{p}_0$ and $W^u(\mathbf{p}_0, f) = \bigcup_{j=0}^\infty f^j(V)$. Let $N = \mathbb{T}^2 \setminus V$. Then, N is a trapping region because $f(V) \supset V$. Let $\Lambda = \bigcap_{j=0}^\infty f^j(N)$.

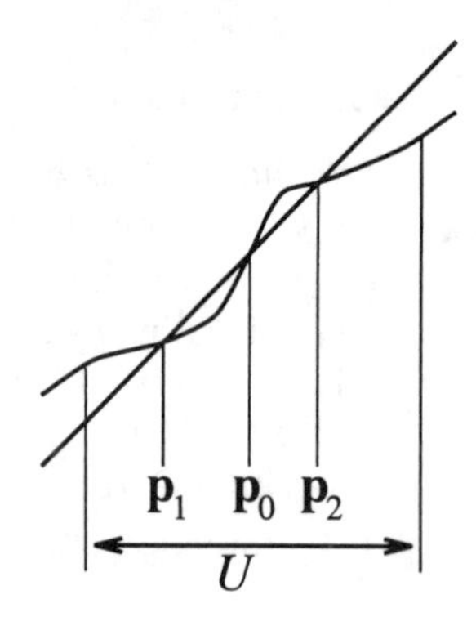

FIGURE 8.1. Graph of $f|W^s(\mathbf{p}_0, g)$

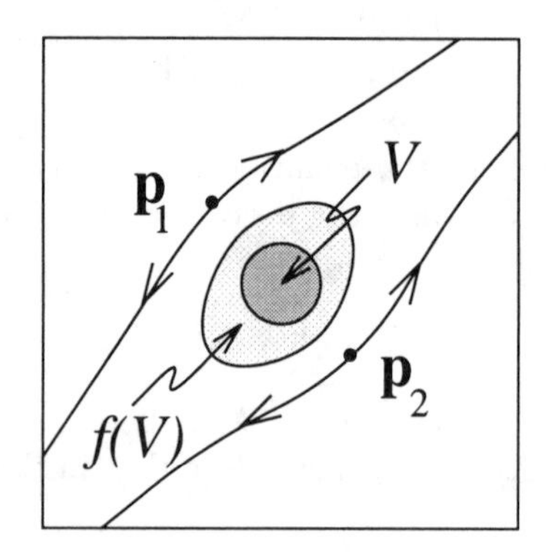

FIGURE 8.2. Image of Open Set V

This is an attracting set, and $\Lambda = \mathbb{T}^2 \setminus W^u(\mathbf{p}_0, f)$. The unstable manifold of $\mathbf{p}_0$ for f, $W^u(\mathbf{p}_0, f)$, is a "thickened" version of the unstable manifold for g. In Step 4 below, we prove that $W^u(\mathbf{p}_0, f)$ is still dense in $\mathbb{T}^2$, so Λ has empty interior.

We proceed to prove Theorem 8.1 through a series of steps.

Step 1. *The map f has a hyperbolic structure on Λ.*

PROOF. In terms of the splitting $\mathbb{E}^u_{\mathbf{q}}(g) \oplus \mathbb{E}^s_{\mathbf{q}}(g)$, the derivative of f, $Df_{\mathbf{q}} = (a_{ij})$, is lower triangular in U and diagonal outside U ($a_{12} = 0$ everywhere and $a_{21} = 0$ outside U). The unstable term $a_{11} = \lambda_u > 1$ everywhere and $0 < a_{22} < 1$ outside $f(V)$ so on Λ. Because of the form of the derivative, $\mathbb{E}^s_{\mathbf{q}}(f) = \mathbb{E}^s_{\mathbf{q}}(g)$ is an invariant bundle and every vector in this bundle is contracted by $Df_{\mathbf{q}}$ for $\mathbf{q} \in \Lambda$. Therefore, this is the stable bundle on Λ. Let C be a bound on $|a_{21}|$ everywhere, define $L = C(\lambda_u - \lambda_s)^{-1}$ and take the cones

$$C^u_{\mathbf{q}} = \{(\mathbf{v}_1, \mathbf{v}_2) \in \mathbb{E}^u_{\mathbf{q}}(g) \oplus \mathbb{E}^s_{\mathbf{q}}(g) : |\mathbf{v}_2| \le L|\mathbf{v}_1|\}.$$

Then, it can be checked using the lower triangular nature of the derivative of f that the cones are invariant and

$$\mathbb{E}^u_{\mathbf{q}}(f) = \bigcap_{j=0}^{\infty} Df^j_{f^{-j}(\mathbf{q})} C^u_{f^{-j}(\mathbf{q})}$$

is an invariant bundle on which the derivative is an expansion for points $\mathbf{q} \in \Lambda$. Thus, this gives the unstable bundle on Λ and so the hyperbolic splitting. $\square$

Step 2. *For $j = 1, 2$, $\mathbf{p}_1, \mathbf{p}_2 \in \Lambda$ and $W^u(\mathbf{p}_j, f) \subset \Lambda$.*

PROOF. The fact that $\mathbf{p}_1, \mathbf{p}_2 \in \Lambda$ follows because $\mathbf{p}_1, \mathbf{p}_2 \notin V$ and they are fixed points, so

$$\mathbf{p}_1, \mathbf{p}_2 \notin \bigcup_{j=0}^{\infty} f^j(V) = \mathbb{T}^2 \setminus \Lambda.$$

The fact about the unstable manifolds was proved for general attracting sets. □

Step 3. *The stable manifolds of f satisfy the following:*

$$W^s(\mathbf{p}_1, f) \cup W^s(\mathbf{p}_2, f) = W^s(\mathbf{p}_0, g) \setminus \{\mathbf{p}_0\}$$

and $W^s(\mathbf{q}, f) = W^s(\mathbf{q}, g)$ for $\mathbf{q} \notin W^s(\mathbf{p}_0, g)$. Thus, $W^s(\mathbf{q}, f)$ is dense in $\mathbb{T}^2$ for all $\mathbf{q} \in \Lambda$.

PROOF. $f(W^s(\mathbf{q}, g)) = W^s(f(\mathbf{q}), g)$ and $W^s(\mathbf{q}, g)$ is tangent to $\mathbb{E}^s(f)$. Also, for $\mathbf{q} \in \Lambda$, f is a contraction on $W^s_{\text{loc}}(\mathbf{q}, g)$. Therefore, $W^s_{\text{loc}}(\mathbf{q}, f) = W^s_{\text{loc}}(\mathbf{q}, g)$, and $W^s(\mathbf{q}, f) \subset W^s(\mathbf{q}, g)$. A line segment I in $W^s(\mathbf{q}, g)$ which does not end in V (goes all the way across V if it intersects it) is lengthened by f^{-1}. Any line segment whose end stays in V for all inverse images must be a subset of $W^s(\mathbf{p}_0, g)$ and have an end in $W^s_{\text{loc}}(\mathbf{p}_0, g)$. Thus, if $\mathbf{q} \notin W^s(\mathbf{p}_0, g)$, then $W^s(\mathbf{q}, f) = W^s(\mathbf{q}, g)$. Also, this implies $W^s(\mathbf{p}_j, f)$ is one component of $W^s(\mathbf{p}_0, g) \setminus \{\mathbf{p}_0\}$ for $j = 1, 2$. The fact that the $W^s(\mathbf{q}, f)$ are dense in $\mathbb{T}^2$ follows because these are lines with irrational slope. □

Step 4. *The unstable manifold of f at $\mathbf{p}_0$, $W^u(\mathbf{p}_0, f)$, is an open dense set in $\mathbb{T}^2$.*

PROOF. By construction, $\mathbb{T}^2 = \Lambda \cup \bigcup_{j=0}^{\infty} f^j(V)$, so we need only prove that $W^u(\mathbf{p}_0, f)$ accumulates on Λ. Let $\mathbf{p} \in \Lambda$, and $Z_{\mathbf{p}}$ be an arbitrarily small neighborhood of $\mathbf{p}$ in $\mathbb{T}^2$. Let $I = \text{comp}_{\mathbf{p}}(W^s(\mathbf{p}, f) \cap Z_{\mathbf{p}})$. As long as $f^{-j}(I)$ does not intersect $f(V)$, it is lengthened by f^{-1} by a uniform amount. But there is a uniform bound on the length of $\text{comp}_{\mathbf{z}}[W^s(\mathbf{z}, g) \setminus f(V)]$. Therefore, for large j,

$$\begin{aligned} &f^{-j}(I) \cap f(V) \neq \emptyset, \\ &f^{-j}(Z_{\mathbf{p}}) \cap f(V) \neq \emptyset, \\ &Z_{\mathbf{p}} \cap f^{j+1}(V) \neq \emptyset, \qquad \text{and} \\ &Z_{\mathbf{p}} \cap W^u(\mathbf{p}_0, f) \neq \emptyset. \end{aligned}$$

Since $Z_{\mathbf{p}}$ is an arbitrarily small neighborhood, it follows that $W^u(\mathbf{p}_0, f)$ is dense at $\mathbf{p}$. Since $\mathbf{p}$ is arbitrary, $W^u(\mathbf{p}_0, f)$ is dense in $\mathbb{T}^2$. □

Step 5. *For $j = 1, 2$, $W^u(\mathbf{p}_j, f)$ is dense in Λ.*

PROOF. Let $\mathbf{p} \in \Lambda \cap W^s(\mathbf{q}, g)$, where $\mathbf{q}$ has period k for g. By Step 4, $\mathbf{p} \in \partial(W^u(\mathbf{p}_0, f))$, since $\mathbf{p} \in \text{cl}(W^u(\mathbf{p}_0, f)) \setminus W^u(\mathbf{p}_0, f)$. In the proof of Step 4, when $f^{-jk}(I)$ intersects V, it must cross $W^u(\mathbf{p}_1, f) \cup W^u(\mathbf{p}_2, f)$. (It crosses from one side to the other.) Therefore,

$$\begin{aligned} &[W^u(\mathbf{p}_1, f) \cup W^u(\mathbf{p}_2, f)] \cap f^{-jk}(Z_{\mathbf{p}}) \neq \emptyset \qquad \text{and} \\ &[W^u(\mathbf{p}_1, f) \cup W^u(\mathbf{p}_2, f)] \cap Z_p \neq \emptyset \end{aligned}$$

because the unstable manifolds are invariant by f. This shows that the union of the two unstable manifolds $W^u(\mathbf{p}_1, f) \cup W^u(\mathbf{p}_2, f)$ is dense in Λ.

We have shown that the union of the two manifolds is dense in Λ, and we need to show that each manifold is dense by itself. $W^s(\mathbf{p}_1, f)$ is dense in $\mathbb{T}^2$ so it must intersect $W^u(\mathbf{p}_2, f)$. Because these are tangent to the bundles $\mathbb{E}^s$ and $\mathbb{E}^u$, the intersections (which are on Λ) are transverse. By the Inclination Lemma, it follows that $W^u(\mathbf{p}_2, f)$ accumulates on $W^u(\mathbf{p}_1, f)$, $\text{cl}(W^u(\mathbf{p}_2, f) \supset W^u(\mathbf{p}_1, f)$, and $\text{cl}(W^u(\mathbf{p}_2, f) = \Lambda$. Similarly, $\text{cl}(W^u(\mathbf{p}_1, f) = \Lambda$. □

Step 6. *The topological dimension of Λ is one.*

PROOF. By Step 4, $W^u(\mathbf{p}_0, f)$ is dense in $\mathbb{T}^2$, so Λ has empty interior and must have topological dimension at most one. The manifolds $W^u(\mathbf{p}_j, f)$ for $j = 1, 2$ are contained in Λ, so it must have topological dimension at least one. □

Step 7. *For $j = 1, 2$, $\{\mathbf{q} \in \Lambda : \mathbf{q}$ is a transverse homoclinic point for $\mathbf{p}_j\}$ is dense in Λ.*

PROOF. If $\mathbf{x} \in \Lambda$, Steps 3 and 5 imply that both $W^u(\mathbf{p}_j, f)$ and $W^s(\mathbf{p}_j, f)$ come arbitrarily near $\mathbf{x}$ for j equal to either 1 or 2. The existence of a hyperbolic structure in Step 1 implies that $W^u(\mathbf{p}_j, f)$ and $W^s(\mathbf{p}_j, f)$ intersect transversally arbitrarily near $\mathbf{x}$ for j equal either to 1 or 2. □

Step 8. *The set Λ is transitive.*

PROOF. This follows from Step 7 and the Birkhoff Transitivity Theorem. □

Step 9. *The periodic points of f are dense in Λ.*

PROOF. This follows from Step 7 and the horseshoe theorem for transverse homoclinic points. □

Together, all these steps prove the theorem. □

8.8.1 The Branched Manifold

A Markov partition for the Anosov automorphism g is given in Figure 8.3. The map f pushes outward from $\mathbf{p}_0$ in the stable direction. If we form equivalence classes of points in $\text{comp}_{\mathbf{z}}(W^s(\mathbf{z}, f) \setminus V)$ and collapse these to points, we get the *branched manifold*, K, indicated in Figure 8.4. This quotient space has the differential structure of a one-dimensional manifold except there are branch points. The fact that there is a C^1 structure on the quotient space is reflected in the picture by the fact that the three curves coming into a branch point all have the same tangent line. There is a map defined on the quotient space (the branched manifold), $g_* : K \to K$. See Williams (1967) for the definition of a one-dimensional branched manifold or Williams (1974) for the definition in any dimension. This map is an expanding map (because we quotiented out the contracting directions and left the expanding directions), and has the following images:

$$\begin{aligned} g_*(A) &= B \\ g_*(B) &= BCB \\ g_*(C) &= CAC. \end{aligned}$$

In fact, these line segments A, B, and C can be oriented so the map preserves the orientation. This map takes the role of the doubling map for the solenoid. It can be proved that f on Λ is topologically conjugate to the inverse limit of g_* on K. See Williams (1970a). We leave the details to the reader and references.

8.9 Plykin Attractors in the Plane

Let Λ be a hyperbolic attractor in the plane with trapping region N. Thus, N must be diffeomorphic to a disk with some (or no) holes removed. Plykin (1974) proved that if Λ is not just a periodic orbit, then N must have at least three holes removed (four holes on the two sphere S^2).

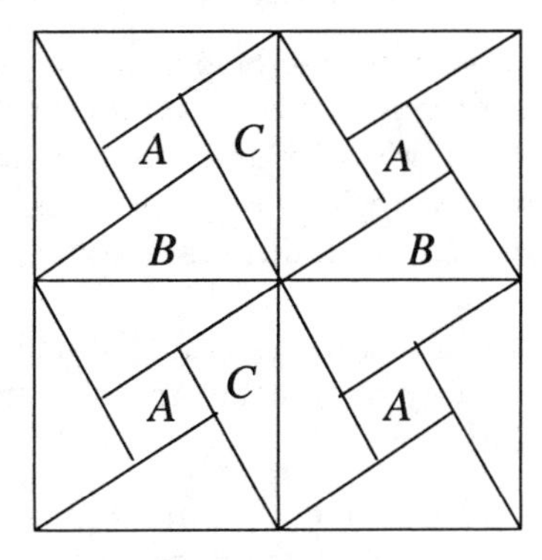

FIGURE 8.3. Markov Partition for DA-Diffeomorphism

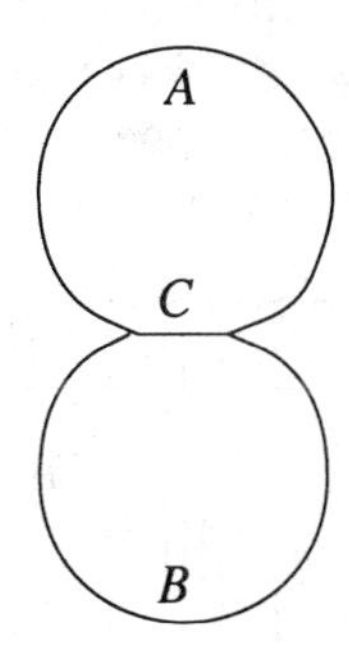

FIGURE 8.4. Branched Manifold for the DA-Diffeomorphism

Theorem 9.1. *Let $N \subset \mathbb{R}^2$ be a trapping region for f. Assume the associated attracting set, $\Lambda = \bigcap_{k=0}^{\infty} f^k(N)$, has a uniform hyperbolic structure for which the expanding bundle has dimension one, $\dim(\mathbb{E}^u_{\mathbf{p}}) = 1$ for $\mathbf{p} \in \Lambda$. (So Λ is not just the orbit of a periodic sink.) Then, N must have at least three holes.*

REMARK 9.1. For the general proof, see Plykin (1974). We give a sketch of a proof that N must have at least two holes. Because of this theorem, any nontrivial attractor (not a periodic orbit) in the plane or sphere is called a *Plykin attractor.*

PROOF. The hyperbolic splitting on Λ can be extended to a small neighborhood U of Λ. (This extension is not hard if it is not assumed that the splitting is invariant off Λ. It is possible to extend it so it is invariant, but this is harder and we do not need this property.) For large k, $f^k(N) \subset U$, so there is a splitting on $f^k(N)$. The neighborhood $f^k(N)$ has the same topological type as N, so we can assume that the splitting is on the entire trapping neighborhood N.

Assume that the extension of the bundle $\mathbb{E}^u_{\mathbf{p}}$ to all points $\mathbf{p} \in N$ is orientable on N. Then, it is possible to take $X(\mathbf{p}) \in \mathbb{E}^u_{\mathbf{p}}$ that is a nonvanishing vector field. Take $\mathbf{p} \in \Lambda$. Then, the integral curve of $\mathbf{p}$ is one side of $W^u(\mathbf{p})$, $W^u(\mathbf{p})^+$. By the Poincaré Bendixson Theorem, $W^u(\mathbf{p})^+$ accumulates on a closed orbit γ for X (because $\omega(\mathbf{p}, X)$ has no fixed points since X is nonvanishing). But $\omega(\mathbf{p}, X) \subset \Lambda$ because Λ is closed. Thus, $\gamma \subset \Lambda$ is a closed curve which is an unstable manifold. But unstable manifolds cannot be closed curves (they are immersed lines). This contradiction shows that the extension $\mathbb{E}^u_{\mathbf{p}}$ cannot be orientable on N.

If N has no holes (and so is a disk), then the extended bundle $\mathbb{E}^u$ must be orientable on N. The above argument shows that this is impossible, so N must have at least one hole.

Next assume that N is a disk with one hole removed (an annular region). By the above argument the extension $\mathbb{E}^u$ must not be orientable on N. In this case, it is possible to take a double cover $\bar{N}$ of N on which there is an orientable bundle $\bar{\mathbb{E}}^u$ which covers $\mathbb{E}^u$ on N. Again, $\bar{N}$ is an annular region. It is also possible to define a map $\bar{f}$ on $\bar{N}$ which covers f. But this leads to a contradiction as above, so N must have at least two holes.

As stated above, Plykin has an argument that N cannot have just two holes. This can also be proved using the theory of "pseudo-Anosov diffeomorphisms" of Thurston. We do not give these arguments. □

Example 9.1. It is possible to describe a geometric model of a map f which has a planar region with three holes, N, as a trapping region. See Figure 9.1. Consider the map f for which the image of N is as indicated in Figure 9.2. This map takes each of the line segments drawn in Figure 9.1 into (subsets of) another one of these line segments. These line segments are pieces of the stable manifolds. The map f stretches in the direction across the line segments. Also, $f(N) \subset N$. The attracting set $\Lambda = \bigcap_{k=0}^{\infty} f^k(N)$ has a hyperbolic structure.

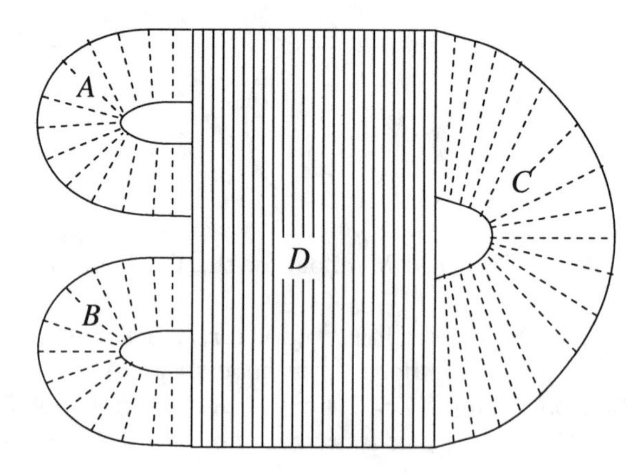

FIGURE 9.1. Neighborhood N

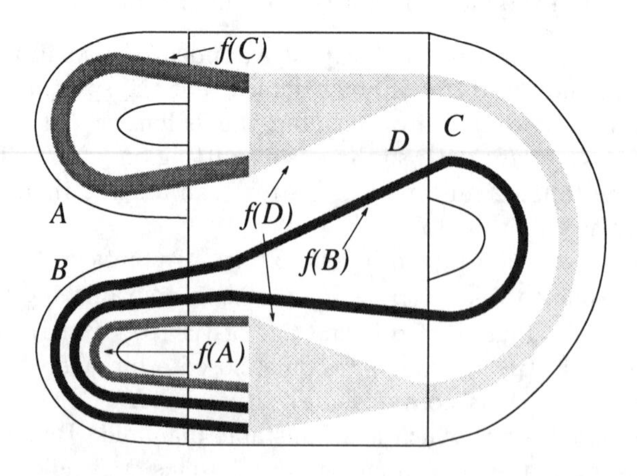

FIGURE 9.2. Image of N inside of N

If we make equivalence classes of points which are in the same components of stable manifolds in N, $\mathbf{q} \sim \mathbf{p}$ if $\mathbf{q} \in \text{comp}_{\mathbf{p}}(W^s(\mathbf{p}) \cap N)$, then we can form the quotient space

$K = N/\sim$. For this example, the quotient is indicated in Figure 9.3. Williams showed that to an expanding attractor there is associated a *branched manifold*, Williams (1970a, 1974). In Section 12.2, we show that the tangent lines, $\mathbb{E}^s_{\mathbf{x}}$ to the various $T_{\mathbf{x}}W^s(\mathbf{p})$, depend in a C^1 fashion on $\mathbf{x}$. This differentiability can be used to show that the quotient space can be given a smooth structure. There is a map defined on the quotient space, $g : K \to K$. This map is an expanding map (because we quotiented out the contracting directions). This map takes the role of the doubling map for the solenoid. In the example being discussed, $g(C) \supset A$, $g(A) \supset B$, $g(B) \supset C$, and $g(D) \supset C$. It can be proved that f on Λ is topologically conjugate to the inverse limit of g on K. See Williams (1967) and (1970a). Also see Barge (1988) for the connection between inverse limits and attractors for diffeomorphisms in the plane.

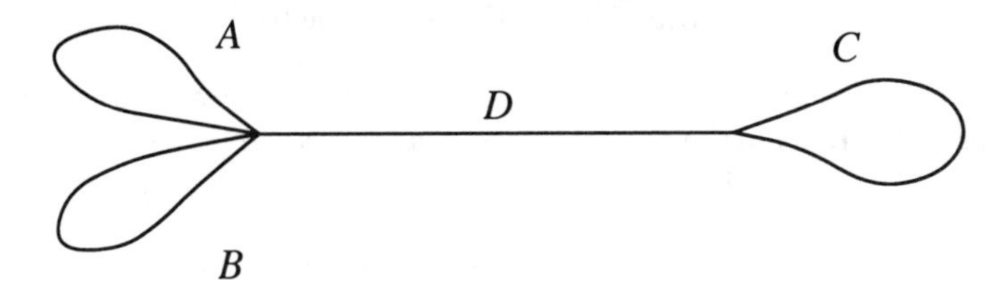

FIGURE 9.3. Branched Manifold for Plykin Example

REMARK 9.2. Some progress has been made to understand hyperbolic attractors which are not expanding attractors, i.e., hyperbolic attractors for which the topological dimension is greater than the dimension of the unstable manifolds. See Wen (1992).

8.10. Attractor for the Hénon Map

Again we consider the Hénon map, $F_{A,B}(x, y) = (A - By - x^2, x)$. In the earlier section, we showed that for large values of A, $F_{A,B}$ has a horseshoe, e.g., $B = \pm 0.3$ and $A = 5$. In this section, we consider smaller values of A for which $F_{A,B}$ has a trapping region. Hénon introduced this map and discussed this map for $B = -0.3$ and $A = 1.4$ for which there is a trapping region N which is topologically a disk (a region with no holes). See Hénon (1976). For the following discussion, let $F = F_{1.4,-0.3}$. Since it is a trapping region, $\Lambda = \bigcap_{k=0}^{\infty} F^k(N)$ is an attracting set. Numerical iteration indicates that F is topologically transitive on Λ because the iteration of a (generic) point appears to have a dense orbit in Λ. By the discussion of Plykin theory in the last section, F cannot have a (uniform) hyperbolic structure on Λ because Λ has arbitrarily small neighborhoods which have no hole (topologically disks), $F^k(N)$. It is still possible that there exists a point with a dense orbit in Λ. This point also might have a positive Liapunov exponent. (That is, there might be some point $\mathbf{p}$ and a vector $\mathbf{v}$ for which $|DF^k_{\mathbf{p}}\mathbf{v}|$ grows at an exponential rate, $\liminf_{k\to\infty}(1/k)\log(|DF^k_{\mathbf{p}}\mathbf{v}|) > 0$.) In spite of the lack of rigorous proof, an attracting set Λ for any map $F_{A,B}$ in the Hénon family such that $F_{A,B}|\Lambda$ appears to be transitive is called a *Hénon attractor*. See Figure 10.1 for $A = 1.4$ and $B = -0.3$. (Also see the comments below about the results of Benedicks and Carleson.)

Since Λ is an attracting set, Λ must contain all the unstable manifolds of hyperbolic periodic points in Λ. The following proposition shows that Λ is the closure of the unstable manifold of the fixed point for the Hénon map for many $B < 0$.

Proposition 10.1. *(a) Let $f : \mathbb{R}^2 \to \mathbb{R}^2$ be a diffeomorphism with a fixed point $\mathbf{p}$. Assume there is a bounded region $\Omega \subset \mathbb{R}^2$ which is positively invariant and $\partial(\Omega) \subset$*

FIGURE 10.1. Hénon Attractor

$L^u \cup L^s$ where L^u is contained in a compact piece of $W^u(\mathbf{p})$ and L^s is contained in a compact piece of $W^s(\mathbf{p})$. Further assume f decreases area on Ω, i.e., there is a $0 < \rho < 1$ such that $|\det(Df_{\mathbf{x}})| \leq \rho$. Then,

$$\Lambda = \text{cl}(\bigcup_{n\geq 0} f^n(L^u)).$$

If $\mathbf{p}$ is in the interior of the L^u as a subset of $W^u(\mathbf{p})$, then

$$\Lambda = \text{cl}(W^u(\mathbf{p}).$$

(b) For the Hénon map $F_{A,B}$, there is a set $\mathcal{S}$ of values (A, B) with $B < 0$ for which part (a) applies. This set includes $(1.4, -0.3)$ as well as values with $1.4 < A < 2$ and B small enough.

REMARK 10.1. The region Ω is not a trapping region since the part of the boundary in $W^u(\mathbf{p})$ is usually in the image $f(\Omega)$. See Figure 10.2. In the case of the Hénon map, the region can be enlarged to make it a trapping region U, but then $W^u(\mathbf{p})$ is in the interior of U.

PROOF. (b) We do not give the proof of part (b) but merely indicate a region that works. A choice of Ω is the shaded region in Figure 10.2 whose boundary is made up of the pieces of $W^u(\mathbf{p})$ and $W^s(\mathbf{p})$. This region is not a trapping region because the part of the boundary contained in $W^u(\mathbf{p})$ is on the boundary of the image of Ω. By slightly enlarging Ω along the part of the boundary contained in $W^u(\mathbf{p})$, it is possible to make a trapping region.

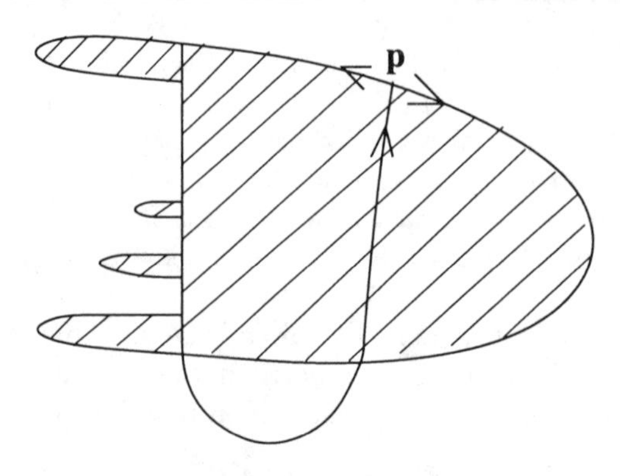

FIGURE 10.2. Invariant Region Ω for the Hénon Attractor is Shaded

(a) Because f decreases area by ρ, the absolute value of the stable eigenvalue at $\mathbf{p}$ is less than ρ, $|\lambda_s| < \rho$. Since L^s is contained in a compact piece of $W^s(\mathbf{p})$, there is a $k > 0$ for which $f^k(L^s) \subset W^s_\epsilon(\mathbf{p})$. For $n \geq k$, the diameter of $f^n(L^s)$ is less than $2\epsilon\rho^{n-k}$. Given $\eta > 0$, there is an n_1 such that this diameter is less than $\eta/2$ for $n \geq n_1$.

Now, take $\mathbf{q} \in \Lambda$. Take $\eta > 0$ as above. Since the area$(f^n(\Omega)) \leq \rho^n$area$(\Omega)$, there is an $n_2 \geq n_1$ such that $D(\mathbf{q}, \eta/2) \not\subset f^{n_2}(\Omega)$, $D(\mathbf{q}, \eta/2) \cap \partial(f^{n_2}(\Omega)) \neq \emptyset$, and $D(\mathbf{q}, \eta/2) \cap (f^{n_2}(L^u) \cup f^{n_2}(L^s)) \neq \emptyset$. Because the diameter of $f^{n_2}(L^s)$ is less than $\eta/2$, $d(\mathbf{q}, f^{n_2}(L^u)) \leq \eta$. (The ends of $\partial(f^{n_2}(\Omega)) \cap f^{n_2}(L^s)$ lie in $f^{n_2}(L^u)$.) Because η is arbitrary, $\mathbf{q}$ is in the closure of the union of the forward iterates of L^u, and Λ is the closure of $\bigcup_{n\geq 0} f^n(L^u)$ as claimed.

If $\mathbf{p}$ is in the interior of the L^u, then $\bigcup_{n\geq 0} f^n(L^u) = W^u(\mathbf{p})$, so Λ is the closure of $W^u(\mathbf{p})$ as claimed. □

Even though the attracting set for the Hénon map is the closure of the unstable manifold, it is not necessarily topologically transitive. In fact, for some parameter values, we argue below that it contains some periodic sinks. However, when we look more closely at the attracting set Λ, the invariant set looks like a Cantor set of curves which seem to be the unstable manifolds of points in Λ. See Figure 10.3. If we look at the attracting set in a smaller box (at a smaller scale), the set still looks like a Cantor set of curves. However, between the curves which reach all the way across the box, there are curves which turn around part way across the box and come out the same edge they entered. These latter curves look like hooks among the other curves which are relatively straight. If all points in Λ had stable and unstable manifolds, then these hooks in the unstable manifolds would be tangent to the stable manifold of some other point in the attracting set. As the parameter A varies, these hooks move in the attracting set. For many parameter values, it would seem likely that there are homoclinic tangencies (tangencies of the stable and unstable manifolds of the fixed point or some periodic point). At other parameter values, the tangencies of stable and unstable manifolds may only be for nonperiodic points. In any case, these tangencies prevent Λ from having a uniform hyperbolic structure.

A numerical study by means of computer graphics seems to indicate that there is a tangency for $B = -0.3$ and A about 1.392. However, it is known that for parameter values near a homoclinic tangency, the attracting set is not transitive but contains infinitely many periodic sinks. This follows from the work of Newhouse (1979) on infinitely many periodic sinks. Also see Robinson (1983). It is still conceivable that the placement of the hooks could be controlled enough to avoid all homoclinic tangencies. It would be hoped that for such a parameter value that most points could be proved to have a positive Liapunov exponent.

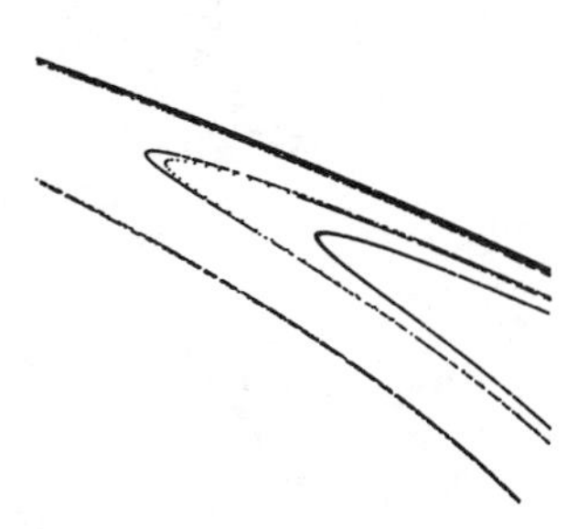

FIGURE 10.3. Enlargement of Piece of Hénon Attractor

Recently, Benedicks and Carleson (1991) have shown that there are other parameter values for which $F_{A,B}$ has a transitive attractor with positive Liapunov exponent. (The map $F_{A,B}$ still cannot have a uniform hyperbolic structure.) In fact, there is a set $\mathcal{S} \subset \{A : 1.0 < A < 2.0\}$ of positive measure such that for $A \in \mathcal{S}$ and $B < 0$ small enough, the attracting set is topologically transitive and has a positive Liapunov exponent. Their proof uses a perturbation argument from the one-dimensional case of the quadratic map, i.e., from $F_{A,0}$. For the one-dimensional map, the "hooks" are the images of the critical point, and can be controlled well enough to make the map transitive with an invariant ergodic measure. This one-dimensional result was first proved by Jakobson (1981). There have been many refinements of the proof, including Benedicks and Carleson (1985). Benedicks and Carleson were able to use the knowledge about the images of the critical point for the one-dimensional map to control the location of the hooks for the two-dimensional map for small values of B and even prove that there is a point with a dense orbit which has a positive Liapunov exponent. (See Section 9.2 for the definition of Liapunov exponents for a two-dimensional map.)

This result for the Hénon map for very small B by Benedicks and Carleson gives plausibility to the conjecture that this is true for $A = 1.4$ and $B = -0.3$, but this is still unproven. Mora and Viana (1993) have shown how this theorem of Benedicks and Carleson applies to maps which are not quadratic maps.

8.11 Lorenz Attractor

As stated in Chapter I, Lorenz (1963) introduced the equations

$$\begin{aligned}\dot{x} &= -\sigma x + \sigma y,\\ \dot{y} &= \rho x - y - xz,\\ \dot{z} &= -\beta z + xy\end{aligned}$$

as a model for fluid flow of the atmosphere (weather). See Guckenheimer and Holmes (1983) or Sparrow (1982) for further discussion of the way these equations model atmospheric movement. The parameter values which have some physical significance occur near $\rho = 1$, but Lorenz discovered some very unusual dynamics for the parameter values $\sigma = 10$, $\rho = 28$, and $\beta = 8/3$. In our treatment of the dynamics, we fix $\sigma = 10$ and $\beta = 8/3$ for the rest of the section and discuss the situation for various values of ρ, but always taking $\rho > 1$.

Before turning to the detailed discussion, note that the equations are invariant under the substitution of $(-x, -y, z)$ for (x, y, z). Therefore, the solutions have this type of symmetry: if $(x(t), y(t), z(t))$ is a solution, then $(-x(t), -y(t), z(t))$ is also a solution.

The fixed points of this system of equations are easy to find, and are at $\mathbf{0} = (0,0,0)$,

$$\begin{aligned}\mathbf{p}^+ &= ([8(\rho-1)/3]^{1/2}, [8(\rho-1)/3]^{1/2}, \rho - 1), \text{ and}\\ \mathbf{p}^- &= (-[8(\rho-1)/3]^{1/2}, -[8(\rho-1)/3]^{1/2}, \rho - 1).\end{aligned}$$

The eigenvalues at the origin are all real, $-8/3$, $-11/2 \pm [121 + 40(\rho - 1)]^{1/2}/2$. Thus, there is one unstable eigenvalue

$$\lambda_u = -11/2 + [121 + 40(\rho - 1)]^{1/2}/2$$

and two stable eigenvalues

$$\begin{aligned}\lambda_s &= -8/3 \quad \text{and}\\ \lambda_{ss} &= -11/2 - [121 + 40(\rho - 1)]^{1/2}/2.\end{aligned}$$

For $\rho = 28$, $\lambda_u \approx 11.83$ and $\lambda_{ss} \approx -22.83$. The unstable manifold of the origin is one-dimensional and has two branches $W^u(\mathbf{0})^{\pm}$. These two branches are related to each other by the symmetry noted above. For small values of ρ, the positive branch of the unstable manifold, $W^u(\mathbf{0})^+$ stays on one side of the stable manifold $W^s(\mathbf{0})$. In fact, numerical integration indicates that it goes to $\mathbf{p}^+$. See Figure 11.1. For $\rho = \rho_0 \approx 13.93$, the unstable manifold for the origin is seen to connect to the stable manifold of the origin forming a homoclinic loop, $W^u(\mathbf{0}) \subset W^s(\mathbf{0})$. (Remember, if one branch forms a homoclinic loop for a parameter value, then the other branch also forms a homoclinic loop by the symmetry.) See Figure 11.2. For $\rho > \rho_0$, each half of the unstable manifold for the origin (going out only one direction from the origin), $W^u(\mathbf{0})^{\pm}$, crosses from one side of $W^s(\mathbf{0})$ to the other side. See Figure 11.3. For ρ near ρ_0, $W^u(\mathbf{0})^+$ falls into $W^s(\mathbf{p}^-)$, and $W^u(\mathbf{0})^-$ falls into $W^s(\mathbf{p}^+)$. For $\rho > \rho_1 = 470/19 \approx 24.74$, this is no longer the case, as we discuss further below.

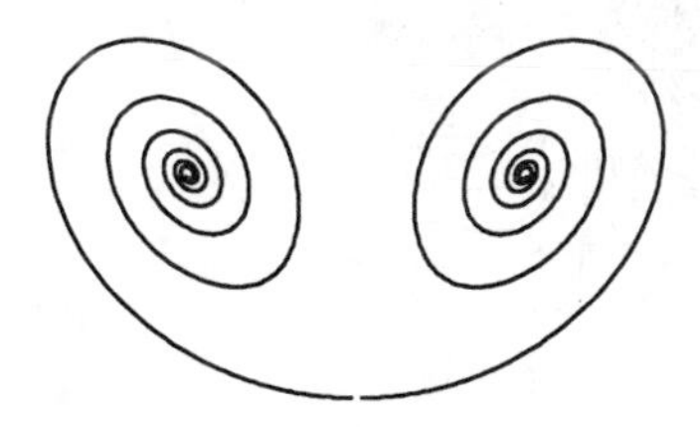

FIGURE 11.1. Unstable Manifold of the Origin, $1 < \rho < \rho_0$

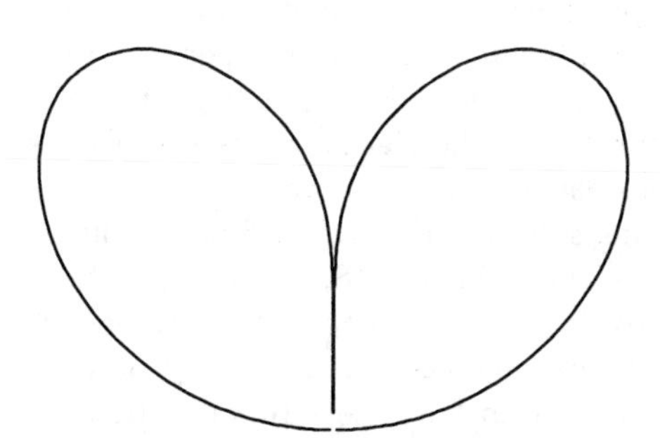

FIGURE 11.2. Unstable Manifold of the Origin, $\rho = \rho_0$

We have already referred to the stable manifolds of the fixed points $\mathbf{p}^{\pm}$. We now turn to a discussion of the stability type of these fixed points. As is verified in Exercise 7.10, the characteristic equation for the fixed points $\mathbf{p}^{\pm}$ is

$$p(\lambda) = \lambda^3 + (41/3)\lambda^2 + \frac{8}{3}(\rho + 10)\lambda + \frac{160}{3}(\rho - 1) = 0.$$

For $\rho \geq 14$, $p(\lambda)$ has one real negative root and two complex roots. There is a bifurcation value at $\rho_1 = 470/19 \approx 24.74$. The two complex roots have a negative real part for

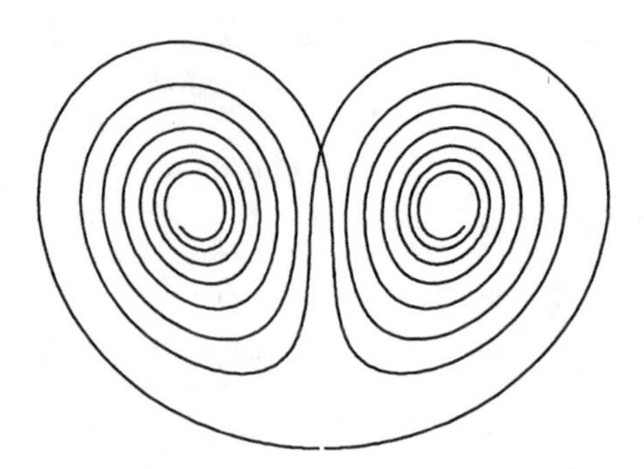

FIGURE 11.3. Unstable Manifold of the Origin, $\rho_0 < \rho < \rho_1$

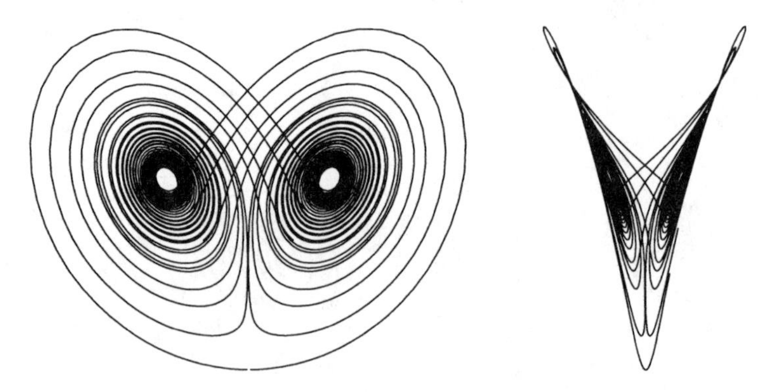

FIGURE 11.4. Two Views of the Unstable Manifold of the Origin for $\rho = 28$

$14 \leq \rho < \rho_1$, and a positive real part $\rho > \rho_1$. Thus, these fixed points are stable for $14 \leq \rho < \rho_1$, and they become unstable at $\rho_1 = 470/19 \approx 24.74$. In fact, a Hopf bifurcation takes place for this parameter value. See Sparrow (1982). It is a subcritical Hopf bifurcation where two unstable periodic orbits disappear at ρ_1. For $\rho < \rho_1$, the fixed points $\mathbf{p}^{\pm}$ are sinks, while for $\rho > \rho_1$, these fixed points push outward in a two-dimensional subspace. The eigenvalue is complex, so the trajectories spiral around in this two-dimensional surface as they move outward. The existence of this two-dimensional expanding subspace for the fixed points pushes the unstable manifolds $W^u(\mathbf{0})^{\pm}$ away from $\mathbf{p}^{\pm}$. In fact, for $\rho = 28$, which is greater than ρ_1, $W^u(\mathbf{0})^+$ crosses from one side of $W^s(\mathbf{0})$ to the other and back again, as indicated in Figure 11.4. (None of the facts stated above are obvious, but follow by detailed calculations. Some of these calculations are contained in Exercise 7.10, and others are referred to in Sparrow (1982).)

From now on we focus our attention on the behavior for $\rho = 28$ and fix this value. There is a trapping region N containing the origin and not containing the other two fixed points. In fact, there are two holes in the region where these two fixed points are located. See Figure 11.5. Let Λ be the maximal invariant set in N,

$$\Lambda = \bigcap_{t \geq 0} \varphi^t(N).$$

Thus, Λ is an attracting set. The unstable manifold of the origin, $W^u(\mathbf{0})$, must be completely contained in the trapping region N, and so in Λ.

The flow cannot have a hyperbolic structure on Λ in the usual sense because Λ contains a fixed point at $\mathbf{0}$: at the fixed point $\mathbf{0}$, $\mathbb{E}^s_{\mathbf{0}}$ has dimension two and $\mathbb{E}^u_{\mathbf{0}}$ has

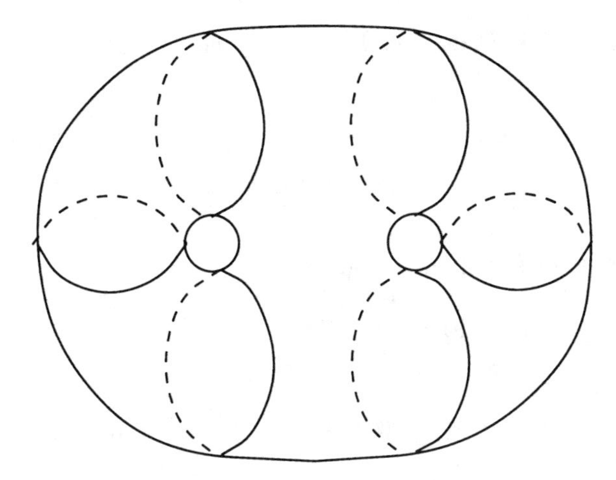

FIGURE 11.5. Trapping Region

dimension one, while at other points $\mathbf{x} \in \Lambda$ the splitting would be of the form

$$\mathbb{E}^u_{\mathbf{x}} \oplus \mathbb{E}^s_{\mathbf{x}} \oplus \mathbb{E}^c_{\mathbf{x}}$$

where each of the subspaces would have dimension one and $\mathbb{E}^c_{\mathbf{x}}$ would be spanned by the vector field at $\mathbf{x}$. However, we could consider a *generalized hyperbolic structure* where the splitting varies as above and the two-dimensional subspace $\mathbb{E}^s_{\mathbf{x}} \oplus \mathbb{E}^c_{\mathbf{x}}$ for $\mathbf{x} \in \Lambda \setminus \{\mathbf{0}\}$ converges to $\mathbb{E}^s_{\mathbf{0}}$ as $\mathbf{x}$ converges to $\mathbf{0}$. Numerical integration indicates that the equations have this type of generalized hyperbolic structure, but this is a global question and has not been verified analytically.

8.11.1 Geometric Model for the Lorenz Equations

Guckenheimer (1976) introduced a *geometric model of the Lorenz equations* which is compatible with the observed numerical integration of the actual equations. This model has been analyzed in Williams (1977, 1979, 1980), Guckenheimer and Williams (1980), Rand (1978), and Robinson (1989, 1992). See Sparrow (1982) and Guckenheimer and Holmes (1983) for a more complete discussion of the model than we give in this section.

To understand the model, first consider the flow of the actual equations near the origin. By a nonlinear change of coordinates, the equations are differentiably conjugate to the linearized equations in a neighborhood of the origin,

$$\begin{aligned}\dot{x} &= ax\\ \dot{y} &= -by\\ \dot{z} &= -cz\end{aligned}$$

where $a = \lambda_u \approx 11.83$, $b = -\lambda_{ss} \approx 22.83$, and $c = -\lambda_s = 8/3$. (The differentiable conjugacy follows by the result of Sternberg (1958). Also see Hartman (1964).) Therefore, $0 < c < a < b$. The solution of the linearized equations is given by $x(t) = e^{at}x_0$, $y(t) = e^{-bt}y_0$, and $z(t) = e^{-ct}z_0$. We want to follow the solutions as they flow past the fixed point, so from the time when $z(t)$ equals to some fixed z_0 until $x(t)$ equals to some fixed $\pm x_1$. Consider the two cross sections

$$\begin{aligned}\Sigma &= \{(x, y, z_0) : |x|, |y| \leq \alpha\},\\ \Sigma' &= \Sigma \setminus \{(x, y, z_0) : x = 0\}, \qquad \text{and}\\ S^{\pm} &= \{(\pm x_1, y, z) : |y|, |z| \leq \beta\}.\end{aligned}$$

Then, for $(x, y, z_0) \in \Sigma'$, the time τ such that $\varphi^\tau(x, y, z_0) \in S = S^+ \cup S^-$ is determined by

$$e^{a\tau}|x| = x_1$$
$$e^\tau = \left(\frac{x_1}{x}\right)^{1/a}.$$

Then, the Poincaré map $P_1 : \Sigma' \to S$ is given by

$$\begin{aligned} P_1(x, y) &= (y(\tau), z(\tau)) \\ &= (e^{-b\tau} y, e^{-c\tau} z_0) \\ &= (y x^{b/a} x_1^{-b/a}, x^{c/a} z_0 x_1^{-c/a}). \end{aligned}$$

Notice that if $x > 0$, then $P_1(x, y) \in S^+$, and if $x < 0$, then $P_1(x, y) \in S^-$. For eigenvalues at the fixed point equal to those of the real Lorenz equations, $b/a = |\lambda_{ss}/\lambda_u| \approx 1.93 > 1$ and $c/a = |\lambda_s/\lambda_u| \approx 0.23 < 1$. Therefore, a square region $\{(x, y, z_0) : 0 < x \leq \alpha, |y| \leq \alpha\}$ in Σ' comes out in a cusp shaped region in S^+. See Figure 11.6. The geometric model assumes that the Poincaré map P_2 from $S^\pm$ back to Σ takes the horizontal lines $z = z_1$ into lines $x = x_1$. This compatibility insures that there is a coherent set of contracting directions. More specifically, we assume that

$$D(P_2)_{(y,z)} = \begin{pmatrix} 0 & \zeta \\ \pm 1 & 0 \end{pmatrix}.$$

Let $P = P_2 \circ P_1$. Then, $P : \Sigma' \to \Sigma$ and the image of Σ' by P is as in Figure 11.7.

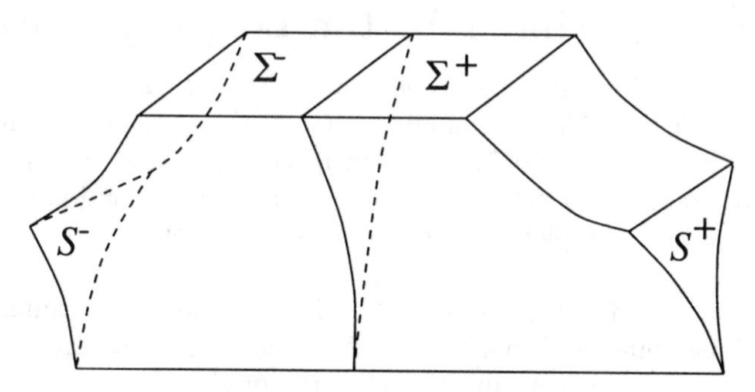

FIGURE 11.6. Flow of Σ Past Fixed Point

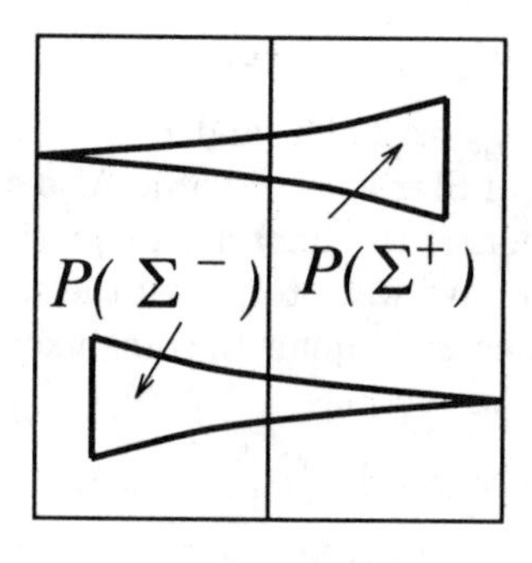

FIGURE 11.7. Image of Σ' by Poincaré Map

Because (1) P_1 takes a line segment with all the same x values to a line segment with all the same z values and (2) P_2 takes a line segment with all the same z values back to a line segment with all the same x values, the x value of the image of a point by P is determined solely by its x value. Therefore, P is of the form

$$P(x, y) = (f(x), g(x, y)).$$

For a fixed x_0, the map $g(x_0, y)$ is a contraction in the y-direction:

$$|g(x_0, y_1) - g(x_0, y_2)| \leq \mu |y_1 - y_2|$$

for $0 < \mu < 1$, and $|f'(x)| \geq \sqrt{2} > 1$ for all x with $|x| \leq \alpha$. Thus, for the Poincaré map, there is a hyperbolic splitting $\mathbb{E}^u_{\mathbf{p}} \oplus \mathbb{E}^s_{\mathbf{p}}$, with $\mathbb{E}^s_{\mathbf{p}} = \{(0, v_2)\}$ and $\mathbb{E}^u_{\mathbf{p}}$ mainly in the x-direction. Because of the form of the Poincaré map, it has an invariant stable foliation $W^s(\mathbf{q}, P)$ made up of curves with constant value of x on Σ. One of these line segments, $W^s(\mathbf{q}, P)$ for $\mathbf{q} \in \Sigma'$, is taken into another such line segment, $W^s(P(\mathbf{q}), P)$, with most likely a different value of x. We make an equivalence class of points on Σ that lie on the same stable line segment, $W^s(\mathbf{q}, P)$. By collapsing equivalence classes to points, we get a map $\pi : \Sigma \to [-\alpha, \alpha]$. (In the above situation, $\pi(\mathbf{q})$ just gives the x-value of $\mathbf{q}$.) Because P takes an equivalence class into an equivalence class, P and π induce a map $f : [-\alpha, \alpha] \setminus \{0\} \to \mathbb{R}$. This description of the map f is more coordinate free than given above where we wrote $P(x, y) = (f(x), g(x, y))$, but it represents the same function.

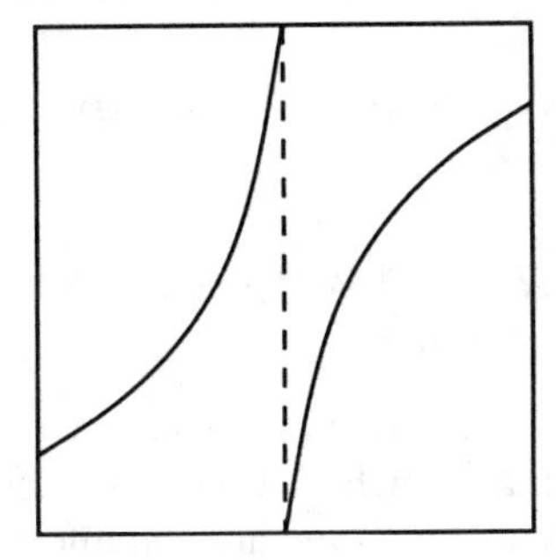

FIGURE 11.8. Graph of f

The assumptions on the map f are more specifically as follows.

(1) The symmetry of the differential equations implies that $f(-x) = -f(x)$.
(2) The map f has a single discontinuity at $x = 0$.
(3) The limit of $f(x)$ as x approaches 0 from the left side is $A > 0$, $f(0-) = A$; and the limit of $f(x)$ as x approaches 0 from the right side is $-A$, $f(0+) = -A < 0$. Also, $0 < f^2(A) < f(A) < A$ and so $0 > f^2(-A) > f(-A) > -A$. Thus,

$$f : [-A, A] \setminus \{0\} \to [-A, A].$$

(4) f is nonuniformly continuously differentiable on $[-A, A] - \{0\}$, with $f'(x) > \sqrt{2}$ for all $x \neq 0$.
(5) The limit of $f'(x)$ is infinity as x approaches 0 from either side.
(6) Each of the two branches of the inverse of f extends to a $C^{1+\alpha}$ function for some $\alpha > 0$ on $[f(-A), A]$ or $[-A, f(A)]$. (Thus, the derivative of the extension of the inverse is α-Hölder for some $\alpha > 0$.)

The graph of f is given in Figure 11.8. The properties of f follow mainly from the form of the Poincaré map of the flow past the fixed point.

Let φ^t be the flow for the geometric model. Because $P(\Sigma') \subset \Sigma$, φ^t has an attracting set Λ. The dynamics of the flow φ^t on Λ are determined by the two-dimensional Poincaré map P. It can be shown that the dynamics of the two-dimensional Poincaré map are determined by the one-dimensional Poincaré map f.

The fact that P has a coherent set of contracting directions can be used to show that there is a bundle $\mathbb{E}^{ss}_{\mathbf{p}}$ for $\mathbf{p} \in \Lambda$ and a complementary plane of directions $\mathbb{E}^{c}_{\mathbf{p}}$ that are taken into themselves by $D(\varphi^t)_{\mathbf{p}}$,

$$\begin{aligned} D(\varphi^t)_{\mathbf{p}}\mathbb{E}^{ss}_{\mathbf{p}} &= \mathbb{E}^{ss}_{\varphi^t(\mathbf{p})} \\ D(\varphi^t)_{\mathbf{p}}\mathbb{E}^{c}_{\mathbf{p}} &= \mathbb{E}^{c}_{\varphi^t(\mathbf{p})} \end{aligned}$$

and the $\mathbb{E}^{ss}_{\mathbf{p}}$ is contracted more strongly than anything in the $\mathbb{E}^{c}_{\mathbf{p}}$ directions,

$$\|D(\varphi^t)_{\mathbf{p}}|\mathbb{E}^{ss}_{\mathbf{p}}\| \le m(D(\varphi^t)_{\mathbf{p}}|\mathbb{E}^{c}_{\mathbf{p}}).$$

This last condition implies that cones about the $\mathbb{E}^{ss}_{\mathbf{p}}$ are taken into themselves by the derivative of the flow, $D(\varphi^t)_{\mathbf{p}}$. The vector field $X(\mathbf{p})$ for the differential equation is in the center direction $\mathbb{E}^{c}_{\mathbf{p}}$. The center direction $\mathbb{E}^{c}_{\mathbf{p}}$ is also more or less "tangent" to the "sheets" in Λ, while the strong stable direction $\mathbb{E}^{ss}_{\mathbf{p}}$ points transverse to the attracting set Λ. There is then a stable manifold theorem which says that there are curves $W^{ss}_{\epsilon}(\mathbf{p}, \varphi^t)$ that are tangent to the $\mathbb{E}^{ss}_{\mathbf{p}}$ directions which are taken into themselves by the flow,

$$\varphi^t(W^{ss}_{\epsilon}(\mathbf{p}, \varphi)) \subset W^{ss}_{\epsilon}(\varphi^t(\mathbf{p}), \varphi).$$

For small $\epsilon > 0$,

$$N' \equiv \bigcup_{\mathbf{p}\in\Lambda} W^{ss}_{\epsilon}(\mathbf{p}, \varphi) \subset N.$$

We can form equivalence classes of points in the same strong stable manifold $W^{ss}_{\epsilon}(\mathbf{p}, \varphi)$, and get a projection from N' to a branched manifold L. See Figure 11.9. See Williams (1974) for the general definition of a branched manifold or Williams (1977) for the branched manifold of the Geometric Model of the Lorenz attractor. The flow on N' induces a semi-flow ψ^t on L. Only a semi-flow and not a flow is induced on L because there are two choices of the backward trajectory at the branch set. The Poincaré map from $\Sigma' = \pi(\Sigma)$ to itself for ψ is f. (See the references for details.)

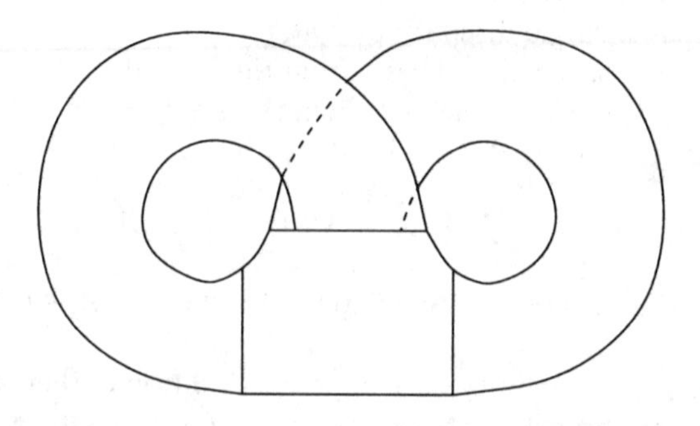

FIGURE 11.9. The Branched Manifold

The fact that the flow φ^t has the above properties is preserved under C^2 perturbations. That is the content of the papers by Robinson (1981, 1984).

Theorem 11.1. *Let φ^t be a flow with the properties of the Geometric Model of the Lorenz equations as given above. In particular, assume the one-dimensional Poincaré map f satisfies properties (1) – (6). Then, there is a neighborhood $\mathcal{N}$ in the C^2 topology such that if $\tilde{\varphi}^t$ is a flow in $\mathcal{N}$ with the symmetry properties of φ^t, then $\tilde{\varphi}^t$ has a bundle of strong stable directions (and so do the Poincaré maps). Forming the quotient by the strong stable manifolds for $\tilde{\varphi}^t$ on the trapping region N', there is induced a semi-flow on the quotient space $\tilde{L}$ which is a branched manifold. The semi-flow $\tilde{\varphi}^t$ has a Poincaré map*

$$\tilde{P} : \Sigma \setminus W^s(\mathbf{0}, \tilde{\varphi}) \to \Sigma.$$

The quotient map taking $W^{ss}(\mathbf{p}, \tilde{\varphi})$ to points, induces a one-dimensional Poincaré map $\tilde{f}$ that satisfies all the properties (1) – (6).

In particular, Williams was able to show $\tilde{f}$ is transitive by the following argument. We first of all make the following definition for one-dimensional maps.

Definition. Let $I \subset \mathbb{R}$ be an interval and $f : I \to I$ be a continuous map. We say that f is *locally eventually onto* provided for any nonempty (small) open interval $K \subset I$, there is an $n > 0$ such that $f^n(K) \supset I$. If f is locally eventually onto, then it is transitive on I by the Birkhoff Transitivity Theorem.

Theorem 11.2. *Assume $f : [-A, 0) \cup (0, A] \to [-A, A]$ satisfies assumptions (1) – (6) given above.*
(a) Then, f is locally eventually onto for the interval $(-A, A)$.
(b) Therefore, f is transitive on $(-A, A)$ and $\Omega(f) = [-A, A]$.

PROOF. Let $I = [-A, A]$. In the proof, we repeatedly throw away points whose iterates hit 0 and thus do not have well-defined forward orbits.

Given an interval $K \subset I$, define

$$K_0 = \begin{cases} K & \text{if } 0 \notin K \\ \text{the longest component of } K \setminus \{0\} & \text{if } 0 \in K. \end{cases}$$

By induction, define K_i for $i \geq 0$ by

$$K_{i+1} = \begin{cases} f(K_i) & \text{if } 0 \notin f(K_i) \\ \text{the longest component of } f(K_i) \setminus \{0\} & \text{if } 0 \in f(K_i). \end{cases}$$

Let

$$\lambda = \inf_{x \in I \setminus \{0\}} \{f'(x)\}.$$

By the assumptions, $\lambda > \sqrt{2}$. Note that by construction, $0 \notin \text{int}(K_i)$. Therefore, $\ell(f(K_i)) \geq \lambda \ell(K_i)$, where $\ell(J)$ is the length of an interval J. Because of the choice of K_{i+1},

$$\ell(K_{i+1}) \geq \begin{cases} \lambda \ell(K_i) & \text{if } 0 \notin f(K_i) \\ \frac{\lambda}{2} \ell(K_i) & \text{if } 0 \in f(K_i). \end{cases}$$

Similarly,

$$\ell(K_{i+2}) \geq \begin{cases} \lambda \ell(K_{i+1}) & \text{if } 0 \notin f(K_{i+1}) \\ \frac{\lambda}{2} \ell(K_{i+1}) & \text{if } 0 \in f(K_{i+1}). \end{cases}$$

Thus, if $0 \notin f(K_i)$ or $0 \notin f(K_{i+1})$ $(0 \notin f(K_i) \cap f(K_{i+1}))$, then

$$\ell(K_{i+2}) \geq \frac{\lambda^2}{2} \ell(K_i).$$

Thus, every two iterates of the interval makes the length of K_i increase by a factor $\lambda^2/2 > 1$ until there is some n for which $0 \in f(K_{n-2}) \cap f(K_{n-1})$. Thus, we have proved that $0 \in f(K_{n-2}) \cap f(K_{n-1})$ for some $n \geq 2$.

Claim. *For the above choice of n, $K_n = (-A, 0]$ or $[0, A)$.*

PROOF. The point $0 \in f(K_{n-2})$ so $0 \in \partial(K_{n-1})$, i.e., K_{n-1} abuts on 0. Let $b > 0$ be such that $f(\pm b) = 0$. Then, the fact that $0 \in f(K_{n-1})$ implies that b or $-b$ are in K_{n-1}, so $K_{n-1} \supset [-b, 0)$ or $(0, b]$. Thus, $f(K_{n-1})$ contains either $f([-b, 0)) = [0, A)$ or $f((0, b]) = (-A, 0]$. □

Let $c = f(A)$. From the claim, it follows that $f(K_n) = (-c, A) = (-c, 0] \cup [0, A)$ or $(-A, c) = (-A, 0] \cup [0, c)$. Note that since $f(c) = f^2(A) > 0 = f(b)$, it follows that $c > b$. Then, in the first case,

$$\begin{aligned} f^2(K_n) &\supset f((-c, 0]) \cup f([0, A)) \\ &\supset f((-b, 0]) \cup f([0, A)) \\ &\supset [0, A) \cup (-A, c) \\ &\supset (-A, A). \end{aligned}$$

A similar argument holds when $f(K_n) = (-A, 0] \cup [0, c)$. This completes the proof of (a).

As mentioned before the theorem, the Birkhoff Transitivity Theorem implies that f is transitive and so all points are nonwandering. □

The next theorem makes some connections between f and P.

Theorem 11.3. *(a) There is a one to one correspondence between the periodic points of f and the periodic points of P.*
(b) Let $\Lambda_P = \bigcap_{n \geq 0} P^n(\Sigma')$. Then, $\Lambda_P = \Omega(P)$.

PROOF. If $P^k(x_0, y_0) = (x_0, y_0)$, then clearly $f^k(x_0) = x_0$. Conversely, assume that $f^k(x_0) = x_0$. Then $P(x, y) = (f(x), g(x, y))$, so $P^k(x_0, y) \in \{x_0\} \times I$ for the interval I of values of y. The map P is a contraction by a factor $\mu < 1$ on fibers $W^{ss}(\mathbf{q}, P)$, so

$$|P^j(x, y_1) - P^j(x, y_2)| \leq \mu^j |y_1 - y_2|,$$

or $P^k(x_0, \cdot) : I \to I$ is a contraction by μ^k. Therefore, $P^k(x_0, y)$ has a unique fixed point y_0 in I, and P has a unique point (x_0, y_0) of period k corresponding to the point x_0 of period k for f. This proves part (a).

For part (b), note that Σ has the property that $P(\Sigma) \subset \text{int}(\Sigma)$. (If this is not the case, then enlarging Σ slightly in the x-direction makes this true.) Then, Λ_P is an attracting set, so $\Omega(P) \subset \Lambda_P$.

We need to show that $\Omega(P) \supset \Lambda_P$. Take $(x_0, y_0) \in \Lambda_P$ and let U be a neighborhood in Σ. Then, there is a small interval J containing x_0 and $K = [y_0 - \epsilon, y_0 + \epsilon]$ containing y_0 such that $J \times K \subset U$. Take $m > 0$ such that $\mu^m \ell(I) < \epsilon$. There exists a point $(u_0, v_0) \in \Lambda_P$ such that $P^m(u_0, v_0) = (x_0, y_0)$. Since f is locally eventually onto, there is a point $x_1 \in J$ such that $f^n(x_1) = u_0$. Take any $y_1 \in K$, so $(x_1, y_1) \in U$, and let $P^n(x_1, y_1) = (u_0, v_1)$. Then,

$$\begin{aligned} |P^{n+m}(x_1, y_1) - (x_0, y_0)| &= |P^m(u_0, v_1) - P^m(u_0, v_0)| \\ &\leq \mu^m |v_1 - v_0| \\ &\leq \epsilon. \end{aligned}$$

Therefore, $(x_1, y_1), P^{n+m}(x_1, y_1) \in U$. This is true for any neighborhood of (x_0, y_0), so $(x_0, y_0) \in \Omega(P)$. This shows $\Lambda_P \subset \Omega(P)$, completing the proof. □

The reduction to the one-dimensional Poincaré map can also be used to prove that the flow is not structurally stable. Small changes in the flow can make it have a homoclinic orbit or not, i.e., $f^n(A) = 0$ for some n, or $f^j(A) \neq 0$ for all $j > 0$. These two different types of flows are not conjugate or even flow equivalent. However, Guckenheimer and Williams have analyzed much of the topological structure of the attracting set for the Geometric Model of the Lorenz equations. See Guckenheimer (1976), Guckenheimer and Williams (1980), and Williams (1977, 1979, 1980).

Birman and Williams (1983a, 1983b) and Williams (1983) have studied the type of knots that occur as periodic orbits for the Geometric Model of the Lorenz equations. The branched manifold plays an important part of this analysis. The branch manifold is modified by removing the fixed point, getting what is called a *template*. In particular, the periodic orbits on the template correspond to the periodic orbits in $\mathbb{R}^3$. The references cited above show that the knots which appear are prime. Also see Holmes (1988) for a good introduction into the theory of knots in Dynamical Systems.

8.11.2 Homoclinic Bifurcation to a Lorenz Attractor

More recently, Rychlik (1990) proved that an *attractor of Lorenz type* could occur for specific cubic differential equations in $\mathbb{R}^3$. The papers by Robinson (1989, 1992) contain a further discussion of this type of bifurcation and connections with stable manifold theory. Also see Ushiki, Oka, and Kokubu (1984). Recently, Dumortier, Kokubu, and Oka (1992) have determined the various codimension two bifurcations of a homoclinic connection for a fixed point of a differential equations in $\mathbb{R}^3$. For the Lorenz equations, after the homoclinic bifurcation at ρ_0, there is an invariant suspension of a horseshoe and not an attractor. Later, at the Hopf bifurcation value, it appears that the gap of the horseshoe disappears and an attractor is formed. In the equations studied by Rychlik and Robinson, there are more parameters than in the Lorenz equations so these two bifurcations can be compressed into one. With certain (codimension two) assumptions on the parameters, it is possible to bifurcate directly from the homoclinic connection to an attractor. Since the homoclinic connection involves only two orbits, it is possible to analyze completely the properties of the flow at this parameter value and prove that a strong stable direction is preserved after the bifurcation. By controlling the expansion rates in comparison to the distance of the unstable manifold to the stable manifold, it is possible to show that the one-dimensional Poincaré map is like the one given for the Geometric Model of the Lorenz equations. Again, this control of the expansion rates requires the extra parameter of the equations studied in these papers which is not present in the Lorenz equations.

This work for the Lorenz equations is somewhat analogous to the comparison of the results of Benedicks and Carleson (1991) for the Hénon map for small B to the observed results for the parameter values $A = 1.4$ and $B = -0.3$.

8.12 Morse-Smale Systems

The examples of the horseshoe, toral Anosov, and solenoid all have infinitely many periodic orbits. In this section we consider a class of systems with only finitely many periodic orbits and no other chain recurrent points (or no other nonwandering points). If we let $\mathcal{R}(f)$ be the set of chain recurrent points, and $\text{Per}(f)$ be the set of periodic points, then we are assuming that $\text{Per}(f) = \mathcal{R}(f)$ and that there are finitely many orbits in $\text{Per}(f)$. We also want this system to be structurally stable. Therefore, we require that all the periodic points are hyperbolic, so they persist under small perturbations of f and the dynamics near the periodic points do not change.

Assume f is a diffeomorphism with $\text{Per}(f) = \mathcal{R}(f)$. For any $\mathbf{q}$, $\alpha(\mathbf{q}), \omega(\mathbf{q}) \subset \text{Per}(f)$, so there must be two periodic points $\mathbf{p}_1, \mathbf{p}_2$ with $\alpha(\mathbf{q}) = \mathcal{O}(\mathbf{p}_1)$ and $\omega(\mathbf{q}) = \mathcal{O}(\mathbf{p}_2)$. Thus, for any $\mathbf{q}$, there must be $\mathbf{p}_1, \mathbf{p}_2 \in \text{Per}(f)$ with $\mathbf{q} \in W^u(\mathbf{p}_1) \cap W^s(\mathbf{p}_2)$. Since the periodic points are hyperbolic, for g sufficiently near to f, there will be nearby periodic points $\mathbf{p}_1(g), \mathbf{p}_2(g) \in \text{Per}(g)$. For the system f to be structurally stable, it is necessary that for g sufficiently near to f, $W^u(\mathbf{p}_1(g), g) \cap W^s(\mathbf{p}_2(g), g) \neq \emptyset$. Thus, we need that any intersection of stable and unstable manifolds cannot be destroyed. The property which assures that these intersections cannot be broken is transversality, which we now define.

Definition. Two submanifolds V and W in M are *transverse* (in M) provided for any point $\mathbf{q} \in V \cap W$, we have that $T_\mathbf{q}V + T_\mathbf{q}W = T_\mathbf{q}M$. (This allows for the possibility that $V \cap W = \emptyset$.)

Notice that for the two submanifolds V, W to intersect transversally at a point $\mathbf{q}$, it is necessary for $\dim T_\mathbf{q}V + \dim T_\mathbf{q}W \geq \dim T_\mathbf{q}M$. In $\mathbb{R}^2$, if V and W are two curves, then V being transverse to W at $\mathbf{q}$ means that the two tangent lines $T_\mathbf{q}V$ and $T_\mathbf{q}W$ are not colinear. In $\mathbb{R}^3$, two planes are transverse at $\mathbf{q}$ if they intersect in a line through $\mathbf{q}$.

The definition of a Morse-Smale system can now be given. Notice that the definition involves conditions on the whole phase portrait and how orbits go between various periodic points. For this reason we only define it when the phase space is compact. Instead of $\mathbb{R}^n$, we need to add the point at infinity to get a system on S^n, or work with some other compact phase space such as $\mathbb{T}^n$.

Definition. A diffeomorphism f (or a flow φ^t) on a compact manifold M is called *Morse-Smale* provided

(1) the chain recurrent set is a finite set of periodic orbits, each of which is hyperbolic, and
(2) each pair of stable and unstable manifolds of periodic points is transverse, i.e., if $\mathbf{p}_1, \mathbf{p}_2 \in \text{Per}(f)$, then $W^u(\mathbf{p}_1)$ is transverse to $W^s(\mathbf{p}_2)$. Notice that in the case of flows, we allow periodic orbits and not just fixed points.

We now give a number of examples of Morse-Smale diffeomorphisms. At the end of the section, we return to some examples of flows, and highlight some of the special aspects of Morse-Smale flows.

Example 12.1. On S^1, we consider the system

$$f(\theta) = \theta + \epsilon \sin(2\pi k \theta) \qquad \text{mod } 1,$$

for $0 < 2\pi k \epsilon < 1$. The lift F of f has $F'(\theta) = 1 + \epsilon \cos(2\pi k \theta)$. This derivative is positive with the assumption on ϵ, so f is a diffeomorphism. This has $2k$ fixed points, $\{\frac{j}{2k}\}_{j=0}^{2k-1}$. Half of the fixed points are attracting, $\{\frac{(2j+1)}{2k}\}_{j=0}^{k-1}$, and the other half are repelling, $\{\frac{j}{k}\}_{j=0}^{k-1}$. All other trajectories are in the stable manifold of one of these sinks and the unstable manifold of one of the sources. Thus, the system is Morse-Smale. See Figure 12.1.

Example 12.2. Again on S^1, we consider the system

$$f(\theta) = \theta + \frac{1}{k} + \epsilon \sin(2\pi k \theta) \qquad \text{mod } 1,$$

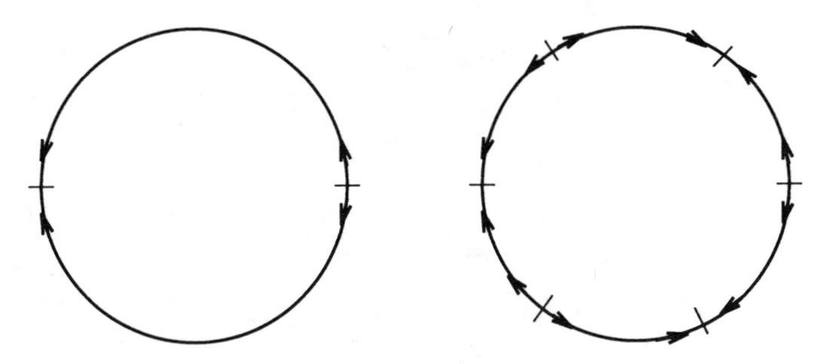

FIGURE 12.1. (a) $k = 1$ (b) $k = 3$

for $0 < 2\pi k\epsilon < 1$. With the assumption on ϵ, f is a diffeomorphism as in the previous example. This system has two periodic orbits of period k,

$$\{\frac{j}{k}\}_{j=0}^{k-1} \quad \text{and} \quad \{\frac{(2j+1)}{2k}\}_{j=0}^{k-1}.$$

The first orbit is repelling and the second is attracting.

Example 12.3. On $\mathbb{T}^2$, take θ_1 and θ_2 as variable modulo 1. Let

$$f\begin{pmatrix}\theta_1\\ \theta_2\end{pmatrix} = \begin{pmatrix}\theta_1 + \epsilon\sin(2\pi\theta_1)\\ \theta_2 + \epsilon\sin(2\pi\theta_2)\end{pmatrix}.$$

We assume that $0 < 2\pi\epsilon < 1$, so that both coordinate functions are one to one and f is a diffeomorphism. This diffeomorphism has fixed points at $\mathbf{p}_1 = (0,0)$, $\mathbf{p}_2 = (1/2,0)$, $\mathbf{p}_3 = (0,1/2)$, and $\mathbf{p}_4 = (1/2,1/2)$. The point $\mathbf{p}_1$ is a source, $\mathbf{p}_2$ and $\mathbf{p}_3$ are saddles, and $\mathbf{p}_4$ is a sink. Then, $W^s(\mathbf{p}_j) \subset W^u(\mathbf{p}_1)$ and $W^u(\mathbf{p}_j) \subset W^s(\mathbf{p}_4)$ for $j = 2, 3$,

$$W^u(\mathbf{p}_1) \subset W^s(\mathbf{p}_2) \cup W^s(\mathbf{p}_3) \cup W^s(\mathbf{p}_4), \quad \text{and}$$
$$W^s(\mathbf{p}_4) \subset W^u(\mathbf{p}_1) \cup W^u(\mathbf{p}_2) \cup W^u(\mathbf{p}_3).$$

It is easily checked that all these intersections are transverse and the diffeomorphism is Morse-Smale. See Figure 12.2.

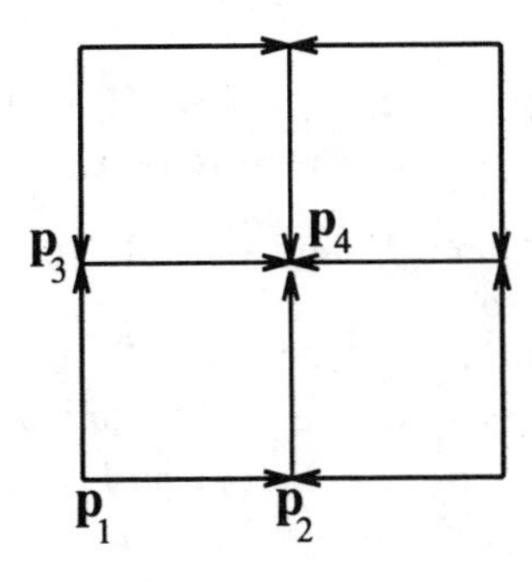

FIGURE 12.2. Example 12.3 on Torus

Example 12.4. On S^2, it is possible to have a Morse-Smale diffeomorphism with one fixed point source at the north pole and one fixed point sink at the south pole. See Figure 12.3.

FIGURE 12.3. North Pole – South Pole Diffeomorphism on S^2

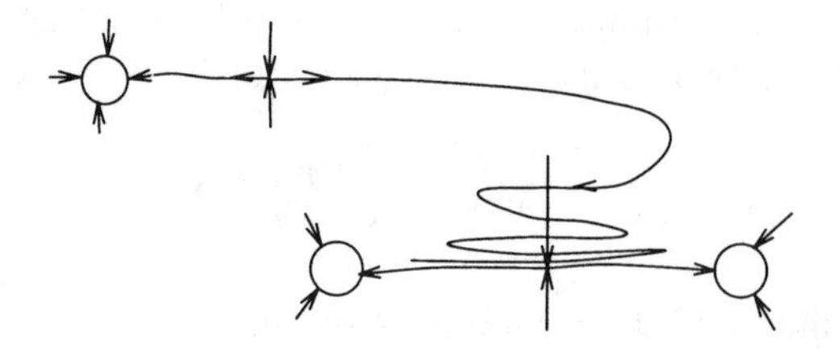

FIGURE 12.4. Example 12.5

Example 12.5. On S^2, it is possible to have a Morse-Smale diffeomorphism with one source at the north pole (infinity of the plane), two saddles, and three sinks, all fixed points. See Figure 12.4.

The next lemma proves that for a Morse-Smale diffeomorphism, there are restrictions on the manner in which the stable and unstable manifolds can intersect. We first define a cycle of periodic points. The lemma then states that a Morse-Smale system can have no cycles among the periodic orbits.

Definition. If $\mathbf{p}$ is a periodic point of f, for $\sigma = u, s$, let

$$W^\sigma(\mathcal{O}(\mathbf{p})) = \bigcup_j W^\sigma(f^j(\mathbf{p})), \qquad \text{and}$$

$$\hat{W}^\sigma(\mathcal{O}(\mathbf{p})) = W^\sigma(\mathcal{O}(\mathbf{p})) \setminus \mathcal{O}(\mathbf{p}).$$

A collection of periodic points $\mathbf{p}_0, \ldots, \mathbf{p}_{k-1} \in Per(f)$ is a *k-cycle* provided $\hat{W}^u(\mathcal{O}(\mathbf{p}_j)) \cap \hat{W}^s(\mathcal{O}(\mathbf{p}_{j+1})) \neq \emptyset$ for $j = 0, \ldots, k-1$, where we let $\mathbf{p}_k = \mathbf{p}_0$.

Lemma 12.1. *(a) If $\mathbf{p}_1, \mathbf{p}_2 \in \operatorname{Per}(f)$ and $\mathbf{q} \in \hat{W}^u(\mathbf{p}_1) \cap \hat{W}^s(\mathbf{p}_2)$, then there is an ϵ-chain from $\mathbf{p}_1$ to $\mathbf{q}$ and then to $\mathbf{p}_2$.*

(b) If $\mathbf{p}_0, \ldots \mathbf{p}_k \in \operatorname{Per}(f)$ and $\mathbf{q}_j \in \hat{W}^u(\mathbf{p}_j) \cap \hat{W}^s(\mathbf{p}_{j+1})$ for $j = 0, \ldots, k-1$, then there is an ϵ-chain from $\mathbf{p}_0$ to $\mathbf{p}_k$ which passes through all the $\mathbf{p}_j$ and the $\mathbf{q}_j$.

(c) If there is a k-cycle $\mathbf{p}_0, \ldots \mathbf{p}_{k-1} \in \operatorname{Per}(f)$ and $\mathbf{q}_j \in \hat{W}^u(\mathcal{O}(\mathbf{p}_j)) \cap \hat{W}^s(\mathcal{O}(\mathbf{p}_{j+1}))$ for $j = 0, \ldots, k-1$, then $\mathbf{q}_j \in \mathcal{R}(f)$ for $j = 0, \ldots, k-1$ and there are chain recurrent points which are not periodic points.

(d) If f is Morse-Smale, then there can be no cycles among the period points of f.

We leave the proof as an exercise for the reader. See Exercise 8.51.

Next we state the theorem about the structural stability of Morse-Smale diffeomorphisms and flows.

Theorem 12.2. *Let M be a compact manifold, and f a C^1 Morse-Smale diffeomorphism (or flow). Then, f is C^1 structurally stable.*

The proof in the case of a diffeomorphism on S^1 is simple because there can only be periodic sources and sinks and no periodic saddle points. The proof in this case is similar to the examples treated in Section 2.6. For the proof in the general case, see Palis and de Melo (1982). The original proof is found in Palis and Smale (1970).

Morse-Smale Flows

We now turn to flows. The definition of a *Morse-Smale flow* is the same as for a diffeomorphism but with periodic orbits replaced by either fixed points or periodic orbits or some of each. There are some differences in the implications of the assumptions of transversality. If $\mathbf{q} \in W^u(\gamma_1) \cap W^s(\gamma_2)$ where γ_1 and γ_2 are either fixed points or periodic orbits, then the whole orbit $\mathcal{O}(\mathbf{q}) \subset W^u(\gamma_1) \cap W^s(\gamma_2)$. Thus, for transversality, we need that $\dim(W^u(\gamma_1)) + \dim(W^s(\gamma_2)) \geq \dim(M) + 1$. In particular, on a surface (dimension two), if γ_1 and γ_2 are each fixed point saddles for a Morse-Smale flow, then $W^u(\gamma_1) \cap W^s(\gamma_2) = \emptyset$.

Example 12.6. On $\mathbb{T}^2$, consider the equations (written on $\mathbb{R}^2$)

$$\begin{aligned} \dot{\theta}_1 &= \epsilon_1 \sin(2\pi\theta_1) \\ \dot{\theta}_2 &= \epsilon_2 \sin(2\pi\theta_2) \end{aligned}$$

for $0 < \epsilon_1, \epsilon_2 < 1/(2\pi)$. Taking all the variables modulo 1, there are four fixed points: $(0,0)$, $(1/2,0)$, $(0,1/2)$, and $(1/2,1/2)$. This is very much like Example 12.3 given above for a diffeomorphism and has one source, two saddles, and one sink. See Figure 12.5. This example is a Morse-Smale flow.

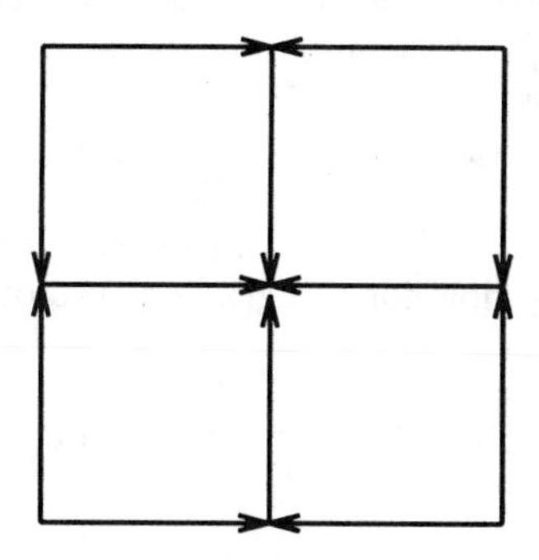

FIGURE 12.5. Example 12.6: Differential Equations on Torus

If φ^t is a Morse-Smale flow with fixed points but no periodic orbits, then the time one map, φ^1, is a Morse-Smale diffeomorphism. (See Exercise 8.44.) Examples 12.1 – 12.4 of Morse-Smale diffeomorphisms discussed above are of this type, but Example 12.5 is not.

One way to get examples of Morse-Smale flows with fixed points but no periodic orbits is as the gradient of a real valued function, which we now define.

Definition. A flow φ^t on $\mathbb{R}^n$ is called a *gradient flow* provided there is a real valued function $V : \mathbb{R}^n \to \mathbb{R}$ such that

$$\frac{d}{dt}\varphi^t(\mathbf{p}) = -\nabla V_{\varphi^t(\mathbf{p})}.$$

We also say that $X(\mathbf{p}) = -\nabla V_{\varphi^t(\mathbf{p})}$ is a *gradient vector field.*

We want to define a gradient flow on a surface or a manifold. A surface can be embedded in $\mathbb{R}^3$. In general, a manifold M can be embedded in a large dimensional Euclidean space $\mathbb{R}^n$. If this is done, then a real valued function $V : M \to \mathbb{R}$ is called differentiable if it can be extended to a neighborhood U of M in $\mathbb{R}^n$, $\bar{V} : U \to \mathbb{R}$ with $\bar{V}|M = V$. With these facts we can define a gradient system on a manifold M.

Definition. Let a manifold M be embedded in a Euclidean space $\mathbb{R}^n$. A vector field $X(\mathbf{p})$ on M is called a *gradient vector field* provided there is a real valued function $V : U \to \mathbb{R}$ where U is a neighborhood of M in $\mathbb{R}^n$ such that

$$X(\mathbf{p}) = -\pi_{\mathbf{p}} \nabla V_{\mathbf{p}},$$

where $\pi_{\mathbf{p}} : T_{\mathbf{p}}\mathbb{R}^n \to T_{\mathbf{p}}M$ is the orthogonal projection of vectors at $\mathbf{p}$ in $\mathbb{R}^n$ to tangent vectors to M. Also, a flow φ^t on M is called a *gradient flow* provided there is a gradient vector field $X(\mathbf{p})$ such that $\frac{d}{dt}\varphi^t(\mathbf{p}) = X(\varphi^t(\mathbf{p}))$.

As an alternative definition of a gradient vector field on a manifold (and more intrinsic to the manifold itself), we can assume that there is a Riemannian metric on M, i.e., for each $\mathbf{p} \in M$ there is an inner product on $T_{\mathbf{p}}M$, a positive definite symmetric bilinear form $\langle\ ,\ \rangle_{\mathbf{p}}$, which varies smoothly with $\mathbf{p}$. (An embedding of M into a Euclidean space is merely one way to get such an inner product.) Given a tangent vector $\mathbf{v}_{\mathbf{p}}$ at $\mathbf{p}$, $\langle \mathbf{v}_{\mathbf{p}},\ \rangle_{\mathbf{p}}$ is something which acts on a tangent vector at $\mathbf{p}$ and gives a real number, and the map sending $\mathbf{w}_{\mathbf{p}}$ to $\langle \mathbf{v}_{\mathbf{p}}, \mathbf{w}_{\mathbf{p}}\rangle_{\mathbf{p}}$ is linear in $\mathbf{w}_{\mathbf{p}}$, so $\langle \mathbf{v}_{\mathbf{p}},\ \rangle_{\mathbf{p}}$ is in the dual space to $T_{\mathbf{p}}M$. Thus, using the inner product, for each $\mathbf{p} \in M$, there is a map from the tangent vectors at $\mathbf{p}$ to the dual space, $J_{\mathbf{p}} : T_{\mathbf{p}}M \to (T_{\mathbf{p}}M)^*$. The fact that the inner product is positive definite implies that this map is an isomorphism. At each point $\mathbf{p}$, the derivative for V, $DV_{\mathbf{p}}$, is a linear map waiting to act on a vector and the outcome is a real number, thus $DV_{\mathbf{p}} \in \mathbf{L}(T_{\mathbf{p}}M, \mathbb{R}) \approx (T_{\mathbf{p}}M)^*$. Combining these two constructions, we define the *gradient of* V by

$$\text{grad}(V)_{\mathbf{p}} = J_{\mathbf{p}}^{-1} DV_{\mathbf{p}}.$$

Finally, a flow φ^t is called a *gradient flow* provided there is a real valued function $V : M \to \mathbb{R}$ such that φ^t is the flow for the gradient vector field $-\text{grad}(V)$,

$$\frac{d}{dt}\varphi^t(\mathbf{p}) = -\text{grad}(V)_{\varphi^t(\mathbf{p})}.$$

Theorem 12.3. *Let $V : M \to \mathbb{R}$ be a C^2 function such that each critical point is nondegenerate, i.e., at each point $\mathbf{x}$ where grad$(V)_{\mathbf{x}} = \mathbf{0}$, the matrix of second partial derivatives, $\left(\dfrac{\partial^2 V}{\partial x_i \partial x_j}(\mathbf{x})\right)$, has nonzero determinant. Let $X(\mathbf{x}) = -$grad$(V)_{\mathbf{x}}$ be the gradient vector field for V. Then, (a) all the fixed points are hyperbolic and (b) the chain recurrent set for X equals the set of fixed points for X (X has no periodic points).*

PROOF. We leave the proof of part (a) to the exercises. See Exercise 8.42. Since the function V is strictly decreasing off the fixed points and all the fixed points are isolated, all points which are not fixed are not in the chain recurrent set. This proves part (b). □

REMARK 12.1. Let $V : M \to \mathbb{R}$ be a C^2 function such that each critical point is nondegenerate. By Theorem 12.3, then $-\text{grad}(V)$ is a Morse-Smale vector field if and only if the stable and unstable manifolds are transverse.

Example 12.7. For a gradient vector field on $\mathbb{T}^2$, let $V : \mathbb{T}^2 \to \mathbb{R}$ be the height function when $\mathbb{T}^2$ is stood up on its end. This defines a gradient vector field with four fixed points: a source at the maximum $\mathbf{p}_{max}$, a sink at the minimum $\mathbf{p}_{min}$, and two saddles $\mathbf{p}_1$ and $\mathbf{p}_2$ with $V(\mathbf{p}_2) > V(\mathbf{p}_1)$. See Figure 12.6. If the maximum, minimum, and saddles are nondegenerate, then the fixed points are hyperbolic. If the torus is "straight up and down," then there is a saddle connection from $\mathbf{p}_2$ to $\mathbf{p}_1$, $W^u(\mathbf{p}_2) \cap W^s(\mathbf{p}_1) \neq \emptyset$. Thus, this example is not Morse-Smale. If the torus is slightly tipped so that $W^u(\mathbf{p}_2) \cap W^s(\mathbf{p}_1) = \emptyset$, then the vector field becomes Morse-Smale.

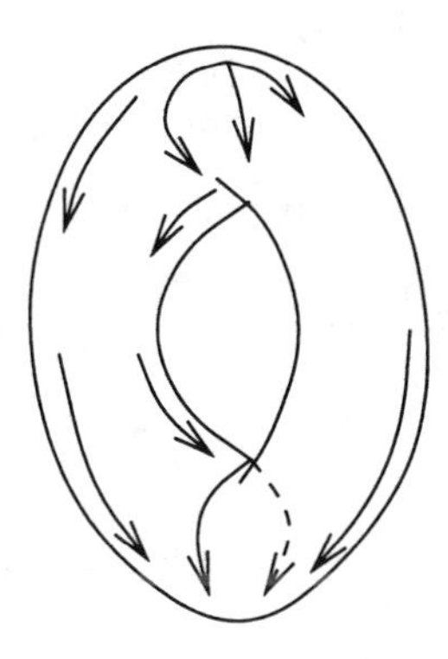

FIGURE 12.6. Gradient Flow on $\mathbb{T}^2$

We want to make the connection between gradient systems and Liapunov functions, so we first define (global) Liapunov functions and then state and prove the theorem. Also see Conley (1978).

Definition. A C^1 function $L : M \to \mathbb{R}$ is called a *weak Liapunov function* for a flow φ^t provided

$$L \circ \varphi^t(\mathbf{x}) \leq L(\mathbf{x}) \quad \text{for all } \mathbf{x} \in M \text{ and all } t \geq 0.$$

(Maybe we should call this a weak, global Liapunov function, but we do not use the adjective global when we use this definition.) Thus, if we let $\dot{L}(\mathbf{x}) \equiv \frac{d}{dt} L(\varphi^t(\mathbf{x}))|_{t=0}$, then L is a weak Liapunov function if and only if $\dot{L}(\mathbf{x}) \leq 0$ for all points $\mathbf{x} \in M$. A real valued function L on M is called a *Liapunov function* (or strong or strict or complete Liapunov function) provided it is a weak Liapunov function with

$$L \circ \varphi^t(\mathbf{x}) < L(\mathbf{x}) \quad \text{for all } \mathbf{x} \notin \mathcal{R}(\varphi^t) \text{ and all } t > 0,$$

i.e., $\dot{L}(\mathbf{x}) < 0$ for all $\mathbf{x} \notin \mathcal{R}(\varphi^t)$. (Or for $\mathbf{x} \notin S$ for some other set S such as $\text{Fix}(\varphi^t)$, or the set $\mathcal{P}$ defined in Theorem X.1.3. If necessary, the set off which the Liapunov function is decreasing under the flow is identified.)

The reader might want to check that it is necessary that a Liapunov function L is constant on any orbit in $\text{Per}(\varphi^t)$ or, in fact, in $\mathcal{R}(\varphi^t)$.

Sometimes a flow is not a gradient flow but has a Liapunov function which is decreasing off the fixed points. We identify such flows by the following definition.

Definition. A flow φ^t is called *gradient-like* provided there is a strict Liapunov function which is strictly decreasing off $\text{Fix}(\varphi^t)$.

Theorem 12.4. *Let $X(\mathbf{p}) = -grad(L)_{\mathbf{p}}$ be a gradient vector field on a manifold M with flow $\varphi^t(\mathbf{p})$. Then, $\dot{L}(\mathbf{p}) = -|grad(L)_{\mathbf{p}}|^2 \le 0$ for all $\mathbf{p} \in M$, and $\dot{L}(\mathbf{p}) = 0$ only at the fixed points of X. Thus, L is a Liapunov function of the flow of X, and φ^t is gradient-like.*

PROOF.

$$\begin{aligned}\dot{L}(\mathbf{p}) &= \frac{d}{dt} L(\varphi^t(\mathbf{p}))|_{t=0} \\ &= DL_{\varphi^t(\mathbf{p})} \frac{d}{dt}\varphi^t(\mathbf{p})|_{t=0} \\ &= DL_{\mathbf{p}} X(\mathbf{p}).\end{aligned}$$

But $DL_{\mathbf{p}}\mathbf{v}_{\mathbf{p}} = \langle \text{grad}(L)_{\mathbf{p}}, \mathbf{v}_{\mathbf{p}}\rangle$ for any tangent vector $\mathbf{v}_{\mathbf{p}} \in T_{\mathbf{p}}M$, so

$$\begin{aligned}\dot{L}(\mathbf{p}) &= \langle \text{grad}(L)_{\mathbf{p}}, X(\mathbf{p})\rangle \\ &= \langle \text{grad}(L)_{\mathbf{p}}, -\text{grad}(L)_{\mathbf{p}}\rangle \\ &= -|\text{grad}(L)_{\mathbf{p}}|^2.\end{aligned}$$

Thus, $\dot{L}(\mathbf{p}) \le 0$. It equals 0 if and only if $\text{grad}(L)_{\mathbf{p}} = \mathbf{0}$, if and only if $DL_{\mathbf{p}} = \mathbf{0}$, if and only if $\mathbf{p}$ is a fixed point of the flow. □

Corollary 12.5. *The minima of L are asymptotically stable fixed points, and the saddles and maxima of L are not Liapunov stable.*

Corollary 12.6. *The trajectories of a gradient system for $L : M \to \mathbb{R}$ cross the level sets of L orthogonally at points which are not fixed points.*

If the real valued function L on M has only finitely many nondegenerate critical points, then the gradient vector field for L, X, has finitely many fixed points all of which are hyperbolic and no other periodic orbits. Thus, $\mathcal{R}(X) = \text{Fix}(X) = \text{Per}(X)$. It can be shown that by perturbing any such gradient vector field slightly to Y, all the stable and unstable manifolds of Y can be made transverse, and hence Y would be a Morse-Smale vector field. Example 12.7 with the torus straight on the end is an example where $\mathcal{R}(X) = \text{Fix}(X) = \text{Per}(X)$, but the system is not Morse-Smale. The perturbation of the gradient system caused by slightly tipping the torus is an example of how such a system can be made Morse-Smale.

There are constraints on the possible dynamics in terms of the topology of the phase space. We give the result for gradient flows or gradient-like flows. If $\mathbf{p}$ is a fixed point, then the *index at* $\mathbf{p}$ is defined to be the dimension of the unstable subspace, $\text{index}(\mathbf{p}) = \dim(\mathbb{E}^u_{\mathbf{p}})$. Let $c_k = \#($ critical points of index $k)$ and $\beta_k = \dim H_k(M, \mathbb{R})$. Then, the *Morse inequalities* are as follows:

$$c_k - c_{k-1} + \cdots \pm\ c_0 \ge \beta_k - \beta_{k-1} + \cdots \pm\ \beta_0$$

for $k = 0, \ldots, \dim(M)$. Also, for $k = \dim(M)$,

$$\sum_{j=0}^{\dim(M)} (-1)^{\dim(M)-j} c_j = \sum_{j=0}^{\dim(M)} (-1)^{\dim(M)-j} \beta_j = \chi(M),$$

where $\chi(M)$ is the Euler characteristic of M. See Franks (1982) for more details, including some results for diffeomorphisms. Also see Palis and de Melo (1982) for further extension of the index to more complicated invariant sets.

8.13 Exercises

Manifolds

8.1. Prove that S^n has the structure of a C^∞ n-manifold. Hint: Use the graphs given in Example 1.3.

8.2. Let $F : \mathbb{R}^{n+1} \to \mathbb{R}$ be a C^r function for some $r \geq 1$. Assume $c \in \mathbb{R}$ is a value such that for each $\mathbf{p} \in F^{-1}(c)$, $DF_{\mathbf{p}} \neq \mathbf{0}$, i.e., $DF_{\mathbf{x}}$ has rank one, or some partial derivative of F is nonzero at $\mathbf{p}$. Prove that $F^{-1}(c)$ has the structure of a C^r n-manifold.

8.3. Let $F : \mathbb{R}^{n+k} \to \mathbb{R}^k$ be a C^r function for some $r \geq 1$. Assume $\mathbf{c} \in \mathbb{R}^k$ is a value such that for each $\mathbf{p} \in F^{-1}(\mathbf{c})$, $DF_{\mathbf{p}}$ has rank k. Prove that $F^{-1}(\mathbf{c})$ has the structure of a C^r n-manifold. Hint: Use Theorem V.2.2.

8.4. Let M and N be manifolds and $f : M \to N$ a C^1 diffeomorphism (a C^1 homeomorphism with a C^1 inverse). Prove that the map $Df_p : T_pM \to T_{f(p)}M$ is an isomorphism for every $p \in M$.

8.5. Let $X = \mathbb{R}$ and consider the two coordinate charts $\varphi_1, \varphi_2 : \mathbb{R} \to X$ given by $\varphi_1(x) = x$ and $\varphi_2(x) = x^3$. For $i = 1, 2$, let M_i denote the manifold consisting of the metric space X together with the one coordinate chart φ_i.
   (a) Prove that the identity map on X is not a diffeomorphism.
   (b) Prove that there is a diffeomorphism from M_1 to M_2.

8.6. Let M be a manifold and $X : M \to TM$ be a vector field. Prove that X is C^r for $r \geq 1$ if and only if for each C^{r+1} function $f : M \to \mathbb{R}$ the function $X(f)$ defined by $X(f)(\mathbf{p}) = Df_{\mathbf{p}}X(\mathbf{p})$ is C^r.

Transitivity Theorems

8.7. A homeomorphism $f : X \to X$ is called topologically mixing if for any two open sets $U, V \subset X$ there is an $N > 0$ such that $n \geq N$ implies $f^n(U) \cap V \neq \emptyset$.
   (a) Prove that no homeomorphism of $I = [0, 1]$ is mixing.
   (b) Give an example of a homeomorphism of a compact connected metric space which has a point with a dense orbit but which is not mixing.

8.8. Let

$$T(x) = \begin{cases} 2x & \text{for } x \leq 1/2 \\ 2 - 2x & \text{for } x \geq 1/2 \end{cases}$$

be the tent map.
   (a) Using the Birkhoff Transitivity Theorem, prove that T has a point with a dense orbit.
   (b) Using the fact that $F_4(x) = 4x(1-x)$ is conjugate to T (see Example II.6.2), prove that F_4 has a point with a dense orbit.

8.9. (Poincaré Recurrence Theorem) Assume μ is a finite measure on a space X, $\mu(X) < \infty$. Assume $f : X \to X$ is a one to one map which preserves the measure μ, i.e., $\mu(f^{-1}(A)) = \mu(A)$ for every measurable set A. Assume S is a measurable set. Let $S_0 = S$,

$$\begin{aligned} S_n &= \{\mathbf{x} \in S_{n-1} : f^j(\mathbf{x}) \in S_{n-1} \text{ for some } j \geq 1\} \\ &= S_{n-1} \cap \bigcup_{j \geq 1} f^{-j}(S_{n-1}) \end{aligned}$$

for $n \geq 1$, and $S_\infty = \bigcap_{n \geq 0} S_n$. Prove that S_∞ is measurable, $\mu(S_\infty) = \mu(S)$, and for $\mathbf{x} \in S_\infty$, $f^j(\mathbf{x})$ returns to S an infinite number of times. Hint: Prove that $\mu(S_n) = \mu(S)$ for $n \geq 1$.

8.10. Assume X is a separable metric space and μ is a finite Borel measure on X (so all open sets are measurable). Assume that $f : X \to X$ is a homeomorphism on X which preserves the measure μ. Using the previous exercise, prove that

$$Y = \{\mathbf{x} \in X : \mathbf{x} \in \alpha(\mathbf{x}) \text{ and } \mathbf{x} \in \omega(\mathbf{x})\}$$

has full measure, i.e., $\mu(Y) = \mu(X)$. Hint: Let $\{\mathbf{x}_k\}_{k=1}^{\infty}$ be a countable dense set in X and consider the open balls $B(\mathbf{x}_k, \delta)$ for $\delta > 0$.

8.11. Assume X is a complete separable metric space and μ is a finite Borel measure on X which is positive on every open set. Assume that $f : X \to X$ is a homeomorphism on X which preserves the measure μ. Prove that

$$Y = \{\mathbf{x} \in X : \mathbf{x} \in \alpha(\mathbf{x}) \text{ and } \mathbf{x} \in \omega(\mathbf{x})\}$$

is a residual set in X and so is dense in X. Hint: Let $\{\mathbf{x}_k\}_{k=1}^{\infty}$ be a countable dense set in X. For $\delta > 0$, let $U(\mathbf{x}_k, \delta, 0) = B(\mathbf{x}_k, \delta)$ and

$$U(\mathbf{x}_k, \delta, n) = \{\mathbf{x} \in U(\mathbf{x}_k, \delta, n-1) : f^j(\mathbf{x}) \in U(\mathbf{x}_k, \delta, n-1) \text{ for some } j \geq 1\}$$

for $n \geq 1$, Prove that each $U(\mathbf{x}_k, \delta, n)$ is dense and open in $B(\mathbf{x}_k, \delta)$.

Two-Sided Shift Spaces

8.12. Let Σ_N be the full two-sided shift space on N symbols with shift map σ_N. For $\lambda > 1$, let ρ_λ be the metric on Σ_N defined by

$$\rho_\lambda(\mathbf{s}, \mathbf{t}) = \sum_{j=-\infty}^{\infty} \frac{\delta(s_j, t_j)}{\lambda^{|j|}}$$

where

$$\delta(s_j, t_j) = \begin{cases} 0 & \text{when } s_j = t_j \\ 1 & \text{when } s_j \neq t_j. \end{cases}$$

Given $\mathbf{t} \in \Sigma_N$ and $k \geq 0$, prove that

$$\{\mathbf{s} : s_j = t_j \text{ for } |j| \leq k\}$$

is an open ball in terms of the metric ρ_λ if and only if $\lambda > 3$.

8.13. Let Σ_N be the full two-sided shift space on N symbols with shift map σ_N. Let A be an $N \times N$ transition matrix and $\Sigma_A \subset \Sigma_N$ be the subshift of finite type determined by A and $\sigma_A = \sigma_N|\Sigma_A$.

(a) Prove that σ_A is topologically transitive on Σ_A using the Birkhoff Transitivity Theorem.

(b) Describe a symbol sequence $\mathbf{s}^* \in \Sigma_A$ that has both a dense forward orbit and a dense backward orbit in Σ_A. Remark: Compare with Theorem II.5.3.

8.14. Let $\sigma : \Sigma_2 \to \Sigma_2$ be the full two-sided two-shift. Define $r : \Sigma_2 \to \Sigma_2$ by $r(\mathbf{a}) = \mathbf{b}$ where $b_j = a_{-j-1}$, and let $s = \sigma \circ r$. Prove that $r \circ r = id$, $s \circ s = id$, and $\sigma = s \circ r$.

Subshifts for Nonnegative Matrices

8.15. Let A be an $n \times n$ adjacency matrix and $N = \sum_{i,j} a_{ij}$. Let T be the $N \times N$ transition matrix on the edges induced by A as defined in Section 8.3.1.

(a) Prove that there are $(A^k)_{ij}$ T-allowable words $\mathbf{w}$ of length $k+1$ with $b(\mathbf{w}) = i$ and $e(\mathbf{w}) = j$.

(b) Prove that A is irreducible if and only if T is irreducible.

(c) Prove that $\#(\mathrm{Fix}(\sigma_T^k)) = \mathrm{tr}(A^k)$.

8.16. Let A be an $n \times n$ adjacency matrix and $N = \sum_{i,j} a_{ij}$. Assume $a_{ij} \in \{0,1\}$ for all $1 \le i, j \le n$. Let T be the $N \times N$ transition matrix on the edges induced by A as defined in Section 8.3.1. Prove that the vertex subshift $\sigma_A : \Sigma_A \to \Sigma_A$ is topologically conjugate to the edge subshift formed from A, $\sigma_T : \Sigma_T \to \Sigma_T$. Hint: Define $h : \Sigma_T \to \Sigma_A$ by $h(\mathbf{s}) = \mathbf{a}$, where $s_j = b(a_j)$ for all j, $a_j \in \mathcal{E}$ and $s_j \in \mathcal{V}$. Prove h is a conjugacy.

Horseshoes

8.17. (Hyperbolic structure for an invariant set) Assume $f : \mathbb{R}^2 \to \mathbb{R}^2$ is a diffeomorphism with a compact hyperbolic invariant set Λ. Assume the dimension of each of the subspaces in the splitting has dimension one. Prove that the splitting is continuous. Hint: See Lemma V.10.9 for the meaning of a continuous splitting.

8.18. Consider the map $f : S^2 \to S^2$ which gives the Geometric Horseshoe. Let S be the square as given in the chapter. Draw the inverse image of S by f, $f^{-1}(S)$.

8.19. Consider the Geometric Horseshoe Map. Let $\mathbf{p}_1$ be the fixed point in $H_1 \cap V_1$, $L^s = \mathrm{comp}_{\mathbf{p}_1}(W^s(\mathbf{p}_1) \cap S)$, and $L^u = \mathrm{comp}_{\mathbf{p}_1}(W^u(\mathbf{p}_1) \cap S)$. Draw the images of L^u by f^3 and L^s by f^{-3}, $f^3(L^u)$ and $f^{-3}(L^s)$.

8.20. Let $f, k_\epsilon, g_\epsilon : \mathbb{R}^2 \to \mathbb{R}^2$ be diffeomorphisms as in Example 4.1 (Section 8.4.2). Let U be the neighborhood of the nontransverse homoclinic point $\mathbf{x}^*$ for f. Prove that the following statements are true for $\epsilon > 0$ small enough:

(a) g_ϵ is a diffeomorphism,

(b) $\mathbf{0}$ is a saddle fixed point for g_ϵ,

(c) letting $I = W^s(\mathbf{0}, f) \cap U$,

$$\bigcup_{j=0}^{\infty} f^j(I) \subset W^s(\mathbf{0}, g_\epsilon),$$

$$\bigcup_{j=-1}^{-\infty} f^j(I) \subset W^u(\mathbf{0}, g_\epsilon), \qquad \text{and}$$

$$k_\epsilon(I) \subset W^u(\mathbf{0}, g_\epsilon),$$

and

(d) $\mathbf{x}^*$ is a transverse homoclinic point for g_ϵ.

8.21. When $B = 0$, show that the family of Hénon maps $F_{A,0}$ is conjugate to the family of quadratic maps F_μ on $\mathbb{R}$. (Unfortunately, the labeling of the two families of functions is very similar.) By a conjugacy in this case, we mean a continuous map $h : \mathbb{R} \to \mathbb{R}^2$ which is a homeomorphism onto the image of $F_{A,0}$ and such that $h \circ F_\mu = F_{A,0} \circ h$.

8.22. Assume $f : \mathbb{R}^2 \to \mathbb{R}^2$ is a diffeomorphism with a compact hyperbolic invariant set Λ. Prove that if V is a small enough neighborhood of Λ, then the maximal invariant set in V,

$$\Lambda_V \equiv \bigcap_{n \in \mathbb{Z}} f^n(V),$$

has a hyperbolic structure. Hint: See the proof of Theorem 4.5.

8.23. Let $f : \mathbb{R} \to \mathbb{R}$ be C^1 function. Assume that $f(0) = 0$ and $f'(0) > 1$. Assume that $q \neq 0$ is a point such that (i) $f(q) = 0$, (ii) there is a choice of a backward orbit $\{q_k\}_{k \le 0}$ with $q_0 = q$, $f(q_{k-1}) = q_k$ for $k \le 0$, $f'(q_k) \neq 0$ for $k \le 0$, and q_k converges to 0 as k goes to $-\infty$, i.e., $q \in W^u(0)$. Prove that there is an invariant set containing q and 0 which is conjugate to a one-sided subshift of finite type. Hint: Since f is not a diffeomrophism, Theorem 4.5 does not apply. Find intervals which work correctly as the boxes do in the proof of Theorem 4.5.

Anosov Diffeomorphisms

8.24. Let $f : \mathbb{T}^2 \to \mathbb{T}^2$ be the map on the two torus induced by the matrix

$$A = \begin{pmatrix} 2 & 0 \\ 0 & 2 \end{pmatrix}$$

acting on $\mathbb{R}^2$. (Note that f is not invertible.) Prove that the periodic points are dense.

8.25. Let f_A be a hyperbolic toral automorphism on $\mathbb{T}^n$ with lift L_A to $\mathbb{R}^n$. Let g be a small C^1 perturbation of f_A with lift G to $\mathbb{R}^n$. Finally, let $\hat{G} = G - L_A$. Let $C^0_{b,per}(\mathbb{R}^n)$, $C^1_{b,per}(\mathbb{R}^n)$, and $\Theta(\hat{G}, v)$ be defined as in the proof of Theorem 5.1.
(a) Prove that $\hat{G} \in C^1_{b,per}(\mathbb{R}^n)$.
(b) Prove that $\Theta(\hat{G}, \cdot)$ preserves $C^0_{b,per}(\mathbb{R}^n)$.

8.26. (a) Give an example of a hyperbolic toral automorphism on the three torus $\mathbb{T}^3$.
(b) Give an example of a hyperbolic toral automorphism on the n-torus $\mathbb{T}^n$ for $n > 3$.

8.27. Let $f : \mathbb{T}^2 \to \mathbb{T}^2$ be the hyperbolic toral automorphism with matrix

$$\begin{pmatrix} 2 & 1 \\ 1 & 1 \end{pmatrix}.$$

Prove that f has sensitive dependence on initial conditions.

8.28. Let f_A be a hyperbolic toral automorphism on $\mathbb{T}^n$ with lift L_A to $\mathbb{R}^n$. Let g be a map of $\mathbb{T}^n$ which is homotopic to f_A. Use Problem 5.30 to prove that g is semi-conjugate to f_A.

Markov Partions for Hyperbolic Toral Automorphisms

8.29. (A horseshoe as a subsystems of a hyperbolic toral automorphism.) Let $f_{A_2} : \mathbb{T}^2 \to \mathbb{T}^2$ be the diffeomorphism induced by the matrix

$$A_2 = \begin{pmatrix} 2 & 1 \\ 1 & 1 \end{pmatrix}$$

discussed in Example 5.4. Let R_{1a} be the rectangle used in the Markov partition for this diffeomorphism. Let $g = f^2_{A_2}$ and

$$\Lambda = \bigcap_{j=-\infty}^{\infty} g^j(R_{1a}).$$

Prove that $g : \Lambda \to \Lambda$ is topologically conjugate to the two-sided full two-shift $\sigma : \Sigma_2 \to \Sigma_2$. Hint: R_{1a} plays the role that S played in the construction of the geometric horseshoe. Prove that $g(R_{1a}) \cap R_{1a}$ is made up of two disjoint rectangles. (These rectangles are similar to V_1 and V_2 in the geometric horseshoe.) Looking at the transition matrix for the Markov partition for f may help.

8.30. Let $f : \mathbb{T}^2 \to \mathbb{T}^2$ be the diffeomorphism induced by the matrix

$$A = \begin{pmatrix} 1 & 1 \\ 1 & 0 \end{pmatrix}.$$

Form a Markov partition with three rectangles by using the line segment $[\mathbf{a}, \mathbf{b}]_s$ as in the text, and an unstable line segment $[\mathbf{g}, \mathbf{c}]_u$ where g is determined by extending the unstable manifold of the origin through the origin so that it terminates at a point $\mathbf{g} \in [\mathbf{a}, \mathbf{0}]_s$. Thus, $\mathbf{0}$ is within the unstable segment $[\mathbf{g}, \mathbf{c}]_u$. Determine the transition matrix B for this partition. Determine the three eigenvalues for the transition matrix. How do the eigenvalues compare with the eigenvalues for A?

8.31. Let $f : \mathbb{T}^2 \to \mathbb{T}^2$ be a hyperbolic toral automorphism and A the transition matrix for a Markov partition. Let $h : \Sigma_A \to \mathbb{T}^2$ be the semi-conjugacy. Prove that $\mathbf{s}$ is periodic for σ_A if and only if $h(\mathbf{s})$ is periodic for f_A. Explain why the periods could be different.

8.32. Let $f_A : \mathbb{T}^2 \to \mathbb{T}^2$ be a hyperbolic toral automorphism and B the transition matrix for a Markov partition. Using the fact that f_A is topologically transitive, prove that B is irreducible.

Zeta Function for a Hyperbolic Toral Automorphism

8.33. Let $f : \mathbb{T}^n \to \mathbb{T}^n$ be a C^1 diffeomorphism with a hyperbolic fixed point $\mathbf{p}$. Prove that the Lefschetz index of f at $\mathbf{p}$ is given as follows:

$$I_{\mathbf{p}}(f) = (-1)^u \Delta$$

where $u = \dim(\mathbb{E}^u_{\mathbf{p}})$ and $\Delta = 1$ provided $Df_{\mathbf{p}}|\mathbb{E}^u_{\mathbf{p}} \to \mathbb{E}^u_{\mathbf{p}}$ preserves orientation and $\Delta = -1$ provided this linear map reverses orientation.

8.34. (Zeta function via Markov partitions) Let $f : \mathbb{T}^2 \to \mathbb{T}^2$ be a hyperbolic toral automorphism and A the transition matrix for a Markov partition $\mathcal{M}$. Let $h : \Sigma_A \to \mathbb{T}^2$ be the semi-conjugacy. The zeta function of σ_A is rational by Theorem III.3.2. A. Manning (1971) proved that $\zeta_f(t)$ is a rational function by relating it to ζ_{σ_A}. Bowen (1978a) sketched the following modification of Manning's proof. Let $\mathcal{P}_k$ be the collection of families of k indices of distinct rectangles $\{i_1, \dots, i_k\}$ such that each $R_{i_j} \in \mathcal{M}$ and $R_{i_1} \cap \cdots \cap R_{i_k} \neq \emptyset$. For each such family, fix an ordering $\mathbf{i} = (i_1, \dots, i_k)$. For $\mathbf{i}, \mathbf{j} \in \mathcal{P}_k$, write $\mathbf{i} \to \mathbf{j}$ provided there is a permutation τ of $\{1, \dots, k\}$ so that $A_{i_\ell, j_{\tau(\ell)}} = 1$ for $1 \leq \ell \leq k$. Define the $\#(\mathcal{P}_k) \times \#(\mathcal{P}_k)$ matrix by

$$A(k)_{\mathbf{i},\mathbf{j}} = \begin{cases} 1 & \text{if } \mathbf{i} \to \mathbf{j} \quad \text{and } \tau \text{ is an even permutation,} \\ -1 & \text{if } \mathbf{i} \to \mathbf{j} \quad \text{and } \tau \text{ is an odd permutation, and} \\ 0 & \text{otherwise.} \end{cases}$$

Notice that $A(1) = A$ and $A(k) = 0$ for $k > 4$ because f is a diffeomorphism on a two-dimensional manifold. The parts of the problem below ask you to prove that

$$N_j(f) = \sum_{k \geq 1} (-1)^{k+1} \operatorname{tr}(A(k)^j). \qquad (*)$$

(a) Prove that if $\mathbf{p} \in \operatorname{int}(R_i)$ for some $R_i \in \mathcal{M}$ has period ℓ for f, then $h^{-1}(\mathbf{p})$ has period ℓ for σ_A. Prove that it does not correspond to a periodic point for $\sigma_{A(k)}$ for any $k > 1$. (Alternatively, $h^{-1}(\mathbf{p})$ does not contribute to the trace of $A(k)^j$ for any $k > 1$ and $j \geq 1$.) Prove that in this case, $h^{-1}(\mathbf{p})$ contributes 1 to the right-hand side of $(*)$ for the j which are multiples of ℓ and 0 for other j.

(b) Assume $\mathbf{p} \in R_{i_1} \cap R_{i_2}$, $\mathbf{p}$ is not in any combination of three rectangles in $\mathcal{M}$, and $\mathbf{p}$ is a fixed point for f. Prove that $\mathbf{i} \to \mathbf{i}$ where $\mathbf{i} = (i_1, i_2)$. (i) If the permutation τ for $\mathbf{i}$ is even, prove that $h^{-1}(\mathbf{p})$ contains exactly two points, each of which is fixed by σ_A. Prove that $A(2)_{\mathbf{i},\mathbf{i}} = 1$ and $\mathbf{p}$ does not correspond to a periodic point of $\sigma_{A(k)}$ for a $k > 2$. In this case, prove that the points in $h^{-1}(\mathbf{p})$ contribute 1 to the right-hand side of $(*)$ for all j. (ii) If the permutation τ is odd, prove that $h^{-1}(\mathbf{p})$ contains exactly two points, which is an orbit of period two for σ_A. Prove that $A(2)^j_{\mathbf{i},\mathbf{i}} = (-1)^j$ and $\mathbf{p}$ does not correspond to a periodic point for $\sigma_{A(k)}$ for a $k > 2$. Prove that the points in $h^{-1}(\mathbf{p})$ contribute 1 to the right-hand side of $(*)$ for all j.

(c) Assume that $\mathbf{p}$ is a fixed point for f. Prove that the points in $h^{-1}(\mathbf{p})$ contribute 1 to the right-hand side of $(*)$ for all j.

(d) Assume that $\mathbf{p}$ has period ℓ for f. Prove that the points in $h^{-1}(\mathbf{p})$ contribute 1 to the right-hand side of $(*)$ for the j which are multiples of ℓ and 0 for other j. Conclude that $(*)$ is true for all j.

(e) Using part (d), prove that $\zeta_f(t)$ is rational.

Attractors

8.35. (Theorem 6.1) Let $f : M \to M$ be a diffeomorphism on a finite-dimensional manifold M. Let A be a compact invariant set for f. Prove that A is an attracting set if and only if there are arbitrarily small neighborhoods V of A such that (i) V is positively invariant and (ii) $\omega(\mathbf{p}) \subset A$ for all $\mathbf{p} \in V$. (Note that the neighborhoods V can be taken to be compact since both M and A are compact.)

8.36. (Theorem 6.2) Let Λ be an attracting set for f. Assume either that (i) $\mathbf{p} \in \Lambda$ is a hyperbolic periodic point or (ii) Λ has a hyperbolic structure and $\mathbf{p} \in \Lambda$. Prove that $W^u(\mathbf{p}, f) \subset \Lambda$.

Solenoids

8.37. Construct a Markov partition for the map $g : S^1 \to S^1$ defined by $g(\theta) = 2\theta \mod 1$. Note in this case there is no contracting direction so $W^u(p, R_i) = R_i$ and $W^s(p, R_i) = \{p\}$. Thus, the fourth condition for a Markov partition becomes the following: if $\text{int}(R_i) \cap g^{-1}(\text{int}(R_j)) \neq \emptyset$ and $p \in \text{int}(R_i) \cap g^{-1}(\text{int}(R_j))$, then $g(\text{int}(R_i)) \cap \text{int}(R_j) = \text{int}(R_j)$ and $g^{-1}(g(p)) \cap \text{int}(R_i) = \{p\}$.

8.38. (a) Construct a Markov partition for the solenoid.

(b) Let $h : \Sigma_A \to \Lambda$ be the map from the subshift of finite type determined by the transition matrix A for the Markov partition to the attractor Λ for the solenoid diffeomorphism defined by

$$h(\mathbf{s}) = \bigcap_{n=0}^{\infty} \text{cl}\,\Big(\bigcap_{j=-n}^{n} f^{-j}(\text{int}(R_{s_j}))\Big).$$

Prove that h is at most two to one.

8.39. Let $\text{comp}_{\mathbf{p}}(S)$ denote the connected component of S containing $\mathbf{p}$. Let $N = S^1 \times D^2$ and $f : N \to N$ be the solenoid map. For $\mathbf{p} = (\theta, z)$, show that $\text{comp}_{\mathbf{p}}(W^s(\mathbf{p}, f) \cap N) = D(\theta)$, and for $\theta_1 < \theta < \theta_2$ that

$$\text{comp}_{\mathbf{p}}(W^u(\mathbf{p}, f) \cap D([\theta_1, \theta_2])) = \bigcap_{n \geq 1} \text{comp}_{\mathbf{p}}(D([\theta_1, \theta_2]) \cap f^n(N)).$$

Here $D([\theta_1, \theta_2]) = (\theta_1, \theta_2) \times D^2$. Hint: Show the inclusion both ways. You may assume that the left side is the one-to-one differentiable image of an interval.

8.40. Consider the map given in polar coordinates by

$$F\begin{pmatrix} \theta \\ r \end{pmatrix} = \begin{pmatrix} 8\,\theta - \dfrac{\pi}{8} \\ 2 + \dfrac{1}{16}r + \dfrac{1}{16}\sin(\theta) \end{pmatrix}$$

for $0 \leq \theta \leq \dfrac{\pi}{2}$ and $1 \leq r \leq 3$. (Notice this definition is only for part of the plane.) Let

$$V_L = \{(\theta, r) : 0 \leq \theta \leq 3\,\pi/32,\ 1 \leq r \leq 3\}$$
$$V_R = \{(\theta, r) : \pi/4 \leq \theta \leq 11\,\pi/32,\ 1 \leq r \leq 3\}.$$

Prove that the maximal invariant set in $V_L \cup V_R$ is hyperbolic.

DA Attractor

8.41. Prove that these exists a neighborhood V for the DA-diffeomorphism as specified in the proof of Theorem 8.1: $f(V) \supset V$, $a_{22} > 1$ for $\mathbf{q} \in V$, and $a_{22} < 1$ for $\mathbf{q} \notin f(V)$.

Morse-Smale Diffeomorphisms

8.42. Let $V : \mathbb{T}^n \to \mathbb{T}^n$ be a C^2 function such that each critical point is nondegenerate, i.e., at each point $\mathbf{x}$ where $\text{grad}(V)_{\mathbf{x}} = \mathbf{0}$, the matrix of second partial derivatives, $\dfrac{\partial^2 V}{\partial x_i \partial x_j}(\mathbf{x})$, has nonzero determinant. Prove that all the fixed points of the gradient vector field of V are hyperbolic.

8.43. Let $L : M \to \mathbb{R}$ be weak Liapunov function for the flow φ^t. If $\mathbf{x} \in \mathcal{R}(\varphi^t)$, prove that $L \circ \varphi^t(\mathbf{x})$ is a constant function of time.

8.44. (a) Let φ^t be a Morse-Smale flow with only fixed points and no periodic orbits. Prove that $f(\mathbf{x}) = \varphi^1(\mathbf{x})$ is a Morse-Smale diffeomorphism.
(b) Let φ^t be a Morse-Smale flow that has a periodic orbit. Prove that $f(\mathbf{x}) = \varphi^1(\mathbf{x})$ is not a Morse-Smale diffeomorphism.
(c) Let $f : M \to M$ be a Morse-Smale diffeomorphism. Let φ^t be the suspension of f. Prove that φ^t is a Morse-Smale flow without any fixed points and only periodic orbits.

8.45. Consider the set of C^1 diffeomorphisms on S^1 with the C^1 topology, $\text{Diff}^1(S^1)$. Prove that any $f \in \text{Diff}^1(S^1)$ can be approximated by a Morse-Smale diffeomorphism, i.e., prove that the set of Morse-Smale diffeomorphisms is dense in $\text{Diff}^1(S^1)$.

8.46. Prove that the set of Morse-Smale diffeomorphisms is not dense in $\text{Diff}^1(\mathbb{T}^2)$ or $\text{Diff}^1(S^2)$ where S^2 is the two sphere. Hint: Consider the examples of diffeomorphisms we have given with infinitely many periodic points.

8.47. Consider the two torus, $\mathbb{T}^2$. The Beti numbers of $\mathbb{T}^2$ are $\beta_0 = \beta_2 = 1$, and $\beta_1 = 2$. Assume that there is a diffeomorphism f on $\mathbb{R}^2$ with one source, $c_2 = 1$, and two sinks, $c_0 = 2$. Using the Morse inequalities, determine how many saddles f must have.

8.48. Which of the following diffeomorphisms are structurally stable? Prove your answer.
(a) $f : S^1 \to S^1$ defined by $f(\theta) = \theta + \alpha \bmod(2\pi)$, where α/π is irrational.

(b) $g : S^1 \to S^1$ defined by $f(\theta) = \theta + \alpha \bmod(2\pi)$, where α/π is rational.
(c) $h_\alpha : S^1 \to S^1$ defined by $h_\alpha(\theta) = \theta + \alpha\sin(\theta) \bmod(2\pi)$, where $\alpha \in (0, 0.2)$. (Is h_α structurally stable for all these values of α, some, or none?)

8.49. Assume that for each $t \in [a, b]$, $f_t : M \to M$ is a diffeomorphism which is structurally stable. Prove that f_a and f_b are topologically conjugate.

8.50. Let $f : \mathbb{R} \to \mathbb{R}$ be defined by $f(x) = x + 1$. Show there is an $\epsilon > 0$ such that whenever g is a C^1 map which satisfies

$$|f(x) - g(x)| < \epsilon \qquad \text{and}$$
$$|f'(x) - g'(x)| < \epsilon$$

for all $x \in \mathbb{R}$, then f and g are *differentiably conjugate.* That is, there is a C^1 diffeomorphism $h : \mathbb{R} \to \mathbb{R}$ such that $f \circ h = h \circ g$.

8.51. Prove Lemma 12.1.

8.52. Assume the flow φ^t is gradient-like with Liapunov function L.
(a) Prove that $\Omega(\varphi^t) = \text{Fix}(\varphi^t)$.
(b) Assume that $L(\text{Fix}(\varphi^t))$ is a totally diconnected subset of $\mathbb{R}$. Prove that $\mathcal{R}((\varphi^t) = \text{Fix}(\varphi^t)$.

CHAPTER IX
Measurement of Chaos in Higher Dimensions

In this chapter we continue the discussion of measurements of chaos started in Chapter III. As mentioned there, topological entropy is a quantity which is used in the mathematical study of dynamical systems. A newcomer to the subject often has difficulty understanding exactly what is being measured from the definition of topological entropy itself. We calculate the entropy of a few simple situations. We also relate the entropy to horseshoes caused by homoclinic points and Markov partitions for Anosov diffeomorphisms. Hopefully, after seeing these examples, the reader will gain an understanding of the concept.

The next section returns to the concept of Liapunov exponents. These are defined in one dimension where the situation is fairly simple. In this chapter we give the definition in higher dimensions. In this discussion, we merely give the definitions and statements of theorems. The goal is for the reader to be aware of the concept. We leave a full treatment to the references.

The next section discusses the Sinai-Ruelle-Bowen measure for an attractor. Our treatment in this book of invariant measures in general and this measure in particular is very sketchy. Again, we defer to the references for a more complete treatment.

The last measurement of chaos relates to the geometry of the invariant set. We define fractal dimension (box dimension, or Hausdorff dimension) and calculate it for a few examples. Again the goal is for the reader to be aware of the concept without giving all the theory or details.

9.1 Topological Entropy

As stated when discussing chaos in Section 3.5, topological entropy is a quantitative measurement of how chaotic the map is. In fact, it is determined by how many "different orbits" there are for a given map (or flow). To get an intuitive idea of the concept, assume that you cannot distinguish points which are closer together than a given resolution ϵ. Then, two orbits of length n can be distinguished provided there is some iterate between 0 and n for which they are distance greater than ϵ apart. Let $r(n, \epsilon, f)$ be the number of such orbits of length n that can be so distinguished. The entropy for a given ϵ, $h(\epsilon, f)$, is the growth rate of $r(n, \epsilon, f)$ as n goes to infinity. The limit of $h(\epsilon, f)$ as ϵ goes to 0 is the entropy of f, $h(f)$. The key idea in this sequence of limits is the growth rate of the number of orbits of length n that are at least ϵ apart. Now, we give a more careful definition.

Definition. Let $f : X \to X$ be a continuous map on the space X with metric d. A set $S \subset X$ is called *(n, ϵ)-separated for f* for n a positive integer and $\epsilon > 0$ provided that for every pair of distinct points $\mathbf{x}, \mathbf{y} \in S$, $\mathbf{x} \neq \mathbf{y}$, there is at least one k with $0 \leq k < n$ such that $d(f^k(\mathbf{x}), f^k(\mathbf{y})) > \epsilon$. Another way of expressing this concept is to introduce the distance

$$d_{n,f}(\mathbf{x}, \mathbf{y}) = \sup_{0 \leq j < n} d(f^j(\mathbf{x}), f^j(\mathbf{y})).$$

Using this distance, a set $S \subset X$ is (n, ϵ)-separated for f provided $d_{n,f}(\mathbf{x}, \mathbf{y}) > \epsilon$ for every pair of distinct points $\mathbf{x}, \mathbf{y} \in S$, $\mathbf{x} \neq \mathbf{y}$.

The *number of different orbits of length n (as measured by ϵ)* is defined by

$$r(n, \epsilon, f) = \max\{\#(S) : S \subset X \text{ is a } (n, \epsilon)\text{-separated set for } f\},$$

where $\#(S)$ is the number (cardinality) of elements in S.

We want to measure the growth rate of $r(n, \epsilon, f)$ as n increases, so we define

$$h(\epsilon, f) = \limsup_{n \to \infty} \frac{\log(r(n, \epsilon, f))}{n}.$$

If $r(n, \epsilon, f) = e^{n\tau}$, then $h(\epsilon, f) = \tau$; thus, $h(\epsilon, f)$ measures the "exponent" of the manner in which $r(n, \epsilon, f)$ grows with respect to n.

Note that $r(n, \epsilon, f) \geq 1$ for any pair (n, ϵ), so $0 \leq h(\epsilon, f) \leq \infty$. We show in Lemma 1.10 that on a compact metric space, $h(\epsilon, f) < \infty$.

Finally, we consider the way that $h(\epsilon, f)$ varies as ϵ goes to 0, and define the *topological entropy of f* as

$$h(f) = \lim_{\epsilon \to 0, \epsilon > 0} h(\epsilon, f).$$

In the next subsection we introduce another way of counting orbits, so we distinguish the notation for the number of (n, ϵ)-separated orbits by a subscript, i.e., we use the notation $r_{sep}(n, \epsilon, f)$ for $r(n, \epsilon, f)$, $h_{sep}(\epsilon, f)$ for $h(\epsilon, f)$, and $h_{sep}(f)$ for $h(f)$

REMARK 1.1. Note for $0 < \epsilon_2 < \epsilon_1$, $r(n, \epsilon_2, f) \geq r(n, \epsilon_1, f)$, so $h(\epsilon, f)$ is a monotone function of ϵ, $h(\epsilon_2, f) \geq h(\epsilon_1, f)$, the limit defining $h(f)$ exists, and $0 \leq h(\epsilon, f) \leq h(f) \leq \infty$ for all $\epsilon > 0$. If f is C^1 on a compact space, then it has been proved that $h(f) < \infty$. See Bowen (1971, 1978a).

REMARK 1.2. The concept of topological entropy was originally introduced by Adler, Konheim, and McAndrew (1965) using a very different definition involving covers of the set by open sets. This definition is given in the next subsection. The definition we use was introduced by Bowen (1970b). Good general references for topological entropy are Bowen (1978a, 1970b), Walters (1982), and Alsedà, Llibre, and Misiurewicz (1993). This last reference treats the case of the entropy of a one-dimensional map quite extensively. It also treats both the definition we have given with separated sets and the original definition using refinements of open sets.

REMARK 1.3. It is also possible to define a measure theoretic entropy $h_\mu(f)$ for an invariant measure μ. Then, under the correct hypothesis, $h(f) = \sup\{h_\mu(f)\}$, where the supremum is taken over all invariant measures μ. See Bowen (1975b) or Walters (1982) for a discussion of this type of entropy and its connection with topological entropy.

It is hard to get a very good idea of what entropy means directly from the above definition. Throughout the section, we determine the entropy of a few examples which helps give some feeling for its meaning. The first example is the "doubling map" which is easy to show that it has entropy equal to log(2). We state this result as a proposition.

Proposition 1.1. *Let $f : S^1 \to S^1$ have a covering map $F : \mathbb{R} \to \mathbb{R}$ given by*

$$F(x) = 2x.$$

This map is called the doubling map. The distance on S^1 is the one inherited from $\mathbb{R}$ by taking x to x mod 1. Thus, points near 1 are close to points near 0. (This can also be *considered as the map $g(z) = z^2$ on the circle, where we use complex notation. If this representation is used, the map is often called the squaring map.) Then, $h(f) = \log(2)$.*

PROOF. Two points x and y stay within ϵ of each other for $n-1$ iterates of f if and only if $|x-y| \le \epsilon 2^{-n+1}$ (as points in $\mathbb{R}$). If we put points exactly distance $\epsilon 2^{-n+1}$ apart in $[0,1)$, then there are at most $[\epsilon^{-1}2^{n-1}]$ points. However, the last point to the right is close to the first point on the left when considered modulo 1 in S^1, so there are $[\epsilon^{-1}2^{n-1}] - 1$ points in S^1. These points can be spread apart slightly to make these points actually (n,ϵ)-separated. Therefore, $r(n,\epsilon,f) = [\epsilon^{-1}2^{n-1}] - 1$, where $[a]$ is the integer part of a. Then,

$$\begin{aligned} h(\epsilon, f) &= \limsup_{n\to\infty} \frac{\log([\epsilon^{-1}2^{n-1}] - 1)}{n} \\ &= \limsup_{n\to\infty} \frac{\log(\epsilon^{-1}) + (n-1)\log(2)}{n} \\ &= \log(2) \end{aligned}$$

for any $\epsilon > 0$, so $h(f) = \log(2)$ as claimed. □

We give a few basic results which are used to help calculate the entropy for specific maps. The first such result relates the entropy of a map f with a power f^k of f.

Theorem 1.2. *Assume f is uniformly continuous (or X is compact) and k is an integer with $k \ge 1$. Then, the entropy of f^k is equal to k times the entropy of f, $h(f^k) = k\, h(f)$.*

PROOF. The points of the orbits considered for $r(n,\delta,f^k)$ constitute a subset of those considered for $r(nk,\delta,f)$,

$$\{f^{ki}(\mathbf{y}) : 0 \le i < n\} \subset \{f^i(\mathbf{y}) : 0 \le i < nk\},$$

so any (n,δ)-separated set for f is also an (nk,δ)-separated set for f^k, and we have that $r(n,\delta,f) \le r(nk,\delta,f^k)$. By uniform continuity, given $\epsilon > 0$, there is $\delta_\epsilon > 0$ such that if $d(\mathbf{x},\mathbf{y}) \le \delta_\epsilon$, then $d(f^j(\mathbf{x}), f^j(\mathbf{y})) \le \epsilon$ for $0 \le j < k$. Therefore, any (n,δ_ϵ)-separated set for f^k is also a (nk,ϵ)-separated set for f, or $r(n,\delta_\epsilon,f^k) \le r(nk,\epsilon,f)$, where δ_ϵ is uniform in n. Combining these two inequalities,

$$\frac{1}{n} r(nk,\delta_\epsilon,f) \le \frac{1}{n} r(n,\delta_\epsilon,f^k) \le \frac{1}{n} r(nk,\epsilon,f),$$

and taking the limits in n and then ϵ

$$\begin{aligned} &k\, h(\delta_\epsilon, f) \le h(\delta_\epsilon, f^k) \le k\, h(\epsilon, f), \\ &k\, h(f) \le h(f^k) \le k\, h(f). \end{aligned}$$

This proves the theorem. □

REMARK 1.4. We leave to the exercises to prove that if f is a homeomorphism, then $h(f^{-1}) = h(f)$. See Exercise 9.9. From equality it follows that $h(f^k) = |k|\, h(f)$ for any integer k.

The next two results relate the entropy of a map with the entropy on invariant subsets: the first result is in terms of disjoint invariant sets and the second one is in terms of the nonwandering set.

Theorem 1.3. *Let f be a continuous map on X. Assume $X = X_1 \cup \cdots \cup X_k$ is a decomposition into disjoint closed invariant subsets which are a positive distance apart. Then,*

$$h(f) = \max_i h(f|X_i).$$

PROOF. If ϵ is smaller than the distance between the subsets X_i, then

$$r(n, \epsilon, f) = \sum_i r(n, \epsilon, f|X_i).$$

Thus, for each n and each j, we must have

$$r(n, \epsilon, f|X_j) \le r(n, \epsilon, f) \le k \max_i r(n, \epsilon, f|X_i).$$

In passing to the limit in calculating $h(\epsilon, f)$, for each j we have

$$\begin{aligned} h(\epsilon, f|X_j) &\le h(\epsilon, f) \\ &= \limsup_{n\to\infty} \frac{\log(r(n, \epsilon, f))}{n} \\ &\le \limsup_{n\to\infty} \frac{\log(k \max_i r(n, \epsilon, f|X_i))}{n} \\ &\le \max_i h(\epsilon, f|X_i), \end{aligned}$$

so $h(\epsilon, f) = \max_j h(\epsilon, f|X_j)$. By taking a countable number of $\epsilon_i > 0$ converging to 0, one j_0 must have

$$h(\epsilon_i, f) = h(\epsilon_i, f|X_{j_0})$$

for infinitely many of the ϵ_i, so

$$h(f) = h(f|X_{j_0}).$$

Since $h(f|X_j) \le h(f)$ for all j, this proves the theorem. □

The next result of Bowen (1970b) says that all the entropy is contained in the nonwandering set, i.e., the wandering orbits do not contribute to the entropy.

Theorem 1.4. *Let $f : X \to X$ be a continuous map on a compact metric space X. Let $\Omega \subset X$ be the nonwandering points of f. Then, the entropy of f equals the entropy of f restricted to its nonwandering set, $h(f) = h(f|\Omega)$.*

The proof of this theorem is delayed until the next subsection because of its length and the relative greater complexity of its argument; it also uses a slightly different definition of topological entropy in addition to that of separated sets. We mention that this result shows that any Morse-Smale diffeomorphism, or any map whose nonwandering set is a finite set of points, has zero entropy as stated in the following result.

Theorem 1.5. *Let X be a compact metric space and $f : X \to X$ be a continuous map for which $\Omega(f)$ is a finite number of periodic orbits. (For example, f could be a Morse-Smale diffeomorphism.) Then, the entropy of f is zero, $h(f) = 0$.*

PROOF. First, $h(f) = h(f|\Omega)$ by Theorem 1.4. Then, by Theorem 1.3, $h(f|\Omega)$ is the maximum of the entropy on the individual periodic orbits. However, the entropy of a single periodic orbit is zero. □

REMARK 1.5. Let $F_\mu(x) = \mu x(1-x)$, and μ_k be taken in the range where F_{μ_k} has one attracting periodic orbit of period 2^k and repelling periodic orbits of periods 2^j for $0 \le j < k$. Since $\mu_k < 4$, F_{μ_k} preserves the interval $I = [0,1]$. Theorem 1.5 implies that the entropy $h(F_{\mu_k}|I) = 0$. In Theorem 1.6 below, we give a direct proof of this fact for $1 < \mu < 3$ without using Theorem 1.5 (and so without the proof of Theorem 1.4). We include the proof of this special case in addition to the proof of the general case given in Theorem 1.4 because its proof is more concrete. In the calculation of entropy, it is necessary to calculate the number of (n,ϵ)-separated wandering orbits which start near the repelling fixed point 0 and end up near the attracting fixed point p_μ. Because these orbits can be partitioned by the iterate when they leave a neighborhood of 0, the number of these orbits grows linearly in n. Because linear growth adds nothing to the entropy, $h(F_\mu) = 0$.

REMARK 1.6. In the proof of Theorem 1.4 we show that the wandering orbits contribute at most a factor which grows in a polynomial fashion in the length n of the orbits considered. A term which grows polynomially in the length n does not contribute to the entropy, so this proves the theorem. The reason that the growth is possibly polynomial rather than linear is that an orbit can make several transitions from a neighborhood of the nonwandering set. For example, for the horseshoe on S^2, an orbit could start near the source at ∞ and proceed near the horseshoe itself and finally leave and go near the fixed point sink. Because there are two times at which these orbits leave a neighborhood of the nonwandering set (the time it leaves a neighborhood of ∞ and the time it leaves a neighborhood of the horseshoe), the number of such orbits can grow quadratically in n. In the general proof, we do not use any decomposition of the nonwandering set. (However, see Theorems X.1.3 and X.4.4.) Because we proceed without any special knowledge of the decomposition of the nonwandering set, we must consider wandering orbits which make several transitions between a neighborhood of the nonwandering set. The proof shows that each of these transitions contributes a possible factor of n to the growth rate of $r(n,\epsilon,f)$ and so the the total number of these (n,ϵ)-separated grows at a polynomial rate. Such a growth rate of $r(n,\epsilon,f)$ contributes nothing to $h(f)$.

REMARK 1.7. On the whole real line $r(n,\epsilon,F_\mu) = \infty$ (because if $x, y < 0$, then the orbits $F^j(x)$ and $F^j(y)$ diverge as j goes to infinity) so $h(F_\mu) = \infty$. This illustrates the fact that the entropy is not a good measurement of the chaotic nature of a map on a noncompact set. In fact, let $id : \mathbb{R} \to \mathbb{R}$ be the identity map, $id(x) = x$ for all x. The identity map is certainly not chaotic, but $r(n,\epsilon,id) = \infty$ for all n and $\epsilon > 0$, so $h(id) = \infty$.

As stated above, we give a direct proof that $h(F_\mu) = 0$ in the special case when F_μ when it has only two fixed points and no other nonwandering points. This proof is more concrete and less complex than the general proof of Theorem 1.4, and we also obtain a better upper bound on the number of (n,ϵ)-separated orbits.

Proposition 1.6. *Let $F_\mu(x) = \mu x(1-x)$. Then, the entropy $h(F_\mu|[0,1]) = 0$ for $1 < \mu < 3$.*

PROOF. For $1 < \mu < 3$, F_μ has a single repelling fixed point at 0, a single attracting fixed point $p_\mu = 1 - 1/\mu$, and no other periodic points or nonwandering points. We write F for F_μ, and p for p_μ.

Let $a = F(0.5)$ be the image of the critical point and $J = [0,a]$. Notice that $F([0,1]) = J$. The reader can easily verify that $h(F|[0,1]) = h(F|J)$. (See Exercise 9.13.) We use $F|J$ because it has unique inverse images of points near 0. Let $H = F|J$ to simplify notation.

The point p is attracting, so there exists $\epsilon_0 > 0$ such that if x and y are two points within a distance ϵ_0 of p, then for all $j \geq 0$, (i) $H^j(x)$ and $H^j(y)$ stay within a distance ϵ_0 of p, and (ii) $|H^j(x) - H^j(y)| \leq |x - y|$. (Take $\epsilon_0 > 0$ such that $|H'(z)| < 1$ for all points $z \in (p - \epsilon_0, p + \epsilon_0)$, and apply the Mean Value Theorem.) For $\epsilon_0 > 0$ possibly smaller, if x and y are within ϵ_0 of 0, then for all $j \leq 0$, (i) $H^j(x)$ and $H^j(y)$ stay within a distance ϵ_0 of 0, and (ii) $|H^j(x) - H^j(y)| \leq |x - y|$. (Notice that this would not be true as stated for $F|[0,1]$.)

The idea of the proof is that as n grows, the (n, ϵ_0)-separated orbits which make the transition from near 0 to near p can be counted by the iterate at which they start to make the transition. Therefore, the number of these orbits grows at most like a multiple of n. Since the number of orbits which remain near either 0 or p is bounded, $r(n, \epsilon_0, H)$ grows linearly in n and $h(\epsilon_0, H) = 0$.

We proceed to make the above idea precise. Now, fix a $0 < \epsilon \leq \epsilon_0$. Let $\mathcal{N}_a = H((p - \epsilon/2, p + \epsilon/2))$ be the open interval about the attracting fixed point p, and let $\mathcal{N}_r = H^{-1}([0, \epsilon/2))$ be the open (in J) interval about the repelling fixed point 0. Let $U = \mathcal{N}_r \cup \mathcal{N}_a$ and $U^c = J \setminus U$. Then, $\mathcal{D} = [0, \epsilon/2] \setminus \mathcal{N}_r$ is a fundamental domain of the unstable manifold of 0.

Next, there is a positive integer n_0 such that if $H^j(x) \in U^c$ for $0 \leq j < n$, then $n \leq n_0$. In particular, if $x \in \mathcal{D}$, then $H^j(x) \in \mathcal{N}_a$ for $j \geq n_0$, and $\bigcup_{i=0}^{n_0-1} H^i(\mathcal{D}) \supset U^c$.

Because U^c is compact and all points are wandering, there is a $\beta > 0$ with $2\beta \leq \epsilon$ such that $H^j([y - \beta, y + \beta]) \cap [y - \beta, y + \beta] = \emptyset$ for all $y \in U^c$ and all $j \geq 1$.

Let $E_{dense}(\beta, \mathcal{D}) \subset \mathcal{D}$ be a set such that

$$E_{dense}(\beta, U^c) = \bigcup_{i=0}^{n_0-1} H^i(E_{dense}(\beta, \mathcal{D}))$$

(i) is β-dense in U^c and (ii) for any $x \in U^c$, there is a $y \in E_{dense}(\beta, U^c)$ such that $d_{n_0,H}(x, y) < \beta$. Thus, for these x and y, $|G^i(x) - G^i(y)| < \beta$ as long as $G^i(x) \in U^c$. Finally, let

$$G = \{0, p\} \cup E_{dense}(\beta, U^c).$$

To estimate $r(n, \epsilon, J)$, we define a map

$$\varphi_n : J \to G^n$$

as follows.

Case (i): If $x \in \mathcal{N}_a$, then $H^i(x) \in \mathcal{N}_a$ for all $i \geq 0$, so we let $\varphi_n(x) = (p, \ldots, p)$.

Case(ii): If $x \in \mathcal{N}_r$, then there is a $j \leq n$ such that $H^i(x) \in \mathcal{N}_r$ for $0 \leq i < j$ and $H^j(x) \in \mathcal{D}$ if $j < n$. If $j < n$, let y_j be a choice of a point in $E_{dense}(\beta, \mathcal{D})$ such that $d_{n_0,H}(H^j(x), y_j) < \beta$. Let $\varphi_n(x) = (y_0, \ldots, y_{n-1})$ where

$$y_i = \begin{cases} 0 & \text{for } 0 \leq i < j \\ H^{i-j}(y_j) & \text{for } j \leq i < \min\{n, j + n_0\} \\ p & \text{for } \min\{n, j + n_0\} \leq i < n. \end{cases}$$

Case(iii): If $x \in U^c$, then let y_0 be a choice of a point in $E_{dense}(\beta, U^c)$ such that $d_{n_0,H}(x, y_0) < \beta$. Let $\varphi_n(x) = (y_0, \ldots, y_{n-1})$, where

$$y_i = \begin{cases} H^i(y_0) & \text{for } 0 \leq i < n_0 \\ p & \text{for } n_0 \leq i < n. \end{cases}$$

We leave to the reader to check the following claim.

Claim 1. *If $x \in J$ and $\varphi_n(x) = (y_0, \dots, y_{n-1})$, then $|H^i(x) - y_i| < \epsilon/2$ for $0 \le i < n$.*

Take an n with $n > n_0$. Let $E(n, \epsilon)$ be a set with the maximal number of (n, ϵ)-separated points in J, $\#(E(n, \epsilon)) = r(n, \epsilon, J, H)$.

Claim 2. *The map $\varphi_n | E(n, \epsilon)$ is one to one.*

PROOF. Assume $\varphi_n(x) = \varphi_n(z) = (y_0, \dots, y_{n-1})$ for $x, z \in E(n, \epsilon)$. Then,

$$|H^i(x) - H^i(z)| \le |H^i(x) - y_i| + |y_i - H^i(z)| < \epsilon$$

for $0 \le i < n$. Since $E(n, \epsilon)$ is (n, ϵ)-separated, $x = z$. □

Claim 3. *The cardinality of $\varphi_n(E(n, \epsilon))$ is less than $(n + 1)\, \#(E_{dense}(\beta, U^c))$.*

PROOF. An n-tuple $(y_0, \dots, y_{n-1}) \in \varphi_n(E(n, \epsilon))$ has the form

$$y_i = \begin{cases} 0 & \text{for } 0 \le i < j \\ y_j & \text{for } j \le i < j + k \\ p & \text{for } j + k \le i < n \end{cases}$$

where $k = \min\{n - j, n_0\}$ if $j \ge 1$, and $k = \min\{n - j, n_0\} = n_0$ or 0 if $j = 0$.

To count these n-tuples, we divide them up by the number k. If $k = \min\{n, j + n_0\} > 0$, then there are at most $\#(E_{dense}(\beta, U^c))$ choices for y_j. As j varies, $0 \le j < n$, we have at most $n\, \#(E_{dense}(\beta, U^c))$ such n-tuples. The only other cases are when $j = n$ or $j + k = 0$. These contribute two more n-tuples, $(0, \dots, 0)$ and $(p, \dots, p)$. Thus,

$$\begin{aligned} \#(\varphi_n(E(n, \epsilon))) &\le n\, \#(E_{dense}(\beta, U^c)) + 2 \\ &\le (n + 1) \#(E_{dense}(\beta, U^c)) \end{aligned}$$

as claimed. □

To calculate the entropy,

$$\begin{aligned} r(n, \epsilon, J, H) &= \#(E(n, \epsilon)) \\ &= \#(\varphi_n(E(n, \epsilon))) \\ &\le (n + 1) \#(E_{dense}(\beta, U^c)). \end{aligned}$$

This quantity grows linearly with n, so does not contribute to the entropy:

$$\begin{aligned} h(\epsilon, H) &\le \limsup_{n \to \infty} \frac{\log(n + 1) + \log(\#(E_{dense}(\beta, U^c)))}{n} \\ &= 0. \end{aligned}$$

Therefore, $h(\epsilon, H) = 0$ and $h(F) = h(H) = 0$, as claimed in the theorem. □

The next two theorems show that the topological entropy of two maps which are topologically conjugate are equal. In fact, the second result is for maps which are only semi-conjugate by a map which is uniformly finite to one.

Theorem 1.7. *Let X and Y be metric spaces with metrics d and d', respectively. Let $F : X \to X$ and $f : Y \to Y$ be semi-conjugate by $k : X \to Y$.*
(a) Assume X and Y are compact and the semi-conjugacy k is onto. Then, $h(F) \geq h(f)$.
(b) If k is one to one (but not necessarily onto), then $h(F) \leq h(f)$.

PROOF. By uniform continuity of k, given $\epsilon > 0$ there is a $\delta > 0$ such that $d(\mathbf{x}_1, \mathbf{x}_2) \geq \delta$ whenever $d'(k(\mathbf{x}_1), k(\mathbf{x}_1)) \geq \epsilon$. Let $E(n, \epsilon, f) \subset Y$ be a maximal (n, ϵ)-separated set for f, i.e., one with $\#(E(n, \epsilon, f)) = r(n, \epsilon, f)$. Form the set $E(n, \delta, F) \subset X$ by taking one $\mathbf{x} \in k^{-1}(\mathbf{y})$ for each $\mathbf{y} \in E(n, \epsilon, f)$. Thus, $\#(E(n, \delta, F)) = \#(E(n, \epsilon, f))$. Then, $E(n, \delta, F)$ is a (n, δ)-separated set for F by the property of uniform continuity of k mentioned above. Therefore,

$$r(n, \delta, F) \geq \#(E(n, \delta, F)) = \#(E(n, \epsilon, f)) = r(n, \epsilon, f).$$

From this it follows that $h(\delta, F) \geq h(\epsilon, f)$ and $h(F) \geq h(f)$, as desired. This proves part (a). We leave part (b) to the reader. □

REMARK 1.8. If two maps f and g are conjugate on invariant compact subsets (or conjugate by a uniformly continuous homeomorphism), then their topological entropies on these subsets are equal. Thus, the topological entropy of the shift map σ_2 on Σ_2 is equal to that of F_μ on Λ_μ for $\mu > 4$. We show below that $h(\sigma_2) = \log(2)$, so we can deduce the value of the entropy for the quadratic map on Λ_μ for $\mu > 4$.

The next result gives a criterion for entropies of F and f to be equal when F is semi-conjugate to f by a map k. We say that k is *uniformly finite to one* provided $k^{-1}(\mathbf{y})$ has a finite number of points for each $\mathbf{y}$ and there is a bound C on the number of elements in $k^{-1}(\mathbf{y})$ which is independent of $\mathbf{y}$. The theorem says that if k is a uniformly finite to one semi-conjugacy from F to f, each of which are defined on compact sets, then the entropies of F and f are equal. This result can be used to calculate the entropy of F_4. This theorem is due to Bowen (1971). See de Melo and Van Strien (1993).

Theorem 1.8. *Assume $F : X \to X$ and $f : Y \to Y$ are continuous maps, where X and Y are compact metric spaces with metrics d and d', respectively. Assume $k : X \to Y$ is a semi-conjugacy from F to f that is onto and uniformly finite to one. Then, $h(F) = h(f)$.*

We delay the proof to the next subsection, and at this time apply it calculate the entropy of $F_4|[0,1]$.

Example 1.1. There is a semi-conjugacy k from the doubling map $D(y) = 2y \bmod 1$ to $F_4|[0,1]$ which is two to one. (This is shown in Example II.6.2.) The entropy of D is $\log(2)$ by Proposition 1.1, so by the above theorem the entropy of $F_4|[0,1]$ is also $\log(2)$.

We end the section by determining the entropy of a subshift of finite type. The first part of the theorem expresses the entropy of any subshift (and not just a subshift of finite type) in terms of the growth rate of the number of words of length n as n goes to infinity.

Theorem 1.9. *(a) Let $\sigma : \Sigma_N \to \Sigma_N$ be the full shift on N symbols (either one- or two-sided). Assume $X \subset \Sigma_N$ is a closed invariant subset, so $\sigma|X$ is a subshift. Let w_n be the number of words of length n in X, i.e.,*

$$w_n = \#\{(s_0, \ldots s_{n-1}) : s_j = x_j \text{ for } 0 \leq j < n \text{ for some } \mathbf{x} \in X\}.$$

Then,

$$h(\sigma|X) = \limsup_{n \to \infty} \frac{\log(w_n)}{n}.$$

(b) Let A be a transition matrix on N symbols, so A is $N \times N$. Let $\sigma_A : \Sigma_A \to \Sigma_A$ be the associated subshift of finite type (either one or two-sided). Then, $h(\sigma_A) = \log(\lambda_1)$, where λ_1 is the real eigenvalue of A such that $\lambda_1 \geq |\lambda_j|$ for all the other eigenvalues λ_j of A.

PROOF. (a) We need to consider the number of (n, ϵ)-separated points for various ϵ. First, take $\epsilon = 2^{-1}$. Two points $\mathbf{s}, \mathbf{t} \in X$ are within 2^{-1} if and only if $s_0 = t_0$. For the first $n-1$ iterates, $\sigma^j(\mathbf{s})$ is within 2^{-1} of $\sigma^j(\mathbf{t})$ for $0 \leq j < n$ if and only if $s_j = t_j$ for $0 \leq j < n$. There are w_n choices of blocks $(s_0, \ldots, s_{n-1})$ (by the definition of w_n), so $r(n, 2^{-1}, \sigma|X) = w_n$. Thus,

$$h(2^{-1}, \sigma|X) = \limsup_{n\to\infty} \frac{\log(w_n)}{n}.$$

Next, we need to consider other values of ϵ. Since $h(\epsilon, \sigma|X)$ is monotonically increasing as ϵ decreases, it is enough to calculate the value for $\epsilon = 2^{-1}3^{-k}$. By Exercise 2.12, $d(\mathbf{s}, \mathbf{t}) \leq 2^{-1}3^{-k}$ if and only if $s_j = t_j$ for $0 \leq j \leq k$. Thus,

$$d(\sigma^i(\mathbf{s}), \sigma^i(\mathbf{t})) > 2^{-1}3^{-k}$$

for some $0 \leq i < n$ if and only if $s_j \neq t_j$ for some $0 \leq j < n + k$. Therefore,

$$r(n, 2^{-1}3^{-k}, \sigma|X) = r(n + k, 2^{-1}, \sigma|X),$$

and

$$\begin{aligned} h(2^{-1}3^{-k}, \sigma|X) &= \limsup_{n\to\infty} \frac{\log(r(n, 2^{-1}3^{-k}, \sigma|X))}{n} \\ &= \limsup_{n\to\infty} \frac{\log(r(n+k, 2^{-1}, \sigma|X))}{n} \\ &= \limsup_{n\to\infty} \left(\frac{n+k}{n}\right) \frac{\log(r(n+k, 2^{-1}, \sigma|X))}{n+k} \\ &= h(2^{-1}, \sigma|X). \end{aligned}$$

Since we have shown that $h(2^{-1}3^{-k}, \sigma|X) = h(2^{-1}, \sigma|X)$ for any positive k and $h(\epsilon, \sigma|X)$ is monotone in ϵ,

$$\begin{aligned} h(\sigma|X) &= h(2^{-1}, \sigma|X) \\ &= \limsup_{n\to\infty} \frac{\log(w_n)}{n}. \end{aligned}$$

This proves part (a).

(b) We prove the case for A irreducible using the Perron-Frobenius Theorem and leave to the exercises the proof of the general case. See Exercise 9.14. (The general case uses Proposition III.2.10.)

We first take the case where A is eventually positive, A^j is positive for $j \geq m$. Later, we discuss the proof of the general irreducible case.

By Lemma III.2.2, for a subshift of finite type with transition matrix A, w_n is the sum of all the entries in A^{n-1} which we denote by $\#(A^{n-1})$,

$$\begin{aligned} w_n &= \sum_{1\leq i\leq N, 1\leq j\leq N} (A^{n-1})_{i,j} \\ &= \#(A^{n-1}). \end{aligned}$$

Therefore, to calculate the entropy, we need to estimate $\#(A^{n-1})$. Letting $\mathbf{e}$ be the column vector with all entries being one, $\mathbf{e} = (1, \dots, 1)^{tr}$,

$$A^{n-1}\mathbf{e} = \left(\textstyle\sum_j (A^{n-1})_{i,j} \right),$$

so

$$\#(A^{n-1}) = \sum_i (A^{n-1}\mathbf{e})_i.$$

Applying the case of the Perron-Frobenius Theorem given in Theorem IV.9.10, (i) there is a positive real eigenvalue λ_1 and corresponding eigenvector $\mathbf{v}^1$ with all positive entries such that $\lambda_1 > |\lambda_j|$ for all the other eigenvalues λ_j of A, (ii) $A^{n-1}\mathbf{e}/|A^{n-1}\mathbf{e}|$ converges to $\mathbf{v}^1/|\mathbf{v}^1|$ as n goes to infinity, and (iii) there are positive constants C_1, C_2 such that $C_1\lambda_1^{n-1} \leq (A^{n-1}\mathbf{e})_i \leq C_2\lambda_1^{n-1}$ for $1 \leq i \leq N$ and $n > m$. (Remember that A^j is positive for $j \geq m$.) Summing on i, we get the estimate

$$NC_1\lambda_1^{n-1} \leq \#(A^{n-1}) = w_n \leq NC_2\lambda_1^{n-1}.$$

Because constant multiples do not affect the exponential growth rate,

$$\begin{aligned}
\log(\lambda_1) &= \lim_{n\to\infty} \frac{\log(N) + \log(C_1) + (n-1)\log(\lambda_1)}{n} \\
&= \lim_{n\to\infty} \frac{\log(NC_1\lambda_1^{n-1})}{n} \\
&\leq \limsup_{n\to\infty} \frac{\log(w_n)}{n} \\
&\leq \lim_{n\to\infty} \frac{\log(NC_2\lambda_1^{n-1})}{n} \\
&\leq \log(\lambda_1),
\end{aligned}$$

so

$$\begin{aligned}
h(\sigma_A) &= \limsup_{n\to\infty} \frac{\log(w_n)}{n} \\
&= \log(\lambda_1).
\end{aligned}$$

This completes the proof of the theorem in the case when A is eventually positive

Finally, consider the case when A is merely irreducible. It is proved in Gantmacher (1959) that by means of a permutation, A can be put into the following "cyclic" form:

$$A = \begin{pmatrix} 0 & A_{12} & 0 & \dots & 0 \\ 0 & 0 & A_{23} & \dots & 0 \\ \dots & \dots & \dots & \dots & \dots \\ 0 & 0 & 0 & \dots & A_{k-1,k} \\ A_{k1} & 0 & 0 & \dots & 0 \end{pmatrix}$$

where the blocks along the diagonal are square. By taking the kth power (where k is the number of blocks),

$$A^k = \operatorname{diag}(B_1, \dots, B_k),$$

where $B_j = A_{j,j+1} \cdots A_{n-1,n}A_{n,1} \cdots A_{j-1,j}$ is eventually positive for each j. Thus, a general irreducible matrix is a "combination" of a cyclic permutation of blocks of symbols and an eventually positive return map A^k on each of these blocks of symbols. All the B_j have the same real eigenvalue λ_1^k, such that $\lambda_1^k > |\mu_{\ell,j}|$ for the other eigenvalues of B_j. In fact, $\lambda_1 e^{2\pi \ell/k}$ for $0 \le \ell < k$ are simple eigenvalues of A; in particular, $\lambda_1 \ge |\lambda_j|$ for the other eigenvalues λ_j of A. (See Gantmacher (1959) for details.) From this form, it follows that Σ_A is the union of k subsets which are invariant under σ_A^k, and σ_A^k on the j subset is σ_{B_j}. By Theorem 1.3, $h(\sigma_A^k) = \max_j h(\sigma_{B_j}) = \log(\lambda_1^k)$. Then, the entropy of σ_A is as follows:

$$\begin{aligned} h(\sigma_A) &= \frac{1}{k}\, h(\sigma_A^k) \\ &= \frac{1}{k}\, \log(\lambda_1^k) \\ &= \log(\lambda_1). \end{aligned}$$

This shows how the result for a general irreducible subshift follows from the general Perron-Frobenius Theorem. □

REMARK 1.9. In the exercises, we ask the reader to use this last theorem to show that the entropy of the full shift on N symbols is $\log(N)$. See Exercise 9.4. We also use it to calculate the entropy of several subshifts of finite type.

9.1.1 Proof of Two Theorems on Topological Entropy

This subsection contains the proofs of Theorems 1.4 and 1.8. We apply Theorem 1.8 to toral automorphisms using their Markov partitions in the next subsection. In Chapter X we apply it to general hyperbolic invariant sets. The main use of Theorem 1.4 is for Morse-Smale systems, which we also discuss in the next subsection.

The proofs of Theorems 1.4 and 1.8 use another method of counting orbits in addition to (n, ϵ)-separated sets called (n, ϵ)-spanning sets. We give the definition in a slightly more general situation where we allow the initial points to be restricted to a subset K that is not necessarily invariant. The following definition makes these ideas precise.

Definition. Let $f : X \to X$ be a continuous map on the space X with metric d. Let $K \subset X$ be a subset. For a positive integer q, let

$$d_{q,f}(\mathbf{w}, \mathbf{z}) = \sup_{0 \le j < q} d(f^j(\mathbf{w}), f^j(\mathbf{z}))$$

as we defined earlier. A set $S \subset K$ is said to (n, ϵ)-*span* K for n a positive integer and $\epsilon > 0$ provided for each $\mathbf{x} \in K$ there exists a $\mathbf{y} \in S$ such that $d_{n,f}(\mathbf{x}, \mathbf{y}) \le \epsilon$. Then, the number $r_{span}(n, \epsilon, K, f)$ is defined to be the smallest number of elements in any set S which (n, ϵ)-span K, and

$$h_{span}(\epsilon, K, f) = \limsup_{n \to \infty} \frac{\log(r_{span}(n, \epsilon, K, f))}{n}.$$

It is easily checked that $h_{span}(\epsilon, K, f)$ is monotonically decreasing in ϵ ($0 < \epsilon_1 < \epsilon_2$ implies that $h_{span}(\epsilon_1, K, f) \ge h_{span}(\epsilon_2, K, f)$), so the limit as ϵ goes to 0 if $h_{span}(\epsilon, K, f)$ exists,

$$h_{span}(K, f) = \lim_{\epsilon \to 0,\ \epsilon > 0} h_{span}(\epsilon, K, f),$$

and for any $\epsilon > 0$, $h_{span}(\epsilon, K, f) \le h_{span}(K, f)$.

For any integer n, $\epsilon > 0$, and subset $K \supset X$, we let $E_{span}(n, \epsilon, K)$ be a minimal (n, ϵ)-spanning set, so

$$\#(E_{span}(m, \epsilon, S)) = r_{span}(m, \epsilon, S, f).$$

As in the last section, $r_{sep}(n,\epsilon,K,f)$ is the maximal number of elements in any set $S \subset K$ which are (n,ϵ)-separated. The definitions for $h_{sep}(\epsilon,K,f)$ and $h_{sep}(K,f)$ are similar. Again, we let $E_{sep}(m,\epsilon,S)$ be a maximal (m,ϵ)-separated set for K, so $\#(E_{sep}(m,\epsilon,S)) = r_{sep}(m,\epsilon,S,f)$.

If $K = X$, then we usually drop the specification of the set K in the notation.

The following lemma shows that $h_{sep}(K,f) = h_{span}(K,f)$, so the limit is the same for separating and spanning sets (and either one can be used to define entropy). After proving the lemma, we denote either quantity by $h(K,f)$.

Lemma 1.10. *Let K be a subset of X. For $\epsilon > 0$ and n a positive integer,*

$$r_{sep}(n,2\epsilon,K,f) \le r_{span}(n,\epsilon,K,f) \le r_{sep}(n,\epsilon,K,f)$$

and

$$h_{sep}(2\epsilon,K,f) \le h_{span}(\epsilon,K,f) \le h_{sep}(\epsilon,K,f).$$

Therefore, $h_{sep}(K,f) = h_{span}(K,f)$. If we further assume that the space X is compact, then $h_{sep}(\epsilon,K,f) < \infty$.

PROOF. Let $E_{sep}(n,\epsilon,K)$ be a maximal (n,ϵ)-separated set for K and let $\mathbf{x} \in K$. There is some $\mathbf{y} \in E_{sep}(n,\epsilon,K)$ such that $d_{n,f}(\mathbf{x},\mathbf{y}) \le \epsilon$, because otherwise $E_{sep}(n,\epsilon,K) \cup \{\mathbf{x}\}$ would be an (n,ϵ)-separated set for K and $E_{sep}(n,\epsilon,K)$ would not be maximal. Therefore, $E_{sep}(n,\epsilon,K)$ (n,ϵ)-spans K, and

$$r_{sep}(n,\epsilon,K,f) = \#(E_{sep}(n,\epsilon,K)) \ge r_{span}(n,\epsilon,K,f).$$

Let $E_{sep}(n,2\epsilon,K)$ be a maximal $(n,2\epsilon)$-separated set for K, and $E_{span}(n,\epsilon,K)$ be a minimal (n,ϵ)-spanning set for K. Using the fact that $E_{span}(n,\epsilon,K)$ spans, we are going to define a map $T : E_{sep}(n,2\epsilon,K) \to E_{span}(n,\epsilon,K)$. For $\mathbf{x} \in E_{sep}(n,2\epsilon,K)$ there is a $\mathbf{y} = T(\mathbf{x}) \in E_{span}(n,\epsilon,K)$ with $d_{n,f}(\mathbf{x},\mathbf{y}) \le \epsilon$ because $E_{span}(n,\epsilon,K)$ spans. If $T(\mathbf{x}_1) = T(\mathbf{x}_2)$ for $\mathbf{x}_1, \mathbf{x}_2 \in E_{sep}(n,2\epsilon,K)$, then

$$d_{n,f}(\mathbf{x}_1,\mathbf{x}_2) \le d_{n,f}(\mathbf{x}_1,\mathbf{y}) + d_{n,f}(\mathbf{y},\mathbf{x}_2) \le 2\epsilon.$$

Because $E_{sep}(n,2\epsilon,K)$ is an $(n,2\epsilon)$-separated set, $\mathbf{x}_1 = \mathbf{x}_2$. This shows that T is one to one, and so

$$\begin{aligned} r_{sep}(n,2\epsilon,K,f) &= \#(E_{sep}(n,2\epsilon,K)) \\ &\le \#(E_{span}(n,\epsilon,K)) \\ &= r_{span}(n,\epsilon,K,f). \end{aligned}$$

By taking the growth rates as n goes to infinity, we get the inequalities

$$h_{sep}(2\epsilon,K,f) \le h_{span}(\epsilon,K,f) \le h_{sep}(\epsilon,K,f).$$

By letting ϵ go to 0, $h_{sep}(K,f) = h_{sep}(K,f)$.

If X is compact, there is a finite number, N_ϵ, such that N_ϵ is the maximal number of disjoint balls of radius ϵ. It follows from this that the maximal number of elements of an (n,ϵ)-separated set is bounded by N_ϵ^n. (There cannot be two orbits with $d_{n,f}(\mathbf{x},\mathbf{y}) \ge \epsilon$ and $f^j(\mathbf{x})$ and $f^j(\mathbf{y})$ in the same ϵ-balls for $0 \le j < n$.) Therefore, $r_{sep}(n,\epsilon,K,f) \le N_\epsilon^n$, and $h_{sep}(\epsilon,K,f) \le \log(N_\epsilon) < \infty$. □

We next give the definition of topological entropy in terms of open covers. We do not use this definition, so it can be skipped. However, the reader may note some similarity between this definition and a construction in the proof of Theorem 1.4.

Definition. A collection $\mathcal{A}$ is called an *open cover of* X provided (i) each $A \in \mathcal{A}$ is an open subset of X and (ii) $\bigcup\{A \in \mathcal{A}\} = X$. A subcollection $\mathcal{B} \subset \mathcal{A}$ is called a *subcover* provided $\bigcup\{A \in \mathcal{B}\} = X$.

For an open cover $\mathcal{A}$, let

$$\mathcal{A}^n = \{\bigcap_{j=0}^{n-1} f^{-j}(A_j) : A_j \in \mathcal{A} \text{ and } \bigcap_{j=0}^{n-1} f^{-j}(A_j) \neq \emptyset\}.$$

Let $N(\mathcal{A})$ be the minimal cardinality of a subcover $\mathcal{B} \subset \mathcal{A}$. Denote the growth rate of the number of elements in (a minimal subcover of) $\mathcal{A}^n$ by

$$h(\mathcal{A}, f) = \limsup_{n\to\infty} \frac{\log(N(\mathcal{A}^n))}{n}.$$

Finally, the entropy of f is given by

$$h(f) = \sup\{h(\mathcal{A}, f) : \mathcal{A} \text{ is an open cover of } X\}.$$

Note that this definition is also given in terms of the growth rate of a number of objects determined by iterates of f. We do not prove this fact, but the above definition is equivalent to the definitions in terms of (n, ϵ)-separated or spanning sets.

PROOF OF THEOREM 1.4. Because $\Omega \subset X$, we always have that $h(f|\Omega) \leq h(f)$. What we need to prove is the reverse inequality. We use an (m, ϵ)-spanning set of Ω to estimate the size of an $(n, 2\epsilon)$-separated set on all of X.

We fix an integer $m \geq 1$ and $\epsilon > 0$ for quite a while in the proof. Take the set $E_{span}(m, \epsilon, \Omega)$ to be a minimal (m, ϵ)-spanning set for $f|\Omega$. Let

$$U = \{\mathbf{x} \in X : d_{m,f}(\mathbf{x}, \mathbf{y}) < \epsilon \text{ for some } \mathbf{y} \in E_m(\epsilon, \Omega)\}.$$

Since the orbits in $E_{span}(m, \epsilon, \Omega)$ also span orbits of points near Ω in X, U is an open neighborhood of Ω in X. Since $U^c = X \backslash U$ is compact and all points in U^c are wandering, there exists a uniform β with $0 < \beta \leq \epsilon$ such that the forward orbit of the ball of radius β about any $\mathbf{y} \in U^c$, $B(\mathbf{y}, \beta)$, never intersects itself, $f^j(B(\mathbf{y}, \beta)) \cap B(\mathbf{y}, \beta) = \emptyset$ for all $j \geq 1$. Now, take a set $E_{span}(m, \beta, U^c)$ which is a minimal (m, β)-spanning set for f with points starting in U^c, so

$$\#(E_{span}(m, \beta, U^c)) = r_{span}(m, \beta, U^c, f).$$

Let $G_{span}(m) = E_{span}(m, \epsilon, \Omega) \cup E_{span}(m, \beta, U^c)$. The set $G_{span}(m)$ is clearly an (m, ϵ)-spanning set for X, so $\#(G_{span}(m)) \geq r_{span}(m, \epsilon, X, f)$.

Let ℓ be a positive integer. To estimate $r_{sep}(n, 2\epsilon, X, f)$, we define a map

$$\varphi_\ell : X \to G_{span}(m)^\ell$$

by $\varphi_\ell(\mathbf{x}) = (\mathbf{y}_0, \ldots, \mathbf{y}_{\ell-1})$, where (i) $\mathbf{y}_s \in E_{span}(m, \epsilon, \Omega)$ and $d_{m,f}(f^{sm}(\mathbf{x}), \mathbf{y}_s) < \epsilon$ if $f^{sm}(\mathbf{x}) \in U$, and (ii) $\mathbf{y}_s \in E_{span}(m, \beta, U^c)$ and $d_{m,f}(f^{sm}(\mathbf{x}), \mathbf{y}_s) < \beta$ if $f^{sm}(\mathbf{x}) \in U^c$. Because $E_{span}(m, \epsilon, \Omega)$ is an (m, ϵ)-spanning set for U and $E_{span}(m, \beta, U^c)$ is an (m, β)-spanning set for U^c, it is always possible to make these choices to define φ_ℓ.

Claim 1. *Assume* $(\mathbf{y}_0, \dots, \mathbf{y}_{\ell-1}) = \varphi_\ell(\mathbf{x})$ *for some* $\mathbf{x} \in X$. *Then, a point* $\mathbf{y}_s \in E_{span}(m, \beta, U^c)$ *cannot be repeated in this* ℓ*-tuple.*

PROOF. This claim follows because the balls $B(\mathbf{y}_s, \beta)$ are wandering for and choice of $\mathbf{y}_s \in E_{span}(m, \beta, U^c)$. □

Now, we take $n > m\, r_{span}(m, \beta, U^c, f)$ to be an integer. Let $E_{sep}(n, 2\epsilon, X)$ be a maximal $(n, 2\epsilon)$-separated set. For such an n, let ℓ be the positive integer with $(\ell - 1)m < n \le \ell m$. The following claim shows that φ_ℓ is one to one on $E_{sep}(n, 2\epsilon, X)$, so we can estimate $r_{sep}(n, 2\epsilon, X, f)$ by means of $\#(\varphi_\ell(E_{sep}(n, 2\epsilon, X)))$.

Claim 2. *The map* φ_ℓ *is one to one on* $E_{sep}(n, 2\epsilon, X)$.

PROOF. Assume that $\varphi_\ell(\mathbf{x}) = \varphi_\ell(\mathbf{z}) = (\mathbf{y}_0, \dots, \mathbf{y}_{\ell-1})$ for $\mathbf{x}, \mathbf{z} \in E_{sep}(n, 2\epsilon, X)$. For $0 \le t < m$ and $0 \le s < \ell$,

$$\begin{aligned} d(f^{sm+t}(\mathbf{x}), f^{sm+t}(\mathbf{z})) &\le d_{m,f}(f^{sm}(\mathbf{x}), \mathbf{y}_s) + d_{m,f}(\mathbf{y}_s, f^{sm}(\mathbf{z})) \\ &< \epsilon + \epsilon = 2\epsilon. \end{aligned}$$

The integer ℓ is chosen such that $\ell m \ge n$, so we get that $d_{n,f}(\mathbf{x}, \mathbf{z}) < 2\epsilon$. Since the set $E_{sep}(n, 2\epsilon, X)$ is $(n, 2\epsilon)$-separated, $\mathbf{x} = \mathbf{z}$. □

Claim 3. *Let* $q = r_{span}(m, \beta, U^c, f)$ *and* $p = r_{span}(m, \epsilon, \Omega, f)$. *Then,*

$$\#(\varphi_\ell(E_{sep}(n, 2\epsilon, X))) \le (q+1)!\,\ell^q p^\ell.$$

PROOF. Let $\mathcal{I}_j$ be the subset of ℓ-tuples in $\varphi_\ell(E_{sep}(n, 2\epsilon, X))$ such that there are exactly j of the $\mathbf{y}_s$ that are in $E_{span}(m, \beta, U^c)$. Because the $\mathbf{y}_s \in E_{span}(m, \beta, U^c)$ cannot be repeated in $\varphi_\ell(\mathbf{x})$, we must have $j \le q$. (Notice that this bound is independent of n or ℓ. Also, $n > mq$ so $\ell > q$.) For $\mathcal{I}_j$, there are $\binom{q}{j}$ ways of picking these j points $\mathbf{y}_s \in E_{span}(m, \beta, U^c)$; there are

$$\ell \cdot (\ell - 1) \cdots (\ell - j + 1) = \frac{\ell!}{(\ell - j)!}$$

ways of arranging these choices among the positions in the ordered ℓ-tuples; finally, there are at most

$$r_{span}(m, \epsilon, \Omega, f)^{\ell - j} = p^{\ell - j} \le p^\ell$$

ways of picking the remaining $\mathbf{y}_s$ from $E_{span}(m, \epsilon, \Omega)$. Thus,

$$\#(\mathcal{I}_j) \le \binom{q}{j} \frac{\ell!}{(\ell - j)!}\, p^\ell,$$

and

$$\begin{aligned} \#(\varphi_\ell(E_{sep}(n, 2\epsilon, X))) &= \sum_{j=0}^{q} \#(\mathcal{I}_j) \\ &\le \sum_{j=0}^{q} \binom{q}{j} \frac{\ell!}{(\ell - j)!}\, p^\ell. \end{aligned}$$

To estimate this summation, note that $\binom{q}{j} \leq q!$ and

$$\frac{\ell!}{(\ell-j)!} = \ell \cdot (\ell-1) \cdots (\ell-j+1) \leq \ell^j \leq \ell^q.$$

Thus,

$$\begin{aligned}\#(\varphi_\ell(E_{sep}(n, 2\epsilon, X))) &\leq \sum_{j=0}^{q} q! \ell^q \, p^\ell \\ &\leq (q+1)! \ell^q \, p^\ell\end{aligned}$$

as claimed. □

REMARK 1.10. Notice how crudely we made the count of possible ℓ-tuples in the set $\varphi_\ell(E_{sep}(n, 2\epsilon, X))$. The estimate $\sum_{j=0}^{q} q!\ell^q \leq (q+1)!\ell^q$ gives a bound on the number of wandering orbits. Because this quantity only grows polynomially in ℓ, it does not contribute to the entropy. A better bound is possible by paying attention to the dynamics of points which wander, but it would still have polynomial growth contributed by the wandering orbits.

We could also form the open cover of sets $\mathcal{V}$ made up of the open sets

$$\{\mathbf{x} \in X : d_{m,f}(\mathbf{x}, \mathbf{y}_s) < \epsilon\}$$

for $\mathbf{y}_s \in E_{span}(m, \epsilon, \Omega)$ and

$$\{\mathbf{x} \in X : d_{m,f}(\mathbf{x}, \mathbf{y}_s) < \beta\}$$

for $\mathbf{y}_s \in E_{span}(m, \beta, U^c)$. Let $\mathcal{V}^\ell$ be the refinement as in the definition of the entropy by open sets. Then, there is a very close connection between $\mathcal{V}^\ell$ and $\varphi_\ell(E_{sep}(n, 2\epsilon, X))$. Since the growth rate of the number of open sets in $\mathcal{V}^\ell$ as ℓ goes to infinity gives the entropy for this open cover, it is not surprising that it can be used to get a bound on the growth rate of $(n, 2\epsilon)$-separated orbits. See Alsedà, Llibre, and Misiurewicz (1993) for a proof of Theorem 1.4 using the definition in terms of open covers.

By Claims 2 and 3,

$$\begin{aligned}r_{sep}(n, 2\epsilon, X, f) &= \#(\varphi_\ell(E_{sep}(n, 2\epsilon, X))) \\ &\leq (q+1)! \ell^q \, p^\ell,\end{aligned}$$

where $q = r_{span}(m, \beta, U^c, f)$ and $p = r_{span}(m, \epsilon, \Omega, f)$. Then,

$$\begin{aligned}h_{sep}(2\epsilon, X, f) &= \limsup_{n\to\infty} \frac{1}{n} \log(r_{sep}(n, 2\epsilon, X, f)) \\ &\leq \limsup_{\ell\to\infty} \frac{\log((q+1)!) + q\log(\ell) + \ell\log(p)}{(\ell-1)m} \\ &\leq \frac{\log(p)}{m} \\ &= \frac{\log(r_{span}(m, \epsilon, \Omega, f))}{m}.\end{aligned}$$

Next, we let m vary and go to infinity, and we obtain the bound

$$\begin{aligned} h_{sep}(2\epsilon, X, f) &\le h_{span}(\epsilon, \Omega, f) \\ &\le h(\Omega, f). \end{aligned}$$

Finally, letting ϵ go to 0, we get $h(X, f) \le h(\Omega, f)$ as desired. □

PROOF OF THEOREM 1.8. Because k is onto, $h(F) \ge h(f)$. Thus, what we need to show is that $h(F) \le h(f)$.

The obvious attempts at proofs do not work. Using the uniform continuity of k, it can be shown that given $\epsilon > 0$ there is a $\delta > 0$ such that if $E_{sep}(n, \delta, f) \subset Y$ is (n, δ)-separated for f, then $k^{-1}(E_{sep}(n, \delta, f))$ is (n, ϵ)-separated for F. However, $k^{-1}(E_{sep}(n, \delta, f))$ is not necessarily the maximal (n, ϵ)-separated set for F, so this fact does not give an upper bound for $r_{sep}(n, \epsilon, F)$ in terms of $r_{sep}(n, \delta, f)$. On the other hand, the inverse image of a spanning set for f is not necessarily a spanning set for F. The proof below shows that given $\epsilon > 0$, there is a $\beta > 0$ such that there is an upper bound on the maximal size of a $(n, 2\epsilon)$-separating set for F, $r_{sep}(n, 2\epsilon, F)$, in terms of the minimal size of a (n, β)-spanning set for f, $r_{span}(n, \beta, f)$.

Fix an $\epsilon > 0$. Let $C \ge 1$ be a bound on the number of points in $k^{-1}(\mathbf{y})$,

$$\#(k^{-1}(\mathbf{y})) \le C$$

for all $\mathbf{y} \in Y$. Let $m \ge 1$ be an integer. With some work we show that the following bound for the size of an $(n, 2\epsilon)$-separating set for F is true:

$$h_{sep}(2\epsilon, F) \le h_{span}(\beta, f) + \frac{1}{m}\log(C) \le h(f) + \frac{1}{m}\log(C),$$

for correctly chosen $\beta > 0$. Since m is an arbitrarily large integer, this implies that $h_{sep}(2\epsilon, F) \le h(f)$ and $h(F) \le h(f)$.

To start this proof, for $\mathbf{y} \in Y$, let

$$U_{\mathbf{y}} = U_{\mathbf{y},m,\epsilon} = \{\mathbf{w} \in X : d_{m,F}(\mathbf{w}, \mathbf{z}) < \epsilon \text{ for some } \mathbf{z} \in k^{-1}(\mathbf{y})\}.$$

By the continuity of F, $U_{\mathbf{y}}$ is an open neighborhood of $k^{-1}(\mathbf{y})$. By the continuity of k and compactness of X, there is an open neighborhood $W_{\mathbf{y}} \subset Y$ of $\mathbf{y}$ such that $k^{-1}(W_{\mathbf{y}}) \subset U_{\mathbf{y}}$. The set Y is compact, so there exists a finite cover $\{W_{\mathbf{y}_1}, \ldots, W_{\mathbf{y}_p}\}$. Let $\beta > 0$ be the Lebesgue number of this finite cover, i.e., if $\mathbf{y} \in Y$ there is some $W_{\mathbf{y}_j}$ such that the closed β ball about $\mathbf{y}$ is contained in $W_{\mathbf{y}_j}$, $\text{cl}(B(\mathbf{y}, \beta)) \subset W_{\mathbf{y}_j}$. Note that given a small $\epsilon > 0$, we have determined a small $\beta > 0$. This is the β that works for the (n, β)-spanning set for f mentioned above. We want to show that

$$\frac{1}{n}\log(r_{sep}(n, 2\epsilon, F)) \le \frac{1}{n}\log(r_{span}(n, \beta, f)) + \frac{1}{m}\log(C) + \frac{1}{n}\log(C).$$

In the proof below, we need to break up an orbit of F into segments of length m. To this end, for an integer n let ℓ be the integer such that

$$(\ell - 1)m < n \le \ell m.$$

Let $E_{sep}(n, 2\epsilon, F) \subset X$ be a maximal $(n, 2\epsilon)$-separating set for F with

$$\#(E_{sep}(n, 2\epsilon, F)) = r_{sep}(n, 2\epsilon, F),$$

and $E_{span}(n, \beta, f) \subset Y$ be a minimal (n, β)-spanning set for f with

$$\#(E_{span}(n, \beta, f)) = r_{span}(n, \beta, f).$$

For $\mathbf{y} \in E_{span}(n, \beta, f)$, let $\mathbf{q}(j, \mathbf{y}) \in \{\mathbf{y}_1, \dots \mathbf{y}_p\}$ be chosen so that

$$\text{cl}(B(f^j(\mathbf{y}), \beta)) \subset W_{\mathbf{q}(j,\mathbf{y})}.$$

To get an estimate on $r_{sep}(n, 2\epsilon, F)$ in terms of $r_{span}(n, \beta, F)$, we define the map

$$\varphi_\ell : E_{sep}(n, 2\epsilon, F) \to E_{span}(n, \beta, f) \times X^\ell$$

by $\varphi_\ell(\mathbf{x}) = (\mathbf{y}; \mathbf{x}_0, \dots, \mathbf{x}_{\ell-1})$, where (i) $d'_{n,f}(\mathbf{y}, k(\mathbf{x})) \leq \beta$ and $\mathbf{y} \in E_{span}(n, \beta, f)$, and (ii) $\mathbf{x}_s \in k^{-1}(\mathbf{q}(sm, \mathbf{y}))$ satisfies $d_{m,F}(F^{sm}(\mathbf{x}), \mathbf{x}_s) < \epsilon$ for $0 \leq s < \ell$. Because

$$\begin{aligned} &k \circ F^{sm}(\mathbf{x}) = f^{sm} \circ k(\mathbf{x}) \in \text{cl}(B(f^{sm}(\mathbf{y}), \beta)) \subset W_{\mathbf{q}(sm,\mathbf{y})}, \\ &F^{sm}(\mathbf{x}) \in k^{-1}(W_{\mathbf{q}(sm,\mathbf{y})}) \subset U_{\mathbf{q}(sm,\mathbf{y}),m,\epsilon} \end{aligned}$$

and it is always possible to choose the $\mathbf{x}_s$.

Claim. *The map φ_ℓ is one to one on $E_{sep}(n, 2\epsilon, F)$.*

PROOF. If $\varphi_\ell(\mathbf{w}) = \varphi_\ell(\mathbf{z}) = (\mathbf{y}; \mathbf{x}_0, \dots, \mathbf{x}_{\ell-1})$, then

$$\begin{aligned} d(F^{sm+t}(\mathbf{w}), F^{sm+t}(\mathbf{z})) &\leq d_{m,F}(F^{sm}(\mathbf{w}), \mathbf{x}_s)) + d_{m,F}(\mathbf{x}_s, F^{sm}(\mathbf{z})) \\ &\leq \epsilon + \epsilon = 2\epsilon. \end{aligned}$$

for $0 \leq t < m$, and $0 \leq s \leq \ell$. Since $m\ell \geq n$, we get that $d_{n,F}(\mathbf{z}, \mathbf{w}) \leq 2\epsilon$. Since $E_{sep}(n, 2\epsilon, F)$ is $(n, 2\epsilon)$-separated, we get that $\mathbf{w} = \mathbf{z}$. □

Now, we use the claim to finish the proof of the theorem. For $\mathbf{y} \in E_{span}(n, \beta, f)$ fixed,

$$\begin{aligned} \#(\varphi_\ell(E_{sep}(n, 2\epsilon, F)) \cap (\{\mathbf{y}\} \times X^\ell)) &\leq \prod_{s=0}^{\ell-1} \#(k^{-1}(\mathbf{q}(sm, \mathbf{y})) \\ &\leq C^\ell. \end{aligned}$$

Because there are $r_{span}(n, \beta, f)$ choices of $\mathbf{y} \in E_{span}(n, \beta, f)$,

$$\begin{aligned} r_{sep}(n, 2\epsilon, F) &= \#(\varphi_\ell(E_{sep}(n, 2\epsilon, F))) \\ &\leq r_{span}(n, \beta, f)\, C^\ell. \end{aligned}$$

Therefore,

$$\begin{aligned} \frac{1}{n} \log(r_{sep}(n, 2\epsilon, F)) &\leq \frac{1}{n} \log(r_{span}(n, \beta, f)) + \frac{1}{n} \log(C^\ell) \\ &= \frac{1}{n} \log(r_{span}(n, \beta, f)) + \frac{\ell m}{nm} \log(C) \\ &\leq \frac{1}{n} \log(r_{span}(n, \beta, f)) + \frac{n+m}{nm} \log(C) \\ &\leq \frac{1}{n} \log(r_{span}(n, \beta, f)) + \frac{1}{m} \log(C) + \frac{1}{n} \log(C). \end{aligned}$$

The second to last inequality is obtained using the definition of ℓ. Taking the limsup as n goes to ∞, we get

$$h_{sep}(2\epsilon, F) \leq h_{span}(\beta, f) + \frac{1}{m}\log(C) \leq h(f) + \frac{1}{m}\log(C).$$

Since this last inequality is true for all $m \geq 1$, we get that $h_{sep}(2\epsilon, F) \leq h(f)$. Letting $\epsilon > 0$ go to 0, we get the desired inequality, $h(F) \leq h(f)$. (Notice that the bound on the number of points in the inverse image of a point contributed $\log(C)/m$ to the bound on the entropy, which is nonzero if $C > 1$. It is only as m goes to infinity that this bound can be neglected.) □

REMARK 1.11. There is a generalization of Theorem 1.8 that does not assume k is uniformly finite to one. In this case,

$$h(f) \leq h(F) \leq h(f) + \sup_{\mathbf{y}\in Y} h(k^{-1}(\mathbf{y}), F).$$

Notice with the assumptions of Theorem 1.8, $k^{-1}(\mathbf{y})$ is finite so that the second term is 0, $\sup_{\mathbf{y}\in Y} h(k^{-1}(\mathbf{y}), F) = 0$. Therefore, the specific result of Theorem 1.8 follows from this more general result. The proof of the more general result is basically the same as that given above but the number m varies with the point $\mathbf{y} \in Y$. This form was also proved by Bowen (1971). See de Melo and Van Strien (1993) for a proof.

9.1.2 Entropy of Higher Dimensional Examples

In this subsection, we apply the theorems about entropy to some of the higher-dimensional examples discussed in Chapter VIII. We start with Morse-Smale diffeomorphisms. Because such a diffeomorphism has only finitely many points in its chain recurrent set, the dynamics are simple and it has zero topological entropy.

Theorem 1.11. *Let M be a compact manifold, and f a C^1 Morse-Smale diffeomorphism. Then, the topological entropy of f is zero, $h(f) = 0$.*

PROOF. Theorem 1.4 shows that $h(f) = h(f|\Omega)$, where $\Omega = \Omega(f)$ is the nonwandering set of f. Since f is Morse-Smale, $\Omega(f)$ is a finite set of points. Therefore, $h(f) = h(f|\Omega) = 0$. □

As a second example, we consider an Anosov diffeomorphism and connect the entropy to the entropy of the subshift of finite type induced by a Markov partition.

Theorem 1.12. *Let f_A be a hyperbolic toral automorphism on $\mathbb{T}^2$. Assume $\mathcal{R} = \{R_j\}_{j=1}^m$ is a Markov partition with transition matrix B. Let λ_1 be the eigenvalue of the transition matrix B with largest absolute value. (This is also the absolute value of the largest eigenvalue of the original matrix A.) Then, the topological entropy of f_A is $\log(\lambda_1)$.*

PROOF. By Theorem VIII.5.3, there is a finite to one semiconjugacy $h : \Sigma_B \to \mathbb{T}^2$. Since h is a finite to one semi-conjugacy, the entropies of f and σ_B are equal by Theorem 1.8. The topological entropy of a two-sided shift is the logarithm of the largest eigenvalue by Theorem 1.9. Combining these facts gives the result. □

REMARK 1.10. Bowen (1971) has also related the topological entropy of a hyperbolic toral automorphism on $\mathbb{T}^n$ to a summation involving all the expanding eigenvalues:

$$h(f_A) = \sum_{|\lambda_j|>1} \log(|\lambda_j|),$$

where λ_j are the eigenvalues of A. Also see Bowen (1978a).

Example 1.2. As a last example, consider a transverse homoclinic point for a diffeomorphism f. By Theorem VIII.4.5(a), f^n has an invariant set Λ which is conjugate to σ_2 on the full two-shift. The orbit of Λ by f is just n sets which are homeomorphic to Λ and for $\mathbf{x} \in \Lambda$, $f^j(\mathbf{x})$ returns to Λ every n iterates. The entropy of $f^n|\Lambda$ is $\log(2)$ and the entropy of f on the orbit of Λ is $(1/n)\log(2)$. On the other hand, by Theorem VIII.4.5(b), f has an invariant set Λ' which is conjugate to a subshift of finite type for a matrix B, where B is given in the proof. As noted in Remark VIII.4.6, the characteristic polynomial of B is $p(\lambda) = \lambda^n - \lambda^{n-1} - 1$. Since $p(2^{1/n}) = 2 - 2^{(n-1)/n} - 1 < 0$, the largest eigenvalue of B is larger than $2^{1/n}$, i.e., the entropy of the subshift of finite type $\sigma_B|\Sigma_B$ and so $f|\Lambda'$ is greater than the entropy of $f|\mathcal{O}(\Lambda)$.

REMARK 1.11. Assume $f : M \to M$ is a C^1 diffeomorphism on a compact manifold M and f has a hyperbolic chain recurrent set (or hyperbolic nonwandering set). Let $N_n(f) = \#(\text{Fix}(f^k))$ be the number of points whose least period divides n. Bowen (1970b) proved that

$$h(f) = \limsup_{n\to\infty} \frac{\log(N_n(f))}{n},$$

i.e., the entropy is equal to the growth rate of the number periodic points. Also see Bowen (1978a).

Bowen (1978a) also introduces the connection between the entropy and the induced maps on the homology groups. See Yomdin (1987) for a proof of the conjecture.

9.2 Liapunov Exponents

Most of this book concerns examples of systems with uniformly hyperbolic invariant sets. In this section we define Liapunov exponents for diffeomorphisms and flows in all dimensions, extending the treatment in Section 3.6. We give only a brief introduction to the ideas. For more details see Ruelle (1989a), Mañé (1987a), Katok (1980), and Walters (1982). For a proof see Ruelle (1979).

Just as in one dimension, the exponents exist almost everywhere in terms of an invariant measure. If these exponents are nonzero almost everywhere on an invariant set, then it has a *nonuniformly hyperbolic structure* (which may in fact be a uniformly hyperbolic structure in some cases). There are examples which have a nonuniformly hyperbolic structure; in fact, the Hénon map for $A < 2$ and $|B|$ small cannot have a uniform hyperbolic structure by the theorem of Plykin, but Benedicks and Carleson (1991) proved that it does have nonzero Liapunov exponents. Therefore, the Hénon map for these parameter values is an example with a nonuniformly hyperbolic structure. For $A = 1.4$ and $B = -0.3$, numerical simulation indicates that it has an invariant set with a nonuniformly hyperbolic structure but this is unproven.

With this introduction we turn to the definitions for a diffeomorphism. Those for flows are similar, but we leave the small differences to the reader.

Definition. Let $f : M \to M$ be a diffeomorphism on a manifold of dimension m. Let $|\cdot|$ be the norm on tangent vectors induced by a Riemannian metric (inner product on tangent vectors) on M. For each $\mathbf{x} \in M$ and $\mathbf{v} \in T_{\mathbf{x}}M$, let

$$\lambda(\mathbf{x}, \mathbf{v}) = \lim_{k\to\infty} \frac{1}{k} \log(|Df^k_{\mathbf{x}}\mathbf{v}|)$$

whenever this limit exists. Note that

$$\limsup_{k\to\infty} \frac{1}{k} \log(|Df^k_{\mathbf{x}}\mathbf{v}|)$$

always exists, so it would be possible to use this as the definition. We restrict to the limit existing because it seems to make some of the statements below easier to make. Thus, $\lambda(\mathbf{x}, \mathbf{v})$ is the exponential growth rate of the vector transported by the linearized equations along the orbit (or transported by the first variation equation for flows). Exercise 9.23 asks the reader to prove that the Liapunov exponent does not depend on the length of the vector. The Multiplicative Ergodic Theorem of Oseledec, which is stated below, says for almost all points $\mathbf{x}$ that (i) the limit exists for all tangent vectors $\mathbf{v} \in T_{\mathbf{x}}M$ and (ii) that there are at most m distinct values of $\lambda(\mathbf{x}, \mathbf{v})$ for one point $\mathbf{x}$. Let $s(\mathbf{x})$ be the number of distinct values of $\lambda(\mathbf{x}, \mathbf{v})$ at $\mathbf{x}$ for $\mathbf{v} \in T_{\mathbf{x}}M$, with tangent vectors $\mathbf{v}^j \in T_{\mathbf{x}}M$ for $1 \leq j \leq s(\mathbf{x})$ giving distinct values,

$$\lambda_j(\mathbf{x}) = \lambda(\mathbf{x}, \mathbf{v}^j),$$

with

$$\lambda_1(\mathbf{x}) < \lambda_2(\mathbf{x}) < \cdots < \lambda_{s(\mathbf{x})}(\mathbf{x}).$$

These $s(\mathbf{x})$ distinct values, $\lambda_j(\mathbf{x})$ for $1 \leq j \leq s(\mathbf{x})$, are called the *Liapunov exponents at* $\mathbf{x}$. The Liapunov exponents are constant along an orbit by the definition. If the diffeomorphism f is ergodic for an invariant measure μ, then the Liapunov exponents are constant μ-almost everywhere, and we can speak of the *Liapunov exponents of the diffeomorphism.* (A diffeomorphism is called *ergodic* for an invariant probability measure μ provided the only invariant sets A with $\mu(A) > 0$ have full measure, $\mu(A) = 1$. Thus, for any set of positive measure B, the orbit of B, $\mathcal{O}(B)$, has full measure. Thus, the concept of ergodicity is like transitivity in a measure theoretic sense.)

The quantities called Liapunov exponents above were introduced by Liapunov (1907). Much of the recent interest and use of these quantities has been the consequence of the Multiplicative Ergodic Theorem of Oseledec (1968), who proved that the limit exists for almost all points and all the quantities vary measurably. This theorem allows the application of ergodic theory methods to such diffeomorphisms.

The Multiplicative Ergodic Theorem is stated in terms of measures on the σ-algebra of Borel subsets of the manifold M. The Borel subsets of M can be defined by either (i) using local coordinates on an open cover of M, (ii) considering M as a subset of a larger dimensional Euclidean space $\mathbb{R}^N$ (M can always be embedded in some $\mathbb{R}^N$), or (iii) using the exponential map (of geodesics of the Riemannian metric) from each tangent space $T_{\mathbf{x}}M$ to a neighborhood in M of $\mathbf{x}$. We let $\mathcal{B}$ be the σ-algebra generated by the Borel subsets of M.

For a diffeomorphism $f : M \to M$, a measure μ on M is said to be a *Borel probability measure on* M provided (i) μ is defined on $\mathcal{B}$, and (ii) $\mu(M) = 1$. The measure μ is said to be an *invariant Borel probability measure for* f provided in addition that (iii) $\mu(f^{-1}(B)) = \mu(B)$ for all $B \in \mathcal{B}$. We let $\mathcal{M}(f)$ be the set of all invariant Borel probability measures for f.

We can now state the theorem.

Theorem 2.1 (Multiplicative Ergodic Theorem). *Let M be a compact manifold of dimension m, $\mathcal{B}$ be the σ-algebra generates by the Borel subsets of M, and $f : M \to M$ be a C^2 diffeomorphism. Then, there is an invariant set $B_f \in \mathcal{B}$ of full measure for every $\mu \in \mathcal{M}(f)$ such that the Liapunov exponents exist for all points $\mathbf{x} \in B_f$.*

More precisely, the following properties are true.

(a) *The set B_f is (i) invariant, $f(B_f) = B_f$, and (ii) of full measure, $\mu(B_f) = 1$ for all $\mu \in \mathcal{M}(f)$.*

(b) *For each $\mathbf{x} \in B_f$, the tangent space at $\mathbf{x}$ can be written as an increasing set of subspaces*

$$\{\mathbf{0}\} = V_{\mathbf{x}}^0 \subset V_{\mathbf{x}}^1 \subset \cdots \subset V_{\mathbf{x}}^{s(\mathbf{x})} = T_{\mathbf{x}}M$$

such that (i) for $\mathbf{v} \in V_{\mathbf{x}}^j \setminus V_{\mathbf{x}}^{j-1}$ *the limit defining* $\lambda(\mathbf{x}, \mathbf{v})$ *exists and* $\lambda_j(\mathbf{x}) = \lambda(\mathbf{x}, \mathbf{v})$ *is the same value for all such* $\mathbf{v}$*, and (ii) the bundle of subspaces*

$$\{V_{\mathbf{x}}^j : \mathbf{x} \in B_f \text{ and } s(\mathbf{x}) \geq j\}$$

are invariant in the sense that $Df_{\mathbf{x}} V_{\mathbf{x}}^j = V_{f(\mathbf{x})}^j$ *for all* $1 \leq j \leq s(\mathbf{x})$.

(c) *The function* $s : B_f \to \{1, \ldots, m\}$ *is a measurable function and invariant,* $s \circ f = s$.

(d) *If* $\mathbf{x} \in B_f$*, the exponents satisfy*

$$-\infty \leq \lambda_1(\mathbf{x}) < \lambda_2(\mathbf{x}) < \cdots < \lambda_{s(\mathbf{x})}(\mathbf{x}).$$

(Note that we allow $\lambda_1(\mathbf{x}) = -\infty$*.) For* $1 \leq j \leq m$*, the function* $\lambda_j(\cdot)$ *is (i) defined and measurable on the set*

$$\{\mathbf{x} \in B_f : s(\mathbf{x}) \geq j\},$$

and (ii) is invariant, $\lambda_j \circ f = \lambda_j$.

REMARK 2.1. The statement of the theorem may not seem like an ergodic theorem except for the involvement of invariant measures. However, the proof definitely involves the use of ideas from ergodic theory. When we defined Liapunov exponents for a map g on the reals in Section 3.6, it involved the product of the derivatives along an orbit,

$$(g^k)'(x) = g'(x_{k-1}) \cdots g'(x_0)$$

where $x_j = g^j(x)$. This product does not depend on the order of multiplication, so the sum

$$\frac{1}{k} \log(|(g^k)'(x)|) = \frac{1}{k} \sum_{j=0}^{k-1} \log(|g'(x_j)|)$$

is the average of numbers $\log(|g'(x_j)|)$ where the order of summation does not matter. Therefore, the Birkhoff Ergodic Theorem can be applied to prove that the limit defining the Liapunov exponents exists for almost all points x. In higher dimensions, the product of the derivatives along the orbit,

$$Df_{\mathbf{x}}^k = Df_{\mathbf{x}_{k-1}} \cdots Df_{\mathbf{x}_0}$$

where $\mathbf{x}_j = f^j(\mathbf{x})$, depends on the order of multiplying the linear maps. If we take norms of the linear maps, then the product is again commutative, but we get an inequality and not an equality,

$$\|Df_{\mathbf{x}}^k\| \leq \|Df_{\mathbf{x}_{k-1}}\| \cdots \|Df_{\mathbf{x}_0}\|,$$

or

$$\frac{1}{k} \log \|Df_{\mathbf{x}}^k\| \leq \frac{1}{k} \sum_{j=0}^{k-1} \log \|Df_{\mathbf{x}_j}\|.$$

Even though this quantity is subadditive rather than additive, there is a generalization of the Birkhoff Ergodic Theorem (called the Subadditive Ergodic Theorem) which says that $\lim_{k\to\infty} \frac{1}{k} \log \|Df_{\mathbf{x}}^k\|$ exists for almost all points $\mathbf{x}$. This limit turns out to be the largest Liapunov exponent which exists almost everywhere as claimed. Thus, the existence of the limits defining the Liapunov exponents is related to ergodic theory.

Another aspect of the proof relates to understanding how $Df^k_{\mathbf{x}}$ changes lengths of all vectors. Even on a manifold it is possible to reduce the proof to considering a square matrix, so we consider $Df^k_{\mathbf{x}}$ as a matrix. The square of the length of the image of a vector $\mathbf{v}$ by $Df^k_{\mathbf{x}}$ is

$$\begin{aligned}\|Df^k_{\mathbf{x}}\mathbf{v}\|^2 &= \left(Df^k_{\mathbf{x}}\mathbf{v}\right)^T Df^k_{\mathbf{x}}\mathbf{v} \\ &= \mathbf{v}^T\left[\left(Df^k_{\mathbf{x}}\right)^T Df^k_{\mathbf{x}}\right]\mathbf{v}.\end{aligned}$$

The matrix $\left(Df^k_{\mathbf{x}}\right)^T Df^k_{\mathbf{x}}$ is symmetric and positive definite. Therefore, it is possible to take roots: $\left[\left(Df^k_{\mathbf{x}}\right)^T Df^k_{\mathbf{x}}\right]^{1/2}$ measures how much lengths are changed by $Df^k_{\mathbf{x}}$, and $\left[\left(Df^k_{\mathbf{x}}\right)^T Df^k_{\mathbf{x}}\right]^{1/2k}$ measures the average amount vectors are stretched. It can be shown that for almost all $\mathbf{x}$, the limit

$$\lim_{k\to\infty}\left[\left(Df^k_{\mathbf{x}}\right)^T Df^k_{\mathbf{x}}\right]^{1/2k} = \Lambda_{\mathbf{x}}$$

exists. The logarithm of the eigenvalues of $\Lambda_{\mathbf{x}}$ are the Liapunov exponents. See Ruelle (1979) for a development of these ideas into a complete proof of the Multiplicative Ergodic Theorem.

REMARK 2.2. The Multiplicative Ergodic Theorem states that the Liapunov exponents exist almost everywhere, but it may not be obvious how to calculate these quantities. We now describe a process that can be used to calculate them. In this discussion, we assume for simplicity that there are m distinct exponents. The general case is really not any more complicated but the notation is more complicated.

Remember that for a single linear map L on $\mathbb{R}^m$, most vectors $\mathbf{v}$ have some component in the direction of the eigenvector corresponding to the largest eigenvalue μ_m, so

$$\lim_{k\to\infty}\frac{1}{k}\log(|L^k\mathbf{v}|) = \log(|\mu_m|).$$

In the present situation, most tangent vectors $\mathbf{v}$ at $\mathbf{x}$ lie in $V^m_{\mathbf{x}} \setminus V^{m-1}_{\mathbf{x}}$, so

$$\lambda_m(\mathbf{x}) = \lim_{k\to\infty}\frac{1}{k}\log(|Df^k_{\mathbf{x}}\mathbf{v}|).$$

Thus, we can calculate $\lambda_m(\mathbf{x})$ by taking an arbitrary tangent vector $\mathbf{v} \in T_{\mathbf{x}}M$ (and assume that it does not lie in $V^{m-1}_{\mathbf{x}}$) and determine $\lambda_m(\mathbf{x})$ by means of $\lambda(\mathbf{x},\mathbf{v})$ (if the limit converges).

We cannot easily find a vector $\mathbf{v} \in V^{m-1}_{\mathbf{x}}$ to calculate $\lambda_{m-1}(\mathbf{x})$. However, for most pairs of vectors $\mathbf{v}^m, \mathbf{v}^{m-1} \in T_{\mathbf{x}}M$, the plane spanned by $\mathbf{v}^m$ and $\mathbf{v}^{m-1}$ is complementary to $V^{m-2}_{\mathbf{x}}$. Therefore, there is a vector $\mathbf{w}^{m-1} \in V^{m-1}_{\mathbf{x}}$ such that $\mathbf{w}^{m-1} = \mathbf{v}^{m-1} + \alpha\mathbf{v}^m$ for some scalar α. Under application of $Df^k_{\mathbf{x}}$, the area of the parallelogram determined by $Df^k_{\mathbf{x}}\mathbf{v}^m$ and $Df^k_{\mathbf{x}}\mathbf{v}^{m-1}$ equals the area of the parallelogram determined by $Df^k_{\mathbf{x}}\mathbf{v}^m$ and $Df^k_{\mathbf{x}}\mathbf{w}^{m-1}$, and so grows at a rate like $e^{k(\lambda_m(\mathbf{x})+\lambda_{m-1}(\mathbf{x}))}$. (This requires some justification and consideration of angles between $V^m_{f^k(\mathbf{x})}$ and $V^{m-1}_{f^k(\mathbf{x})}$.) Let $|\mathbf{v}\wedge\mathbf{w}|$ be the area of the parallelogram determined by $\mathbf{v}$ and $\mathbf{w}$. Then, for most $\mathbf{v}^m, \mathbf{v}^{m-1} \in T_{\mathbf{x}}M$,

$$\lambda_m(\mathbf{x}) + \lambda_{m-1}(\mathbf{x}) = \lim_{k\to\infty}\frac{1}{k}\log(|(Df^k_{\mathbf{x}}\mathbf{v}^m)\wedge(Df^k_{\mathbf{x}}\mathbf{v}^{m-1})|).$$

This gives a means of calculating $\lambda_{m-1}(\mathbf{x})$ once we know $\lambda_m(\mathbf{x})$.

Taking j-tuples of tangent vectors $\mathbf{v}^m, \dots, \mathbf{v}^{m-j+1} \in T_{\mathbf{x}}M$, most choices span a subspace which is complementary to $V_{\mathbf{x}}^j$, so

$$\lambda_m(\mathbf{x}) + \cdots + \lambda_{m-j+1}(\mathbf{x}) = \lim_{k\to\infty} \frac{1}{k} \log(|(Df_{\mathbf{x}}^k \mathbf{v}^m) \wedge \cdots \wedge (Df_{\mathbf{x}}^k \mathbf{v}^{m-j+1})|),$$

where $|\mathbf{v}^m \wedge \cdots \wedge \mathbf{v}^{m-j+1}|$ is the j-dimensional volume of the j-dimensional parallelepiped determined by $\mathbf{v}^m, \dots, \mathbf{v}^{m-j+1}$. Thus, it is possible to determine all the Liapunov exponents by induction.

In interpreting the Liapunov exponents, $\lambda_j(\mathbf{x})$ is like the logarithm of the eigenvalue of a fixed point of a map. Therefore, $\lambda_j(\mathbf{x}) > 0$ corresponds to an expanding direction, $\lambda_j(\mathbf{x}) < 0$ corresponds to a contracting direction, and $\lambda_j(\mathbf{x}) = 0$ corresponds to a neutral direction (at least as far as exponential growth rates are concerned).

For the rest of the section, we assume that $\lambda_j(\mathbf{x}) \neq 0$ for all j and almost all points $\mathbf{x} \in \Lambda \cap B_f$, where Λ is an invariant set for f and B_f is defined by Theorem 2.1. This assumption is analogous to a hyperbolic structure on Λ, and is what we called above a nonuniform hyperbolic structure on Λ. Let $r(\mathbf{x})$ be the integer function such that $\lambda_j(\mathbf{x}) < 0$ for $1 \leq j \leq r(\mathbf{x})$ and $\lambda_j(\mathbf{x}) > 0$ for $r(\mathbf{x}) < j \leq s(\mathbf{x})$. Let $\mathbb{E}_{\mathbf{x}}^s = V_{\mathbf{x}}^{r(\mathbf{x})}$. This is the *stable subspace at* $\mathbf{x}$. The *unstable subspace at* $\mathbf{x}$ can be determined using f^{-1}, $\mathbb{E}_{\mathbf{x}}^u$. The splitting of the tangent space by means of the stable and unstable subspaces, $T_{\mathbf{x}}M = \mathbb{E}_{\mathbf{x}}^s \oplus \mathbb{E}_{\mathbf{x}}^u$ for $\mathbf{x} \in \Lambda \cap B_f$, is measurable in $\mathbf{x}$ but not necessarily continuous. This type of structure is called a *nonuniformly hyperbolic structure on* Λ.

Pesin (1976) proved that a nonlinear map f on a nonuniformly hyperbolic invariant set has local stable and unstable manifolds for points $\mathbf{x} \in \Lambda$ which are invariant and tangent to $\mathbb{E}_{\mathbf{x}}^s$ and $\mathbb{E}_{\mathbf{x}}^u$, respectively. In this case, the diameter of the local manifolds varies with the base point $\mathbf{x}$, so the situation is much more complicated than the uniformly hyperbolic case. Also see Pugh and Shub (1989) for a proof more like the one given in this book for uniformly hyperbolic sets. Ruelle (1979) has another proof.

Katok (1980) used these ideas to prove the following theorem which gives a connection between positive Liapunov exponents and uniformly hyperbolic sets which have positive topological entropy.

Theorem 2.2. *Let $f : M \to M$ be a C^2 diffeomorphism on a compact manifold M. Assume that (i) f is ergodic for some invariant Borel probability measure $\mu_0 \in \mathcal{M}(f)$ where μ_0 is not concentrated on a single periodic orbit (μ_0 is a non-atomic measure), and (ii) the Liapunov exponents are nonzero μ_0-almost everywhere. Then, f has a closed invariant uniformly hyperbolic subset, Λ, (a "horseshoe") such that (i) $f|\Lambda$ is topologically conjugate to a subshift of finite type and (ii) the topological entropy of f on Λ is positive, $h(f|\Lambda) > 0$.*

REMARK 2.3. With the assumptions of this theorem, it follows that there must be at least one positive Liapunov exponent μ_0-almost everywhere. Thus, this theorem proves that a positive Liapunov exponent implies positive topological entropy and so chaos.

This theorem also says that the condition on individual orbits about Liapunov exponents implies an aggregate condition on a hyperbolic invariant set. In principle, it is easier to show the existence of nonzero Liapunov exponents than the existence of a uniform hyperbolic structure. The theorem implies that there is not as much difference between these two conditions as one might think or fear.

An earlier theorem of Pesin (1977) implies that if a diffeomorphism f (i) preserves a measure $\mu_0 \in \mathcal{M}(f)$ which is equivalent to the volume for the Riemannian metric, and (ii) at least one of the Liapunov exponents is positive on a set of positive μ_0 measure, then the topological entropy is positive. In fact, there is a lower bound on the topological entropy in terms of an integral of the sum of the positive Liapunov exponents. Let $k_j(\mathbf{x})$ be the multiplicity of $\lambda_j(\mathbf{x})$, $k_j(\mathbf{x}) = \dim(V_{\mathbf{x}}^j) - \dim(V_{\mathbf{x}}^{j-1})$, and

$$\chi^u(\mathbf{x}) = \sum_{j=r(\mathbf{x})+1}^{s(\mathbf{x})} k_j(\mathbf{x})\lambda_j(\mathbf{x}),$$

then

$$h(f) \geq \int_M \chi^u(\mathbf{x})\, d\mu_0.$$

This theorem requires the measure to be smooth, but does not assume that all the exponents are nonzero. In fact, what Pesin proved was that the measure theoretic entropy with respect to μ_0 is greater than or equal to the integral,

$$h_{\mu_0}(f) \geq \int_M \chi^u(\mathbf{x})\, d\mu_0.$$

Earlier, Margulis (1969) proved the other inequality

$$h_{\mu_0}(f) \leq \int_M \chi^u(\mathbf{x})\, d\mu_0,$$

so

$$h_{\mu_0}(f) = \int_M \chi^u(\mathbf{x})\, d\mu_0.$$

See Katok (1980) or Mañé (1987a) for further discussion of this type of result.

Several recent papers have used a field of invariant cones to prove the existence of a positive Liapunov exponent and the ergodicity of the system. See Wojtkowski (1985), Burns and Gerber (1989), and Katok and Burns (1994) for references dealing with the ergodicity of geodesic flows. See Sinai (1970), Chernov and Sinai (1987), Katok and Strelcyn (1986), and Liverani and Wojtkowski (1993) for references dealing with the ergodicity of billiards problems. (The last reference gives a good introduction and many other references.)

9.3 Sinai-Ruelle-Bowen Measure for an Attractor

In this section, $f : M \to M$ is a C^2 diffeomorphism, and Λ is a uniformly hyperbolic attractor. (Since we assume that all the points of an attractor are chain recurrent and have a hyperbolic structure, the periodic points are dense in Λ. See Theorem X.4.1.) Our goal in this section is to explain why the forward orbit of most points in the basin of attraction of Λ tend to be dense in Λ: the computer picture generated by the forward orbit of most points $\mathbf{x}$ is a picture of the attractor.

We denote the basin of attraction of Λ by $W^s(\Lambda)$. Let U be a compact neighborhood of Λ in $W^s(\Lambda)$. For $\mathbf{x} \in U$, a probability measure, $\nu_{\mathbf{x}}^n$, can be associated to the partial forward orbit of length n, $\{\mathbf{x}, f(\mathbf{x}), \ldots, f^{n-1}(\mathbf{x})\}$:

$$\nu_{\mathbf{x}}^n = \frac{1}{n}\sum_{i=0}^{n-1} \delta_{f^i(\mathbf{x})}$$

where $\delta_{\mathbf{y}}$ is the atomic measure at the point $\mathbf{y}$. The Sinai-Ruelle-Bowen Theory says that there exist (i) a subset $U' \subset U$ of full Lebesgue measure and (ii) a measure μ with support on Λ such that for any $\mathbf{x} \in U'$ the measures $\nu_{\mathbf{x}}^n$ converge weakly to μ, i.e., for any continuous function $\varphi : U \to \mathbb{R}$, $\frac{1}{n}\sum_{i=0}^{n-1} \varphi \circ f^i(\mathbf{x})$ converges to $\int \varphi(\mathbf{y})\, d\mu(\mathbf{y})$. This measure μ is called the *Sinai-Ruelle-Bowen measure of the attractor*, or just the *SRB measure of the attractor*. The fact that the average of the evaluation of the function φ along the forward orbit of $\mathbf{x}$ converges to the integral means that the forward orbit is uniformly spread out over Λ in terms of the measure μ. The measure μ is characterized by the fact that (i) there is a positive Liapunov exponent μ-a.e. and (ii) μ has absolutely continuous conditional measures on the unstable manifolds for points of Λ relative to the Riemannian measure on the unstable manifolds. See Young (1993) for an introduction to this theory and other related measure theoretic results. Also see Sinai (1972), Bowen (1975b), Ruelle (1976), and Bowen and Ruelle (1975).

We mentioned above that a uniformly hyperbolic attractor always has an SRB measure. The only situation where an SRB measure has been shown to exist for a nonuniformly hyperbolic invariant set is the Hénon attractor for B very small and A just less than 2, Benedicks and Young (1993). This is the same situation where Benedicks and Carleson (1991) proved there is a transitive attractor. See Young (1993) for a discussion of this result. For $A = 1.4$ and $B = -0.3$, the Hénon map appears to have an SRB measure (the orbits of most points appear to be uniformly dense in the whole attracting set), but this result has not been verified mathematically. The same remark about an apparent SRB measure that has not been verified mathematically applies to the Lorenz attractor for $\rho = 28$.

9.4 Fractal Dimension

Another measurement of the complicated or chaotic nature of the dynamics is the dimension of the invariant set. In Section 8.6 we have defined the topological dimension of a set. This dimension is always an integer and is useful in some considerations but does not relate to the chaotic nature of the system. In this section, we discuss some dimensions which are nonnegative real numbers and do not have to be integers. The definitions of these concepts go back to Hausdorff, but recent interest has been stimulated by Mandelbrot. Generally, these types of dimensions are called fractal dimensions. We mainly consider the box dimension (also known as capacity dimension), but define Hausdorff dimension at the end of the section. We give references to other sources which introduce some of the other types of dimension: information dimension, correlation dimension, and Liapunov dimension.

Chapter 5 in Broer, Dumortier, van Strien, and Takens (1991) discusses the sense in which the box dimension is related to the chaotic nature of the system. In doing this it also defines both the box dimension and the entropy for a time series of a system. Because the time series can either be generated by an explicit map (or differential equation) or by an experiment, this indicates how the ideas can be applied to experimental situations.

Definition. In our discussion of box dimension, we only consider compact subsets A of some Euclidean space $\mathbb{R}^n$. (These definitions also make sense in a metric space. Since a manifold M can be embedded in some Euclidean space $\mathbb{R}^n$, our definitions apply to compact manifolds.) For $\epsilon > 0$, consider the subdivision of $\mathbb{R}^n$ into boxes or cubes of sides of length ϵ: for $(j_1, \ldots, j_n) \in \mathbb{Z}^n$, let

$$R_{j_1,\ldots,j_n} = \{(x_1, \ldots x_n) : j_i\epsilon \le x_i < (j_i + 1)\epsilon \text{ for } 1 \le i \le n\}.$$

A box of this kind is said to be a *box from the ϵ-grid.* Let $N(\epsilon, A)$ be the number of boxes $R_{\mathbf{j}}$ among all the choices of $\mathbf{j} \in \mathbb{Z}^n$ such that $A \cap R_{\mathbf{j}} \neq \emptyset$.

To motivate the definition of box dimension, we consider the number of boxes from the ϵ-grid, $N(\epsilon, A)$, that are needed to cover various objects. For a line segment, $N(\epsilon, A)$ is roughly ϵ^{-1} times the length. For a rectangle in a plane, $N(\epsilon, A)$ is roughly ϵ^{-2} times the area. It is not as obvious, but for a curve, $N(\epsilon, A)$ also is roughly ϵ^{-1} times the length, and for a piece of surface, $N(\epsilon, A)$ is roughly ϵ^{-2} times the area. Next, consider a compact submanifold A of $\mathbb{R}^n$ of dimension d. (The dimension is here used in the sense of the number of variables in local coordinate charts.) For such a manifold, $N(\epsilon, A)\epsilon^d$ is roughly equal to the d-dimensional volume, or $N(\epsilon, A)$ is proportional to ϵ^{-d}; more precisely, there are two constants $C_1, C_2 > 0$ such that $C_1 \leq N(\epsilon, A)\epsilon^d \leq C_2$ or $C_1\epsilon^{-d} \leq N(\epsilon, A) \leq C_2\epsilon^{-d}$. Taking logarithms of the inequality, we get that

$$\log(C_1) \leq \log(N(\epsilon, A)) - d\log(\epsilon^{-1}) \leq \log(C_2),$$

so

$$\frac{\log(N(\epsilon, A)) - \log(C_2)}{\log(\epsilon^{-1})} \leq d \leq \frac{\log(N(\epsilon, A)) - \log(C_1)}{\log(\epsilon^{-1})} \qquad \text{and}$$

$$d = \lim_{\epsilon \to 0} \frac{\log(N(\epsilon, A))}{\log(\epsilon^{-1})}.$$

Notice that for real numbers $0 \leq p < d < q$ and $0 < \epsilon < 1$, $\epsilon^p > \epsilon^d > \epsilon^q$ so

$$\lim_{\epsilon \to 0} N(\epsilon, A)\epsilon^p = \infty \qquad \text{and}$$

$$\lim_{\epsilon \to 0} N(\epsilon, A)\epsilon^q = 0.$$

Thus, for a compact submanifold A of dimension d and $0 \leq p < d < q$, the growth rate of the p-dimensional volume of boxes which cover A is infinite, the growth rate of the q-dimensional volume of boxes which cover A is zero, and the growth rate of the d-dimensional volume of the boxes which cover A is a finite number. Thus, the dimension d is characterized as that number p at which the $\lim_{\epsilon \to 0} N(\epsilon, A)\epsilon^p$ changes from being infinite to being zero. If the limit exists for $p = d$, then this limit is the d-dimensional measure of A.

Thus, the dimensions can both be defined in terms of the growth rate of $N(\epsilon, A)$ in terms of ϵ^{-1}, and as the number for which the d-dimensional measure makes sense. We use the first characterization in our definition of box dimension and the second in our definition of Hausdorff dimension. We now turn to organizing these ideas into precise definitions.

Definition. For a general compact subset $A \subset \mathbb{R}^n$, we define the *box dimension of A*, $\dim_b(A)$, by

$$\dim_b(A) = \liminf_{\epsilon \to 0} \frac{\log(N(\epsilon, A))}{\log(\epsilon^{-1})}.$$

This dimension is often called the *capacity dimension of A* or *limit capacity of A*. Because the word capacity has other meanings in certain discussions of complex dynamics, the term "box dimension" is becoming the standard term for this measurement. If we use the limsup instead of the liminf, we get the *upper box dimension of A*, which is denoted by $\dim_B(A)$,

$$\dim_B(A) = \limsup_{\epsilon \to 0} \frac{\log(N(\epsilon, A))}{\log(\epsilon^{-1})}.$$

REMARK 4.1. Clearly, $\dim_b(A) \leq \dim_B(A) \leq n$, when $A \subset \mathbb{R}^n$.

REMARK 4.2. The box dimension of a set A is the same as the box dimension of its closure. This is the reason we restrict to closed sets. We take the sets to be compact so there are at most a finite number of boxes from the ϵ-grid which intersect A.

REMARK 4.3. We use an exercise to show that

$$\liminf_{\epsilon \to 0} N(\epsilon, A)\epsilon^p = \begin{cases} \infty & \text{for } 0 \le p < \dim_{\mathrm{b}}(A) \\ 0 & \text{for } \dim_{\mathrm{b}}(A) < p < \infty \end{cases}$$

and

$$\limsup_{\epsilon \to 0} N(\epsilon, A)\epsilon^p = \begin{cases} \infty & \text{for } 0 \le p < \dim_{\mathrm{B}}(A) \\ 0 & \text{for } \dim_{\mathrm{B}}(A) < p < \infty. \end{cases}$$

See Exercise 9.30.

Instead of using a fixed set of boxes from the ϵ-grid, it is possible to cover the set A by a finite set of closed cubes with length ϵ on a side and sides parallel to the axes (without fixing the grid of hyperplanes which form the surfaces). Let $N'(\epsilon, A)$ be the minimum number of such cubes of size ϵ which cover A. The following result shows that $N'(\epsilon, A)$ can be used to calculate the box dimension instead of $N(\epsilon, A)$.

Proposition 4.1. *Let $A \subset \mathbb{R}^n$ be a compact subset. The box dimension of A can be calculated by the following limit:*

$$\dim_{\mathrm{b}}(A) = \liminf_{\epsilon \to 0} \frac{\log(N'(\epsilon, A))}{\log(\epsilon^{-1})},$$

where $N'(\epsilon, A)$ is defined above as the minimum number of boxes without fixing the grid.

PROOF. Clearly, $N'(\epsilon, A) \le N(\epsilon, A)$ since it is possible to choose the cover from boxes from the ϵ-grid. Also, any cube of size ϵ is contained in 2^n boxes from the ϵ-grid, so $N(\epsilon, A) \le 2^n N'(\epsilon, A)$. Taking logarithms and taking the limit, we get the following result:

$$\begin{aligned} \dim_{\mathrm{b}}(A) &= \liminf_{\epsilon \to 0} \frac{\log(N(\epsilon, A))}{\log(\epsilon^{-1})} \\ &\le \liminf_{\epsilon \to 0} \frac{n\log(2) + \log(N'(\epsilon, A))}{\log(\epsilon^{-1})} \\ &= \liminf_{\epsilon \to 0} \frac{\log(N'(\epsilon, A))}{\log(\epsilon^{-1})} \\ &\le \liminf_{\epsilon \to 0} \frac{\log(N(\epsilon, A))}{\log(\epsilon^{-1})} \\ &= \dim_{\mathrm{b}}(A). \end{aligned}$$

□

It is also possible to cover the set A with a ball of diameter ϵ in terms of the Euclidean metric or any metric equivalent to the Euclidean metric. A metric d is said to be equivalent to the Euclidean metric provided there are constants $C_1, C_2 > 0$ such that

$$C_1 |\mathbf{x} - \mathbf{y}| \le d(\mathbf{x}, \mathbf{y}) \le C_2 |\mathbf{x} - \mathbf{y}|$$

for all $\mathbf{x}, \mathbf{y} \in \mathbb{R}^n$. Using a metric equivalent to the Euclidean metric, let $N''_\epsilon(A)$ be the minimum number of closed balls of diameter ϵ which cover A. We defer to Exercise 9.31 to verify the following proposition.

Proposition 4.2. *Let $A \subset \mathbb{R}^n$ be a compact subset. The box dimension of A is given by the following limit:*

$$\dim_b(A) = \liminf_{\epsilon \to 0} \frac{\log(N''(\epsilon, A))}{\log(\epsilon^{-1})},$$

where $N''(\epsilon, A)$ is defined above as the minimum number of closed balls of diameter ϵ which cover A.

It is also useful to calculate the box dimension using a sequence of diameters going to zero and not having to use all possible ϵ. To that end, we have the following result.

Proposition 4.3. *Let $A \subset \mathbb{R}^n$ be a compact subset. Assume $0 < r < 1$. Then*

$$\dim_b(A) = \liminf_{j \to \infty} \frac{\log(N'(r^j, A))}{j \log(r^{-1})}.$$

PROOF. For any $\epsilon > 0$, there is a j such that $r^{j+1} < \epsilon \leq r^j$. In particular,

$$\begin{aligned}
&r^{-1} r^{-j} > \epsilon^{-1} \geq r\, r^{-j-1}, \\
&\log(r^{-1}) + j \log(r^{-1}) > \log(\epsilon^{-1}) \geq \log(r) + (j+1)\log(r^{-1}), \quad \text{and} \\
&[\log(r^{-1}) + j \log(r^{-1})]^{-1} < [\log(\epsilon^{-1})]^{-1} \leq [\log(r) + (j+1)\log(r^{-1})]^{-1}.
\end{aligned}$$

Since $N'(\epsilon, A) \geq N'(\delta, A)$ whenever $\epsilon < \delta$,

$$N'(r^j, A) \leq N'(\epsilon, A) \leq N'(r^{j+1}, A).$$

Therefore,

$$\begin{aligned}
\liminf_{j \to \infty} \frac{\log(N'(r^j, A))}{j \log(r^{-1})} &= \liminf_{j \to \infty} \frac{\log(N'(r^j, A))}{\log(r^{-1}) + j \log(r^{-1})} \\
&\leq \liminf_{\epsilon \to 0} \frac{\log(N'(\epsilon, A))}{\log(\epsilon^{-1})} \\
&= \dim_b(A) \\
&\leq \liminf_{j \to \infty} \frac{\log(N'(r^{j+1}, A))}{\log(r) + (j+1)\log(r^{-1})} \\
&= \liminf_{j \to \infty} \frac{\log(N'(r^{j+1}, A))}{(j+1)\log(r^{-1})} \\
&= \liminf_{j \to \infty} \frac{\log(N'(r^j, A))}{j \log(r^{-1})}.
\end{aligned}$$

Because the first and last entries are equal, they must equal the box dimension. □

Example 4.1. Let C be the middle-α Cantor set in the line. Let β be chosen such that $2\beta + \alpha = 1$, so that $0 < \beta < 1/2$. In the formation of the Cantor set, there are 2^j intervals of length β^j which cover C, so $N'(\beta^j) = 2^j$. Therefore,

$$\begin{aligned}
\dim_b(C) &= \liminf_{j \to \infty} \frac{\log(N'(\beta^j, C))}{j \log(\beta^{-1})} \\
&= \liminf_{j \to \infty} \frac{\log(2^j)}{j \log(\beta^{-1})} \\
&= \frac{\log(2)}{\log(\beta^{-1})}.
\end{aligned}$$

First of all, note that $0 < \dim_b(C) < 1$. Thus, these Cantor sets have nonintegral box dimension. Also, the dimension depends on β and hence α. In fact, any number between 0 and 1 can be realized as the box dimension by the proper choice of α.

It is also possible to construct a Cantor set in the line (which is not a middle-α Cantor set) with positive Lebesgue measure so $\dim_b(C) = 1$.

Example 4.2. For $0 < \beta < 1/2$, a solenoid can be formed using the map

$$f(t, z) = (g(t), \beta\, z + \frac{1}{2}e^{2\pi t i}),$$

where $g(t) = 2t \mod 1$. This map f takes the neighborhood $N = S^1 \times D^2$ into itself. Let $\Lambda = \bigcap_{k\geq 0} f^k(N)$ be the attractor as in the usual construction of the solenoid. The set $\mathcal{S}_k = f^k(N)$ is the finite intersection. Let $D(t) = \{t\} \times D^2$ be a single fiber as before. For each t, $\mathcal{S}_k \cap D(t)$ is the union of 2^k disks of diameter $2\beta^k$. Therefore,

$$\begin{aligned}
\dim_b(\Lambda \cap D(t)) &= \liminf_{k\to\infty} \frac{\log(N'(2\beta^k, \Lambda \cap D(t)))}{k \log(\beta^{-1})} \\
&= \liminf_{k\to\infty} \frac{k \log(2)}{k \log(\beta^{-1})} \\
&= \frac{\log(2)}{\log(\beta^{-1})}.
\end{aligned}$$

We do not give the details but

$$\dim_b(\Lambda) = 1 + \frac{\log(2)}{\log(\beta^{-1})}$$

with the other dimension coming from the expanding direction. Notice that by picking β close to $1/2$, $\dim_b(\Lambda \cap D(t))$ can be made almost equal to one and $\dim_b(\Lambda)$ can be made almost equal to two.

REMARK 4.4. If we define Cantor sets of the type of $\Lambda \cap D(t)$ in the above example but with different rates of contraction in different directions, it is often impossible to calculate the box dimension exactly.

Hausdorff Dimension

We mentioned above for the box dimension that

$$\liminf_{\epsilon\to 0} N(\epsilon, A)\epsilon^p = \begin{cases} \infty & \text{for } 0 \leq p < \dim_b(A) \\ 0 & \text{for } \dim_b(A) < p < \infty. \end{cases}$$

The definition of Hausdorff dimension uses a characterization like this one.

Definition. If U is a nonempty subset of $\mathbb{R}^n$, let

$$|U| = \sup\{|\mathbf{x} - \mathbf{y}| : \mathbf{x}, \mathbf{y} \in U\}$$

denote the diameter of U. If $A \subset \bigcup_i U_i$, where each U_i is a ball of diameter less than or equal to δ, then $\{U_i\}$ is called a *δ-cover of A*.

Let $0 \leq p$. We want to define a p measure of a Borel set A. To do this, first take $\delta > 0$ and define

$$\mathcal{H}^p_\delta(A) = \inf \sum_{i=1}^{\infty} |U_i|^p,$$

where the infimum is taken over all countable δ-covers. Notice in this summation, all the diameters of the different sets do not have to be equal; this contrasts with the definition for box dimension. Next let δ go to zero and define

$$\mathcal{H}^p(A) = \lim_{\delta \to 0} \mathcal{H}^p_\delta(A).$$

Because $\mathcal{H}^p_\delta(A)$ increases as δ decreases, the limit exists but can equal infinity. This quantity, $\mathcal{H}^p(A)$, is called the *Hausdorff p-dimensional measure of A*. If $A \subset \mathbb{R}^n$, then $\mathcal{H}^n(A)$ is the Lebesgue measure of A. The *Hausdorff dimension of A* is defined to be d if

$$\mathcal{H}^p(A) = \begin{cases} \infty & \text{for } 0 \le p < d \\ 0 & \text{for } d < p \le \infty. \end{cases}$$

The Hausdorff dimension of A is denoted by $\dim_H(A)$. If a set A has Hausdorff dimension d, then the Hausdorff d-dimensional measure of A can be zero, infinity, or a finite number.

REMARK 4.5. For a compact set $A \subset \mathbb{R}^n$,

$$\dim_H(A) \le \dim_{\mathrm{b}}(A) \le \dim_{\mathrm{B}}(A) \le n.$$

In many ways the Hausdorff dimension is defined in a manner more similar to Lebesgue measure than the box dimension.

REMARK 4.6. In Palis and Takens (1993), Cantor subsets of $\mathbb{R}$ which are defined in terms of a map in a manner like Λ_μ is defined by $F_\mu(x) = \mu x(1-x)$ are called *dynamically defined.* These sets include the middle-α Cantor sets, C_α, as well as less uniform ones like Λ_μ. They prove that if $C \subset \mathbb{R}$ is a dynamically defined Cantor set, then the Hausdorff dimension equals the upper box dimension and so also the box dimension, $\dim_H(A) = \dim_{\mathrm{b}}(A) = \dim_{\mathrm{B}}(A) \le 1$. In particular, $\dim_H(C_\alpha) = \dim_{\mathrm{b}}(C_\alpha) = \log(2)/\log(2/(1-\alpha))$ for the middle-α Cantor set. See Palis and Takens (1993), Takens (1988), or Manning and McCluskey (1983).

For a further discussion of fractal dimensions, see Edgar (1990) or Falconer (1990). For an introduction to dimension in terms of dynamical systems, see Farmer, Ott, and Yorke (1983) or Palis and Takens (1993). For a systematic treatment of different fractal dimensions in terms of dynamical systems, see Pesin (1997). For a connection with measures and entropy, see Young (1982).

9.5 Exercises

Topological Entropy

9.1. Let $f_\mu(x) = \mu x \bmod 1$, for $\mu > 1$. Calculate the entropy of f_μ, $h(f_\mu)$.

9.2. Let f be a diffeomorphism on the circle, S^1. Prove that the topological entropy of f is zero, $h(f) = 0$.

9.3. Let $d > 1$ be an integer. Let $f : S^1 \to S^1$ have a covering map $F : \mathbb{R} \to \mathbb{R}$ given by

$$F(x) = dx.$$

Prove that the entropy of f is $\log(d)$, $h(f) = \log(d)$.

9.4. Let $\sigma_N : \Sigma_N \to \Sigma_N$ be the full shift on N symbols. Prove that the entropy of σ_N is $\log(N)$, $h(\sigma_N) = \log(N)$.

9.5. Let $A = \begin{pmatrix} 1 & 1 \\ 1 & 0 \end{pmatrix}$. Find the entropy of the subshift of finite type with transition matrix A.

9.6. Use Theorem 1.9(a) (but not Theorem 1.9(b)) to calculate the entropy of the subshifts of finite type with the following transition matrices A, $h(\sigma_A)$. Note: These examples illustrate again that (i) the growth rate of the number of the wandering orbits does not contribute to the entropy and (ii) the entropy is the maximum of the entropy on disjoint invariant pieces of the nonwandering set.

(a) Let $A = \begin{pmatrix} 1 & 1 \\ 0 & 1 \end{pmatrix}$.

(b) Let $A = \begin{pmatrix} 1 & 1 & 1 \\ 0 & 1 & 1 \\ 0 & 0 & 1 \end{pmatrix}$.

(c) Let $A = \begin{pmatrix} 1 & 1 & 1 & 1 & 1 \\ 1 & 1 & 1 & 1 & 1 \\ 1 & 1 & 1 & 1 & 1 \\ 0 & 0 & 0 & 1 & 1 \\ 0 & 0 & 0 & 1 & 1 \end{pmatrix}$.

9.7. Let A be an N by N transition matrix and $\Sigma_A \subset \Sigma_N$ be the corresponding subshift of finite type. Prove that the entropy $h(\sigma_A) \leq \log(N)$.

9.8. Let $f : M \to M$ and $g : N \to N$ be two maps on compact metric spaces. Consider the map $f \times g : M \times N \to M \times N$ defined by $f \times g(\mathbf{x}, \mathbf{y}) = (f(\mathbf{x}), g(\mathbf{y}))$. Prove that $h(f \times g) = h(f) + h(g)$.

9.9. Assume $f : X \to X$ is a homeomorphism. Prove that $h(f^{-1}) = h(f)$.

9.10. Let $f_\omega : S^1 \to S^1$ be a rotation by ω. Prove that the entropy $h(f_\omega) = 0$.

9.11. Assume $f : [0,1] \to [0,1]$ is continuous and has two subinterval $I_1, I_2 \subset [0,1]$ such that $f(I_1) \supset I_1 \cup I_2$ and $f(I_2) \supset I_1 \cup I_2$. Prove that the entropy of f is at least $\log(2)$, $h(f) \geq \log(2)$.

9.12. Assume $f : [0,1] \to [0,1]$ is continuous and has a periodic point of a period which is not a power of 2. Prove that f has positive entropy. Hint: Use the Stefan cycle to get intervals which cover each other.

9.13. Assume $f : X \to X$ is a continuous map on a metric space X and $f(X) = Y$. Prove that $h(f) = h(f|Y)$.

9.14. Use Proposition III.2.10 to prove Theorem 1.9(b) in the case when A is reducible.

9.15. Let $A : \Sigma_2 \to \Sigma_2$ be the adding machine map:

$$A(s_0 s_1 s_2 \ldots) = (s_0 s_1 s_2 \ldots) + (1000\ldots) \bmod 2,$$

i.e., $(1000\ldots)$ is added to $(s_0 s_1 s_2 \ldots)$ mod 2 with carrying.

(a) Prove that $h(A) = 0$. Hint: Take $\epsilon = 3^{-k+1}2^{-1}$ so the closed balls of radius of radius ϵ are cylinder sets given in Exercise 2.15. Then, show that A permutes these cylinder sets.

(b) Let $f_\infty : [0,1] \to [0,1]$ be the map defined in Example III.1.3. Prove that $h(f_\infty) = 0$.

9.16. Let f be the solenoid diffeomorphism given in Section 8.7. Using a Markov partition, prove that $h(f) = \log(2)$.

9.17. Let g be the hyperbolic toral automorphism for the matrix $\begin{pmatrix} 2 & 1 \\ 1 & 1 \end{pmatrix}$. Prove that the the entropy $h(g) = (3 + \sqrt{5})/2$.

9.18. Let f be the DA-diffeomorphism formed from the hyperbolic toral automorphism g for the matrix $\begin{pmatrix} 2 & 1 \\ 1 & 1 \end{pmatrix}$. Using the result of Bowen given in Remark 1.11, prove that the entropies of f and g are equal, $h(f) = h(g)$.

9.19. Let $f : \mathbb{T}^2 \to \mathbb{T}^2$ be the noninvertible toral map induced by the matrix

$$\begin{pmatrix} 1 & -1 \\ 1 & 1 \end{pmatrix}.$$

Calculate the entropy of f. Hint: Find a power f^k for which it is easy to calculate the entropy and use Theorem 1.2 relating the entropy of f^k to the entropy of f.

9.20. Let $f : [0, 3] \to [0, 3]$ be the piecewise linear function such that $f(0) = 3$, $f(1) = 2$, $f(2) = 3$, and $f(3) = 0$. (The map f is linear between each pair of adjacent integers.)
  (a) Find a Markov partition for f and its transition matrix.
  (b) Find the topological entropy of f.

Liapunov Exponents

9.21. Let f be the solenoid map given in Section 8.7. Prove that the Liapunov exponents satisfy $\lambda_2 = \lambda_3 = \log(4^{-1})$ and $\lambda_1 \geq \log(2)$.

9.22. (Generalized Baker's map) Assume $0 < \mu_1 < \mu_2 < 1$ satisfy $\mu_1 + \mu_2 < 1$. Define the function

$$f_{\mu_1,\mu_2}(x, y) = \begin{cases} (2x, \mu_1 y) & \text{for } 0 \leq x < 0.5,\ 0 \leq y \leq 1 \\ (2x - 1, 1 - \mu_2 + \mu_2 y) & \text{for } 0.5 \leq x \leq 1,\ 0 \leq y \leq 1 \end{cases}$$

on the square $S = [0, 1] \times [0, 1]$.
  (a) Show that the Liapunov exponents satisfy $\lambda_1(x, y) = \log(2)$ and

$$\log(\mu_1) \leq \lambda_2(x, y) \leq \log(\mu_2)$$

  for all $(x, y) \in S$.
  (b) Notice that the first coordinate function of f_{μ_1,μ_2} is essentially (except for $x = 1$) the doubling map $D(x) = 2x \mod 1$ and so is ergodic with respect to Lebesgue measure on $[0, 1]$. (You do not need to prove this fact.) Using the Birkhoff Ergodic Theorem, prove that $\lambda_2(x, y) = 0.5[\log(\mu_1) + \log(\mu_2)]$ almost everywhere with respect to Lebesgue measure on S.
  (c) Find three different periodic orbits (of any period) for which $\lambda_2(x, y)$ equals $\log(\mu_1)$, $\log(\mu_2)$, and $[\log(\mu_1) + \log(\mu_2)]/2$, respectively.

9.23. Let $f : M \to M$ be a C^1 diffeomorphism. Assume $\mathbf{x} \in M$ and $\mathbf{v} \in T_{\mathbf{x}}M$ are a point and a vector for which the Liapunov exponent $\lambda(\mathbf{x}, \mathbf{v})$ exists.
  (a) Let α be a real number. Prove that

$$\lambda(\mathbf{x}, \alpha\mathbf{v}) = \lambda(\mathbf{x}, \mathbf{v}).$$

  (b) Prove that

$$\lambda(f(\mathbf{x}), Df_{\mathbf{x}}\mathbf{v}) = \lambda(\mathbf{x}, \mathbf{v}).$$

Fractal Dimension

9.24. Let $S = \{0\} \cup \{1/k : k \text{ is a positive integer }\}$.
  (a) Prove that $\dim_b(S) = 1/2$.
  (b) Prove that $\dim_H(S) = 0$.

9.25. Let $S = \{0\} \cup \{2^{-k} : k \text{ is a positive integer }\}$.
(a) Prove that $\dim_{\mathrm{B}}(S) = 0$.
(b) Prove that $\dim_H(S) = 0$.

9.26. Let $F_\mu(x) = \mu x(1-x)$ and $\Lambda_\mu = \bigcap_{n\geq 0} F_\mu^n([0,1])$. Let $\lambda_\mu = (\mu^2 - 4\mu)^{1/2}$. Prove that

$$\dim_{\mathrm{b}}(\Lambda_\mu) \leq \frac{\log(2)}{\log(\lambda_\mu)}$$

for $\mu > 2(1+\sqrt{2})$.

9.27. Construct a Cantor set in the line with box dimension equal to one.

9.28. Let $N = S^1 \times D^2$ and

$$f(t,z) = (g(t), \frac{1}{2}z + \beta\, e^{2\pi t i})$$

where $g(t) = 4t \bmod 1$. Let $\Lambda = \bigcap_{k\geq 0} f^k(N)$.
(a) Prove for $0 < \beta < 2^{-1/2}$, that f is an embedding of N into itself.
(b) Let $D(t) = \{t\} \times D^2$. Prove that

$$\dim_{\mathrm{b}}(\Lambda \cap D(t)) = \frac{\log(4)}{\log(\beta^{-1})}.$$

Also prove for correction choice of β, that $\dim_{\mathrm{b}}(\Lambda \cap D(t)) > 1$. Note that $\Lambda \cap D(t)$ is a totally disconnected set that has box dimension greater than one.
(c) Prove that

$$\dim_{\mathrm{b}}(\Lambda) = 1 + \frac{\log(4)}{\log(\beta^{-1})}.$$

9.29. Calculate the box dimension of the invariant set Λ for the Geometric horseshoe.

9.30. Let $A \subset \mathbb{R}^n$ be a compact set. Prove that

$$\liminf_{\epsilon\to 0} N(\epsilon, A)\epsilon^p = \begin{cases} \infty & \text{for } 0 \leq p < \dim_{\mathrm{b}}(A) \\ 0 & \text{for } \dim_{\mathrm{b}}(A) < p < \infty \end{cases}$$

and

$$\limsup_{\epsilon\to 0} N(\epsilon, A)\epsilon^p = \begin{cases} \infty & \text{for } 0 \leq p < \dim_{\mathrm{B}}(A) \\ 0 & \text{for } \dim_{\mathrm{B}}(A) < p < \infty. \end{cases}$$

9.31. Let $A \subset \mathbb{R}^n$ be a compact subset. Using a metric equivalent to the Euclidean metric, let $N''_\epsilon(A)$ be the minimum number of balls of diameter ϵ which cover A. Prove that

$$\dim_{\mathrm{b}}(A) = \liminf_{\epsilon\to 0} \frac{\log(N''(\epsilon, A))}{\log(\epsilon^{-1})}.$$

9.32. Let f_{μ_1,μ_2} be the generalized Baker's map defined in Exercise 9.21.
(a) Prove there is a unique positive number d such that $1 = \mu_1^d + \mu_2^d$.
(b) Let d be the number given in part (a). Let

$$\Lambda = \bigcap_{n=0}^{\infty} f_{\mu_1,\mu_2}^n([0,1] \times [0,1]).$$

Prove that the Hausdorff dimension of $\Lambda \cap (\{0\} \times [0,1])$ is less than or equal to d. Remark: Theorem 6.3.12 in Edgar (1990) proves that the Hausdorff dimension of $\Lambda \cap (\{0\} \times [0,1])$ is actually equal to d. The Hausdorff dimension of Λ is then equal to $1+d$. Because the set is dynamically defined, the box dimension of $\Lambda \cap (\{0\} \times [0,1])$ is also d, so the box dimension of Λ is $1+d$.

CHAPTER X
Global Theory of Hyperbolic Systems

In this chapter, we take the ideas introduced in the last chapter and make them into a more complete theory. The first section shows that in some sense any map can be decomposed into a map on chain recurrent pieces and a gradient-like map (or flow) between the pieces. This is a very general theorem of Conley and can be proved without using much of the material introduced earlier in this book except the definition of chain recurrent. The next section indicates how the proof of the stable manifold theorem for a fixed point can be modified to prove the case for a hyperbolic invariant set. Using these results, we prove the possibility of shadowing near a hyperbolic invariant set, the Ω-stability of diffeomorphisms with a hyperbolic chain recurrent set, and the structural stability of diffeomorphisms which satisfy a "transversality" condition in addition to having a hyperbolic chain recurrent set. These theorems form the heart of the theory of hyperbolic diffeomorphisms (ones for which the chain recurrent set is hyperbolic) which was articulated by Smale (1967) and carried out in the following years by Smale and other researchers.

10.1 Fundamental Theorem of Dynamical Systems

In this section, we study a flow φ^t on a metric space M. Usually, M is compact. Once we give the results for flows, we indicate how they can be carried over to homeomorphisms (or diffeomorphisms) in the next subsection.

The main tool is that there is a Liapunov function which is decreasing off the chain recurrent set, i.e., off the set of points on which complicated dynamics takes place. Different parts of this chain recurrent set may lie at different levels of the Liapunov function and so there is no way to move from one of these pieces to another piece and then back to the first piece (since the Liapunov function is strictly decreasing off the chain recurrent set). Therefore, if the pieces of the chain recurrent set are collapsed to points, the flow on the quotient space is a gradient-like system, so the original flow can be thought of as a gradient-like extension of a flow that is transitive on disjoint pieces.

The main reference and original reference for this section is Conley (1978). Franks (1988) has a proof of many of the theorems of this section with proof directly for diffeomorphisms. Hurley (1991, 1992) has extended some of these results to noncompact sets.

In Section 2.3 we defined the chain recurrent set for a map and in Section 5.4 we gave the modifications for a flow. In particular, we defined an ϵ-chain. For a subset $Y \subset M$ we defined the ϵ-chain limit set of Y, $\Omega_\epsilon^+(Y)$, and the chain limit set of Y, $\Omega^+(Y)$. Similarly, backward sets $\Omega_\epsilon^-(Y)$ and $\Omega^-(Y)$. These definitions play an important role in this section and should be reviewed. In particular, the *chain recurrent set* is given by

$$\begin{aligned}\mathcal{R}(\varphi^t) &= \{\mathbf{x} : \text{ there is an } \epsilon\text{-chain from } \mathbf{x} \text{ to } \mathbf{x} \text{ of length greater than } T \\ &\qquad\qquad \text{for all } \epsilon > 0 \text{ and for all } T > 1\} \\ &= \{\mathbf{x} : \mathbf{x} \in \Omega^+(\mathbf{x})\} \\ &= \{\mathbf{x} : \mathbf{x} \in \Omega^-(\mathbf{x})\}.\end{aligned}$$

The definitions for a flow are the same once an ϵ-chain is defined, which is done in Section 5.4.

In order to give an interpretation of the fundamental theorem below, we form equivalence classes of points in $\mathcal{R}(\varphi^t)$ for which there are chains from one point to another and back to the the first point. This concept is made precise in the following definition.

Definition. We define a relation $\sim$ on $\mathcal{R}(\varphi^t)$ by $\mathbf{x} \sim \mathbf{y}$ if $\mathbf{y} \in \Omega^+(\mathbf{x})$ and $\mathbf{x} \in \Omega^+(\mathbf{y})$. It is clear that this is an equivalence relation. The equivalence classes are called the *chain components* of $\mathcal{R}(\varphi^t)$. It is not hard to prove that for a flow, the chain components are indeed the connected components of $\mathcal{R}(\varphi^t)$.

In Section 8.12 on Morse-Smale Systems, the definition of a Liapunov function in a global sense is given, as opposed to near a fixed point. Also, a gradient vector field and gradient-like vector fields and flows on Euclidean spaces and manifolds are defined in that section. The following definition of Conley seems very different than the definition of gradient-like flow but it is compatible in the sense that a corollary of the Fundamental Theorem of Dynamical Systems (given below) says that a strongly gradient-like flow is gradient-like.

Definition (Conley). A flow φ^t is called *strongly gradient-like* provided $\mathcal{R}(\varphi^t)$ is totally disconnected (and consequently made up entirely of fixed points, $\mathcal{R}(\varphi^t) = \text{Fix}(\varphi^t)$).

Using these definitions, we can state the main theorem.

Theorem 1.1 (Conley's Fundamental Theorem of Dynamical Systems). *A flow φ^t on a compact metric space has a Liapunov function L which is strictly decreasing off $\mathcal{R}(\varphi^t)$ and such that $L(\mathcal{R}(\varphi^t))$ is a nowhere dense subset of $\mathbb{R}$.*

REMARK 1.1. Conley interprets his Fundamental Theorem by looking at the induced flow after collapsing points in the same chain component to points and stating that the resulting flow is strongly gradient-like. More specifically, let $\mathbf{x} \sim \mathbf{y}$ if $\mathbf{x}$ and $\mathbf{y}$ are in the same chain component, i.e., $\mathbf{y} \in \Omega^+(\mathbf{x})$ and $\mathbf{x} \in \Omega^+(\mathbf{y})$. The flow φ^t induces a flow Φ^t on the quotient space $M^* = M/\{\mathbf{x} \sim \mathbf{y}\}$ in a natural way. Then, $\mathcal{R}(\Phi^t)$ is clearly totally disconnected. The theorem says that there is a strong Liapunov function L for the flow φ^t. This function is constant on chain components so it induces a strong Liapunov function $\bar{L}$ for Φ^t, and Φ^t is gradient-like. In particular, if φ^t is strongly gradient-like (the chain recurrent set of φ^t is already totally disconnected), then it is gradient-like.

The proof of the theorem requires proving the connection between the chain recurrent set and the collection of all pairs of attracting and repelling sets. We start by giving these definitions for a flow; they are much the same as for a diffeomorphism except that the "time" is continuous rather than discrete.

Definition. Let φ^t be a flow on a metric space M. A set U is called a *trapping region* (or *isolating neighborhood* by Conley) provided U is positively invariant and there is a $T > 0$ such that $\varphi^T(\text{cl}(U)) \subset \text{int}(U)$. A set A is called an *attracting set for a flow* φ^t provided there exists a trapping region U such that $A = \bigcap_{t \geq 0} \varphi^t(U)$. A set A^* is called a *repelling set* provided there exists a trapping region U such that $A^* = \bigcap_{t \leq 0} \varphi^t(M \setminus U)$. The set A^* is also called the *dual repelling set* for the attracting set A. The ordered sets (A, A^*) are called the *attracting-repelling pair* for the trapping region U. The collection of all attracting-repelling pairs is denoted by

$$\mathcal{A} = \{(A, A^*) : U \text{ is a trapping region for the attracting set } A \\ \text{and the repelling set } A^*\}.$$

A set V is called a *weak trapping region* for the flow φ^t provided there is some $T > 0$ such that $\text{cl}(\varphi^T(V)) \subset \text{int}(V)$, i.e., V is not necessarily positively invariant.

REMARK 1.2. The sets A and A^*, which we call an attracting set and a repelling set, respectively, Conley calls an attractor and a dual repeller, respectively. We reserve the word "attractor" for an attracting set which is transitive.

REMARK 1.3. Proposition 1.10 proves that if V is a weak trapping region with $A = \bigcap_{t\geq 0} \varphi^t(V)$ and $A^* = \bigcap_{t\leq 0} \varphi^t(M \setminus V)$, then there is a trapping region with the same attracting-repelling pair, (A, A^*). Proposition 1.9 proves that an attracting-repelling pair has a trapping region which is strongly positively invariant in the sense that $\varphi^t(\text{cl}(U)) \subset \text{int}(U)$ for all $t > 0$. Thus, these two propositions give two different ways in which the definition of a trapping region can be changed.

The following proposition gives a couple of useful properties of attracting sets.

Proposition 1.2. *(a) Any attracting set or repelling set is closed.*
(b) Both A and A^* are positively and negatively invariant.

PROOF. (a) Because of the property that $\varphi^T(\text{cl}(U)) \subset \text{int}(U)$, it follows that A is also the intersection of closed sets and so is closed: $A = \bigcap_{t\geq 0} \varphi^t(\text{cl}(U))$. The proof for A^* is similar.

(b) The proofs that attracting sets and repelling sets are invariant are similar, so we only look at A. Let $\mathbf{x} \in A$ and fix any real s. Then, $\mathbf{x} \in \varphi^t(U)$ for all $t \geq 0$, so $\varphi^s(\mathbf{x}) \in \varphi^{t+s}(U)$. Therefore, $\varphi^s(\mathbf{x}) \in \bigcap_{t\geq|s|} \varphi^{t+s}(U) = A$. Since s is arbitrary, A is both positively and negatively invariant. □

The following theorem gives the connection between the chain recurrent set and the set of all attracting-repelling pairs, and is used to prove the existence of a Liapunov function which is strictly decreasing off $\mathcal{R}(\varphi^t)$, as given in Theorem 1.1.

Theorem 1.3. *Let φ^t be a flow on a compact metric space M. Let*

$$\mathcal{P} = \bigcap \{A \cup A^* : (A, A^*) \in \mathcal{A}\}.$$

Then, $\mathcal{P} = \mathcal{R}(\varphi^t)$.

Before starting the proof, we give an example.

Example 1.1. Let φ^t be the gradient flow on T^2 with one sink $\mathbf{p}_0$, two saddles $\mathbf{p}_1$ and $\mathbf{p}_2$, and one source $\mathbf{p}_3$. If $U_0 = \emptyset$, then $A_0 = \emptyset$ and $A_0^* = T^2$. Next, let U_1 be an attracting neighborhood of the sink $\mathbf{p}_0$. Then, $A_1 = \{\mathbf{p}_0\}$ and $A_1^* = \{\mathbf{p}_3\} \cup W^s(\mathbf{p}_2) \cup W^s(\mathbf{p}_1)$. If U_2 is a positively invariant set that contains $\mathbf{p}_0$ and $\mathbf{p}_1$ but not $\mathbf{p}_2$ or $\mathbf{p}_3$, then $A_2 = \{\mathbf{p}_0\} \cup W^u(\mathbf{p}_1)$ and $A_2^* = \{\mathbf{p}_3\} \cup W^s(\mathbf{p}_2)$. If U_3 is a positively invariant set that contains $\mathbf{p}_0$ and $\mathbf{p}_2$ but not $\mathbf{p}_1$ or $\mathbf{p}_3$, then $A_3 = \{\mathbf{p}_0\} \cup W^u(\mathbf{p}_2)$ and $A_3^* = \{\mathbf{p}_3\} \cup W^s(\mathbf{p}_1)$. If U_4 is a set such that $T^2 \setminus U_4$ is a repelling neighborhood of $\mathbf{p}_3$, then $A_4 = \{\mathbf{p}_0\} \cup W^u(\mathbf{p}_1) \cup W^u(\mathbf{p}_2)$ and $A_4^* = \{\mathbf{p}_3\}$. Finally, if $U_5 = T^2$, then $A_5 = T^2$ and $A_5^* = \emptyset$. By taking the intersection of these sets, $\mathcal{P} \supset \bigcap_{0\leq j\leq 5} A_j \cup A_j^* = \{\mathbf{p}_j : 0 \leq j \leq 3\}$. By the remarks preceding the example, we see that the other inclusion is also true and so we have equality (as stated in Theorem 1.3).

The following lemma proves $\mathcal{R}(\varphi^t) \supset \mathcal{P}$, which is one direction of the inclusion of the two sets in Theorem 1.3.

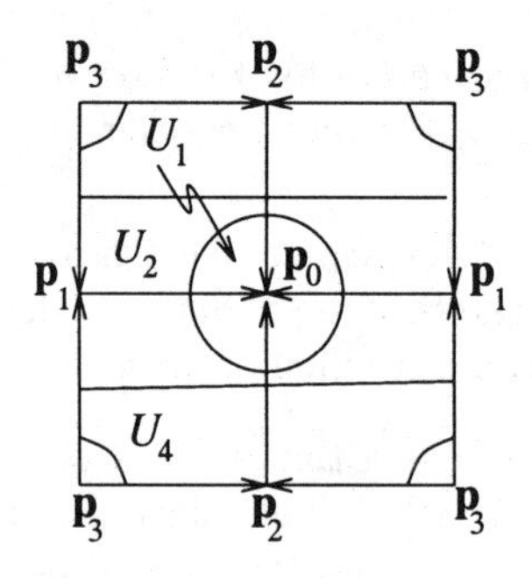

FIGURE 1.1. Neighborhoods U_1, U_2, and U_4 for Example 1.1

Lemma 1.4. *If* $\mathbf{y} \notin \mathcal{R}(\varphi^t)$, *then there exists an* $(A, A^*) \in \mathcal{A}$ *such that* $\mathbf{y} \notin A \cup A^*$. *Therefore,* $\mathcal{R}(\varphi^t) \supset \mathcal{P}$.

PROOF. Take $\mathbf{y} \notin \mathcal{R}(\varphi^t)$. Then, there is an $\epsilon > 0$ such that $\mathbf{y} \notin \Omega^+_\epsilon(\mathbf{y})$. Let $U = \Omega^+_\epsilon(\mathbf{y})$. Then, $\varphi^1(\mathrm{cl}(U)) \subset \mathrm{int}(U)$ (or else it would be possible to ϵ-chain out of U and the set would be bigger). Let $A = \bigcap_{t\geq 0} \varphi^t(U)$ be the attracting set for U and $A^* = \bigcap_{t\leq 0} \varphi^t(M \setminus U)$ be the dual repelling set. Since $\mathbf{y} \notin U$, $\mathbf{y} \notin A$. From the definition of Ω^+_ϵ, it follows that there is a $T > 0$ such that $\varphi^T(\mathbf{y}) \in U$ so $\varphi^T(\mathbf{y}) \notin A^*$. Because A^* is negatively invariant, $\mathbf{y} \notin A^*$. This proves $\mathbf{y} \notin A \cup A^*$. □

REMARK 1.4. To prove the second inclusion, that $\mathcal{R}(\varphi^t) \subset \mathcal{P}$, first note that for any attracting-repelling pair (A, A^*), all the fixed points and periodic points must be contained in $A \cup A^*$. Thus, $\mathrm{Fix}(\varphi^t) \cup \mathrm{Per}(\varphi^t) \subset \mathcal{P}$. In the following lemma, we prove that given any attracting-repelling pair (A, A^*), there is a Liapunov function which is strictly decreasing off $A \cup A^*$. Then, in Lemma 1.6 below, we use this Liapunov function to show that $\mathcal{R}(\varphi^t) \subset \mathcal{P}$.

Lemma 1.5. *Let* $(A, A^*) \in \mathcal{A}$. *Then, there is a continuous function* $L : M \to \mathbb{R}$ *such that* $L|A = 0$, $L|A^* = 1$, $L(M \setminus (A \cup A^*)) \subset (0, 1)$, *and* L *is strictly decreasing off* $A \cup A^*$.

PROOF. Since A and A^* are disjoint closed sets in a metric space, the function $V : M \to [0, 1]$ given by

$$V(\mathbf{x}) = \frac{d(\mathbf{x}, A)}{d(\mathbf{x}, A) + d(\mathbf{x}, A^*)}$$

is continuous, $V|A = 0$, $V|A^* = 1$, and $V(M \setminus (A \cup A^*)) \subset (0, 1)$. (The existence of such a function in a normal space which is not a metric space is given by the Tietze-Urysohn Theorem.) This function V is not necessarily decreasing or even non-increasing.

Let

$$V^*(\mathbf{x}) = \sup\{V(\varphi^t(\mathbf{x})) : t \geq 0\}.$$

For $\mathbf{x} \in A$, $\varphi^t(\mathbf{x}) \in A$ so $V^*(\mathbf{x}) = 0$. Similarly, if $\mathbf{x} \in A^*$, then $V^*(\mathbf{x}) = 1$. If $\mathbf{x} \notin A \cup A^*$, then $\varphi^t(\mathbf{x})$ approaches A as t goes to plus infinity. (See Exercise 10.7.) Therefore, if $\mathbf{x} \notin A^*$, then $V(\varphi^t(\mathbf{x}))$ goes to zero and the maximum of $V(\varphi^t(\mathbf{x}))$ is attained. Moreover, there is a neighborhood W and a bounded time interval $[0, t_1]$ such that for $\mathbf{y} \in W$, the maximum of $V(\varphi^t(\mathbf{y}))$ for $t \geq 0$ is attained for some $t \in [0, t_1]$; it follows that V^* is continuous. Finally, $V^*(\varphi^t(\mathbf{x})) \leq V^*(\mathbf{x})$ for $t \geq 0$ by construction. Note that the inequality is not necessarily strict.

To make a strictly decreasing function, define L to be a weighted average of V^* over the forward orbit:

$$L(\mathbf{x}) = \int_0^\infty e^{-s}\, V^* \circ \varphi^s(\mathbf{x})\, ds.$$

This function is easily seen to be continuous. Then, for $t \geq 0$,

$$\begin{aligned} L(\varphi^t(\mathbf{x})) &= \int_0^\infty e^{-s}\, V^*(\varphi^{s+t}(\mathbf{x}))\, ds \\ &\leq \int_0^\infty e^{-s}\, V^*(\varphi^{s}(\mathbf{x}))\, ds \\ &\leq L(\mathbf{x}). \end{aligned}$$

The second inequality follows because $V^*(\varphi^{s+t}(\mathbf{x})) \leq V^*(\varphi^s(\mathbf{x}))$ for $t \geq 0$. Thus, L is non-increasing along orbits.

Now, take $t > 0$. If $L(\varphi^t(\mathbf{x})) = L(\mathbf{x})$, then $V^*(\varphi^{s+t}(\mathbf{x})) = V^*(\varphi^s(\mathbf{x}))$ for all $s > 0$. In particular, taking $s = nt$, we get that $V^*(\varphi^{nt}(\mathbf{x})) = V^*(\mathbf{x})$ for all $n > 0$. However, if $\mathbf{x} \notin A \cup A^*$, then $\varphi^{nt}(\mathbf{x})$ goes to A so $V^*(\varphi^{nt}(\mathbf{x}))$ goes to zero. Thus, it is impossible since $V^*(\varphi^{s+t}(\mathbf{x})) = V^*(\varphi^s(\mathbf{x}))$ for all $s > 0$. Therefore, we have shown that $L(\mathbf{x})$ is strictly decreasing off $A \cup A^*$ as required. □

Lemma 1.6. *(a) Let $(A, A^*) \in \mathcal{A}$ be an attracting-repelling pair. Then, $\mathcal{R}(\varphi^t) \subset A \cup A^*$. (b) Further, $\mathcal{R}(\varphi^t) \subset \mathcal{P}$, so $\mathcal{R}(\varphi^t) = \mathcal{P}$. (c) Let $\mathbf{x}, \mathbf{y} \in \mathcal{R}(\varphi^t)$. Then, $\mathbf{x}$ is chain equivalent to $\mathbf{y}$ if and only if for each attracting-repelling pair (A, A^*) both points are in A or both are in A^*.*

PROOF. Part (b) easily follows from part (a). This, together with Lemma 1.4, proves $\mathcal{R}(\varphi^t) = \mathcal{P}$, Theorem 1.3.

To prove part (a), we take $\mathbf{p} \notin A \cup A^*$ and show $\mathbf{p} \notin \mathcal{R}(\varphi^t)$. Let L be the Liapunov function decreasing off $A \cup A^*$, which is given by Lemma 1.5. Let $c_0 = L(\mathbf{p})$ and $c_1 = L(\varphi^1(\mathbf{p}))$. Since L is strictly decreasing at all points of $L^{-1}([c_1, c_0])$, there is a $\delta > 0$ with $\delta < (c_0 - c_1)/2$ such that if $\mathbf{q} \in L^{-1}([c_1, c_1 + \delta])$, then $\varphi^1(\mathbf{q}) \in L^{-1}([0, c_1])$. For each $\mathbf{q} \in L^{-1}([0, c_1])$, there is an $\epsilon > 0$ such that for $\mathbf{q}'$ within ϵ of $\mathbf{q}$, $L(\mathbf{q}') < c_1 + \delta$. By compactness, there is one ϵ that works for all points at once. Now, if $\{\mathbf{p} = \mathbf{p}_0, \ldots, \mathbf{p}_n; t_1, \ldots, t_n\}$ is an ϵ-chain with $t_k \geq 1$, then $L(\varphi^{t_1}(\mathbf{p}_0)) \leq L(\varphi^1(\mathbf{p}_0)) = c_1$. Then, $L(\mathbf{p}_1) \leq c_1 + \delta$ by the choice of ϵ. Next, $L(\varphi^{t_2}(\mathbf{p}_1)) \leq L(\varphi^1(\mathbf{p}_1)) \leq c_1$. Again, $L(\mathbf{p}_2) \leq c_1 + \delta$ by the choice of ϵ. Continuing by induction $L(\mathbf{p}_k) \leq c_1 + \delta$ for $1 \leq k \leq n$. Thus, the chain can never get back to $\mathbf{p}$. This proves that $\mathbf{p}$ is not chain recurrent.

For part (c), if $\mathbf{x}$ and $\mathbf{y}$ are in the same chain component, then by an argument like that for part (b) both points have to be in A or both in A^*. Conversely, assume $\mathbf{x}$ and $\mathbf{y}$ are in different chain components, and in particular that it is not possible to ϵ-chain from $\mathbf{x}$ to $\mathbf{y}$ for small $\epsilon > 0$, $\mathbf{y} \notin \Omega^+_\epsilon(\mathbf{x})$. Then $\mathbf{x}$ is in the attracting set for $\Omega^+_\epsilon(\mathbf{x})$, and $\mathbf{y}$ is not in the attracting set and so is in the repelling set. Thus, if $\mathbf{x}$ and $\mathbf{y}$ are in different chain components, then they are in different halves of an attracting-repelling pair. □

This proves Theorem 1.3. To prove Theorem 1.1, we combine the Liapunov functions for different attracting-repelling pairs to prove that there is a Liapunov function which is strictly decreasing off $\mathcal{P}$ (which equals $\mathcal{R}(\varphi^t)$). To carry out this construction, we need to know that the number of such pairs is at most countable.

Lemma 1.7. *The set $\mathcal{A}$ is at most countable (i.e., finite or countable).*

PROOF. We want a neighborhood to uniquely determine the attracting set. Since there are often proper subsets of attracting sets which are attracting sets, we use the pair $A \times A^*$, which is the unique pair that is contained in $\text{int}(U) \times \text{int}(M \setminus U)$. Since $M \times M$ is a compact metric space, it has a countable basis. Therefore, there is at most a countable number of such pairs $A \times A^*$ and $\mathcal{A}$ is countable. □

Example 1.2. The Morse-Smale examples all have finitely many attracting-repelling pairs. The following is an example of a function on the one torus (or circle), having countably many attracting-repelling pairs. Let $\mathbb{T}^1 = \{x \mod 1\}$ be the one torus. Let the equations on $\mathbb{T}^1$ be given by

$$\dot{x} = f(x) = x^2 \sin(1/x) \qquad \text{for } -2/(3\pi) \leq x \leq 2/\pi.$$

Then, $f(x)$ can be extended outside this interval to have no more zeroes and be periodic. The fixed points are $x = 0, 1/(n\pi)$ for all integers n. Since

$$f'(x) = 2x \sin(1/x) - \cos(1/x),$$

$f'(1/(n\pi)) = (-1)^{n+1}$. Therefore, for any integer k, $x = 1/(2k\pi)$ is a fixed point sink, and $x = 1/((2k+1)\pi)$ is a source. Thus, there are countably many fixed point attracting sets. There are other attracting sets made up of intervals of the form $A = [1/(2j\pi), 1/(2k\pi)]$, where j and k are integers and either $j < 0 < k$, or j and k have the same sign and $0 < |k| < |j|$. The dual repelling set is also an interval; if $j < 0 < k$, then $A^* = \mathbb{T}^1 \setminus (1/((2j-1)\pi), 1/((2k-1)\pi)$. (There are other attracting sets made up of intervals which do not contain 0.) This gives the different types of attracting sets. Notice that the set $\{0\}$ is not an attracting set, but is the intersection of attracting sets.

Theorem 1.8. *Assume φ^t is a flow on a compact metric space M. Then, there is a Liapunov function $L : M \to \mathbb{R}$ which is strictly decreasing off $\mathcal{P} = \mathcal{R}(\varphi^t)$, such that $L(\mathcal{P})$ is a nowhere dense subset of $\mathbb{R}$. Also, for $\mathbf{x}, \mathbf{y} \in \mathcal{R}(\varphi^t)$, $L(\mathbf{x}) = L(\mathbf{y})$ if and only if $\mathbf{x}$ and $\mathbf{y}$ are in the same chain component.*

PROOF. By Lemma 1.7, $\mathcal{A}$ is at most countable, i.e., $\mathcal{A} = \{(A_j, A_j^*)\}_{j=1}^{\infty}$. (The case when $\mathcal{A}$ is finite requires small modifications.) For each j, there is a weak Liapunov function L_j, given by Lemma 1.5, which is strictly decreasing off $A_j \cup A_j^*$. Also, $L_j(M) \subset [0,1]$. Let

$$L(\mathbf{x}) = 2 \sum_{j=1}^{\infty} 3^{-j} L_j(\mathbf{x}).$$

The sum is absolutely convergent, so L is continuous. Also, $L(M) \subset [0,1]$. The function is a combination of non-increasing functions and so is non-increasing along trajectories. Next, if $\mathbf{x} \notin \mathcal{P}$, then $\mathbf{x} \notin A_k \cup A_k^*$ for some k. Then, for $t > 0$, $L_j(\varphi^t(\mathbf{x})) \leq L_j(\mathbf{x})$ for all j and $L_k(\varphi^t(\mathbf{x})) < L_k(\mathbf{x})$, so $L(\varphi^t(\mathbf{x})) < L(\mathbf{x})$. This proves that L is strictly decreasing off $\mathcal{P}$ as desired. Next, for any point $\mathbf{p} \in \mathcal{P}$, each $L_j(\mathbf{p})$ is equal to either 0 or 1, so $L(\mathbf{p})$ is a number whose ternary expansion has only 0's or 2's. (Note the factor of 2 in the definition of L.) Therefore, $L(\mathcal{P})$ is contained in the nowhere dense Cantor set, made up of points whose ternary expansion has only 0's or 2's.

The final claim follows from Lemma 1.6(c) and the fact that the ternary expansion of a point in the Cantor set is unique. Let $\mathbf{x}, \mathbf{y} \in \mathcal{R}(\varphi^t)$. They are chain equivalent if and only if for each attracting-repelling pair (A_j, A_j^*) they are both in A_j or both in A_j^*, so $L_j(\mathbf{x}) = L_j(\mathbf{y})$ for all j. Therefore, $L(\mathbf{x}) = L(\mathbf{y})$. If $\mathbf{x}, \mathbf{y} \in \mathcal{R}(\varphi^t)$ but are not chain equivalent, then there is some j_1 such that $L_{j_1}(\mathbf{y}) \neq L_{j_1}(\mathbf{x})$. Let j_0 be the smallest index such that $L_{j_0}(\mathbf{x}) \neq L_{j_0}(\mathbf{y})$ and assume $L_{j_0}(\mathbf{x}) = 0$ and $L_{j_0}(\mathbf{y}) = 1$. Note that $L_j(\mathbf{x})$ and $L_j(\mathbf{y})$ can only equal 0 or 1 for all j. Then,

$$\begin{aligned} L(\mathbf{y}) - L(\mathbf{x}) &\geq \frac{2}{3^{j_0}} + \sum_{j=j_0+1}^{\infty} \frac{2}{3^j} \\ &\geq \frac{2}{3^{j_0}} + \frac{2}{3^{j_0+1}} (\frac{1}{1-\frac{1}{3}}) \\ &\geq \frac{2}{3^{j_0}} (1 - \frac{1}{3}\frac{3}{2}) \\ &> 0. \end{aligned}$$

□

Theorems 1.3 and 1.8 combine to prove Theorem 1.1.

We end this section with three results which follow from the above results or are closely related.

Proposition 1.9. *Any attracting set A has a neighborhood U which is a trapping region for A and which is positively invariant in the strong sense that $\varphi^t(\mathrm{cl}(U)) \subset \mathrm{int}(U)$ for all $t > 0$.*

PROOF. Let L be the Liapunov function given by Lemma 1.5 and let $U = L^{-1}([0, \epsilon))$ for small $\epsilon > 0$. Then, $\mathrm{cl}(U) \subset L^{-1}([0, \epsilon])$ and $\varphi^t(\mathrm{cl}(U)) \subset \varphi^t(L^{-1}[0, \epsilon]) \subset L^{-1}([0, \epsilon))$ for $t > 0$. □

The following proposition proves that it is enough to assume there is a weak trapping region.

Proposition 1.10. *Let V be a weak trapping region for the pair (A, A^*) of attracting and repelling sets. Then, there is an trapping region U such that $A = \bigcap_{t \geq 0} \varphi^t(U)$ and $A^* = \bigcap_{t \leq 0} \varphi^t(M \setminus U)$.*

PROOF. Let V be a weak trapping region and T the time such that $\mathrm{cl}(\varphi^T(V)) \subset \mathrm{int}(V)$. Let

$$U = \bigcap_{0 \leq t \leq T} \varphi^t(\mathrm{int}(V)).$$

We leave to Exercise 10.5 the verification that U is open using the continuity of the flow on initial conditions. For $0 \leq s = jT + s'$ with $0 \leq s' < T$,

$$\begin{aligned} \varphi^s(U) &= [\bigcap_{0 \leq t < T - s'} \varphi^{t+jT+s'}(\mathrm{int}(V))] \cap [\bigcap_{T-s' \leq t < T} \varphi^{t+jT+s'}(\mathrm{int}(V))] \\ &= [\bigcap_{s' \leq \tau < T} \varphi^{\tau} \circ \varphi^{jT}(\mathrm{int}(V))] \cap [\bigcap_{0 \leq \tau < s'} \varphi^{\tau} \circ \varphi^{(j+1)T}(\mathrm{int}(V))] \\ &\subset \bigcap_{0 \leq \tau < T} \varphi^{\tau}(\mathrm{int}(V)) \\ &\subset U. \end{aligned}$$

For $s = T$,

$$\begin{aligned} \varphi^T(\mathrm{cl}(U)) &\subset \bigcap_{0 \leq t < T} \varphi^t \circ \varphi^T(\mathrm{cl}(V)) \\ &\subset \bigcap_{0 \leq t < T} \varphi^t(\mathrm{int}(V)) \\ &\subset U. \end{aligned}$$

This proves that U is a trapping region. It is also easily checked that the attracting-repelling pair for U is still equal to (A, A^*). □

The next proposition says that the chains from a point $\mathbf{x} \in \mathcal{R}(\varphi^t)$ to itself can actually be taken inside $\mathcal{R}(\varphi^t)$.

Proposition 1.11. *Let φ^t be a flow on a compact metric space M. Then, the map restricted to the chain recurrent set has all points chain recurrent, $\mathcal{R}(\varphi^t|\mathcal{R}(\varphi^t)) = \mathcal{R}(\varphi^t)$.*

REMARK 1.5. This result is not true if the ambient space is not compact. Note also that the analogous theorem is not true in general for the nonwandering set, i.e., there exist examples where $\Omega(\varphi^t|\Omega(\varphi^t)) \neq \Omega(\varphi^t)$. Remark 4.2 later in this chapter gives one such example.

PROOF. Let $\mathbf{p} \in \mathcal{R}(\varphi^t)$. For each $n > 0$, there is a periodic $1/n$-chain through $\mathbf{p}$: $\{\mathbf{p} = \mathbf{p}_0^n, \dots, \mathbf{p}_k^n = \mathbf{p}; t_1 = 1, \dots, t_k = 1\}$. It is easily seen to be possible to take all these $t_j = 1$. Let $C_n = \{\mathbf{p}_j^n\}_{j \in \mathbb{Z}}$ be a periodic $1/n$-chain through $\mathbf{p}$. Then, the set C_n is a compact subset of M. In the Hausdorff metric on compact subsets of M, there is a subsequence C_{n_j} that converges to some compact set $C \subset M$. We show below that for any $\mathbf{q} \in C$ and $\epsilon > 0$, there is a periodic ϵ-chain through $\mathbf{q}$, $\{\mathbf{q} = \mathbf{x}_0, \dots, \mathbf{x}_k; 1, \dots, 1\}$ with $\mathbf{x}_j \in C$. It follows that $\mathbf{q} \in \mathcal{R}(\varphi^t|C) \subset \mathcal{R}(\varphi^t)$. This is true for any $\mathbf{q} \in C$, so $C \subset \mathcal{R}(\varphi^t)$. Next, $\mathbf{p} \in C$, so $\mathbf{p} \in \mathcal{R}(\varphi^t|C) \subset \mathcal{R}(\varphi^t)$. This argument applies to any $\mathbf{p} \in \mathcal{R}(\varphi^t)$, so $\mathcal{R}(\varphi^t) \subset \mathcal{R}(\varphi^t|C)$ and so they are equal.

We have only to show that there is a periodic ϵ-chain through $\mathbf{q}$ with $\mathbf{x}_j \in C$. By uniform continuity of the flow φ^t on M, there is a $\delta = \delta(\epsilon/3) > 0$ such that for $d(\mathbf{a}, \mathbf{b}) < \delta$, $d(\varphi^1(\mathbf{a}), \varphi^1(\mathbf{b})) < \epsilon/3$. We can also take $\delta < \epsilon/3$. Because the C_{n_j} converge to C, we can find an $n = n_k$ such that $1/n < \epsilon/3$ and the distance from C_n to C in the Hausdorff metric is less than δ. Assume $\{\mathbf{p}_i^n\}$ has period j so $\mathbf{p}_j^n = \mathbf{p}_0^n = \mathbf{p}$. We can extend $\mathbf{p}_i^n$ periodically to all $i \in \mathbb{Z}$, so $\mathbf{p}_{i+j}^n = \mathbf{p}_i^n$ for all i. For each $\mathbf{p}_i^n$, take $\mathbf{x}_i \in C$ with $d(\mathbf{x}_i, \mathbf{p}_i^n) < \delta$, $\mathbf{x}_{i+j} = \mathbf{x}_i$ for all i, and $\mathbf{x}_i = \mathbf{q}$ for some i. Then, $d(\varphi^1(\mathbf{x}_i), \mathbf{x}_{i+1}) < d(\varphi^1(\mathbf{x}_i), \varphi^1(\mathbf{p}_i^n)) + d(\varphi^1(\mathbf{p}_i^n), \mathbf{p}_{i+1}^n) + d(\mathbf{p}_{i+1}^n, \mathbf{x}_{i+1}) < \epsilon$. Therefore, $\mathbf{x}_i$ is a periodic ϵ-chain through $\mathbf{q}$. □

10.1.1 Fundamental Theorem for a Homeomorphism

Let $f : N \to N$ be a homeomorphism on a compact metric space. We can form the suspension of f and obtain a flow φ^t on $M = (N \times \mathbb{R})/\sim$. By the Fundamental Theorem for flows, there is a Liapunov function $L : M \to \mathbb{R}$ which is strictly decreasing off the chain recurrent set of φ^t. By restricting L to $(N \times \{0\})/\sim$, which we identify with N, we get a Liapunov function for f which is strictly decreasing off the chain recurrent set of f. This proves the following theorem.

Theorem 1.12 (Conley's Fundamental Theorem for a Homeomorphism). *A homeomorphism f on a compact metric space has a Liapunov function L which is strictly decreasing off $\mathcal{R}(f)$ and such that $L(\mathcal{R}(f))$ is a nowhere dense subset of $\mathbb{R}$.*

There is the following corollary just as in the case for flows.

Corollary 1.13. *Let $f : N \to N$ be a homeomorphism on a compact metric space. Then, $\mathcal{R}(f|\mathcal{R}(f)) = \mathcal{R}(f)$.*

10.2 Stable Manifold Theorem for a Hyperbolic Invariant Set

The previous section gave some of the basic results about the chain recurrent set using only ideas from point set topology. The other results which we give use the hyperbolic structure on the chain recurrent set, or at least on the closure of the periodic points. One of the key results about systems with a hyperbolic structure on the chain recurrent set is the existence of stable and unstable manifolds as stated in Section 8.1.3. In this section, we restate the theorem and sketch the proof.

In the statement of the theorem, we use coordinates near points in the invariant set Λ. In a Euclidean space, a neighborhood $\mathcal{V}_{\mathbf{p}}(\epsilon)$ of $\mathbf{p}$ can be taken as $\{\mathbf{p}\}+\mathbb{E}^u_{\mathbf{p}}(\epsilon)\times\mathbb{E}^s_{\mathbf{p}}(\epsilon)$. In a manifold, a neighborhood can be taken of the same form if we use local coordinates. However, it is not always possible to take one set of coordinates for all the different points of Λ. Another approach is to use the exponential map from tangent vectors at $\mathbf{p}$ into the manifold:

$$\exp_{\mathbf{p}} : T_{\mathbf{p}}M \to M.$$

In a Euclidean space, $\exp_{\mathbf{p}}(\mathbf{v_p}) = \mathbf{p}+\mathbf{v_p}$. In a manifold, the point $\exp_{\mathbf{p}}(\mathbf{v_p})$ is obtained by going along a geodesic (distance minimizing curve) starting at $\mathbf{p}$ in the direction $\mathbf{v_p}$. If the manifold is embedded in some larger Euclidean space $\mathbb{R}^N$, then $\exp_{\mathbf{p}}(\mathbf{v_p})$ can be visualized as taking the point $\mathbf{p}+\mathbf{v_p}$ which is probably not in the manifold M and then projecting this point into M along the subspace $(T_{\mathbf{p}}M)^{\perp}$ of vectors perpendicular to $T_{\mathbf{p}}M$. See Franks (1979) for a more complete explanation of this latter method. In any case, the neighborhood of $\mathbf{p}$ can be taken as

$$\mathcal{V}_{\mathbf{p}}(\epsilon) = \exp_{\mathbf{p}}(\mathcal{B}_{\mathbf{p}}(\epsilon))$$

where

$$\mathcal{B}_{\mathbf{p}}(\epsilon) = \mathbb{E}^u_{\mathbf{p}}(\epsilon) \times \mathbb{E}^s_{\mathbf{p}}(\epsilon).$$

In the proofs below, we use the fact that the $D(\exp_{\mathbf{p}})_{\mathbf{0_p}} = id$, so $\|D(\exp_{\mathbf{p}})_{\mathbf{v_p}} - id\|$ is small for $\mathbf{v_p} \in \mathcal{B}_{\mathbf{p}}(\epsilon)$, although we do not explicitly isolate the fact that $D(\exp_{\mathbf{p}})_{\mathbf{0_p}}$ is different from the identity.

Theorem 2.1 (Stable Manifold Theorem for a Hyperbolic Set). *Let $f : M \to M$ be a C^k diffeomorphism. Let Λ be a compact invariant set with a hyperbolic structure for f with hyperbolic constants $0 < \mu < 1 < \lambda$ and $C \geq 1$. Then, there is an $\epsilon > 0$ such that for each $\mathbf{p} \in \Lambda$, there are two C^k embedded disks $W^s_\epsilon(\mathbf{p}, f)$ and $W^u_\epsilon(\mathbf{p}, f)$ which are tangent to $\mathbb{E}^s_{\mathbf{p}}$ and $\mathbb{E}^u_{\mathbf{p}}$, respectively.*

Using the exponential map discussed above, $W^s_\epsilon(\mathbf{p}, f)$ can be represented as the graph of a C^k function $\sigma^s_{\mathbf{p}} : \mathbb{E}^s_{\mathbf{p}}(\epsilon) \to \mathbb{E}^u_{\mathbf{p}}(\epsilon)$ with $\sigma^s_{\mathbf{p}}(\mathbf{0_p}) = \mathbf{0_p}$ and $D(\sigma^s_{\mathbf{p}})_{\mathbf{0_p}} = \mathbf{0}$:

$$W^s_\epsilon(\mathbf{p}, f) = \exp_{\mathbf{p}}(\{(\sigma^s_{\mathbf{p}}(\mathbf{y}), \mathbf{y}) : \mathbf{y} \in \mathbb{E}^s_{\mathbf{p}}(\epsilon)\}).$$

Also, the function $\sigma^s_{\mathbf{p}}$ and its first k derivatives vary continuously as $\mathbf{p}$ varies. Similarly, there is a C^k function $\sigma^u_{\mathbf{p}} : \mathbb{E}^u_{\mathbf{p}}(\epsilon) \to \mathbb{E}^s_{\mathbf{p}}(\epsilon)$ with $\sigma^u_{\mathbf{p}}(\mathbf{0_p}) = \mathbf{0_p}$ and $D(\sigma^u_{\mathbf{p}})_{\mathbf{0_p}} = \mathbf{0}$ and with the function $\sigma^u_{\mathbf{p}}$ and its first k derivatives varying continuously as $\mathbf{p}$ varies such that

$$W^u_\epsilon(\mathbf{p}, f) = \exp_{\mathbf{p}}(\{(\mathbf{x}, \sigma^u_{\mathbf{p}}(\mathbf{x})) : \mathbf{x} \in \mathbb{E}^u_{\mathbf{p}}(\epsilon)\}).$$

Moreover, for $\epsilon > 0$ small enough and using the neighborhoods $\mathcal{V}_{\mathbf{p}}(\epsilon)$ defined above,

$$\begin{aligned} W^s_\epsilon(\mathbf{p}, f) &= \{\mathbf{q} \in \mathcal{V}_{\mathbf{p}}(\epsilon) : f^j(\mathbf{q}) \in \mathcal{V}_{f^j(\mathbf{p})}(\epsilon) \text{ for } j \geq 0\} \\ &= \{\mathbf{q} \in \mathcal{V}_{\mathbf{p}}(\epsilon) : f^j(\mathbf{q}) \in \mathcal{V}_{f^j(\mathbf{p})}(\epsilon) \text{ for } j \geq 0 \text{ and} \\ &\qquad d(f^j(\mathbf{q}), f^j(\mathbf{p})) \leq C\mu^j d(\mathbf{q}, \mathbf{p}) \text{ for all } j \geq 0\}. \end{aligned}$$

Similarly,

$$\begin{aligned} W^u_\epsilon(\mathbf{p}, f) &= \{\mathbf{q} \in \mathcal{V}_{\mathbf{p}}(\epsilon) : f^{-j}(\mathbf{q}) \in \mathcal{V}_{f^{-j}(\mathbf{p})}(\epsilon) \text{ for } j \geq 0\} \\ &= \{\mathbf{q} \in \mathcal{V}_{\mathbf{p}}(\epsilon) : f^{-j}(\mathbf{q}) \in \mathcal{V}_{f^{-j}(\mathbf{p})}(\epsilon) \text{ for } j \geq 0 \text{ and} \\ &\qquad d(f^{-j}(\mathbf{q}), f^{-j}(\mathbf{p})) \leq C\lambda^{-j} d(\mathbf{q}, \mathbf{p}) \text{ for all } j \geq 0\}. \end{aligned}$$

REMARK 2.1. The idea of the proof we sketch is to repeat the argument for a single fixed point, adding the fact that points near $\mathbf{p}$ get mapped to points near $f(\mathbf{p})$.

It is possible to prove the Stable Manifold Theorem for a hyperbolic set as a corollary of the Stable Manifold Theorem for a fixed point in a Banach space. See Hirsch and Pugh (1970) or Shub (1987). These references also contain a complete proof (rather than a sketch of a proof which we give for our approach).

PROOF. We also assume that the constant $C \geq 1$ in the definition of hyperbolic structure is taken to be 1. This is always possible by taking an adapted norm as we proved in Theorem VIII.1.1 of Section 8.1.3.

For each point $\mathbf{p} \in \Lambda$, we take $\mathcal{B}_{\mathbf{p}}(\epsilon) \subset T_{\mathbf{p}}M$ and $\mathcal{V}_{\mathbf{p}}(\epsilon) = \exp_{\mathbf{p}}(\mathcal{B}_{\mathbf{p}}(\epsilon))$, as defined above. The map from a neighborhood of $\mathbf{p}$ to a neighborhood of $f(\mathbf{p})$ induces a C^k map $F_{\mathbf{p}} : \mathcal{B}_{\mathbf{p}}(r) \to \mathcal{B}_{f(\mathbf{p})}(C_0 r)$ for some $C_0 \geq \lambda$ defined by

$$F_{\mathbf{p}}(\mathbf{v}_{\mathbf{p}}) = \exp^{-1}_{f(\mathbf{p})} \circ f \circ \exp_{\mathbf{p}}(\mathbf{v}_{\mathbf{p}}).$$

For a small $\mathbf{v}_{\mathbf{p}}$, $\exp_{\mathbf{p}}(\mathbf{v}_{\mathbf{p}})$ is a point in M near $\mathbf{p}$, $f \circ \exp_{\mathbf{p}}(\mathbf{v}_{\mathbf{p}})$ is a point in M near $f(\mathbf{p})$, and $\exp^{-1}_{f(\mathbf{p})} \circ f \circ \exp_{\mathbf{p}}(\mathbf{v}_{\mathbf{p}})$ is a relatively small vector in $T_{f(\mathbf{p})}M$. Notice that

$$\begin{aligned} F_{\mathbf{p}}(\mathbf{0}_{\mathbf{p}}) &= \exp^{-1}_{f(\mathbf{p})} \circ f(\mathbf{p}) \\ &= \mathbf{0}_{f(\mathbf{p})}. \end{aligned}$$

Also, $D(F_{\mathbf{p}})_{\mathbf{0}_{\mathbf{p}}} = Df_{\mathbf{p}}$ because $D(\exp_{\mathbf{p}})_{\mathbf{0}_{\mathbf{p}}} = id$ and $D(\exp_{f(\mathbf{p})})_{\mathbf{0}_{\mathbf{p}}} = id$. Therefore, $Df_{\mathbf{p}}$ is a linear approximation for the nonlinear map $F_{\mathbf{p}}$ in $\mathcal{B}_{\mathbf{p}}(\epsilon)$.

In order to consider all these maps for different $\mathbf{p}$ at once, we use the bundle structure of $T_\Lambda M$ to keep the different neighborhoods disjoint. The set

$$\mathcal{B}_\Lambda(r) = \{\mathbf{v}_{\mathbf{p}} \in \mathcal{B}_{\mathbf{p}}(r) : \mathbf{p} \in \Lambda\} \subset T_\Lambda M$$

is a bundle over Λ with all the sets $\mathcal{B}_{\mathbf{p}}(r)$ contained in distinct tangent spaces $T_{\mathbf{p}}M$ and so do not intersect. (The neighborhoods $\mathcal{V}_{\mathbf{q}}(\epsilon)$ and $\mathcal{V}_{\mathbf{p}}(\epsilon)$ do intersect for $\mathbf{q}$ near $\mathbf{p}$.) Since Λ is often not a manifold (e.g., a Cantor set), this space is a metric space but not a manifold. The map

$$F : \mathcal{B}_\Lambda(r) \to \mathcal{B}_\Lambda(C_0 r)$$

is continuous and C^k on each fixed fiber $\mathcal{B}_{\mathbf{p}}(r)$. Note that since Λ is compact, C_0 can be taken independent of $\mathbf{p}$.

Let $0 < \mu < 1 < \lambda$ be bounds on the derivatives on $\mathbb{E}^s_{\mathbf{p}}$ and $\mathbb{E}^u_{\mathbf{p}}$: $\|Df_{\mathbf{p}}|\mathbb{E}^s_{\mathbf{p}}\| < \mu$ and $m(Df_{\mathbf{p}}|\mathbb{E}^u_{\mathbf{p}}) > \lambda$ for all $\mathbf{p} \in \Lambda$. (Remember that for a linear map A, $m(A)$ is the minimum norm of A. See Section 4.1.) Take $\epsilon > 0$ small enough so that $\mu + 2\epsilon < 1$ and $\lambda - 2\epsilon > 1$. Given such an $\epsilon > 0$, there is an $r > 0$ small enough so that for $\mathbf{q} \in \mathcal{B}_{\mathbf{p}}(r)$,

$$D(F_{\mathbf{p}})_{\mathbf{q}} = \begin{pmatrix} A^{ss}_{\mathbf{p}}(\mathbf{q}) & A^{su}_{\mathbf{p}}(\mathbf{q}) \\ A^{us}_{\mathbf{p}}(\mathbf{q}) & A^{uu}_{\mathbf{p}}(\mathbf{q}) \end{pmatrix}$$

with $\|A^{ss}_{\mathbf{p}}(\mathbf{q})\| < \mu$, $m(A^{uu}_{\mathbf{p}}(\mathbf{q})) > \lambda$, $\|A^{su}_{\mathbf{p}}(\mathbf{q})\| < \epsilon$, and $\|A^{us}_{\mathbf{p}}(\mathbf{q})\| < \epsilon$. In this expression, the derivative of $F_{\mathbf{p}}$ is taken along the fiber with $\mathbf{p}$ fixed. (We have used the fact that $D(\exp_{\mathbf{p}})_{\mathbf{q}}$ is near the identity.)

We let

$$\tilde{W}^s_r(\mathbf{p}) = \bigcap_{j=0}^{\infty} F^j_{f^j(\mathbf{p})}(\mathcal{B}_{f^j(\mathbf{p})}(r)) \qquad \text{and}$$
$$W^s_r(\mathbf{p}) = \exp_{\mathbf{p}}(\tilde{W}^s_r(\mathbf{p})),$$

where we think of the local stable manifold $\tilde{W}^s_r(\mathbf{p})$ as being represented in the local coordinates given by the linear space $T_{\mathbf{p}}M$. We want to show that it is 1-Lipschitz. As before, we consider the stable and unstable cones (but now at different points):

$$\begin{aligned} C^s_{\mathbf{p}} &= \{(\mathbf{v}^s_{\mathbf{p}}, \mathbf{v}^u_{\mathbf{p}}) \in \mathbb{E}^s_{\mathbf{p}} \times \mathbb{E}^u_{\mathbf{p}} : |\mathbf{v}^u_{\mathbf{p}}| \le |\mathbf{v}^s_{\mathbf{p}}|\} \\ &= \{\mathbf{v}_{\mathbf{p}} \in T_{\mathbf{p}}M : |\pi^u_{\mathbf{p}}\mathbf{v}_{\mathbf{p}}| \le |\pi^s_{\mathbf{p}}\mathbf{v}_{\mathbf{p}}|\}, \qquad \text{and} \\ C^u_{\mathbf{p}} &= \{\mathbf{v}_{\mathbf{p}} \in T_{\mathbf{p}}M : |\pi^u_{\mathbf{p}}\mathbf{v}_{\mathbf{p}}| \ge |\pi^s_{\mathbf{p}}\mathbf{v}_{\mathbf{p}}|\}. \end{aligned}$$

As before, the condition that

$$\tilde{W}^s_r(\mathbf{p}) \cap [\{\mathbf{q}\} + C^u_{\mathbf{p}}] = \{\mathbf{q}\}$$

for all points $\mathbf{q} \in \tilde{W}^s_r(\mathbf{p})$ is equivalent to the graph being 1-Lipschitz for each fixed $\mathbf{p}$.

The following facts are proved just as in the proof of the stable manifold of a single point. (Cf. Section 5.10.1.)

1. For $\mathbf{q} \in \mathcal{B}_{\mathbf{p}}(r)$, $D(F_{\mathbf{p}})_{\mathbf{q}} C^u_{\mathbf{p}} \subset C^u_{f(\mathbf{p})}$.
2. Let $\mathbf{q}_1, \mathbf{q}_2 \in \mathcal{B}_{\mathbf{p}}(r)$ with $\mathbf{q}_2 \in \{\mathbf{q}_1\} + C^u_{\mathbf{p}}$. Then, $F_{\mathbf{p}}(\mathbf{q}_2) \in \{F_{\mathbf{p}}(\mathbf{q}_2)\} + C^u_{f(\mathbf{p})}$, and $|\pi^u_{f(\mathbf{p})} F_{\mathbf{p}}(\mathbf{q}_2) - \pi^u_{f(\mathbf{p})} F_{\mathbf{p}}(\mathbf{q}_1)| \ge (\lambda - \epsilon)|\pi^u_{\mathbf{p}}(\mathbf{q}_2 - \mathbf{q}_1)|$.
3. Let $\mathbf{q}_1, \mathbf{q}_2 \in \mathcal{B}_{\mathbf{p}}(r)$ with $\mathbf{q}_2 \in \{\mathbf{q}_1\} + C^s_{\mathbf{p}}$. Then, $F^{-1}_{\mathbf{p}}(\mathbf{q}_2) \in \{F^{-1}_{\mathbf{p}}(\mathbf{q}_2)\} + C^s_{f^{-1}(\mathbf{p})}$, and $|\pi^s_{f^{-1}(\mathbf{p})} F^{-1}_{\mathbf{p}}(\mathbf{q}_2) - \pi^s_{f^{-1}(\mathbf{p})} F^{-1}_{\mathbf{p}}(\mathbf{q}_1)| \ge (\mu + \epsilon)^{-1}|\pi^s_{\mathbf{p}}(\mathbf{q}_2 - \mathbf{q}_1)|$.
4. Let $D^u_{0,\mathbf{p}}$ be an unstable disk in $\mathcal{B}_{\mathbf{p}}(r)$, i.e., the image of a C^1 function $\psi : \mathbb{E}^u_{\mathbf{p}}(r) \to \mathbb{E}^s_{\mathbf{p}}(r)$ with $\mathrm{Lip}(\psi) \le 1$. Let

$$D^u_{n,f^n(\mathbf{p})} = F_{f^{n-1}(\mathbf{p})}(D^u_{n-1,f^{n-1}(\mathbf{p})}) \cap \mathcal{B}_{f^n(\mathbf{p})}(\mathbf{p})$$

be defined by induction. Then, for $n \ge 1$, $D^u_{n,f^n(\mathbf{p})}$ is an unstable disk in $\mathcal{B}_{f^n(\mathbf{p})}(r)$ and

$$F^{-n}(D^u_{n,f^n(\mathbf{p})}) \subset F^{-n+1}(D^u_{n-1,f^{n-1}(\mathbf{p})}) \subset \cdots \subset D^u_{0,\mathbf{p}}$$

is a nested set of unstable disks in $\mathcal{B}_{\mathbf{p}}(r)$ with

$$\mathrm{diam}[\bigcap_{j=0}^{n} F^{-j}(D^u_{j,f^j(\mathbf{p})})] \le (\lambda - \epsilon)^{-n} 2r.$$

5. The manifold $\tilde{W}^s_r(\mathbf{p})$ is a graph of a 1-Lipschitz function $\varphi^s_{\mathbf{p}} : \mathbb{E}^s_{\mathbf{p}}(r) \to \mathbb{E}^u_{\mathbf{p}}(r)$ with $\varphi^s_{\mathbf{p}}(\mathbf{0}_{\mathbf{p}}) = \mathbf{0}_{\mathbf{p}}$.
6. If $\mathbf{q} \in \tilde{W}^s_r(\mathbf{p})$, then $|F^j_{\mathbf{p}}(\mathbf{q}) - \mathbf{0}_{f^j(\mathbf{p})}| \le (\mu+\epsilon)^j |\mathbf{q} - \mathbf{0}_{\mathbf{p}}|$ for $j \ge 0$. If $\mathbf{q} \in W^s_r(\mathbf{p}) \subset M$ is a point in the stable manifold in M, then $d(f^j(\mathbf{q}), f^j(\mathbf{p})) \le (\mu + \epsilon)^j d(\mathbf{q}, \mathbf{p})$ converges to zero at an exponential rate.

7. For $\mathbf{p}$ fixed, the manifolds $\tilde{W}^s_r(\mathbf{p})$ and $W^s_r(\mathbf{p})$ are C^1; $\tilde{W}^s_r(\mathbf{p})$ is tangent to $\mathbb{E}^s_{\mathbf{p}}$ at $\mathbf{0}_{\mathbf{p}}$ and $W^s_r(\mathbf{p})$ is tangent to $\mathbb{E}^s_{\mathbf{p}}$ at $\mathbf{p}$.
8. For $\mathbf{p}$ fixed, the manifolds $\tilde{W}^s_r(\mathbf{p})$ and $W^s_r(\mathbf{p})$ are C^k if f is C^k.

The fact that the function $\sigma^s_{\mathbf{p}}$ and its derivatives vary continuously as $\mathbf{p}$ varies follows easily from the fact that this is true about $F_{\mathbf{p}}$ and its derivative along fibers. Thus, the construction for $\mathbf{p}$ and that for $\mathbf{p}'$ are close to each other if $\mathbf{p}$ is near $\mathbf{p}'$.

Since we are assuming that f is invertible (a diffeomorphism), the corresponding facts about $W^u_r(\mathbf{p})$ follow from looking at f^{-1}. □

In the next section, we outline a modification of the above proof which shows that a δ-chain can be ϵ-shadowed near a hyperbolic invariant set. We later give another variation which proves the structural stability of Anosov diffeomorphisms.

10.3 Shadowing and Expansiveness

In this section, we prove that it is possible to shadow in a neighborhood of a hyperbolic invariant set and that a diffeomorphism is expansive on a hyperbolic invariant set. R. Bowen carried over the idea of shadowing to hyperbolic invariant sets, based on the work of D. Anosov (1967) and Ya. Sinai (1972) for Anosov systems. See Bowen (1975a, 1975b). In the next section, we show that if $\mathcal{R}(f)$ has a hyperbolic structure, then it breaks up into a finite number of pieces, and the results of this section show that we can ϵ-shadow an δ-chain by an orbit in one of these pieces.

We start by defining shadowing precisely. We also want a condition on the invariant set which allows us to conclude that the orbit which shadows a δ-chain lies in a given invariant set. The necessary assumption is that the invariant set is isolated, which we define next. After these definitions, we state the result about existence of shadowing.

Definition. Let $f : M \to M$ be a homeomorphism. Let $\{\mathbf{x}_j\}_{j=j_1}^{j_2}$ be a δ-chain for f. (In this section, we often add the requirement that either $j_1 = -\infty$ or $j_2 = \infty$ or both.) A point $\mathbf{y} \in M$ *ϵ-shadows* $\{\mathbf{x}_j\}_{j=j_1}^{j_2}$ provided $d(f^j(\mathbf{y}), \mathbf{x}_j) < \epsilon$ for $j_1 \le j \le j_2$.

Definition. A closed invariant set Λ is said to be *isolated* provided there is a neighborhood U of Λ such that $\Lambda \subset \operatorname{int}(U)$ and

$$\Lambda = \bigcap_{n \in \mathbb{Z}} f^n(U).$$

The neighborhood U is called the *isolating neighborhood.* Notice that the isolated invariant set is the *maximal invariant set* contained in its isolating neighborhood, i.e., any invariant set Λ contained entirely inside the neighborhood U is a subset of Λ.

The geometric horseshoe given in Section 8.3 is an example of an isolated invariant set which is not an attractor. The set S given in that section is an isolating neighborhood.

When we defined an isolating neighborhood (trapping region) for an attracting set, we required that U is positively invariant. With this assumption on U, the intersection of $f^n(U)$ for all integers n (given above) is the same as the intersection for all positive integers n (used for attracting sets).

For an invariant set which is not an attracting set, we only require that the set Λ be locally isolated in U. There can be other points $\mathbf{x}$ in the neighborhood U such that the orbit of $\mathbf{x}$ leaves U and later returns to U. These points could be either periodic, nonwandering, or chain recurrent.

Example 3.1. We give an example of a compact hyperbolic invariant set which is not isolated and which is a subset of the symbol space Σ_2. Such an example can be realized as a subset of a horseshoe by using the conjugacy. Let $\Lambda \subset \Sigma_2$ be the subshift that is not of finite type given in Example III.2.1: between any two 2's in the string $\mathbf{s}$, there are an even number of 1's. As noted in Chapter III, Λ is closed and invariant by the shift map σ. It is enough to consider neighborhoods U_k of all points $\mathbf{s}$ such that there is a point $\mathbf{t} \in \Lambda$ with $s_j = t_j$ for $-k \leq j \leq k$. Given such a k, there is a point $\mathbf{s}^0 \in \Sigma_2 \setminus \Lambda$ that has $2k+3$ consecutive 1's between two 2's but is still inside U_k. We leave it to the reader to check that such a $\mathbf{s}^0 \in \bigcap_{n\in\mathbb{Z}} f^n(U_k) \setminus \Lambda$. This shows that Λ is not isolated.

Example 3.2. For a second example of a non-isolated invariant set, let $\mathbf{q}$ be a transverse homoclinic point for a fixed point $\mathbf{p}$. Let $\Lambda = \{\mathbf{p}\} \cup \mathcal{O}(\mathbf{q})$. It follows from the theorem on the the existence of a horseshoe, Theorem VIII.4.5, that Λ is not isolated.

Theorem 3.1 (Shadowing). *Let Λ be a compact hyperbolic invariant set. Given $\epsilon > 0$, there exist $\delta > 0$ and $\eta > 0$ such that if $\{\mathbf{x}_j\}_{j=j_1}^{j_2}$ is a δ-chain for f with $d(\mathbf{x}_j, \Lambda) < \eta$ for $j_1 \leq j \leq j_2$, then there is a $\mathbf{y}$ which ϵ-shadows $\{\mathbf{x}_j\}_{j=j_1}^{j_2}$. If the δ-chain is periodic, then $\mathbf{y}$ is periodic. Moreover, if $j_1 = -\infty$ and $j_2 = \infty$ for the δ-chain, then $\mathbf{y}$ is unique. If $j_2 = -j_1 = \infty$ and Λ is an isolated invariant set (or has a local product structure), then the unique point $\mathbf{y} \in \Lambda$.*

REMARK 3.1. R. Bowen's proof of this theorem assumes that the set has a local product structure and uses the conclusion of the stable manifold theorem and then uses point set topology ideas. Instead, we use the proof of the stable manifold theorem as given in the last section to prove the result directly.

REMARK 3.2. Meyer (1987) and Meyer and Sell (1989) give a proof of the shadowing theorem using the implicit function theorem. At some general level, their proof is really the same as the one given here but it appears very different. We present the ideas geometrically, while their proof is cleaner analytically. Meyer and Hall (1992) give a complete exposition of this proof in the case when the invariant set is a subset of a Euclidean space. Also see Palmer (1984) and Chow, Lin, and Palmer (1989) for another proof.

REMARK 3.3. Grebogi, Hammel, and Yorke (1988) have extended this result to show that systems without a uniform hyperbolic structure can often shadow δ-chains for long intervals of time.

PROOF. Throughout the proof, we take $0 < r < \epsilon$. First, extend the splitting $\mathbb{E}^u_{\mathbf{p}} \times \mathbb{E}^s_{\mathbf{p}}$ from on Λ to a neighborhood $V \subset M$. Take η small enough so the η-neighborhood of Λ is inside V. Let $\mathcal{B}_{\mathbf{p}}(r) = \mathbb{E}^u_{\mathbf{p}}(r) \times \mathbb{E}^s_{\mathbf{p}}(r) \subset T_{\mathbf{p}}M$ and $\mathcal{V}_{\mathbf{p}}(r) = \exp_{\mathbf{p}}(\mathcal{B}_{\mathbf{p}}(r))$ be as in the last section. The box $\mathcal{B}_{\mathbf{p}}(r)$ is a subset of $T_{\mathbf{p}}M$ while $\mathcal{V}_{\mathbf{p}}(r)$ is the comparable neighborhood of $\mathbf{p}$ in M. Let $\{\mathbf{x}_j\}_{j=j_1}^{j_2}$ be a δ-chain for f. If $\delta > 0$ and $\eta > 0$ are small enough, then $f(\mathcal{V}_{\mathbf{x}_j}(r)) \subset \mathcal{V}_{\mathbf{x}_{j+1}}(C_0 r)$ for some $C_0 > \lambda$. We introduce the map $F_j : \mathcal{B}_{\mathbf{x}_j}(r) \to \mathcal{B}_{\mathbf{x}_{j+1}}(C_0 r)$, which should be considered as the map f represented in local coordinates at $\mathbf{x}_j$ and $\mathbf{x}_{j+1}$

$$F_j(\mathbf{y}) = \exp^{-1}_{\mathbf{x}_{j+1}} \circ f \circ \exp_{\mathbf{x}_j}.$$

Note since $f(\mathbf{x}_j)$ is not necessarily equal to $\mathbf{x}_{j+1}$, $|F_j(\mathbf{0}_{\mathbf{x}_j}) - \mathbf{0}_{\mathbf{x}_{j+1}}|$ is small, say less than δ, but is not necessarily equal to zero. However, for small enough δ, the estimates for this sequence of maps are similar to those in the last section. Here we use $\nu > 0$ instead of ϵ to measure the change in the derivatives because we are using ϵ for something else. One difference between the two proofs is that in order to know that an unstable disk has an image which is an unstable disk, we need to take into consideration the fact that $F_j(\mathbf{0}_{\mathbf{x}_j}) \neq \mathbf{0}_{\mathbf{x}_{j+1}}$: the requirement becomes $(\lambda-\nu)r-\delta \geq r$ or $(\lambda-\nu-1)r \geq \delta$, so that the jumps measured by δ are bounded in terms of quantities determined by the hyperbolicity. Similarly in the consideration of stable disks, we need that $(\mu+\nu)^{-1}r - \delta \geq r$ or $[(\mu+\nu)^{-1} - 1]r \geq \delta$. Thus, we first take the neighborhood V of Λ in M and $0 < r < \epsilon$ small enough so that the estimates on the derivative of

$$F_{\mathbf{p}} = \exp_{f(\mathbf{p})}^{-1} \circ f \circ \exp_{\mathbf{p}} : \mathcal{B}_{\mathbf{p}}(r) \to \mathcal{B}_{f(\mathbf{p})}(C_0 r)$$

are true for all $\mathbf{p} \in V$ in terms of the extended splitting; for $\mathbf{q} \in \mathcal{B}_{\mathbf{p}}(r)$,

$$D(F_{\mathbf{p}})_{\mathbf{q}} = \begin{pmatrix} A_{\mathbf{p}}^{ss}(\mathbf{q}) & A_{\mathbf{p}}^{su}(\mathbf{q}) \\ A_{\mathbf{p}}^{us}(\mathbf{q}) & A_{\mathbf{p}}^{uu}(\mathbf{q}) \end{pmatrix}$$

the entries satisfy the following estimates: $\|A_{\mathbf{p}}^{ss}(\mathbf{q})\| < \mu$, $m(A_{\mathbf{p}}^{uu}(\mathbf{q})) > \lambda$, $\|A_{\mathbf{p}}^{su}(\mathbf{q})\| < \nu/2$, and $\|A_{\mathbf{p}}^{us}(\mathbf{q})\| < \nu/2$. Next take $\delta > 0$ such that the same estimates are true for F_j constructed from a δ-chain $\{\mathbf{x}_j\}_{j=j_1}^{j_2}$ provided $\mathbf{x}_j \in V$ for all j.

Assume that $j_1 = -\infty$. To shorten the notation, for $k > 0$ write F_m^k for $F_{m+k-1} \circ \cdots \circ F_m$. Using this notation,

$$\tilde{D}_j^u(r) = \bigcap_{k=0}^{\infty} F_{j-k}^k(\mathcal{B}_{\mathbf{x}_{j-k}}(r))$$

is an unstable disk in $\mathcal{B}_{\mathbf{x}_j}(r)$ and $D_j^u(r) = \exp_{\mathbf{x}_j}(\tilde{D}_j^u(r))$ is an unstable disk in M near $\mathbf{x}_j$. (The fact that $F_j(\mathbf{0}_{\mathbf{x}_j}) \neq \mathbf{0}_{\mathbf{x}_{j+1}}$ means that the disks $\tilde{D}_j^u(r)$ do not necessarily go through $\mathbf{0}_{\mathbf{x}_j}$ and the $D_j^u(r)$ do not necessarily go through $\mathbf{x}_j$ but are only nearby.) If $\mathbf{q} \in \tilde{D}_0^u(r)$ and $k \geq 0$, then $\mathbf{q} \in F_{-k}^k(\mathcal{B}_{\mathbf{x}_{-k}}(r))$, so $(F_{-k}^k)^{-1}(\mathbf{q}) \in \mathcal{B}_{\mathbf{x}_{-k}}(r)$. In terms of $D_0^u(r)$, if $\mathbf{y} \in D_0^u(r)$, then $f^{-k}(\mathbf{y}) \in \mathcal{V}_{\mathbf{x}_{-k}}(r)$ and the backward orbit of $\mathbf{y}$ stays within r of the δ-chain.

Similarly, if $j_2 = \infty$, we can find stable disks. For k negative, let $F_m^k = F_{m+k}^{-1} \circ \cdots \circ F_{m-1}^{-1}$. Then,

$$\tilde{D}_j^s(r) = \bigcap_{k=-\infty}^{0} F_{j-k}^k(\mathcal{B}_{\mathbf{x}_{j-k}}(r))$$

is a stable disk in $\mathcal{B}_{\mathbf{x}_j}(r)$ and $D_j^s(r) = \exp_{\mathbf{x}_j}(\tilde{D}_j^s(r))$ is a stable disk in M near $\mathbf{x}_j$. If $\mathbf{q} \in \tilde{D}_0^s(r)$, then for all $k \leq 0$, $\mathbf{q} \in F_{-k}^k(\mathcal{B}_{\mathbf{x}_{-k}}(r))$ so $F_{-k}^{-k}(\mathbf{q}) \in \mathcal{B}_{\mathbf{x}_{-k}}(r)$. In terms of $D_0^s(r)$, if $\mathbf{y} \in D_0^s(r)$, then $f^{|k|}(\mathbf{y}) \in \mathcal{V}_{\mathbf{x}_{|k|}}(r)$ and the forward orbit of $\mathbf{y}$ stays within r of the chain.

Because of the slopes of these two disks,

$$\bigcap_{k=-\infty}^{\infty} F_{j-k}^k(\mathcal{B}_{\mathbf{x}_{j-k}}(r))$$

is a single point $\mathbf{q}_j \in \mathcal{B}_{\mathbf{x}_j}(r)$ and $\exp_{\mathbf{x}_j}(\mathbf{q}_j) = \mathbf{y}_j \in \mathcal{V}_{\mathbf{x}_j}(r)$. Because of the nature of the intersection defining $\mathbf{y}_0$, $f^j(\mathbf{y}_0)$ stays within r of $\mathbf{x}_j$ for each j. Since $r < \epsilon$, $\mathbf{y}_0$ ϵ-shadows the chain. The fact that $\mathbf{y}_0$ is a single point shows that the point which ϵ-shadows the chain is unique when $j_2 = -j_1 = \infty$. Also, the uniqueness shows that $F_j(\mathbf{q}_j) = \mathbf{q}_{j+1}$, or $f(\mathbf{y}_j) = \mathbf{y}_{j+1}$ for the points of M.

If the chain is k periodic, then the uniqueness shows that $F_0^k(\mathbf{q}_0) = \mathbf{q}_k = \mathbf{q}_0$, so $\mathbf{q}_0$ is a periodic point for F, or $\mathbf{y}_0 \in M$ is a periodic point for f.

If the chain is not infinite in one direction or the other, then the intersection is a nonempty strip but not a point. This gives the existence of shadowing but not the uniqueness.

Now, in the case that Λ is also an isolated invariant set, let U be an isolating neighborhood and take V and r small enough so that $\mathcal{V}_{\mathbf{p}}(r) \subset U$ for all $\mathbf{p} \in V$. Then $f^j(\mathbf{y}_0) = \mathbf{y}_j \in \mathcal{V}_{\mathbf{x}_j}(r) \subset U$. Therefore, $f^j(\mathbf{y}_0) \in U$ for all j, and so $\mathbf{y}_0 \in \Lambda$ since Λ is the maximal invariant set in U. □

The next result compares the points with ω-limit sets in Λ with the stable manifolds of points. It shows that if a point approaches the whole set Λ, then it has to approach the orbit of some point in Λ. First, we give a definition of the stable manifold of the invariant set.

Definition. If Λ is an invariant set, the *stable manifold of* Λ, $W^s(\Lambda)$, is defined to be all points $\mathbf{q}$ such that $\omega(\mathbf{q}) \subset \Lambda$. Notice that often the stable manifold is not a manifold. Even for the horseshoe, it is a Cantor set of curves. Similarly, the *unstable manifold of* Λ, $W^u(\Lambda)$, is defined to be all points $\mathbf{q}$ such that $\alpha(\mathbf{q}) \subset \Lambda$.

A point $\mathbf{q}$ in the stable manifold of a single point $\mathbf{p}$ in an invariant set has a forward orbit which approaches the forward orbit of $\mathbf{p}$ in phase, $d(f^k(\mathbf{q}), f^k(\mathbf{p}))$ goes to zero as k goes to infinity. If the invariant set is not hyperbolic, then a point $\mathbf{q}$ might approach the invariant set and be in the stable manifold of the invariant set, without being in phase to any one point in the invariant set. The following theorem proves that a point which approaches a compact isolated hyperbolic invariant set necessarily is in phase with some point in the invariant set.

Corollary 3.2 (In Phase). *Let Λ be a compact isolated hyperbolic invariant set. Then,*

$$W^s(\Lambda) = \bigcup_{\mathbf{x}\in\Lambda} W^s(\mathbf{x}) \quad \textit{and}$$

$$W^u(\Lambda) = \bigcup_{\mathbf{x}\in\Lambda} W^u(\mathbf{x}).$$

PROOF. Let $\mathbf{y} \in W^s(\Lambda)$. Then, there is an $N > 0$ such that $d(f^j(\mathbf{y}), \Lambda) < \nu$ for $j \geq N$. Take $\mathbf{x}_j \in \Lambda$ such that $d(f^j(\mathbf{y}), \mathbf{x}_j) < \nu$ for $j \geq N$. By uniform continuity of f, $\mathbf{x}_j$ is a δ-chain if ν is small enough. Let $\mathbf{x}_j = f^{j-N}(\mathbf{x}_N) \in \Lambda$ for $j \leq N$. This δ-chain can be uniquely ϵ-shadowed by a point $\mathbf{x} \in \Lambda$. Thus, for $j \geq N$,

$$\begin{aligned} d(f^j(\mathbf{y}), f^j(\mathbf{x})) &\leq d(f^j(\mathbf{y}), \mathbf{x}_j) + d(\mathbf{x}_j, f^j(\mathbf{x})) \\ &\leq \nu + \epsilon. \end{aligned}$$

Since the forward orbit of $\mathbf{y}$ stays near the forward orbit of $\mathbf{x}$, the Stable Manifold Theorem implies that $\mathbf{y} \in W^s(\mathbf{x})$.

The proof for $W^u(\Lambda)$ is similar. □

Example 3.3. Let $\Lambda \subset \Sigma_2$ be the subshift of Example 3.1. Let $\mathbf{t} \in \Sigma_2$ be the sequence with $t_i = 0$ for $i < 0$, $t_0 = 1$, $t_1 = 0$, $t_2 = 1$, $t_3 = t_4 = t_5 = 0$, $t_6 = 1$, etc. After each 1 in the sequence, there are odd number of zeroes, with each time one more zero than the previous time. The point t is in $W^s(\Lambda)$ because for any $\epsilon > 0$, there is an $n > 0$ such that $\sigma^n(\mathbf{t})$ is within ϵ of $\bar{0}1\bar{0} \in \Lambda$. On the other hand, $\mathbf{t}$ does not converge to a single orbit in Λ.

Corollary 3.3 (Expansiveness). *Let Λ be a compact hyperbolic invariant set. There exists a $\delta > 0$ such that if $\mathbf{x}, \mathbf{y} \in \Lambda$ with $d(f^j(\mathbf{x}), f^j(\mathbf{y})) \le \delta$ for all $j \in \mathbb{Z}$, then $\mathbf{x} = \mathbf{y}$.*

REMARK 3.4. A diffeomorphism with the property of this corollary is called *expansive.*

Any diffeomorphism that is expansive also has sensitive dependence on initial conditions. (See Section 3.5 for the definition.) Therefore, a diffeomorphism restricted to a hyperbolic invariant set has sensitive dependence on initial conditions.

PROOF. Let $\mathbf{x}_j = f^j(\mathbf{x})$ and $\mathbf{y}_j = f^j(\mathbf{y})$. Both of these are infinite δ-chains. Each is an orbit which δ-shadows the other. By uniqueness of shadowing, $\mathbf{x} = \mathbf{y}$ if δ is small enough. □

10.4 Anosov Closing Lemma

In this section we prove that if Λ is either the set limit set $L(f)$ or the chain recurrent set $\mathcal{R}(f)$ and Λ has a hyperbolic structure, then $\Lambda = \text{cl}(\text{Per}(f))$. Thus, if (i) Λ is one of these two sets and is hyperbolic, and (ii) there are transverse intersections of stable and unstable manifolds for some points in Λ, then we can conclude that there are transverse intersections of stable and unstable manifolds for periodic points. Sometimes we can even get a homoclinic point for the same periodic point. Since a transverse homoclinic intersection for a periodic point implies the existence of a horseshoe, this result can be used to get an invariant set nearby with additional periodic points.

We proceed with the statement and proof of the Anosov Closing Lemma.

Theorem 4.1 (Anosov Closing Lemma). *Assume $f : M \to M$ is a C^1 diffeomorphism (or flow) on a compact manifold M.*

(a) Assume that the chain recurrent set of f, $\mathcal{R}(f)$, has a hyperbolic structure. Then, the periodic points are dense in the chain recurrent set, and $\text{cl}(\text{Per}(f)) = \mathcal{R}(f) = L(f) = \Omega(f)$.

(b) Assume that the limit set of f, $L(f)$, has a hyperbolic structure. Then, the periodic points are dense in the limit set, $\text{cl}(\text{Per}(f)) = L(f)$.

(c) Assume that the nonwandering set $\Omega(f)$ is hyperbolic. Then, the periodic points are dense in the nonwandering set of the map restricted to the nonwandering set, $\text{cl}(\text{Per}(f)) = \Omega(f|\Omega(f))$, i.e., $\text{cl}(\text{Per}(f)) = \Omega(F)$, where $F = f|\Omega(f)$.

REMARK 4.1. In part (c), if $\Omega(f)$ is hyperbolic, it is not true that $\text{cl}(\text{Per}(f)) = \Omega(f)$. A. Dankner (1978) gives an example of a diffeomorphism with a hyperbolic nonwandering set for which $\text{cl}(\text{Per}(f)) \neq \Omega(f)$. The problem is that there are a point $\mathbf{x} \in \Omega(f)$ and a neighborhood U of $\mathbf{x}$ such that if $\mathbf{z}, f^k(\mathbf{z}) \in U$ with $k \ge 1$, then $\{f^j(\mathbf{z}) : 0 \le j \le k\}$ is not contained in a small neighborhood of $\Omega(f)$. Thus, the orbit segment $\{f^j(\mathbf{z}) : 0 \le j \le k\}$ does not stay in the region of M where f is hyperbolic and so the periodic ϵ-chain generated by $\{f^j(\mathbf{z}) : 0 \le j \le k\}$ cannot be shadowed by a periodic orbit.

REMARK 4.2. There are examples where $L(f)$ is hyperbolic but $L(f) \neq \Omega(f) \subset \mathcal{R}(f)$. Figure 4.1 gives the phase portrait of an example where the limit set is hyperbolic, but the nonwandering set is not hyperbolic. The point $\mathbf{q}$ of non-transverse intersection is nonwandering but not a limit point, $\mathbf{q} \in \Omega(f) \setminus L(f)$. It follows that the whole orbit of $\mathbf{q}$ are nonwandering points which are not limit points. Also, $\mathbf{q} \in \Omega(f) \setminus \Omega(f|\Omega(f))$. Exercise 10.14 asks the reader to prove these facts about $\mathbf{q}$.

PROOF. (a) We showed in the section on Conley's Theorem that $\mathcal{R}(f|\mathcal{R}(f)) = \mathcal{R}(f)$. This means that given $\mathbf{x} \in \mathcal{R}(f)$, there is a periodic δ-chain $\{\mathbf{x}_j\}$ with $\mathbf{x}_0 = \mathbf{x}$, $\mathbf{x}_{j+k} = \mathbf{x}_j$ for all j, and $\mathbf{x}_j \in \mathcal{R}(f)$. By the Shadowing Theorem, there is a periodic point $\mathbf{y} \in \mathcal{R}(f)$

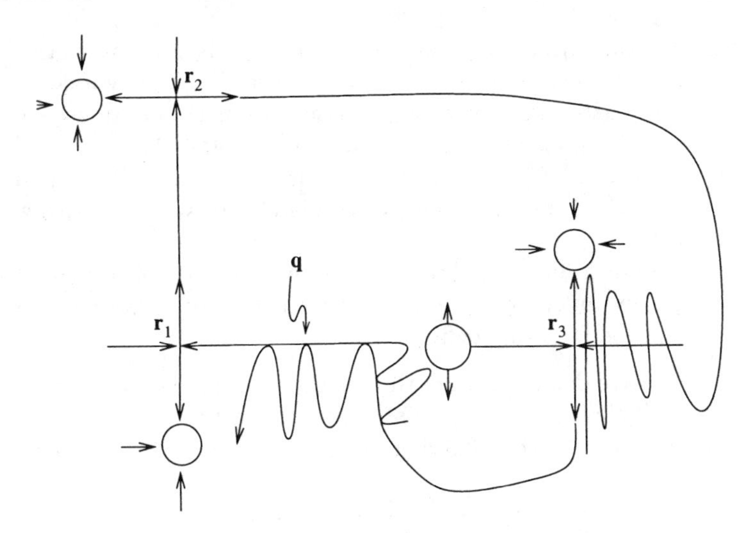

FIGURE 4.1. Phase Portrait for Remark 4.2

which ϵ-shadows the chain. The point $\mathbf{y}$ is within $\delta + \epsilon$ of $\mathbf{x}$. Since this is arbitrarily small, $\mathbf{x}$ is in the closure of the periodic points.

Since $\text{cl}(\text{Per}(f)) \subset L(f) \subset \Omega(f) \subset \mathcal{R}(f)$, and the outer two are equal, we have that $\text{cl}(\text{Per}(f)) = L(f) = \Omega(f) = \mathcal{R}(f)$.

(b) Since $L(f) = \text{cl}\left(\bigcup_{\mathbf{z}} \omega(\mathbf{z})\right)$, it is sufficient to prove that the periodic points are dense in an arbitrary $\omega(\mathbf{z})$. Let $\mathbf{x} \in \omega(\mathbf{z})$. Since M is compact, $\omega(\mathbf{z})$ is compact. In this circumstance, we proved earlier that $d(f^j(\mathbf{z}), \omega(\mathbf{z}))$ goes to zero as j goes to infinity. Therefore, given $\delta > 0$, we can find k and n such that $d(f^k(\mathbf{z}), f^{k+n}(\mathbf{z})) \leq \delta$, $d(f^k(\mathbf{z}), \mathbf{x}) < \delta$, and $d(f^i(\mathbf{z}), \omega(\mathbf{z})) \leq \delta$ for $k \leq i \leq k+n$. Let $\mathbf{z}_j$ be the n-periodic δ-chain with $\mathbf{z}_j = f^j(\mathbf{z})$ for $k \leq j < k+n$ and $\mathbf{z}_{k+n} = \mathbf{z}_k$. By construction, this whole periodic δ-chain is within δ of $\omega(\mathbf{z})$ and so of $L(f)$. This chain can be ϵ-shadowed by a periodic point $\mathbf{y}$. Then, $\mathbf{y}$ is within $\delta + \epsilon$ of $\mathbf{x}$. The sum $\delta + \epsilon$ can be made arbitrarily small, so $\mathbf{x}$ is in the closure of the periodic points.

We leave the proof of part (c) to the exercises. See Exercise 10.15. □

10.5 Decomposition of Hyperbolic Recurrent Points

Conley's Fundamental Theorem of Dynamical Systems gives a decomposition of the chain recurrent set into invariant sets. In this section we show that if f has a hyperbolic structure on $\mathcal{R}(f)$, then $\mathcal{R}(f)$ has a finite number of chain components, each of which is an isolated invariant set. This conclusion is not true without the added assumption of hyperbolicity.

In other books, it is often assumed that f satisfies a condition in terms of the nonwandering set called Axiom A. A diffeomorphism of flow f is said to satisfy *Axiom A* provided it has a hyperbolic structure on $\Omega(f)$ and $\text{cl}(\text{Per}(f)) = \Omega(f)$. See Smale (1967). We mentioned in the last section that the second condition of Axiom A does not follow from the first condition.

Rather than assume $\mathcal{R}(f)$ has a hyperbolic structure or f satisfies Axiom A, we often only assume that f has a hyperbolic structure $\text{cl}(\text{Per}(f))$. This latter condition is a weaker assumption than either of the other two assumptions and isolates what is actually necessary to make the theorem true. A second part of the main theorem assumes the limit set $L(f)$ has a hyperbolic structure. Again, this assumption is implied by the assumption that either (i) $\mathcal{R}(f)$ has a hyperbolic structure or (ii) f satisfies Axiom A and so is a weaker assumption. This weaker assumption is exactly what is needed to make the theorem work. The treatment given in this section closely follows the lecture notes by Newhouse (1980), who first isolated the actual assumptions which we use. Many of the original theorems are due to Smale. See Smale (1967).

Throughout this section, f is a C^1 diffeomorphism on a compact manifold M. The results are also true for a flow on a compact manifold but we do not state the results using this terminology.

As indicated in the introduction above, we use several types of recurrent sets in this section: limit set, nonwandering set, and chain recurrent set. We review the notation for these concepts in the following definition.

Definition. As we have defined before, $\Omega(f)$ is the set of all nonwandering points and $\mathcal{R}(f)$ is the set of all chain recurrent points. As a third type of recurrent points, the *limit set of* f is defined to be the following set:

$$L(f) = \text{cl}\,\Big(\bigcup_{\mathbf{p}\in M} \omega(\mathbf{p}) \cup \alpha(\mathbf{p}) \Big).$$

Using the definitions, it is easy to check that $\text{Per}(f) \subset L(f) \subset \Omega(f) \subset \mathcal{R}(f)$. In the theorems of this section, we often assume that cl(Per(f)), $L(f)$, or $\mathcal{R}(f)$ has a hyperbolic structure. We could also assume that $\Omega(f)$ has a hyperbolic structure, but then we have to add the assumption that $\text{cl}(\text{Per}(f)) = \Omega(f)$, i.e., that f satisfies Axiom A.

To break up the periodic points into pieces, we form equivalence classes of periodic points using the relation given in the following definition.

Definition. We let $H(f)$ be the set of all hyperbolic periodic points of f. If $\mathbf{q}, \mathbf{p} \in H(f)$, then we say that $\mathbf{p}$ is *heteroclinically related to* $\mathbf{q}$, or $\mathbf{p}$ *is h-related to* $\mathbf{q}$, provided $W^u(\mathcal{O}(\mathbf{p}))$ has a nonempty transverse intersection with $W^s(\mathcal{O}(\mathbf{q}))$ and $W^u(\mathcal{O}(\mathbf{q}))$ has a nonempty transverse intersection with $W^s(\mathcal{O}(\mathbf{p}))$. (Newhouse calls this property *homoclinically related.*) We form equivalence classes of h-related points and write $\mathbf{p} \sim \mathbf{q}$ if $\mathbf{p}$ is h-related to $\mathbf{q}$. For $\mathbf{p} \in H(f)$, let

$$H_{\mathbf{p}} = \{\mathbf{q} \in H(f) : \mathbf{p} \sim \mathbf{q}\}.$$

The set $H_{\mathbf{p}}$ is called the *h-class of* $\mathbf{p}$. Note if $\mathbf{p}$ is h-related to $\mathbf{q}$, then $\dim W^u(\mathbf{p}) = \dim W^u(\mathbf{q})$ since there is a transverse intersection of the stable manifold of $\mathbf{p}$ and the unstable manifold $\mathbf{q}$ and vice versa.

Proposition 5.1. *Being h-related is an equivalence relation on* $H(f)$.

PROOF. It is clear that $\mathbf{p} \sim \mathbf{p}$ and that $\mathbf{p} \sim \mathbf{q}$ if and only if $\mathbf{q} \sim \mathbf{p}$.

What we need to check is transitivity: assume $\mathbf{p} \sim \mathbf{q}$ and $\mathbf{q} \sim \mathbf{r}$ and show that $\mathbf{p} \sim \mathbf{r}$. Let n_1, n_2, and n_3 be the periods of $\mathbf{p}$, $\mathbf{q}$, and $\mathbf{r}$, respectively. Since the manifolds for the orbits are transverse, we can find $\mathbf{p}' \in \mathcal{O}(\mathbf{p})$ such that $W^u(\mathbf{p}')$ intersects $W^s(\mathbf{q})$ transversally at a point $\mathbf{x}$. By replacing $\mathbf{x}$ by $f^{jn_1n_2}(\mathbf{x})$ for some $j \geq 1$, we can assume that $\mathbf{x}$ lies in the local stable manifold of $\mathbf{q}$. Let Δ^u be a small disk in $W^u(\mathbf{p}')$ through $\mathbf{x}$. By the Inclination Lemma, $f^{jn_2}(\Delta^u)$ accumulates on $W^u_{loc}(\mathbf{q})$ in a C^1 manner.

Next, because $\mathbf{q}$ is h-related to $\mathbf{r}$, there is an $\mathbf{r}' \in \mathcal{O}(\mathbf{r})$ such that $W^u(\mathbf{q}')$ intersects $W^s(\mathbf{r}')$ transversally at some point $\mathbf{y}$. By replacing $\mathbf{y}$ by $f^{-jn_2n_3}(\mathbf{y})$ for some $j \geq 1$, we can assume that $\mathbf{y}$ lies in the local unstable manifold of $\mathbf{q}$. Let Δ^s be a small disk in $W^s(\mathbf{r}')$ through $\mathbf{y}$. See Figure 5.1. By the Inclination Lemma, $f^{-jn_2n_3}(\Delta^s)$ accumulates on $W^s_{loc}(\mathbf{q})$ in a C^1 manner. Since $W^u_{loc}(\mathbf{q})$ and $W^s_{loc}(\mathbf{q})$ cross transversally at $\mathbf{q}$, and $f^{jn_1n_2}(\Delta^u)$ and $f^{-jn_2n_3}(\Delta^s)$ accumulate on them in a C^1 manner, it follows that

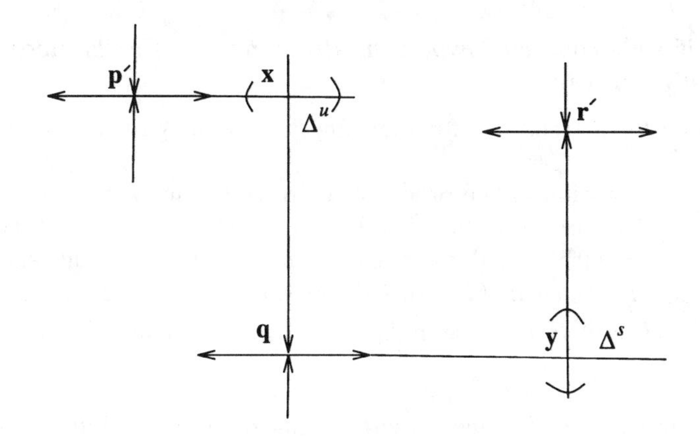

FIGURE 5.1. Disks Δ^u and Δ^s

$f^{jn_1n_2}(\Delta^u)$ and $f^{-jn_2n_3}(\Delta^s)$ cross transversally for large enough j. Thus, $W^u(\mathcal{O}(\mathbf{p}))$ has a nonempty transverse intersection with $W^s(\mathcal{O}(\mathbf{r}))$. A similar argument shows that $W^u(\mathcal{O}(\mathbf{r}))$ has a nonempty transverse intersection with $W^s(\mathcal{O}(\mathbf{p}))$. Combining, $\mathbf{p}$ is h-related to $\mathbf{r}$. □

Next, we show that the closure of a single class $H_{\mathbf{p}}$, $\text{cl}(H_{\mathbf{p}})$, is topologically transitive using the Birkhoff Transitivity Theorem.

Proposition 5.2. *For any $\mathbf{p} \in H(f)$, the set $\text{cl}(H_{\mathbf{p}})$ is closed, invariant under f, and topologically transitive for f.*

PROOF. Let $X = \text{cl}(H_{\mathbf{p}})$. Clearly, X is closed. It is the closure of an invariant set and so is invariant. (See Exercise 10.4.)

To show that f is topologically transitive on X, we need to show that the Birkhoff Transitivity Theorem applies. The set X is a complete metric space with countable basis. Take any two open sets U_1 and U_2 in the ambient space such that $\hat{U}_i = X \cap U_i \neq \emptyset$. It is sufficient to show that $\mathcal{O}(\hat{U}_1) \cap \hat{U}_2 \neq \emptyset$. Take any $\mathbf{r}_j \in U_j \cap H_p$ for $j = 1, 2$. The orbit $\mathcal{O}(\mathbf{r}_1) \subset \mathcal{O}^+(U_1)$, so $W^u_\epsilon(\mathcal{O}(\mathbf{r}_1)) \subset \mathcal{O}^+(U_1)$. By iteration, we get that $W^u(\mathcal{O}(\mathbf{r}_1)) \subset \mathcal{O}^+(U_1)$. Because $\mathbf{r}_1$ is h-related to $\mathbf{r}_2$ (both are in $H_{\mathbf{p}}$), $W^u(\mathcal{O}(\mathbf{r}_1))$ intersects $W^s(\mathcal{O}(\mathbf{r}_2))$ transversally at some point $\mathbf{y}_1$ and $W^u(\mathcal{O}(\mathbf{r}_2))$ intersects $W^s(\mathcal{O}(\mathbf{r}_1))$ transversally at some point $\mathbf{y}_2$. These transverse heteroclinic intersections imply the existence of a horseshoe by Theorem VIII.4.5; therefore, $\mathbf{y}_1$ is in the closure of periodic points which must be heteroclinically related to $\mathbf{r}_1$ and $\mathbf{r}_2$, and $\mathbf{y}_1 \in X$. We have shown that $\mathbf{y}_1 \in \mathcal{O}^+(\hat{U}_1)$. By an argument as above applied to $\mathbf{r}_2$, $\mathbf{y}_1 \in \mathcal{O}^-(\hat{U}_2)$ and so there is some $k > 0$ such that $f^k(\mathbf{y}_1) \in \hat{U}_2$. Because $\mathcal{O}^+(\hat{U}_1)$ is positively invariant, $f^k(\mathbf{y}_1) \in \mathcal{O}^+(\hat{U}_1)$. Thus, $\mathcal{O}^+(\hat{U}_1) \cap \hat{U}_2 \neq \emptyset$. □

Example 5.1. Let f be the map on the standard horseshoe Λ. Any two periodic points are h-related. Also, $\Lambda = \text{cl}(H_{\mathbf{p}})$. By the above theorem it follows that f is topologically transitive on Λ. (The fact that f is topologically transitive on Λ also follows from the conjugacy to the two-sided shift.)

Example 5.2. The fact that the solenoid, or any other connected hyperbolic attracting set, is topologically transitive follows from the following proposition.

Proposition 5.3. *(a) Let N be compact and connected and $f : N \to N$ be a C^1 diffeomorphism into N. Assume that N is a trapping region, so that $f(\text{cl}(N)) \subset \text{int}(N)$. Let $\Lambda = \bigcap_{j \geq 0} f^j(N)$ be the attracting set. Assume that f has a hyperbolic structure on Λ and the periodic points are dense in Λ. Then, f is topologically transitive on Λ.*

(b) In fact, if Λ is a connected hyperbolic attracting set for a diffeomorphism f, then f is topologically transitive on Λ.

PROOF. A connected attracting set has a connected trapping region, so part (b) follows from part (a).

To simplify the notation in the proof, we write H to mean the hyperbolic periodic points in Λ, $H(f) \cap \Lambda$. Because there is a hyperbolic structure on Λ, there is an $\epsilon > 0$ such that if $\mathbf{p}, \mathbf{q} \in H$ with $d(\mathbf{p}, \mathbf{q}) < \epsilon$, then $\mathbf{p}$ is h-related to $\mathbf{q}$. Thus, each $H_{\mathbf{p}}$ is open in H. Take $U_{\mathbf{p}} \supset H_{\mathbf{p}}$ open in M with $U_{\mathbf{p}}$ contained within a distance of $\epsilon/3$ of $H_{\mathbf{p}}$. Since $\bigcup_{\mathbf{p} \in H} H_{\mathbf{p}} = H$, and H is dense in Λ, $\bigcup_{\mathbf{p} \in H} U_{\mathbf{p}} \supset \Lambda$. Therefore, we have an open cover of Λ.

Next, we claim that two $U_{\mathbf{p}}$ and $U_{\mathbf{q}}$ either coincide or are disjoint. If $\mathbf{x} \in U_{\mathbf{p}} \cap U_{\mathbf{q}}$, there is a $\mathbf{p}' \in H_{\mathbf{p}}$ and $\mathbf{q}' \in H_{\mathbf{q}}$ such that $d(\mathbf{x}, \mathbf{p}') < \epsilon/3$ and $d(\mathbf{x}, \mathbf{q}') < \epsilon/3$. Then, $d(\mathbf{p}', \mathbf{q}') \leq d(\mathbf{p}', \mathbf{x}) + d(\mathbf{x}, \mathbf{q}') < (2/3)\epsilon < \epsilon$. By the choice of ϵ, it follows that $\mathbf{p}'$ is h-related to $\mathbf{q}'$, and so $\mathbf{p}$ is h-related to $\mathbf{q}$. Thus, $H_{\mathbf{p}} = H_{\mathbf{q}}$ and $U_{\mathbf{p}} = U_{\mathbf{q}}$. Therefore, the only time that $U_{\mathbf{p}}$ and $U_{\mathbf{q}}$ can intersect is for them to coincide.

We have shown that the $U_{\mathbf{p}}$ form a cover of the set Λ by disjoint open sets. Since each $f^k(N)$ is connected, Λ is the intersection of the connected sets $f^k(N)$, and Λ is connected. (See Exercise 5.11.) Therefore, there is only one set $U_{\mathbf{p}}$ and so only one equivalence class $H_{\mathbf{p}}$, $H = H_{\mathbf{p}}$, and $\Lambda = \text{cl}(H_{\mathbf{p}})$. By Proposition 5.2, it follows that f is topologically transitive on Λ. □

The next theorem shows that the closure of the periodic points can be split up into a finite union of invariant sets that are topologically transitive if cl(Per(f)) has a hyperbolic structure. In some sense, Proposition 5.3 is a special case of this theorem where there is one piece. The theorem also proves that each invariant set has a local product structure which is defined next.

Definition. Let Λ be a hyperbolic invariant set for a diffeomorphism f. We say that Λ has a *local product structure* provided there is an $r > 0$ such that for every $\mathbf{p}, \mathbf{q} \in \Lambda$, $W_r^u(\mathbf{p}) \cap W_r^s(\mathbf{q}) \subset \Lambda$.

Let Λ be a hyperbolic invariant set. The continuity of the stable and unstable manifolds for points of Λ implies that there is an $r_0 > 0$ such that for any $0 < r \leq r_0$, there is an $\epsilon = \epsilon(r) > 0$ for which $W_r^u(\mathbf{p}) \cap W_r^s(\mathbf{q})$ is a single point for any $\mathbf{p}, \mathbf{q} \in \Lambda$ with $d(\mathbf{p}, \mathbf{q}) \leq \epsilon$, i.e., this intersection is nonempty and a single point. Thus, if Λ is a hyperbolic invariant set with a local product structure and $d(\mathbf{p}, \mathbf{q}) < \epsilon$, then $W_r^u(\mathbf{p}) \cap W_r^s(\mathbf{q}) = \{\mathbf{z}\} \subset \Lambda$. Using this fact, it can be shown that given $\mathbf{p} \in \Lambda$, for small $r > 0$ a neighborhood of $\mathbf{p}$ in Λ is homeomorphic to $[W_r^u(\mathbf{p}) \cap \Lambda] \times [W_r^s(\mathbf{p}) \cap \Lambda]$ by the map

$$h : [W_r^u(\mathbf{p}) \cap \Lambda] \times [W_r^s(\mathbf{p}) \cap \Lambda] \to \Lambda$$
$$h(\mathbf{x}, \mathbf{y}) = W_r^u(\mathbf{x}) \cap W_r^s(\mathbf{y}).$$

Thus, locally the invariant set is homeomorphic to a product, hence the name local product structure. Now, we give the Spectral Decomposition Theorem. (The name of the theorem was introduced by Smale and has been retained even though the theorem does not refer to the spectra of anything.)

Theorem 5.4 (Spectral Decomposition). *Let M be compact and $f : M \to M$ be a C^1 diffeomorphism.*

(a) Assume that f has a hyperbolic structure on cl(Per(f)). Then, there are a finite number of sets, $\Lambda_1, \ldots, \Lambda_N$, each of which is the closure of disjoint h-classes, such that

$\text{cl}(\text{Per}(f)) = \Lambda_1 \cup \cdots \cup \Lambda_N$. *Thus, each* Λ_j *is closed, invariant by* f*, the periodic points are dense, and* f *is topologically transitive on* Λ_j. *Further, each* Λ_j *has a local product structure.*

(b) *If we assume that* $L(f)$ *has a hyperbolic structure, then* $M = \bigcup_j W^s(\Lambda_j) = \bigcup_j W^u(\Lambda_j)$, *where* $W^u(\Lambda_j) = \{\mathbf{q} : \alpha(\mathbf{q}) \subset \Lambda_j\}$ *and* $W^s(\Lambda_j) = \{\mathbf{q} : \omega(\mathbf{q}) \subset \Lambda_j\}$.

Definition. Assume that $\text{cl}(\text{Per}(f))$ has a hyperbolic structure for f. Then, by the Spectral Decomposition Theorem, $\text{cl}(\text{Per}(f)) = \Lambda_1 \cup \cdots \cup \Lambda_N$. The sets Λ_j are called *basic sets.* Such a basic set is (i) a closed invariant isolated hyperbolic invariant set, and (ii) $f|Lam_i$ is topologically transitive. Some authors take these properties as the definition, but we use the more global assumptions for the definition.

REMARK 5.1. By Theorem 4.1, if we assume either (i) $\Lambda = \mathcal{R}(f)$ has a hyperbolic structure, (ii) $\Lambda = L(f)$ has a hyperbolic structure, or (iii) $\Lambda = \Omega(f)$ and f satisfies Axiom A, then $\Lambda = \text{cl}(\text{Per}(f))$. Therefore, both parts of the Spectral Decomposition Theorem are true if we assume that f satisfies any of the above three assumptions. Since it is possible for $\Omega(f)$ to be hyperbolic and $\Omega(f) \neq L(f) = \text{cl}(\text{Per}(f))$, if we state the theorem in terms of the nonwandering set, we need to assume Axiom A and not just that $\Omega(f)$ has a hyperbolic structure.

REMARK 5.2. It is not hard to prove that each basic set Λ_j can be further split up into pieces for which a power of f is topologically mixing: $\Lambda_j = \bigcup_{i=1}^{n_j} X_{j,i}$ with (i) the sets $X_{j,i} = \text{cl}(W^u(\mathbf{p}) \cap W^s(\mathbf{p}))$ for some periodic point $\mathbf{p} \in X_{j,i}$, (ii) the sets $X_{j,i}$ are pairwise disjoint, (iii) $f(X_{j,i}) = X_{j,i+1}$ for $1 \le i < n_j$ and $f(X_{j,n_j}) = X_{j,1}$, and (iv) the n_j-power of f restricted to each $X_{j,i}$ is topologically mixing, $f^{n_j}|X_{j,i}$ is topologically mixing for each j and i. See Bowen (1975b). (See Section 2.5 for the definition of topologically mixing and the difference from topologically transitive.)

PROOF. As in the proof of Proposition 5.3, there is an $\epsilon > 0$ such that if $\mathbf{p}, \mathbf{q} \in H(f)$ with $d(\mathbf{p}, \mathbf{q}) < \epsilon$, then $\mathbf{p}$ is h-related to $\mathbf{q}$. Therefore, if $\text{cl}(H_{\mathbf{p}}) \cap \text{cl}(H_{\mathbf{q}}) \neq \emptyset$, then $H_{\mathbf{p}} = H_{\mathbf{q}}$. That is, the closure of h-classes either agree or are disjoint. Also, $\text{cl}(H_{\mathbf{p}})$ is closed, invariant, and topologically transitive by Proposition 5.2.

As in the proof of Proposition 5.3, we can take sets $U_{\mathbf{p}} \supset H_{\mathbf{p}}$ which are open in M with $U_{\mathbf{p}}$ contained within $\epsilon/3$ of $H_{\mathbf{p}}$. These open sets are disjoint (or coincide) and cover $\text{cl}(\text{Per}(f))$. Since $\text{cl}(\text{Per}(f))$ is compact, a finite number of these $U_{\mathbf{p}}$ cover. Therefore, there are only a finite number of distinct classes $H_{\mathbf{p}}$.

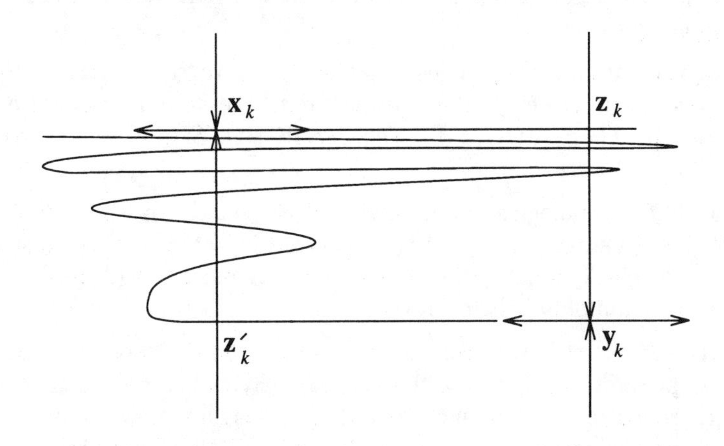

FIGURE 5.2. Local Product Structure: $W_r^s(\mathbf{x}_k)$, $W_r^u(\mathbf{x}_k)$, $W_r^s(\mathbf{y}_k)$, and $W^u(\mathbf{y}_k)$

Finally, we prove the local product structure of each $\Lambda_j = \text{cl}(H_{\mathbf{p}_j})$. Let $\mathbf{x}, \mathbf{y} \in \Lambda_j$, and choose $\mathbf{x}_k, \mathbf{y}_k \in H_{\mathbf{p}_j} \cap \text{Per}(f)$ so that $\mathbf{x}_k$ converges to $\mathbf{x}$ and $\mathbf{y}_k$ converges to $\mathbf{y}$. By the continuity of the local stable and unstable manifolds, both $W^u_r(\mathbf{x}_k) \cap W^s_r(\mathbf{y}_k) = \{\mathbf{z}_k\}$ and $W^u_r(\mathbf{y}_k) \cap W^s_r(\mathbf{x}_k) = \{\mathbf{z}'_k\}$ are a single point at a transverse intersection. By the Inclination Lemma, $W^u(\mathbf{y}_k)$ accumulates on $W^u_r(\mathbf{x}_k)$, and so $W^u(\mathbf{y}_k)$ has a transverse homoclinic intersection with $W^s_r(\mathbf{y}_k)$ arbitrarily near the point $\mathbf{z}_k$. See Figure 5.2. Each of these transverse homoclinic intersections for a periodic point is in the closure of the periodic points, so $\mathbf{z}_k \in \text{cl}(\text{Per}(f)) = \Lambda_1 \cup \cdots \cup \Lambda_N$. Because the Λ_i are a finite distance apart, $\mathbf{z}_k \in \Lambda_j$ if ϵ is small enough. As $\mathbf{x}_k$ approaches $\mathbf{x}$ and $\mathbf{y}_k$ approaches $\mathbf{y}$, $\mathbf{z}_k$ approaches some $\mathbf{z} \in W^u_r(\mathbf{x}) \cap W^s_r(\mathbf{y})$. Since Λ_j is closed, it follows that $\mathbf{z} \in \Lambda_j$ as was to be proved. This completes the proof of part (a).

For the proof of part (b) of Theorem 5.4, let $\mathbf{p} \in M$. Then, $\omega(\mathbf{p}) \subset L(f) = \Lambda_1 \cup \cdots \cup \Lambda_N$. The Λ_j are disjoint and a bounded distance apart, so there exists an $\nu > 0$ such that they are a distance apart greater than ν. We want to show that $\omega(\mathbf{p})$ must be contained in a single Λ_j. We assume the opposite and get a contradiction, i.e., we assume that $\omega(\mathbf{p}) \cap \Lambda_j \neq \emptyset$ and $\omega(\mathbf{p}) \cap \Lambda_k \neq \emptyset$ for $j \neq k$. Let $D(\Lambda_i, \nu/3)$ be the $\nu/3$ neighborhood of Λ_i. Because the sets Λ_i are invariant, there are neighborhoods U_i of each Λ_i, such that $\text{cl}(U_i) \subset D(\Lambda_i, \nu/3)$ and $f(\text{cl}(U_i)) \cap D(\Lambda_t, \nu/3) = \emptyset$ for $t \neq i$. With the above assumptions on $\mathbf{p}$, there is an increasing sequence of iterates n_i with $f^{n_{2i}}(\mathbf{p}) \in U_j$ and $f^{n_{2i+1}}(\mathbf{p}) \in U_k$. Let m_i be the largest integer m such that $n_{2i} < m < n_{2i+1}$ with $f^{m-1}(\mathbf{p}) \in \text{cl}(U_j)$. Then, $f^{m_i}(\mathbf{p}) \notin U_j$ by the choice of m_i. On the other hand, $f^{m_i-1}(\mathbf{p}) \in \text{cl}(U_j)$, so $f^{m_i}(\mathbf{p}) = f \circ f^{m_i-1}(\mathbf{p}) \notin D(\Lambda_t, \nu/3) \supset U_t$ for $t \neq j$. Therefore, $f^{m_i}(\mathbf{p}) \in M \setminus \bigcup_t U_t$. By taking a subsequence of the m_i, we get that $f^{m_i}(\mathbf{p})$ accumulates on a point $\mathbf{q}$ that is not in any of the Λ_t. We have a contradiction because $\mathbf{q}$ is in $\omega(\mathbf{p})$. Thus, we have shown that $\omega(\mathbf{p})$ is contained in a single Λ_j. Therefore, $\mathbf{p} \in W^s(\Lambda_j)$.

The proof that $\alpha(\mathbf{p}) \subset \Lambda_k$ for some k is similar, so $\mathbf{p} \in W^u(\Lambda_k)$. This completes the proof of the theorem. □

Corollary 5.5. *If $\mathcal{R}(f)$ has a hyperbolic structure, then f has a finite number of chain components.*

PROOF. The Spectral Decomposition Theorem implies that $\mathcal{R}(f) = \Lambda_1 \cup \cdots \cup \Lambda_N$ is a finite decomposition. Since f is topologically transitive on each of the Λ_j, each of these Λ_j is a chain component. Therefore, there are a finite number of chain components. □

The following proposition shows that the stable manifold of a single periodic orbit is dense in the stable manifold of the whole invariant set. This fact clarifies the meaning of a cycle as defined below.

Proposition 5.6. *Assume Λ is a basic set for f. (We could assume that Λ is an isolated invariant set with* Per(f) *dense in Λ and all the period points of Λ h-related.) Let $\mathbf{p} \in \Lambda \cap$* Per$(f)$. *Then, $W^s(\mathcal{O}(\mathbf{p}))$ is dense in $W^s(\Lambda)$. Similarly, $W^u(\mathcal{O}(\mathbf{p}))$ is dense in $W^u(\Lambda)$.*

REMARK 5.3. If f is topologically mixing on a basic set and $\mathbf{p} \in \Lambda \cap \text{Per}(f)$, then the stable manifold of the single point $\mathbf{p}$, $W^s(\mathbf{p})$, is dense in $W^s(\Lambda)$ and not just $W^u(\mathcal{O}(\mathbf{p}))$.

If f is not topologically mixing, then $W^s(\mathbf{p})$ is not dense in $W^s(\Lambda)$ and it is necessary to take the stable manifold of the orbit of $\mathbf{p}$.

PROOF. Let $\mathbf{p} \in \Lambda \cap \text{Per}(f)$ be as in the statement of the theorem have period n. Let $\mathbf{q}$ be any other periodic point in Λ with period m. By the assumptions, $\mathbf{q}$ is h-related to $\mathbf{p}$. Thus, there is a $\mathbf{p}_1 \in \mathcal{O}(\mathbf{p})$ such that $W^s(\mathbf{p}_1)$ has a transverse intersection with $W^u(\mathbf{q})$. Using the Inclination Lemma applied to f^{mn}, $W^s(\mathbf{p}_1)$ accumulates on the local stable manifold of $\mathbf{q}$, and so by iteration of f^{mn} it accumulates on all of $W^s(\mathbf{q})$. Thus, $W^s(\mathcal{O}(\mathbf{p}))$ accumulates on the stable manifold of any periodic point.

Because the periodic points are dense in Λ, it follows from the continuous dependence of the stable manifolds on the point that $\bigcup\{W^s(\mathbf{q}) : \mathbf{q} \in \Lambda \cap \text{Per}(f)\}$ is dense in $\bigcup\{W^s(\mathbf{q}) : \mathbf{q} \in \Lambda\} = W^s(\Lambda)$. Since $W^s(\mathcal{O}(\mathbf{p}))$ is dense in $\bigcup\{W^s(\mathbf{q}) : \mathbf{q} \in \Lambda \cap \text{Per}(f)\}$, it follows that $W^s(\mathcal{O}(\mathbf{p}))$ is dense in $W^s(\Lambda)$. This completes the proof of the theorem for stable manifolds. The proof for unstable manifolds is similar. □

Definition. Assume that cl(Per(f)), $L(f)$, or $\mathcal{R}(f)$ has a hyperbolic structure for f. (If we assume that $\Omega(f)$ has a hyperbolic structure, then we also have to assume that $\text{cl}(\text{Per}(f)) = \Omega(f)$.) Let $\Lambda_1 \cup \cdots \cup \Lambda_N$ be the basic sets given by the Spectral Decomposition Theorem, and define $\hat{W}^u(\Lambda_j) = W^u(\Lambda_j) \setminus \Lambda_j$. Define a partial ordering on the basic sets by declaring $\Lambda_j << \Lambda_k$ if $\hat{W}^u(\Lambda_j) \cap \hat{W}^s(\Lambda_k) \neq \emptyset$. A *k-cycle* is a sequence of basic sets $\Lambda_{j_1}, \ldots, \Lambda_{j_k}$ with $\Lambda_{j_1} << \Lambda_{j_2} << \cdots << \Lambda_{j_k} << \Lambda_{j_1}$. Thus, a 1-cycle is a basic set Λ_j with $\hat{W}^u(\Lambda_j) \cap \hat{W}^s(\Lambda_j) \neq \emptyset$, i.e., the stable and unstable manifolds intersect off Λ_j.

REMARK 5.4. If $L(f)$ (or $\Omega(f)$ or $\mathcal{R}(f)$) is hyperbolic, then a cycle contains some non-transverse intersections because otherwise all the periodic points are h-related and intersections are all in $L(f)$. The example given above with $L(f) \neq \Omega(f)$ is an example for which the limit set is hyperbolic with a 3-cycle, but the nonwandering set is not hyperbolic, i.e., the points of non-transverse intersection are nonwandering but not limit points.

Definition. Assume that $L(f)$ or $\mathcal{R}(f)$ has a hyperbolic structure. (Alternatively, assume that $\Omega(f)$ has a hyperbolic structure and $\text{cl}(\text{Per}(f)) = \Omega(f)$.) Then, f is said to have *no cycles* or *satisfy the no cycle property* provided there are no cycles among the basic sets formed from $L(f)$ (or Ω or $\mathcal{R}(f)$).

Theorem 5.7. *If $\mathcal{R}(f)$ is hyperbolic, then f satisfies the no cycle property.*

PROOF. Assume $\{\Lambda_{j_i}\}_{i=1}^k$ is a k-cycle for f. It is easy to check that all the points of the intersections $\hat{W}^u(\Lambda_{j_i}) \cap \hat{W}^s(\Lambda_{j_{i+1}})$ are in $\mathcal{R}(f)$. (See Exercise 10.2 dealing with periodic points. Also compare with Exercise 10.18.) Thus, these points are not outside the basic sets, so there is no cycle but merely one larger basic set. □

REMARK 5.5. If we assume that $L(f)$ or $\Omega(f)$ has a hyperbolic structure, then the fact that f has no cycles among the basic sets is an additional assumption.

10.6 Markov Partitions for a Hyperbolic Invariant Set

Throughout this section, $f : M \to M$ is a C^1 diffeomorphism on a manifold M, Λ is a compact, isolated, hyperbolic invariant set for f with a local product structure. The set Λ could be a basic set from the spectral decomposition but this is not necessary. We take an adapted norm on tangent vectors at points of Λ, and d a compatible distance on M. With this distance, two points in a stable manifold get closer together under forward iteration and two points in an unstable manifold get closer together under backward iteration. Therefore, $f^{-1}(W^s_\epsilon(f(\mathbf{p}))) \supset W^s_\epsilon(\mathbf{p})$ and $f(W^u_\epsilon(f^{-1}(\mathbf{p}))) \supset W^u_\epsilon(\mathbf{p})$. Since Λ has a local product structure, there are $\epsilon \geq \alpha > 0$ such that $W^s_\epsilon(\mathbf{p}) \cap W^u_\epsilon(\mathbf{q})$ a single point in Λ whenever $\mathbf{p}, \mathbf{q} \in \Lambda$ satisfy $d(\mathbf{p}, \mathbf{q}) \leq \alpha$. However, by taking $\epsilon \geq \alpha > 0$ smaller, we can make sure that

$$f^{-1}(W^s_\epsilon(f(\mathbf{p}))) \cap f(W^u_\epsilon(f^{-1}(\mathbf{q}))) = W^s_\epsilon(\mathbf{p}) \cap W^u_\epsilon(\mathbf{q}),$$

whenever $\mathbf{p}, \mathbf{q} \in \Lambda$ satisfy $d(\mathbf{p}, \mathbf{q}) \leq \alpha$, and so the intersection of the manifolds on the left-hand side is also a single point. Note that the the two sets which are intersected on the left-hand side of the equality contain the respective sets on the right-hand side. We fix ϵ and α with these properties.

For this general situation, we need to define a rectangle and then the stable and unstable manifolds in a rectangle. Note that throughout this section the interiors of all subsets of Λ are taken as subsets of Λ and not as a subset of the ambient manifold M; thus, $\text{int}(R)$ means the interior of R in Λ.

Definition. A set $R \subset \Lambda$ is a *rectangle* provided
(i) it has diameter less than α, where α is as above, and
(ii) $\mathbf{p}, \mathbf{q} \in R$ implies that $W^s_\epsilon(\mathbf{p}) \cap W^u_\epsilon(\mathbf{q}) \in R$.
The rectangle is call *proper* if, in addition,
(iii) $R = \text{cl}(\text{int}(R))$ so that it is closed.
If R is a rectangle, then for $\mathbf{p} \in R$ let

$$W^s(\mathbf{p}, R) = W^s_\epsilon(\mathbf{p}) \cap R \qquad \text{and}$$
$$W^u(\mathbf{p}, R) = W^u_\epsilon(\mathbf{p}) \cap R.$$

Note the comparison of the definition of $W^s(\mathbf{p}, R)$ with what is used for hyperbolic toral automorphisms in Chapter VIII.

With these definitions we can define a Markov partition as before.

Definition. Assume that Λ has a local product structure for f as above. A *Markov partition of* Λ for f is a finite collection of proper rectangles, $\mathcal{R} = \{R_j\}_{j=1}^m$, that satisfy the following four conditions:
(i) $\Lambda = \bigcup_{j=1}^m R_j$,
(ii) if $i \neq j$, then $\text{int}(R_i) \cap \text{int}(R_j) = \emptyset$, (so $\text{int}(R_i) \cap R_j = \emptyset$),
(iii) if $\mathbf{z} \in \text{int}(R_i) \cap f^{-1}(\text{int}(R_j))$, then

$$f(W^u(\mathbf{z}, R_i)) \supset W^u(f(\mathbf{z}), R_j) \qquad \text{and}$$
$$f(W^s(\mathbf{z}, R_i)) \subset W^s(f(\mathbf{z}), R_j),$$

and
(iv) if $\mathbf{z} \in \text{int}(R_i) \cap f^{-1}(\text{int}(R_j))$, then

$$\text{int}(R_j) \cap f(W^u(\mathbf{z}, R_i) \cap \text{int}(R_i)) = W^u(f(\mathbf{z}), R_j) \cap \text{int}(R_j) \qquad \text{and}$$
$$\text{int}(R_i) \cap f^{-1}(W^s(R_j) \cap \text{int}(R_j)) = W^s(\mathbf{z}, R_i) \cap \text{int}(R_i).$$

REMARK 6.1. In the context of the toral Anosov automorphisms, condition (iv) is needed to insure that the image of a rectangle only crosses another rectangle once; this fact enables large rectangles to be used and still to be able to get a single orbit which passes through the prescribed sequence of rectangles, i.e., to get symbolic dynamics. In the present context, the rectangles found are small. Their small size is used to show that condition (iii) implies condition (iv). (Note the added condition on the size of the local stable manifolds and iterates of the map which we imposed above.) Because the small rectangles automatically satisfy condition (iv), Bowen and other authors do not add a condition like this one to the definition of a Markov partition.

We can now state the main result of this section which is due to Bowen (1970a). We follow the treatment of Bowen (1975).

Theorem 6.1. *Let Λ be a hyperbolic invariant set with a local product structure for a diffeomorphism f. Then, there exists a Markov partition of Λ for f with rectangles arbitrarily small (diameter less than α).*

REMARK 6.2. Once there is a Markov partition, then it is possible to define a subshift of finite type Σ_A as for a toral Anosov automorphism and a semi-conjugacy $h : \Sigma_A \to \Lambda$ that is finite to one. By Theorem IX.1.8, the entropy of $f|\Lambda$, $h(f|\Lambda)$, is equal to the entropy of σ_A, $h(\sigma|\Sigma_A)$. By Theorem IX.1.9, $h(\sigma|\Sigma_A) = \log(\lambda_1)$, where λ_1 is the largest eigenvalue of A.

PROOF. Let $\epsilon > 0$ and $\alpha > 0$ be small as above. Let $\beta > 0$ be such that any β-chain can be $(\alpha/2)$-shadowed in Λ because Λ is isolated. Next, take $\gamma > 0$ with $\gamma \leq \min\{\beta/2, \alpha/2\}$, so that if $d(\mathbf{x}, \mathbf{y}) < \gamma$, then $d(f(\mathbf{x}), f(\mathbf{y})) < \beta/2$ and $d(f^{-1}(\mathbf{x}), f^{-1}(\mathbf{y})) < \beta/2$.

Let $\mathcal{P} = \{\mathbf{p}_1, \ldots, \mathbf{p}_r\}$ be a finite set of γ-dense points in Λ. Let $B = (b_{ij})$ be the transition matrix with

$$b_{ij} = \begin{cases} 1 & d(f(\mathbf{p}_i), \mathbf{p}_j) < \beta \\ 0 & d(f(\mathbf{p}_i), \mathbf{p}_j) \geq \beta. \end{cases}$$

Because of the choice of γ, for each i, there is at least one j such that $b_{ij} = 1$. Let Σ_B be the two-sided subshift of finite type determined by B. Then, the cylinder sets

$$C_j = \{\mathbf{s} \in \Sigma_B : s_0 = j\}$$

form a Markov partition of Σ_B. Remember that for $\mathbf{s} \in \Sigma_B$,

$$\begin{aligned} W^s_{\text{loc}}(\mathbf{s}, \sigma_B) &= \{\mathbf{t} \in \Sigma_B : t_i = s_i \text{ for } i \geq 0\} \qquad \text{and} \\ W^u_{\text{loc}}(\mathbf{s}, \sigma_B) &= \{\mathbf{t} \in \Sigma_B : t_i = s_i \text{ for } i \leq 0\}. \end{aligned}$$

If $\mathbf{s}, \mathbf{s}' \in \Sigma_B$ with $s_0 = s'_0$, then

$$W^s_{\text{loc}}(\mathbf{s}, \sigma_B) \cap W^u_{\text{loc}}(\mathbf{s}', \sigma_B) = \{\mathbf{s}^*\}$$

with $\mathbf{s}^* \in \Sigma_B$, where

$$s^*_i = \begin{cases} s_i & \text{for } i \geq 0 \text{ and} \\ s'_i & \text{for } i \leq 0. \end{cases}$$

We use this partition for Σ_B to construct a partition for Λ. We define a map

$$\theta : \Sigma_B \to \Lambda$$

which we use to take a rectangle in Σ_B to a rectangle in Λ. For each $\mathbf{s} \in \Sigma_B$, let $\theta(\mathbf{s})$ be the point $\mathbf{z} \in \Lambda$ which $(\alpha/2)$-shadows the β-chain $\{\mathbf{p}_{s_j}\}_{j=-\infty}^{\infty}$. The main properties of θ are contained in the following lemma.

Lemma 6.2. *The map $\theta : \Sigma_B \to \Lambda$ is continuous and onto. Further, if $\mathbf{s}, \mathbf{s}' \in \Sigma_B$ have $s_0 = s'_0$ (are in the same cylinder set of the partition of Σ_B), then*

$$\begin{aligned} &d(\theta(\mathbf{s}), \theta(\mathbf{s}')) < \alpha \\ &\theta(W^s_{\text{loc}}(\mathbf{s}, \sigma_B)) \subset W^s_\alpha(\theta(\mathbf{s}), f), \\ &\theta(W^u_{\text{loc}}(\mathbf{s}', \sigma_B)) \subset W^u_\alpha(\theta(\mathbf{s}'), f), \qquad \text{and} \\ &\theta(W^s_{\text{loc}}(\mathbf{s}, \sigma_B) \cap W^u_{\text{loc}}(\mathbf{s}', \sigma_B)) = W^s_\epsilon(\theta(\mathbf{s}), f) \cap W^u_\epsilon(\theta(\mathbf{s}'), f). \end{aligned}$$

PROOF. To show θ is onto, let $\mathbf{z} \in \Lambda$. For each j, let $\mathbf{p}_{s_j}$ be chosen within γ of $f^j(\mathbf{z})$. Then,

$$\begin{aligned} d(f(\mathbf{p}_{s_j}), \mathbf{p}_{s_{j+1}}) &\leq d(f(\mathbf{p}_{s_j}), f^{j+1}(\mathbf{z})) + d(f^{j+1}(\mathbf{z}), \mathbf{p}_{s_{j+1}}) \\ &\leq \beta/2 + \gamma \\ &\leq \beta, \end{aligned}$$

so $\{\mathbf{p}_{s_j}\}$ is a β-chain. The orbit of $\mathbf{z}$ γ-shadows, so it $(\alpha/2)$-shadows this β-chain; thus, $\theta(\mathbf{s}) = \mathbf{z}$ and θ is onto.

If two sequences $\mathbf{s}, \mathbf{s}' \in \Sigma_B$ have $s_0 = s_0'$, then $\theta(\mathbf{s})$ and $\theta(\mathbf{s}')$ are both within $\alpha/2$ of $\mathbf{p}_{s_0}$, so are within α of each other.

Note that two symbol sequences $\mathbf{s}$ and $\mathbf{s}'$ are close if $s_j = s_j'$ for $-n \leq j \leq n$ for some large n. Thus, $\mathbf{s}$ and $\mathbf{s}'$ correspond to two β-chains which agree for a large number of points. By an argument like we have given in earlier sections, $\theta(\mathbf{s})$ and $\theta(\mathbf{s}')$ are nearby points. This shows that θ is continuous.

For the properties about stable manifolds stated in the lemma, take $\mathbf{s}^* \in W^s_{\text{loc}}(\mathbf{s}, \sigma_B)$. Then, $s_j^* = s_j$ for $j \geq 0$. Then, both $\theta(\mathbf{s})$ and $\theta(\mathbf{s}^*)$ $(\alpha/2)$-shadow the same forward β-chain, so

$$d(f^j \circ \theta(\mathbf{s}^*), f^j \circ \theta(\mathbf{s})) \leq \alpha \leq \epsilon$$

for $j \geq 0$. It follows that

$$\begin{aligned} &\theta(\mathbf{s}^*) \in W^s_\alpha(\theta(\mathbf{s}), f) \subset W^s_\epsilon(\theta(\mathbf{s}), f) \qquad \text{or} \\ &\theta(W^s_{\text{loc}}(\mathbf{s}, \sigma_B)) \subset W^s_\alpha(\theta(\mathbf{s}), f). \end{aligned}$$

The proof that

$$\theta(W^u_{\text{loc}}(\mathbf{s}, \sigma_B)) \subset W^u_\alpha(\theta(\mathbf{s}), f)$$

is similar using $j \leq 0$.

Finally, assume let $\mathbf{s}, \mathbf{s}' \in \Sigma_B$ with $s_0 = s_0'$. Then,

$$W^s_{\text{loc}}(\mathbf{s}, \sigma_B) \cap W^u_{\text{loc}}(\mathbf{s}', \sigma_B) = \{\mathbf{s}^*\}$$

where

$$s_i^* = \begin{cases} s_j & \text{for } j \geq 0 \quad \text{and} \\ s_j' & \text{for } j \leq 0. \end{cases}$$

By above

$$\begin{aligned} &\theta(\mathbf{s}^*) \in W^s_\alpha(\theta(\mathbf{s}), f) \cap W^u_\alpha(\theta(\mathbf{s}'), f) = W^s_\epsilon(\theta(\mathbf{s}), f) \cap W^u_\epsilon(\theta(\mathbf{s}'), f), \qquad \text{or} \\ &\theta(W^s_{\text{loc}}(\mathbf{s}, \sigma_B) \cap W^u_{\text{loc}}(\mathbf{s}', \sigma_B)) = W^s_\epsilon(\theta(\mathbf{s}), f) \cap W^u_\epsilon(\theta(\mathbf{s}'), f) \end{aligned}$$

as claimed. □

We check in Lemma 6.3 below that the sets

$$T_j = \theta(C_j) = \{\theta(\mathbf{s}) : \mathbf{s} \in \Sigma_B \text{ and } s_0 = j\}$$

are rectangles in Λ for $1 \leq j \leq r$. Since θ is continuous, each of the T_j is closed. We do not know that these rectangles are proper. Also, the collection of these rectangles might not have disjoint interiors. On the other hand, they do form a cover of Λ, and Lemma 6.3 also checks the first condition on the stable and unstable manifolds in the definition of a Markov partition.

Lemma 6.3. *The collection of sets* $\{T_j : 1 \leq j \leq r\}$ *satisfy the following conditions.*

(a) Each T_j *is a rectangle in* Λ.

(b) The $\{T_j\}$ *cover* Λ, $\Lambda = \bigcup_{j=1}^{r} T_j$.

(c) If $\mathbf{x} = \theta(\mathbf{s})$ *for* $\mathbf{s} \in \Sigma_B$, *then*

$$f(W^s(\mathbf{x}, T_{s_0})) \subset W^s(f(\mathbf{x}), T_{s_1}) \qquad \text{and}$$
$$f(W^u(\mathbf{x}, T_{s_0})) \supset W^u(f(\mathbf{x}), T_{s_1}).$$

PROOF. (a) The diameter of T_j is less than α, because any two points in a T_j are within $\alpha/2$ of the same point $\mathbf{p}_j$. Thus, condition (i) in the definition of a rectangle is true. Let $\mathbf{x}, \mathbf{y} \in T_j$ with $\theta(\mathbf{s}) = \mathbf{x}$ and $\theta(\mathbf{s}') = \mathbf{y}$. Then, $\mathbf{s}$ and $\mathbf{s}'$ are in the same cylinder set C_j,

$$\mathbf{s}^* = W^s_{\text{loc}}(\mathbf{s}, \sigma_B) \cap W^u_{\text{loc}}(\mathbf{s}', \sigma_B) \in C_j, \qquad \text{and}$$
$$W^s_\epsilon(\theta(\mathbf{s}), f) \cap W^u_\epsilon(\theta(\mathbf{s}'), f) = \theta(\mathbf{s}^*) \in T_j.$$

Thus, condition (ii) in the definition of a rectangle is true, and the T_j are rectangles.

(b) Since θ is onto, the $\{T_j\}$ cover Λ.

(c) If $\mathbf{y} \in W^s(\mathbf{x}, T_{s_0}) \subset W^s_\epsilon(\mathbf{x}, f)$, then $\mathbf{y} = W^s_\epsilon(\mathbf{x}, f) \cap W^u_\epsilon(\mathbf{y}, f)$. On the other hand, if $\mathbf{y} = \theta(\mathbf{s}')$, define

$$\mathbf{s}^* = W^s_{\text{loc}}(\mathbf{s}, \sigma_B) \cap W^u_{\text{loc}}(\mathbf{s}', \sigma_B),$$
$$s^*_j = \begin{cases} s_j & \text{for } j \geq 0 \\ s'_j & \text{for } j \leq 0. \end{cases} \qquad \text{and}$$

By Lemma 6.2,

$$\theta(\mathbf{s}^*) = W^s_\epsilon(\theta(\mathbf{s}), f) \cap W^u_\epsilon(\theta(\mathbf{s}'), f),$$

so $\theta(\mathbf{s}^*) = \mathbf{y}$. Also, (i) $\sigma_B(\mathbf{s}^*)_j = \sigma_B(\mathbf{s})_j = s_{j+1}$ for $j \geq 0$ and $\theta(\sigma_B(\mathbf{s})) = f(\mathbf{x})$ so $\theta(\sigma_B(\mathbf{s}^*)) \in W^s_\epsilon(f(\mathbf{x}), f)$, and (ii) $\sigma_B(\mathbf{s}^*)_j = \sigma_B(\mathbf{s}')_j = s'_{j+1}$ for $j \leq 0$ and $\theta(\sigma_B(\mathbf{s}')) = f(\mathbf{y})$ so $\theta(\sigma_B(\mathbf{s}^*)) \in W^u_\epsilon(f(\mathbf{y}), f)$:

$$\theta(\sigma_B(\mathbf{s}^*)) \in W^s_\epsilon(f(\mathbf{x}), f) \cap W^u_\epsilon(f(\mathbf{y}), f) = \{f(\mathbf{y})\},$$
$$\theta(\sigma_B(\mathbf{s}^*)) = f(\mathbf{y}).$$

But $\theta(\sigma_B(\mathbf{s}^*)) \in T_{s_1}$, so $f(\mathbf{y}) \in W^s(f(\mathbf{x}), T_{s_1})$, $f(W^s(\mathbf{x}, T_{s_0})) \subset W^s(f(\mathbf{x}), T_{s_1})$, and we are done.

A similar argument proves the last inclusion,

$$f^{-1}(W^u(f(\mathbf{x}), T_{s_1}) \subset W^u(\mathbf{x}, T_{s_0}).$$

□

To get the other two properties of the Markov partition, we need to subdivide the T_j's. The subdivision uses the different parts of the boundary of the rectangles. Therefore, before making the refinements of the covering, we characterize the different parts of the boundary of a rectangle in the following lemma.

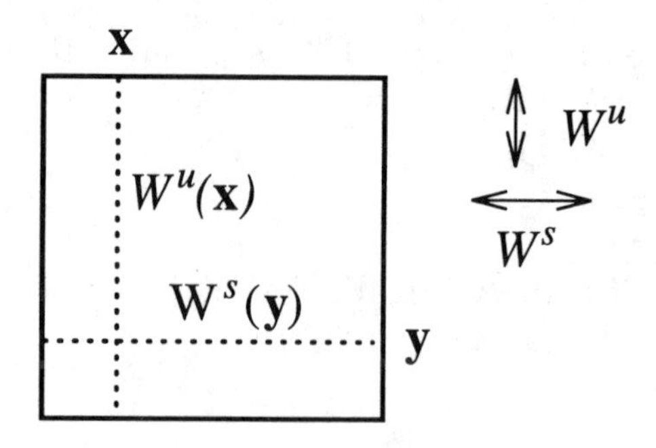

FIGURE 6.1. The Points $\mathbf{x} \in \partial^s(R)$ and $\mathbf{y} \in \partial^u(R)$

Lemma 6.4. *Let R be a closed rectangle of Λ. The boundary of R (relative to Λ) can be written as the union of two subsets, $\partial(R) = \partial^s(R) \cup \partial^u(R)$, where*

$$\partial^s(R) = \{\mathbf{x} \in R : \mathbf{x} \notin \operatorname{int}(W^u(\mathbf{x}, R))\} \quad \text{and}$$
$$\partial^u(R) = \{\mathbf{x} \in R : \mathbf{x} \notin \operatorname{int}(W^s(\mathbf{x}, R))\}.$$

(The interiors of $W^s(\mathbf{x}, R)$ and $W^u(\mathbf{x}, R)$ in the above definitions of $\partial^s(R)$ and $\partial^u(R)$ are as subsets of $W^s_\epsilon(\mathbf{x}) \cap \Lambda$ and $W^u_\epsilon(\mathbf{x}) \cap \Lambda$, respectively.)

REMARK 6.3. The notation is chosen so that $\partial^s(R)$ is made up of pieces of stable manifolds and $\partial^u(R)$ is made up of pieces of unstable manifolds. See Figure 6.1.

PROOF. It is sufficient to prove that $\operatorname{int}(R)$ is equal to the set $R \setminus (\partial^s(R) \cup \partial^u(R))$.

If $\mathbf{x} \in \operatorname{int}(R)$, then $W^\sigma(\mathbf{x}, R) = R \cap W^\sigma_\epsilon(\mathbf{x})$ is a neighborhood of $\mathbf{x}$ in $W^\sigma_\epsilon(\mathbf{x}) \cap \Lambda$ for $\sigma = s, u$. This proves that $\operatorname{int}(R) \subset R \setminus (\partial^s(R) \cup \partial^u(R))$.

Conversely, assume

$$\mathbf{x} \in R \setminus (\partial^s(R) \cup \partial^u(R)) \qquad \text{or}$$
$$\mathbf{x} \in \operatorname{int}(W^s(\mathbf{x}, R)) \cap \operatorname{int}(W^u(\mathbf{x}, R)).$$

For $\mathbf{y} \in \Lambda$ near to $\mathbf{x}$,

$$\mathbf{y}^s \equiv W^s_\epsilon(\mathbf{x}) \cap W^u_\epsilon(\mathbf{y}) \in \operatorname{int}(W^s(\mathbf{x}, R)) \quad \text{and}$$
$$\mathbf{y}^u \equiv W^s_\epsilon(\mathbf{y}) \cap W^u_\epsilon(\mathbf{x}) \in \operatorname{int}(W^u(\mathbf{x}, R))$$

since the intersections depend continuously on $\mathbf{y}$. See Figure 6.2. Thus,

$$\mathbf{y} = W^s_\epsilon(\mathbf{y}^u) \cap W^u_\epsilon(\mathbf{y}^s)$$

for $\mathbf{y}^s \in \operatorname{int}(W^s(\mathbf{x}, R))$ and $\mathbf{y}^u \in \operatorname{int}(W^u(\mathbf{x}, R))$. The set of such points forms a neighborhood of $\mathbf{x}$ in Λ, so $\mathbf{x} \in \operatorname{int}(R)$. □

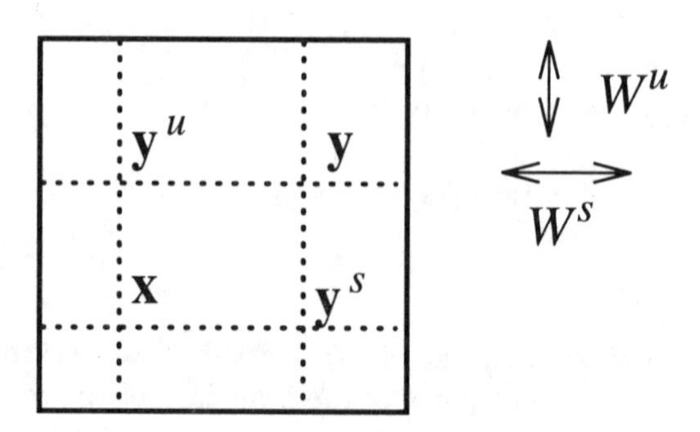

FIGURE 6.2. Determination of Points $\mathbf{y}^s$ and $\mathbf{y}^u$

For $\mathbf{x} \in \Lambda$, we associate elements of the cover $\mathcal{T} = \{T_1, \ldots, T_r\}$ of Λ by letting

$$\mathcal{T}(\mathbf{x}) = \{T_j \in \mathcal{T} : \mathbf{x} \in T_j\} \qquad \text{and}$$
$$\mathcal{T}^*(\mathbf{x}) = \{T_k \in \mathcal{T} : T_k \cap T_j \neq \emptyset \text{ for some } T_j \in \mathcal{T}(\mathbf{x})\}.$$

Since $\mathcal{T}$ is a cover of Λ,

$$Z = \Lambda \setminus \bigcup_{j=1}^{r} \partial(T_j)$$

is open and dense in Λ. We also need to remove the extensions of the $\partial^s(T_j)$ by stable manifolds and the extensions of the $\partial^u(T_j)$ by unstable manifolds, so we define the set

$$Z^* = \{\mathbf{x} \in \Lambda : W^s_\epsilon(\mathbf{x}) \cap \partial^s(T_k) = \emptyset \text{ and}$$
$$W^u_\epsilon(\mathbf{x}) \cap \partial^u(T_k) = \emptyset \text{ for all } T_k \in \mathcal{T}^*(\mathbf{x})\}.$$

This set is also open and dense by an argument like Lemma 6.4.

Now, we fix $T_j \in \mathcal{T}$ and subdivide it for T_k with $T_j \cap T_k \neq \emptyset$ as follows:

$$\begin{aligned}
T^1_{j,k} &= T_j \cap T_k \\
&= \{\mathbf{x} \in T_j : W^u(\mathbf{x}, T_j) \cap T_k \neq \emptyset \text{ and } W^s(\mathbf{x}, T_j) \cap T_k \neq \emptyset\} \\
T^2_{j,k} &= \{\mathbf{x} \in T_j : W^u(\mathbf{x}, T_j) \cap T_k \neq \emptyset \text{ and } W^s(\mathbf{x}, T_j) \cap T_k = \emptyset\} \\
T^3_{j,k} &= \{\mathbf{x} \in T_j : W^u(\mathbf{x}, T_j) \cap T_k = \emptyset \text{ and } W^s(\mathbf{x}, T_j) \cap T_k \neq \emptyset\} \\
T^4_{j,k} &= \{\mathbf{x} \in T_j : W^u(\mathbf{x}, T_j) \cap T_k = \emptyset \text{ and } W^s(\mathbf{x}, T_j) \cap T_k = \emptyset\}.
\end{aligned}$$

See Figure 6.3. With these definitions, $T^1_{j,j} = T_j$ and $T^2_{j,j} = T^3_{j,j} = T^4_{j,j} = \emptyset$. Each of the $T^n_{j,k}$ is a rectangle as the following observation implies: if $\mathbf{x}, \mathbf{y} \in T_j$ and $\mathbf{z} = W^s(\mathbf{x}, T_j) \cap W^u(\mathbf{y}, T_j)$, then $W^s(\mathbf{z}, T_j) = W^s(\mathbf{x}, T_j)$ and $W^u(\mathbf{z}, T_j) = W^u(\mathbf{y}, T_j)$.

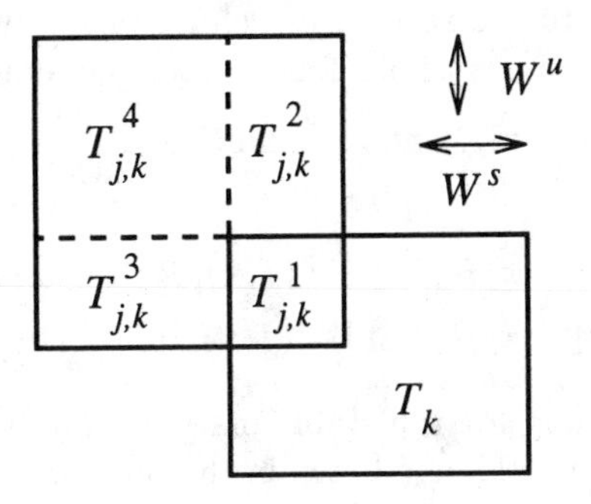

FIGURE 6.3. Subdivision of Rectangles

For $\mathbf{x} \in Z^*$, we take intersections of the $T^n_{j,k}$ to define a rectangle at $\mathbf{x}$,

$$R(\mathbf{x}) = \bigcap \{\text{int}(T^n_{j,k}) : T_j \in \mathcal{T}(\mathbf{x}), T_j \cap T_k \neq \emptyset, \text{ and } \mathbf{x} \in T^n_{j,k}\}.$$

Since there are only a finite number of sets involved, $R(\mathbf{x})$ is open, and nonempty since $\mathbf{x} \in R(\mathbf{x})$. Since each of the $T^n_{j,k}$ is a rectangle, each $\text{int}(T^n_{j,k})$ is a rectangle, so $R(\mathbf{x})$ is a rectangle.

A finite collection of the $\text{cl}(R(\mathbf{x}))$, $\mathcal{R}$, form the Markov partition claimed in the theorem. The following lemma proves properties of the $R(\mathbf{x})$'s used to verify the conditions of a Markov partition in the sequence of lemmas which follow. Notice that $\mathcal{R}$ is a refinement of the cover $\mathcal{T}$.

Lemma 6.5. *(a) For* $\mathbf{x}, \mathbf{y} \in Z^*$, *either* $R(\mathbf{x}) = R(\mathbf{y})$ *or* $R(\mathbf{x}) \cap R(\mathbf{y}) = \emptyset$.
(b) For $\mathbf{x} \in Z^*$, $\partial(R(\mathbf{x})) \cap Z^* = \emptyset$.
(c) For $\mathbf{x} \in Z^*$, $\text{int}(\text{cl}(R(\mathbf{x}))) = R(\mathbf{x})$, *i.e., the rectangles* $\text{cl}(R(\mathbf{x}))$ *are proper.*

PROOF. Assume that $R(\mathbf{x}) \cap R(\mathbf{y}) \neq \emptyset$ for $\mathbf{x}, \mathbf{y} \in Z^*$. There is a $\mathbf{z} \in R(\mathbf{x}) \cap R(\mathbf{y}) \cap Z^*$. By the definition of $R(\mathbf{x})$, if $T_j \in \mathcal{T}(\mathbf{x})$, then $\mathbf{z} \in T^1_{j,j} = T_j$, so $T_j \in \mathcal{T}(\mathbf{z})$ and $\mathcal{T}(\mathbf{z}) \supset \mathcal{T}(\mathbf{x})$. On the other hand, if $T_j \notin \mathcal{T}(\mathbf{x})$, then $T_j \cap R(\mathbf{x}) = \emptyset$, so $\mathbf{z} \notin T_j$ and $\mathcal{T}(\mathbf{z}) \subset \mathcal{T}(\mathbf{x})$. Combining, we get $\mathcal{T}(\mathbf{z}) = \mathcal{T}(\mathbf{x})$. Similarly, $\mathcal{T}(\mathbf{y}) = \mathcal{T}(\mathbf{z})$, so $\mathcal{T}(\mathbf{y}) = \mathcal{T}(\mathbf{x})$. Thus, if $R(\mathbf{x}) \cap R(\mathbf{y}) \neq \emptyset$, then $R(\mathbf{x}) = R(\mathbf{y})$. This proves part (a).

Assume $\mathbf{y} \in \partial(R(\mathbf{x})) \cap Z^*$. Since $R(\mathbf{y})$ is a neighborhood of $\mathbf{y}$ in Λ and $\mathbf{y} \in \partial(R(\mathbf{x}))$, it must be that $R(\mathbf{y}) \cap R(\mathbf{x}) = \emptyset$ or else $R(\mathbf{x}) = R(\mathbf{y})$ and $\mathbf{y}$ is not a boundary point. But also, this neighborhood $R(\mathbf{y})$ must not intersect $R(\mathbf{x})$ so $\mathbf{y}$ cannot be a boundary point. This is a contradiction and shows that $\partial(R(\mathbf{x})) \cap Z^* = \emptyset$. This proves part (b).

Since Z^* is dense in Λ and $\partial(R(\mathbf{x})) \cap Z^* = \emptyset$, $\text{int}(\partial(R(\mathbf{x}))) = \emptyset$, and so $\text{int}(\text{cl}(R(\mathbf{x}))) = R(\mathbf{x})$. □

There are only a finite number of different $R(\mathbf{x})$ because there are only finitely many T_j and $T^n_{j,k}$. Let

$$\mathcal{R} = \{\text{cl}(R(\mathbf{x})) : \mathbf{x} \in Z^*\} = \{R_1, \dots, R_m\}$$

be an enumeration of this finite collection of rectangles. Lemma 6.5(c) proves that the rectangles are proper. Since the collection of the interiors,

$$\{\text{int}(R_1), \dots, \text{int}(R_m)\},$$

cover Z^* which is dense in Λ, the collection $\mathcal{R}$ satisfies the first condition for a Markov partition, $\bigcup_{j=1}^{m} R_j = \Lambda$. By Lemma 6.5(a,c), condition (ii) for a Markov partition is true: $\text{int}(R_j) \cap \text{int}(R_k) = \emptyset$ for $j \neq k$. We are only left to show that the stable and unstable manifolds relative to the rectangles behave properly, conditions (iii) and (iv) for a Markov partition. These conditions follow from the lemmas which follow.

Lemma 6.6. *The rectangles in* $\mathcal{R}$ *satisfy condition (iii) in the definition of a Markov Partition: if* $\mathbf{x} \in \text{int}(R_i) \cap f^{-1}(\text{int}(R_j))$, *then*

$$f(W^s(\mathbf{x}, R_i)) \subset W^s(f(\mathbf{x}), R_j) \qquad \textit{and}$$
$$f(W^u(\mathbf{x}, R_i)) \supset W^u(f(\mathbf{x}), R_j).$$

We prove only the inclusion for the stable manifolds, and remark how the case for the unstable manifolds follows. The first step in the proof of this lemma is the following sublemma.

Sublemma 6.7. *Assume* $\mathbf{x}, \mathbf{y} \in Z^* \cap f^{-1}(Z^*)$, $R(\mathbf{x}) = R(\mathbf{y})$, *and* $\mathbf{y} \in W^s_\epsilon(\mathbf{x})$. *Then,* $R(f(\mathbf{x})) = R(f(\mathbf{y}))$.

PROOF. The first step is to show that $\mathcal{T}(f(\mathbf{x})) = \mathcal{T}(f(\mathbf{y}))$. Assume that $f(\mathbf{x}) \in T_j$. Then, there is a $\mathbf{s} \in \Sigma_B$ with $\mathbf{x} = \theta(\mathbf{s})$ and $s_1 = j$. Let $\mathbf{s}'$ be the point in Σ_B with $s'_0 = s_0$ and $\mathbf{y} = \theta(\mathbf{s}')$. If we let

$$s^*_i = \begin{cases} s_i & \text{for } i \geq 0 \quad \text{and} \\ s'_i & \text{for } i \leq 0, \end{cases}$$

then $\mathbf{y} = \theta(\mathbf{s}^*)$ as we argued before. Since $\mathbf{y} = \theta(\mathbf{s}^*)$ and $s^*_1 = j$, $f(\mathbf{y}) \in T_j$. Thus, $\mathcal{T}(f(\mathbf{x})) \subset \mathcal{T}(f(\mathbf{y}))$. The other inclusion is proved by reversing the roles of $\mathbf{x}$ and $\mathbf{y}$, so $\mathcal{T}(f(\mathbf{x})) = \mathcal{T}(f(\mathbf{y}))$.

Now, assume $f(\mathbf{x}), f(\mathbf{y}) \in T_j$ and $T_j \cap T_k \neq \emptyset$. We need to show that $f(\mathbf{x})$ and $f(\mathbf{y})$ are in the same $T^n_{j,k}$. Note that $W^s(f(\mathbf{x}), T_j) = W^s(f(\mathbf{y}), T_j)$, so both these stable manifolds either intersect T_k or neither intersects T_k. If $f(\mathbf{x})$ and $f(\mathbf{y})$ are not in the same $T^n_{j,k}$, then one of the unstable manifolds $W^u(f(\mathbf{x}), T_j)$ and $W^u(f(\mathbf{y}), T_j)$ intersects T_k and the other does not. Assume that

$$f(\mathbf{z}) \in W^u(f(\mathbf{x}), T_j) \cap T_k \neq \emptyset \qquad \text{and}$$
$$W^u(f(\mathbf{y}), T_j) \cap T_k = \emptyset.$$

See Figure 6.4. Let $\mathbf{s} \in \Sigma_B$ be such that $f(\mathbf{x}) = \theta \circ \sigma(\mathbf{s})$ with $s_1 = j$. In Lemma 6.3 we proved that it follows that $W^u(f(\mathbf{x}), T_j) \subset f(W^u(\mathbf{x}, T_{s_0}))$, so $f(\mathbf{z}) \in f(W^u(\mathbf{x}, T_{s_0}))$ and $\mathbf{z} \in W^u(\mathbf{x}, T_{s_0})$.

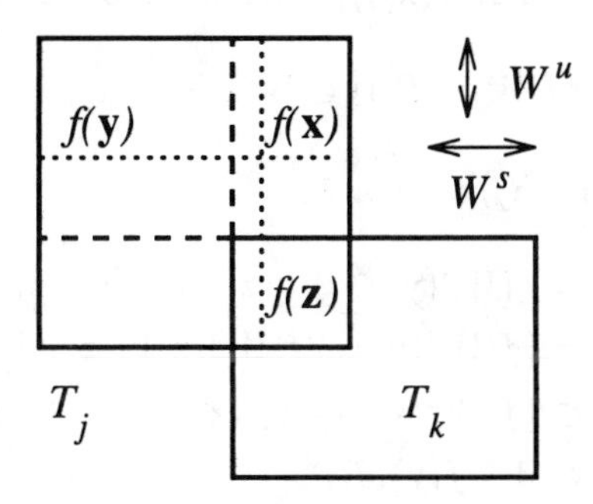

FIGURE 6.4

Let $\mathbf{s}^* \in \Sigma_B$ be such that $\mathbf{z} = \theta(\mathbf{s}^*)$ with $s_1^* = k$. Thus, $T_{s_0} \in \mathcal{T}(\mathbf{x}) = \mathcal{T}(\mathbf{y})$ and $\mathbf{z} \in T_{s_0} \cap T_{s_0^*} \neq \emptyset$. Since $\mathbf{z} \in W^u(\mathbf{x}, T_{s_0}) \cap T_{s_0^*}$ and $R(\mathbf{x}) = R(\mathbf{y})$ (so they are in the same $T^n_{s_0, s_0^*}$), the intersection $W^u(\mathbf{y}, T_{s_0}) \cap T_{s_0^*} \neq \emptyset$. In fact, there is a point

$$\begin{aligned} \mathbf{z}' &= W^s_\epsilon(\mathbf{z}) \cap W^u_\epsilon(\mathbf{y}) \\ &= W^s(\mathbf{z}, T_{s_0}) \cap W^u(\mathbf{y}, T_{s_0}). \end{aligned}$$

See Figure 6.5. Therefore,

$$\begin{aligned} f(\mathbf{z}') &= W^s_\epsilon(f(\mathbf{z})) \cap W^u_\epsilon(f(\mathbf{y})) \\ &= W^s(f(\mathbf{z}), T_k) \cap W^u(f(\mathbf{y}), T_j). \end{aligned}$$

This contradicts the fact that $W^u(f(\mathbf{y}), T_j) \cap T_k = \emptyset$. Therefore, $f(\mathbf{x})$ and $f(\mathbf{y})$ are in the same $T^n_{j,k}$. This is true for an arbitrary T_k, so $R(f(\mathbf{x})) = R(f(\mathbf{y}))$. □

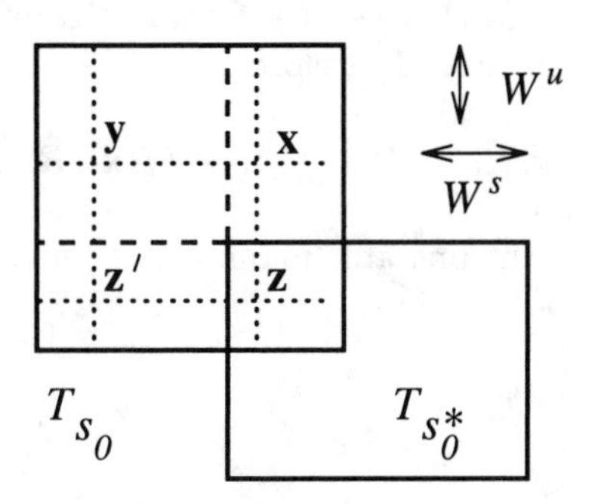

FIGURE 6.5

Proof of Lemma 6.6. Let $\mathbf{x} \in Z^* \cap f^{-1}(Z^*)$. The set

$$W^s(\mathbf{x}, R(\mathbf{x})) \cap Z^* \cap f^{-1}(Z^*)$$

is open and dense in $W^s(\mathbf{x}, \mathrm{cl}(R(\mathbf{x}))$ by an argument like Lemma 6.4. By Lemma 6.7 and continuity,

$$\begin{aligned} f(W^s(\mathbf{x}, \mathrm{cl}(R(\mathbf{x})) &\subset \mathrm{cl}(R(f(\mathbf{x}))) \cap W^s_\epsilon(f(\mathbf{x})) \\ &\subset W^s(f(\mathbf{x}), \mathrm{cl}(R(f(\mathbf{x}))). \end{aligned}$$

If $\mathrm{int}(R_i) \cap f^{-1}(\mathrm{int}(R_j)) \neq \emptyset$, then this open subset of Λ contains an $\mathbf{x} \in Z^* \cap f^{-1}(Z^*)$ with $R_i = \mathrm{cl}(R(\mathbf{x}))$ and $R_j = \mathrm{cl}(R(f(\mathbf{x})))$. Therefore, for such $\mathbf{x}$,

$$f(W^s(\mathbf{x}, R_i)) \subset W^s(f(\mathbf{x}), R_j).$$

For any $\mathbf{y} \in \mathrm{int}(R_i) \cap f^{-1}(\mathrm{int}(R_j))$,

$$\begin{aligned} f(W^s(\mathbf{y}, R_i)) &= f\{W^s_\epsilon(\mathbf{y}) \cap W^u_\epsilon(\mathbf{z}) : \mathbf{z} \in W^s(\mathbf{x}, R_i)\} \\ &= \{f(W^s_\epsilon(\mathbf{y})) \cap f(W^u_\epsilon(\mathbf{z})) : \mathbf{z} \in W^s(\mathbf{x}, R_i)\} \\ &\subset \{f(W^s_\epsilon(\mathbf{y})) \cap W^u_\epsilon(\mathbf{z}') : \mathbf{z}' \in W^s(f(\mathbf{x}), R_j)\} \\ &\subset W^s(f(\mathbf{y}), R_j). \end{aligned}$$

This proves the lemma for the stable manifolds.

The result for the unstable manifolds follows by applying the above argument to f^{-1} (or modifying it for the unstable manifolds directly). □

Lemma 6.8. *The rectangles in $\mathcal{R}$ satisfy a strong version of condition (iv) in the definition of a Markov Partition: if $\mathbf{x} \in \mathrm{int}(R_i) \cap f^{-1}(\mathrm{int}(R_j))$, then*

$$\begin{aligned} W^s(\mathbf{x}, R_i) &= R_i \cap f^{-1}(W^s(f(\mathbf{x}), R_j)) \qquad \text{and} \\ W^u(f(\mathbf{x}), R_j) &= f(W^u(\mathbf{x}, R_i)) \cap R_j. \end{aligned}$$

Proof. By Lemma 6.6,

$$\begin{aligned} W^s(\mathbf{x}, R_i) &\subset R_i \cap f^{-1}(W^s(f(\mathbf{x}), R_j)) \\ &\subset R_i \cap f^{-1}(W^s_\epsilon(f(\mathbf{x}))) = W^s(\mathbf{x}, R_i). \end{aligned}$$

The last equality follows by the choices of ϵ for the size of the stable and unstable manifolds made early in the section. Therefore,

$$W^s(\mathbf{x}, R_i) = R_i \cap f^{-1}(W^s(f(\mathbf{x}), R_j))$$

as claimed. The argument for the unstable manifolds is similar. This proves the lemma and completes the proof of the theorem. □

10.7 Local Stability and Stability of Anosov Diffeomorphisms

In this section we prove the semi-stability and structural stability of Anosov diffeomorphisms and also the stability of a single basic set. The proofs given are in the spirit of those of Anosov, and repeat the construction of stable and unstable manifolds. We start with an Anosov diffeomorphism. Throughout the section, M is a compact manifold (or at least the invariant set Λ is a compact invariant set).

Definition. Consider the set of homeomorphisms on a compact manifold M. If we use the usual C^0-sup topology on functions, then the set of homeomorphisms are not open and so not a complete space. Therefore, to study perturbations g of a homeomorphism f which we want to remain a homeomorphism, we require that both $d_0(f,g)$ and $d_0(f^{-1},g^{-1})$ to be small, i.e., we use the metric

$$\begin{aligned} d_{homeo}(f,g) &= \max\{d_0(f,g), d_0(f^{-1},g^{-1})\} \\ &= \sup\{d(f(\mathbf{x}),g(\mathbf{x})), d(f^{-1}(\mathbf{x}),g^{-1}(\mathbf{x})) : \mathbf{x} \in M\}. \end{aligned}$$

The space of homeomorphisms is complete in terms of the metric d_{homeo}.

Let $f : M \to M$ be a homeomorphism (or diffeomorphism). We say that f is *semi-stable* provided there exists a $\epsilon > 0$ such that for every homeomorphism $g : M \to M$ with the C^0 distance from f to g and from f^{-1} to g^{-1} less than ϵ, $d_{homeo}(f,g) < \epsilon$, there exists a $h : M \to M$ which is a semi-conjugacy from g to f, with h continuous, onto, and $h \circ g = f \circ h$. These conditions imply that the dynamics of g are at least as complicated as those of f.

An example of two diffeomorphisms which are semi-conjugate is where f is a hyperbolic toral automorphism on $\mathbb{T}^2$ and g the DA-diffeomorphism constructed from f. In fact, this f is semi-stable as the following theorem of Walters (1970) proves.

Theorem 7.1 (Semi-Stability of Anosov Diffeomorphisms). *Assume M is a compact manifold and that $f : M \to M$ is an Anosov diffeomorphism (f has a hyperbolic structure on all of M). Then, f is semi-stable.*

REMARK 7.1. The proof we give is in the spirit of that given by Anosov, although the ideas are filtered through the concept of shadowing. A different type of proof was given by Moser (1969). This latter proof solves a functional equation using a contraction mapping.

PROOF. Let g be a homeomorphism within $\epsilon > 0$ of f, $d_{homeo}(f,g) < \epsilon$. Then, for each $\mathbf{x} \in M$, $\{g^j(\mathbf{x})\}_{j=-\infty}^{\infty}$ is an ϵ-chain for f. Given $\eta > 0$, there is an $\epsilon > 0$ such that all ϵ-chains can be uniquely η-shadowed. Let $\mathbf{y} = h(\mathbf{x})$ be the point that η-shadows $g^j(\mathbf{x})$, $d(f^j \circ h(\mathbf{x}), g^j(\mathbf{x})) < \eta$. We now check the properties of h. Because the point which shadows is unique, h is a well-defined function.

Claim 1. *The map h is continuous.*

PROOF. Let $\mathcal{V}_{\mathbf{p}}(r) = \exp_{\mathbf{p}}(\mathcal{B}_{\mathbf{p}}(r))$ be the neighborhoods of $\mathbf{p}$ in M defined before in the proof of shadowing. The proof of the shadowing result shows that

$$h(\mathbf{x}) = \bigcap_{j=-\infty}^{\infty} f^j(\mathcal{V}_{g^{-j}(\mathbf{x})}(r)).$$

Given $\eta > 0$, there is an N such that all points in the finite intersection

$$\bigcap_{j=-N}^{N} f^j(\mathcal{V}_{g^{-j}(\mathbf{x})}(r))$$

are within $\eta/2$ of $h(\mathbf{x})$. Then, for $\mathbf{y}$ near enough to $\mathbf{x}$, all points in the finite intersection

$$\bigcap_{j=-N}^{N} f^j(\mathcal{V}_{g^{-j}(\mathbf{y})}(r))$$

are within η of $h(\mathbf{x})$. Since

$$h(\mathbf{y}) = \bigcap_{j=-\infty}^{\infty} f^j(\mathcal{V}_{g^{-j}(\mathbf{y})}(r)) \subset \bigcap_{j=-N}^{N} f^j(\mathcal{V}_{g^{-j}(\mathbf{y})}(r)),$$

we have that $d(h(\mathbf{x}), h(\mathbf{y})) < \eta$. This proves the continuity of h. □

Claim 2. $h \circ g = f \circ h$.

PROOF. Using the point $g(\mathbf{x})$ to η-shadow,

$$d(g^j \circ g(\mathbf{x}), f^j \circ h \circ g(\mathbf{x})) = d(g^{j+1}(\mathbf{x}), f^{j+1} \circ f^{-1} \circ h \circ g(\mathbf{x})) < \eta.$$

But also,

$$d(g^{j+1}(\mathbf{x}), f^{j+1} \circ h(\mathbf{x})) < \eta.$$

By the uniqueness of $h(\mathbf{x})$, $f^{-1} \circ h \circ g(\mathbf{x}) = h(\mathbf{x})$, or $h \circ g(\mathbf{x}) = f \circ h(\mathbf{x})$. □

Claim 3. *The map h is onto.*

PROOF. We argue that h is onto using ideas from algebraic topology. (In the proof of Theorem 7.3 below, h is one to one and there is another proof which uses the Invariance of Domain Theorem.) The map h is within η of the identity map and can be continuously deformed into the identity (h is homotopic to the identity). The top homology group of M, $H_n(M)$ where $n = \dim(M)$, is nontrivial and the identity induces an isomorphism on this group. Because h is homotopic to the identity, it also induces an isomorphism on $H_n(M)$. Since any map which induces an isomorphism on $H_n(M)$ is onto, this is true for h. □

These claims combine to prove Theorem 7.1. □

REMARK 7.2. Let g be the DA-diffeomorphism constructed from the hyperbolic toral automorphism f. Then, the semi-conjugacy h from g to f is not one to one. In fact, h takes the line segments in $W^s(q, f)$ between $W^s(p_1, g)$ and $W^s(p_2, g)$ and collapses them to points.

Theorem 7.2 (Openness of Anosov Diffeomorphisms). *Assume M is a compact manifold. The set of Anosov diffeomorphisms on M is open in the C^1 topology.*

We leave the proof of this theorem as an exercise. See Exercise 10.27. (The proof uses cones.)

Theorem 7.3 (Structural Stability of Anosov Diffeomorphisms). *Assume that M is a compact manifold and $f : M \to M$ is an Anosov diffeomorphism. (The map f has a hyperbolic structure on all of M.) Then, f is structurally stable (under all small C^1 perturbations).*

REMARK 7.3. This result was first proved by Anosov (1967). Also see Moser (1969) and the appendix by Mather in Smale (1967).

PROOF. The previous result showed there is a semi-conjugacy, h, with $h \circ g = f \circ h$. We only need to show that h is one to one. Assume that $h(\mathbf{x}) = h(\mathbf{y})$. Then, $h \circ g^j(\mathbf{x}) =$

$f^j \circ h(\mathbf{x}) = f^j \circ h(\mathbf{y}) = h \circ g^j(\mathbf{y})$. Therefore, $d(g^j(\mathbf{x}), g^j(\mathbf{y})) < 2\eta$. Since g is Anosov, it is expansive. Therefore, if η is small enough, $\mathbf{x} = \mathbf{y}$. (The expansive constant can be shown to be uniform for a neighborhood of f.)

In this case there are alternative proofs that h is onto that do not use the induced maps on the top homology group. One such proof is given below in the case of a hyperbolic invariant set. Another proof uses the fact that h is one to one. By the Invariance of Domain Theorem, h is an open map, so $h(M)$ is open in M. Because M is compact, $h(M)$ is compact and so closed. If M is connected, then the image $h(M)$ is both open and closed in M and so is all of M, and h is onto. (This is the same type of proof that we used for the Hartman-Grobman Theorem, Lemma V.7.5.) If M is not connected, then the above argument can be applied to the connected components and still shows that h is onto. □

Finally, we want to give a version of this last theorem for a hyperbolic isolated invariant set. See Hirsch and Pugh (1970).

Theorem 7.4 (Stability of an Hyperbolic Invariant Set). *Let Λ_f be a compact hyperbolic isolated invariant set for $f : M \to M$ with isolating neighborhood U. There is an $\epsilon > 0$ such that if g is a C^1 map within ϵ of f, then g has a hyperbolic structure on $\Lambda_g = \bigcap\{g^j(U) : j \in \mathbb{Z}\}$, and there exists a homeomorphism $h : \Lambda_g \to \Lambda_f$ (onto) that gives a topological conjugacy.*

PROOF. Given $\eta > 0$, there exists a $\delta > 0$ such that any δ-chain within δ of Λ_f can be uniquely η-shadowed by an orbit in Λ_f. Take N such that

$$\bigcap_{j=-N}^{N} f^j(U) \subset \{\mathbf{q} : d(\mathbf{q}, \Lambda_f) < \delta/2\}.$$

There exists a C^0 neighborhood $\mathcal{N}'$ of f such that for g in $\mathcal{N}'$

$$\bigcap_{j=-N}^{N} g^j(U) \subset \{\mathbf{q} : d(\mathbf{q}, \Lambda_f) < \delta/2\},$$

and $g^j(\mathbf{p})$ is a δ-chain for f for any $\mathbf{p} \in \bigcap\{g^j(U) : -N \leq j \leq N\}$. Let $\Lambda_g = \bigcap\{g^j(U) : j \in \mathbb{Z}\}$. Using cones, it can be shown that if $\mathcal{N} \subset \mathcal{N}'$ is a small enough neighborhood of f in the C^1 topology, then for $g \in \mathcal{N}$, g has a hyperbolic structure on Λ_g. (See Exercise 10.28.)

Take $g \in \mathcal{N}$. For $\mathbf{p} \in \Lambda_g$, $g^j(\mathbf{p})$ is a δ-chain for f. Therefore, there is a unique $\mathbf{q} = h(\mathbf{p}) \in \Lambda_f$ such that $d(f^j \circ h(\mathbf{p}), g^j(\mathbf{p})) < \eta$. This defines $h : \Lambda_g \to \Lambda_f$. Just as in the case of an Anosov diffeomorphism, the fact that the shadowing is unique proves that $h \circ g = f \circ h$. The map h is continuous, just as for an Anosov diffeomorphism.

Because g has a hyperbolic structure on Λ_g, we can define a map $k : \Lambda_f \to \Lambda_g$ such that $k \circ f = g \circ k$. In fact, if $\mathbf{y} \in \Lambda_f$, then $f^j(\mathbf{y})$ is an δ-chain for g. Because g has a hyperbolic structure on Λ_g, this chain can be uniquely shadowed by $g^j(\mathbf{p_y})$ for $\mathbf{p_y} = k(\mathbf{y})$ with $d(g^j(\mathbf{p_y}), f^j(\mathbf{y}))$ small. Just as for h, k is continuous and satisfies $k \circ f = g \circ k$.

The following claim proves that k is the inverse for h, so h is one to one and onto. This claim completes the proof of Theorem 7.4. □

Claim. *The map h is a homeomorphism between Λ_f and Λ_g and so is a conjugacy. In fact, k is the inverse of h as a map between Λ_f and Λ_g.*

PROOF. For $\mathbf{p} \in \Lambda_g$, $d(f^j \circ h(\mathbf{p}), g^j(\mathbf{p})) < \eta$ is small for all j, and the g orbit of $\mathbf{p}$ shadows the f orbit of $h(\mathbf{p})$, so $k \circ h(\mathbf{p}) = \mathbf{p}$. Thus, if $h(\mathbf{p}_1) = h(\mathbf{p}_2)$, then $\mathbf{p}_1 = k \circ h(\mathbf{p}_1) = k \circ h(\mathbf{p}_2) = \mathbf{p}_2$, so h is one to one.

Next, for $\mathbf{y} \in \Lambda_f$, $d(g^j \circ k(\mathbf{y}), f^j(\mathbf{y}))$ is small for all j, and the f orbit of $\mathbf{y}$ shadows the g orbit of $k(\mathbf{y})$, so $h \circ k(\mathbf{y}) = \mathbf{y}$. Thus, h is onto Λ_f. We showed above that h is continuous. □

10.8 Stability of Anosov Flows

The results of the preceding section are also true for flows. We restrict ourselves to proving the structural stability of Anosov flows. We use the proof of this theorem to discuss expansiveness for a flow (flow expansiveness) and to introduce some constructions which are used in the proof of the global structural stability theorem for flows. (The global structural stability theorem for diffeomorphisms is stated in the next section.) The reader should also notice that we solve a slightly different functional equation in this section than that which is implicitly used in the last section by means of shadowing. The statement of the theorem is similar to before.

Theorem 8.1 (Structural Stability of Anosov Flows). *Let M be a compact manifold and φ^t be a C^1 Anosov flow on M, i.e., φ^t has a hyperbolic structure on all of M. Then, φ^t is structurally stable, i.e., φ^t is topologically equivalent to any flow ψ^t which is C^1 near to φ^t for $-2 \le t \le 2$.*

REMARK 8.1. Remember that for φ^t to be topologically equivalent to ψ^t, it is permissible to reparameterize either ψ^t or φ^t.

REMARK 8.2. One example of an Anosov flow is obtained by suspension of a hyperbolic toral automorphism. The other standard example is the geodesic flow on a manifold with (constant) negative curvature. See Katok and Hasselblatt (1994).

PROOF. The main difference in the proof is that the flow does not expand or contract along the direction of the flow. To keep the trajectory of the perturbation from running ahead of the trajectory for the original flow, we construct a reparameterization of ψ^t that keeps its trajectory in a transversal of $\varphi^t(\mathbf{x})$.

For each point $\mathbf{x} \in M$, let $\Sigma(\mathbf{x})$ be a small transversal at $\mathbf{x}$. These can be taken so that they vary differentiably with $\mathbf{x}$. For $\eta > 0$, let $\Sigma(\mathbf{x}, \eta)$ be the subset of $\Sigma(\mathbf{x})$ made up of points within η of $\mathbf{x}$. As is done to define the Poincaré map, there are $\eta > 0$ and a differentiable function $\tau(t, \mathbf{x}, \mathbf{y})$ with $\tau(0, \mathbf{x}, \mathbf{y}) = 0$ and such that for $-2 \le t \le 2$ and $\mathbf{y} \in \Sigma(\mathbf{x}, \eta)$,

$$\psi^{\tau(t,\mathbf{x},\mathbf{y})}(\mathbf{y}) \in \Sigma(\varphi^t(\mathbf{x})).$$

Then, we can use this "reparameterization" to define

$$F^t(\mathbf{x}, \mathbf{y}) = (\varphi^t(\mathbf{x}), \psi^{\tau(t,\mathbf{x},\mathbf{y})}(\mathbf{y})).$$

The flow F^t is on a subset of $M \times M$ which can be thought of as a bundle over M. (We could let $\Sigma(\mathbf{x}) = \exp_{\mathbf{x}}(\tilde{\Sigma}(\mathbf{x}))$, where $\tilde{\Sigma}(\mathbf{x}) \subset T_{\mathbf{x}}M$ is a disk in a subspace which is a complement to the line spanned by the vector field for φ^t. In the construction of the stable and unstable disks below, we would first construct disks in $\tilde{\Sigma}(\mathbf{x})$ and then exponentiate them to get disks in M. In this discussion we do not include these steps explicitly. See Robinson (1975a).) The first point $\mathbf{x}$ gives the base point in M and the second point $\mathbf{y}$ gives the point in the fiber (which is the transversal at $\mathbf{x}$). Thus, for $-2 \le t \le 2$,

$$F^t : \bigcup_{\mathbf{x} \in M} \{\mathbf{x}\} \times \Sigma(\mathbf{x}, \eta) \to \bigcup_{\mathbf{x} \in M} \{\mathbf{x}\} \times \Sigma(\mathbf{x}).$$

This flow can be extended for all times for which $\psi^{\tau(t,\mathbf{x},\mathbf{y})}(\mathbf{y})$ stays in the transversal $\Sigma(\varphi^t(\mathbf{x}))$. As in the case for diffeomorphisms,

$$D^u(\mathbf{x},\eta) = \bigcap_{t\geq 0} F^t(\varphi^{-t}(\mathbf{x}),\Sigma(\varphi^{-t}(\mathbf{x}),\eta))$$

is an unstable disk at $\mathbf{x}$;

$$D^s(\mathbf{x},\eta) = \bigcap_{t\leq 0} F^t(\varphi^{-t}(\mathbf{x}),\Sigma(\varphi^{-t}(\mathbf{x}),\eta))$$

is a stable disk at $\mathbf{x}$; and

$$\begin{aligned} D^u(\mathbf{x},\eta)\cap D^s(\mathbf{x},\eta) &= \bigcap_{t\in\mathbb{R}} F^t(\varphi^{-t}(\mathbf{x}),\Sigma(\varphi^{-t}(\mathbf{x}),\eta)) \\ &\equiv (\mathbf{x},h(\mathbf{x})) \end{aligned}$$

is a single point. By construction of the disks and the definition of F^t,

$$(\varphi^t(\mathbf{x}),\psi^{\tau(t,\mathbf{x},h(\mathbf{x}))}(h(\mathbf{x}))) = F^t(\mathbf{x},h(\mathbf{x})) \in D^u(\varphi^t(\mathbf{x}),\eta)\cap D^s(\varphi^t(\mathbf{x}),\eta),$$

but also,

$$(\varphi^t(\mathbf{x}),h\circ\varphi^t(\mathbf{x})) \in D^u(\varphi^t(\mathbf{x}),\eta)\cap D^s(\varphi^t(\mathbf{x}),\eta);$$

so by uniqueness of this point,

$$h\circ\varphi^t(\mathbf{x}) = \psi^{\tau(t,\mathbf{x},h(\mathbf{x}))}(h(\mathbf{x})).$$

This formula has built into it the reparameterization of ψ. The function $\tau(t,\mathbf{x},h(\mathbf{x}))$ is a monotone function of t, so it has an inverse $\sigma(s,\mathbf{x})$. Then, σ can be used to reparameterize φ^t,

$$h\circ\varphi^{\sigma(s,\mathbf{x})}(\mathbf{x}) = \psi^s(h(\mathbf{x})).$$

The fact that h is onto and continuous is the same as for diffeomorphisms. The reparameterization is continuous by the fact that both h and τ (and so σ) are continuous.

Lastly, we need to check that h is one to one. Assume $h(\mathbf{x}_1) = h(\mathbf{x}_2)$. Then,

$$\begin{aligned} h\circ\varphi^{\sigma(s,\mathbf{x}_1)}(\mathbf{x}_1) &= \psi^s(h(\mathbf{x}_1)) \\ &= \psi^s(h(\mathbf{x}_2)) \\ &= h\circ\varphi^{\sigma(s,\mathbf{x}_2)}(\mathbf{x}_2), \end{aligned}$$

so,

$$d(\varphi^{\sigma(s,\mathbf{x}_1)}(\mathbf{x}_1),\varphi^{\sigma(s,\mathbf{x}_2)}(\mathbf{x}_2)) < 2\eta$$

for all t. In fact, by taking ψ^t nearer to φ^t, we can make this as small as we desire.

The above property is closely related to expansiveness of flows. A flow φ^t is *flow expansive* on an invariant set Λ provided given any $\epsilon > 0$, there exists a $\delta > 0$ such that if $\mathbf{x}_1,\mathbf{x}_2 \in \Lambda$ with

$$d(\varphi^t(\mathbf{x}_1),\varphi^{\sigma(t)}(\mathbf{x}_2)) < \delta$$

for all t where

$$\lim_{t\to\pm\infty}\sigma(t) = \pm\infty,$$

then $\mathbf{x}_2 = \varphi^s(\mathbf{x}_1)$ for some $|s|\leq\epsilon$.

Proposition 8.2 (Flow Expansiveness). *Let Λ be a compact hyperbolic invariant set for a flow φ^t. Then, φ^t is flow exansive on Λ.*

We defer the proof of the proposition to Exercise 10.30.

RETURNING TO THE PROOF OF THEOREM 8.1. By taking ψ^t near enough to φ^t, we can insure that the two trajectories are within δ. By flow expansiveness, $\mathbf{x}_2 = \varphi^s(\mathbf{x}_1)$ for some $|s| \le \epsilon$, and the points are on the same trajectory. Because transversals for nearby points on the same trajectory are disjoint, we must have that $\mathbf{x}_1 = \mathbf{x}_2$, and h is one to one. This completes the proof of Theorem 8.1. □

The above proof uses the flow F^t on a bundle of transversals. This construction works fine away from fixed points of the flow. Since we are considering Anosov flows (or possibly flows on a hyperbolic invariant set without fixed points), this causes no problem in the present situation. However, in the proof of the global stability theorem for flows, it is necessary to allow fixed points. As other points limit on a fixed point, the transversals do not have a continuous extension to the fixed point. In the proof of the Hartman-Grobman Theorem near a fixed point for a flow, no reparameterization is needed and the variation of the nonlinear flow from the linear flow in all directions (not just those transverse to the flow lines) is used to construct the conjugacy. In the proof of the global stability theorem for flows, it is necessary to make a transition from (i) reparameterizations of the flow and the conjugating function taking values in a transversal for most points to (ii) no reparameterization of the flow and the conjugating function allowing displacements in all directions near fixed points. One way to accomplish this is given in Robinson (1975a). Although in this section we only consider flows without fixed points, we introduce some of the ideas used in the more general situation.

It is possible to extend the flow F^t used above so that it is defined for all pairs $(\mathbf{x}, \mathbf{y})$ with $\mathbf{y}$ near $\mathbf{x}$ (not just in the transversal). As before, there are $\eta > 0$ and a differentiable function $\tau(t, \mathbf{x}, \mathbf{y})$ such that for $-2 \le t \le 2$ and $d(\mathbf{x}, \mathbf{y}) < \eta$,

$$\psi^{\tau(t,\mathbf{x},\mathbf{y})}(\mathbf{y}) \in \Sigma(\varphi^t(\mathbf{x})).$$

In this context, $\tau(0, \mathbf{x}, \mathbf{y})$ is not necessarily equal to 0. Next, let

$$\mu(t, \mathbf{x}, \mathbf{y}) = \tau(t, \mathbf{x}, \mathbf{y}) - e^{-\alpha t}\tau(0, \mathbf{x}, \mathbf{y})$$

where $\alpha > 0$ is small enough so that

$$\mu'(t, \mathbf{x}, \mathbf{y}) = \tau'(t, \mathbf{x}, \mathbf{y}) + \alpha e^{-\alpha t}\tau(0, \mathbf{x}, \mathbf{y}) > 0.$$

(This last condition gives monotonicity of the reparameterization by μ.) Notice that $\mu(0, \mathbf{x}, \mathbf{y}) = 0$. Let

$$F^t(\mathbf{x}, \mathbf{y}) = (\varphi^t(\mathbf{x}), \psi^{\tau(t,\mathbf{x},\mathbf{y})}(\mathbf{y}))$$

as before, but is defined on a larger space. It can be checked that F^t satisfies the group property, $F^t \circ F^s = F^{t+s}$. The flow F^t preserves the bundle of transversals on which we previously defined the flow. We have altered the flow so that F^t is hyperbolic in all the "fiber" directions: it contracts in the direction pointing off the transversal (in the flow direction) in the fiber. Applying the construction as before, we get $D^u(\mathbf{x}, \eta)$ and $D^s(\mathbf{x}, \eta)$ with $D^u(\mathbf{x}, \eta) \subset \Sigma(\mathbf{x})$ so $h(\mathbf{x}) \in \Sigma(\mathbf{x})$. (The disk $D^s(\mathbf{x}, \eta)$ includes the direction along the flow lines of ψ^t in each fiber.) We refer the reader to Robinson (1975a) for details.

In the present context, there is no real advantage to extending the flow F^t as indicated above. However, this construction allows the style of proof discussed above to be used to apply to the global stability theorem where fixed points are allowed. We present these ideas in the present context where they are somewhat simple and can be understood without the complicated induction construction needed for the general global stability theorem.

10.9 Global Stability Theorems

In this section we give the global stability theorems. The first result gives the conjugacy only on the chain recurrent set. Because the original version of this theorem gave the conjugacy only on the nonwandering set, $\Omega(f)$, we call the result the Ω-stability theorem. Remember that in the last section we proved the existence of a conjugacy on each basic set.

Definition. Let $f : M \to M$ be a C^1 diffeomorphism on a compact manifold M. Then, f is *$\mathcal{R}$-stable* provided there exists a neighborhood $\mathcal{N}$ of f in the C^1 topology such that for $g \in \mathcal{N}$ there is a homeomorphism $h : \mathcal{R}(f) \to \mathcal{R}(g)$ (onto $\mathcal{R}(g)$) such that $h \circ f = g \circ h$. Similarly, f is called *Ω-stable* provided there is a homeomorphism h from $\Omega(f)$ onto $\Omega(g)$ with $h \circ f = g \circ h$.

Theorem 9.1 (Ω-Stability Theorem). *Assume that $f : M \to M$ for M compact is a C^1 diffeomorphism for which $\mathcal{R}(f)$ has a hyperbolic structure. Then, f is $\mathcal{R}$-stable.*

Alternatively, using the nonwandering set to express the assumptions, assume that $\Omega(f)$ has a hyperbolic structure, $\Omega(f) = \text{cl}(\text{Per}(f))$, and f has no cycles. Then, f is Ω-stable.

REMARK 9.1. This theorem was originally given in Smale (1970).

REMARK 9.2. We give the proof with the assumptions on the chain recurrent set. With the assumptions on the nonwandering set, it can be proved that f has a filtration. The rest of the proof is similar to the one given. See Shub (1987).

PROOF. By the assumptions, $\mathcal{R}(f) = \text{cl}(\text{Per}(f))$; so by the Spectral Decomposition Theorem, $\mathcal{R}(f) = \Lambda_1 \cup \cdots \cup \Lambda_N$ is the finite union of basic sets. Since we are using the chain recurrent set, each Λ_k has an isolating neighborhood U_k such that $\Lambda_k = \bigcap_{j=-\infty}^{\infty} f^j(U_k)$. In this context, there are only a finite number of attracting-repelling pairs. See Exercise 10.16.

By Conley's Fundamental Theorem, there is a Liapunov function $V : M \to \mathbb{R}$ that is strictly decreasing off of $\mathcal{R}(f)$. Because each pair of Λ_j and Λ_k can be put in different attracting-repelling pairs, V has different values on each of the Λ_j. In fact, the Λ_j can be renumbered and V can be modified so that $V(\Lambda_j) = j$. For this modified V, $V : M \to [1, N]$. Exercise 10.32 asks the reader to carry out the details of the above modification of the Liapunov function V. This puts a total ordering on the Λ_j that is compatible with the partial ordering with $\Lambda_j << \Lambda_k$ if $W^s(\Lambda_j) \cap W^u(\Lambda_k) \neq \emptyset$. (In this ordering, the orbits flow down the ordering. In many of my papers, I used the ordering for which the index increases along a forward orbit.)

Using the modified Liapunov function V, we can define subsets of M by

$$M_j = V^{-1}((-\infty, j + \frac{1}{2}]).$$

These sets form what is called a *filtration*. They have the following properties which characterize a filtration: for each j such that $1 \le j \le N$,

(1) $M = M_N \supset M_{N-1} \supset \cdots \supset M_1 \supset M_0 = \emptyset$,
(2) $f(M_j) \subset \text{int}(M_j)$, so each M_j is a trapping region,
(3) $\Lambda_j \subset \text{int}(M_j \setminus M_{j-1})$,
(4) $\Lambda_j = \bigcap_{k=-\infty}^{\infty} f^k(M_j \setminus M_{j-1})$, and

(5)

$$\bigcap_{k=0}^{\infty} f^k(M_j) = \bigcup_{i \le j} W^u(\Lambda_i)$$
$$= \bigcup_{i \le j} \mathrm{cl}(W^u(\Lambda_i)).$$

We leave it to Exercise 10.33 to check these properties. Note that we used the existence of a Liapunov function to prove the existence of a filtration. From now on we use the filtration, and no longer use the Liapunov function. It is possible to prove the existence of the filtration without using the Liapunov function. Properties (1) – (4) are used in the proof of the Ω-Stability Theorem, while Property (5) is also used in the proof of the Structural Stability Theorem. Note in Property (5) that $\mathrm{cl}(W^u(\Lambda_j))$ is not usually equal to $W^u(\Lambda_i)$ but can have points in unstable manifolds of basic sets lower in the filtration.

Given the filtration, using the Local Stability Theorem of the last section, it is possible to take the isolating neighborhoods $U_j = M_j \setminus M_{j-1}$. Then, there exists a neighborhood $\mathcal{N}$ of f such that for $g \in \mathcal{N}$, (i) each of the $M_j \setminus M_{j-1}$ is an isolating neighborhood for $\Lambda_j(g) = \bigcap_{k=-\infty}^{\infty} g^k(M_j \setminus M_{j-1})$, (ii) there exists $h_j : \Lambda_j(f) \to \Lambda_j(g)$ that is a conjugacy, and (iii) $g(M_j) \subset \mathrm{int}(M_j)$. We define $h : \bigcup_{j=1}^N \Lambda_j(f) \to \bigcup_{j=1}^N \Lambda_j(g)$. This gives a conjugacy on these sets. We show below that for $g \in \mathcal{N}$, $\mathcal{R}(g) = \bigcup_{j=1}^N \Lambda_j(g)$. Therefore, h is a $\mathcal{R}$-conjugacy from f to g. It remains to show that $\mathcal{R}(g) = \bigcup_{j=1}^N \Lambda_j(g)$, which we do by means of the following two claims.

Claim 1. $\mathcal{R}(g) \supset \bigcup_{j=1}^N \Lambda_j(g)$.

PROOF. The periodic points of f are dense in $\mathcal{R}(f)$ and so in each $\Lambda_j(f)$. The conjugacy h_j takes these periodic points of f into periodic points for g. Therefore, the periodic points of g are dense in each $\Lambda_j(g)$, and $\mathcal{R}(g) \supset \mathrm{cl}(\mathrm{Per}(g)) \supset \Lambda_j(g)$. Taking the union over j, we get the result. □

Claim 2. $\mathcal{R}(g) \subset \bigcup_{j=1}^N \Lambda_j(g)$.

PROOF. Take $\mathbf{y} \in \mathcal{R}(g)$. Then, $\mathbf{y} \in M_j \setminus M_{j-1}$ for some j. We want to show that $\mathbf{y} \in \Lambda_j$.

The first step is to show that $g^i(\mathbf{y}) \notin M_{j-1}$ for all $i > 0$. Assume to the contrary that $g^k(\mathbf{y}) \in M_{j-1}$ for some $k > 0$. Since $g(M_{j-1}) \subset \mathrm{int}(M_{j-1})$, $g^i(\mathbf{y}) \in \mathrm{int}(M_{j-1})$ for all $i \ge k$. Also, for $\epsilon > 0$ small enough, any ϵ-chain $\{\mathbf{y}_i\}$ with $\mathbf{y}_0 = \mathbf{y}$ has $\mathbf{y}_k \in M_{j-1}$ by the continuity of g. Next, $g(\mathbf{y}_k) \in g(M_{j-1}) \subset \mathrm{int}(M_{j-1})$. For $\epsilon > 0$ small enough, $\mathbf{y}_{k+1} \in M_{j-1}$ since $d(\mathbf{y}_{k+1}, g(\mathbf{y}_k)) < \epsilon$. Continuing by induction, $\mathbf{y}_i \in M_{j-1}$ for $i \ge k$. This contradicts the fact that $\mathbf{y}$ is chain recurrent. Therefore, $g^i(\mathbf{y}) \notin M_{j-1}$ for all $i \ge 0$.

Because $g(M_j) \subset M_j$ and $\mathbf{y} \in M_j$, $g^i(\mathbf{y}) \in g^i(M_j) \subset M_j$ for all $i \ge 0$. Combining, $g^i(\mathbf{y}) \in M_j \setminus M_{j-1}$ for all $i \ge 0$.

A similar argument applied to backward iterates shows that $g^i(\mathbf{y}) \in M_j \setminus M_{j-1}$ for all $i \le 0$. Combining the results for positive and negative i, $g^i(\mathbf{y}) \in M_j \setminus M_{j-1}$ for all $i \in \mathbb{Z}$, i.e., $\mathbf{y} \in \Lambda_j(g)$. □

Combining the claims, we have completed the proof of Theorem 9.1. □

In the exercises, we ask the reader to give a direct proof of the Ω-Stability Theorem for a flow on the two sphere S^2 with one source, one sink, and no other chain recurrent points. See Exercise 10.31.

Definition. Assume f is a diffeomorphism with a hyperbolic structure on $\mathcal{R}(f)$ and $\mathcal{R}(f) = \Lambda_1 \cup \cdots \cup \Lambda_N$, where the Λ_j are basic sets. We say that f *satisfies the transversality condition* provided for every $\mathbf{p}, \mathbf{q} \in \mathcal{R}(f)$, $W^u(\mathbf{p}, f)$ is transverse to $W^s(\mathbf{q}, f)$. (This condition allows the fact that some unstable manifolds do not intersect other stable manifolds.) Note that the condition is that the stable and unstable manifolds of points are transverse and not just that the stable and unstable manifolds of basic sets are transverse. Alternatively, we could assume that $L(f)$ (resp. $\Omega(f)$) has a hyperbolic structure, and f satisfies the transversality condition with respect to $L(f)$ (resp. $\Omega(f)$), i.e., $W^u(\mathbf{p}, f)$ and $W^s(\mathbf{q}, f)$ are transverse for all $\mathbf{p}, \mathbf{q} \in L(f)$ (resp. $\Omega(f)$). Note that if $L(f)$ (resp. $\Omega(f)$) has a hyperbolic structure and f satisfies the transversality condition with respect to $L(f)$ (resp. $\Omega(f)$), then f can be shown to have no cycles. (Remember that we proved that if $\mathcal{R}(f)$ has a hyperbolic structure, then f has no cycles, so certainly it has no cycles if $\mathcal{R}(f)$ has a hyperbolic structure and f satisfies the transversality condition.)

Theorem 9.2 (Structural Stability Theorem). *Assume that M is a compact manifold, $f : M \to M$ is a C^1 diffeomorphism, (i) f has a hyperbolic structure on $\mathcal{R}(f)$, and (ii) f satisfies the transversality condition with respect to $\mathcal{R}(f)$. Then, f is structurally stable. That is, there exists a neighborhood $\mathcal{N}$ of f in the set of C^1 diffeomorphisms such that if $g \in \mathcal{N}$, then g is topologically conjugate to f on all of M.*

Assumptions (i) and (ii) and be replaced with either of the following alternatives:

(a) *(i) f has a hyperbolic structure on $L(f)$ and (ii) f satisfies the transversality condition with respect to $L(f)$, or*

(b) *(i) f has a hyperbolic structure on $\Omega(f)$ and $\Omega(f) = \mathrm{cl}(\mathrm{Per}(f))$ and (ii) f satisfies the transversality condition with respect to $\Omega(f)$.*

REMARK 9.3. This theorem was proved for several special cases before the general proof was given.

The case when $\mathcal{R}(f) = M$ was proved by Anosov (1967) and Moser (1969). (This case is the Anosov Stability Theorem, Theorems 7.3 and VIII.5.1(e).) These proofs applied to both diffeomorphisms or flows (although Moser's proof for flows had to be somewhat modified from what he gave).

The case when $\mathcal{R}(f)$ is a finite number of (periodic) points (so f is Morse–Smale) was proved by Palis and Smale (1970). This proof applies to either diffeomorphisms or flows. Also see Palis and de Melo (1982).

The case when f is a C^2 diffeomorphism (but the neighborhood is still in the C^1 topology) was proved by Robbin (1971). This is the first proof of the general theorem stated above.

The case when f is a C^1 diffeomorphism was first proved in Robinson (1976a). (This result has weaker hypothesis than the result of Robbin referred to above.)

The general case of the theorem when f^t is a C^1 flow was proved by Robinson (1975a) after first proving it for C^2 vector fields in Robinson (1974).

REMARK 9.4. Besides the original proofs referred to above, also see Robinson (1975b, 1976b, and 1977) for sketch of the proof and discussion of the ideas for the full theorem.

REMARK 9.5. The proof in one dimension is especially easy because there can only be sources and sinks. In Section 2.6, we treated some examples on the line. The one complication comes from the fact that the line is not compact. We also considered some examples, which have critical points which is more complicated. These methods can be applied to show that Morse-Smale diffeomorphisms on S^1 are structurally stable. See Exercises 10.38 and 10.39 for some special cases.

We conclude by giving a few comments about the construction. The conjugacy is built up in neighborhoods of successive basic sets. It is necessary to extend the map onto the stable and unstable manifolds. We give the following definition.

Definition. Let Λ be a hyperbolic basic set. A *fundamental domain for the stable manifold of* Λ is a closed set $D^s \subset W^s(\Lambda) \setminus \Lambda$ such that there exists a set $D^{s\prime}$ with $D^s = \text{cl}(D^{s\prime})$ and $f^j(D^{s\prime}) \cap D^{s\prime} = \emptyset$ for all integers $j \neq 0$. Note that $D^s \cap \Lambda = \emptyset$.

A *fundamental domain for the unstable manifold of* Λ is defined similarly.

REMARK 9.6. One way to show the existence of a fundamental domain is to use a Liapunov function. Let $V : M \to \mathbb{R}$ be a Liapunov function with $\Lambda \subset V^{-1}(i)$ and $W^s(\Lambda) \subset V^{-1}([i, \infty))$. Let $S = V^{-1}([0, i + \epsilon]) \cap W^s(\Lambda)$ be the "local stable manifold" of Λ for some choice of ϵ, $0 < \epsilon < 0.5$. Let $D^{s\prime} = S \setminus f(S)$ and and $D^s = \text{cl}(D^{s\prime})$. This is a fundamental domain for the stable manifold. See Exercise 10.38.

The proof is not that difficult in the case when f has only one repeller and one attractor. We consider the even easier case of a north pole - south pole diffeomorphism of S^n, i.e., f has a single fixed point source $\mathbf{x}_2$ and a single fixed point sink $\mathbf{x}_1$ and no other chain recurrent points. Let D^s be a fundamental domain for the sink $\mathbf{x}_1$ of f. Also let D^s be constructed as above with the upper edge equal to $V^{-1}(1.5)$. Let g be a small C^1 perturbation which is $\mathcal{R}$-conjugate (so it has only two fixed points, $\mathbf{y}_1$ and $\mathbf{y}_2$). Assume g is near enough to f so that $g(V^{-1}(1.5)) \subset V^{-1}([1, 1.5))$, i.e., g still moves this level set down in terms of V. Let $h_0(\mathbf{x}) = \mathbf{x}$ on $V^{-1}(1.5)$. On the image of $f(V^{-1}(1.5))$, define h_0 by $h_0(\mathbf{x}) = g \circ h_0 \circ f^{-1}(\mathbf{x})$. Using a bump function h_0 can be filled in to define a function $h_0 : D^s \to S^n$. (This is similar to the construction in Section 2.6.) Since D^s is a fundamental domain, h_0 can be extended to a function h defined on $S^n \setminus \{\mathbf{x}_1, \mathbf{x}_2\}$ by $h(\mathbf{x}) = g^j \circ h_0 \circ f^{-j}(\mathbf{x})$ where j is chosen so that $f^{-j}(\mathbf{x}) \in D^s$. Since h_0 is continuous on D^s, this extension is continuous on $S^n \setminus \{\mathbf{x}_1, \mathbf{x}_2\}$. Also define $h(\mathbf{x}_1) = \mathbf{y}_1$ and $h(\mathbf{x}_2) = \mathbf{y}_2$. A little checking shows that h is continuous at these points as well. This completes the sketch of why f is structurally stable in this case.

10.10 Exercises

Fundamental Theorem

10.1. Let $f : S^2 \to S^2$ be the diffeomorphism with one source, one sink, and a hyperbolic (saddle) invariant set that is a Cantor set, i.e., f is the horseshoe map on S^2. Find enough pairs of attracting-repelling pairs to show that their intersection as in Theorem 1.3 is equal to $\mathcal{R}(f)$.

10.2. Let f be a diffeomorphism, and all the $\mathbf{p}_j$ listed below are periodic points for f.
  (a) Assume that $\mathbf{q} \in W^u(\mathcal{O}(\mathbf{p}_1)) \cap W^s(\mathcal{O}(\mathbf{p}_2)) \neq \emptyset$. Show that for all $\epsilon > 0$, there is an ϵ-chain from $\mathbf{p}_1$ to $\mathbf{q}$ and then to $\mathbf{p}_2$.
  (b) Assume that $\mathbf{q}_j \in \hat{W}^u(\mathcal{O}(\mathbf{p}_j)) \cap \hat{W}^s(\mathcal{O}(\mathbf{p}_{j+1})) \neq \emptyset$ for $j = 0, \ldots, n$ and $\mathbf{p}_{n+1} = \mathbf{p}_0$. Show that all the $\mathbf{q}_j$ are chain recurrent.

10.3. Let $\mathbf{x}, \mathbf{y} \in \mathcal{R}$ and assume $\mathbf{y} \notin \Omega^+(\mathbf{x})$. Prove there is a $(A, A^*) \in \mathcal{A}$ such that $\mathbf{x} \in A$ and $\mathbf{y} \in A^*$. Hint: Let $U = \Omega^+_\epsilon(\mathbf{x})$ for small ϵ, and for the corresponding attracting-repelling pair, $(A, A^*) \in \mathcal{A}$, show that $\mathbf{x} \in A$ and $\mathbf{y} \in A^*$.

10.4. Prove that if Y is an invariant set for f and $X = \text{cl}(Y)$, then X is an invariant set for f.

10.5. Let φ^t be a continuous flow that is defined on a space X for all t, e.g., X is compact. Assume V is an open set in X and let $U = \bigcap_{0 \le t \le T} \varphi^t(V)$. Assume $\mathbf{x} \in U$.
(a) Prove that $\varphi^{-t}(\mathbf{x}) \in V$ for $0 \le t \le T$.
(b) Prove that U is open.

10.6. Assume that the set Y is positively invariant for the flow φ^t. Prove that Y^c is negatively invariant.

10.7. Let φ^t be a continuous flow on a compact metric space M. Let (A, A^*) be an attracting-repelling pair for a trapping neighborhood U. If $\mathbf{x} \notin A \cup A^*$, without using a Liapunov function prove that $\omega(\mathbf{x}) \subset A$ and $\alpha(\mathbf{x}) \subset A^*$.

10.8. Let φ^t be a continuous flow on a compact metric space M. Let A be an attracting set. Prove its dual repelling set A^* does not depend on which trapping neighborhood U is used (for which $A = \bigcap_{t \ge 0} \varphi^t(U)$) but only depends on A.

Shadowing and Expansiveness

10.9. Let $D(x) = 2x \bmod 1$ be the doubling map on S^1. Let $\{x_j\}_{j=0}^{\infty}$ be a δ-chain for D.
(a) Show that the inverse iterates $D^{-i}(x_j)$ can be chosen so that

$$d(D^{-i}(x_j), x_{j-i}) \le \delta(\frac{1}{2} + \cdots + \frac{1}{2^i}) \le \delta.$$

(b) Choosing the inverse iterates as in part (a), let $y_k = \lim_{j \to \infty} D^{-j+k}(x_j)$. Prove that $D(y_k) = y_{k+1}$.
(c) Prove that y_0 δ-shadows the δ-chain $\{x_j\}_{j=0}^{\infty}$.

10.10. Let Σ_A be a one-sided subshift of finite type with metric d as defined in Chapter II. Let $\delta = 0.5$. Assume $\{\mathbf{s}^{(j)} \in \Sigma_A\}_{j=0}^{\infty}$ is a 0.5-chain for σ_A on Σ_A. Specify the point $\mathbf{t} \in \Sigma_A$ which 0.5-shadows the 0.5-chain.

10.11. Prove that a hyperbolic toral automorphism on $\mathbb{T}^2$ is expansive without using the result about shadowing.

10.12. Assume Λ is a compact hyperbolic invariant set. Prove that Λ is isolated if and only if it has a local product structure.

10.13. Assume Λ is a compact hyperbolic invariant set that is not isolated. Prove that if V is a small enough neighborhood of Λ, then the maximal invariant set in V, $\Lambda_V = \bigcap_{n \in \mathbb{Z}} f^i(V)$, has a hyperbolic structure. Hint: See the proof of Theorem VIII.4.5, the existence of a horseshoe for a transverse homoclinic point.

Anosov Closing Lemma

10.14. Consider the example given in Remark 4.2 and Figure 4.1.
(a) Explain why the point labeled $\mathbf{q}$ is in $\Omega(f)$ but not $L(f)$.
(b) Explain why the point labeled $\mathbf{q}$ is in $\Omega(f)$ but not in $\Omega(f|\Omega(f))$. Conclude that $\Omega(f|\Omega(f)) \ne \Omega(f)$ for this example.

10.15. Prove Theorem 4.1(c), i.e., assume that the nonwandering set $\Omega(f)$ is hyperbolic, and prove that $\mathrm{cl}(\mathrm{Per}(f)) = \Omega(f|\Omega(f))$.

Decomposition of Recurrent Points

10.16. Assume M is a compact manifold and $f : M \to M$ is a diffeomorphism with a hyperbolic structure on the chain recurrent set, $\mathcal{R}(f)$. Let $\mathcal{A}$ be the set of attracting-repelling pairs for f. Let $\{\Lambda_1, \dots, \Lambda_N\}$ be the collection of basic sets.

(a) Let $(A, A^*) \in \mathcal{A}$. If $\Lambda_j \cap A \neq \emptyset$, prove that $\Lambda_j \subset A$ and $W^u(\Lambda_j) \subset A$. If $\Lambda_j \cap A^* \neq \emptyset$, prove that $\Lambda_j \subset A^*$ and $W^s(\Lambda_j) \subset A^*$.

(b) Let $(A, A^*) \in \mathcal{A}$. Prove that

$$A = \bigcup\{W^u(\Lambda_j) : \Lambda_j \cap A \neq \emptyset\} \qquad \text{and}$$
$$A^* = \bigcup\{W^s(\Lambda_j) : \Lambda_j \cap A^* \neq \emptyset\}.$$

(c) For a diffeomorphism with a hyperbolic structure on the chain recurrent set, prove that there are a finite number of distinct attracting-repelling pairs in $\mathcal{A}$.

10.17. Give an example of an attracting set which is not an attractor.

10.18. Assume that (i) the periodic points are dense in two hyperbolic basic sets Λ_{j_1} and Λ_{j_2}, (ii) there exist $\mathbf{q}_1 \in \Lambda_{j_1}$ and $\mathbf{q}_2 \in \Lambda_{j_2}$ such that $W^u(\mathbf{q}_1)$ has a non-empty transverse intersection with $W^s(\mathbf{q}_2)$, and (iii) there exist $\mathbf{q}_1' \in \Lambda_{j_1}$ and $\mathbf{q}_2' \in \Lambda_{j_2}$ such that $W^u(\mathbf{q}_2')$ has a non-empty transverse intersection with $W^s(\mathbf{q}_1')$. Prove that all the points of intersection, $\hat{W}^u(\Lambda_{j_1}) \cap \hat{W}^s(\Lambda_{j_2})$ and $\hat{W}^u(\Lambda_{j_2}) \cap \hat{W}^s(\Lambda_{j_1})$, are in $\text{cl}(\text{Per}(f))$ and so in $\Omega(f)$.

10.19. Assume that M is compact, $f : M \to M$ has a hyperbolic structure on the nonwandering set $\Omega(f)$, and f has a cycle. Prove that the nonwandering set is not equal to the chain recurrent set, $\Omega(f) \neq \mathcal{R}(f)$.

10.20. Assume that $f : M \to M$ has a hyperbolic structure on the chair recurrent set $\mathcal{R}(f)$ and M is compact. Assume Λ_j is a basic set for which $\text{int}(W^s(\Lambda_j)) \neq \emptyset$. Prove that Λ_j is an attractor.

10.21. Assume that $f : M \to M$ has a hyperbolic structure on the chair recurrent set $\mathcal{R}(f)$ and M is compact. Let

$$S = \bigcup\{W^s(\Lambda_j) : \Lambda_j \text{ is an attractor }\}.$$

Prove that S is open and dense in M.

10.22. Let $f : M \to M$ be an Anosov diffeomorphism on a connected manifold. (It is not assumed that f is a hyperbolic toral automorphism.)

(a) Let $\mathbf{p} \in \text{Per}(f)$ be a periodic point. Prove that $W^s(\mathbf{p})$ is dense in M. Note that this is the stable manifold of $\mathbf{p}$ and not the stable manifold of the orbit of $\mathbf{p}$. Hint: Prove that $\text{cl}(W^s(\mathbf{p}))$ is open in M.

(b) Let $\mathbf{q} \in M$ be any point. Prove that $W^s(\mathbf{p})$ is dense in M.

(c) Prove that f is topologically mixing.

10.23. Show on any compact two-dimensional manifold M that there exists a diffeomorphism f for which (i) $\mathcal{R}(f)$ has a hyperbolic structure and (ii) f has infinitely many periodic points. Hint: Take a Morse-Smale diffeomorphism on M with a fixed point sink. Replace a neighborhood of the sink with the Smale horseshoe on the disk N (in S^2).

10.24. Assume that $f : M \to M$ has a hyperbolic structure on the chair recurrent set $\mathcal{R}(f)$ and M is compact. Let $\mathcal{R}(f) = \Lambda_1 \cup \cdots \cup \Lambda_N$ be the spectral decomposition into basic sets. Prove that each basic set Λ_j can be decomposed into subsets

$$\Lambda_j = \bigcup_{i=1}^{n_j} X_{j,i}$$

with the following properties.

(i) The sets $X_{j,i} = \mathrm{cl}(W^u(\mathbf{p}) \cap W^s(\mathbf{p}))$ for some periodic point $\mathbf{p} \in X_{j,i}$.

(ii) The sets $X_{j,i}$ are pairwise disjoint.

(iii) The sets $X_{j,i}$ are permuted, $f(X_{j,i}) = X_{j,i+1}$ for $1 \leq i < n_j$ and $f(X_{j,n_j}) = X_{j,1}$.

(iv) The n_j-power of f restricted to each $X_{j,i}$ is topologically mixing, $f^{n_j}|X_{j,i}$ is topologically mixing for each j and i.

Markov Partitions

10.25. Let Σ_B be a two-sided subshift of finite type with shift map σ_B. Find Markov partitions of arbitrarily small diameter. (Note that there is no differential structure, so this map does not have a true hyperbolic structure, but it does have stable and unstable manifolds which is all that is needed.)

10.26. Let $f : \mathbb{R}^2 \to \mathbb{R}^2$ be the diffeomorphism which has the geometric horseshoe Λ as an invariant set. Find Markov partitions of arbitrarily small diameter.

Local Stability

10.27. Prove that the set of Anosov diffeomorphisms is open by proving the following steps. Let f be an Anosov diffeomorphism and define cones C^u_p using the splitting $\mathbb{E}^u_{\mathbf{p}} \oplus \mathbb{E}^s_{\mathbf{p}}$ such that $Df_{f^{-1}(\mathbf{p})} C^u_{f^{-1}(\mathbf{p})} \subset C^u_{\mathbf{p}}$.

(a) Show that if g is C^1 near enough, then $Dg_{g^{-1}(\mathbf{p})} C^u_{g^{-1}(\mathbf{p})} \subset C^u_{\mathbf{p}} \subset T_{\mathbf{p}}M$.

(b) Prove that

$$\bigcap_{n \geq 0} Dg^n_{g^{-n}(\mathbf{p})} C^u_{g^{-n}(\mathbf{p})} \subset T_{\mathbf{p}}M$$

is a subspace $\mathbb{E}^{u,g}_{\mathbf{p}}$ that is near to the subspace $\mathbb{E}^u_{\mathbf{p}}$ for f. Similarly, get the subspace

$$\mathbb{E}^{s,g}_{\mathbf{p}} = \bigcap_{n \geq 0} Dg^{-n}_{g^n(\mathbf{p})} C^s_{g^n(\mathbf{p})} \subset T_{\mathbf{p}}M,$$

where $C^s_{\mathbf{q}} = \mathrm{cl}(T_{\mathbf{q}} \setminus C^u_{\mathbf{q}})$.

(c) Since the subspaces $\mathbb{E}^{u,g}_{\mathbf{p}}$ and $\mathbb{E}^{s,g}_{\mathbf{p}}$ are near the subspaces $\mathbb{E}^u_{\mathbf{p}}$ and $\mathbb{E}^s_{\mathbf{p}}$, argue that $T_{\mathbf{p}}M = \mathbb{E}^{u,g}_{\mathbf{p}} \oplus \mathbb{E}^{s,g}_{\mathbf{p}}$ is a direct sum decomposition.

(d) Since the decomposition for g is near the decomposition for f, argue that this is a hyperbolic structure: $Dg_{\mathbf{p}}$ expands vectors in $\mathbb{E}^{u,g}_{\mathbf{p}}$ and contracts vectors in $\mathbb{E}^{s,g}_{\mathbf{p}}$.

10.28. Assume that Λ_f is a hyperbolic isolated invariant set for $f : M \to M$ with isolating neighborhood U. Prove that if $\epsilon > 0$ is small enough and g is a C^1 diffeomorphism within ϵ of f, then g has a hyperbolic structure on $\Lambda_g = \bigcap\{g^j(U) : j \in \mathbb{Z}\}$.

10.29. Assume that $f : M \to M$ has a hyperbolic chain recurrent set on a compact manifold M. Prove that each of the basic sets Λ_j is an isolated invariant set.

Anosov Flows

10.30. Prove Proposition 8.2 on Flow Expansiveness.

Global Stability Theorems

10.31. Let φ^t be the flow on the two sphere, S^2, with one sink and one source and no other recurrent points. This flow is often called the north pole - south pole flow. Prove that φ^t is $\mathcal{R}$-stable using only the local stability near the fixed points (i.e., without using the Ω-Stability Theorem).

10.32. Assume that $f : M \to M$ is a C^1 diffeomorphism for which M is compact and $\mathcal{R}(f)$ has a hyperbolic structure. Prove that the Λ_j can be renumbered and there exists a modified Liapunov function $V : M \to [1, N]$ which has the following properties:
   (i) V is decreasing off $\mathcal{R}(f)$, and
   (ii) $V(\Lambda_j) = j$.

10.33. Assume that $f : M \to M$ is a C^1 diffeomorphism for which M is compact and $\mathcal{R}(f)$ has a hyperbolic structure. Let Λ_j for $1 \le j \le N$. Let V be a Liapunov function that is strictly decreasing off of $\mathcal{R}(f)$ with $V(\Lambda_j) = j$. Let

$$M_j = V^{-1}((-\infty, j + \frac{1}{2}]).$$

Prove these sets have the five properties of a filtration listed in Section 10.9.

10.34. Assume that $f : M \to M$ is a C^1 diffeomorphism for which M is compact and $\mathcal{R}(f)$ has a hyperbolic structure. Prove there is at least one basic set which is an attractor and at least one basic set which is a repeller.

10.35. Assume that $f : M \to M$ is a C^1 diffeomorphism for which M is compact. Also assume that f has finitely many periodic orbits, all of which are hyperbolic. Prove that there can be no cycle between these periodic points for which all the intersections of stable and unstable manifolds are transverse.

10.36. Assume that $f_j : M_j \to M_j$ is a C^1 diffeomorphism on a compact manifold M_j for $j = 1, 2$. Let $F : M_1 \times M_2 \to M_1 \times M_2$ be defined by $F(\mathbf{x}, \mathbf{y}) = (f_1(\mathbf{x}), f_2(\mathbf{y}))$.
   (a) If f_j has a hyperbolic structure on the chain recurrent set $\mathcal{R}(f_j)$ for $j = 1, 2$, prove that F has a hyperbolic structure on $\mathcal{R}(F)$.
   (b) If in addition to the assumptions of part (a), if f_j satisfies the transversality condition for $j = 1, 2$, prove that F satisfies the transversality condition and so is structurally stable.

10.37. Let Λ be a basic set for a diffeomorphism f with a hyperbolic chain recurrent set on a compact manifold M. Let $V : M \to \mathbb{R}$ be a Liapunov function with $\Lambda \subset V^{-1}(i)$. Let $S = V^{-1}([0, i + \epsilon)) \cap W^s(\Lambda)$ for some choice of $\epsilon > 0$, and $D^s = \text{cl}(S \setminus f(S))$. Prove for good choices of ϵ that D^s is a fundamental domain.

10.38. Let f be the diffeomorphism on S^1 whose lift $F : \mathbb{R} \to \mathbb{R}$ is given by

$$F(\theta) = \theta + \epsilon \sin(2\pi k \theta) \qquad \text{mod } 1,$$

for $0 < 2\pi k \epsilon < 1$. Prove that f is structurally stable.

10.39. Let f be the diffeomorphism on S^1 whose lift $F : \mathbb{R} \to \mathbb{R}$ is given by

$$F(t) = t + 1/n + \epsilon \sin(2\pi n t)$$

for n a positive integer and $0 < \epsilon < 1/(2\pi n)$. Prove that f is structurally stable.

10.40. Prove that the set constructed in Remark 9.6 is a fundamental domain for the basic set.

CHAPTER XI
Generic Properties

In this chapter, systems with hyperbolic chain recurrent sets are shown to be $\mathcal{R}$-stable. Certain systems are also shown to be structurally stable: (i) Anosov systems, (ii) Morse-Smale systems, and (iii) systems with a hyperbolic chain recurrent set which also satisfy the transversality condition. Although we have given examples of such systems, we have not discussed the prevalence of such systems. At one time it was hoped that any system could be approximated by a structurally stable system. By now, many counter examples have been constructed. For these examples, there is a whole open set of systems that are not structurally stable or even Ω-stable. However, it is possible to prove that there are certain properties which are generic in the sense of Baire category, i.e., any system can be approximated by another for which these properties are true and the condition is open at least in a weak sense. This chapter considers several of the basic generic properties. It also gives a counter-example to the density of structurally stable systems. The proofs of the genericity use methods from transversality theory, so a section develops these ideas.

11.1 Kupka-Smale Theorem

A generic property is one which holds for most functions in the function space under consideration. We make this precise in the following definition.

Definition. Let $\mathcal{F}$ be a topological space. A subset $\mathcal{R} \subset \mathcal{F}$ is called a *residual subset* (in the sense of Baire category) provided $\mathcal{R}$ contains the countable intersection of dense open subsets, more precisely, $\mathcal{R} \supset \bigcap_{j=1}^{\infty} S_j$ where each S_j is open and dense in $\mathcal{F}$. In a complete metric space, a residual subset of X is always dense. A topological space X is called a *Baire space* provided any residual subset of X is dense in X. A property is *generic* in a function space $\mathcal{F}$ which is a Baire space provided the property is true for functions in a residual subset of $\mathcal{F}$.

If M is a compact manifold and $1 \leq k \leq \infty$, then $C^k(M, M)$, $\text{Diff}^k(M)$, and the set of C^k vector fields $\mathcal{X}^k(M)$ are all Baire spaces. See Hirsch (1976). If M is noncompact and these function spaces are given the Whitney topology (or strong topology as defined in Hirsch (1976)), then they are also Baire spaces.

The first generic property which we consider is the hyperbolicity of periodic points and the transversality of the stable and unstable manifolds. Because the statement and proof of the theorem can be expressed in terms of the set of all periodic points and the set of hyperbolic periodic points, we give some notation for these sets. For any diffeomorphism f on M, let $\text{Per}(k, f)$ be the set of all periodic points with period less than or equal to n,

$$\text{Per}(n, f) = \{\mathbf{p} \in M : f^j(\mathbf{p}) = \mathbf{p} \text{ for some } j \leq n\},$$

$\text{Per}(f)$ be the set of all periodic points,

$$\text{Per}(f) = \bigcup_{n=1}^{\infty} \text{Per}(n, f),$$

$\text{Per}_h(n, f)$ be the set of all hyperbolic periodic points with period less than or equal to n,

$$\text{Per}_h(n, f) = \{\mathbf{p} \in \text{Per}(n, f) : \mathbf{p} \text{ is a hyperbolic periodic point }\},$$

and $\text{Per}_h(f)$ be the set of all hyperbolic periodic points,

$$\text{Per}_h(f) = \{\mathbf{p} \in \text{Per}(f) : \mathbf{p} \text{ is a hyperbolic periodic point }\}.$$

(Notice that this usage of the notation for $\text{Per}(n, f)$ does not agree with how it is used in the rest of the book where it is the set of all periodic points with least period exactly n.)

Note that all the periodic points of period less than or equal to n are hyperbolic if and only if $\text{Per}(n, f) = \text{Per}_h(n, f)$. Using this fact, we define

$$\mathcal{H}_n = \{f \in \text{Diff}^k(M) : \text{Per}(n, f) = \text{Per}_h(n, f)\}, \qquad \text{and}$$

$$\mathcal{H} = \bigcap_{n=1}^{\infty} \mathcal{H}_n.$$

Therefore, $f \in \mathcal{H}$ if and only if all the periodic points of f are hyperbolic.

The second half of the theorem deals with the transversality of the stable and unstable manifolds. We let

$$\text{KS}(M) = \{f \in \mathcal{H} : W^s(\mathbf{p}, f) \text{ is transverse to } W^u(\mathbf{q}, f) \text{ for all } \mathbf{p}, \mathbf{q} \in \text{Per}(f)\}.$$

Theorem 1.1 (Kupka-Smale). *Assume M is a compact manifold and $1 \leq k \leq \infty$.*

(a) The set $\mathcal{H}_n$ defined above is dense and open in $\text{Diff}^k(M)$*, and $\mathcal{H}$ is a residual subset of* $\text{Diff}^k(M)$.

(b) The set $\text{KS}(M)$ *is a residual subset of* $\text{Diff}^k(M)$.

REMARK 1.1. This theorem is also true for vector fields (or flows). We require that both all the fixed points and all the periodic orbits are hyperbolic in the definition of $\mathcal{H}(X)$. In the definition of $\text{KS}(X)$, we use the stable and unstable manifolds of periodic orbits and not just the stable and unstable manifolds of individual points in the periodic orbits: we require that $W^s(\gamma_1)$ is transverse to $W^u(\gamma_2)$ where γ_1 and γ vary over all fixed points and periodic orbits.

There are two aspects which are different about flows or vector fields than diffeomorphisms. First, there are both fixed points and closed orbits. This difference is minor. Second, the periods of the periodic orbits can be any positive real number so the induction on the period is slightly more cumbersome to implement. Even with these differences, the main ideas of the proof are the same.

REMARK 1.2. The Kupka-Smale Theorem was proved independently by Kupka (1963) and Smale (1963). A nice proof for the case of vector fields is given in Peixoto (1966) which includes the case of noncompact manifolds. Palis and de Melo (1982) and Abraham and Robbin (1967) also give proofs for vector fields.

REMARK 1.3. We delay the proof (for diffeomorphisms) until Section 11.3 because it uses the transversality theorems which we discuss in Section 11.2. The idea of the proof is that a periodic point can be approximated by a hyperbolic periodic point. Since the hyperbolicity of a single periodic point is an open condition, the set $\mathcal{H}_n$ is both dense and open. Similarly, a nontransverse intersection of stable and unstable manifolds can be approximated by a transverse one which implies that $\text{KS}(M)$ is dense. The transversality of the intersections on compact pieces is an open condition and the manifolds can be represented as the countable union of compact subsets, so it can be shown that $\text{KS}(M)$ is residual.

Definition. A diffeomorphism (respectively flow) which satisfies the properties of the set KS(M) in Theorem 1.1(b) is called a *Kupka-Smale* diffeomorphism (respectively flow).

The next set of results concerns the genericity of the condition that the closure of the periodic points equals the nonwandering set. The first step is the possibility of approximating a nonwandering point by a periodic orbit, the Closing Lemma. This lemma was first thought to be obvious (hence the title of a lemma), which it is in the C^0 topology. Its proof is very difficult in the C^1 topology and unknown in the C^2 topology. In other words, the approximating diffeomorphism g with the periodic orbit can be taken to be C^∞ but is only C^1 near to the original f. The reason that the proof is complicated is that one localized perturbation is not enough to change an orbit which returns near to itself into a periodic orbit. The proof for an approximation in the C^1 topology uses many localized perturbations to accomplish the feat. A proof for an approximation in the C^2 topology (if and when it is given) will probably not use a localized perturbation, but will have to control the effects of an orbit passing several times through a single perturbation.

Theorem 1.2 (Closing Lemma of Pugh). *Assume M is a compact manifold, f a C^1 diffeomorphism, $\mathcal{N}$ a neighborhood of f in* $\text{Diff}^1(M)$, *and* $\mathbf{p} \in \Omega(f)$. *Then, there is a $g \in \mathcal{N}$ such that $\mathbf{p}$ is a periodic point for g.*

REMARK 1.4. This result is also true in the spaces of C^1 flows or C^1 vector fields. It was originally proved in Pugh (1967a) for flows. That paper claims to prove the result for C^1 vector fields but there is a technical difficulty in the smoothness of the perturbed vector field (which do not arise when considering the space of flows). These difficulties were discovered by Pugh and are corrected (by Pugh) in Pugh and Robinson (1983). The theorem is also proved for many other function spaces in this latter paper: Hamiltonian and volume preserving diffeomorphisms and flows. Further papers on this result include Liao (1979), Mai (1986), and Wen (1991). For an intuitive discussion of the proof and its difficulties, see Robinson (1978).

The next lemma is much easier than the Closing Lemma. It shows that once a periodic orbit has been produced, the diffeomorphism (vector field, or flow) can be approximated by a new diffeomorphism (vector field, or flow) with a hyperbolic periodic point.

Lemma 1.3. *Let $\mathbf{p}$ be a periodic point of period j for a C^k diffeomorphism $g : M \to M$, for $1 \le k \le \infty$. Then, g can be approximated arbitrarily closely in the C^k topology by a diffeomorphism g' such that $\mathbf{p}$ is a hyperbolic periodic point of the same period.*

PROOF. Let $\varphi : V \to U$ be a coordinate chart at $\mathbf{p}$ with $\varphi(\mathbf{0}) = \mathbf{p}$. Let $r > 0$ be small enough so that $g^i(B(\mathbf{p}, r)) \cap B(\mathbf{p}, r) = \emptyset$ for $1 \le i < j$, where $B(\mathbf{p}, r) = \varphi(\{\mathbf{x} : |\mathbf{x}| \le r\})$. Let $\beta : \mathbb{R}^n \to \mathbb{R}$ be a C^∞ bump function such that $\beta(\mathbf{x}) = 1$ for $|\mathbf{x}| \le r/2$ and $\beta(\mathbf{x}) = 0$ for $|\mathbf{x}| \ge r$. Finally, let $g_\epsilon(\mathbf{q}) = g(\mathbf{q})$ for $\mathbf{q} \notin B(\mathbf{p}, r)$ and

$$g_\epsilon \circ \varphi(\mathbf{x}) = g \circ \varphi(\mathbf{x} + \epsilon\beta(\mathbf{x})\mathbf{x})$$

for $\mathbf{x} \in \varphi^{-1}(B(\mathbf{p}, r)) = \{\mathbf{x} : |\mathbf{x}| \le r\}$. Then,

$$D(\varphi^{-1} \circ g_\epsilon^j \circ \varphi)_\mathbf{0} = D(\varphi^{-1} \circ g^j \circ \varphi)_\mathbf{0}(1 + \epsilon)I.$$

Because of the form of this derivative, g_ϵ has a hyperbolic periodic point at $\mathbf{p}$ for arbitrarily small $\epsilon > 0$. Clearly, g_ϵ converges to g in the C^k topology as ϵ goes to 0. □

A corollary of the Closing Lemma is the General Density Theorem which was also originally proved by Pugh (1967b). This result is only true in the C^1 topology because it uses the Closing Lemma.

Theorem 1.4 (General Density Theorem). *Assume M is a compact manifold. Let $\mathcal{G} = \{f \in \text{Diff}^1(M) : \text{cl}(\text{Per}_h(f)) = \Omega(f)\}$. Then, $\mathcal{G}$ is residual in* $\text{Diff}^1(M)$.

Before starting the proof of the General Density Theorem, we give some definitions and results about semi-continuous set valued functions which are used in the proof.

Let M be a complete metric space, and $\mathcal{C}_M$ be the collection of all compact subsets of M. The *Hausdorff metric* on $\mathcal{C}_M$ is defined as follows: for $A, B \in \mathcal{C}_M$,

$$d(A,B) \equiv \sup\{d(a,B),\ d(b,A) : a \in A,\ b \in B\},$$

where

$$d(b,A) = \inf\{d(b,a) : a \in A\}.$$

(Note: This metric has nothing to do with the Hausdorff property for a general topological space.) If M is a complete metric space, then $\mathcal{C}_M$ is a complete metric space with the metric d defined above. (This result is an exercise in many of the books on topology.)

The proof also uses the concept of a semi-continuous set valued function. (The functions we use take f to $\text{Per}_h(n,f)$.) Let $A_n \in \mathcal{C}_M$ for $n \geq 1$. Define

$$\liminf_{n\to\infty} A_n = \{\mathbf{y} \in M : \text{ there exist } \mathbf{y}_n \in A_n \text{ for } n \geq 1 \text{ such that } \mathbf{y} = \lim_{n\to\infty} \mathbf{y}_n\}.$$

If a point $\mathbf{y}$ is contained in all the A_n for $n \geq N$ for some N, then $\mathbf{y} \in \liminf_{n\to\infty} A_n$. Thus, the set $\liminf_{n\to\infty} A_n$ is the set of points which are "essentially" contained by each of the A_n for large n. (In fact, $\mathbf{y}$ only has to be the limit of points in the A_n and not actually lie in the sets.) It is easy to check that if $A_n \in \mathcal{C}_M$, then $\liminf_{n\to\infty} A_n$ is compact, so is in $\mathcal{C}_M$.

Let X be a topological space and M a complete metric space. A set valued function $\Gamma : X \to \mathcal{C}_M$ is called *lower semi-continuous at* $\mathbf{x}$ provided

$$\Gamma(\mathbf{x}) \subset \liminf_{n\to\infty} \Gamma(\mathbf{x}_n)$$

for every sequence $\mathbf{x}_n \in X$ converging to $\mathbf{x}$. This means that any point in $\Gamma(\mathbf{x})$ can be approached by a sequence of points $\mathbf{y}_n \in \Gamma(\mathbf{x}_n)$. In an intuitive sense, the set $\Gamma(\mathbf{x})$ can be smaller than the $\Gamma(\mathbf{x}_n)$ for nearby $\mathbf{x}_n$ but cannot be bigger. The set valued function Γ is called *lower semi-continuous* provided it is lower semi-continuous at all points $\mathbf{x} \in X$.

Lower semi-continuity can also be expressed in terms of a nonreflexive "semi-metric" on $\mathcal{C}_M$ defined by

$$d_{sub}(A,B) \equiv \sup\{d(a,B) : a \in A\}.$$

For B compact, $d_{sub}(A,B) = 0$ if and only if $A \subset B$. (In the Hausdorff metric, $d(A,B) = 0$ if and only if $A = B$.) Then, $\Gamma : X \to \mathcal{C}_M$ is lower semi-continuous at $\mathbf{x}$ provided

$$\lim\{d_{sub}(\Gamma(\mathbf{x}), \Gamma(\mathbf{y})) : \mathbf{y} \text{ converges to } \mathbf{x} \in X\} = 0.$$

Exercise 11.3 asks the reader to prove the following result. Assume $\Gamma_n : X \to \mathcal{C}_M$ are lower semi-continuous set valued functions for $n \geq 1$, and define $\Gamma : X \to \mathcal{C}_M$ by $\Gamma(\mathbf{x}) = \text{cl}(\bigcup_n \Gamma_n(\mathbf{x}))$. Then, Γ is a lower semi-continuous set valued function.

Finally, assume $\Gamma : X \to \mathcal{C}_M$ is a lower semi-continuous set valued function. Let $\mathcal{R} \subset X$ be the points of continuity of Γ. Then, $\mathcal{R}$ is residual, i.e., a semi-continuous function is continuous at a residual subset. See Choquet (1969).

PROOF OF THEOREM 1.4 (GENERAL DENSITY THEOREM). Define $\Gamma_n, \Gamma : \mathrm{Diff}^1(M) \to \mathcal{C}_M$ by

$$\Gamma_n(f) = \mathrm{Per}_h(n, f),$$
$$\Gamma(f) = \mathrm{cl}(\mathrm{Per}_h(f)),$$

where $\mathrm{Per}_h(n, f)$ and $\mathrm{Per}_h(f)$ are the set of hyperbolic periodic points of period less than or equal to n and of all periods, respectively. The sets $\mathrm{Per}_h(n, f)$ can easily seen to be closed and thus compact. For $n \geq 1$, the map Γ_n is lower semi-continuous because a hyperbolic periodic point persists under perturbations. (See Theorem V.6.4.) These functions are not continuous at all f: if f_0 is a diffeomorphism with a periodic point $\mathbf{x}_0$ with eigenvalue of absolute value one, then under arbitrarily small perturbations of f_0 to g the periodic point $\mathbf{x}_0(g)$ can become hyperbolic (or disappear). Thus, the set $\Gamma_n(g)$ can get bigger for small C^1 perturbations of an f ($\mathbf{x}_0(g) \in \Gamma_n(g)$) but cannot get smaller, so Γ_n is lower semi-continuous but not continuous at f_0. Because each of the set valued functions Γ_n are lower semi-continuous, the function $\Gamma(\cdot) = \mathrm{cl}(\bigcup_n \Gamma_n(\cdot))$ is also lower semi-continuous by Exercise 11.3.

Let $\mathcal{R} \subset \mathrm{Diff}^1(M)$ be the points of continuity of Γ. As mentioned above, a semi-continuous set valued function is continuous at a residual subset, so $\mathcal{R}$ is residual in $\mathrm{Diff}^1(M)$.

Claim. *The set $\mathcal{R} \subset \mathcal{G}$ where $\mathcal{G}$ is the set defined in the theorem, so $\mathcal{G}$ is residual.*

PROOF. Suppose $f \in \mathcal{R} \setminus \mathcal{G}$. Therefore, Γ is continuous at f, but there is a point $\mathbf{p} \in \Omega(f) \setminus \mathrm{cl}(\mathrm{Per}(f))$. By the Closing Lemma, there are $g_i \in \mathrm{Diff}^1(M)$ converging to f such that $\mathbf{p} \in \mathrm{Per}(g_i)$. By Lemma 1.3, each g_i can be approximated by $g_i' \in \mathrm{Diff}^1(M)$ such that $\mathbf{p} \in \mathrm{Per}_h(g_i')$ and the g_i' still converge to f. Therefore, $\mathbf{p} \in \Gamma(g_i')$ and

$$\lim\{d(\Gamma(g_i'), \Gamma(f)) : i \to \infty\} \geq d(\mathbf{p}, \Gamma(f)) \neq 0,$$

which contradicts the fact that f is a continuity point of Γ. This contradiction proves that $\mathcal{R} \subset \mathcal{G}$, which proves the claim and finishes the proof of the theorem. □

11.2 Transversality

We have already defined what it means for two submanifolds to be transverse in an ambient manifold. In this chapter we need to consider functions which are transverse to a submanifold in the space where the function takes its values. After giving the definition of this concept, we present several transversality theorems which state that most functions are transverse to a given submanifold.

Definition. Let M and N be differentiable manifolds, $V \subset N$ be a submanifold, and $K \subset M$ be a subset. A C^1 function $f : M \to N$ is *transverse to* V *at* $\mathbf{p} \in M$ provided that either (i) $f(\mathbf{p}) \notin V$, or (ii) $T_{f(\mathbf{p})}N$ is spanned by the two subspaces $Df_{\mathbf{p}}T_{\mathbf{p}}M$ and $T_{f(\mathbf{p})}V$ whenever $f(\mathbf{p}) \in V$, i.e.,

$$T_{f(\mathbf{p})}N = Df_{\mathbf{p}}T_{\mathbf{p}}M + T_{f(\mathbf{p})}V.$$

A C^1 function $f : M \to N$ is *transverse to* V *along* K provided it is transverse to V at all points $\mathbf{p} \in K$. We use the notation $f \pitchfork_K V$ to indicate that f is transverse to V along K. If $K = M$, then we say that f is *transverse to* V, and denote it by $f \pitchfork V$.

We also use the notion of the codimension of a submanifold. Assume N is a manifold and $V \subset N$ a submanifold. The *codimension of* V *in* N is defined to be $\dim(N) - \dim(V)$, and is denoted by $\mathrm{codim}(V)$.

The following theorem gives a generalization of the fact that the inverse image of a regular value is a submanifold.

Theorem 2.1. *Assume M and N manifolds and $V \subset N$ is a submanifold. (In our use of the terminology, a submanifold is an embedded submanifold.) Assume that $f : M \to N$ is a C^r map for $r \geq 1$, and it is transverse to V. Then, $f^{-1}(V)$ is a C^r submanifold of M. Moreover, the codimension of $f^{-1}(V)$ in M is the same as the codimension of V in N.*

See Hirsch (1976) for a proof.

Theorem 2.2 (Thom Transversality Theorem). *Let M and N be manifolds with M compact, and let $V \subset N$ be a closed submanifold. Assume that $k \geq 1$. The set of functions $\mathcal{T}^k(M, V) = \{f \in C^k(M, N) : f \pitchfork V\}$ is dense and open in $C^k(M, N)$.*

Again, see Hirsch (1976) for a proof.

The difficulty in using the Thom Transversality Theorem is that we do not always have the use of all the functions in $C^k(M, N)$. In the proof of the Kupka-Smale Theorem for $f \in \text{Diff}^1(M)$, we consider the maps $\rho_n(f) = (id, f^n) : M \to M \times M$, where n is a positive integer. We want to conclude that most diffeomorphisms f have $\rho_n(f)$ transverse to the diagonal in $M \times M$. Thus, we do not have all functions in $C^1(M, M \times M)$ at our disposal but only those arising as $\rho_n(f)$. The Thom Transversality Theorem certainly implies that there is an open set of f which have this property. (See Theorem 2.4 below.) To prove the density of the f which satisfy the property, we must know that the function space $\text{Diff}^1(M)$ is large enough to be able to make a perturbation of f for which $\rho_n(f)$ is transverse to the diagonal. There are various ways around this difficulty. Palis and de Melo (1982) and Peixoto (1966) use only the Thom Transversality Theorem to prove density. They both build into the proof the necessary step to make the Thom Transversality Theorem applicable to construct the perturbation of the diffeomorphism f itself and not just of $\rho_n(f)$. Abraham and Robbin (1967) proves a very general form of the transversality theorem which is directly applicable to representations like ρ_n. The difficulty is that this proof of the Kupka-Smale Theorem requires proving that $\text{Diff}^1(M)$ is a Banach manifold which we do not want to verify. (It is true that $\text{Diff}^k(M)$ is a Banach manifold and $\rho_n^{ev} : \text{Diff}^k(M) \times M \to M \times M$ is C^k. See Franks (1979), or Hirsch (1976).) We give a proof more like Abraham and Robbin but only require that $\text{Diff}^1(M)$ contains finite-dimensional subsets of functions on which ρ_n is differentiable. Thus, we use the Parametric Transversality Theorem stated below. Also see Abraham and Robbin (1967) for many related results.

First we give one definition which we use repeatedly.

Definition. Let M and N be manifolds, $\mathcal{F}$ a topological space, and $\rho : \mathcal{F} \to C^k(M, N)$ a continuous map for some $k \geq 0$. The *evaluation of* ρ is defined to be the map $\rho^{ev} : \mathcal{F} \times M \to N$ given by

$$\rho^{ev}(f, \mathbf{x}) = \rho(f)(\mathbf{x}).$$

Theorem 2.3 (Parametric Transversality Theorem). *Assume D^q is an open subset of some $\mathbb{R}^q$, M and N are manifolds, $K \subset M$ compact, $V \subset N$ is a closed submanifold, and $k > \max\{0, \dim(M) - \text{codim}(V)\}$. Assume $\rho : D^q \to C^k(M, N)$ satisfies the following two conditions:*

(i) *ρ is continuous,*
(ii) *the evaluation map $\rho^{ev} : D^q \times M \to N$ is C^k, and*
(iii) *ρ^{ev} is transverse to V along $D^q \times K$.*

Then, $\mathcal{T}(\rho, K, V) \equiv \{\mathbf{t} \in D^q : \rho(\mathbf{t}) \pitchfork_K V\}$ is dense and open in D^q.

Remark 2.1. The differentiability assumption on ρ^{ev} is that it is C^k, where k is greater than the dimension of M minus the codimension of V in N. Note that there is a misprint in this assumption in the Parametric Transversality Theorem given in Hirsch (1976).

REMARK 2.2. The transversality assumption on the evaluation map means that there are enough parameters with which to make the necessary perturbations at one point at a time. The conclusion of the theorem is that the function can be approximated by one which is transverse at all points.

REMARK 2.3. See Hirsch (1976) for a proof.

The following theorem proves the openness of the set of maps which are transverse.

Theorem 2.4. *Assume that $\mathcal{F}$ is a topological space, M and N are manifolds, $K \subset M$ compact, $V \subset N$ is a closed submanifold, and $1 \le k \le \infty$. Assume the map $\rho : \mathcal{F} \to C^k(M, N)$ is continuous where $C^k(M, N)$ is given the compact open topology on the first k derivatives. (If M is compact, then the sup topology on the first k derivatives can be used. Hirsch calls the compact open topology the weak topology.) Then, $\mathcal{T}(\rho, K, V) = \{f \in \mathcal{F} : \rho(f) \pitchfork_K V\}$ is open in $\mathcal{F}$.*

SKETCH OF THE PROOF. The idea is that $\mathcal{T}^k(K, V) = \{f \in C^k(M, N) : f \pitchfork_K V\}$ is open in $C^k(M, N)$ and ρ is continuous. Therefore, $\rho^{-1}(\mathcal{T}^k(K, V)) = \mathcal{T}(\rho, K, V)$ is open. See Hirsch (1976) for details. □

REMARK 2.4. See Abraham and Robbin (1967) or Hirsch (1976) for a more detailed discussion of transversality theorems and their applications.

11.3 Proof of the Kupka–Smale Theorem

In the proof of the theorem, we need to consider not only whether a periodic point is hyperbolic, but also whether it satisfies a condition connected with a transversality condition (or the Implicit Function Theorem). A periodic point $\mathbf{p}$ of f of period n is called *elementary* provided 1 is not an eigenvalue of $Df^n_{\mathbf{p}}$. Thus, a hyperbolic periodic point is elementary, but not all elementary periodic points are hyperbolic. The condition of being elementary is the natural first step in the proof below.

As mentioned in the last section on transversality, we consider the maps

$$\rho_n : \mathrm{Diff}^k(M) \to C^k(M, M \times M)$$

defined by

$$\rho_n(f)(\mathbf{x}) = (\mathbf{x}, f^n(\mathbf{x})).$$

Let Δ be the diagonal in $M \times M$, $\Delta = \{(\mathbf{y}, \mathbf{y}) : \mathbf{y} \in M\}$. Clearly, Δ is a closed submanifold of $M \times M$. Also, $\rho_n(f)(\mathbf{p}) \in \Delta$ if and only if $\mathbf{p}$ is fixed by f^n, i.e., $\mathbf{p}$ has period n in the weak sense of the word. The following lemma characterizes the condition that $\rho_n(f)$ is transverse to Δ as being equivalent to the condition that all fixed points of f^n are elementary.

Lemma 3.1. *(a) The point $\mathbf{p}$ is a fixed point of f if and only if $\rho_1(f)(\mathbf{p}) \in \Delta$.*

(b) Assume $\rho_1(f)(\mathbf{p}) \in \Delta$. Then, $\rho_1(f) \pitchfork_{\mathbf{p}} \Delta$ if and only if $\mathbf{p}$ is an elementary fixed point.

(c) The point $\mathbf{p}$ is a fixed point of f^n if and only if $\rho_n(f)(\mathbf{p}) \in \Delta$.

(d) Assume $\rho_n(f)(\mathbf{p}) \in \Delta$. Then, $\rho_n(f) \pitchfork_{\mathbf{p}} \Delta$ if and only if $\mathbf{p}$ is an elementary fixed point of f^n.

PROOF. (a) and (c) These statements follow easily from the definitions.

(b) Assume $\rho_1(f)(\mathbf{p}) \in \Delta$. For $\mathbf{v} \in T_{\mathbf{p}}M$,

$$D(\rho_1(f))_{\mathbf{p}}\mathbf{v} = (\mathbf{v}, Df_{\mathbf{p}}\mathbf{v}).$$

The tangent space to the diagonal is clearly given as follows:

$$T_{(\mathbf{p},\mathbf{p})}\Delta = \{(\mathbf{w},\mathbf{w}) : \mathbf{w} \in T_{\mathbf{p}}M\}.$$

If $\rho_1(f) \pitchfork_{\mathbf{p}} \Delta$, then for any $\mathbf{u}_1, \mathbf{u}_2 \in T_{\mathbf{p}}M$, we can solve the set of equations

$$\begin{aligned} \mathbf{v} + \mathbf{w} &= \mathbf{u}_1 \\ Df_{\mathbf{p}}\mathbf{v} + \mathbf{w} &= \mathbf{u}_2 \end{aligned}$$

for $\mathbf{v}, \mathbf{w} \in T_{\mathbf{p}}M$. But this means $\mathbf{w} = \mathbf{u}_1 - \mathbf{v}$, so we can solve $Df_{\mathbf{p}}\mathbf{v} - \mathbf{v} = \mathbf{u}_2 - \mathbf{u}_1$ for $\mathbf{v}$, $Df_{\mathbf{p}} - I$ is onto, and 1 is not an eigenvalue of $Df_{\mathbf{p}}$.

Conversely, assume 1 is not an eigenvalue of $Df_{\mathbf{p}}$. Then, for any $\mathbf{u}_1, \mathbf{u}_2 \in T_{\mathbf{p}}M$, we can solve $Df_{\mathbf{p}}\mathbf{v} - \mathbf{v} = \mathbf{u}_2 - \mathbf{u}_1$ for $\mathbf{v}$, and set $\mathbf{w} = \mathbf{u}_1 - \mathbf{v}$. Thus, we can solve the set of equations

$$\begin{aligned} \mathbf{v} + \mathbf{w} &= \mathbf{u}_1 \\ Df_{\mathbf{p}}\mathbf{v} + \mathbf{w} &= \mathbf{u}_2 \end{aligned}$$

for $\mathbf{v}, \mathbf{w} \in T_{\mathbf{p}}M$, and $\rho_1(f) \pitchfork_{\mathbf{p}} \Delta$.

(d) The case for higher period is proved similarly to part (b). □

We use the following notation for a disk in the proof: for $r > 0$ and a positive integer J, $D^J(r)$ is the open ball of radius r centered at $\mathbf{0}$ in $\mathbb{R}^J$, $D^J(r) = \{\mathbf{x} \in \mathbb{R}^J : |\mathbf{x}| < r\}$.

The first step in the proof of the theorem is to prove that the set of diffeomorphisms all of whose fixed points are elementary is open and dense.

Lemma 3.2. *The set*

$$\mathcal{T}(\rho_1, \Delta) \equiv \{f \in \mathrm{Diff}^k(M) : \rho_1(f) \pitchfork \Delta\}$$

is open and dense in $\mathrm{Diff}^k(M)$.

Proof. The openness of $\mathcal{T}(\rho_1, \Delta)$ follows from Theorem 2.2 since ρ_1 is clearly continuous. (The openness of $\mathcal{T}(\rho_1, \Delta)$ is also related to Theorem V.6.4.)

We fix a $f \in \mathrm{Diff}^k(M)$ for the rest of the proof. To prove that f is in the closure of $\mathcal{T}(\rho_1, \Delta)$ in $\mathrm{Diff}^k(M)$, we apply the Parametric Transversality Theorem 2.3 with a finite-dimensional subspace of perturbations. We construct a map $\zeta : D^{Lm}(r) \to \mathrm{Diff}^k(M)$ and apply Theorem 2.3 to $\rho_1 \circ \zeta$. We have assumed that M is compact. Clearly, $\Delta \subset M \times M$ is a closed submanifold. The differentiability required to apply Theorem 2.3 is

$$k > \dim(M) - \dim(M \times M) + \dim(\Delta) = m - 2m + m = 0,$$

or $k \geq 1$. Finally, we must construct a large enough space of perturbations of the given f so that $(\rho_1 \circ \zeta)^{ev}$ is transverse to Δ.

Fix $f \in \mathrm{Diff}^k(M)$. To construct the perturbations, we use local coordinates (although it is possible to use more global constructions). Cover M by a finite number of open coordinate charts U_i, $\{\varphi_i : V_i \subset \mathbb{R}^m \to U_i \subset M\}_{i=1}^I$, with compact subsets $K_i \subset U_i$ which cover M, $\bigcup_{i=1}^I K_i = M$. Let $\hat{U}_i$ be open sets in M with $K_{i,1} \subset \hat{U}_i \subset \mathrm{cl}(\hat{U}_i) \subset U_i$. Let $K_{i,0} = \{\mathbf{x} \in K_i : f(\mathbf{x}) \notin \hat{U}_i\}$, and $K_{i,1} = \mathrm{cl}(K_i \backslash K_{i,0})$. Thus, each K_i is union of two compact subsets, $K_i = K_{i,0} \cup K_{i,1}$, such that $f(K_{i,0}) \cap K_i = \emptyset$ and $f(K_{i,1}) \subset \mathrm{cl}(\hat{U}_i) \subset U_i$. With this decomposition, there cannot possibly be any fixed points in any of the sets $K_{i,0}$: $\rho_1(f)(K_{i,0}) \cap \Delta = \emptyset$ so $\rho_1(f) \pitchfork_{K_{i,0}} \Delta$. On the other hand, for each i, we need to construct a finite-dimensional set of perturbations of f, $\zeta_i : D^m(r_i) \to \mathrm{Diff}^k(M)$, so that $(\rho_1 \circ \zeta_i)^{ev}$ is transverse to Δ along $\{\mathbf{0}\} \times K_{i,1} \subset \mathbb{R}^m \times M$. The transversality means that this finite-dimensional subset of diffeomorphisms, $\{\zeta_i(D^m(r_i))\}$, is large enough to perturb f so that it makes all the fixed points in $K_{i,1}$ elementary.

We proceed to construct the perturbations along $K_{i,1}$. Let U_i' be open sets in M with $K_{i,1} \subset U_i' \subset \text{cl}(U_i') \subset U_i$. Let $\beta_i : M \to \mathbb{R}$ be a C^∞ bump function with $\beta_i | U_i' \equiv 1$ and $\text{cl}(\{\mathbf{x} : \beta_i(\mathbf{x}) \neq 0\}) \subset U_i$. The set $\text{cl}(\{\mathbf{x} : \beta_i(\mathbf{x}) \neq 0\})$ is called the *support of* β and is denoted by $\text{supp}(\beta)$. For $r_i > 0$, define $\hat{\zeta}_i, \zeta_i : D^m(r_i) \to \text{Diff}^k(M)$ by

$$\hat{\zeta}_i(\mathbf{v})(\mathbf{x}) = \begin{cases} \mathbf{x} & \text{for } \mathbf{x} \notin U_i \\ \varphi_i(\varphi_i^{-1}(\mathbf{x}) + \beta_i(\mathbf{x})\mathbf{v}) & \text{for } \mathbf{x} \in U_i, \end{cases} \qquad \text{and}$$
$$\zeta_i(\mathbf{v})(\mathbf{x}) = \hat{\zeta}_i(\mathbf{v}) \circ f(\mathbf{x}).$$

Because the set of diffeomorphisms is open, for small enough $r_i > 0$, the image of $D^m(r_i)$ by ζ is in $\text{Diff}^k(M)$. Clearly, $(\rho_1 \circ \zeta_i)^{ev} : D^m(r_i) \times M \to M \times M$ is C^k for $k \geq 1$ as required. For $\mathbf{p} \in K_{i,1} \cap (\rho_1(f))^{-1}(\Delta)$, $\mathbf{p} = f(\mathbf{p}) \in K_{i,1} \subset U_i'$, $\beta_i \circ f(\mathbf{p}) = 1$, and

$$(\rho_1 \circ \zeta_i)^{ev}(\mathbf{v}, \mathbf{p}) = \big(\mathbf{p}, \varphi_i(\varphi_i^{-1}(\mathbf{p}) + \mathbf{v})\big).$$

Differentiating with respect to $\mathbf{v}$ for such a $\mathbf{p}$,

$$D(\rho_1 \circ \zeta_i)^{ev}_{(\mathbf{0},\mathbf{p})}(\mathbf{v}, \mathbf{0_p}) = \big(\mathbf{0_p}, D(\varphi_i)_{\varphi_i^{-1}(\mathbf{p})}\mathbf{v}\big) \qquad \text{and}$$
$$D(\rho_1 \circ \zeta_i)^{ev}_{(\mathbf{0},\mathbf{p})}(\mathbb{R}^m \times \{\mathbf{0_p}\}) = \{\mathbf{0_p}\} \times T_{\mathbf{p}}M.$$

As noted above,

$$T_{(\mathbf{p},\mathbf{p})}\Delta = \{(\mathbf{w}, \mathbf{w}) : \mathbf{w} \in T_{\mathbf{p}}M\}.$$

Together $\{\mathbf{0_p}\} \times T_{\mathbf{p}}M$ and $\{(\mathbf{w}, \mathbf{w}) : \mathbf{w} \in T_{\mathbf{p}}M\}$ span $T_{(\mathbf{p},\mathbf{p})}(M \times M) = T_{\mathbf{p}}M \times T_{\mathbf{p}}M$. Thus,

$$(\rho_1 \circ \zeta_i)^{ev} \pitchfork_{\{\mathbf{0}\} \times K_{i,1}} \Delta.$$

Before applying Theorem 2.3, we combine all the ζ_i into a single map to get transversality along all of M in one step. Let

$$P = D^m(r_1) \times \cdots \times D^m(r_I)$$

and $\zeta : P \to \text{Diff}^k(M)$ be given by

$$\zeta(\mathbf{v}_1, \ldots, \mathbf{v}_I) = \hat{\zeta}_1(\mathbf{v}_1) \circ \cdots \circ \hat{\zeta}_I(\mathbf{v}_I) \circ f.$$

Then,

$$D(\rho_1 \circ \zeta)^{ev}_{(\mathbf{0},\mathbf{p})}(\mathbf{0}, \ldots, \mathbf{0}, \mathbf{v}_i, \mathbf{0}, \ldots, \mathbf{0}, \ldots, \mathbf{0}, \mathbf{0_p}) = D(\rho_1 \circ \zeta_i)^{ev}_{(\mathbf{0},\mathbf{p})}(\mathbf{v}_i, \mathbf{0_p}),$$

and $(\rho_1 \circ \zeta)^{ev}$ is transverse to Δ along $\{\mathbf{0}\} \times \bigcup_{i=1}^I K_{i,1}$. As mentioned above, $\rho_1(f)$ is transverse to Δ along $\bigcup_{i=1}^I K_{i,0}$, so $(\rho_1 \circ \zeta)^{ev}$ is transverse to Δ along $\{\mathbf{0}\} \times M$. By openness of transverse intersection, there is some open neighborhood $P' \subset P$ of $\mathbf{0}$ such that $(\rho_1 \circ \zeta)^{ev}$ is transverse to Δ along $P' \times M$. By Theorem 2.3,

$$\mathcal{T}(\rho_1 \circ \zeta, \Delta) \equiv \{\mathbf{t} \in P' : \rho_1 \circ \zeta(\mathbf{t}) \pitchfork \Delta\}$$

is dense in P'. Because ζ is continuous, $f = \zeta(\mathbf{0})$ is in the closure of $\zeta(\mathcal{T}(\rho_1 \circ \zeta, \Delta))$ in $\text{Diff}^k(M)$. Since

$$\zeta(\mathcal{T}(\rho_1 \circ \zeta, \Delta)) \subset \mathcal{T}(\rho_1, \Delta),$$

f is in the closure of $\mathcal{T}(\rho_1, \Delta)$ in $\text{Diff}^k(M)$. This completes the proof of the lemma. □

PROOF THAT $\mathcal{H}_1$ IS OPEN AND DENSE. For $f \in \mathcal{T}(\rho_1, \Delta)$, Lemma 3.1 shows that if $\rho_1(f)(\mathbf{p}) \in \Delta$, then $\mathbf{p}$ is a elementary fixed point. Because the elementary fixed points are isolated, each $f \in \mathcal{T}(\rho_1, \Delta)$ has only finitely many fixed points. (The set $(\rho_1(f))^{-1}(\Delta)$ is a manifold of the same codimension as Δ, which is n, so it is a 0-dimensional manifold, i.e., isolated points.) By Lemma 1.3, each $f \in \mathcal{T}(\rho_1, \Delta)$ can be approximated by a diffeomorphism g for which each fixed point of f becomes a hyperbolic fixed point of g. In fact, there are an open neighborhood U of all the fixed points of f and a perturbation g of f such that $\text{Per}(1, g|U) = \text{Per}(1, f|U)$. Also, the nonexistence of fixed points on the compact set $M \setminus U$ is an open condition, i.e.,

$$\{g \in \text{Diff}^k(M) : \rho_1(g)(M \setminus U) \cap \Delta = \emptyset\}$$

is open in $\text{Diff}^k(M)$. Combining the consideration on and off U, the g which approximates f can be taken so that $\text{Per}(1, g) = \text{Per}(1, f)$ and each fixed point of g is hyperbolic, so $g \in \mathcal{H}_1$. This proves that $\mathcal{H}_1$ is dense in $\text{Diff}^k(M)$.

By the openness of $\mathcal{T}(\rho_1, \Delta)$, and the openness of the hyperbolicity of one particular fixed point, it follows that $\mathcal{H}_1$ is open. □

To prove that $\mathcal{H}_n$ is dense in $\text{Diff}^k(M)$, we cannot just take $\rho_n : \text{Diff}^k(M) \to C^k(M, M \times M)$ and proceed to construct perturbations for an arbitrary $f \in \text{Diff}^k(M)$. The difficulty is that if $\mathbf{p}$ is a fixed point for f for which -1 is an eigenvalue with eigenvector $\mathbf{v}^1$ for $Df_{\mathbf{p}}$, then for any perturbations of f, $\zeta : D^J(r) \to \text{Diff}^k(M)$,

$$D(\rho_2 \circ \zeta)_{(\mathbf{0},\mathbf{p})}(\mathbf{v}, \mathbf{0})$$

does not span the direction corresponding $(\mathbf{0}, \mathbf{v}^1)$. This difficulty is overlooked by many books giving the proof. The correct proof proceeds by induction on n and considers $\rho_n : \mathcal{H}_{n-1} \to C^k(M, M \times M)$, i.e., only constructs perturbations for $f \in \mathcal{H}_{n-1}$. The idea is that for $f \in \mathcal{H}_{n-1}$, if $\mathbf{p}$ has period less than n and $f^n(\mathbf{p}) = \mathbf{p}$ (so $\mathbf{p}$ has period n/j for some integer j), then $D(f^n)_{\mathbf{p}}$ is hyperbolic so we do not need to construct any perturbations.

Lemma 3.3. *Assume $\mathcal{H}_{n-1}$ is dense and open in* $\text{Diff}^k(M)$. *Then,*

$$\mathcal{T}(\rho_n, \Delta) \cap \mathcal{H}_{n-1} \equiv \{f \in \mathcal{H}_{n-1} : \rho_1(f) \pitchfork \Delta\}$$

is open and dense in $\text{Diff}^k(M)$.

PROOF. By the openness of transverse intersection, $\mathcal{T}(\rho_n, \Delta) \cap \mathcal{H}_{n-1}$ is open in $\mathcal{H}_{n-1}$, and so in $\text{Diff}^k(M)$.

To prove the density in $\text{Diff}^k(M)$, it is enough to prove density in $\mathcal{H}_{n-1}$, so we fix $f \in \mathcal{H}_{n-1}$. Let $\{\varphi_i : V_i \to U_i \subset M\}_{i=1}^I$ be open coordinate charts which cover M as before. Let U be an open neighborhood of $\text{Per}(n-1, f)$ in M and $\mathcal{N}$ be an open neighborhood of f in $\mathcal{H}_{n-1}$ such that for $g \in \mathcal{N}$, $\text{Per}(n, g) \cap \text{cl}(U) = \text{Per}(n-1, g) \subset U$ and all the periodic points in $\text{Per}(n-1, g)$ are hyperbolic. If $\mathbf{p} \in \text{cl}(U)$ and $\rho_n(f)(\mathbf{p}) \in \Delta$, then $\mathbf{p}$ has least period less than n, $\mathbf{p}$ is a hyperbolic periodic point, $\mathbf{p}$ is a hyperbolic fixed point of f^n, and $\rho_n(f) \pitchfork_{\mathbf{p}} \Delta$. Therefore, $\rho_n(f) \pitchfork_{\text{cl}(U)} \Delta$.

Let $K_i \subset U_i \setminus U$ be compact subsets such that $\bigcup_{i=1}^I K_i = M \setminus U$. (Note that $K_i = \emptyset$ is allowed.) Also, $K_i \cap \text{Per}(n-1, f) \subset K_i \cap U = \emptyset$. To proceed as in the case for $n = 1$, we need to divide K_i into subsets; however, for $n > 1$, we may need more than two subsets because the orbit of a point $\mathbf{x} \in K_i$ can pass through U_i for some intermediate iterate, and we must construct a perturbation g of f such that the orbit of a point in K_i goes only once through the set $\{\mathbf{x} : g(\mathbf{x}) \neq f(\mathbf{x})\}$. In particular, each K_i can be written as the union of a finite number of compact subsets, $K_i = \bigcup_{j=0}^{L_i} K_{i,j}$, such that (i) $f^n(K_{i,0}) \cap K_i = \emptyset$, (ii) $f^n(K_{i,j}) \subset U_i$ for $1 \leq j \leq L_i$, and (iii) $f^\ell(K_{i,j}) \cap K_{i,j} = \emptyset$ for $0 < \ell < n$ and $1 \leq j \leq L_i$. (Note that $L_i = 0$ is allowed.) For $1 \leq j \leq L_i$ and $1 \leq i \leq I$, let $U'_{i,j}$ and $U''_{i,j}$ be open sets of M such that (i) $K_{i,j} \subset U'_{i,j} \subset \mathrm{cl}(U'_{i,j}) \subset U''_{i,j} \subset \mathrm{cl}(U''_{i,j}) \subset U_i$, (ii) $f^n(U'_{i,j}) \subset U_i$, and (iii) $f^\ell(U'_{i,j}) \cap U''_{i,j} = \emptyset$ for $0 < \ell < n$. For $1 \leq j \leq L_i$ and $1 \leq i \leq I$, let $\beta_{i,j} : M \to \mathbb{R}$ be a C^∞ bump function with $\beta_{i,j}|U'_{i,j} \equiv 1$ and $\mathrm{supp}(\beta_{i,j}) \subset U''_{i,j}$. For $r_{i,j} > 0$, define $\hat{\zeta}_{i,j}, \zeta_{i,j} : D^m(r_{i,j}) \to \mathrm{Diff}^k(M)$ by

$$\hat{\zeta}_{i,j}(\mathbf{v})(\mathbf{x}) = \begin{cases} \mathbf{x} & \text{for } \mathbf{x} \notin U''_{i,j} \\ \varphi_i\big(\varphi_i^{-1}(\mathbf{x}) + \beta_{i,j}(\mathbf{x})\mathbf{v}\big) & \text{for } \mathbf{x} \in U''_{i,j}, \quad \text{and} \end{cases}$$

$$\zeta_{i,j}(\mathbf{v})(\mathbf{x}) = \hat{\zeta}_{i,j}(\mathbf{v}) \circ f(\mathbf{x}).$$

For $r_{i,j} > 0$ small enough, the image $\hat{\zeta}_{i,j}(D^m(r_{i,j}))$ is in $\mathrm{Diff}^k(M)$. Fix $\mathbf{p} \in K_{i,j}$ with $\rho_n(f)(\mathbf{p}) \in \Delta$. Let $f_\mathbf{v} = \zeta_{i,j}(\mathbf{v})$. For $1 \leq \ell < n$, $f^\ell_\mathbf{v}(\mathbf{p}) \notin U''_{i,j}$ so $f^\ell_\mathbf{v}(\mathbf{p}) = f^\ell(\mathbf{p})$. The nth iterate,

$$\begin{aligned} f^n_\mathbf{v}(\mathbf{p}) &= \varphi_i\big(\varphi_i^{-1} \circ f^n(\mathbf{p}) + \beta_{i,j} \circ f^n(\mathbf{p})\mathbf{v}\big) \\ &= \varphi_i\big(\varphi_i^{-1}(\mathbf{p}) + \mathbf{v}\big), \end{aligned}$$

or

$$(\rho_n \circ \zeta_{i,j})^{ev}(\mathbf{v}, \mathbf{p}) = \big(\mathbf{p}, \varphi_i\big(\varphi_i^{-1}(\mathbf{p}) + \mathbf{v}\big)\big).$$

As for the case $n = 1$,

$$(\rho_n \circ \zeta_{i,j})^{ev} \pitchfork_{\{\mathbf{0}\} \times K_{i,j}} \Delta.$$

Let $L = \sum_{i=1}^{I} L_i$, $P_i = D^m(r_{i,1}) \times \cdots \times D^m(r_{i,L_i})$, and $P = P_1 \times \cdots \times P_I \subset \mathcal{R}^{Lm}$. Define $\zeta : P \to \mathrm{Diff}^k(M)$ by

$$\zeta(\mathbf{v}_{1,1}, \cdots, \mathbf{v}_{I,L_I}) = \hat{\zeta}_{1,1}(\mathbf{v}_{1,1}) \circ \cdots \circ \hat{\zeta}_{I,L_I}(\mathbf{v}_{I,L_I}) \circ f.$$

As for the case $n = 1$, $(\rho_n \circ \zeta)^{ev}$ is transverse to Δ along $\{\mathbf{0}\} \times \bigcup_{i=1}^{I} \bigcup_{j=1}^{L_i} K_{i,j}$. Also, $\rho_n(f)$ is transverse to Δ along $\mathrm{cl}(U) \cup \bigcup_{i=1}^{I} K_{i,0}$. Because

$$M = \mathrm{cl}(U) \cup \bigcup_{i=1}^{I} \big(K_{i,0} \cup \bigcup_{j=1}^{L_i} K_{i,j}\big),$$

$(\rho_n \circ \zeta)^{ev}$ is transverse to Δ along $\{\mathbf{0}\} \times M$. By the openness of transverse intersection, there is an open neighborhood $P' \subset P$ of $\mathbf{0}$ in $\mathbb{R}^{Lm}$ such that $(\rho_n \circ \zeta)^{ev}$ is transverse to Δ along $P' \times M$. By Theorem 2.3,

$$\mathcal{T}(\rho_n \circ \zeta, \Delta) = \{\mathbf{t} \in P' : \rho_n \circ \zeta(\mathbf{t}) \pitchfork \Delta\}$$

is dense in P'. Because ζ is continuous, f is in the closure of $\zeta(\mathcal{T}(\rho_n \circ \zeta, \Delta))$ in $\mathrm{Diff}^k(M)$. Since

$$\begin{aligned} \zeta(\mathcal{T}(\rho_n \circ \zeta, \Delta)) &\subset \mathcal{T}(\rho_n, \Delta) \cap \mathcal{H}_{n-1} \\ &\equiv \{g \in \mathcal{H}_{n-1} : \rho_n(g) \pitchfork \Delta\}, \end{aligned}$$

f is in the closure of $\mathcal{T}(\rho_n, \Delta) \cap \mathcal{H}_{n-1}$ in $\text{Diff}^k(M)$. This completes the proof of the lemma. $\square$

PROOF THAT $\mathcal{H}_n$ IS OPEN AND DENSE. For $f \in \mathcal{T}(\rho_n, \Delta)$, each point of least period n can be made hyperbolic using Lemma 1.3, so $\mathcal{H}_n$ is dense in $\mathcal{T}(\rho_n, \Delta)$. The set of diffeomorphisms for which a particular periodic point is hyperbolic is open, so $\mathcal{H}_n$ is open in $\text{Diff}^k(M)$. $\square$

$\mathcal{H}$ IS A RESIDUAL SUBSET. Since $\mathcal{H} = \bigcap_{n=1}^{\infty} \mathcal{H}_n$ is the countable intersection of open dense subsets of $\text{Diff}^k(M)$, it is residual. $\square$

KS(M) IS A RESIDUAL SUBSET. We define a countable collection of sets of diffeomorphisms $\mathcal{K}(n, R_i)$ such that (i) the intersection of all the $\mathcal{K}(n, R_i)$ is equal to KS(M) and (ii) we can prove each set $\mathcal{K}(n, R_i)$ is open and dense in $\text{Diff}^k(M)$. For a hyperbolic fixed point $\mathbf{p} \in \text{Per}_h(f)$, let $W^s_R(\mathbf{p}, f)$ be the points $\mathbf{q}$ in $W^s(\mathbf{p}, f)$ for which there is a curve $\{\gamma(t) : 0 \leq t \leq 1\}$ such that (i) the length of γ is less than or equal to R, (ii) $\gamma(0) = \mathbf{q}$ and $\gamma(1) = \mathbf{p}$, and (iii) $\gamma(t) \in W^s(\mathbf{p}, f)$ for $0 \leq t \leq 1$. Similarly, define $W^u_R(\mathbf{p}, f)$. We say that $W^u_R(\mathbf{p}_1, f)$ is transverse to $W^s_R(\mathbf{p}_2, f)$ as a shorter way of saying that $W^u(\mathbf{p}_1, f)$ is transverse to $W^s(\mathbf{p}_2, f)$ at points of $W^u_R(\mathbf{p}_1, f) \cap W^s_R(\mathbf{p}_2, f)$. For a positive integer n and $R > 0$, let

$$\mathcal{K}(n, R) = \{f \in \mathcal{H}_n : W^u_R(\mathbf{p}_1, f) \text{ is transverse to } W^s_R(\mathbf{p}_2, f) \text{ for all } \mathbf{p}_1, \mathbf{p}_2 \in \text{Per}(n, f)\}.$$

Then,

$$\text{KS}(M) = \bigcap_{1 \leq n < \infty} \bigcap_{1 \leq R < \infty} \mathcal{K}(n, R).$$

If each $\mathcal{K}(n, R)$ is dense and open in $\mathcal{H}_n$, then KS(M) is residual in $\text{Diff}^k(M)$. (It is possible to take only a countable number of $R > 0$ which go to infinity.)

The fact that $\mathcal{K}(n, R)$ is open is not difficult. For $f \in \mathcal{H}_n$, there are only finitely many points in $\text{Per}(n, f)$. By the arguments which prove the openness of $\mathcal{H}_n$, there is a neighborhood $\mathcal{N}$ of f such that for $g \in \mathcal{N}$, (i) the cardinality of $\text{Per}(n, g)$ is the same as the cardinality of $\text{Per}(n, f)$ and (ii) each periodic point of g in $\text{Per}(n, g)$ is hyperbolic, so $\mathcal{N} \subset \mathcal{H}_n$. For $g \in \mathcal{N}$, let $\{\mathbf{p}_i(g)\}_{i=1}^{I}$ be the periodic points in $\text{Per}(n, g)$. The stable and unstable maps depend continuously in the C^k topology on compact subsets. For each $1 \leq i \leq I$ and $g \in \mathcal{N}$, there are maps $\sigma^s_i(g) : \mathbb{R}^{s_i} \to M$ such that (i) the image of $\sigma^s_i(g)$ is $W^s(\mathbf{p}_i(g), g)$ and (ii) the image of the close ball $\bar{D}^{s_i}(R)$ is $W^s_R(\mathbf{p}_i(g), g)$, $\sigma^s_i(g)(\bar{D}^{s_i}(R)) = W^s_R(\mathbf{p}_i(g), g)$. Similarly, $\sigma^u_i(g) : \mathbb{R}^{u_i} \to M$ gives $W^u(\mathbf{p}_i(g), g)$. The continuity of the stable and unstable manifolds can be expressed by saying that the maps $\sigma^s_i : \mathcal{N} \to C^k(\mathbb{R}^{s_i}, M)$ and $\sigma^u_i : \mathcal{N} \to C^k(\mathbb{R}^{u_i}, M)$ are continuous if $C^k(\mathbb{R}^{s_i}, M)$ and $C^k(\mathbb{R}^{u_i}, M)$ are given the compact open topology on the first k derivatives. For i and j fixed, we can combine these into a map $\sigma_{i,j} : \mathcal{N} \to C^k(\mathbb{R}^{s_i} \times \mathbb{R}^{u_j}, M \times M)$ by

$$\sigma_{i,j}(g)(\mathbf{x}, \mathbf{y}) = (\sigma^s_i(g)(\mathbf{x}), \sigma^u_j(g)(\mathbf{y})).$$

The reader can check that $W^s_R(\mathbf{p}_i(g), g)$ is transverse to $W^u_R(\mathbf{p}_j(g), g)$ if and only if $\sigma_{i,j}(g)$ is transverse to Δ along $\bar{D}^{s_i}(R) \times \bar{D}^{u_j}(R)$. The openness of the transverse intersection of $W^s_R(\mathbf{p}_i(g), g)$ and $W^u_R(\mathbf{p}_j(g), g)$ follows from Theorem 2.4. By using all pairs $1 \leq i, j \leq I$, it follows that $\mathcal{K}(n, R)$ is open in $\mathcal{N}$, and so in $\mathcal{H}_n$ and $\text{Diff}^k(M)$.

To complete the proof, we only need to prove that $\mathcal{K}(n, R)$ is dense at functions $f \in \mathcal{H}_n$. We prove this density by applying Theorem 2.3 to $\sigma_{i,j}$. If we can show for each pair (i, j), that the set of g for which $\sigma_{i,j}(g)$ is transverse to Δ is dense in $\mathcal{N}$, then it clearly follows that $\mathcal{K}(n, R)$ is dense in $\mathcal{N}$. We show that we can make $W^s_R(\mathbf{p}_i(g), g)$ transverse to $W^u_R(\mathbf{p}_j(g), g)$ at a single point $\mathbf{q} \in W^s_R(\mathbf{p}_i(g), g) \cap W^u_R(\mathbf{p}_j(g), g)$. The transversality at all points can then be obtained by combining these perturbations in a way similar to that used to prove that $\mathcal{H}_1$ and $\mathcal{H}_n$ are dense. (This argument uses the fact that $\bar{D}^{s_i}(R) \times \bar{D}^{u_j}(R)$ is compact.)

Assume

$$\sigma_{i,j}(g)(\mathbf{x}_0, \mathbf{y}_0) = (\mathbf{q}, \mathbf{q}) \in \Delta.$$

Since the stable and unstable manifolds are transverse at the periodic points, $\mathbf{q} \neq \mathbf{p}_i, \mathbf{p}_j$. Let $\varphi : V \to U$ be a coordinate chart at $\mathbf{q}$ with $\mathbf{p}_i, \mathbf{p}_j \notin U$. Then, $\omega(\mathbf{q}) = \mathcal{O}(\mathbf{p}_i)$ and $\alpha(\mathbf{q}) = \mathcal{O}(\mathbf{p}_j)$; so for a small enough neighborhood $U'' \subset U$ of $\mathbf{q}$, $g^n(\mathbf{q}) \notin U''$ for $n \neq 0$. Let $U' \subset U''$ be a smaller neighborhood of $\mathbf{q}$, and $\beta : M \to \mathbb{R}$ be a bump function with $\beta|U' \equiv 1$ and $\text{supp}(\beta) \subset U''$. Finally, define $\hat{\zeta}_{i,j}, \zeta_{i,j} : D^m(r_{i,j}) \to \text{Diff}^k(M)$ by

$$\hat{\zeta}_{i,j}(\mathbf{v})(\mathbf{x}) = \begin{cases} \mathbf{x} & \text{for } \mathbf{x} \notin U'' \\ \varphi(\varphi^{-1}(\mathbf{x}) + \beta(\mathbf{x})\mathbf{v}) & \text{for } \mathbf{x} \in U'', \end{cases}$$
$$\zeta_{i,j}(\mathbf{v})(\mathbf{x}) = \hat{\zeta}_{i,j}(\mathbf{v}) \circ g(\mathbf{x}).$$

The two periodic points $\mathbf{p}_i, \mathbf{p}_j \notin U''$, so they remain hyperbolic periodic points. Also, $g(\mathbf{q}) \notin U''$, so $\zeta_{i,j}(\mathbf{v})(\mathbf{q}) = g(\mathbf{q})$. Similarly, $g^n(\mathbf{q}) \notin U''$ for $n > 0$, so $\zeta_{i,j}(\mathbf{v})^n(\mathbf{q}) = g^n(\mathbf{q})$, $\mathbf{q} \in W^s(\mathbf{p}_i(\zeta_{i,j}(\mathbf{v})), \zeta_{i,j}(\mathbf{v}))$, and

$$\sigma^s_i \circ \zeta_{i,j}(\mathbf{v})(\mathbf{x}_0) = \mathbf{q}$$

is still the same point in M. For the unstable manifold, a similar argument shows that $g^{-1}(\mathbf{q}) \in W^u(\mathbf{p}_j(\zeta_{i,j}(\mathbf{v})), \zeta_{i,j}(\mathbf{v}))$. However,

$$\begin{aligned} \zeta_{i,j}(\mathbf{v})(g^{-1}(\mathbf{q})) &= \hat{\zeta}_{i,j}(\mathbf{v})(\mathbf{q}) \\ &= \varphi(\varphi^{-1}(\mathbf{q}) + \mathbf{v}), \end{aligned}$$

so

$$\sigma^u_j \circ \zeta_{i,j}(\mathbf{v})(\mathbf{y}_0) = \varphi(\varphi^{-1}(\mathbf{q}) + \mathbf{v}).$$

Therefore, the perturbation $\zeta_{i,j}(\mathbf{v})$ of g changes the unstable manifold at $\mathbf{q}$ but not the stable manifold, i.e., the perturbation can move $W^u(\mathbf{p}_j, \zeta_{i,j}(\mathbf{v}))$ off $W^s(\mathbf{p}_i, \zeta_{i,j}(\mathbf{v}))$. Taking the derivative with respect to $\mathbf{v}$,

$$D(\sigma_{i,j} \circ \zeta_{i,j})^{ev}_{(\mathbf{0},\mathbf{x}_0,\mathbf{y}_0)}(\mathbb{R}^m \times \{\mathbf{0},\} \times \{\mathbf{0}\}) = \{\mathbf{0}_\mathbf{q}\} \times T_\mathbf{q}M.$$

Because this image and $T_{(\mathbf{q},\mathbf{q})}\Delta$ span $T_\mathbf{q}M \times T_\mathbf{q}M$, $(\sigma_{i,j} \circ \zeta_{i,j})^{ev}$ is transverse to Δ at $(\mathbf{0}, \mathbf{x}_0, \mathbf{y}_0)$. A finite number of the sets of the type of U' cover $W^s_R(\mathbf{p}_i, g) \cap W^u_R(\mathbf{p}_j, g)$, so perturbations in these various U' can be combined to define a $\zeta_{i,j} : D^J(r) \to \text{Diff}^k(M)$ such that $(\sigma \circ \zeta_{i,j})^{ev}$ is transverse to Δ along $\{\mathbf{0}\} \times \bar{D}^{s_i}(R) \times \bar{D}^{u_j}(R)$. By Theorem 2.3, g is in the closure of $\mathcal{K}(n, R)$. This completes the proof of Theorem 1.1. □

11.4 Necessary Conditions for Structural Stability

Chapter X gives sufficient conditions for $\mathcal{R}$-stability and structural stability. In this section, we consider the necessity of some of these conditions. After several people made contributions to this question, Mañé (1978b, 1987b) and Liao (1980) eventually proved that if f is a C^1 $\mathcal{R}$-stable diffeomorphism on a compact manifold, then f has a hyperbolic structure on $\mathcal{R}(f)$. The proof of this result is very difficult and beyond the scope of this book, but we discuss some of the easier aspects. We start with the hyperbolicity of fixed points.

Theorem 4.1. *Let M be a compact manifold and $f : M \to M$ a diffeomorphism. Assume f is C^1 $\mathcal{R}$-stable. Then, all the periodic points of f are hyperbolic.*

REMARK 4.1. This theorem was proved by Franks (1971) and Pliss (1971, 1972). The papers by Pliss also obtained some further results related to the theorem of Mañé and Liao.

REMARK 4.2. Note that if f is structurally stable, then it satisfies the assumptions of the theorem.

REMARK 4.3. We prove this theorem assuming that f is C^1 $\mathcal{R}$-stable, but it is true if f is C^r structurally stable for any $r \geq 1$. See Robinson (1973).

The first step in the proof of Theorem 4.1 is to show that a diffeomorphism can be C^1 approximated by a new diffeomorphism which is equal to its linear part in a neighborhood of a periodic point.

Lemma 4.2. *Assume $\mathbf{p}$ is a periodic point for f of period n. Let $\varphi : V \to U$ be a coordinate chart at $\mathbf{p}$ with $\varphi(\mathbf{0}) = \mathbf{p}$ and $A = D(\varphi^{-1} \circ f^n \circ \varphi)_{\mathbf{0}}$. Let $\mathcal{N}$ be a neighborhood of f in the C^1 topology.*

(a) Then, there are $r > 0$ and $g \in \mathcal{N}$ such that (i) $\mathbf{p}$ has period n for g and (ii) $\varphi^{-1} \circ g^n \circ \varphi(\mathbf{x}) = A\mathbf{x}$ for $\mathbf{x} \in V \cap \{\mathbf{x} \in \mathbb{R}^n : |\mathbf{x}| \leq r\}$.

(b) If $\|B - A\|$ is small enough, then there are $r > 0$ and $g \in \mathcal{N}$ such that (i) $\mathbf{p}$ has period n for g and (ii) $\varphi^{-1} \circ g^n \circ \varphi(\mathbf{x}) = B\mathbf{x}$ for $\mathbf{x} \in V \cap \{\mathbf{x} \in \mathbb{R}^n : |\mathbf{x}| \leq r\}$.

REMARK 4.4. Lemma 4.2 is not true in the C^2 topology.

PROOF. (a) We only consider the case of a fixed point. The reader can supply the changes for higher period. We also leave the proof of part (b) to the reader.

Let $\beta : \mathbb{R}^n \to \mathbb{R}$ be a C^∞ bump function with (i) $0 \leq \beta(\mathbf{x}) \leq 1$ for all $\mathbf{x}$, (ii) $\beta(\mathbf{x}) = 1$ for $|\mathbf{x}| \leq 1$, and (iii) $\beta(\mathbf{x}) = 0$ for $|\mathbf{x}| \geq 2$. For $r > 0$, let $\beta_r : \mathbb{R}^n \to \mathbb{R}$ be defined by $\beta_r(\mathbf{x}) = \beta(\mathbf{x}/r)$. Thus, β_r is a bump function which is equal to 1 for $|\mathbf{x}| \leq r$ and is equal to 0 for $|\mathbf{x}| \geq 2r$. The estimates on β_r and its derivative depend on r in the following fashion: $\sup\{\beta_r(\mathbf{x})\} = \sup\{\beta(\mathbf{x})\} = 1$, and

$$\sup\{\|D(\beta_r)_{\mathbf{x}}\|\} = \frac{C_0}{r}$$

where $C_0 = \sup\{\|D(\beta)_{\mathbf{x}}\|\}$.

Using the bump function β_r, we can define a perturbation g_r. First we write f in terms of its linear and nonlinear terms: let

$$\varphi^{-1} \circ f \circ \varphi(\mathbf{x}) = A\mathbf{x} + \hat{f}(\mathbf{x})$$

where $\hat{f}(\mathbf{0}) = \mathbf{0}$ and $D(\hat{f})_{\mathbf{0}} = \mathbf{0}$. Thus, $\|D(\hat{f})_{\mathbf{x}}\| = o(|\mathbf{x}|^0)$ and $|\hat{f}(\mathbf{x})| = o(|\mathbf{x}|)$, i.e., $\|D(\hat{f})_{\mathbf{x}}\|$ and $|\hat{f}(\mathbf{x})|/|\mathbf{x}|$ both go to zero as $|\mathbf{x}|$ goes to zero. We only consider $r > 0$ for which $\operatorname{supp}(\beta_r) \subset \{\mathbf{x} : |\mathbf{x}| \leq 2r\} \subset V$. Let g_r be equal to f outside of U, and

$$\begin{aligned}\varphi^{-1} \circ g_r \circ \varphi(\mathbf{x}) &= \beta_r(\mathbf{x})A\mathbf{x} + (1 - \beta_r(\mathbf{x}))\varphi^{-1} \circ f \circ \varphi(\mathbf{x}) \\ &= A\mathbf{x} + (1 - \beta_r(\mathbf{x}))\hat{f}(\mathbf{x}).\end{aligned}$$

for $\mathbf{x} \in V$. Note that $\varphi^{-1} \circ g_r \circ \varphi(\mathbf{x}) = \varphi^{-1} \circ f \circ \varphi(\mathbf{x})$ for $|\mathbf{x}| \geq 2r$. On the other hand, for $|\mathbf{x}| \leq r$, $\beta_r(\mathbf{x}) = 1$ and $\varphi^{-1} \circ g_r \circ \varphi(\mathbf{x}) = A\mathbf{x}$.

To check that g_r is near f for small r, we need to calculate the derivative of g_r:

$$\begin{aligned} D(\varphi^{-1} \circ g_r \circ \varphi)_{\mathbf{x}} &= A + (1 - \beta_r(\mathbf{x}))D\hat{f}_{\mathbf{x}} - \hat{f}(\mathbf{x})D(\beta_r)_{\mathbf{x}} \\ &= D(\varphi^{-1} \circ f \circ \varphi)_{\mathbf{x}} - \beta_r(\mathbf{x})D\hat{f}_{\mathbf{x}} - \hat{f}(\mathbf{x})D(\beta_r)_{\mathbf{x}}. \end{aligned}$$

For $|\mathbf{x}| \geq 2r$, $D(\varphi^{-1} \circ g_r \circ \varphi)_{\mathbf{x}} = D(\varphi^{-1} \circ f \circ \varphi)_{\mathbf{x}}$, so we only need to consider $\mathbf{x}$ with $|\mathbf{x}| \leq 2r$. For $|\mathbf{x}| \leq 2r$,

$$\begin{aligned} \|D(\varphi^{-1} \circ g_r \circ \varphi)_{\mathbf{x}} - &D(\varphi^{-1} \circ f \circ \varphi)_{\mathbf{x}}\| \\ &\leq \beta_r(\mathbf{x})\|D(\hat{f})_{\mathbf{x}}\| + |\hat{f}(\mathbf{x})| \cdot \|D(\beta_r)_{\mathbf{x}}\| \\ &= o(r^0) + o(r)\left(\frac{C_0}{r}\right) \\ &= o(r^0). \end{aligned}$$

In this last calculation, we used the estimates given above for $\hat{f}(\mathbf{x})$ and $D\hat{f}_{\mathbf{x}}$ and that $\sup\{\|D(\beta_r)_{\mathbf{x}}\|\} = C_0/r$. From the estimate, the derivative of g_r approaches that of f as r goes to 0. Since $g_r(\mathbf{p}) = \mathbf{p} = f(\mathbf{p})$, the Mean Value Theorem proves that the C^0 distance from g_r to f goes to zero also. Therefore, for $r > 0$ small enough, $g_r \in \mathcal{N}$, the C^1 neighborhood of f. □

PROOF OF THEOREM 4.1. Since f is $\mathcal{R}$-stable, there is an open neighborhood $\mathcal{N}$ of f such that any $g \in \mathcal{N}$ is $\mathcal{R}$-conjugate to f and so g is also $\mathcal{R}$-stable itself. Assume that $\mathbf{p}$ is a nonhyperbolic periodic point of period n. By Lemma 4.2(a), there is a $g_1 \in \mathcal{N}$ such that in local coordinates g_1^n is linear, $\varphi^{-1} \circ g_1^n \circ \varphi(\mathbf{x}) = A\mathbf{x}$ for $|\mathbf{x}| < r$. Since A is nonhyperbolic, it can be approximated by another matrix B that has an eigenvalue λ which is a jth root of unity with eigenvector $\mathbf{v}$. By Lemma 4.2(b), g_1 can be approximated by $g_2 \in \mathcal{N}$ such that in local coordinates $\varphi^{-1} \circ g_2^n \circ \varphi(\mathbf{x}) = B\mathbf{x}$ for $|\mathbf{x}| < r$. Then,

$$\varphi^{-1} \circ g_2^{jn} \circ \varphi(s\mathbf{v}) = B^j s\mathbf{v} = \lambda^j s\mathbf{v} = s\mathbf{v},$$

for $|s\mathbf{v}| < r$, so g_2^{jn} has a curve of fixed points which are nonisolated, i.e., g_2 has nonisolated periodic points of a given period. By the Kupka-Smale Theorem, there is a $g_3 \in \mathcal{N}$ with only hyperbolic periodic points, so the periodic points of any given period are isolated. But the two diffeomorphisms g_3 and g_2 cannot be $\mathcal{R}$-conjugate. This contradicts the assumptions on f, and so all the periodic points of f must be hyperbolic. □

We end the section by stating a result about the transversality of stable and unstable manifolds. We give the result in two dimensions because part (a) is easier to state in this case. Parts (b) and (c) are true in any dimension. We leave the proof of this theorem to the exercises. See Exercise 11.11.

Theorem 4.3. *Assume M is a compact surface (two-dimensional manifold) and $f, g : M \to M$ are C^1 diffeomorphisms.*

(a) Assume f is a Kupka-Smale diffeomorphism. Assume g has two periodic points $\mathbf{p}$ and $\mathbf{q}$ such that $W^u(\mathbf{p}, g)$ is tangent to $W^s(\mathbf{q}, g)$ at $\mathbf{z}$ and $W^u(\mathbf{p}, g)$ is on one side of $W^s(\mathbf{q}, g)$ locally near $\mathbf{z}$. (The two manifolds are not topologically transverse at $\mathbf{z}$.) Then, f and g are not conjugate.

(b) If f is structurally stable, then f is Kupka-Smale.

(c) Assume f is structurally stable and that $\mathcal{R}(f)$ has a hyperbolic structure. Then, f satisfies the transversality condition with respect to $\mathcal{R}(f)$.

REMARK 4.5. As stated in the opening paragraph of this section, Mañé (1978b, 1987b) and Liao (1980) proved a much stronger theorem. If f is a C^1 $\mathcal{R}$-stable diffeomorphism on a compact manifold, then f has a hyperbolic structure on $\mathcal{R}(f)$. Hayashi (1992, 1997) has some improvements of the proof. Also see Aoki (1991, 1992) and Wen (1996).

Assume f is C^1 structurally stable. Then, by Theorem 4.3 (or Exercise 11.11(c)), f also satisfies the transversality condition with respect to $\mathcal{R}(f)$.

The proof of the result of Mañé and Liao for diffeomorphisms implies the same result for flows without fixed points. Hu (1994) proved a comparable theorem for flows on three-dimensional compact manifolds even when fixed points are allowed. Finally Hayashi (1997) proved the result for flows with fixed points on manifolds of dimension greater than 3.

11.5 Nondensity of Structural Stability

As we have remarked elsewhere, the set of structurally stable diffeomorphisms (or Ω-stable diffeomorphisms) are not dense in $\text{Diff}^1(M)$ (unless $M = S^1$). There was a sequence of papers dealing with this problem, including Smale (1966), Abraham and Smale (1970), Williams (1970a), Newhouse (1970a), and Simon (1972). Later examples contradicted further conjectures of genericity of various forms of stability. Below, we describe the the example in Williams (1970a). This example is simple once the the DA-diffeomorphism is understood. See Section 8.7 for the description of the DA-diffeomorphism.

Theorem 5.1. *There is an open set of diffeomorphisms $\mathcal{N} \subset \text{Diff}^1(\mathbb{T}^2)$ such that no $g \in \mathcal{N}$ is structurally stable.*

PROOF. We start with the DA-diffeomorphism f_1 on $\mathbb{T}^2$ with a fixed point source $\mathbf{p}_0$ and a hyperbolic invariant set Λ. The DA-diffeomorphism f_1 can be constructed from a hyperbolic toral automorphism g with $f_1|(\mathbb{T}^2 \setminus U) = g|(\mathbb{T}^2 \setminus U)$, where $U \subset \mathbb{T}^2$ is an open set. As we saw in Section 8.7,

$$\Lambda = \bigcap_{n=0}^{\infty} f_1^n(\mathbb{T}^2 \setminus U).$$

We now modify f_1 to form a f_2 with $f_1|(\mathbb{T}^2 \setminus U) = f_2|(\mathbb{T}^2 \setminus U)$, so f_2 still has Λ as a hyperbolic invariant set. Inside U, we replace the single fixed point source with two fixed point sources $\mathbf{q}_1$ and $\mathbf{q}_2$ and a fixed point saddle $\mathbf{p}_0$. The construction can be made so that

$$W^u(\mathbf{p}_0, f_2) \subset W^s(\mathbf{p}_1, f_2) \cup W^s(\mathbf{p}_2, f_2).$$

See Figure 5.1.

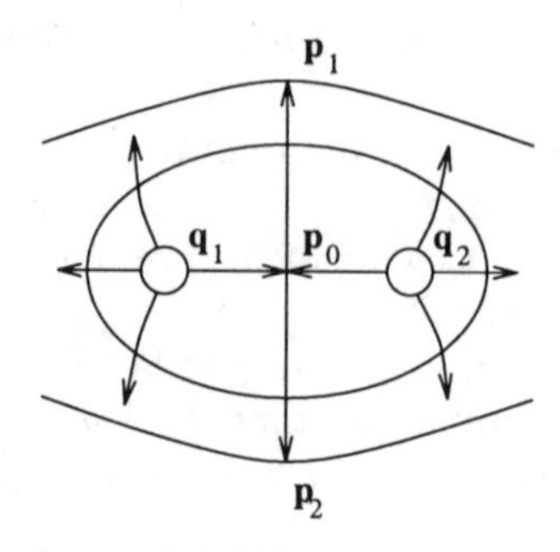

FIGURE 5.1

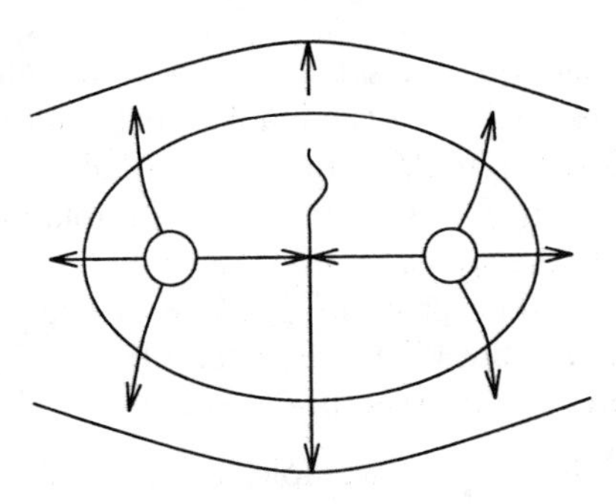

FIGURE 5.2

Finally, f_2 can be modified a third time to obtain a diffeomorphism f_3. Let

$$[\mathbf{a}, \mathbf{b}] \subset W^u(\mathbf{p}_0, f_2) \cap W^s(\mathbf{p}_1, f_2)$$

be a fundamental domain with $f_2(\mathbf{a}) = \mathbf{b}$, and let $\mathbf{c} = f_2(\mathbf{b})$. Let $\beta(\mathbf{x})$ be a bump function such that (i) it is zero away from $[\mathbf{b}, \mathbf{c}]$ and on the part of $W^u(\mathbf{p}_0, f_2)$ from $\mathbf{p}_0$ to $\mathbf{b}$, (ii) it equals one in a neighborhood of the midpoint of $[\mathbf{b}, \mathbf{c}]$, and (iii) the forward orbit by f_2 of $\text{supp}(\beta)$ does not intersect $\text{supp}(\beta)$,

$$\text{supp}(\beta) \cap \bigcup_{n=1}^{\infty} f_2^n(\text{supp}(\beta)) = \emptyset.$$

Let $\mathbf{v}$ be a vector transverse to $W^u(\mathbf{p}_0, f_2)$. For small $\epsilon > 0$, let

$$f_3(\mathbf{x}) = \begin{cases} f_2(\mathbf{x}) & \text{for } \beta \circ f_2(\mathbf{x}) = 0 \\ f_2(\mathbf{x}) + \epsilon \beta \circ f_2(\mathbf{x}) \mathbf{v} & \text{for } \beta \circ f_2(\mathbf{x}) \neq 0. \end{cases}$$

(Here we write the sum as if it makes sense on $\mathbb{T}^2$. This should really be written in the coordinates of the covering space $\mathbb{R}^2$.) Because of the support of β, the part of the unstable manifold $W^u(\mathbf{p}_0, f_2)$ from $\mathbf{p}_0$ to $\mathbf{b}$ remains part of $W^u(\mathbf{p}_0, f_3)$; in particular, $[\mathbf{a}, \mathbf{b}] \subset W^u(\mathbf{p}_0, f_3)$, so also $f_3([\mathbf{a}, \mathbf{b}]) \subset W^u(\mathbf{p}_0, f_3)$. However, $f_3([\mathbf{a}, \mathbf{b}])$ is no longer equal to the line segment $[\mathbf{b}, \mathbf{c}]$, but it bends outward from the stable manifold $W^s(\mathbf{p}_1, f_2)$. See Figure 5.2. Because the perturbation from f_2 to f_3 does not affect the forward orbits of points in $\text{supp}(\beta)$, these points stay on the stable manifolds of the same points in Λ, and f_3 has a tangency between $W^u(\mathbf{p}_0, f_3)$ and $W^s(\mathbf{z}, f_3)$ for some $\mathbf{z}(f_3) \in \Lambda$. For a small enough C^1 neighborhood $\mathcal{N}$ of f_3, every $f \in \mathcal{N}$ has (i) a hyperbolic invariant set $\Lambda(f)$, (ii) a saddle fixed point $\mathbf{p}_0(f)$, (iii) two fixed point sources $\mathbf{q}_1(f)$ and $\mathbf{q}_2(f)$, (iv) $\mathcal{R}(f) = \Lambda(f) \cup \{\mathbf{p}_0(f), \mathbf{q}_1(f), \mathbf{q}_2(f)\}$, and (v) a tangency between $W^u(\mathbf{p}_0(f), f)$ and $W^s(\mathbf{z}(f), f)$ for some $\mathbf{z}(f) \in \Lambda(f)$. In the C^1 topology (as opposed to the C^2 topology), f may have more than one tangency. However, if we use the extremal tangency, then the stable manifold $W^s(\mathbf{z}(f), f)$ is well defined. (Note that the point $\mathbf{z}(f)$ is not itself well defined.) This neighborhood $\mathcal{N}$ is the one indicated in the statement of the theorem.

By means of a bump function near the point of tangency, the stable manifold on which the tangency occurs, $W^s(\mathbf{z}(f), f)$, can be moved within $W^s(\Lambda, f)$. The periodic points of f in $\Lambda(f)$ are dense, so the union of stable manifolds of periodic points,

$$\bigcup_{\mathbf{x} \in \text{Per}(f)} W^s(\mathbf{x}, f),$$

is dense in $W^s(\Lambda(f), f)$. Therefore, the set $\mathcal{N}_1$ of $f \in \mathcal{N}$ for which $W^s(\mathbf{z}(f), f)$ contains a periodic point in Λ is dense in $\mathcal{N}$. By the Kupka-Smale Theorem, the set $\mathcal{N}_2 = \{f' \in \mathcal{N} : f \text{ is Kupka-Smale }\}$ is dense in $\mathcal{N}$. For $f \in \mathcal{N}_1$ and $f' \in \mathcal{N}_2$, f is not conjugate to f' by Theorem 4.3, since (i) f has a tangency of stable and unstable manifolds of periodic points where locally the stable manifold is on one side of the unstable manifold and (ii) f' is Kupka-Smale. (The proof of Theorem 4.3 is an exercise. See Exercise 11.11.) Since (i) $\mathcal{N}_1$ and $\mathcal{N}_2$ are both dense in $\mathcal{N}$ and (ii) no $f \in \mathcal{N}_1$ is conjugate to any $f' \in \mathcal{N}_2$, no diffeomorphism in $\mathcal{N}$ is structurally stable. □

REMARK 5.1. The idea of the counter-examples to Ω-stability is to make the tangency between the stable and unstable manifolds take place at a point which is nonwandering. Abraham and Smale (1970) do this for an example on $S^2 \times \mathbb{T}^2$. Simon (1972) does this for a three-dimensional manifold. Newhouse (1970a) does this on a two-dimensional manifold but needs to use the C^2 topology.

11.6 Exercises

Transversality

11.1. Let M and N be compact submanifolds of $\mathbb{R}^k$ with $\dim(M) + \dim(N) < k$. Prove that $\{\mathbf{a} \in \mathbb{R}^k : (M + \mathbf{a}) \cap N = \emptyset\}$ is dense and open.

11.2. Let M be a submanifold of $\mathbb{R}^n$. Fix $k < n$ and let $\mathcal{B}$ be the set of all $n \times k$ matrices A such that A has rank k. Note that for $A \in \mathcal{B}$, $A(\mathbb{R}^k)$ is a k-plane in $\mathbb{R}^n$. The set $\mathcal{B}$ is an open subset of all $n \times k$ matrices, so $\mathcal{B}$ is a manifold. (You do not need to prove this fact.) Show that for a generic set of $A \in \mathcal{B}$, the k-plane $A(\mathbb{R}^k)$ is transverse to M.

Kupka-Smale Theorem

11.3. Let $\mathcal{F}$ be a topological space, M a compact metric space, and $\mathcal{C}$ be the collection of all compact subsets of M with the Hausdorff metric. Assume $\Gamma_n : \mathcal{F} \to \mathcal{C}$ is lower semi-continuous for all $n \geq 1$. Define $\Gamma(f) = \text{cl}(\bigcup_n \Gamma_n(f)$. Prove that $\Gamma : \mathcal{F} \to \mathcal{C}$ is lower semi-continuous.

11.4. Let $MS(S^1)$ be the set of Morse-Smale diffeomorphisms on S^1.
- (a) Prove that $MS(S^1)$ is open and dense in $\text{Diff}^1(S^1)$.
- (b) Prove that if $f \in MS(S^1)$, then there are positive integers n and k such that f has j periodic sinks of period n, and j periodic sources of period n, and no other periodic points.

11.5. Assume M is a compact manifold and $L : M \to \mathbb{R}$ is a C^2 Morse function, i.e., all the critical points of L are nondegenerate (or $\nabla(L)$ has only hyperbolic fixed points). Assume X to be a C^1 vector field such that L is constant along the trajectories of X (e.g., X could be the Hamiltonian vector field for L).
- (a) Prove that X can be C^1 approximated by a vector field Y for which $\mathcal{R}(Y)$ is a finite set of hyperbolic fixed points.
- (b) Prove that X can be C^1 approximated by a vector field Y that is Morse-Smale and has no periodic orbits.

11.6. Assume M is a compact manifold. Let $\mathcal{X}^k(M)$ be the set of C^k vector fields on M. Fix $k \geq 1$. Let

$$\mathcal{H}_0^k = \{X \in \mathcal{X}^k(M) : \text{ all the fixed points of } X \text{ are hyperbolic }\}.$$

Let $Z = \{\mathbf{0_p} : \mathbf{p} \in M\}$ be the zero section of TM.
- (a) Let $X \in \mathcal{X}^k(M)$. Prove that $X : M \to TM$ is transverse to Z if and only if each fixed point of X does not have 0 as an eigenvalue.

(b) Prove that $\mathcal{T}^k(M, Z) = \{X \in \mathcal{X}^k(M) : X \pitchfork Z\}$ is dense and open in $\mathcal{X}^k(M)$.

(c) Prove that $\mathcal{H}_0^k$ is dense and open in $\mathcal{X}^k(M)$.

11.7. Let $KS(S^2)$ be the set of Kupka-Smale vector fields on the two sphere. Prove that $KS(S^2)$ is open in $\mathcal{X}^1(S^2)$.

11.8. Prove that the suspension of a Kupka-Smale diffeomorphism is a Kupka-Smale flow.

11.9. Prove that the set of Kupka-Smale diffeomorphisms is not open in $\text{Diff}^1(M)$. Hint: Given an example with $\Omega(f) \neq \text{cl}(\text{Per}(f))$ for which f is Kupka-Smale.

Necessary Conditions for Structural Stability

11.10. Let M be a compact manifold and $X \in \mathcal{X}^1(M)$ a structurally stable vector field on M. Prove that all the fixed points of X are hyperbolic.

11.11. (Proof of Theorem 4.3.) Assume M is a compact surface and $f, g : M \to M$ are C^1 diffeomorphisms.

(a) Assume f is a Kupka-Smale diffeomorphism and g has two periodic points $\mathbf{p}$ and $\mathbf{q}$ such that $W^u(\mathbf{p}, g)$ has a point of nontransverse intersection with $W^s(\mathbf{q}, g)$ where $W^u(\mathbf{p}, g)$ locally lies on one side of $W^s(\mathbf{q}, g)$. Prove that f and g are not conjugate.

(b) Prove that if f is a structurally stable, then f is Kupka-Smale.

(c) Assume f is structurally stable and that $\mathcal{R}(f)$ has a hyperbolic structure. Prove that f satisfies the transversality condition.

CHAPTER XII

Smoothness of Stable Manifolds and Applications

In Chapters V and IX, we proved various results about stable manifolds for a fixed point and a hyperbolic invariant set. In this chapter we present some general results about the existence and differentiability of an invariant section for a fiber contraction. These results generalize the parameterized version of the Contraction Mapping Theorem which is given in Exercise 5.4. Later in the chapter, we give applications of this theory to prove (i) the differentiability of the Center Manifold, (ii) the differentiability of the hyperbolic splitting for an Anosov diffeomorphism on $\mathbb{T}^2$, and (iii) the persistence and differentiability of a "normally contracting" invariant submanifold.

12.1 Differentiable Invariant Sections for Fiber Contractions

In this section, we consider maps $F : X \times Y_0 \to X \times Y_0$ of the form $F(\mathbf{x}, \mathbf{y}) = (f(\mathbf{x}), g(\mathbf{x}, \mathbf{y}))$, where X is a metric space and Y_0 is a closed ball in a Banach space Y. Such a map is called a *bundle map*, X is called the *base space*, and Y_0 or Y is called the *fiber*. In applications, the set of points in the base space that we want to consider is not always invariant. A subset $X_0 \subset X$ is called *overflowing for* f provided $f(X_0) \supset X_0$. The bundle map $F(\cdot,\cdot) = (f(\cdot), g(\cdot,\cdot))$ is called a *fiber contraction over* X_0 provided (i) X_0 is overflowing for f, and (ii) there is a κ with $0 < \kappa < 1$ such that $g(\mathbf{x}, \cdot) : Y_0 \to Y_0$ is Lipschitz and $\text{Lip}(g(\mathbf{x}, \cdot)) \leq \kappa$ for each $\mathbf{x} \in X_0$, i.e.,

$$|g(\mathbf{x}, \mathbf{y}_1) - g(\mathbf{x}, \mathbf{y}_2)| \leq \kappa |\mathbf{y}_1 - \mathbf{y}_2|$$

for all $\mathbf{x} \in X_0$ and $\mathbf{y}_1, \mathbf{y}_2 \in Y_0$. An *invariant section over* X_0 of such a map is a function $\sigma^* : X_0 \to Y_0$ such that $F(\mathbf{x}, \sigma^*(\mathbf{x})) = (f(\mathbf{x}), \sigma^* \circ f(\mathbf{x}))$ for all $\mathbf{x} \in X_0 \cap f^{-1}(X_0)$, i.e., the graph of σ^* is invariant (or overflowing) by F.

If F is a fiber contraction, then Theorem 1.1 proves that there is a continuous invariant section. After proving this result, we give conditions on F which imply that the invariant section is differentiable. In the case when F is a fiber contraction and $f(\mathbf{x}) = \mathbf{x}$ for all $\mathbf{x}$, the first theorem is merely the result about fixed points for a parameterized contraction mapping. See Exercise 5.4.

The results not only apply to product spaces of the form $X \times Y_0 \subset X \times Y$, but also to subsets of vector bundles. In the applications, several of the spaces are of the form $\mathbb{E} = \bigcup_{\mathbf{x} \in X} \{\mathbf{x}\} \times \mathbb{E}_{\mathbf{x}}$, where the space $\mathbb{E}_{\mathbf{x}}$ is a vector space that varies with the point $\mathbf{x}$, and the space X is at least a metric space and often a manifold. With a few assumptions on how $\mathbb{E}_{\mathbf{x}}$ varies, such a space $\mathbb{E}$ is a *vector bundle over* X. (In the proof of Theorem 1.2, the induction step uses a bundle map on a vector bundle even if the original space is a product.) The space X is called the *base space* and $\mathbb{E}_{\mathbf{x}}$ is called the *fiber over* $\mathbf{x}$. The map $\pi : \mathbb{E} \to X$ taking $\{\mathbf{x}\} \times \mathbb{E}_{\mathbf{x}}$ to $\mathbf{x}$ is called the *projection*. The tangent space of a manifold is an example of a vector bundle. See Hirsch (1976) for a formal definition of a vector bundle.

We say that a metric d on a vector bundle $\mathbb{E}$ is an *admissible metric* provided
(1) it induces a norm on each fiber $\mathbb{E}_{\mathbf{x}}$ for $\mathbf{x} \in X$ (so we write $|\mathbf{v}_{\mathbf{x}} - \mathbf{w}_{\mathbf{x}}|$ for $d(\mathbf{v}_{\mathbf{x}}, \mathbf{w}_{\mathbf{x}})$),
(2) there is a complementary bundle $\mathbb{E}'$ over X such that (i) the sum of the two bundles

$$\mathbb{E} \oplus \mathbb{E}' = \bigcup_{X} \{\mathbf{x}\} \times \mathbb{E}_{\mathbf{x}} \times \mathbb{E}'_{\mathbf{x}}$$

is isomorphic to a product bundle $X \times Y$, where Y is a Banach space, (ii) the product metric on $X \times Y$ induces the metric d on $\mathbb{E}$, and (iii) the projection $\{\mathbf{x}\} \times Y$ onto $\mathbb{E}_{\mathbf{x}}$ along $\mathbb{E}'_{\mathbf{x}}$ is of norm 1.

In a vector bundle with an admissible metric, the *disk bundle of radius* R is the subset

$$\mathbb{E}(R) = \bigcup_{\mathbf{x} \in X} \{\mathbf{x}\} \times \mathbb{E}_{\mathbf{x}}(R)$$

where $\mathbb{E}_{\mathbf{x}}(R)$ is the closed ball (or disk) of radius R in $\mathbb{E}_{\mathbf{x}}$.

A map $F : \mathbb{E}(R) \to \mathbb{E}(R)$ is called a *bundle map* provided $\pi \circ F(\{\mathbf{x}\} \times \mathbb{E}_{\mathbf{x}}(R))$ takes on a unique value $f(\mathbf{x})$. Therefore, a bundle map F induces a map $f : X \to X$ such that $\pi \circ F = f \circ \pi$. A bundle map on a vector space $\mathbb{E}$ with an admissible metric induces a map $\tilde{F} : X \times Y(R) \to X \times Y(R)$ on the associated product space by means of the composition of (i) the isomorphism from $X \times Y(R)$ to $(\mathbb{E} \oplus \mathbb{E}')(R)$, (ii) the projection from $(\mathbb{E} \oplus \mathbb{E}')(R)$ to $\mathbb{E}(R)$, (iii) the map F from $\mathbb{E}(R)$ to $\mathbb{E}(R)$, (iv) the inclusion of $\mathbb{E}(R)$ in $(\mathbb{E} \oplus \mathbb{E}')(R)$, and finally (v) the isomorphism from $(\mathbb{E} \oplus \mathbb{E}')(R)$ to $X \times Y(R)$. By using this construction, the theorems which we state for a map on a product spaces are valid for a bundle map on a disk bundle. See Hirsch and Pugh (1970) or Shub (1987) for details.

Now, we can state the first result.

Theorem 1.1 (Invariant Section Theorem). *Assume X is a metric space, $X_0 \subset X$, and Y_0 is a closed bounded ball in a Banach space Y. Assume $F : X_0 \times Y_0 \to X \times Y_0$ is a continuous fiber contraction over X_0 with $F(\mathbf{x}, \mathbf{y}) = (f(\mathbf{x}), g(\mathbf{x}, \mathbf{y}))$. (Or F is a bundle map on a disk bundle over X_0 and is a contraction on each fiber.) Assume that (i) f is overflowing on X_0, and (ii) $f|X_0$ is one to one. Then, there is a unique invariant section over X_0, $\sigma^* : X_0 \to Y_0$, and σ^* is continuous.*

PROOF. Consider the spaces of bounded functions and continuous functions given by

$$\mathcal{G} = \{\sigma : X_0 \to Y_0\} \qquad \text{and}$$
$$\mathcal{G}^0 = \{\sigma : X_0 \to Y_0 : \sigma \text{ is continuous}\}.$$

Note that because Y_0 is a bounded ball, $\mathcal{G}$ is the space of bounded functions. On $\mathcal{G}$ and $\mathcal{G}^0$, put the C^0-sup topology

$$\|\sigma_1 - \sigma_2\|_0 = \sup_{\mathbf{x} \in X_0} |\sigma_1(\mathbf{x}) - \sigma_2(\mathbf{x})|.$$

This norm makes both $\mathcal{G}$ and $\mathcal{G}^0$ complete spaces (closed balls in normed linear spaces) since Y_0 is complete and the sections are bounded. See Dieudonne (1960), 7.1.3 and 7.2.1.

There is an induced map on the sections $\Gamma_F : \mathcal{G} \to \mathcal{G}$ such that the graph of $\Gamma_F(\sigma)$ is a subset of the image by F of the graph of σ. (These two graphs are not necessarily equal because X_0 is overflowing but not necessarily invariant for f.) This map Γ_F is called the *graph transform of* F. In fact, it is possible to give a formula for $\Gamma_F(\sigma)$. Let

$\mathbf{x} \in X_0$. Then, $f^{-1}(\mathbf{x}) \in X_0$ is well defined because f is one to one on X_0; the value of σ at $f^{-1}(\mathbf{x})$ is given by $\sigma \circ f^{-1}(\mathbf{x})$ and is in the fiber over $f^{-1}(\mathbf{x})$; the image of $\sigma \circ f^{-1}(\mathbf{x})$ by F must be $(\mathbf{x}, \Gamma_F(\sigma)(\mathbf{x}))$, so

$$\begin{aligned}(\mathbf{x}, \Gamma_F(\sigma)(\mathbf{x})) &= F(f^{-1}(\mathbf{x}), \sigma \circ f^{-1}(\mathbf{x})) \\ &= (\mathbf{x}, g(f^{-1}(\mathbf{x}), \sigma \circ f^{-1}(\mathbf{x}))) \\ &= (\mathbf{x}, g \circ \hat{\sigma} \circ f^{-1}(\mathbf{x}))\end{aligned}$$

where $\hat{\sigma}(\mathbf{x}) = (\mathbf{x}, \sigma(\mathbf{x}))$. Thus,

$$\Gamma_F(\sigma) = g \circ \hat{\sigma} \circ f^{-1}.$$

We claim that Γ_F is Lipschitz on $\mathcal{G}$ with $\text{Lip}(\Gamma_F) \leq \kappa$, where $0 < \kappa < 1$ is the fiber contraction constant for F. Let $\sigma_1, \sigma_2 \in \mathcal{G}$, and $\hat{\sigma}_1, \hat{\sigma}_2 : X_0 \to X_0 \times Y_0$ be induced as above. Then,

$$\begin{aligned}||\Gamma_F(\sigma_1) - \Gamma_F(\sigma_2)||_0 &= \sup_{\mathbf{x}\in X_0} |g \circ \hat{\sigma}_1 \circ f^{-1}(\mathbf{x}) - g \circ \hat{\sigma}_2 \circ f^{-1}(\mathbf{x})| \\ &\leq \sup_{\mathbf{p}\in X_0} |g \circ \hat{\sigma}_1(\mathbf{p}) - g \circ \hat{\sigma}_2(\mathbf{p})| \\ &\leq \sup_{\mathbf{p}\in X_0} \kappa\, |\sigma_1(\mathbf{p}) - \sigma_2(\mathbf{p})| \\ &\leq \kappa\, ||\sigma_1 - \sigma_2||_0.\end{aligned}$$

Because Γ_F is a contraction on the complete metric space $\mathcal{G}$, it has a unique fixed point σ^*, or F has a unique bounded invariant section over X_0. Because Γ_F preserves the closed subspace $\mathcal{G}^0$, $\sigma^* \in \mathcal{G}^0$ and the invariant section is continuous. □

We want conditions which imply that the invariant section is differentiable. It is not enough for F to be a fiber contraction. The following example gives a function which contracts more along the base space X than along the fibers Y and there is no differentiable invariant section.

Example 1.1. We take the base space as $S^1 = \{x \in \mathbb{R} \bmod 1\}$ and fiber space $\mathbb{R}$. Let $f : S^1 \to S^1$ be a C^∞ diffeomorphism with two fixed points, an attracting fixed point at 0 and a repelling fixed point at 1/2. Further, assume that $f(x) = x/3$ for $-1/3 \leq x \leq 1/3$. (This is really the form of the function on the covering space $\mathbb{R}$.) To define g, let $\beta : [0,1] \to [0,1] \subset \mathbb{R}$ be a C^∞ bump function with $\text{supp}(\beta) = [1/9, 1/3]$ and $\beta(2/9) = 1$. The map $g : S^1 \times \mathbb{R} \to \mathbb{R}$ is defined by $g(x,y) = \beta(x) + y/2$. Notice that f contracts more strongly on the base S^1 (by a factor of 1/3) than g contracts on the fiber (by a factor of 1/2). Let $F = (f,g)$ be the bundle map on $S^1 \times \mathbb{R}$, which takes $S^1 \times [-3,3]$ into its interior.

Since F is a fiber contraction, it has a unique continuous section σ^*, $g \circ \hat{\sigma}^* f^{-1} = \sigma^*$. By the iterative process used in the proof of Theorem 1.1, $\sigma^*(x) = 0$ for $1/3 \leq x \leq 1$. For $1/9 \leq x \leq 1/3$, $f^{-1}(x) \in [1/3, 1/2]$ (because $f(1/2) = 1/2$ and f is monotone), so $\sigma^* \circ f^{-1}(x) = 0$ and

$$\sigma^*(x) = g \circ \hat{\sigma}^* \circ f^{-1}(x) = g(f^{-1}(x), 0) = \beta \circ f^{-1}(x) = 0.$$

However, for $1/3^3 \leq x \leq 1/3^2$, $f^{-1}(x) = 3x \in [1/9, 1/3]$ and

$$\sigma^*(x) = g \circ \hat{\sigma}^*(3x) = g(3x, 0) = \beta(3x).$$

In particular, $\sigma^*(2/3^3) = 1$ and $\sigma^*(1/3^3) = \sigma^*(1/3^2) = 0$. Next for $1/3^{3+1} \leq x \leq 1/3^{2+1}$, $f^{-1}(x) = 3x \in [1/3^3, 1/3^2]$ and

$$\begin{aligned}\sigma^*(x) &= g \circ \hat{\sigma}^*(3x) \\ &= g(3x, \beta(3^2 x)) \\ &= \frac{1}{2}\beta(3^2 x).\end{aligned}$$

In particular, $\sigma^*(2/3^4) = 1/2$ and $\sigma^*(1/3^4) = \sigma^*(1/3^3) = 0$. By induction on n, for $1/3^{3+n} \leq x \leq 1/3^{2+n}$ with $n \geq 1$, $f^{-1}(x) = 3x \in [1/3^{2+n}, 1/3^{1+n}]$ and

$$\begin{aligned}\sigma^*(x) &= g \circ \hat{\sigma}^*(3x) \\ &= g(3x, \frac{1}{2^{n-1}}\beta(3^{n+1}x)) \\ &= \frac{1}{2^n}\beta(3^{n+1}x).\end{aligned}$$

In particular, $\sigma^*(2/3^{3+n}) = 1/2^n$ and $\sigma^*(1/3^{3+n}) = \sigma^*(1/3^{2+n}) = 0$. Since $\sigma^*(0) = 0$ and $\sigma^*(2/3^{3+n}) = 1/2^n$, the difference quotient

$$\frac{\sigma^*(2/3^{3+n}) - \sigma^*(0)}{(2/3^{3+n}) - 0} = \frac{3^{3+n}}{2^{n+1}}$$

does not converge as n goes to infinity. Therefore, σ^* is not differentiable at $x = 0$. By the calculation of the form of σ^* for $x \neq 0$, it is differentiable at all other points. The reader can check using the Mean Value Theorem that there are points in the interval $[1/3^{3+n}, 2/3^{3+n}]$ at which the derivative is at least $3^{n+3}/2^n$, and this quantity is unbounded as n goes to infinity. This completes the analysis of this example.

To insure that the invariant section is C^1, we need to assume that the contraction on fibers is stronger than the contraction in the base space. The following theorem makes this precise.

Theorem 1.2 (C^r Section Theorem). *Assume X is a manifold, Y_0 is a closed bounded ball in a Banach space Y, and $X_0 \subset X$ is a subset for which $X_0 = \text{cl}(\text{int}(X_0))$ (so differentiation makes sense on X_0). Assume $F : X_0 \times Y_0 \to X \times Y_0$ is a C^1 fiber contraction on X_0 with uniformly bounded derivatives on $X_0 \times Y_0$ and $F(\mathbf{x}, \mathbf{y}) = (f(\mathbf{x}), g(\mathbf{x}, \mathbf{y}))$. (Or F is a bundle map on a disk bundle over X_0.) Assume that (i) f is overflowing on X_0, and (ii) $f|X_0$ is a diffeomorphism from X_0 to its image $f(X_0) \supset X_0$. For each $\mathbf{x} \in X_0$, let $\lambda_{\mathbf{x}} = \|(Df_{\mathbf{x}})^{-1}\| = 1/m(Df_{\mathbf{x}})$. (Note that $\lambda_{\mathbf{x}}$ measures the greatest expansion of f^{-1} at $f(\mathbf{x})$, and $1/\lambda_{\mathbf{x}}$ measures the minimum expansion or greatest contraction of f at $\mathbf{x}$.) Let $D_2 g_{(\mathbf{x},\mathbf{y})} : Y \to Y$ be the derivative with respect to the variables in Y. For each $\mathbf{x} \in X_0$, let*

$$\kappa_{\mathbf{x}} = \sup_{\mathbf{y} \in Y_0} \|D_2 g_{(\mathbf{x},\mathbf{y})}\|.$$

The map F is a fiber contraction, so

$$\sup_{\mathbf{x} \in f^{-1}(X_0)} \kappa_{\mathbf{x}} < 1.$$

(a) Assume further that $1 \leq r \leq \infty$, F is C^r, and

$$\sup_{\mathbf{x} \in f^{-1}(X_0)} \kappa_{\mathbf{x}} \lambda_{\mathbf{x}}^j < 1$$

for $1 \le j \le r$. Then, the unique invariant section is C^r.

(b) Assume that $0 \le r \le \infty$, $\alpha > 0$, F is $C^{r+\alpha}$ with uniformly bounded Hölder constant,

$$\sup_{\mathbf{x} \in f^{-1}(X_0)} \kappa_{\mathbf{x}} \lambda_{\mathbf{x}}^j < 1$$

for $1 \le j \le r$, and

$$\sup_{\mathbf{x} \in f^{-1}(X_0)} \kappa_{\mathbf{x}} \lambda_{\mathbf{x}}^{r+\alpha} < 1.$$

Then, the unique invariant section is $C^{r+\alpha}$, i.e., the rth derivative is α-Hölder.

REMARK 1.1. When $r = 0$ and $\alpha > 0$, the theorem is true when Y_0 is merely a metric space.

REMARK 1.2. This theorem is essentially in Hirsch and Pugh (1970). For part (b), they make the comparisons of derivatives at different points:

$$[\sup_{\mathbf{x} \in f^{-1}(X_0)} \kappa_{\mathbf{x}} \lambda_{\mathbf{x}}^r][\sup_{\mathbf{x} \in f^{-1}(X_0)} \lambda_{\mathbf{x}}^\alpha] < 1.$$

Also see Hirsch, Pugh, and Shub (1977).

REMARK 1.3. Hurder and Katok (1990) prove the result with the purely pointwise assumption that

$$\sup_{\mathbf{x} \in f^{-1}(X_0)} \kappa_{\mathbf{x}} \lambda_{\mathbf{x}}^{r+\alpha} < 1.$$

We refer to these references for the proof in the case of Hölder continuous derivatives. Also see Exercises 12.2 and 12.3.

REMARK 1.4. Below, we let $\mathcal{G}^1$ be the C^1 sections in $\mathcal{G}^0$. The graph transform Γ_F preserves $\mathcal{G}^1$ but it is not a contraction on $\mathcal{G}^1$ with the C^1 topology, so we cannot prove that $\sigma^* \in \mathcal{G}^1$ by that means.

We do not proceed exactly like the proof of the stable manifold, but instead show that the graph transform preserves Lipschitz sections. After this step, we show that the derivative takes a family of appropriate cones of vectors into itself. Note the assumption that $\sup_{\mathbf{x} \in f^{-1}(X_0)} \kappa_{\mathbf{x}} \lambda_{\mathbf{x}} < 1$ is just the condition needed to verify this invariance of cones.

PROOF FOR $r = 1$. Let $\mathcal{G}^0 = \{\sigma : X_0 \to Y_0 : \sigma \text{ is continuous }\}$ and $\Gamma_F : \mathcal{G}^0 \to \mathcal{G}^0$ be as in the proof of Theorem 1.1, $\Gamma_F(\sigma) = g \circ (id, \sigma) \circ f^{-1}$. By Theorem 1.1, there is a unique invariant section $\sigma^* \in \mathcal{G}^0$.

We first prove that the invariant section σ^* is Lipschitz. Let

$$\mathcal{G}^1 = \{\sigma : X_0 \to Y_0 : \sigma \text{ is } C^1\},$$
$$\mathcal{G}^1(L) = \{\sigma \in \mathcal{G}^1 : \sup_{x \in X_0} \|D\sigma_{\mathbf{x}}\| \le L\},$$

and $\mathcal{G}^{Lip}(L)$ be the closure of $\mathcal{G}^1(L)$ in $\mathcal{G}^0$ in terms of the C^0-sup topology. We justify the notation for $\mathcal{G}^{Lip}(L)$ with the following lemma.

Lemma 1.3. *Every $\sigma \in \mathcal{G}^{Lip}(L)$ is Lipschitz with $Lip(\sigma) \le L$,*

PROOF. By the Mean Value Theorem, every $\sigma \in \mathcal{G}^1(L)$ is Lipschitz with $Lip(\sigma) \le L$, If $\sigma_j \in \mathcal{G}^1(L)$ converges to $\sigma \in \mathcal{G}^{Lip}(L)$, then

$$\begin{aligned} |\sigma(\mathbf{x}) - \sigma(\mathbf{y})| &= \lim_{j \to \infty} |\sigma_j(\mathbf{x}) - \sigma_j(\mathbf{y})| \\ &\le \lim_{j \to \infty} L\, d(\mathbf{x}, \mathbf{y}) \\ &\le L\, d(\mathbf{x}, \mathbf{y}), \end{aligned}$$

which proves that σ has a Lipschitz constant as claimed. □

REMARK 1.5. The proof of the above lemma actually shows that the set of maps whose Lipschitz constant is bounded by L is closed in $\mathcal{G}^0$.

Also, $\mathcal{G}^{Lip}(L)$ is actually the set of all Lipschitz sections with Lipschitz constant less than or equal to L, but we do not need this fact. We merely use that $\mathcal{G}^{Lip}(L)$ is closed in $\mathcal{G}^0$ and $\mathcal{G}^1(L)$ is dense in $\mathcal{G}^{Lip}(L)$.

Let

$$\mu = \sup_{\mathbf{x} \in f^{-1}(X_0)} \kappa_{\mathbf{x}} \lambda_{\mathbf{x}}.$$

The derivatives of g are uniformly bounded so we can take $C > 0$ such that

$$\|D_1 g_{(\mathbf{x},\mathbf{y})}\| \, \|(Df_{\mathbf{x}})^{-1}\| \leq C$$

for all $(\mathbf{x},\mathbf{y}) \in X_0 \times Y_0$, where $D_1 g_{(\mathbf{x},\mathbf{y})} : T_{\mathbf{x}}X \to Y$ is the derivative with respect to the variables in X. Take $L_0 = C/(1-\mu) > 0$, i.e., $L_0 > 0$ so that $C + \mu L_0 = L_0$. With these constants we can show that Γ_F preserves $\mathcal{G}^{Lip}(L_0)$.

Lemma 1.4. *With L_0 as chosen above, Γ_F takes $\mathcal{G}^{Lip}(L_0)$ into itself.*

PROOF. For $\sigma \in \mathcal{G}^1(L_0)$ and $\mathbf{x} \in X_0$,

$$\begin{aligned}
\|D(\Gamma_F(\sigma))_{\mathbf{x}}\| &= \|D(g \circ \hat{\sigma} \circ f^{-1})_{\mathbf{x}}\| \\
&\leq \|D_1 g_{\hat{\sigma} \circ f^{-1}(\mathbf{x})} D(f^{-1})_{\mathbf{x}}\| + \|D_2 g_{\hat{\sigma} \circ f^{-1}(\mathbf{x})} D\sigma_{f^{-1}(\mathbf{x})} D(f^{-1})_{\mathbf{x}}\| \\
&\leq C + \kappa_{f^{-1}(\mathbf{x})} L_0 \lambda_{f^{-1}(\mathbf{x})} \\
&\leq C + \mu L_0 \\
&\leq L_0.
\end{aligned}$$

Therefore, $\Gamma_F(\sigma) \in \mathcal{G}^1(L_0)$. Because $\mathcal{G}^1(L_0)$ is dense in $\mathcal{G}^{Lip}(L_0)$ and $\mathcal{G}^{Lip}(L_0)$ is closed,

$$\Gamma_F(\mathcal{G}^{Lip}(L_0)) \subset \mathcal{G}^{Lip}(L_0).$$

□

Take any $\sigma^0 \in \mathcal{G}^1(L_0)$. Then, for $n \geq 1$, $\Gamma_F^n(\sigma^0) \in \mathcal{G}^1(L_0)$, $\Gamma_F^n(\sigma^0)$ converges to σ^* as n goes to infinity, $\mathcal{G}^{Lip}(L_0)$ is closed, and so $\sigma^* \in \mathcal{G}^{Lip}(L_0)$. This proves that the invariant section is Lipschitz.

To prove that the invariant section is C^1, consider the cones

$$C^u_{\mathbf{x}}(2L_0) = \{(\mathbf{v},\mathbf{w}) \in T_{\mathbf{x}}X \times Y : |\mathbf{w}| \leq 2L_0 |\mathbf{v}|\}.$$

The calculation of Lemma 1.4 shows that

$$DF_{(\mathbf{x},\mathbf{y})}[C^u_{\mathbf{x}}(2L_0) \setminus \{\mathbf{0}\}] \subset \operatorname{int}(C^u_{f(\mathbf{x})}(2L_0)).$$

We need to measure the maximum opening of a cone. If $C^u \subset C^u_{\mathbf{x}}(2L_0)$ is a cone over $T_{\mathbf{x}}X$, then the angle of opening of C^u is defined to be

$$\measuredangle(C^u) = \sup\{\frac{|\mathbf{w}-\mathbf{w}'|}{|\mathbf{v}|} : (\mathbf{v},\mathbf{w}),(\mathbf{v},\mathbf{w}') \in C^u \subset T_{\mathbf{x}}X \times Y\}.$$

The following lemma proves the related fact that the angle of opening of the cone decreases by a factor of μ.

Lemma 1.5. *If $C^u \subset C^u_{\mathbf{x}}(2\,L_0)$ has $\measuredangle(C^u) = \beta > 0$, then $\measuredangle(DF_{(\mathbf{x},\mathbf{y})}C^u) \leq \mu\beta$.*

PROOF. Assume

$$(\mathbf{v}, \mathbf{w}), (\mathbf{v}, \mathbf{w}') \in DF_{(\mathbf{x},\mathbf{y})}C^u \subset T_{f(\mathbf{x})}X \times Y,$$

then

$$\begin{aligned}(\mathbf{v}, \mathbf{w}) &= DF_{(\mathbf{x},\mathbf{y})}(\hat{\mathbf{v}}, \hat{\mathbf{w}})\\ (\mathbf{v}, \mathbf{w}') &= DF_{(\mathbf{x},\mathbf{y})}(\hat{\mathbf{v}}, \hat{\mathbf{w}}').\end{aligned}$$

The vector $\mathbf{v} = Df_{\mathbf{x}}\hat{\mathbf{v}}$, so $\hat{\mathbf{v}} = (Df_{\mathbf{x}})^{-1}\mathbf{v}$ (so $\hat{\mathbf{v}}$ is the same for both pairs of vectors), $|\hat{\mathbf{v}}| \leq \lambda_{\mathbf{x}}\,|\mathbf{v}|$, and $|\mathbf{v}|^{-1} \leq \lambda_{\mathbf{x}}\,|\hat{\mathbf{v}}|^{-1}$. To estimate the components in Y,

$$\begin{aligned}\mathbf{w} - \mathbf{w}' &= D_1 g_{(\mathbf{x},\mathbf{y})}\hat{\mathbf{v}} + D_2 g_{(\mathbf{x},\mathbf{y})}\hat{\mathbf{w}} - D_1 g_{(\mathbf{x},\mathbf{y})}\hat{\mathbf{v}} - D_2 g_{(\mathbf{x},\mathbf{y})}\hat{\mathbf{w}}'\\ &= D_2 g_{(\mathbf{x},\mathbf{y})}[\hat{\mathbf{w}} - \hat{\mathbf{w}}'],\end{aligned}$$

so

$$|\mathbf{w} - \mathbf{w}'| \leq \kappa_{\mathbf{x}}\,|\hat{\mathbf{w}} - \hat{\mathbf{w}}'|.$$

Thus, the angle of the opening of the cone $DF_{(\mathbf{x},\mathbf{y})}C^u$ is estimated as follows:

$$\begin{aligned}\measuredangle(DF_{(\mathbf{x},\mathbf{y})}C^u) &= \sup\{\frac{|\mathbf{w} - \mathbf{w}'|}{|\mathbf{v}|} : (\mathbf{v}, \mathbf{w}), (\mathbf{v}, \mathbf{w}') \in DF_{(\mathbf{x},\mathbf{y})}C^u\}\\ &\leq \sup\{\kappa_{\mathbf{x}}\,\lambda_{\mathbf{x}}\frac{|\hat{\mathbf{w}} - \hat{\mathbf{w}}'|}{|\hat{\mathbf{v}}|} : (\hat{\mathbf{v}}, \hat{\mathbf{w}}), (\hat{\mathbf{v}}, \hat{\mathbf{w}}') \in C^u\}\\ &\leq \mu\,\beta.\end{aligned}$$

This proves the bound on the angle of the opening which is claimed. □

By induction on n,

$$\begin{aligned}\measuredangle(DF^n_{\hat{\sigma}^*\circ f^{-n}(\mathbf{x})}C^u_{f^{-n}(\mathbf{x})}(2\,L_0) &\leq \mu^n \measuredangle(C^u_{f^{-n}(\mathbf{x})}(2\,L_0))\\ &= \mu^n 4\,L_0.\end{aligned}$$

As in Lemma V.10.9 (in the proof of the stable manifold), it follows that

$$P_{\mathbf{x}} = \bigcap_{n=0}^{\infty} DF^n_{\hat{\sigma}^*\circ f^{-n}(\mathbf{x})}C^u_{f^{-n}(\mathbf{x})}(2\,L_0)$$

is a linear subspace which is a graph over $T_{\mathbf{x}}X$ of slope less than or equal to L_0. By using an argument like that for Proposition V.10.8, we get that σ^* is C^1. This completes the proof of Theorem 1.2 for $r = 1$. □

PROOF FOR $r \geq 2$. By induction, σ^* is C^{r-1}, so at least C^1. Let $A^*_{\mathbf{x}} = D\sigma^*_{\mathbf{x}}$. Then, $D\sigma^*_{\mathbf{x}} : T_{\mathbf{x}}X \to Y$ is linear for each $\mathbf{x}$, so $D\sigma^*_{\mathbf{x}} \in \mathbf{L}(T_{\mathbf{x}}X, Y)$ where $\mathbf{L}(V_1, V_2)$ is the space of bounded linear maps between Banach spaces. We put the norm on $\mathbf{L}(T_{\mathbf{x}}, Y)$ given by the norm of a linear operator. (Notice that this definition uses a Riemannian norm on TX.) We form a space of possible derivatives of such sections. Let $\mathcal{L}(X)$ be the bundle

$$\mathcal{L}(X) = \bigcup_{\mathbf{x}\in X}\{\mathbf{x}\} \times \mathbf{L}(T_{\mathbf{x}}X, Y).$$

Let $\mathcal{L}(X_0) = \mathcal{L}(X)|X_0$ be the bundle restricted to fibers over points in X_0, and $\mathcal{L}_{\mathbf{x}} = \{\mathbf{x}\} \times \mathbf{L}(T_{\mathbf{x}}X, Y)$.

Next we want to define a bundle map on $\mathcal{L}(X_0)$ that is compatible with the transformations of derivatives by the action of Γ_F on $\mathcal{G}^1$. If $\sigma \in \mathcal{G}^1$, then $\Gamma_F(\sigma)$ is C^1 and

$$D(\Gamma_F(\sigma))_{\mathbf{x}} = Dg_{\hat{\sigma}\circ f^{-1}(\mathbf{x})}(id, D\sigma_{f^{-1}(\mathbf{x})})D(f^{-1})_{\mathbf{x}}$$

by the Chain Rule. Using this a motivation, we define a bundle map $\Psi = (f, \psi) : \mathcal{L}(X_0) \to \mathcal{L}(X)$ by

$$\begin{aligned}\Psi(\mathbf{x}, S) &= (f(\mathbf{x}), \psi(\mathbf{x}, S)) \\ &= (f(\mathbf{x}), Dg_{\hat{\sigma}^*(\mathbf{x})} \circ (id, S) \circ D(f^{-1})_{f(\mathbf{x})}).\end{aligned}$$

Note that in this definition, we consider (id, S) as taking values at tangent vectors at $(\mathbf{x}, \sigma^*(\mathbf{x}))$. Thus, we only consider the space of possible derivatives along $\hat{\sigma}^*$. The map Ψ is C^{r-1} because F is C^r and σ^* is C^{r-1}. Lemma 1.6 shows that Ψ is a fiber contraction over $\mathbf{x}$ by a factor of $\beta_{\mathbf{x}} = \kappa_{\mathbf{x}}\, \lambda_{\mathbf{x}} \le \mu < 1$. By the definitions,

$$\sup_{\mathbf{x}\in X_0} \beta_{\mathbf{x}}\, \lambda_{\mathbf{x}}^j \le \mu < 1$$

for $1 \le j \le r - 1$. The invariant section σ^* is C^1 by the induction hypothesis, so the map

$$\mathbf{x} \mapsto A_{\mathbf{x}}^* = D\sigma_{\mathbf{x}}^*$$

is an invariant section of Ψ, i.e., a fixed point of Γ_Ψ. By applying this theorem for $r-1$ to Ψ, A^* is a C^{r-1} invariant section. The fact that A^* is C^{r-1} implies that σ^* is C^r. This completes the induction step in the proof except for the following lemma. □

Lemma 1.6. *The map $\Psi : \mathcal{L}(X_0) \to \mathcal{L}(X)$ is a contraction on the fiber over $\mathbf{x}$ by $\kappa_{\mathbf{x}}\, \lambda_{\mathbf{x}}$.*

PROOF. The map Ψ is a bundle map by its definition. For $(\mathbf{x}, S), (\mathbf{x}, S') \in \mathcal{L}_{\mathbf{x}}$,

$$\begin{aligned}\|\psi(\mathbf{x}, S) - \psi(\mathbf{x}, S')\| &= \|Dg_{\hat{\sigma}^*(\mathbf{x})}[(id, S) - (id, S')]D(f^{-1})_{f(\mathbf{x})}\| \\ &\le \|D_2 g_{\hat{\sigma}(\mathbf{x})}[S - S']\|\, \|D(f^{-1})_{f(\mathbf{x})}\| \\ &\le \kappa_{\mathbf{x}}\, \lambda_{\mathbf{x}}\, \|S - S'\|\end{aligned}$$

as claimed. □

REMARK 1.6. Note that the proof of Lemma 1.6 is essentially the same as the proof of Lemma 1.5. Also, it is very important that both S and S' take their values at the same point $\hat{\sigma}^*(\mathbf{x})$, so that the derivative of g is calculated at the same point for both terms.

REMARK 1.7. Note that the proof of the induction step (the case $r \ge 2$) does not apply to prove the case for $r = 1$. The reason is that, without using the proof for $r = 1$, we do not know that the invariant section A^* for Ψ is in fact the derivative of the invariant section σ^* for F.

12.2 Differentiability of Invariant Splitting

As an application of the C^r Section Theorem, we consider an Anosov diffeomorphism on $\mathbb{T}^2$ and show that its splitting is C^1.

Theorem 2.1. *Assume $f : \mathbb{T}^2 \to \mathbb{T}^2$ is a C^2 Anosov diffeomorphism. Then, the subbundles $\mathbb{E}^s$ and $\mathbb{E}^u$ in the hyperbolic splitting depend in a C^1 fashion on the base point.*

PROOF. We prove that the unstable bundle is C^1. For this proof, we use that the stable bundle has one-dimensional fibers but not that the unstable bundle is one-dimensional. For this reason, to prove that $\mathbb{E}^u$ is C^1, we assume that f is a C^2 Anosov diffeomorphism on $\mathbb{T}^n$ with $\mathbb{E}^s$ one-dimensional.

In the proof we write the derivative as if f were a map on $\mathbb{R}^n$, i.e., we confuse f with its lift to $\mathbb{R}^n$. By the assumptions, there is a continuous splitting $\mathbb{E}^u \oplus \mathbb{E}^s$. This splitting can be approximated by a C^1 splitting $\mathbb{F}^u \oplus \mathbb{F}^s$ which is almost invariant. In terms of the splitting $\mathbb{F}^u \oplus \mathbb{F}^s$,

$$Df_{\mathbf{x}} = \begin{pmatrix} A_{\mathbf{x}} & B_{\mathbf{x}} \\ C_{\mathbf{x}} & D_{\mathbf{x}} \end{pmatrix},$$

where $A_{\mathbf{x}} : \mathbb{F}^u_{\mathbf{x}} \to \mathbb{F}^u_{f(\mathbf{x})}$, $B_{\mathbf{x}} : \mathbb{F}^s_{\mathbf{x}} \to \mathbb{F}^u_{f(\mathbf{x})}$, etc. Because the splitting is near the hyperbolic invariant splitting, there are $\lambda > 1$, $0 < \mu' < \mu < 1$, and $\epsilon > 0$ such that $\lambda - \epsilon > 1$, $\mu + \epsilon < 1$, $\lambda^{-1} + \epsilon + 2\epsilon/\mu' < 1$, $m(A_{\mathbf{x}}) \geq \lambda$, $\mu' \leq m(D_{\mathbf{x}}) = \|D_{\mathbf{x}}\| \leq \mu$, and $\|B_{\mathbf{x}}\|, \|C_{\mathbf{x}}\| \leq \epsilon$. (Note that the bundle $\mathbb{F}^s$ is one-dimensional, so the norm of $D_{\mathbf{x}}$ is really an absolute value.)

The bundle $\mathbb{E}^u_{\mathbf{x}}$ is the graph of a linear map $\mathbb{F}^u_{\mathbf{x}} \to \mathbb{F}^s_{\mathbf{x}}$. Let $\mathcal{L}_{\mathbf{x}}$ be the space of all linear maps from $\mathbb{F}^u_{\mathbf{x}}$ to $\mathbb{F}^s_{\mathbf{x}}$) with norm less than or equal to 1,

$$\mathcal{L}_{\mathbf{x}} = \{\sigma_{\mathbf{x}} \in \mathbf{L}(\mathbb{F}^u_{\mathbf{x}}, \mathbb{F}^s_{\mathbf{x}}) : \|\sigma_{\mathbf{x}}\| \leq 1\}.$$

Let $\mathcal{L}$ be the disk bundle over $\mathbb{T}^n$ of all the $\mathcal{L}_{\mathbf{x}}$,

$$\mathcal{L} = \bigcup_{\mathbf{x} \in \mathbb{T}^n} \{\mathbf{x}\} \times \mathcal{L}_{\mathbf{x}}.$$

The natural transformation on $\mathcal{L}$ is given in terms of the graph transform induced by the derivative of f, $\Psi = (f, \psi) : \mathcal{L} \to \mathcal{L}$, where $\psi_{\mathbf{x}} : \mathcal{L}_{\mathbf{x}} \to \mathcal{L}_{f(\mathbf{x})}$. To derive the formula for $\psi_{\mathbf{x}}$, let $\sigma_{\mathbf{x}} \in \mathcal{L}_{\mathbf{x}}$ and $\mathbf{v} \in \mathbb{F}^u_{\mathbf{x}}$. Then,

$$\begin{aligned} Df_{\mathbf{x}}(\mathbf{v}, \sigma_{\mathbf{x}}\mathbf{v}) &= ((A_{\mathbf{x}} + B_{\mathbf{x}}\sigma_{\mathbf{x}})\mathbf{v}, (C_{\mathbf{x}} + D_{\mathbf{x}}\sigma_{\mathbf{x}})\mathbf{v}) \\ &= (\mathbf{w}, \psi_{\mathbf{x}}(\sigma_{\mathbf{x}})\mathbf{w}). \end{aligned}$$

So, $\mathbf{v} = (A_{\mathbf{x}} + B_{\mathbf{x}}\sigma_{\mathbf{x}})^{-1}\mathbf{w}$ and

$$\begin{aligned} \psi_{\mathbf{x}}(\sigma_{\mathbf{x}})\mathbf{w} &= (C_{\mathbf{x}} + D_{\mathbf{x}}\sigma_{\mathbf{x}})(A_{\mathbf{x}} + B_{\mathbf{x}}\sigma_{\mathbf{x}})^{-1}\mathbf{w} \qquad \text{or} \\ \psi_{\mathbf{x}}(\sigma_{\mathbf{x}}) &= (C_{\mathbf{x}} + D_{\mathbf{x}}\sigma_{\mathbf{x}})(A_{\mathbf{x}} + B_{\mathbf{x}}\sigma_{\mathbf{x}})^{-1}. \end{aligned}$$

To check that $A_{\mathbf{x}} + B_{\mathbf{x}}\sigma_{\mathbf{x}}$ is invertible, note that

$$\begin{aligned} m(A_{\mathbf{x}} + B_{\mathbf{x}}\sigma_{\mathbf{x}}) &\geq m(A_{\mathbf{x}}) - \|B_{\mathbf{x}}\|\,\|\sigma_{\mathbf{x}}\| \\ &\geq \lambda - \epsilon \\ &> 1. \end{aligned}$$

(See Exercise 12.6.) Since $m(A_{\mathbf{x}} + B_{\mathbf{x}}\sigma_{\mathbf{x}}) > 1 > 0$, $A_{\mathbf{x}} + B_{\mathbf{x}}\sigma_{\mathbf{x}}$ is invertible.

Since f is C^2, $\Psi : \mathcal{L} \to \mathcal{L}$ is C^1. Also, Ψ covers the map $f : \mathbb{T}^n \to \mathbb{T}^n$. The following lemma shows that Ψ is a fiber contraction.

Lemma 2.2. *(a) The map Ψ is a fiber contraction by a factor of $\kappa_{\mathbf{x}} = (\|D_{\mathbf{x}}\|/m(A_{\mathbf{x}})) + 2\epsilon$ on $\mathcal{L}_{\mathbf{x}}$.*

(b) Let $C = \sup\{\|\psi_{\mathbf{x}}(\mathbf{0}_{\mathbf{x}})\| : \mathbf{x} \in \mathbb{T}^n\}$. The bundle map Ψ preserves the disk bundle of radius $L_0 = C/(1-\kappa)$ in $\mathcal{L}$, where $\kappa = \sup_{\mathbf{x}\in\mathbb{T}^n} \kappa_{\mathbf{x}}$.

PROOF. (a) If $\sigma^1_{\mathbf{x}}, \sigma^2_{\mathbf{x}} \in \mathcal{L}_{\mathbf{x}}$, then by adding and subtracting the same term and applying the triangle inequality, we get

$$\begin{aligned}
&\|\psi_{\mathbf{x}}(\sigma^1_{\mathbf{x}}) - \psi_{\mathbf{x}}(\sigma^2_{\mathbf{x}})\| \\
&\quad = \|(C_{\mathbf{x}} + D_{\mathbf{x}}\sigma^1_{\mathbf{x}})(A_{\mathbf{x}} + B_{\mathbf{x}}\sigma^1_{\mathbf{x}})^{-1} - (C_{\mathbf{x}} + D_{\mathbf{x}}\sigma^2_{\mathbf{x}})(A_{\mathbf{x}} + B_{\mathbf{x}}\sigma^2_{\mathbf{x}})^{-1}\| \\
&\quad \le \|(C_{\mathbf{x}} + D_{\mathbf{x}}\sigma^1_{\mathbf{x}}) - (C_{\mathbf{x}} + D_{\mathbf{x}}\sigma^2_{\mathbf{x}})\|\,\|(A_{\mathbf{x}} + B_{\mathbf{x}}\sigma^1_{\mathbf{x}})^{-1}\| \\
&\qquad + \|(C_{\mathbf{x}} + D_{\mathbf{x}}\sigma^2_{\mathbf{x}})\|\,\|(A_{\mathbf{x}} + B_{\mathbf{x}}\sigma^1_{\mathbf{x}})^{-1} - (A_{\mathbf{x}} + B_{\mathbf{x}}\sigma^2_{\mathbf{x}})^{-1}\|.
\end{aligned}$$

Now, we estimate each term on the right-hand side above. The first term has the following estimate:

$$\begin{aligned}
\|(C_{\mathbf{x}} + D_{\mathbf{x}}\sigma^1_{\mathbf{x}}) - (C_{\mathbf{x}} + D_{\mathbf{x}}\sigma^2_{\mathbf{x}})\| &= \|D_{\mathbf{x}}[\sigma^1_{\mathbf{x}} - \sigma^2_{\mathbf{x}}]\| \\
&\le \|D_{\mathbf{x}}\|\,\|\sigma^1_{\mathbf{x}} - \sigma^2_{\mathbf{x}}\|.
\end{aligned}$$

Next,

$$\begin{aligned}
\|(A_{\mathbf{x}} + B_{\mathbf{x}}\sigma^2_{\mathbf{x}})^{-1}\| &= [m(A_{\mathbf{x}} + B_{\mathbf{x}}\sigma^2_{\mathbf{x}})]^{-1} \\
&\le [m(A_{\mathbf{x}}) - \|B_{\mathbf{x}}\| \cdot \|\sigma^2_{\mathbf{x}}\|]^{-1} \\
&\le [m(A_{\mathbf{x}}) - \epsilon]^{-1}
\end{aligned}$$

and

$$\begin{aligned}
\|(C_{\mathbf{x}} + D_{\mathbf{x}}\sigma^2_{\mathbf{x}})\| &\le \|(C_{\mathbf{x}}\| + \|D_{\mathbf{x}}\|\,\|\sigma^2_{\mathbf{x}})\| \\
&\le \epsilon + \mu.
\end{aligned}$$

To estimate the next term, note that

$$\begin{aligned}
\|a^{-1} - b^{-1}\| &= \|a^{-1}bb^{-1} - a^{-1}ab^{-1}\| \\
&\le \|a^{-1}\|\,\|b - a\|\,\|b^{-1}\|,
\end{aligned}$$

so

$$\begin{aligned}
&\|(A_{\mathbf{x}} + B_{\mathbf{x}}\sigma^1_{\mathbf{x}})^{-1} - (A_{\mathbf{x}} + B_{\mathbf{x}}\sigma^2_{\mathbf{x}})^{-1}\| \\
&\qquad \le \|(A_{\mathbf{x}} + B_{\mathbf{x}}\sigma^1_{\mathbf{x}})^{-1}\|\,\|B_{\mathbf{x}}\|\,\|\sigma^1_{\mathbf{x}} - \sigma^2_{\mathbf{x}}\|\,\|(A_{\mathbf{x}} + B_{\mathbf{x}}\sigma^2_{\mathbf{x}})^{-1}\| \\
&\qquad \le \epsilon\,(\lambda - \epsilon)^{-2}\,\|\sigma^1_{\mathbf{x}} - \sigma^2_{\mathbf{x}}\|,
\end{aligned}$$

using the fact that $\|(A_{\mathbf{x}} + B_{\mathbf{x}}\sigma^1_{\mathbf{x}})^{-1}\| \le (\lambda - \epsilon)^{-1}$, $\|(A_{\mathbf{x}} + B_{\mathbf{x}}\sigma^2_{\mathbf{x}})^{-1}\| \le (\lambda - \epsilon)^{-1}$, and $\|B_{\mathbf{x}}\| \le \epsilon$.

Combining these estimates, we get

$$\begin{aligned}
&\|\psi_{\mathbf{x}}(\sigma^1_{\mathbf{x}}) - \psi_{\mathbf{x}}(\sigma^2_{\mathbf{x}})\| \\
&\qquad \le \|D_{\mathbf{x}}\|\,\|\sigma^1_{\mathbf{x}} - \sigma^2_{\mathbf{x}}\|\,[m(A_{\mathbf{x}}) - \epsilon]^{-1} + (\epsilon + \mu)\,\epsilon\,(\lambda - \epsilon)^{-2}\,\|\sigma^1_{\mathbf{x}} - \sigma^2_{\mathbf{x}}\| \\
&\qquad \le \left[\frac{\|D_{\mathbf{x}}\|}{m(A_{\mathbf{x}})} + 2\epsilon\right]\|\sigma^1_{\mathbf{x}} - \sigma^2_{\mathbf{x}}\|
\end{aligned}$$

as claimed. (In the last inequality, we used that $||D_{\mathbf{x}}||[m(A_{\mathbf{x}})-\epsilon]^{-1} \leq [||D_{\mathbf{x}}||/m(A_{\mathbf{x}})]+\epsilon$ and $(\epsilon+\mu)\,\epsilon\,(\lambda-\epsilon)^{-2} \leq \epsilon$.)

(b) Assume $\sigma_{\mathbf{x}} \in \mathcal{L}_{\mathbf{x}}(L_0)$. Then,

$$\begin{aligned}||\psi_{\mathbf{x}}(\sigma_{\mathbf{x}})|| &\leq ||\psi_{\mathbf{x}}(\sigma_{\mathbf{x}}) - \psi_{\mathbf{x}}(\mathbf{0}_{\mathbf{x}})|| + ||\psi_{\mathbf{x}}(\mathbf{0}_{\mathbf{x}})|| \\ &\leq \kappa_{\mathbf{x}}\,||\sigma_{\mathbf{x}}|| + C \\ &\leq \kappa\, L_0 + C \\ &\leq L_0.\end{aligned}$$

Therefore, $\psi_{\mathbf{x}}(\sigma_{\mathbf{x}}) \in \mathcal{L}_{f(\mathbf{x})}(L_0)$. □

Since Ψ covers the map f, we need the estimate on $||(Df_{\mathbf{x}})^{-1}||$: $\lambda_{\mathbf{x}} = ||(Df_{\mathbf{x}})^{-1}|| \leq ||(D_{\mathbf{x}})^{-1}|| + \epsilon = ||D_{\mathbf{x}}||^{-1} + \epsilon$. (This uses the fact that $\mathbb{F}^s_{\mathbf{x}}$ is one-dimensional.) Then, the fiber contraction times the maximum expansion of f^{-1} is less than 1:

$$\begin{aligned}\sup_{\mathbf{x}\in\mathbb{T}^n} \kappa_{\mathbf{x}}\,\lambda_{\mathbf{x}} &\leq \sup_{\mathbf{x}\in\mathbb{T}^n} [(||D_{\mathbf{x}}||\,/m(A_{\mathbf{x}})) + 2\epsilon][||D_{\mathbf{x}}||^{-1} + \epsilon] \\ &\leq m(A_{\mathbf{x}})^{-1} + \epsilon + 2\epsilon/||D_{\mathbf{x}}|| \\ &\leq \lambda^{-1} + \epsilon + 2\epsilon/\mu' \\ &< 1.\end{aligned}$$

It is important that the product of the rates is taken before the supremum is taken so that the factors $||D_{\mathbf{x}}||$ and $||D_{\mathbf{x}}||^{-1}$ multiply to give 1. Thus, the contraction on fibers $\mathcal{L}(L_0)$ is stronger than the contraction within $\mathbb{T}^n$ and the invariant section is C^1. By uniqueness, the invariant section has $\mathbb{E}^u_{\mathbf{x}}$ as a graph and the unstable bundle is C^1.

The proof for the stable bundle is similar provided that the unstable bundle has one-dimensional fiber. □

REMARK 2.1. The bundles are not necessarily C^2 even if the diffeomorphism is C^3. This is because $||A_{\mathbf{x}}||^{-1}[||D_{\mathbf{x}}||^{-1} + \epsilon]$ is not necessarily less than 1.

However, various rigidity results have been proven. If an Anosov diffeomorphism $f : \mathbb{T}^2 \to \mathbb{T}^2$ is C^∞, area preserving, and has C^2 invariant bundles, then the bundles are C^∞. See Hurder and Katok (1990). Also see de la Llave, Marco, and Moriyon (1986) and de la Llave (1987).

REMARK 2.2. In higher dimensions, the bundles are not necessarily C^1 but they are Holder. See Hirsch and Pugh (1970). Their treatment uses the C^α Section Theorem of the last section.

The same type of argument can be used to prove that the stable bundle is C^1 for a hyperbolic attractor with $\dim(\mathbb{E}^u_{\mathbf{x}}) = 1$.

Theorem 2.3. *Assume $f : M \to M$ is C^2 and has a hyperbolic attractor Λ with $\dim(\mathbb{E}^u_{\mathbf{x}}) = 1$. Let $\mathbb{E}^s_{\mathbf{x}} = T_{\mathbf{x}}W^s(\mathbf{x})$ for $\mathbf{x}$ in a neighborhood of Λ. Then, the stable bundle is C^1.*

PROOF. The proof is like above, but the map f^{-1} is overflowing on a neighborhood U of Λ. The estimates similar to those of Theorem 2.1 but applied to $D(f^{-1})_{\mathbf{x}}$ give a fiber contraction on the space of linear maps whose graphs could potentially give $\mathbb{E}^s_{\mathbf{x}}$. See Hirsch and Pugh (1970) for more details. □

12.3 Differentiability of the Center Manifold

The existence of a C^r Center Manifold is stated in Section 5.10.2. In this section, we indicate how this result follows from the C^r Section Theorem. What is needed is to show that the center-unstable manifold, $W^{cu}(\mathbf{0}, f)$, and the center-stable manifold, $W^{cs}(\mathbf{0}, f)$, are C^r. (The center manifold is the intersection of these two manifolds.) We concentrate on showing that $W^{cu}(\mathbf{0}, f)$ is C^r because the proof for $W^{cs}(\mathbf{0}, f)$ is similar. The proof in Section 5.10.2 (using the methods of Section 5.10.1) shows that $W^{cu}(\mathbf{0}, f)$ is C^1. We indicate the induction argument which proves that it is C^r.

Let $2 \leq r < \infty$ be a fixed level of differentiability. (The proof for $r = 1$ is done.) We assume $f : U \subset \mathbb{R}^n \to \mathbb{R}^n$ is a C^r map with $f(\mathbf{0}) = \mathbf{0}$. Also, the tangent space at $\mathbf{0}$ splits, $\mathbb{R}^n = \mathbb{E}^u \oplus \mathbb{E}^c \oplus \mathbb{E}^s$, where these subspaces are labeled in the usual manner. Let $0 < \mu < 1$ be such that $\|Df_{\mathbf{0}}|\mathbb{E}^s\| < \mu$. Let $\lambda > 1$ be chosen so that $\mu\lambda^r < 1$. A basis can be chosen so that in terms of the norm in this basis, $\|Df_{\mathbf{0}}^{-1}|\mathbb{E}^u \oplus \mathbb{E}^c\| < \lambda$. (Notice that for a fixed $\lambda > 1$, it is not possible to satisfy this inequality for all $r \geq 1$. This is essentially the reason the manifold cannot be proven to be C^∞. Also, if $Df_{\mathbf{0}}|\mathbb{E}^c$ is diagonalizable, then it is possible to make $\|Df_{\mathbf{0}}^{-1}|\mathbb{E}^u \oplus \mathbb{E}^c\| \leq 1$.)

Next, we want to get global estimates and not just at $\mathbf{0}$. We do this by extending f using a bump function to all of $\mathbb{R}^n$ so it is uniformly near the derivative at $\mathbf{0}$. Let $\beta : \mathbb{R}^n \to \mathbb{R}$ be a bump function with $\text{supp}(\beta) = \bar{B}(2, \mathbf{0})$ and $\beta|\bar{B}(1, \mathbf{0}) = 1$. Define $\beta_\epsilon(\mathbf{x}) = \beta(\epsilon\mathbf{x})$, so $\text{supp}(\beta_\epsilon) = \bar{B}(2\epsilon, \mathbf{0})$ and $\beta_\epsilon|\bar{B}(\epsilon, \mathbf{0}) = 1$. Let $A = Df_0$ (in the basis indicated above) and

$$F_\epsilon(\mathbf{x}) = \beta_\epsilon(\mathbf{x})f(\mathbf{x}) + (1 - \beta_\epsilon(\mathbf{x}))A\mathbf{x}.$$

As ϵ goes to 0, F_ϵ converges to the linear map $A\mathbf{x}$ in the C^1 topology. (See Proposition V.7.7 and the proof of Lemma XI.4.2.) In particular, if $\epsilon > 0$ is small enough, then $\|D(F_\epsilon)_{\mathbf{x}}|\mathbb{E}^s\| < \mu$ and $\|D(F_\epsilon)_{\mathbf{x}}^{-1}|\mathbb{E}^u \oplus \mathbb{E}^c\| < \lambda$ for all $\mathbf{x} \in \mathbb{R}^n$. We fix this ϵ and write F for F_ϵ. (Notice that even if $Df_{\mathbf{0}}|\mathbb{E}^c$ is diagonalizable, it is not possible to make $\|Df_{\mathbf{x}}^{-1}|\mathbb{E}^u \oplus \mathbb{E}^c\| \leq 1$ for all $\mathbf{x} \in \mathbb{R}^n$.)

Let $\mathcal{L}(\mathbb{R}^n)$ be the bundle and $\Psi : \mathcal{L}(\mathbb{R}^n) \to \mathcal{L}(\mathbb{R}^n)$ be the bundle map as defined in Section 12.1 for the proof for $r \geq 2$. Lemma 1.6 proves that Ψ is a fiber contraction by a factor of $\mu\,\lambda$. The construction above gives that $(\mu\,\lambda)\lambda^{r-1} < 1$, so the the invariant section is C^{r-1}.

Let $\sigma^* : \mathbb{E}^u \oplus \mathbb{E}^c \to \mathbb{E}^s$ be the C^1 invariant section whose graph gives $W^{cu}(\mathbf{0}, F)$. The map $A^*_{\mathbf{x}} = D\sigma^*_{\mathbf{x}}$ is an invariant section for Ψ. It follows that $D\sigma^*_{\mathbf{x}}$ is C^{r-1} and σ^* is C^r. This completes the proof.

12.4 Persistence of Normally Contracting Manifolds

In this section, we assume that we are given a C^r diffeomorphism on a manifold, $f : M \to M$ with an invariant compact C^1 submanifold $V \subset M$, $f(V) = V$. The main theorem gives conditions for an invariant manifold to persist for perturbations of f which are C^1 small. The first step is to define a condition on the invariant submanifold called normally contracting for f at V.

To make the definitions, we need the notion of a normal bundle of a submanifold. At each point $\mathbf{x} \in V$, it is possible to pick a subspace $N_{\mathbf{x}}$ of the tangent space $T_{\mathbf{x}}M$ which is a complementary subspace to $T_{\mathbf{x}}V$, $T_{\mathbf{x}}M = T_{\mathbf{x}}V \oplus N_{\mathbf{x}}$. These subspaces can be chosen such that they vary differentiably on the point $\mathbf{x} \in V$, and together they form a vector bundle over V,

$$\mathcal{N} = \bigcup_{\mathbf{x} \in V} \{\mathbf{x}\} \times N_{\mathbf{x}}.$$

This vector bundle $\mathcal{N}$ is called a *normal bundle to V (in M)*. For each point $\mathbf{x} \in V$, there are two projections defined: $\pi^V_{\mathbf{x}} : T_{\mathbf{x}}M \to T_{\mathbf{x}}V$, the projection along $N_{\mathbf{x}}$ onto $T_{\mathbf{x}}V$, and $\pi^N_{\mathbf{x}} : T_{\mathbf{x}}M \to N_{\mathbf{x}}$, the projection along $T_{\mathbf{x}}V$ onto $N_{\mathbf{x}}$.

Definition. A diffeomorphism $f : M \to M$ is called *normally contracting at V* provided V is a compact invariant submanifold for f, and there are constants $C \geq 1$ and $0 < \mu < 1$ such that

$$\|\pi^N_{f^k(\mathbf{x})} Df^k_{\mathbf{x}}|N_{\mathbf{x}}\| \leq C\mu^k \qquad \text{and}$$
$$\|\pi^N_{f^k(\mathbf{x})} Df^k_{\mathbf{x}}|N_{\mathbf{x}}\| \leq C\mu^k\, m(Df^k_{\mathbf{x}}|T_{\mathbf{x}}V)$$

for all $\mathbf{x} \in V$ and $k \geq 1$. These conditions mean that f contracts toward V and the rate of contraction toward V is stronger than any contraction within V. (The term $m(Df^k_{\mathbf{x}}|T_{\mathbf{x}}V)$ measures any possible contraction within V.)

To get a higher degree of smoothness of the invariant manifold of a perturbation, we need to make further assumptions on the rate of contractions toward V relative to the contractions within V. For $r \geq 1$, a diffeomorphism $f : M \to M$ is called *r-normally contracting at V* provided V is a compact invariant submanifold for f, f is C^r, and there are constants $C \geq 1$ and $0 < \mu < 1$ such that

$$\|\pi^N_{f^k(\mathbf{x})} Df^k_{\mathbf{x}}|N_{\mathbf{x}}\| \leq C\mu^k\, m(Df^k_{\mathbf{x}}|T_{\mathbf{x}}V)^j$$

for all $0 \leq j \leq r$, $\mathbf{x} \in V$, and $k \geq 1$.

REMARK 4.1. There is a generalization of r-normally contracting to r-normally hyperbolic invariant manifolds; see Hirsch, Pugh, and Shub (1977) and Fenichel (1971). This latter condition allows there to be both contracting and expanding directions within the normal bundle.

In our definition, we did not require that the normal bundle be invariant by the derivative map. However, if f is normally contracting along V, then it is possible to choose another normal bundle that is invariant.

Proposition 4.1. *Assume $f : M \to M$ is normally contracting at V.*

(a) There is a continuous choice of the normal bundle that is invariant by the derivative of f, $Df_{\mathbf{x}}(N_{\mathbf{x}}) = N_{f(\mathbf{x})}$.

(b) By changing the Riemannian norm of M, it is possible to take $C = 1$ in the definition of normally contracting at V.

We leave the proof of this proposition to the exercises. See Exercise 12.9. In the rest of this section, we take the invariant normal bundle and adapted norm given by this proposition.

Once the normal bundle is invariant and $C = 1$, we can leave the projection out of the conditions on the rates of contraction, and write that $\|Df_{\mathbf{x}}|N_{\mathbf{x}}\| \leq \mu\, m(Df^k_{\mathbf{x}}|T_{\mathbf{x}}V)^j$ for $0 \leq j \leq r$. This condition that Df contracts vectors in $\mathcal{N}$ more strongly than vectors tangent to V implies that the derivative of f preserves a family of cones of vectors which point more along V than in the normal direction. This fact is crucial in the main theorem of this section, which we state next.

Theorem 4.2. *Let $f : M \to M$ be a C^r diffeomorphism, $r \geq 1$. Assume f is r-normally contracting at V, where V is a compact C^1 submanifold of M. If $g : M \to M$ is a C^r diffeomorphism which is C^1 near f, then g has a C^r invariant r-normally contracting submanifold V_g which is C^1 near V.*

REMARK 4.2. A similar theorem is true for flows with very little change in the definitions, statement, or proof.

REMARK 4.3. If g is C^r near f, then its invariant manifold V_g is C^r near V.

REMARK 4.4. This theorem has a long history. See the remarks in Hale (1969) and Hirsch, Pugh, and Shub (1977). Sacker (1967), Fenichel (1971), and Hirsch, Pugh, and Shub (1977) have recent results in this direction and beyond.

REMARK 4.5. This theorem has many applications in Dynamical Systems. The proof of the Andronov-Hopf Theorem for diffeomorphisms is related to this theorem. For this bifurcation result, the full nonlinear map is considered as a perturbation of a normal form. The normal form of the map trivially has an invariant circle which is normally contracting. As the parameter goes to the bifurcation value, the extent of contraction toward the invariant circle goes to one. However, the perturbation effects of the true nonlinear map from the normal form is small enough so the nonlinear map also has an invariant closed curve. See Ruelle and Takens (1971).

Before starting the proof of the theorem, we discuss some constructions and results related to the theorem. We start by showing that V is necessarily a C^r manifold under the hypothesis of the theorem. This same proposition shows that V_g is C^r once we have shown that it is C^1.

Proposition 4.3. *Assume f is C^r and r-normally contracting at V, where V is a compact invariant C^1 submanifold of M. Then, V is a C^r submanifold of M.*

REMARK 4.6. There are examples of f, which are r-normally contracting at V but not (r+1)-normally contracting, such that V is C^r but not C^{r+1}; in fact, this is the generic situation. See Mañé (1978a).

PROOF. The proof is very similar to that of Theorem 2.1. Again we approximate the invariant splitting $TV \oplus \mathcal{N}$ by a differentiable splitting $\mathbb{F}^V \oplus \mathbb{F}^N$, in terms of which

$$Df_{\mathbf{x}} = \begin{pmatrix} A_{\mathbf{x}} & B_{\mathbf{x}} \\ C_{\mathbf{x}} & D_{\mathbf{x}} \end{pmatrix}.$$

Given $\epsilon > 0$ with $\mu + \epsilon < 1$, because the splitting is near the invariant splitting, $||B_{\mathbf{x}}||, ||C_{\mathbf{x}}|| \leq \epsilon$, and $||D_{\mathbf{x}}|| \leq \mu m(A_{\mathbf{x}}) m(Df_{\mathbf{x}}|T_{\mathbf{x}}V)^j$ for $0 \leq j \leq r-1$.

Let $\mathcal{D}_{\mathbf{x}}$ be the space of all linear maps from $\mathbb{F}^V_{\mathbf{x}}$ to $\mathbb{F}^N_{\mathbf{x}}$ with norm less than or equal to 1,

$$\mathcal{D}_{\mathbf{x}} = \{\sigma_{\mathbf{x}} \in L(\mathbb{F}^V_{\mathbf{x}}, \mathbb{F}^N_{\mathbf{x}}) : ||\sigma_{\mathbf{x}}|| \leq 1\}.$$

Let $\mathcal{D}$ be the disk bundle over V of all the $\mathcal{D}_{\mathbf{x}}$,

$$\mathcal{D} = \bigcup_{\mathbf{x} \in V} \{\mathbf{x}\} \times \mathcal{D}_{\mathbf{x}}.$$

Let $\Psi = (f, \psi) : \mathcal{D} \to \mathcal{D}$ be the graph transform induced by the derivative of f,

$$\psi_{\mathbf{x}}(\sigma_{\mathbf{x}}) = (C_{\mathbf{x}} + D_{\mathbf{x}}\sigma_{\mathbf{x}})(A_{\mathbf{x}} + B_{\mathbf{x}}\sigma_{\mathbf{x}})^{-1}$$

for $\sigma_{\mathbf{x}} \in \mathcal{D}_{\mathbf{x}}$. The proof of Lemma 2.2 shows that Ψ is a fiber contraction by a factor of $||D_{\mathbf{x}}||\, m(A_{\mathbf{x}})^{-1} + 2\epsilon$. The map on the base space V has $\lambda_{\mathbf{x}} = m(Df_{\mathbf{x}}|T_{\mathbf{x}}V)^{-1}$. By the conditions above, Ψ satisfies the assumptions of the C^{r-1} Section Theorem. Thus, the invariant section, whose image is TV, is C^{r-1} and so V is C^r. □

Definition. Next, we introduce the notion of a *tubular neighborhood* of a submanifold. The idea is to identify a point in a neighborhood of V with a point in V and a displacement in the normal direction which is represented by a vector in $\mathcal{N}$. To state this more carefully, we use the disk bundle in $\mathcal{N}$. For $a > 0$, let $N_{\mathbf{x}}(a)$ be the vectors in $N_{\mathbf{x}}$ with length less than or equal to a; thus, $N_{\mathbf{x}}(a)$ is a closed disk in $N_{\mathbf{x}}$. Then, let $\mathcal{N}(a)$ be the bundle of all these disks,

$$\mathcal{N}(a) = \bigcup_{\mathbf{x}\in V} \{\mathbf{x}\} \times N_{\mathbf{x}}(a).$$

The Tubular Neighborhood Theorem says that there is an embedding φ from $\mathcal{N}(a)$ for small a onto a neighborhood of V in M. If M is a Euclidean space, then φ can be taken to be given by $\varphi(\mathbf{v}_{\mathbf{x}}) = \mathbf{x} + \mathbf{v}_{\mathbf{x}}$. In a manifold, $\varphi(\mathbf{v}_{\mathbf{x}}) = \exp_{\mathbf{x}}(\mathbf{v}_{\mathbf{x}})$ works but may be only C^{r-1} if V is C^r. With a little care, φ can be made to be C^r. See Hirsch (1976) for a more complete discussion.

PROOF OF THEOREM 4.2. The bundle $\mathcal{N}$ is defined at all points of the tubular neighborhood $\varphi(\mathcal{N}(a))$ by taking tangent spaces to $\varphi(N_{\mathbf{x}})$ at points in the image of a fiber. We continue to call this bundle $\mathcal{N}$. The tangent space to V can be extended to this neighborhood to be differentiable. (We do not give the details.) We denote the fibers of this bundle by $T_{\mathbf{x}}$.

Rather than use the family of cones of vectors which point more along V than in the normal direction, we use the cones which point more in the normal direction. For $\mathbf{x} \in \varphi(\mathcal{N}(a))$, let

$$C_{\mathbf{x}} = \{\mathbf{v} = (\mathbf{v}_1, \mathbf{v}_2) \in T_{\mathbf{x}} \oplus N_{\mathbf{x}} : |\mathbf{v}_1| \leq |\mathbf{v}_2|\}.$$

Because f is normally contracting at V (with $C = 1$ from an adapted norm), if a is small enough, then these cones are invariant under the action of the derivative of f^{-1}:

$$D(f^{-1})_{\mathbf{x}}(C_{\mathbf{x}}) \subset C_{f^{-1}(\mathbf{x})}.$$

We leave the details to the reader. See Exercise 12.10.

Next let D_0 be a "vertical" disk in the tubular neighborhood $\varphi(\mathcal{N}(a))$ which is the same dimension as a fiber of the normal bundle, $N_{\mathbf{y}}$, and whose tangent space $T_{\mathbf{y}}D_0$ is contained in the cone $C_{\mathbf{y}}$. We also assume that the boundary of D_0 is in the boundary of $\varphi(\mathcal{N}(a))$ and D_0 goes all the way across $\varphi(\mathcal{N}(a))$: in local coordinates we could assume that D_0 is the graph of a function from $N_{\mathbf{x}}(a)$ into $T_{\mathbf{x}}V$ for some point $\mathbf{x} \in V$. Because of the invariance of the bundles under f^{-1}, $f^{-n}(D_0) \cap \varphi(\mathcal{N}(a))$ is a disk with the same properties. As in the proof of the stable manifold theorem,

$$D_n = f^n(f^{-n}(D_0) \cap \varphi(\mathcal{N}(a))) \subset D_0$$

is a nested set of disks which converge to a single point. (Here, $\varphi(\mathcal{N}(a))$ is the tubular neighborhood which is the image of the normal disk bundle over points of V.) This point is the unique point in D_0 which stays in $\varphi(\mathcal{N}(a))$ for all backward iterates.

If $D_0 = \varphi(N_{\mathbf{x}})$ for $\mathbf{x} \in V$, then $\mathbf{x} \in D_0 \cap V$ stays in $\varphi(\mathcal{N}(a))$ for all backward iterates and so in the unique point in the intersections of the D_n. Thus, for these choices of disks, using f we recover V:

$$V = \bigcup_{\mathbf{x}\in V} \bigcap_{n\geq 0} f^n(f^{-n}(\varphi(N_{\mathbf{x}})) \cap \varphi(\mathcal{N}(a)))$$

Now, if g is a C^r diffeomorphism which is C^1 near to f, then g^{-1} will also preserve the above set of cones:

$$D(g^{-1})_{\mathbf{x}}(C_{\mathbf{x}}) \subset C_{g^{-1}(\mathbf{x})}.$$

Again, if D_0 is a "vertical" disk of the same type as above, then

$$\bigcap_{n\geq 0} g^n(g^{-n}(D_0) \cap \varphi(\mathcal{N}(a)))$$

is a single point. Thus,

$$V_g = \bigcup_{\mathbf{x}\in V} \bigcap_{n\geq 0} g^n(g^{-n}(\varphi(N_{\mathbf{x}})) \cap \varphi(\mathcal{N}(a)))$$

is a graph over V in the sense that $\varphi^{-1}(V_g) \subset \mathcal{N}(a)$ can be represented as the graph of a function $\psi_g : V \to \mathcal{N}(a)$ with $\psi_g(\mathbf{x}) \in N_{\mathbf{x}}(a)$. This function is also Lipschitz because it has a unique point in $\varphi(C_{\mathbf{x}})$ for $\mathbf{x} \in V$. (This follows because any vertical disk has a unique point in $\varphi^{-1}(V_g)$ just as in the proof of the Stable Manifold Theorem.)

An argument like that for the stable manifold proves that V_g is a C^1 graph over V. Just as the tangent space to V is an invariant section for the graph transform Ψ as defined in the proof of Proposition 4.3, the tangent bundle to V_g is a fixed section for a similar graph transform Ψ^g induced by the derivative of g. This map Ψ^g is C^0 near Ψ provided g is C^1 near f, so the sections are C^0 near. The fact that the tangent space to V_g is C^0 near the tangent space to V implies that V_g is C^1 near V. We leave the details to the reader.

If g is C^1 near enough to f, then g is r-normally contracting at V_g, so by Proposition 4.3, V_g is C^r. □

12.5 Exercises

Fiber Contractions

12.1. Let $F : X_0 \times Y_0 \to X \times Y_0$ be as in Theorem 1.1

(a) Prove that for each $\mathbf{x} \in X_0$,

$$\bigcap_{n=0}^{\infty} F^n(\{f^{-n}(\mathbf{x})\} \times Y_0)$$

is a unique point $(\mathbf{x}, \sigma^*(\mathbf{x}))$. (Do not use the conclusion of Theorem 1.1.)

(b) Prove that the map $\sigma^* : X_0 \to Y$ defined in part (a) is continuous.

12.2. Consider Theorem 1.2 for $r = 0$ and $0 < \alpha < 1$. Let

$$\mathcal{G}^\alpha(L) = \{\sigma \in \mathcal{G}^0 : |\sigma(\mathbf{x}) - \sigma(\mathbf{x}')| \leq L\, d(\mathbf{x}, \mathbf{x}')^\alpha \text{ for all } \mathbf{x}, \mathbf{x}' \in X_0\}.$$

(a) Prove that $\mathcal{G}^\alpha(L)$ is closed in $\mathcal{G}^0$.

(b) Prove that for $L_0 > 0$ large enough, Γ_F preserves $\mathcal{G}^\alpha(L_0)$.

(c) Prove that the invariant section $\sigma^* \in \mathcal{G}^\alpha(L_0)$.

12.3. Consider Theorem 1.2 for $r = 1$ and $0 < \alpha \le 1$. Let

$$\mathcal{G}^{1+\alpha}(C_1, L) = \{\sigma \in C^1(X_0, Y_0) : |D\sigma_{\mathbf{x}}| \le C_1 \text{ and } |D\sigma_{\mathbf{x}} - D\sigma_{\mathbf{x}'}| \le L\, d(\mathbf{x}, \mathbf{x}')^\alpha \text{ for all } \mathbf{x}, \mathbf{x}' \in X_0\}.$$

Henry (1981) in Lemma 6.1.6 proves that $\mathcal{G}^{1+\alpha}(C_1, L)$ is closed in $\mathcal{G}^0$ for any $0 < \alpha \le 1$. (Note this is not true for $\alpha = 0$.) In fact, he proves with the suitable definition that $\mathcal{G}^{r+\alpha}(C_1, L)$ is closed in $\mathcal{G}^0$ for any integer $r \ge 0$ and $0 < \alpha \le 1$.

(a) Prove that for $C_1, L_0 > 0$ large enough, Γ_F preserves $\mathcal{G}^{1+\alpha}(C_1, L_0)$.

(b) Prove that the invariant section $\sigma^* \in \mathcal{G}^{1+\alpha}(C_1, L_0)$. (You may use the result of Henry.)

12.4. (Fiber Contraction Theorem) Assume X and Y are metric spaces with Y complete. Assume $F : X \times Y \to X \times Y$ is a uniformly continuous fiber contraction on X with $F(\mathbf{x}, \mathbf{y}) = (f(\mathbf{x}), g(\mathbf{x}, \mathbf{y}))$, i.e., there exists a $0 < \kappa < 1$ such that

$$d(g(\mathbf{x}, \mathbf{y}_1), g(\mathbf{x}, \mathbf{y}_2)) \le \kappa\, d(\mathbf{y}_1, \mathbf{y}_2)$$

for all $\mathbf{x} \in X$ and $\mathbf{y}_1, \mathbf{y}_2 \in Y$. Assume there is a $\mathbf{x}^* \in X$ such that for any $\mathbf{x} \in X$, $d(f^n(\mathbf{x}), \mathbf{x}^*)$ goes to zero as n goes to ∞. (Such a fixed point is called *attractive.* Let $\mathbf{y}^*$ be the fixed point of $g(\mathbf{x}^*, \cdot) : Y \to Y$. Let $(\mathbf{x}, \mathbf{y})$ be a point in $X \times Y$. Let $\pi_2 : X \times Y \to Y$ be the projection onto Y. Note that

$$\begin{aligned} d(\pi_2 \circ F^n(\mathbf{x}, \mathbf{y}), \mathbf{y}^*) &\le d(\pi_2 \circ F^n(\mathbf{x}, \mathbf{y}), \pi_2 \circ F^n(\mathbf{x}, \mathbf{y}^*)) \\ &\quad + \sum_{j=0}^{n-1} d(\pi_2 \circ F^{n-1-j}(f^{j+1}(\mathbf{x}), g(f^j(\mathbf{x}), \mathbf{y}^*)), \\ &\qquad \pi_2 \circ F^{n-1-j}(f^{j+1}(\mathbf{x}), \mathbf{y}^*)). \end{aligned}$$

(a) Show that $d(\pi_2 \circ F^n(\mathbf{x}, \mathbf{y}), \pi_2 \circ F^n(\mathbf{x}, \mathbf{y}^*))$ goes to zero as n goes to infinity.

(b) Let $\delta_j = d(g(f^j(\mathbf{x}), \mathbf{y}^*), y^*)$. Prove δ_j goes to zero as j goes to infinity.

(c) Estimate

$$d(\pi_2 \circ F^{n-1-j}(f^{j+1}(\mathbf{x}), g(f^j(\mathbf{x}), \mathbf{y}^*)), \pi_2 \circ F^{n-1-j}(f^{j+1}(\mathbf{x}), y^*)).$$

(d) Splitting the sum from 0 to $n-1$ into the two sums from 0 to $k-1$ and from k to $n-1$, show that $d(\pi_2 \circ F^n(\mathbf{x}, \mathbf{y}), \mathbf{y}^*)$ goes to zero as n goes to infinity. Thus, prove for any $(\mathbf{x}, \mathbf{y}) \in X \times Y$, $d(F^n(\mathbf{x}, \mathbf{y}), (\mathbf{x}^*, \mathbf{y}^*))$ goes to zero as n goes to infinity, and $(\mathbf{x}^*, \mathbf{y}^*)$ is an attractive fixed point.

12.5. Assume $F : X_0 \times Y_0 \to X \times Y_0$ is as in Theorem 1.2 for some $r \ge 1$. Let $\mathcal{L}(X)$ be as defined in the proof. Write $Y_0 \oplus \mathcal{L}(X)$ for

$$\bigcup_{\mathbf{x} \in X} \{\mathbf{x}\} \times Y_0 \times \mathbf{L}(T_{\mathbf{x}}X, Y).$$

Define $\Theta : Y_0 \oplus \mathcal{L}(X_0) \to Y_0 \oplus \mathcal{L}(X)$ by $\Theta(\mathbf{x}, \mathbf{y}, S) = (f(\mathbf{x}), g(\mathbf{x}, \mathbf{y}), \psi(\mathbf{x}, \mathbf{y}, S))$, where

$$\psi(\mathbf{x}, \mathbf{y}, S) = Dg_{(\mathbf{x}, \mathbf{y})} \circ (id, S) \circ D(f^{-1})_{f(\mathbf{x})}.$$

(Note the similarity to ψ defined in Section 12.1.)

(a) Show that Θ is a C^{r-1} fiber contraction over X_0.

(b) Note that the continuous sections of $Y_0 \oplus \mathcal{L}(X_0)$ can be written as $\mathcal{G}^0 \times \mathcal{H}^0$, where $\mathcal{G}^0$ is as before and $\mathcal{H}^0$ are continuous sections of $\mathcal{L}(X_0)$. Let Γ_Θ be the graph transform of Θ on $\mathcal{G}^0 \times \mathcal{H}^0$. Show that $\Gamma_\Theta(\sigma, S) = (\Gamma_F(\sigma), \Gamma_\psi(\sigma, S))$, where

$$\Gamma_\psi(\sigma, S)(\mathbf{x}) = Dg_{\hat{\sigma} \circ f^{-1}(\mathbf{x})} \circ (id, S_{f^{-1}(\mathbf{x})}) \circ D(f^{-1})_{\mathbf{x}}.$$

(c) Using the previous exercise, show that Γ_{Θ} has a unique fixed point (σ^*, A^*). Conclude that if $\sigma \in \mathcal{G}^1$, then $\Gamma_{\Theta}(\sigma, D\sigma)$ converges to (σ^*, A^*) and σ^* is C^1.

(d) Prove that σ^* is C^r.

12.6. Let $A : \mathbb{F}^u_{\mathbf{x}} \to \mathbb{F}^u_{\mathbf{x}'}$, $\sigma : \mathbb{F}^s_{\mathbf{x}} \to \mathbb{F}^u_{\mathbf{x}}$, and $B_{\mathbf{x}} : \mathbb{F}^s_{\mathbf{x}} \to \mathbb{F}^u_{\mathbf{x}'}$ be linear maps between Banach spaces. Prove that $m(A + B\sigma) \geq m(A) - \|B\|\,\|\sigma\|$.

Differentiability of an Invariant Splitting

12.7. Set up the bundle map for the proof of Theorem 2.2. Show that it is overflowing on the base space and a fiber contraction with a stronger contraction on the fiber than the contraction on the base space.

Differentiability of the Center Manifold

12.8. Let f be a C^r diffeomorphism on $\mathbb{R}^n$ for $1 \leq r \leq \infty$. Assume $\mathbf{0}$ is a fixed point that has a center (some eigenvalues with absolute value 1). Assuming that the local unstable manifold $W^u(\mathbf{0}, f)$ is C^1, prove that it is C^r. (Use the theorems of this chapter, but do not use the theorems of Section 5.10.2.)

Normally Contracting Manifolds

12.9. (Proposition 4.1) Assume $f : M \to M$ is normally contracting at V.

(a) Prove that there is a continuous choice of the normal bundle that is invariant by the derivative of f, $Df_{\mathbf{x}}(N_{\mathbf{x}}) = N_{f(\mathbf{x})}$.

(b) Prove that it is possible to take $C = 1$ in the definition of normally contracting at V by changing the Riemannian norm of M.

12.10. Assume $f : M \to M$ is normally contracting at V. Let $\{C_{\mathbf{x}}\}$ be the family of cones as defined in the proof of Theorem 4.2. Prove that these cones are invariant under the action of the derivative of f^{-1}:

$$D(f^{-1})_{\mathbf{x}}(C_{\mathbf{x}}) \subset C_{f^{-1}(\mathbf{x})}.$$

12.11. Assume $f : (-\epsilon_0, \epsilon_0) \times M \to M$ is C^1 and $f(\epsilon, \cdot) : M \to M$ is a diffeomorphism for each $\epsilon \in (-\epsilon_0, \epsilon_0)$. Assume $f(0, \cdot)$ has a 1-normally contracting invariant manifold V_0. Theorem 4.2 proves that there is an $\epsilon_1 > 0$ such that $f(\epsilon, \cdot)$ has a 1-normally contracting invariant manifold V_ϵ which is C^1 for $|\epsilon| \leq \epsilon_1$. Define the map $F : (-\epsilon_0, \epsilon_0) \times M \to (-\epsilon_0, \epsilon_0) \times M$ by $F(\epsilon, \mathbf{x}) = (\epsilon, f(\epsilon, \mathbf{x}))$.

(a) By constructing an invariant set of cones for F, prove that the set

$$\mathcal{V} = \{(\epsilon, \mathbf{x}) : \mathbf{x} \in V_\epsilon \text{ for } |\epsilon| \leq \epsilon_1\}$$

is a Lipschitz manifold in $(-\epsilon_0, \epsilon_0) \times M$. This says that the normally contracting invariant manifold varies in a Lipschitz manner. This manifold $\mathcal{V}$ could be called the "Center Manifold" for the normally contracting invariant manifold V_0.

(b) Prove that $\mathcal{V}$ is a C^1 manifold in $(-\epsilon_0, \epsilon_0) \times M$. This says that the normally contracting invariant manifold varies differentiably.

References

Abraham, R. and Marsden, J. (1978), *Foundation of Mechanics (2nd Edition)*, Benjamin-Cummings Publ. Co., Reading MA.

Abraham, R. and Robbin, J. (1967), *Transversal Mappings and Flows*, Benjamin, Reading MA.

Abraham, R. and Smale, S. (1970), Nongenericity of Ω-stability, *Proc. Symp. in Pure Math., Amer. Math. Soc.* **14**, 5–8.

Adler, R. and Weiss, B. (1970), Similarity of automorphisms of the torus, *Memoirs of Amer. Math. Soc.* **98**.

Alekseev, V. M. (1968a), Quasirandom dynamical systems, I, *Math. USSR-Sb.* **5**, 73–128.

Alekseev, V. M. (1968b), Quasirandom dynamical systems, II, *Math. USSR-Sb.* **6**, 505–560.

Alekseev, V. M. (1969), Quasirandom dynamical systems, III, *Math. USSR-Sb.* **7**, 1–43.

Alekseev, V. M. (1981), Quasirandom oscillations and qualitative questions in celestial mechanics, *Amer. Math. Soc. Translations (Three Papers on Dynamical Systems)* **166 (Series 2)**, 97–169.

Alligood, K. and Yorke, J. (1989), Fractal basin boundaries and chaotic attractors, *Proc. of Symposia in Applied Mathematics* **39**, 41–55.

Alligood, K., Tedeschini-Lalli, L., and Yorke, J. (1991), Metamorphoses: sudden jumps in basin boundaries, *Commun. Math. Physics* **141**, 1–8.

Alsedà, L., Llibre, J., and Misiurewicz, M. (1993), *Combinatorial Dynamics and Entropy in Dimension One, Advanced Series in Nonlinear Dynamics*, vol. 5, World Scientific Publ., River Edge NJ & Singapore.

Andronov, A. A. (1929), Application of Poincaré's theorem on bifurcation points and change in stability to simple autooscillatory systems, *C. R. Acad. Sci. (Paris)* **189**, 559–561.

Andronov, A. A. and Leontovich–Andronova, E. A. (1939), Some cases of the dependence of periodic motions on a parameter, *Ucenye Zapiski Gorki Gosudarstvenny Univ.* **6**, 3.

Andronov, A. A. and Pontryagin, L. (1937), Systèmes Grossiers, *Dokl. Akad. Nauk. USSR* **14**, 247–251.

Andronov, A. A., Leontovich, E. A., Gordon, I. I., and Maier, A. G. (1973), *Theory of Bifurcation of Dynamical Systems on the Plane*, Wiley, New York.

Anosov, D. V. (1967), Geodesic flows on closed Riemannian manifolds of negative curvature, *Proc. Steklov Inst.* **90**, Amer. Math. Soc. (transl. 1969).

Aoki, N. (1991), The set of Axiom A diffeomorphisms with no cycles, in Collection: *Dynamical Systems and Related Topics*, World Scientific Publ., River Edge NJ, pp. 20–35.

Aoki, N. (1992), The set of Axiom A diffeomorphisms with no cycles, *Bol. Soc. Brasil Mat.* **23**, 21–65.

Apostol, T. (1974), *Mathematical Analysis*, Addison-Wesley Publ. Co., New York and Reading MA.

Arnold, V. I. (1961), On the mapping of the circle into itself, *Izvestia Akad. Nauk. USSR Math.* **25**, 21–86.

Arnold, V. I. (1971), *Ordinary Differential Equations*, M.I.T. Press, Cambridge MA.

Arnold, V. I. (1978), *Mathematical Methods of Classical Mechanics*, Springer-Verlag, New York, Heidelberg, Berlin.

Arnold, V. I. (1983), *Geometric Methods in the Theory of Ordinary Differential Equations*, Springer-Verlag, New York, Heidelberg, Berlin.

Arnold, V. I. and Avez, A. (1968), *Ergodic Problems of Classical Mechanics*, Benjamin, New York, Amsterdam.

Arnold, V. I., Kozlov, V. V., and Neishtadti, A. I. (1993), *Mathematical Aspects of Classical and Celestial Mechanics*, Springer-Verlag, New York, Heidelberg, Berlin.

Arrowsmith, D. K. and Place, C. M. (1990), *An Introduction to Dynamical Systems*, Cambridge University Press, Cambridge, New York.

Artin, M. and Mazur B. (1965), On periodic points, *Ann. Math.* **81**, 82 – 89.

Banks, J., Brooks, J., Cairns, G., Davis, G., and Stacey, P. (1992), On Devaney's definition of chaos, *Amer. Math. Monthly* **99**, 332 – 334.

Barge, M. (1986), Horseshoe maps and inverse limits, *Pacific J. Math.* **121**, 29 – 39.

Barge, M. (1988), A method for constructing attractors, *Ergodic Theory Dynamical Systems* **8**, 331 – 349.

Barge, M. and Gillette, R. (1990), Rotation and periodicity in plane separating continua, *Ergodic Theory Dynamical Systems* **11**, 619-631.

Barge, M. and Martin, J. (1990), Construction of global attractors in the plane, *Proc. Amer. Math. Soc.* **110**, 523 – 525.

Belitskii, G. R. (1973), Functional equations and conjugacy of local diffeomorphism of a finite smoothness class, *Functional Analysis and Its Applications* **7**, 268 – 277.

Benedicks, M. and Carleson, L. (1985), On iterates of $x \mapsto 1 - ax^2$ on $(0, 1)$, *Ann. Math.* **122**, 1 – 25.

Benedicks, M. and Carleson, L. (1991), The dynamics of the Hénon map, *Ann. Math.* **133**, 73 – 169.

Benedicks, M. and Young, L. S. (1993), SBR measures for certain Hénon maps, *Inven. Math.* **112**, 541 – 576.

Birkhoff, G. D. (1927), *Dynamical Systems*, Amer. Math. Soc., Providence, RI.

Birman, J. and Williams, R. (1983a), Knotted periodic orbits in dynamical systems I: Lorenz equations, *Topology* **22**, 47 – 82.

Birman, J. and Williams, R. (1983b), Knotted periodic orbits in dynamical systems II: Knot holders and fibered knots, *Contemporary Math.* **20**, 1 – 60.

Blanchard, P. (1984), Complex analytic dynamics on the Riemann sphere, *Bull. Amer. Math. Soc.* **11**, 85 – 141.

Block, L. and Coppel, W. (1992), *Dynamics in One Dimension (Lecture Notes in Mathematics* **1513***)*, Springer-Verlag, New York, Heidelberg, Berlin.

Block, L., Guckenheimer, J., Misiurewicz, M., and Young, L. S. (1980), Periodic points and topological entropy of one dimensional maps, in *Lect. Notes in Math.*, vol. **819**, Springer-Verlag, New York, Heidelberg, Berlin, pp. 18 – 34.

Bowen, R. (1970a), Markov partitions for Axiom A diffeomorphisms, *Amer. J. Math.* **92**, 725 – 747.

Bowen, R. (1970b), Topological entropy and Axiom A, *Global Analysis, Proc. Sympos. Pure Math., Amer. Math. Soc.* **14**, 23 – 42.

Bowen, R. (1971), Entropy for group endomorphisms and homogeneous spaces, *Trans. Amer. Math. Soc.* **153**, 401 – 414.

Bowen, R. (1975a), ω-Limit sets for Axiom A diffeomorphisms, *J. Diff. Eq.* **18**, 333 – 339.

Bowen, R. (1975b), *Equilibrium States and the Ergodic Theory of Anosov Diffeomorphisms (Lecture Notes in Mathematics* **470***)*, Springer-Verlag, New York, Heidelberg, Berlin.

Bowen, R. (1978a), *On Axiom A Diffeomorphisms (Conference Board Math. Science)*, vol. **35**, Amer. Math. Soc., Providence, RI.

Bowen, R. (1978b), Markov partitions are not smooth, *Proc. Amer. Math. Soc.* **71**, 130 – 132.

Bowen, R. and Lanford, O. (1970), Zeta function of restrictions of the shift transformation, *Global Analysis, Proc. Sympos. Pure Math., Amer. Math. Soc.* **14**, 43–49.

Bowen, R. and Ruelle, D. (1975), The ergodic theory of Axiom A flows, *Inven. Math.* **29**, 181–202.

Boyle, M. (1993), Symbolic dynamics and matrices, *IMA Vol. in Math. and Its Appl.* **50**, Springer-Verlag, New York, Heidelberg, Berlin, 1–38.

Braun, M. (1978), *Differential Equations and Their Applications*, Springer-Verlag, New York, Heidelberg, Berlin.

Broer, H., Dumortier, F., van Strien, S., and Takens, F. (1991), *Structures in Dynamics: Finite Dimensional Deterministic Studies*, North-Holland, Elsevier Sci. Publ., Amsterdam.

Brunovsky, P. (1971), On one parameter families of diffeomorphisms II: generic branching in higher dimensions, *Commun. Math. Univ. Carolinae* **12**, 765–784.

Burns, K. and Gerber, M. (1989), Continuous invariant cone families and ergodicity of flows in dimension three, *Ergodic Theory Dynamical Systems* **9**, 19–25.

Burns, K. and Weiss, H. (1995), A geometric criterion for positive topological entropy, *Commun. Math. Phys.* **172**, 95–118.

Carbinatto, M., Kwapisz, J., and Mischaikow, K., (1997), Horseshoes and the Conley index spectrum, *preprint Georgia Institute of Technology.*

Carleson, L. and Gamelin, T. (1993), *Complex Dynamics*, Springer-Verlag, New York, Heidelberg, Berlin.

Carr, J. (1981), *Applications of Centre Manifold Theory*, Springer-Verlag, New York, Heidelberg, Berlin.

Cartwright, M. and Littlewood, J. (1945), On nonlinear differential equations of the second order: I. The equation $\ddot{y} - k(1 - y^2)\dot{y} + y = b\lambda k\cos(\lambda t + \alpha)$, k large, *J. London Math. Soc.* **20**, 180–189.

Cesari, L. (1959), *Asymptotic Behavior and Stability Problems in Ordinary Differential Equations*, Springer-Verlag, New York, Heidelberg, Berlin.

Chernov, N. I. and Sinai, Ya. G. (1987), Ergodic properties of some systems of 2-dimensional discs and 3-dimensional spheres, *Russ. Math. Surveys* **42**, 181–207.

Chillingworth, D. (1976), *Differential Topology with a View to Applications*, Pitman Publ., London, San Francisco, Melbourne.

Choquet, G. (1969), *Lectures on Analysis, I*, Benjamin, New York.

Chow, S.-N. and Hale, J. (1982), *Methods of Bifurcation Theory*, Springer-Verlag, New York, Heidelberg, Berlin.

Chow, S.-N., Lin, X. B., and Palmer, K. (1989), A shadowing lemma with applications to semilinear parabolic equations, *SIAM J. Math. Anal.* **20**, 547–557.

Collet, P. and Eckmann, J. P. (1980), *Iterated Maps on the Interval as Dynamical Systems, Progress on Physics*, vol. 1, Birkhäuser, Boston.

Conley, C. (1978), Isolated Invariant Sets and Morse Index, *Amer. Math. Soc., CBMS* **38**.

Coullet, C. and Tresser, C. (1978), Itération d'endomorphismes et groupe de renormalisation, *J. de Physique Colloque* **39**, C5–C25.

Croom, F. (1989), *Principles of Topology*, Saunders College Pub., Philadelphia.

Curry, J. H. (1979), On the Hénon transformation, *Commun. Math. Phys.* **68**, 129–140.

Dankner, A. (1978), On Smale's Axiom A dynamical systems, *Ann. Math.* **107**. 517–553.

Dankowicz, H. (1997), *Chaotic Dynamics in Hamiltonian Systesm*, World Scientific, Singapore, New Jersey, London, Hong Kong.

Denjoy, A. (1932), Sur les courbes définies par les équations différentielles à la surface du tore, *J. Math. Pure Appl.* **11**, 333–375.

Devaney, R. (1989), *Chaotic Dynamical Systems*, Addison-Wesley Publ. Co., New York & Reading, MA.

Devaney, R. and Nitecki, Z. (1979), Shift automorphism in the Hénon mapping, *Commun. Math. Phys.* **67**, 137–148.

Dieudonné, J. (1960), *Foundations of Modern Analysis*, Academic Press, New York.

Dold, A. (1972), *Lectures on Algebraic Topology*, Springer-Verlag, New York, Berlin, Heidelberg.

Douady, A. and Hubbard, J. H. (1985), On the dynamics of polynomial-like mappings, *Ann. Sc. E. N. S.* **4 serié 18**, 287–343.

Dugundji, J. (1966), *Topology*, Allyn and Bacon, Boston, London, Sydney, Toronto.

Dumortier, F., Kokubu, H., and Oka, H. (1992), On degenerate singularities that generate geometric Lorenz attractors, *Preprint.*

Easton, R. (1986), Trellises formed by stable and unstable manifolds, *Trans. Amer. Math. Soc.* **294**, 719–731.

Easton, R. (1991), Transport through chaos, *Nonlinearity* **4**, 583–590.

Easton, R. (1998), *Geometric Methods for Discrete Dynamical Systems*, Oxford Universtiy Press, New York, Oxford.

Edgar, G. (1990), *Measure, Topology, and Fractal Geometry*, Springer-Verlag, New York, Berlin, Heidelberg.

Edwards, C. H. and Penney, D. E. (1990), *Calculus and Analytical Geometry*, Prentice Hall, Englewood Cliffs, NJ.

Falconer, K. (1990), *Fractal Geometry*, John Wiley & Sons, New York.

Farmer, D., Ott, E., and Yorke, J. (1983), The dimension of chaotic attractors, *Physica D* **3**, 153–180.

Feigenbaum, M. (1978), Quantitative universality for a class of non-linear transformations, *J. Stat. Phys.* **21**, 25–52.

Fenichel, N. (1971), Persistence and smoothness of invariant manifolds for flows, *Indiana Univ. Math. J.* **21**, 193–226.

Fenichel, N. (1974), Asymptotic stability with rate conditions, *Indiana Univ. Math. J.* **23**, 1109–1137.

Fenichel, N. (1977), Asymptotic stability with rate conditions, II, *Indiana Univ. Math. J.* **26**, 81–93.

Franks, J. (1969), Anosov diffeomorphisms on tori, *Trans. Amer. Math. Soc.* **145**, 117–124.

Franks, J. (1970), Anosov diffeomorphisms, *Global Analysis, Proc. Sympos. Pure Math., Amer. Math. Soc.* **14**, 61–94.

Franks, J. (1971), Necessary conditions for stability of diffeomorphisms, *Trans. Amer. Math. Soc.* **158**, 301–308.

Franks, J. (1972), Differentiably Ω-stable diffeomorphisms, *Topology* **11**, 107–114.

Franks, J. (1973), Absolutely structurally stable diffeomorphisms, *Proc. Amer. Math. Soc.* **37**, 293–296.

Franks, J. (1974), Time dependent structural stability, *Inven. Math.* **24**, 163–172.

Franks, J. (1977a), Constructing structurally stable diffeomorphisms, *Ann. Math.* **105**, 343–359.

Franks, J. (1977b), Invariant sets of hyperbolic toral automorphisms, *Amer. J. Math.* **99**, 1089–1095.

Franks, J. (1979), Manifolds of C^r mappings and applications to differentiable dynamical systems, *Adv. Math. Suppl. Studies (Studies in Analysis)* **4**, 271–290.

Franks, J. (1982), Homology and Dynamical Systems, *Amer. Math. Soc., CBMS* **49**, Providence, RI.

Franks, J. (1988), A variation on the Poincaré-Birkhoff Theorem, *Contemporary Math.* **81**, 111–117.

Franks, J. and Shub, M. (1981), The existence of Morse Smale diffeomorphisms, *Topology* **20**, 273–290.

Fried, D. (1982), Geometry of cross-sections to flows, *Topology* **21**, 353–371.

Fried, D. (1987), Rationality for isolated expansive sets, *Adv. Math.* **65**, 35–38.

Frobenius, G. (1912), Über Matrizen aus nicht negativen Elementen, *S.-B. Deutsch. Akad. Wiss. Berlin Math.-Nat. Kl.*, 456–77.

Gantmacher, F.R. (1959), *The Theory of Matrices, Volume I, II*, Chelsea Publ. Co., New York.

Grebogi, C., Hammel, S., and Yorke, J. (1988), Numerical orbits of chaotic processes represent true orbits, *Bull. Amer. Math. Soc.* **19**, 465–469.

Grebogi, C., Ott, E., and Yorke, J. (1987), Basin boundary metamorphoses: changes in accessible boundary orbits, *Physica D* **24**, 243–262.

Gruendler, J. (1985), The existence of homoclinic orbits and the method of Melnikov for systems in $\mathbb{R}^n$, *SIAM Jour. Math. Anal.* **16**, 907–931.

Guckenheimer, J. (1970), Axiom A + no cycles implies $\zeta_f(t)$ rational, *Bull. Amer. Math. Soc.* **76**, 592–595.

Guckenheimer, J. (1972), Absolutely Ω-stable diffeomorphisms, *Topology* **11**, 195–197.

Guckenheimer, J. (1976), A strange, strange attractor, in *The Hopf Bifurcation and Its Applications* (by Marsden and McCracken), Springer-Verlag, New York, Heidelberg, Berlin.

Guckenheimer, J. (1979), Sensitive dependence to initial conditions for one dimensional maps, *Commun. Math. Phys.* **70**, 133–160.

Guckenheimer, J. (1980), Bifurcations of dynamical systems, in *Dynamical Systems, Progress in Math.* **8**, Birkhauser, Boston, Basel, Stuttgart, pp. 115–232.

Guckenheimer, J. and Holmes, P. (1983), *Nonlinear Oscillations, Dynamical Systems and Bifurcations of Vector Fields*, Springer-Verlag, New York, Heidelberg, Berlin.

Guckenheimer, J. and Williams R. (1980), Structural stability of the Lorenz attractor, *Publ. Math. I.H.E.S.* **50**, 73–100.

Guillemin, V. and Pollack A. (1974), *Differential Topology*, Prentice-Hall, Englewood Cliffs, NJ.

Hadamard, J. (1901), Sur l'itération it les solutions asymptotiques des équations différentielles, *Bull. Soc. Math. France* **29**, 224–228.

Hahn, W. (1967), *Stability of Motion*, Springer-Verlag, New York, Heidelberg, Berlin.

Hale, J. (1969), *Ordinary Differential Equations*, Wiley, New York.

Hale, J. and Koçak, H. (1991), *Dynamics and Bifurcation*, Springer-Verlag, New York, Heidelberg, Berlin.

Hammel, S. and Jones, C. (1989), Jumping stable manifolds for dissipative maps of the plane, *Physica D* **35**, 87–106.

Hartley, R. and Hawkes, T. (1970), *Rings, Modules, and Linear Algebra*, Chapman and Hall, New York and London.

Hartman, P. (1960), On local homeomorphisms of Euclidean spaces, *Bol. Soc. Mat. Mexicana* **5**, 220–241.

Hartman, P. (1964), *Ordinary Differential Equations*, Wiley, New York.

Hayashi, S. (1992), Diffeomorphisms in $\mathcal{F}^1(M)$ satisfy Axiom A, *Ergodic Theory Dynamical Systems* **12**, 233–253.

Hayashi, S. (1997), Connecting invariant manifolds and the solution of the C^1 stability and Ω-stability conjectures for flows, *Ann. of Math.* **145**, 81–137.

Hénon, M. (1976), A two-dimensional mapping with a strange attractor, *Commun. Math. Phys.* **50**, 69–77.

Henry, B. R. (1973), Escape from the unit interval under the transformation $x \mapsto \lambda x(1-x)$, *Proc. Amer. Math. Soc.* **41**, 146–150.

Henry, D. (1981), *Geometric Theory of Semilinear Parabolic Equations (Lecture Notes in Math.* **840** *)*, Springer-Verlag, New York, Heidelberg, Berlin.

Herman, M. (1979), Sur la conjugation différentiable des difféomorphisms du cercle à des rotations, *Publ. Math. I.H.E.S.* **49**, 5–233.

Herman, M. (1983), Sur les courbes invariantes par les diffeomorphismes de l'anneau I, *Asterisque* **103**, 3–221.

Herman, M. (1983), Sur les courbes invariantes par les diffeomorphismes de l'anneau II, *Asterisque* **144**, 3–221.

Hille, E. (1962), *Analytic Function Theory*, vol. II, Chelsea Publ. Co., New York.

Hirsch, M. (1976), *Differential Topology*, Springer-Verlag, New York, Heidelberg, Berlin.

Hirsch, M. (1982), Systems of differential equations that are competitive or cooperative, I: limit sets, *SIAM J. Math. Anal.* **13**, 167–79.

Hirsch, M. (1984), The dynamical systems approach to differential equations, *Bull. Amer. Math. Soc.* **11**, 1–64.

Hirsch, M. (1985), Systems of differential equations that are competitive or cooperative, II: convergence almost everywhere, *SIAM J. Math. Anal.* **16**, 423–39.

Hirsch, M. (1988), Systems of differential equations that are competitive or cooperative, III: competitive species, *Nonlinearity* **1**, 51–71.

Hirsch, M. (1990), Systems of differential equations that are competitive or cooperative, IV: structural stability in 3-dimensional systems, *SIAM J. Math. Anal.* **21**, 1225–34.

Hirsch, M. and Pugh, C. (1970), Stable manifolds and hyperbolic sets, *Proc. Symp. Pure Math.* **14**, 133–163.

Hirsch, M., Pugh, C., and Shub, M. (1977), *Invariant Manifolds (Lecture Notes in Math.* **583***)*, Springer-Verlag, New York, Heidelberg, Berlin.

Hirsch, M. and Smale, S. (1974), *Differential Equations, Dynamical Systems, and Linear Algebra*, Academic Press, New York.

Hocking, J. G. and Young, G. S. (1961), *Topology*, Addison-Wesley Publ. Co., New York & Reading, MA.

Hoffman, K. and Kunze, K. (1961), *Linear Algebra*, Prentice-Hall, Englewood Cliffs, NJ.

Holmes, P. (1980), Averaging and chaotic motions in forced oscillations, *SIAM J. Appl. Math.* **38**, 65–80.

Holmes, P. (1984), Bifurcation sequences in horseshoe maps: infinitely many routes to chaos, *Phys. Lett. A* **104**, 299–302.

Holmes, P. (1988), Knots and orbit genealogies in nonlinear oscillations, in *New Directions in Dynamical Systems (London Math. Soc. Lecture Notes* **127**), Cambridge University Press, Cambridge, New York, pp. 150–191.

Holmes, P. and Whitley, D. (1984), Bifurcation in one and two dimensional maps, *Phil. Trans. Roy. Soc. London (Ser. A)* **1515**, 43–102.

Hofbauer, J. and Sigmund, K. (1988), *The Theory of Evolution and Dynamical Systems*, Cambridge University Press, Cambridge, New York.

Hopf, E. (1942), Abzweigung einer periodischen Lösung von einer stationären Lösung eines Differentialsystems, *Ber. Math. Phys. Sächsische Akad. der Wissen. Leipzig* **94**, 1–22 (See Marsden and McCracken (1976) for an English translation).

Hoppensteadt, F. C. (1982), *Mathematical Methods of Population Biology*, Cambridge University Press, Cambridge, New York.

Hu, S. (1994), A proof of C^1 stability conjecture for 3-dimensional flows, *Transact. Amer. Math. Soc.* **342**, 753–772.

Hubbard, J. and West, B. (1992), *MacMath: A Dynamical Systems Software Package for the Macintosh*, Springer-Verlag, New York, Heidelberg, Berlin.

Hurder, S. and Katok, A. (1990), Differentiability, rigidity, and Godbillon-Vey Classes for Anosov flows, *Publ. Math. I.H.E.S.* **72**, 5–61.

Hurewicz, W. (1958), *Lectures on Ordinary Differential Equations*, MIT Press, Cambridge, MA.

Hurewicz, W. and Wallman, H. (1941), *Dimension Theory*, Princeton University Press, Princeton, NJ.

Hurley, M. (1991), Chain recurrence and attraction in noncompact spaces, *Ergodic Theory Dynamical Systems* **11**, 709–729.

Hurley, M. (1992), Chain recurrence and attraction in noncompact spaces, II, *Proc. of the Amer. Math. Soc.* **115**, 1139–1148.

Iooss, G. (1979), *Birfurcation of Maps and Applications*, North-Holland Publ. Co., Amsterdam, New York, and Oxford.

Irwin, M. C. (1970a), A classification of elementary cycles, *Topology* **9**, 35–47.

Irwin, M. C. (1970b), On the stable manifold theorem, *Bull. London Math. Soc.* **2**, 196–198.

Irwin, M. C. (1972), On the smoothness of the composition map, *Q. J. Math. Oxford Ser.* **23**, 113–133.

Irwin, M. C. (1980), *Smooth Dynamical Systems*, Academic Press, New York.

Jakobson, M. (1971), On smooth mappings of the circle into itself, *Math. USSR Sb.* **14**, 161–185.

Jakobson, M. (1981), Absolutely continuous invariant measures for one-parameter families of one-dimensional maps, *Commun. Math. Phys.* **81**, 39–88.

Katok, A. (1980), Lyapunov exponents, entropy, and periodic orbits for diffeomorphisms, *Publ. Math. I.H.E.S.* **51**, 137–173.

Katok, A. and Burns, K. (1994), Infinitesimal Lyapunov functions, invariant cone families and stochastic properties of smooth dynamical systems, *Ergodic Theory Dynamical Systems* **14**, 757–786.

Katok, A. and Hasselblatt, B. (1994), *Introduction to the Modern Theory of Smooth Dynamical Systems*, Cambridge University Press, Cambridge, New York.

Katok, A. and Strelcyn, J.-M. (1986), *Invariant manifolds, entropy, and billiards; smooth maps with singularities (Lecture Notes in Math.* **1986**), Springer-Verlag, New York, Heidelberg, Berlin.

Kelley, A. (1967), The stable, center-stable, center center-unstable, and unstable manifolds (Appendix C), in *Transversal Mappings and Flows* by R. Abraham and J. Robbin, Benjamin, Reading, MA.

Koçak, H. (1986, revised 1989), *Differential and Difference Equations Through Computer Experiments*, Springer-Verlag, New York, Heidelberg, Berlin.

Kraft, R. (1999), Chaos, Cantor sets, and hyperbolicity for the logistic maps, *Amer. Math. Monthly (in press)*.

Kupka, I. (1963), Contribution à la théorie des champs génériques, *Contrib. to Diff. Eq.* **2**, 457–484.

Lanford, O. (1982), A computer-assisted proof of the Feigenbaum conjecture, *Bull. Amer. Math. Soc.* **6**, 427–434.

Lanford, O. (1984), A shorter proof of the existence of Feigenbaum fixed point, *Commun. Math. Phys.* **96**, 521–538.

Lanford, O. (1986), Computer-assisted proofs in analysis, *Proc. Int. Congress of Math.* **1,2**, 1385–1394.

Lang, S. (1967), *Introduction to Differentiable Manifolds*, Interscience Publ. (Wiley & Sons, Inc.), New York/London/ Sydney.

Lang, S. (1968), *Analysis I*, Addison-Wesley Publ. Co., New York and Reading, MA.

Lefever, F. and Nicholis, G. (1971), Chemical instabilities and sustained oscillations, *J. Theoretical Biology* **30**, 267–284.

Levi, M. (1981), Qualitative analysis of the periodically forced relaxation oscillator, *Memoirs Amer. Math. Soc.* **32**.

Levinson, N. (1949), A second order differential equation with singular solutions, *Ann. Math.* **50**, 127–153.

Li, T. and Yorke, J. (1975), Period three implies chaos, *Amer. Math. Monthly* **82**, 985–992.

Liao, S. T. (1979), An extension of the C^1 closing lemma, *Acta Scientiarum Naturalium Universitatis Pekinensis* **3**, 1–14. (Chinese)

Liao, S. T. (1980), Chinese Ann. Math. **1**, 9–30.

Liapunov, A. (1907), Problème général de la stabilité du mouvement, *Ann. Fac. Sci. Univ. Toulouse* **9**, 203–475 [reproduced in Ann. Math. Study (17) Princeton (1947)].

Lichtenberg, A. and Lieberman, M. (1983), *Regular and Stochastic Motion*, Springer-Verlag, New York, Heidelberg, Berlin.

Lind, D. and Marcus, B. (1995), *Symbolic Dynamics and Coding*, Cambridge University Press, Cambridge, New York.

Liverani, C. and Wojtkowski, M. (1995), Ergodicity in Hamiltonian systems, *Dynamics Reported* **N.S. 4**, 130–202.

de la Llavé, R. (1987), Invariants for smooth conjugacy of hyperbolic dynamical systems II, *Commun. Math. Phys.* **109**, 369–378.

de la Llavé, R., Marco, J. M., and Moriyon, R. (1986), Canonical perturbation theory of Anosov systems and regularity results for the Livsic cohomology equation, *Ann. Math.* **123**, 537–611.

Lorenz, E. N. (1963), Deterministic non-periodic flow, *J. Atmos. Sci.* **20**, 130–141.

Mai, J. (1986), A simpler proof of C^1 closing lemma, *Scientia Sinica (Series A)* **29 (10)**, 1021–1031.

Mallet-Paret, J. and Yorke, J. (1982), Snakes: oriented families of periodic orbits, their sources, sinks, and continuation, *J. Diff. Eq.* **43**, 419–450.

Mañé R. (1978a), Persistence manifolds are normally hyperbolic, *Trans. Amer. Math. Soc.* **246**, 261–283.

Mañé R. (1978b), Contributions to the stability conjecture, *Topology* **17**, 383–396.

Mañé R. (1987a), *Ergodic Theory and Differentiable Dynamics*, Springer-Verlag, New York, Heidelberg, Berlin.

Mañé R. (1987b), A proof of the C^1 stability conjecture, *Publ. Math. I.H.E.S.* **66**, 161–210.

Manning, A. (1971), Axiom A diffeomorphisms have rational zeta functions, *Bull. London Math. Soc.* **3**, 215–220.

Manning, A. (1974), There are no new Anosov diffeomorphisms on tori, *Amer. J. Math.* **96**, 422–429.

Manning, A. and McCluskey, H. (1983), Hausdorff dimension for horseshoes, *Ergodic Theory Dynamical Systems* **3**, 251–261.

Marek, M. and Schreiber, I. (1991), *Chaotic Behaviour of Deterministic Dissipative Systems*, Cambridge University Press, Cambridge, New York.

Marsden, J. (1974), *Elementary Classical Analysis*, Freeman and Co., New York.

Marsden, J. (1984), Chaos in dynamical systems by the Poincaré-Melnikov-Arnold method, in *Nonlinear Dynamical Systems* (edited by J. Chandra), SIAM, Philadelphia, pp. 19–31.

Marsden, J. and McCracken, M. (1976), *The Hopf Bifurcation and Its Applications*, Springer-Verlag, New York, Heidelberg, Berlin.

Margulis, G. A. (1969), On some applications of ergodic theory to the study of manifolds of negative curvature, *Funct. Analy. Appl.* **3**, 89–90.

May, R. (1975), *Stability and Complexity in Model Ecosystems, 2nd edn.*, Princeton Univ. Press, Princeton, NJ.

McGehee, R. (1973), A stable manifold theorem for degenerate fixed points with applications to celestial mechanics, *J. Diff. Eq.* **14**, 70–88.

Meiss, J.D. (1002), Symplectic maps, variational principles, and transport, *Reviews of Modern Physics* **64**, 795–848.

Melnikov, V. K. (1963), On the stability of the center for time periodic perturbations, *Trans. Moscow Math. Soc.* **12**, 1–57.

de Melo, W. (1973), Structural stability of diffeomorphisms on two manifolds, *Inven. Math.* **21**, 233–246.

de Melo, W. (1989), *Lectures on One-Dimensional Dynamics*, Instituto de Matematica Pura e Aplicada, Rio de Janeiro, Brazil.

de Melo, W. and Van Strien, S. J. (1993), *One-Dimensional Dynamics*, Springer-Verlag, New York, Heidelberg, Berlin.

Meyer, K. (1987), An analytic proof of the shadowing lemma, *Funkcialaj Ekvacioj* **30**, 127–133.

Meyer, K. and Hall, G. H. (1992), *Introduction to Hamiltonian Dynamical Systems and the N-Body Problem*, Springer-Verlag, New York, Heidelberg, Berlin.

Meyer, K. and Sell, G. (1989), Melnikov transforms, Bernoulli bundles, and almost periodic perturbations, *Trans. Amer. Math. Soc.* **314**, 63–105.

Milnor, J. (1985), On the concept of attractor, *Commun. Math. Phys.* **99**, 177–195.

Misiurewicz, M. (1981), Absolutely continuous measures for certain maps of an interval, *Publ. Math. I.H.E.S.* **53**, 17–51.

Mora, L. and Viana, M. (1993), Abundance of strange attractors, *Acta Math.* **171**, 1–71.

Moser, J. (1962), On invariant curves of area preserving mappings of an annulus, *Nachr. Akad. Wiss. Göttingen, Math. Phys. Kl*, 1–20.

Moser, J. (1968), Lectures on Hamiltonian Systems, *Memoirs of the Amer. Math. Soc.* **81**, 1–60.

Moser, J. (1969), On a theorem of Anosov, *J. Diff. Eq.* **5**, 411–440.

Moser, J. (1973), *Stable and Random Motions in Dynamical Systems*, Princeton University Press, Princeton, NJ.

Munkres, J. (1975), *Topology, A First Course*, Prentice-Hall, Englewood Cliffs, NJ.

Naimark, J. (1967), Motions close to doubly asymptotic motions, *Soviet Math. Dokl.* **8**, 228–231.

Newhouse, S. (1970a), Nondensity of Axiom A(a) on S^2, *Proc. Symp. in Pure Math., Amer. Math. Soc.* **14**, 191–202.

Newhouse, S. (1970b), On codimension one Anosov diffeomorphisms, *Amer. J. Math.* **92**, 761–770.

Newhouse, S. (1972), Hyperbolic limit sets, *Trans. Amer. Math. Soc.* **167**, 125–150.

Newhouse, S. (1979), The abundance of wild hyperbolic sets and non-smooth stable sets for diffeomorphism, *Publ. Math. I.H.E.S.* **50**, 101–151.

Newhouse, S. (1980), Lectures on Dynamical Systems, in *Dynamical Systems, (Progress in Math.* **8**), Birkhauser, Boston, Basel, Stuttgart, pp. 1–114.

Nitecki, Z. (1971), *Differentiable Dynamics*, MIT Press, Cambridge MA.

Nitecki, Z. (1971), On semi-stability for diffeomorphisms, *Inven. Math.* **14**, 83–122.

Oseledec, V. I. (1968), A multiplicative ergodic theorem. Liapunov characteristic numbers for dynamical systems, *Trans. Moscow Math. Soc.* **19**, 197–221.

Ott, E. (1993), *Chaos in Dynamical Systems*, Cambridge University Press, Cambridge, New York.

Palis, J. (1968), On the local structure of hyperbolic fixed points in Banach space, *Anais Acad. Brasil Ciencais* **40**, 263–266.

Palis, J. and de Melo, W. (1982), *Geometric Theory of Dynamical Systems, An Introduction*, Springer-Verlag, New York, Heidelberg, Berlin.

Palis, J. and Smale, S. (1970), Structural Stability Theorems, *Proc. Symp. in Pure Math., Amer. Math. Soc.* **14**, 223–232.

Palis, J. and Takens, F. (1993), *Hyperbolicity and Sensitive Chaotic Dynamics at Homoclinic Bifurcations*, Cambridge University Press, Cambridge, New York.

Palmer, K. (1984), Exponential dichotomies and transversal homoclinic points, *J. Diff. Eq.* **55**, 225–256.

Patterson, S. and Robinson, C. (1988), Basins for general nonlinear Hénon attracting sets, *Proc. Amer. Math. Soc.* **103**, 615–623.

Peixoto, M. (1962), Structurally stability on two dimensional manifolds, *Topology* **1**, 101–120.

Peixoto, M. (1966), On an approximation theorem of Kupka and Smale, *J. Diff. Eq.* **3**, 214–227.

Perron, O. (1907), Über Matrizen, *Math. Ann.* **64**, 248–63.

Perron, O. (1929), Uber stabilität und asymptotishes Verhalten der Lösungen eines Systemes endlicher Differenzengleichungen, *J. Reine Angew. Math.* **161**, 41–64.

Pesin, Ja. (1976), Families of invariant manifolds corresponding to nonzero characteristic exponents, *Math. USSR-Izvestia* **10**, 1261–1305.

Pesin, Ja. (1977), Characteristic Lyapunov exponents and smooth ergodic theory, *Russian Math. Surveys* **32**, 55–114.

Pesin, Ya. (1997), *Dimension Theory in Dynamical Systems*, University of Chicago Press, Chicago.

Pliss, V. A. (1971), Properties of solutions of a sequence of second-order periodic system with small nonlinearities, *Diff. Eq.* **7**, 501–508.

Pliss, V. A. (1972), A hypothesis due to Smale, *Diff. Eq.* **8**, 203–214.

Plykin, R. V. (1974), Sources and sinks for A-diffeomorphisms of surfaces, *Mathematics of the USSR, Sbornik* **23**, 233–253.

Pollicott, M. (1993), *Lectures on Ergodic Theory and Pesin Theory on Compact Manifolds (London Math. Soc. Lecture Note Series No.* **180***)*, Cambridge University Press, Cambridge, New York.

Pöschel, J. (1982), Integrability of Hamiltonian systems on Cantor sets, *Commun. on Pure and Applied Mathematics* **35**, 653–695.

Prigongine, I. and Lefever, R. (1968), Symmetry breaking instabilities in dissipative systems II, *J. Chem. Phys.* **48**, 1695–1700.

Pugh, C. (1967a), The Closing Lemma, *Amer. J. Math.* **89**, 956–1009.

Pugh, C. (1967b), An improved Closing Lemma and a General Density Theorem, *Amer. J. Math.* **89**, 1010–1021.

Pugh, C. (1969), On a theorem of Hartman, *Amer. J. Math.* **91**, 363–367.

Pugh, C. and Robinson, C. (1983), The C^1 Closing Lemma, including Hamiltonians, *Ergodic Theory Dynamical Systems* **3**, 261–313.

Pugh, C. and Shub, M. (1970a), Linearization of normally hyperbolic diffeomorphisms and flows, *Inven. Math.* **10**, 187–198.

Pugh, C. and Shub, M. (1970b), Ω-stability for flows, *Inven. Math.* **11**, 150–158.

Pugh, C. and Shub, M. (1989), Ergodic attractors, *Transact. Amer. Math. Soc.* **312**, 1–54.

Rand, D. (1978), The topological classification of the Lorenz attractor, *Math. Proc. Camb. Phil. Soc.* **83**, 451–460.

Rasband, N. (1990), *Chaotic Dynamics of Nonlinear Systems*, Wiley-Interscience Publ., New York.

Riesz, Z. and Nagy, B. (1955), *Functional Analysis*, Fredrick Ungar Publ. Co., New York.

Robbin, J. (1971), A structural stability theorem, *Ann. Math.* **94**, 447–493.

Robinson, C. (1973), C^r structural stability implies Kupka-Smale, in *Dynamical Systems* (edited by M. M. Peixoto), Academic Press, New York, pp. 443–449.

Robinson, C. (1974), Structural stability of vector fields, *Ann. Math.* **99**, 154–174.

Robinson, C. (1975a), Structural stability of C^1 flows, in *Lecture Notes in Math.* **468**, Springer-Verlag, New York, Heidelberg, Berlin, pp. 262–277.

Robinson, C. (1975b), The geometry of the structural stability proof using unstable disks, *Bol. Soc. Brasil. Math.* **6**, 129–144.

Robinson, C. (1976a), Structural stability of C^1 diffeomorphisms, *J. Diff. Eq.* **22**, 28–73.

Robinson, C. (1976b), Structural stability theorems, in *Dynamical Systems, Vol. II*, Academic Press, New York, pp. 33–36.

Robinson, C. (1977), Stability theorems and hyperbolicity in dynamical systems, *Rocky Mountain J. Math.* **7**, 425–437.

Robinson, C. (1978), Introduction to the Closing Lemma, in *The Structure of Attractors in Dynamical Systems* (edited by Markley, Martin, and Perrizo) *(Lecture Notes in Math.* **668**), Springer-Verlag, New York, Heidelberg, Berlin, pp. 225–230.

Robinson, C. (1981), Differentiability of the stable foliation of the model Lorenz equations, in *Dynamical Systems and Turbulence* (edited by Rand and Young) *Lecture Notes in Math.* **898**), Springer-Verlag, New York, Heidelberg, Berlin, pp. 302–315.

Robinson, C. (1983), Bifurcation to infinitely many sinks, *Commun. Math. Phys.* **90**, 433–459.

Robinson, C. (1984), Transitivity and invariant measures for geometric model of the Lorenz equations, *Ergodic Theory Dynamical Systems* **4**, 605–611.

Robinson, C. (1985), Phase analysis using the Poincaré map, *Nonlinear Anal. Theory Methods Appl.* **9**, 1159–1164.

Robinson, C. (1988), Horseshoes for autonomous Hamiltonian systems using the Melnikov integral, *Ergodic Theory and Dynamical Systems* **8***, 395–409.

Robinson, C. (1989), Homoclinic bifurcation to a transitive attractor of Lorenz type, *Nonlinearity* **2**, 495–518.

Robinson, C. (1992), Homoclinic bifurcation to a transitive attractor of Lorenz type, II, *SIAM J. Math. Anal.* **23**, 1255–1268.

Rudin, W. (1964), *Principles of Mathematical Analysis*, McGraw-Hill, New York.

Ruelle, D. (1976), A measure associated with Axiom A attractor, *Amer. J. Math.* **98**, 619–654.

Ruelle, D. (1979), Ergodic theory of differentiable dynamical systems, *Publ. Math. I.H.E.S.* **50**, 27–58.

Ruelle, D. (1989a), *Chaotic Evolution and Strange Attractors*, Cambridge University Press, Cambridge, New York.

Ruelle, D. (1989b), *Elements of Differentiable Dynamics and Bifurcation Theory*, Academic Press, New York.

Ruelle, D. and Takens, F. (1971), On the nature of turbulence, *Commun. Math. Phys.* **20**, 167–192.

Rychlik, M. (1990), Lorenz attractors through Sil'nikov-type bifurcation, Part I, *Ergodic Theory Dynamical Systems* **10**, 793–822.

Rykken, E. (1993), Markov Partitions and the Expanding Factor for Pseudo-Anosov Homeomorphisms, *Thesis, Northwestern University.*

Sacker, R. (1964), On invariant surfaces and bifurcation of periodic solutions of ordinary differential equations, *New York Univ. IMM-NYU* **333**.

Sacker, R. (1967), A Perturbation for invariant Riemannian manifolds, in *Proc. Symp. at Univ. of Puerto Rico* (edited by J. Hale and J. LaSalle), Academic Press, New York, pp. 43 – 54.

Sharkovskii, A. N. (1964), Coexistence of cycles of a continuous map of a line into itself, *Ukrainian Math. J.* **16**, 61 – 71.

Shub, M. (1987), *Global Stability of Dynamical Systems*, Springer-Verlag, New York, Heidelberg, Berlin.

Siegel, C. L. and Moser, J. (1971), *Lecutres on Celestial Mechanics*, Springer-Verlag, New York, Heidelberg, Berlin.

Simon, C. (1972), Instability in $\text{Diff}^r(T^3)$ and the non-genericity of rational zeta functions, *Trans. Amer. Math. Soc.* **174**, 217 – 242.

Sinai, Ya. (1968), Markov partitions and C-diffeomorphisms, *Func. Anal. Appl.* **2**, 64 – 89.

Sinai, Ya. (1970), Dynamical systems with elastic reflections, *Russ. Math. Surveys* **25**, 137 – 189.

Sinai, Ya. (1972), Gibbs measures in ergodic theory, *Russ. Math. Surveys* **166**, 21 – 69.

Sitnikov, K. (1960), Existence of oscillating motions for the three body problem, *Dokl. Akad. Nauk.* **133**, 303 – 306.

Smale, S. (1963), Stable manifolds for differential equations and diffeomorphisms, *Ann. Scuola Normale Pisa* **18**, 97 – 116.

Smale, S. (1965), Diffeomorphisms with many periodic points, *Differential and Combinatorial Topology*, Princeton University Press, Princeton, NJ, 63 – 80.

Smale, S. (1966), Structurally stable systems are not dense, *Amer. J. Math.* **88**, 491 – 496.

Smale, S. (1967), Differentiable dynamical systems, *Bull. Amer. Math. Soc.* **73**, 747 – 817.

Smale, S. (1970), The Ω-Stability Theorem, *Proc. Symp. Pure Math.* **14**, 289 – 297.

Smale, S. (1980), *The Mathematics of Time: Essays on Dynamical Systems, Economic Processes, and Related Topics*, Springer-Verlag, New York, Heidelberg, Berlin.

Smith, K. (1971), *Primer of Modern Analysis*, Bogden & Quigley, Inc., Publishers, Tarrytown-on-Hudson, New York, Belmont CA.

Snavely M. (1990), Markov Partitions for Hyperbolic Automorphisms of the Two-Dimensional Torus, *Thesis, Northwestern University.*

Sotomayor, J. (1973a), Generic one parameter families of vector fields on two manifolds, *Publ. Math. Inst. Hautes Études Scientifiques* **43**, 5 – 46.

Sotomayor, J. (1973b), Generic bifurcations of dynamical systems, in *Dynamical Systems* (edited by M. Peixoto), Academic Press, New York, pp. 561 – 582.

Sparrow, C. (1982), *The Lorenz Equations: Bifurcations, Chaos, and Strange Attractors*, Springer-Verlag, New York, Heidelberg, Berlin.

Stefan, P. (1977), A theorem of Sarkovskii on the existence of periodic orbits of continuous endomorphisms of the real line, *Commun. Math. Phys.* **54**, 237 – 248.

Sternberg, S. (1958), On the structure of local homeomorphisms of Euclidean n-space, II, *Amer. J. Math.* **81**, 623 – 631.

van Strien, S. (1979), Center manifolds are not C^∞, *Math. Z.* **166**, 143 – 145.

van Strien, S. (1981), On the bifurcation creating horseshoes, in *Dynamical Systems and Turbulence* (edited by Rand and Young) (*Lecture Notes in Math.* **898**), Springer-Verlag, New York, Heidelberg, Berlin, pp. 316 – 351.

Strogatz, S. (1994), *Nonlinear Dynamics and Chaos*, Addison-Wesley Publ. Co., New York & Reading, MA.

Takens, F. (1974), Singularities of vector fields, *Publ. Math. Inst. Hautes Études Scientifiques* **43**, 47–100.

Takens, F. (1975), Tolerance stability, in *Dynamical Systems - Warwick 1974 (Lecture Notes in Math.* **468**), Springer-Verlag, New York, Heidelberg, Berlin, pp. 293–304.

Takens, F. (1988), Limit capacity and Hausdorff dimension of dynamically defined Cantor sets, in *Dynamical Systems (Lecture Notes in Math.* **1331**), Springer-Verlag, New York, Heidelberg, Berlin, pp. 196–212.

Thom, R. (1973), *Stabilité Structurelle et Morphogénèse*, Addison-Wesley Publ. Co., New York and Reading, MA; English (1975).

Ushiki, S., Oka, H., and Kokubu, H. (1984), Existence of strange attractors in the unfolding of a degenerate singularity of a translation invariant vector field, *C. R. Acad. Ac. Paris, Ser. I Math.* **298**, 39–42.

Vinograd, R. E. (1957), The inadequacy of the method of characteristic exponents for the study of nonlinear differential equations, *Mat. Sbornik* **41**, 431–438.

Walters, P. (1970), Anosov diffeomorphisms are topologically stable, *Topology* **9**, 71–78.

Walters, P. (1978), On the pseudo-orbit tracing property and its relationship to stability, in *The Structure of Attractors in Dynamical Systems* (edited by Markley, Martin, and Perrizo), *Lecture Notes in Math.* **668**, Springer-Verlag, New York, Heidelberg, Berlin, pp. 231–244.

Walters, P. (1982), *An Introduction to Ergodic Theory*, Springer-Verlag, New York, Heidelberg, Berlin.

Waltman, P. (1983), *Competition Models in Population Biology*, Society for Industrial and Applied Mathematics, Philadelphia, PA.

Wen, L. (1991), The C^1 closing lemma for non-singular endomorphisms, *Ergodic Theory Dynamical Systems* **11**, 393–412.

Wen, L. (1992), Anosov endomorphisms on branched surfaces, *J. Complexity* **8**, 239–264.

Wen, L. (1996), On the C^1 stability conjecture for flows, *J. Diff. Equat.* **129**, 334–357.

Wiggins, S. (1988), *Global Bifurcation and Chaos*, Springer-Verlag, New York, Heidelberg, Berlin.

Wiggins, S. (1990), *Introduction to Applied Nonlinear Dynamical Systems and Chaos*, Springer-Verlag, New York, Heidelberg, Berlin.

Williams, R. (1967), One-dimensional nonwandering sets, *Topology* **6**, 473–487.

Williams, R. (1968), The zeta function of an attractor, in *Conference on the Topology of Manifolds* (J. C. Hocking, ed.), Prindle Weber and Smith, Boston MA, pp. 155–161.

Williams, R. (1970a), The 'DA' maps of Smale and structural stability, *Global Analysis, Proc. Sympos. Pure Math., Amer. Math. Soc.* **14**, 329–334.

Williams, R. (1970b), The zeta function in global analysis, *Global Analysis, Proc. Sympos. Pure Math., Amer. Math. Soc.* **14**, 335–340.

Williams, R. (1974), Expanding attractors, *Publ. Math. I.H.E.S.* **43**, 169–203.

Williams, R. (1977), The structure of Lorenz attractors, in *Turbulence Seminar* (edited by P. Bernard) *Lect. Notes in Math.*, vol. **615**, Springer-Verlag, New York, Heidelberg, Berlin.

Williams, R. (1979), The bifurcation space of the Lorenz attractor, in *Bifurcation Theory and Its Applications in Scientific Disciplines,* (edited by O. Gurel and O. E. Rössler) Ann. NY Acad. Sci., vol. **316**, pp. 393–399.

Williams, R. (1980), Structure of Lorenz attractors, *Publ. Math. I.H.E.S.* **50**, 59–72.

Williams, R. (1983), Lorenz knots are prime, *Ergodic Theory Dynamical Systems* **4**, 147–163.

Wilson, W. (1967), The structure of the level surfaces of a Lyapunov function, *J. Diff. Eq.* **3**, 323–329.

Wojtkowski, M. (1985), Invariant families of cones and Lyapunov exponents, *Ergodic Theory Dynamical Systems* **5**, 145–161.

Yomdin, Y. (1987), Volume growth and entropy, *Israel J. Math.* **57**, 285–318.

Yorke, J. and Alligood, K. (1985), Period doubling cascades of attractors: a prerequisite for horseshoes, *Commun. Math. Phys.* **101**, 305–321.

Young, L. S. (1982), Dimension, entropy, and Lyapunov exponents, *Ergodic Theory Dynamical Systems* **2**, 109–124.

Young, L. S. (1993), Ergodic theory of chaotic dynamical systems, in *From Topology to Computation: Proceedings of the Smalefest* (editors Hirsch, Marsden, and Shub), Springer-Verlag, New York, Heidelberg, Berlin.

Zeeman, E. C. (1980), Population dynamics from game theory, in *Lecture Notes in Math.* **819**, Springer-Verlag, New York, Heidelberg, Berlin, pp. 471–497.

Index